Crops as Enhancers of Nutrient Use

Crops as Enhancers of Nutrient Use

V. C. BALIGAR
United States Department of Agriculture
Agricultural Research Service
Appalachian Soil and Water Conservation Research Laboratory
Beckley, West Virginia

R. R. DUNCAN
Department of Agronomy
University of Georgia
Griffin, Georgia

ACADEMIC PRESS, INC.
Harcourt Brace Jovanovich, Publishers
San Diego New York Boston London
Sydney Tokyo Toronto

This book is printed on acid-free paper. ∞

Academic Press, Inc.
San Diego, California 92101

United Kingdom Edition published by
Academic Press Limited
24–28 Oval Road, London NW1 7DX

Library of Congress Cataloging-in-Publication Data

Crops as enhancers of nutrient use / edited by V. C. Baligar, R. R. Duncan.
p. cm.
Includes bibliographical references.
ISBN 0-12-077125-X (alk. paper)
1. Crops and soils. 2. Crops--Nutrition. 3. Fertilizers.
I. Baligar, V. C. II. Duncan, R. R.
S596.7.C76 1991
581.1'335--dc20 90-542
CIP

Printed in the United States of America
90 91 92 93 9 8 7 6 5 4 3 2 1

Contents

Contributors

Numbers in parentheses indicate the pages on which the authors' contributions begin.

D. Atkinson (411), School of Agriculture, Aberdeen AB9 1UD, Scotland

V. C. Baligar (3, 351, 475), United States Department of Agriculture, Agricultural Research Service, Appalachian Soil and Water Conservation Research Laboratory, Beckley, West Virginia 25802

David F. Bezdicek (375), Department of Agronomy and Soils, Washington State University, Pullman, Washington 99164

Joe H. Bouton (211), Department of Agronomy, University of Georgia, Athens, Georgia 30602

J. R. Caradus (253), Grasslands Division, Department of Scientific and Industrial Research, Palmerston North, New Zealand

Ralph B. Clark (131), United States Department of Agriculture, Agricultural Research Service, Department of Agronomy, University of Nebraska, Lincoln, Nebraska 68583

Thomas E. Devine (211), Plant Molecular Biology Laboratory, PSI, ARS, BARC, USDA, Beltsville, Maryland 20705

R. R. Duncan (3, 351), Department of Agronomy, University of Georgia, Griffin, Georgia 30223

D. G. Edwards (475), Faculty of Agriculture, University of Queensland, St. Lucia, Queensland 4067, Australia

N. K. Fageria (351, 475), EMBRAPA National Center for Rice and Beans, 74000, Goiania Go, Brazil

Anthony D. M. Glass (37), Department of Botany, The University of British Columbia, Vancouver, British Columbia V6T 2B1, Canada

D. W. James (509), Plant, Soil, and Biometeorology Department, Utah State University, Logan, Utah 84322

Seshadri Kannan (313), Mineral Plant Nutrition Section, Nuclear Agriculture Division, Bhabha Atomic Research Center, Bombay 400085, India

Calvin F. Konzak (81), Department of Crop and Soil Sciences, Washington State University, Pullman, Washington 99164

Tadesse Mabrahtu (211), Virginia State University, Petersburg, Virginia 23803
Ian L. Pepper (375), Department of Soil and Water Science, University of Arizona, Tucson, Arizona 85721
Enrique A. Polle (81), Department of Crop and Soil Sciences, Washington State University, Pullman, Washington 99164
J. F. Power (453), United States Department of Agriculture, Agricultural Research Service, University of Nebraska, Lincoln, Nebraska 68583
P. B. Vose (65), Grange Over Sands Cambria, LA11 7BB, England

Preface

Genetics and physiological components of plants have profound effects on the plant's ability to absorb and utilize nutrients under various environmental and ecological conditions. In this book, we examine the various plant and soil factors that contribute to nutrient use efficiency of plants. Plant–soil interactions are emphasized, particularly those nutritional interactions involving the rhizosphere, microbes, and stress (moisture, low pH, and high pH) on the root system.

Many books have been written with emphasis on the soil, fertilizers, and their interactions. This book deals with plants and how they respond genetically and physiologically to nutrients (1) at the cellular level, (2) on a whole-plant basis, and (3) when subjected to stress.

Comprehensive coverage of genetics and breeding of cereals and legumes as related to nutrient use efficiency is given in Chapter 4 by Polle and Konzak and in Chapter 6 by Devine, Bouton, and Mebrahtu. In Chapter 2, Glass deals extensively with the cellular mechanisms involved in the acquisition and utilization of inorganic nutrients. The intrinsic and extrinsic factors that influence or regulate nutrient acquisition are emphasized. Variations in function and metabolism of roots, root–shoot ratios, and N and P use efficiency on the whole-plant level are covered in Chapter 3 by Vose. The role of root morphology and root actions on efficiency of nutrient use in various types of plant species is given in Chapter 11 by Atkinson. Clark in Chapter 5 and Caradus in Chapter 7 review the effects of plant physiological factors on nutrient efficiency in cereal and legume plant species, respectively. Abiotic and biotic effects of genotypic and environmental interaction are also emphasized. The role of leaves in the uptake of nutrients is covered in Chapter 8 by Kannan. Various mechanisms involving foliar absorption and transport of nutrients are emphasized with respect to plant species and environmental impacts. In Chapter 12, Power deals with various aspects of soil moisture stress and its effect on plant nutrition. Emphasis is also given to the role of soil moisture in biological transformation of plant nutrients. Fertility, nutrient relations, and plant performances in acid and alkaline soils are covered in Chapter 13 by Fageria, Baligar, and Edwards and in Chapter 14 by James. Various chemical constraints that limit plant growth along with management of acid and alkaline soils are emphasized. Root and microbial interaction and rhizosphere nutrient dynamics are covered in Chapter 10 by Pepper and Bezdicek.

This book is a timely addition to the scientific community in light of federally mandated policies regarding Low Input Sustainable Agriculture (LISA), conservation-oriented cropping systems, and reductions in environmental contaminants. We attempt to address these concerns and to present more long-term remedies to some of the inherent problems of high volume applications of expensive fertilizer nutrients. We feel that this book will lead to more cost-effective and judicious nutrient usage of our major crops during the next decade.

The editors gratefully acknowledge the excellent cooperation of all authors who contributed to this book. Without their timely efforts this publication could not have been realized. The editors also express their appreciation to the following reviewers who generously provided their time and talents: M. P. Barbosa Filho, A. B. Bennett, J. H. Bouton, N. Claassen, J. Dunlop, C. D. Foy, D. M. Glenn, A. L. Hart, J. J. Roberts, C. A. Jones, W. R. Kussow, J. W. Maranville, R. H. Miller, J. A. Morgan, I. P. Oliveira, K. D. Ritchey, L. M. Shuman, C. Y. Sullivan, and R. J. Wright.

The editors especially thank their wives and children for their understanding and patience.

V. C. Baligar
R. R. Duncan

Part I

Genetic and Physiological Basis of Nutrient Uptake and Use Efficiency

1

Genetics, Breeding, and Physiological Mechanisms of Nutrient Uptake and Use Efficiency: An Overview

R. R. DUNCAN and V. C. BALIGAR

Plants seldom are grown under the appropriate optimal environmental conditions that can theoretically maximize yield productivity. Physical (extreme moisture and temperature regimes) and chemical (pH, nutritional imbalances, toxic and deficient nutrients) problems limit crop yields in most regions of the world. The increasing worldwide recognition of problems associated with "the greenhouse effect," with environmental pollution, with erosion control, and with stabilizing crop yields and food availability is forcing a restructuring of research and development strategies to deal with these problems. As the world's population continues to increase their demand for food, forcing extension of farming into marginal production areas and requiring the devisement of low-input stress-buffered production systems, the era of

Crops as Enhancers of Nutrient Use

changing the environment to fit the needs of current cultivars has passed. The emerging era of adapting the plant to the natural environment is paramount to stabilizing crop yields and developing low-input, technologically advanced, sustainable production systems for the future. The key to this effort will be the breeding of plants with improved stress tolerance and with improved nutrient uptake and utilization efficiency. Breeders, geneticists, stress physiologists, plant nutritionists, biotechnologists, and the fertilizer industry must all cooperate in this endeavor. Genetic manipulations and breeding strategies for improving nutrient uptake and utilization efficiency among cereal grasses and legumes will hopefully set the tone for continued improvement of crop productivity, i.e., productivity in terms of the crops' ability to efficiently utilize natural resources or limited agronomic inputs when producing a harvestable product. However, we must also improve nutrient recycling strategies while minimizing soil erosion and environmental pollution to parallel the genetic improvements in crop productivity.

I. Gene Implications

A. NUTRITIONAL EFFICIENCY/INEFFICIENCY AND TOLERANCE/SUSCEPTIBILITY CONSIDERATIONS

The study of plant nutritional requirements and associated physiological processes has traditionally concentrated on their response to environmental factors. To exploit the genetic control of these plant processes, independent selection must theoretically be merged with controlled recombination of individual nutritional and physiological characteristics using biotechnological and traditional plant breeding methods. To employ selection techniques, four basic criteria must be available (Mahon, 1983).

1. Genetic variability exhibiting a range of expression is needed to assess the trait.
2. Improvement strategies rely on detailed information about the genetic systems. Genotypic performance stability (broad sense heritability) over a range of spatial and temporal environments is necessary for useful selection.
3. Improvements in nutritional and physiological traits must be related to some feature of agronomic importance, such as yield stability, improved quality of harvestable product, or reduced production costs.
4. Practical exploitation of the traits will be difficult unless phenotypic expression can be expressed in large-scale trials, in large germplasm collections, or in segregating populations.

The genetic diversity necessary for tolerance to toxic or deficient nutritional environments is widely available in many crop species of the plant kingdom. Different genes may be involved for tolerance to different levels of toxicity and to different toxic minerals (Little, 1988). Selection programs should utilize cultivars as source germplasm from specific locations in the world where inherent nutritional stresses exist in the soil. Normal plant capability to adjust to a range of nutritional conditions must be recognized. Many plants perform satisfactorily with wide ranges of nutrient concentrations in their tissues and under variable conditions of soil nutrient availability. Modification of nutrient uptake mechanisms and general plant responses to nutrient stresses are related (Millikan, 1961; Butler *et al.*, 1962; Rorison, 1969; Brown *et al.*, 1972; Barber, 1976; Hovin *et al.*, 1978; Brown, 1979; Stratton and Sleper, 1979; Wegrzyn *et al.*, 1980; Chichester, 1981; Saric, 1981).

Heritability expresses the genetic proportion of the total statistical variance. Heritability estimates for specific traits provide information on available genetic variation which can help identify superior genotypes. More importantly, heritability estimates express the reliability of phenotypic values as a guide to the breeding value (Falconer, 1960).

Recessive genes often block essential processes, causing the loss of enzymatic activity and resulting in deleterious phenotypes (Gerloff and Gabelman, 1983), or may even lead to a secondary response role (of the recessive allele) in crop plants. Relatively high heritability estimates suggest that variation in segregating progeny reflects real genetic differences. Most studies suggest that one gene controls the nutritional efficiency response. Since more efficient segregates can be isolated from small populations in segregating generations, few genes probably contribute to efficiency.

1. Mechanisms for Nutrient Uptake and Use Efficiency

Considerable variation for nutritional efficiency exists between crop species and among cultivars within species. The numerous nutritional differences suggest genetic control of inorganic plant nutrition (Gerloff and Gabelman, 1983). The genotypic nutritional differences expressed in terms of physiological mechanisms and morphological features are listed in Table I. Studies on adaptation of exotic plant species to specific nutritionally stressed environments have usually preceded studies on genetic control of nutrition in crop plants.

2. Plant Adaptability to Low Nutrient Conditions

When soils are limiting in such essential nutrients as nitrogen, phosphorus, and potassium, either nutrient amendments must be added to correct the deficiency or adapted plants grown which can tolerate the low nutrient conditions. Categorization of selected crop plants for response to variable phosphorus conditions is presented in Table II.

Low nutrient adaptability is mostly the result of the inability of crop roots to absorb an essential nutrient, such as phosphorus, from growth media low in that specific nutrient and then to metabolically and physiologically function because of

Table I. Possible Components for Genotypic Variations in Plant Nutrient Efficiency Requirements[a]

I. Acquisition from the environment
1. Morphological root features (root system efficiency)
a. High root/shoot ratio under nutrient deficiency
b. Greater lateral and vertical spread of roots
c. High root density or absorbing surface, more root hairs, especially under stress
2. Physiological efficiency of ion uptake per unit root length
3. Generation of reducing and chelating capacity (i.e., Fe)
4. "Extension" of the root system by mycorrhizae
5. Longevity of roots
6. Ability of the roots to modify rhizosphere and overcome deficient or toxic levels of nutrients
II. Nutrient movement across roots and delivery to the xylem
1. Lateral transfer through the endodermis
2. Release to the xylem
3. Control of ion uptake and distribution by either root or shoot systems, or by both
a. Delivery to root or shoot under deficiency
b. Overall regulation of nutrient uptake in intact plants and use on a whole-plant level
III. Nutrient distribution within plants
1. Degree of retranslocation and reutilization under nutrient stress
2. Release of ions from vacuoles under nutrient deficiency
3. Natural iron chelating compounds in xylem
4. Rate of leaf abscission and rate of hydrolysis (i.e., organic P)
5. Capacity for rapid storage for later use when nutrient is available
IV. Growth and metabolic efficiency under nutrient limitations
1. Capacity for normal functioning at relatively low tissue concentrations
2. Nutrient substitutions (i.e., Na^+ for K^+)
V. Polyploidy and hybridity levels

[a]Sources: Gerloff and Gabelman, 1983; Seetharama *et al.*, 1987.

the low available nutrient concentrations in the plant. Caradus and Snaydon (1986b) found that phosphorus uptake per plant in white clover[1] was determined largely by shoot factors. Among the cereals, rye is especially nutrient efficient with regard to copper, zinc, manganese, and phosphorus (Graham, 1984). Rye is also the most adapted cereal for impoverished acid sandy soils.

Under soil acidity stress where toxicities exist, plant response may be quite different from efficiency adaptations (Table III). However, plants adapted to low nutrient availability conditions and to soil stress are much more likely to be grown in marginal environments with low inputs by subsistence farmers. Nevertheless, some nitrogen, phosphorus, and potassium should be applied to most crop plants to enhance productivity (Table IV).

Soil fertility management is quite important in any crop production program. The efficiency with which a crop produces a harvestable product per unit of available nutrient varies with the species (Table V). Sweet potato, potato, sorghum, and rice are very nitrogen efficient, while the legumes groundnut (peanut), field beans, and

[1]A glossary of scientific plant names appears after Chapter 14.

Table II. Adaptability to Low Phosphorus Conditions[a]

Strong	Medium	Weak
Rice	Wheat	Sugar beet
Maize	Barley	Tomato
Azuki bean	Soybean	Pea
Sorghum	Field beans	Cotton
Sweet potato		
Buckwheat		

[a]Source: Tanaka, 1980.

soybean are inefficient. With phosphorus, potato and sweet potato are very efficient and groundnut is inefficient. Rice is very potassium efficient and wheat is inefficient. However, the distribution of absorbed nutrients in the harvested product versus the residue is critical in multiple cropping, conservation, and subsistence farming programs. In general, percentage distribution of nitrogen and phosphorus is high in the grain and tuber crops. Except for legumes, little nitrogen or phosphorus is returned to the soil. However, the amount of recyclable potassium in the residue varies considerably among crop species. Soybean and potato crops require careful monitoring of potassium fertility while wheat, rice, maize, sorghum, and groundnut contain substantial amounts of potassium in their residue. If the residue is removed and not recycled, soil fertility must be adjusted.

3. Nutrient Insufficiency

Plant response to essential and nonessential nutrients can be depicted by a dose–response curve (Berry and Wallace, 1981) (Figure 1). Three continuous phases—deficiency, tolerance, and toxicity—constitute the response phases of plants to

Table III. Tolerance for High Aluminum Concentrations in Various Field Crops[a]

Very Strong	Strong	Medium	Weak	Very Weak
Buckwheat	Oats	Maize	Onion	Carrot
Rice	Soybeans	Field beans	Barley	Spinach
	Broad bean	Cabbage	Sugar beet	Celery
		Wheat	Cucumber	
		Barnyard millet	Sorghum	
		Pea	Turnip	
		Eggplant		

[a]Source: Tanaka and Hayakawa, 1975.

Table IV. Average Yields of Crops with No Nitrogen, Phosphorus, or Potassium, Relative to Yield with Three Fertilizer Ingredients[a]

Treatment	Maize	Potato	Sugar Beet	Soybean
No nitrogen	8	41	48	112
No phosphorus	87	78	59	79
No potassium	95	24	51	55

[a]Source: Tanaka, 1980.

essential nutrients. Interactions among nutrients occur when plants are exposed to stresses that enhance high or low concentrations of more than one nutrient. These two concentration-dependent phases are separated by a region of zero slope or a tolerance plateau. The entire dose–response curve is a continuum starting at or near zero and ending when the dose reaches lethal levels. An induced deficiency occurs when a yield decrease is associated with an increase in dose due to an imbalance of some other essential nutrient (Berry and Wallace, 1981). When a yield decrease occurs without an imbalance in any other essential element being the cause, the nutrient becomes toxic. Genetic manipulation of plants has targeted the tolerance plateau portion of the curve to enhance nutrient use efficiency. Genetic specificity among and within species for critical levels of nutrients, multiple nutrient interactions, and environmental stresses compound problems involved in breeding and developmental efforts.

Different physiological mechanisms govern nutrient uptake efficiency and nu-

Table V. Efficiency of Nutrient Elements Absorbed by Crops to Produce a Harvested Organ and Nutrient Content in Harvested Organ in Relation to Total Amount Absorbed[a]

Crop	Efficiency of Absorbed Nutrient (g Dry Matter of Harvested Organ per g Nutrient Absorbed)			% Nutrients in Harvested Organ		
	N	P	K	N	P	K
Rice	40	205	44	65	67	11
Wheat	23	151	20	61	70	9
Maize	37	215	29	69	69	14
Sorghum	44	191	38	58	63	17
Soybean	13	154	27	85	88	58
Field bean	18	151	25	78	76	37
Groundnut	11	125	24	56	56	18
Sweet potato	50	321	29	45	54	35
Potato	47	320	33	58	76	65
Sugar beet	30	225	26	35	40	26

[a]Source: Tanaka, 1980.

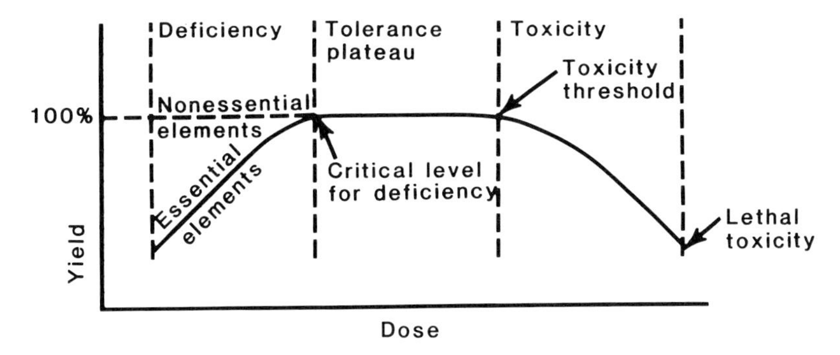

Figure 1. Generalized dose–response curve. Source: Berry and Wallace, 1981.

trient utilization efficiency. Breeding programs designed to select plants based on variations in nutrient uptake and utilization have produced minimal improvements. Complex external factors such as root exudates, symbiotic and nonsymbiotic microbes, soil acidity, or moisture stress and internal factors such as absorption, translocation, assimilation, detoxification, and metabolism hamper progress (Vose, 1963). Knowing the critical concentration ranges for specific nutrients by crop species provides a starting point for the developmental program (Reuter and Robinson, 1986). Crop response to mineral nutrition and growth of plants in soils of high and low nutrient availability is depicted in Figure 2.

Vose (1963) provided additional insight into general mechanisms for nutritional variation. Nutrient absorption depends on the number (root volume) and type of roots (root surface area and degree of branching). However, any improvement in root volume must be accompanied by an improvement in metabolic efficiency. The

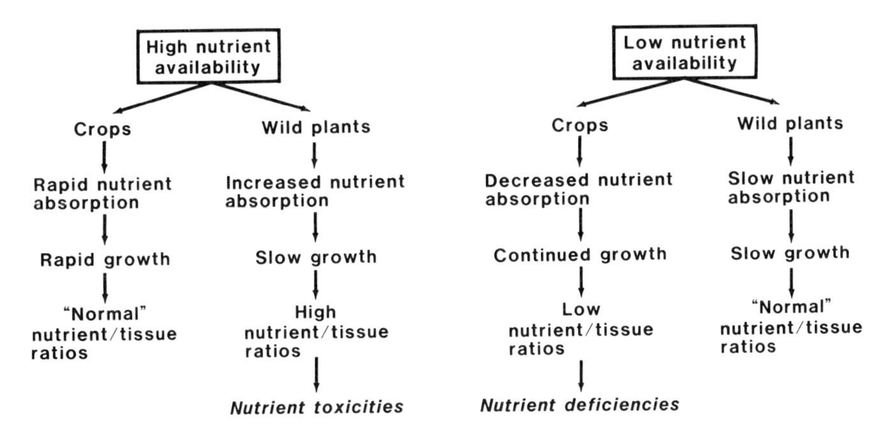

Figure 2. Responses in mineral nutrition and growth of crops adapted to highly fertile soils and of wild plants that are nutrient-stress-tolerant in soils of high and low nutrient availability. Source: Bieleski and Lauchli, 1983.

concept of exchange-absorption and cation-exchange capacity (CEC) of roots has led to studies which show:

1. Plants in the Leguminosae family have approximately twice the root CEC as plants in the Graminae family.
2. Species making the most demands on soil fertility have a higher root CEC than those with lower requirements.
3. Low root CEC favors monovalent ions (K) to divalent ions (Ca, Mg), which may account for some cultivar differences of major cation absorption and their subsequent concentration.
4. The active uptake carrier concept advocates that the entrance of ions into living cells is accompanied by binding to carrier molecules (ribonucleoproteins) of some protoplasmic constituent. The ion–carrier complexes are able to traverse barriers of only limited permeability to free ions. The carrier system is selective to various ions and highly specific regarding competition for absorption sites. Nye (1973) proposed that root surface area is most likely the limiting factor in ion absorption for most crop plants when a nutrient is at a low concentration in the medium. Kramer (1983) proposed that roots undergo genetically determined adaptations to nutritional stress by modification of transport systems. Genetic variability in the plant kingdom for nutrient acquisition and use reflects differences in root morphology and mechanisms that enhance or prevent ion movement into the root (Gabelman *et al.*, 1986). Factors causing variability in ion influx into roots include: (a) chemical nature of the carrier, (b) specificity of the carrier, (c) concentrations of the carrier, (d) speed of turnover of the carrier, (e) multiple carriers for certain ions, (f) carriers with multiple absorption sites for more than one type of ion, (g) relative proportions of carriers, (h) possible adaptive synthesis of carriers, and (i) rate of removal of ions from the sink (Vose, 1963).
5. Translocation factors are inherently affected by plant response to stress factors and interactions with nutrients and with biochemical processes in the plant.
6. Enzyme systems and subsequent plant metabolic processes are sensitive to stress-induced nutritional imbalances in plants. For example, cultivar differences in toxicity tolerance and deficiency susceptibility for plants grown on acid soils are directly related to the effect of Mn and Al on enzyme activity. Hoener and DeTurk (1938) found that differences in N utilization efficiency among high- and low-protein maize cultivars were an inherent lack of adaptive capacity to develop adequate nitrate reductase to fully utilize high nitrate levels. The high-protein plants had a greater enzymatic capacity to reduce nitrate than did the low-protein plants.

Based on available evidence, specific nutritional requirements are apparently under genetic control and crop productivity improvements are feasible through directed selection and breeding programs. As with most breeding programs, genotype × biotic and abiotic environmental interactions on nutrient uptake and utilization have challenged efforts to improve crop species. Exploitation of physiological diversity is a major challenge to plant scientists (Mahon, 1983). However, the most effective progress will occur when mechanisms controlling genotypic variations are better understood.

B. GENETICS OF NUTRITIONAL CHARACTERS

Genetic control of crop plant nutrition is generally complex for macronutrients and relatively simple (single gene control in many cases) for micronutrient efficiency factors (Graham, 1984). With micronutrients, a small increase in the rate of the efficiency process compounded daily into growth in the plant is needed to result in a yield increase during the entire growth cycle (Nye and Tinker, 1977). Many nutrient efficiency factors are specific for a single nutrient. A good historical perspective dealing with the effects of genetic factors on plant nutritional requirements can be found in Vose (1982).

1. Macronutrients

The complicated inheritance of nitrogen-utilization efficiency is governed by four major variables (Vose, 1982): (1) uptake of nitrate or ammonium, (2) level and activity of nitrate reductase, (3) size of the nitrate storage pool, and (4) ability to mobilize and translocate nitrogen to the harvestable product. These processes are all under separate genetic control.

The development of high-yielding cereal cultivars has resulted in important gains in nitrogen retranslocation efficiency from vegetative to reproductive organs during maturation; however, this improvement can be linked with higher harvest indices (as influenced by dwarfing genes with simple major gene inheritance) and to more grain per unit of nitrogen taken up rather than the increased uptake per se (Fisher, 1981). Developing genotypes with a vigorous and efficient juvenile seedling root system and the ability to store nitrogen in tissues for later utilization offers breeders additional challenges (Burns, 1980; Kramer, 1983).

Balko and Russell (1980) recommended avoiding nitrogen deficiency stress for maize parental line development and using intermediate nitrogen fertility levels. Muruli and Paulsen (1981) found that maize cultivars developed for high yield at low nitrogen levels were unresponsive to high nitrogen levels. They suggested selecting unique lines that produce high yields at extreme fertility levels or those that yield highest at a modest nitrogen level. Additional information on genotypes and nitrogen nutrition can be found in Chevalier and Schrader (1977), Burns (1980), and Thibaud and Grignon (1981).

Since both N and P are involved in most plant growth and development processes, separating the direct effects of the nutrient's efficiency from its influence on the genetic determinants of yield has been difficult. The application of *in vivo* techniques in studying metabolic aspects of ion absorption is extremely important (Loughman, 1987).

Gabelman and Gerloff (1983) looked at genetic control of macronutrients in beans, tomatoes, and alfalfa. They found that K efficiency of beans was controlled by a single recessive gene and that differential K remobilization was involved in the mechanism. Progeny tests for P efficiency in beans indicated a wide range of multigenetic controls for P utilization, with additive, dominance, additive $\times$ additive, and dominance $\times$ dominance epistasis generally significant. Potassium efficiency in tomatoes was governed by additive effects, but dominance and epistatic gene effects were important in certain crosses. Calcium efficiency in tomatoes was under multiple genetically controlled mechanisms. Phosphorus, Ca, and Mn accumulation in maize is controlled by genes on chromosome 9 (Naismith *et al.*, 1974). Phosphorus efficiency in rye is controlled by genes on three chromosomes (Graham, 1984). Additional research on phosphorus efficiency in crop plants can be found in Fox (1978) and Lyness (1936).

Caradus (1980) found that grasses were more tolerant of low soil phosphorus conditions than legumes. This difference was attributed to more extensive root systems and their internal ability to regulate phosphorus between the root and shoot.

2. Micronutrients

Micronutrient efficiency characters generally but not always involve single, major gene inheritance. Data in Table VI summarize several micronutrient genetic features. Additional research concerning micronutrient problems can be found in Pope and Munger (1953b), Epstein (1972), Nambiar (1976), and Graham (1984).

3. Salinity

The genetics of salinity tolerance in cereals have been reviewed by Epstein (1980). In general, yield potential and degree of salt tolerance are inversely related (Shannon, 1985). The physiological mechanisms for salt tolerance are not completely understood, and consequently the genetics cannot be resolved satisfactorily. Several genes are involved, but genotype $\times$ environment interactions on yield and vigor confound the picture. Genetic control for plant chloride exclusion in soybeans was due to a single dominant gene (Abel, 1969). Additional information on screening plants for salinity tolerance can be found in Nieman and Shannon (1976).

II. Breeding Techniques

Varying degrees of nutritional adaptation are widespread in nature and the situation suggests that progress for direct selection of differential plant response can be quite

Table VI. Micronutrient Genetic Features of Selected Crops[a]

Micronutrient	Crop	Genetic Explanation	Reference
Iron	Soybean	Single locus, efficiency dominant	Weiss (1943)
Iron	Soybean	Major gene governs uptake, additional genes with quantitative inheritance and additive gene action contribute to efficiency	Fehr (1982) Fehr (1983)
Iron	Tomato	Major gene + minor genes contribute to efficiency	Brown and Wann (1982)
Boron	Celery	Single gene governs efficiency	Pope and Munger (1953a,b)
Copper	Rye	Single gene governs efficiency located on chromosome 5 locus	Graham (1984)
Manganese	Soybean	Multigenic, maternal effects govern toxicity	Brown and Devine (1980)
Manganese	Soybean	Additive gene action, no maternal effects govern toxicity	Devine (1982)

[a]Additional single-gene mutants are cited in Epstein (1972) and Lauchli (1976).

rapid (Antonovics *et al.*, 1967). Consequently, considerable genetic variation must exist on which selection can act, plus strong selection pressures must be available for the evolutionary adaptation to nutritional factors. Both features suggest that breeding for nutritional adaptation is important and is agronomically, economically, and ecologically feasible. Developing cultivars capable of fully exploiting available nutrients or that can be productive at low soil fertility levels and under stress conditions could broaden the cultivation of crops in marginal environments.

A. PLANT GROWTH STAGE EFFECT ON TOLERANCE/EFFICIENCY

As the plant progresses through ontogenetic growth and developmental stages, its sensitivity to nutritional deficiencies and toxicities may change. Sugar beet is highly tolerant to salinity during most of its life cycle, but sensitive during seed germination (Ayers and Hayward, 1948). Other crops tolerate salinity equally well during seed germination and during later growth stages (Bernstein and Hayward, 1958). Salt tolerance of rice (Pearson *et al.*, 1966), tomato (Dumbroff and Cooper, 1974), and wheat and barley (Ayers *et al.*, 1952) diminish after germination, but these crops become more sensitive during the early seedling stage when compared to germination or later vegetative stages. Rice becomes more sensitive again during anthesis (Pearson and Bernstein, 1959), while another study detected no decrease in sensitivity during anthesis and seed set (Kaddah *et al.*, 1975).

Logically, all stages of growth, beginning with germination and progressing through vegetative and into reproductive growth stages, are subjected to shifting priorities within the plant in terms of nutrition, water requirements, and hormonal activity. The shifting priorities may require different mechanisms for handling stress-related problems. Sorghum is thought to undergo three distinct response reactions to acid soil stress based on stage of growth (Duncan, 1988). The impact response phase begins at germination and tolerance to acid soil stress is essentially established by the appearance of the first true leaf from the soil. A preflowering response phase appears to start functioning at panicle initiation (35–45 days postplanting) and a postflowering response phase starts functioning at anthesis and continues during grain filling. Separate tolerance mechanisms are thought to function during each of these response phases. The relationships among mineral nutrients, stress, and plant growth substances are discussed by Moorby and Besford (1983). Climate (humidity, temperature, light, ozone, water) can also influence these relationships (Bates, 1971; Foy, 1983b; Marschner, 1986).

B. SCREENING METHODS

1. Nutrient Cultures

Traditional genetic protocol requires response evaluations of large numbers of segregating progeny in a specified nutritional environment during the developmental program. Used as a preliminary screening, nutrient cultures can rapidly and efficiently identify plants adapted to the specific nutritional constraints of the medium. However, genotypes selected in nutrient culture often have difficulty surviving field stress situations, unless significant effort is made to closely equate greenhouse and growth chamber conditions with field stress conditions (Howeler and Cadavid, 1976; Lafever *et al.*, 1977; Wiersum, 1981; Duncan *et al.*, 1983). Even then, because of the complexity of the stress tolerance mechanisms and the quantitatively inherited nature of overall stress tolerance, field stress survival may still be quite low. Boken (1966) suggested that each genotype should be evaluated over a range of concentrations for the particular limiting factor. Simple solution cultures should rarely be used to select nutrient efficiency factors which operate on some feature of the root–soil interface (Graham, 1984). Efficiency factors operating internal to the root surface may be evaluated in solution, i.e., characteristics of the absorption isotherm, xylem leaching, translocation (mobilization), and nutrient utilization efficiency (carbon fixed per unit of nutrient absorbed). Gerloff and Gabelman (1983) used the solution culture approach for macronutrient efficiency characters in several species. Their technique was relevant to characters controlling efficiency of internal utilization and mobilization of macronutrients rather than uptake processes. They also contend that solution cultures may be used to identify geometrical attributes of the root system. However, root development in solution cultures may be suppressed relative to that in soil and it is doubtful whether this technique can be very effective in selecting useful root system characteristics for breeding programs (Graham, 1984). Solution techniques can identify internally operating micronutrient efficiency factors, elucidate critical nutrient deficiency and toxicity levels, provide information about physiological and genetic mechanisms, and discern relative tolerance to specific mineral toxicities (Clark and Brown, 1980; Foy, 1983a; Gabelman and Gerloff, 1983). Additional discussion of nutrient cultures is presented by Jones (1982), Wilcox (1982), and Wild *et al.* (1987).

2. Soil Cultures

Soil stress requirements should be identical when screening in the field or in pots in greenhouses/growth chambers. Foy (1987) has traditionally utilized Tatum soil in pot experiments to study aluminum toxicity factors in plants. However, pot culture provides uniform soil for each genotype without environmental interference and requires less labor to maintain than solution cultures; however, they also provide an atypically uniform environment. A novel technique whereby root morphogenesis and mineral nutrient capture may be examined in a standardized, nutritionally heterogeneous rooting environment has been described by Campbell and Grime (1989). The method has been used in pot experiments to detect species differences in morphogenetic response to localized nutrient depletion.

With micronutrient efficiency factors, this technique can generally rank genotypes in the same order as field studies (Graham, 1984). To maximize screening pressure, the severity of the deficiency is extremely important and can be manipulated by mixing variable proportions of subsoil and topsoil. One problem associated with assessment of efficiency using seedling performance is that seed nutrient content can significantly affect the growth rate when that specific nutrient is limiting in the soil. For example, manganese concentrations in the seed may vary widely and this variation can be influenced by both environmental and genetic factors.

Caradus and Dunlop (1979) devised a two-stage screening method for phosphorus efficiency in white clover. Stage one consisted of an unreplicated comparison of growth at high and low phosphorus levels. In stage two, replicated phosphorus response curves were obtained for selection of efficient and inefficient genotypes based on stage one data.

3. Quick Tests

Several attempts have been made to devise rapid, short-term juvenile stage (seedling) tests for evaluating stress tolerance. These tests generally involve brief exposure to high concentrations of toxic metals or salts, followed by some quantitative measure of injury or efficiency response. Again, these quick tests target specific nutrients or salts, which probably involve only a small portion of the overall stress tolerance or efficiency mechanisms and correlation with field performance per se has been quite low. These rapid tests are usually conducted at only one specific level of the limiting factor. This approach allows the maximum number of genotypes to be screened with the resources available, but can be counterproductive if large potential yield differences exist among the genotypes (Graham, 1984).

Short-term (2–4 days) bioassay techniques using root development of seedlings have been used to identify soils with Al toxicity and Ca deficiency problems (Karr *et al.*, 1984) and differentiate relative Al tolerance among cultivars and among plant species (Ritchey *et al.*, 1988; Wright *et al.*, 1989). These methods are based on the idea that during juvenile growth stages, plants derive their nutritional needs from the seed and only Al and H^+ toxicity plus Ca and B deficiencies limit seedling root elongation. The Al-pulse technique has been employed by Moore *et al.* (1976). They separated wheat cultivars into four distinct categories of response to Al sensitivity by subjecting each cultivar plus a standard check to several Al concentrations during one 48-hour period, followed by a recovery period of 72 hours in a non-Al toxic solution. The Al concentration that causes irreversible and permanent damage of the root meristem and inhibits further growth was determined. Instead of using the recovery period in normal nutrient solution, a staining procedure with hematoxylin can also be substituted (Howeler, 1987).

4. Cell Cultures

Use of an *in vitro* culture system offers many advantages in the study of plant cell responses to stress (Hasegawa *et al.*, 1984): (1) allows the elimination of all stress responses, except those operating at the cellular level; (2) provides an opportunity to rigorously control the physical environment and nutritional status, parameters which are usually difficult to regulate in intact plants; (3) cells in liquid suspension are fairly uniform ontogenetically and have highly reproducible growth patterns; (4) cells suspended in liquid medium can be exposed uniformly and indefinitely to different stresses; (5) quantitative measurements of stress–response parameters can be made effectively on small cell samples; (6) the cultured cell population is composed of rather uniformly growing cells, and physiological and biochemical changes in response to stress will mimic those same responses associated with growing cells; and (7) provides the ability to obtain cell lines which exhibit varying degrees of tolerance or efficiency (heritable variants that differ in the rate of physiological adjustment to some stress or that differ in their stable level of adjustment). Cell suspension and tissue culture techniques can be used to screen and select genotypes with specific nutritional attributes. Mutagenic agents can be employed to increase the frequency of cells possessing the desired nutritional feature. Desirable cells can then be differentiated into whole plants. The whole plant must have the capability to adjust to the specific stress environment and field screening will provide the ultimate survival or efficiency test. The cell culture environment must equate to the field stress environment (Conner and Meredith, 1985). Aluminum resistance acquired by cultured cells in carrots was retained and inherited in regenerated plants (Ojima and Ohira, 1988).

5. Field Tests

Regardless of which preliminary screening method is used, the final appraisal for stress tolerance and nutrient efficiency evaluations must be conducted under appropriate field conditions. Each tolerant or more efficient plant must cope with environmental pressures and produce a harvestable product. Caradus and Snaydon (1986a) concluded that field tolerance of white clover to soils with low phosphorus status can be accurately identified only by screening in realistic field conditions. Duncan (1988) has devised a dual-phase field screening program for acid soil tolerance improvement in sorghum. Phase one is conducted on a soil with a pH range of 4.4 to 4.8 at 20% Al saturation and suspected manganese toxicity problems. Phase two is conducted on a soil with a pH range of 4.2–4.4 at 50% Al saturation. The dual-phase approach allows for incorporation of desirability, adaptability, and pest resistance characteristics simultaneously as the acid soil tolerance level is improved during the breeding process.

C. CONVENTIONAL TECHNIQUES

Inherent in a breeding program to improve nutritional tolerances or efficiencies is the philosophy that a genetic deficiency exists in the plant rather than a deficiency in the soil (Graham, 1984). Unfortunately, an inverse relationship between yield and efficiency or tolerance to stresses often occurs. Research must then focus on the physiological mechanisms and the accompanying genetics (Wallace *et al.*, 1972) in order to develop meaningful screening and evaluation techniques independent of yield assessment. Nutrient efficiency is considered to be quite low in modern cultivars, probably because breeding programs are conducted in areas of fertile soils with low or zero selection pressure for nutrient efficiency factors. Graham (1984) discusses this "regression hypothesis" (regression from original efficiency in exotic landraces to inefficiency during domestication because of negative selection pressure) which can also be termed "co-evolution during domestication." He argues that improved cultivars have not evolved with efficient mechanisms in the uptake of or the retranslocation of phloem-immobile or variably mobile nutrients (Loneragan *et al.*, 1976; Loneragan, 1978) at the same pace with improvements incurred during domestication, probably because fertile nursery soils were rarely limiting in micronutrients, particularly calcium. The availability of nitrogen and phosphorus and the poor retranslocation of phloem-immobile nutrients such as calcium have become limiting factors as crops have been forced into marginal production areas.

Several criteria can be used to evaluate genetic specificity or efficiency of mineral nutrition in plants (Table VII).

Concentration of nutrients in various plant parts during sequential growth stages becomes important in breeding programs when it influences reproductive growth. Unless the stress-tolerant or efficient genotype produces a harvestable product as competitive or more competitive than existing cultivars, very little progress will have been made. Therefore, breeding programs must be directed toward high organic matter production as well as high nutrient utilization (Bates, 1971; Wallace *et al.*, 1972; Clark and Brown, 1980; Duvick *et al.*, 1981; Saric, 1981, 1983; Foy, 1983a; Gabelman and Gerloff, 1983; Graham, 1984).

Data in Table VIII summarize breeding techniques employed in several programs

Table VII. Criteria for Evaluation of Genetic Specificity of Plant Mineral Nutrition[a]

1. Cytological and anatomical features of cells and subcellular units
2. Plant morphological features
3. Uptake, exudation, translocation (remobilization), distribution, and reutilization of nutrients
4. Nutrient concentration of various plant parts
5. Total content and form of ions
6. Physiological and biochemical processes
7. Total dry or fresh weight
8. Yield and quality parameters

[a]Source: Saric, 1983.

involving nutritional toxicities or efficiencies. Breeding approaches for stress tolerance improvement include backcrossing, progeny testing, and pure line selection (Shannon, 1985). Depending on the breeding system (inbreeding or outbreeding), pedigree, single-seed descent, and population breeding with various selection strategies can be used to develop improved stress-tolerant or nutrient-efficient cultivars. Because of early generation heterogeneity and instability caused by environmental

Table VIII. Breeding Techniques Employed in Selected Breeding Programs for Quantitatively Inherited Nutritional Problems

Crop	Characteristic	Breeding Method	Source
Soybean	Fe-deficiency chlorosis	1. Population improvement—recurrent selection (S_1) 2. Backcrossing of most resistant F_2 progeny	Fehr (1983)
Alfalfa	Al toxicity tolerance	Divergent, recurrent selection	Devine (1977)
Sorghum	Acid soil tolerance complex at 50% Al saturation	1. Pedigree/backcross 2. Genetic male sterile-facilitated recurrent selection	Duncan (1988) Borgonovi *et al.* (1987); Duncan (1988)
Beans	P efficiency	Inbred, backcross line method, evaluation in nutrient solution culture before field testing	Schettini *et al.* (1987)
Alfalfa	Total residual root nitrogen concentration (improvement in N_2 fixation of nondormant genotypes)	Phenotypic recurrent selection	Heichel *et al.* (1989)

interaction, early generation selection may be difficult and could result in loss of segregates with genes for tolerance or efficiency. Environmental effects (genotype × environment interactions) can be minimized by selection under controlled conditions in solution cultures (Blamey *et al.*, 1987) or soil cultures in the greenhouse or growth chamber, by quick screening tests, or by properly stressed field evaluations. Additional practices available to reduce environmental effects during early generations in field situations include family bulking (Duncan, 1988) to assure a maximum number of recombinants, single-seed descent, and use of numerous field sites and stress levels. The acid soil tolerance program in sorghum (Duncan, 1988) involves seven soils at four locations varying in altitude from sea level to 1000 m. Genotypes emerging from the high soil acidity (50% aluminum, less than 5% organic matter) problem soils must survive at least two screenings at each soil site and produce seed for the next generation.

Population breeding with various selection schemes provides an additional effective approach for nutrient toxicity, deficiency, and efficiency improvement (Fehr, 1983; Shannon, 1985; Blum, 1988; Duncan, 1988). Mass selection with various types of progeny testing are designed to accumulate favorable alleles through recurrent selection over generations and should be equally effective if the screening environments are appropriate to allow selection of segregates for tolerance or efficiency. Recurrent selection within families which includes multilocation, replicated evaluations can be used to minimize interactions between stress level and environment as well as those between genotype and phenotype. The reconstitution of advancing generations from donor parents with high evaluation scores for tolerance or efficiency at several locations should lead to quantitative improvement over time.

Standardized Checks

With any breeding program, initial evaluation of various genotypes within species for a specific trait should reveal tolerant and susceptible or efficient and inefficient cultivars which can be used as reference "checks." These checks provide the best method for the breeder to assess progress in any crop improvement endeavor. Adoption of a standard set of checks by breeders working in different environments allows valid comparisons of genotypes for adaptability and stability of performance. This use of standardized checks has been quite beneficial in the monitoring of pathogen pathotype changes and the movement of diseases in specific crops across different environments.

Establishment of reference checks for toxic or deficient nutrient tolerances/susceptibilities and/or nutritional efficiencies/inefficiencies is extremely important when the breeding program is conducted under the proper field stress conditions. Duncan (1988, 1989a) has utilized a tolerant and a susceptible check to monitor progress in breeding sorghum for acid soil tolerance. The checks were selected on the basis of previous research results from Brazil and following genotype evaluations in the southern United States (Duncan, 1981, 1987). Heichel *et al.* (1989) have used non-nitrogen-fixing strains of fall dormant alfalfa to:

1. select for N_2 fixation in agronomically adapted germplasm
2. integrate assessments of N_2 fixation with forage yields in cultivar trials
3. facilitate collection of baseline data on N_2 fixation of cultivars
4. assess the contribution of different N sources to a succeeding crop
5. measure N fertilizer equivalence of legumes and of non-N rotation effects

D. BIOTECHNOLOGICAL TECHNIQUES

The application of genetic engineering techniques to supplement conventional breeding methods offers new opportunities for elucidating quantitatively inherited mechanisms that are involved in nutritional tolerances and efficiencies in crop plants. Data in Figure 3 illustrate how molecular biotechniques can be used to supplement and extend conventional breeding methods (Austin, 1988).

Many molecular biology techniques are now available (Table IX) for systematic studies of tolerance and efficiency relationships. However, the degree of specificity and diversity in the tolerance mechanisms remains to be discerned. These tech-

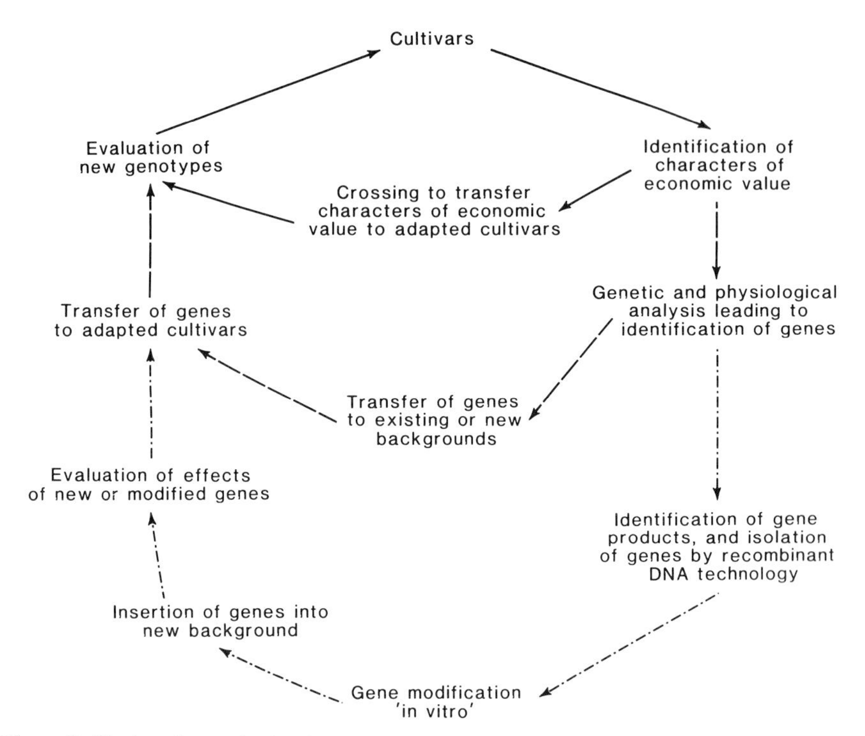

Figure 3. The breeding cycle showing conventional (—), new (---), and proposed (–·–) genetic engineering methods of gene transfer and evaluation. Source: Austin, 1988.

Table IX. Techniques of Molecular Biology[a]

1. Genomic library preparation; cloning isolation of genes
2. Gene detection in genomic libraries
   a. Given the gene protein product
   b. Given the mRNA
   c. Given a dominant gene allele (by transposable elements, if available)
3. Location of genes on chromosomes (mapping)
   a. Given phenotypic gene expression
   b. Given the gene protein product
   c. Given the mRNA
4. Gene sequencing
5. New gene construction, with appropriate control sequences
6. Gene insertion into plant cells; regulation of the plant from these cells
7. Deletion or inactivation of existing genes

[a]Source: Austin, 1988.

niques include isolation and mapping of genes, gene modification *in vitro,* gene reinsertion into plants, gene inactivation, gene cloning at a given locus, and modification of gene expression (Austin, 1988).

With conventional techniques, breeders must rely on recognition of the phenotypic expression of genes. With molecular biology techniques, breeders will be able to recognize particular alleles in segregating progeny by using cDNA probes, by products of the alleles, and by chemical or immunochemical assays. Gene vectors should allow the transfer of genes across species and across generic barriers. Protoplast fusion should provide an avenue for introducing exotic or extrinsic variation via reassociation of nuclei and organelle sequences (gene splicing). New genes can be synthesized. These procedures can be used to investigate the function of modified gene products, especially with proteins and enzymes involved in metabolism (Austin, 1988).

Several genetic engineering techniques which should be useful in plant breeding programs are summarized in the following sections (Gunn and Day, 1986; Day, 1985):

1. Tissue Culture

1. Allows for rapid clonal multiplication of individual genotypes
2. Allows for somaclonal variation (Scowcroft and Larkin, 1988)
3. Several *in vitro* selection agents can be employed following embryo callus regeneration (Duncan, 1989a,b):

Selective Agent	Target Character	Evaluation
Potassium chlorate	Nitrate reductase activity	Hydroponics
Sodium chloride	Salt tolerance	Salt stress in field
Hydroxyproline, polypropaline glycol, polyethylene glycol	Cellular adaptation to osmotic stress	Drought stress in field
Aluminum chloride	Aluminum toxicity	Acid soil field stress

4. Stepwise *in vitro* selection (such as beginning with 100 μg/g $AlCl_3$ and increasing up to 700 μg/g) followed by serial stress screening during the callus passage phases ($AlCl_3 \rightarrow NaCl \rightarrow PPG \rightarrow$ etc.) may be needed before subjecting the regenerants to field evaluation. *In vitro* selective agents can target specific characteristics and, in essence, can be used to "plug in" various pieces of the complex mechanisms involved in stress tolerance per se. The speed and accuracy of using the tissue culture approach compared to making a specific cross in the field with hopes for combining the right parents and the right characters into one package definitely favor the tissue culture system.

2. DNA Probes

1. Uses the ability to detect DNA sequences by their hybridization to probes as a means of classifying plants
2. Useful for detecting the presence of systemic pathogens such as viruses and viroids
3. Useful for detecting the presence of mitochondrial DNA sequences associated with cytoplasmic male sterility

3. Plant Transformations

The extent to which the methods summarized in Table X can be used to effect agronomic changes in plant nutritional tolerances or efficiencies is limited by an inadequate knowledge concerning the physiological and biochemical determinants of the desired traits and the genes involved. Tolerance to specific mineral stresses associated with marginal soils is particularly amenable to cell culture selection (Meredith *et al.*, 1988). Somaclonal variants selected under low phosphorus concentration in suspension cell culture have been used to evaluate red clover for phosphorus efficiency (Bagley and Taylor, 1987). The use of transgenic plants to provide knowledge about basic plant processes will continue to escalate in the future (Hall and DeRose, 1988). The challenge for plant breeders is to capitalize on these advancements. Additional reviews are available (Chaleff, 1983; Larkin, 1985; Duncan and Widholm, 1986; Wilke-Douglas *et al.*, 1986; Beversdorf, 1987; Cohen, 1988; Blum, 1988).

Embryo rescue can be used to overcome abortion of interspecific hybrids be-

Table X. Summary of Methods to Achieve Transformation of Plant Cells[a]

A. Modifying the sexual system
   1. Gametic gene transfer—pollen grains are treated with heavy doses of X-rays to pulverize the chromosomes. The fragmented DNA pairs with homologous egg chromosome segments and becomes incorporated into newly synthesized daughter chromosomes. Success is dependent on occurrence of parthenocarpic development of the embryo.

B. Modifications via asexual methods
   1. Protoplast techniques (parasexual hybridization)
      a. Microinjection of DNA directly into the nucleus of plant protoplasts and fusion of the protoplast with DNA containing liposomes.
      b. Direct gene transfer of DNA into plant protoplasts involving either chemical stimulation of uptake (polyethylene glycol) or a high-voltage pulse (electroporation) which generates transient holes in the protoplast membrane.
      c. Dominant selectable marker genes are quite useful in these techniques to detect a successful transformation.
   2. Introgression via gene transformation vectors
      a. Uses DNA vectors based on the tumor-inducing (Ti) plasmid of the bacterial pathogen *Agrobacterium tumefaciens* and involves transfer of a DNA segment (T-DNA) from the plasmid to the nuclear genome of a susceptible dicotyledonous host cell (insertion of foreign or synthesized genes into a species). The bacterial pathogen infects the wounded plant tissue and causes neoplasia.
      b. Agroinfection or Ti-plasmid-mediated virus infection in which single copies of the virus genomes can escape, replicate, and spread systematically. The assay for successful transfer is extremely sensitive because of amplification of the transferred viral DNA and resulting symptom formation.
      c. Transposons (transposable elements) are segments of DNA which can move around and, when inserted into a particular locus on a chromosome, can disrupt gene expression. Transposon tagging is the principle of using a cloned transposable element as a hybridization probe for the isolation of genes into which the segment has been inserted.
   3. Nonintegrating transformation systems
      a. Geminiviruses replicate by DNA intermediates only and are a potential means for use as plant expression vectors.
      b. Pseudovirions use RNA viruses by the *in vivo* uncoating and efficient expression of foreign mRNAs. Inoculation with pseudovirions may provide a more protective route to introduce selected transcripts directly into the cytoplasm of whole plant cells or isolated protoplasts than inoculation with naked RNAs. They also provide a tool for studying transient gene expression at the translational level in plant molecular biology.
   4. Direction injection of DNA is an alternative approach to achieve cereal transformation without involvement of tissue culture techniques.

[a]Sources: Woolhouse, 1987; Hall and DeRose, 1988.

tween closely related species. Cereal endosperm storage proteins can be evaluated using polyacrylamide gel electrophoresis followed by protein staining. Biosynthetic pathways leading to secondary metabolites (lipids, fatty acids, alkaloids, pigments) can be altered using integrating transformation vectors. Restriction fragment length polymorphisms (RFLPs) can be used in cultivar identification, genetic analysis of economic traits, and breeding methodologies (Beckmann and Soller, 1986). For example, RFLP data can be utilized to formulate graphical genotypes which portray the parental origin and allelic composition throughout the entire genome. This concept would be useful in whole-genome selection for polygenic traits in plant breeding programs (Young and Tanksley, 1989).

III. Physiological Mechanisms

Genotypic variation among plants for nutritional requirements and efficiency of nutrient utilization exists in most crop species. However, identification of specific physiological mechanisms, associated biochemical processes, and resultant anatomical and morphological features is limited. A logical initial step in determining control mechanisms is to ascertain whether inefficiency is associated with acquisition of nutrients from the environment, with distribution within the plant, or with utilization during metabolism (Gerloff and Gabelman, 1983). Good reviews are available for several nutrients (Foy, 1983b; Clark, 1984; Marschner, 1986). Several generalized mechanisms are listed in Table I (see section on acquisition from the environment). Additional specific mechanisms governing nutritional problems in plants are listed in Table XI. Root structure and function (Kramer, 1983), morphological features (particularly with phosphorus) (Gerloff and Gabelman, 1983), and biological processes (Loughman, 1987) responsible for initial transfer of ions across root cell membranes are involved in variability of nutrient acquisition from the soil. Läuchli (1976) investigated ion transport processes as a basis for genotypic differences in nutrient utilization. He concluded that inducible enzyme or protein systems provide plants with a mechanism in terms of nutrient uptake, transport across roots, release to the xylem following initial transfer through cytoplasmic membranes, and specific ion transport capacities in the presence of substrate ions for adapting to fluctuations in ion concentrations in the environment.

Other genotypic differences in nutrient utilization could be attributed to endodermal fine structure, the mechanism of ion transfer from the stele to xylem-conducting nutrients, and enhancement of long-distance transport under nutrient stress (Gerloff and Gabelman, 1983; Marschner, 1986). Variability among genotypes for intensity of nutrient redistribution and reutilization from deficient or senescing plant parts could affect the efficiency of nutrient utilization in growth processes. Weak redistribution could result in nutrient accumulation in old, inactive plant parts rather than utilization in actively growing parts. Excessive redistribution might deplete nutrients in critical plant parts at inopportune times of peak metabolic activity, particularly with phloem-immobile nutrients such as calcium and boron. Heritable differences in efficiency of phosphorus, nitrogen, and potassium can be attributed to utilization of those nutrients in metabolic processes (Gerloff and Gabelman, 1983). Above-average efficiency in utilization during growth and metabolism would be suggested for genotypes producing the highest yields and having the lowest overall nutrient concentrations (dilution effect) (Jarrell and Beverly, 1981) at a given specific amount of absorbed nutrient. Perhaps the greater yields of efficient genotypes resulted from enhanced redistribution of phloem-mobile nutrients such as potassium, phosphorus, and magnesium to terminal growing points and subsequent maintenance of the required nutrient concentrations in the most active regions of the plant (metabolic effectiveness).

Hypotheses regarding mechanisms of toxic metal tolerance (such as aluminum) can be grouped into two major categories based on the site of metal detoxification or immobilization, or on the location of the adaptation to toxic nutrient stress—either in the apoplasm (exclusion or external) or in the symplasm (internal) (Taylor, 1988). Research has failed to produce a clear understanding about the physiological and biochemical basis of metal tolerance in plants. Horst (1983) showed that genotypical manganese toxicity tolerance in cowpea was controlled at the leaf tissue level. Tolerance, or the negative response to a nutrient, probably is due to a combination of both exclusion and internal mechanisms (Table XII) (Taylor, 1988). The plasma membrane and cell wall probably play key exclusionary roles, but their contribution to differential metal tolerance requires additional research. Formation of pH barriers (such as chelating ligands or mucilage) at the root–soil interface are likely involved in the tolerance mechanisms of plants. Internal chelation by carboxylic acids or Al-binding proteins, vacuolar compartmentation, and formation of metal-toxicity-tolerant enzymes are all potentially significant. Until these mechanisms have been determined for each crop species, breeding programs will continue to move forward slowly. Fageria *et al.* (1988) have discussed the aluminum toxicity problem in crop plants in more detail. They categorized plants as:

1. Tolerant and responsive = cultivars that yield well under high levels of Al and respond well to added lime.
2. Tolerant and nonresponsive = cultivars that produce well under high levels of Al, but do not respond to added lime.
3. Susceptible and responsive = cultivars that produce poorly under high Al levels, but respond to added lime.
4. Susceptible and nonresponsive = cultivars that perform poorly under high and low Al levels.

Thung *et al.* (1987) classified beans in similar fashion for phosphorus efficiency and inefficiency.

IV. Summary

The genetics of plant nutritional characteristics are just beginning to be elucidated. Limited knowledge concerning physiological mechanisms, biochemical interactions, chromosome maps, and effective screening techniques is hindering progress.

Table XI. Mechanisms Governing Nutritional Problems in Plants

Nutrient	Crop	Mechanism	Reference
Boron toxicity	Wheat, barley	Exclusion	Nable (1988)
Phosphorus efficiency	White lupine	Releases citrate to free phosphate bound to iron	Gardner *et al.* (1982)
Phosphorus efficiency	Rape	Rhizosphere pH changes	Hedley *et al.* (1982)
Iron efficiency	Rape	Reduce Fe^{3+} to Fe^{2+} at root surface	Brown (1978)
	Sunflower	Proton excretion	Romheld and Marschner (1981)
	White lupine	Release iron-binding ligands such as citrate or other siderophores	Gardner *et al.* (1982); Ripperger and Schreiber (1982)
	White lupine	Resist interferences from other nutrients	Clark (1982)
Salinity tolerance	*Elytrigia*	High affinity for sodium at low concentrations, exclusion at high concentrations	Graham (1984)
	Wheat	Low affinity for sodium, no exclusion at high concentrations	Graham (1984)

Coppery toxicity	*Agrostis*	Ligands (metallothionein proteins) binding	Rauser and Curvetto (1980)
Nitrate uptake	Maize	Hyperpolarization related to functioning of induced NO_3^- uptake system that is a $2NO_3^-/10H^-$ antiport	Thibaud and Grignon (1981)
Calcium efficiency	Tomato	Greater uptake at low concentrations and remobilization	Gabelman and Gerloff (1983)
Phosphorus and potassium efficiency	Beans	Retention in roots and older leaves, remobilization	Gabelman and Gerloff (1983)
Manganese toxicity	Beans	Differential internal tolerance	Kohno and Foy (1983)
Manganese toxicity	Rice, barley, alfalfa, cucumber, pumpkin, tomato	Oxidation of Mn by roots, excretion of oxidized Mn around trichomes (cucumber)	Horiguchi (1987)
Manganese toxicity	Rice	Silicon alleviates toxicity by decreasing uptake and increasing internal tolerance	Horiguchi (1988)
Zinc toxicity	*Agrostis*	Metal inactivation at cell wall	Turner and Marshall (1972)

Table XII. Summary of Aluminum Tolerance Mechanisms[a]

I. Exclusion Mechanisms

A. Cell Wall Immobilization Hypothesis
Aluminum is prevented from entering the symplasm and from reaching sensitive metabolic sites.

1. Selective permeability of the plasma membrane
   a. The membrane acts as a barrier to the movement of Al into the cytosol.
   b. The effectiveness of this barrier is reduced under nonmetabolic conditions.
   c. Differential permeability of the membranes to Al may be involved in Al tolerance.
2. Plant-induced pH barrier at the root–soil interface
   Since Al phytotoxicity is strongly pH dependent, plants which maintain a relatively high rhizosphere or root apoplasm pH could reduce Al toxicity via:
   a. Al hydrolysis or polymerization, resulting in formation of less toxic, soluble ion species.
   b. Formation of sparingly soluble $Al(OH)_3$ at near neutral pH.
   c. Precipitation of $Al(OH)_2 \cdot H_2PO_4$.

   Differences among cultivars for plant-induced pH modifications appear to involve relative uptake of NH_4^+ and NO_3^-. Al-tolerant plants have slower uptake of NH_4^+ and consequently induce a higher rhizosphere pH, whereas Al-sensitive plants rapidly take up NH_4^+ and the rhizosphere pH remains lower.
3. Exudation of chelate ligands
   a. Chelate ligands (citrate, humic acids, organic matter, mucilage at root cap) released into the rhizosphere form stable complexes with Al and reduce or eliminate specific toxic actions of Al in the symplasm.

B. CEC Hypothesis
Cultivar tolerance to Al has been associated with low root cation-exchange capacity via:

1. Selective exclusion of polyvalent cations.;
2. Low CEC reduction of Al binding on exchange sites which may be the first step in ion uptake.
3. Low cation uptake (relative to anions) acidity reduction of the rhizosphere, thereby reducing uptake of Al into the symplasm.

   Concentrations of Al in leaves increased rapidly in some species only after concentrations of Al in roots exceeded root CEC, suggesting that translocation to the shoots increased when root exchange sites were saturated by Al. Since the root tip is a site of growth, of cell wall synthesis, and perhaps the primary site of Al phytotoxicity, synthesis of new cell wall material could provide long-term protection to a sensitive meristem at concentrations of Al sufficient to saturate cell wall binding sites.

II. Internal Mechanisms

A. Chelation Hypothesis (in the cytosol)
Organic acids (such as carboxylic acid or citrate) detoxify Al by:

1. Preventing formation of an Al–ATPase complex.
2. Preventing a direct effect of Al on the ATPase itself.
3. Reduction of Al binding to calmodulin (Haug and Caldwell, 1985).
4. Partially restoring the native structure of calmodulin once an Al–calmodulin complex has been formed.

Table XII. *(Continued)*

5. Preventing the inhibition of a calmodulin-dependent phosphodiesterase.
6. Preventing the inhibition of a calmodulin-stimulated membrane ATPase.
B. Compartmentation Hypothesis (in the vacuole) Isolation of Al in sites which are insensitive to Al toxicity effects.
C. Metal-Binding Protein Hypothesis An inducible, Al-binding protein that detoxifies Al in the cytosol.
D. Evolution of Metal-Tolerant Enzymes Hypothesis Differential enzyme activity (such as metal-tolerant isozymes of acid phosphatase) may function at phytotoxic levels of the metal in intact cells.

[a]Source: Adapted from Taylor, 1988.

The advantages of incorporating nutrient tolerance and efficiency objectives into existing breeding programs were summarized by Graham (1984):

1. Higher yields and lower fertilizer inputs should result in better fertilizer-use efficiency, greater disease resistance, enhanced tolerance to toxicities, better exploitation of subsoil moisture, and elimination of subclinical deficiencies.
2. Higher nutritional quality in terms of essential minerals for both human and animal consumption and hopefully a decline in malnutritional problems worldwide.
3. An increased objectivity and efficiency in breeding programs that are more broadly based in terms of soil constraints and localized problems. A nutritional component contributes to a more integrated approach to problem solving in the field.

As breeding programs move toward the next century under the context of low-input, sustainable agriculture, genetic engineering of plant ideotypes, and nonpollution of the environment, crop nutritional parameters must be addressed. Only mutually beneficial cooperation among many scientific disciplines will answer the difficult questions and lead to sufficient, high-quality food for future generations.

References

Abel, G. H. 1969. Inheritance of the capacity for chloride inclusion and chloride exclusion by soybeans. *Crop Sci.* 9:697–698.

Antonovics, J., J. Lovett, and A. D. Bradshaw. 1967. The evolution of adaptation to nutritional factors in populations of herbage plants. *Isotopes in Plant Nutrition and Physiology. Proc. Symp., IAEA, Vienna, 1966* pp. 549–567.

Austin, R. B. 1988. New opportunities in breeding. *HortScience* 23:41–47.

Ayers, A. D., J. W. Brown, and C. H. Wadleigh. 1952. Salt tolerance of barley and wheat in soil plots receiving several salinization regimes. *Agron. J.* 44:307–310.

Ayers, A. D., and H. E. Hayward. 1948. A method for measuring the effects of soil salinity on seed germination with observations on several crop plants. *Soil Sci. Soc. Am. Proc.* 14:224–226.
Bagley, P. C., and N. L. Taylor. 1987. Evaluation of phosphorus efficiency in somaclones of red clover. *Iowa St. J. Res.* 61:459–480.
Balko, L. G., and W. A. Russell. 1980. Effects of rates of nitrogen fertilizer on maize inbred lines and hybrid progeny. I. Prediction of yield response. *Maydica* 25:65–79.
Barber, S. A. 1976. Efficient fertilizer use. In F. L. Patterson (ed.), Agronomic Research for Food. ASA Spec. Publ. No. 26. Am. Soc. Agron., Madison, Wisconsin. Pp. 13–29.
Bates, T. E. 1971. Factors affecting critical nutrient concentrations in plants and their evaluation: A review. *Soil Sci.* 112:116–130.
Beckmann, J. S., and M. Soller. 1986. Restriction fragment length polymorphisms and genetic improvement of agricultural species. *Euphytica* 35:111–124.
Bernstein, L., and H. E. Hayward. 1958. Physiology of salt tolerance. *Annu. Rev. Plant Physiol.* 9:25–46.
Berry, W. L., and A. Wallace. 1981. Toxicity: The concept and relationship to the dose response curve. *J. Plant Nutr.* 3:13–19.
Beversdorf, W. D. 1987. Application of plant tissue and cell culture to field crops improvement, the up- and down-side. In C. E. Green (ed.), Plant Tissue and Cell Culture. Alan R. Liss, New York. Pp. 359–365.
Bieleski, R. L., and A. Lauchli. 1983. Synthesis and outlook. In A. Lauchli and R. L. Bieleski (eds.), Inorganic Plant Nutrition. Vol. 15B. Springer-Verlag, New York. Pp. 745–755.
Blamey, F. P. C., C. J. Asher, and D. G. Edwards. 1987. Hydrogen and aluminum tolerance. In H. W. Gabelman and B. C. Loughman (eds.), Genetic Aspects of Plant Mineral Nutrition. Nijhoff, Dordrecht, Netherlands. Pp. 173–179.
Blum, A. 1988. Plant Breeding for Stress Environments. CRC Press, Boca Raton, Florida.
Boken, E. 1966. Studies on methods of determining varietal utilization of nutrients. Thesis, Royal Vet. Agric. Coll., Copenhagen.
Borgonovi, R. A., R. E. Schaffert, and G. V. E. Pitta. 1987. Breeding aluminum tolerant sorghums. In L. M. Gourley and J. G. Salinas (eds.), *Sorghum for Acid Soils. Proc. Workshop on Evaluating Sorghum for Tolerance to Al-Toxic Tropical Soils in Latin America, Cali, Colombia, 1984*. Pp. 271–292.
Brown, J. C. 1978. Mechanism of iron uptake by plants. *Plant Cell Environ.* 1:249–257.
Brown, J. C. 1979. Genetic improvement and nutrient uptake in plants. *BioScience* 29:289–292.
Brown, J. C., J. E. Ambler, R. L. Chaney, and C. D. Foy. 1972. Differential responses of plant genotypes to micronutrients. In J. J. Mortvedt, P. M. Giordano, and W. L. Lindsay (eds.), Micronutrients in Agriculture. Soil Sci. Soc. Am., Madison, Wisconsin. Pp. 389–418.
Brown, J. C., and T. E. Devine. 1980. Inheritance of tolerance to manganese toxicity in soybeans. *Agron. J.* 72:898–904.
Brown, J. C., and E. V. Wann. 1982. Breeding for iron efficiency: Use of indicator plants. *J. Plant Nutr.* 5:623–635.
Burns, I. G. 1980. The influence of the spatial distribution of nitrate on the uptake of N by plants: A review and a model for rooting depth. *J. Soil Sci.* 31:155–173.
Butler, G. W., P. C. Barclay, and A. C. Glenday. 1962. Genetic and environmental differences in the mineral composition of ryegrass herbage. *Plant Soil* 16:214–228.
Campbell, B. D., and J. P. Grime. 1989. A new method of exposing developing root systems to controlled patchiness in mineral nutrient supply. *Ann. Bot.* 63:395–400.
Caradus, J. R. 1980. Distinguishing between grass and legume species for efficiency of phosphorus use. *J. Agric. Res.* 23:75–81.
Caradus, J. R., and J. Dunlop. 1979. Screening white clover plants for efficient phosphorus use. In A. R. Ferguson, R. L. Bieleski, and I. S. Ferguson (eds.), Plant Nutrition 1978. Proc. 8th Int. Colloq. on Plant Analysis and Fertilizer Problems, Auckland, 1978. DSIR Inf. Ser. No. 134. Gov. Printer, Wellington, New Zealand. Pp. 75–82.

Caradus, J. R., and R. W. Snaydon. 1986a. Response to phosphorus of populations of white clover. 3. Comparison of experimental techniques. *N.Z. J. Agric. Res.* 29:169–178.

Caradus, J. R., and R. W. Snaydon. 1986b. Plant factors influencing phosphorus uptake by white clover from solution culture. I. Population differences. *Plant Soil* 93:153–163.

Chaleff, R. S. 1983. Isolation of agronomically useful mutants from plant cell cultures. *Science* 219:676–682.

Chevalier, P., and L. E. Schrader. 1977. Genotypic differences in nitrate absorption and partitioning of N among plant parts in maize. *Crop Sci.* 17:897–901.

Chichester, F. W. 1981. Selecting for nutrient use efficiency within forage grass species. I. Development of a screening system. *J. Plant Nutr.* 4:231–236.

Clark, R. B. 1982. Plant genotype differences to uptake, translocation, accumulation, and use of mineral elements. In M. R. Saric (ed.), Genetic Specificity of Mineral Nutrition in Plants. Vol. 13. Serb. Acad. Sci. Arts, Belgrade. Pp. 41–55.

Clark, R. B. 1984. Physiological aspects of calcium, magnesium, and molybdenum deficiencies in plants. In F. Adams (ed.), Soil Acidity and Liming. 2nd Ed., Agron. Monogr. No. 12. Am. Soc. Agron., Madison, Wisconsin. Pp. 99–170.

Clark, R. B., and J. C. Brown. 1980. Role of the plant in mineral nutrition as related to breeding and genetics. In L. S. Murphy, L. F. Welch, and E. C. Doll (eds.), Moving Up the Yield Curve: Advances and Obstacles. ASA Spec. Publ. No. 39. Am. Soc. Agron., Madison, Wisconsin. Pp. 45–70.

Cohen, J. I. 1988. Models for integrating biotech into crop improvement programs. *Biotechnology* 6:387–392.

Conner, A. J., and C. P. Meredith. 1985. Simulating the mineral environment of aluminum toxic soils in plant cell culture. *J. Exp. Bot.* 36:870.

Day, P. R. 1985. Crop improvement: Breeding and genetic engineering. *Philos. Trans. R. Soc. London, Ser. B* 310:193–200.

Devine, T. E. 1977. Aluminum and manganese toxicities in legumes. In M. J. Wright (ed.), Plant Adaptation to Mineral Stress in Problem Soils. Cornell Univ. Press, Ithaca, New York. Pp. 65–72.

Devine, T. E. 1982. Genetic fitting of crops to problem soils. In M. N. Christiansen and C. F. Lewis (eds.), Breeding Plants for Less Favorable Environments. Wiley (Interscience), New York. Pp. 143–173.

Dumbroff, E. B., and A. W. Cooper. 1974. Effects of salt stress applied in balanced nutrient solutions at several stages during growth of tomato. *Bot. Gaz.* 135:219–224.

Duncan, D. R., and J. M. Widholm. 1986. Cell selection for crop improvement. *Plant Breed. Rev.* 4:154–173.

Duncan, R. R. 1981. Variability among sorghum genotypes for uptake of elements under acid soil field conditions. *J. Plant Nutr.* 4:21–32.

Duncan, R. R. 1987. Sorghum genotype comparisons under variable acid soil stress. *J. Plant Nutr.* 10:1079–1088.

Duncan, R. R. 1988. Sequential development of acid soil tolerant sorghum genotypes under field stress conditions. *Commun. Soil Sci. Plant Anal.* 19:1295–1305.

Duncan, R. R. 1989a. Strategies in breeding sorghum for improved soil stress tolerance and plant nutritional traits. *Proc. Bienn. Grain Sorghum Research and Utilization Conf., Sorghum Improvement Conf. of North America and Grain Sorghum Producers Association, Abernathy, Tex.* 16:208–216.

Duncan, R. R. 1989b. Sorghum breeding for stress tolerance: Tissue culture and implications in Africa. In D. Gambourg (ed.), *Proc. 3rd IBP Network Conf., Nairobi.* pp. 25–34.

Duncan, R. R., R. B. Clark, and P. R. Furlani. 1983. Laboratory and field evaluations of sorghum for response to aluminum and acid soil. *Agron. J.* 75:1023–1026.

Duvick, D. N., R. A. Kleese, and N. M. Frey. 1981. Breeding for tolerance of nutrient imbalances and constraints to growth in acid, alkaline and saline soils. *J. Plant Nutr.* 4:111–129.

Epstein, E. 1972. Mineral Nutrition of Plants: Principles and Perspectives. Wiley, New York.

Epstein, E. 1980. Responses of plants to saline environments. In D. W. Raines, R. C. Valentine, and A. Hollaender (eds.), Genetic Engineering of Osmoregulation. Plenum, New York. Pp. 7–21.
Fageria, N. K., V. C. Baligar, and R. J. Wright. 1988. Aluminum toxicity in crop plants. *J. Plant Nutr.* 11:303–319.
Falconer, D. S. 1960. Introduction to Quantitative Genetics. Ronald Press, New York.
Fehr, W. R. 1982. Control of iron deficiency chlorosis in soybeans by plant breeding. *J. Plant Nutr.* 5:611–621.
Fehr, W. R. 1983. Modification of mineral nutrition in soybeans by plant breeding. *Iowa St. J. Res.* 57:393–407.
Fisher, R. A. 1981. Optimizing the use of water and nitrogen through breeding of crops. *Plant Soil* 58:249–278.
Fox, R. H. 1978. Selection for phosphorus efficiency in corn. *Commun. Soil Sci. Plant Anal.* 9:13–37.
Foy, C. D. 1983a. Plant adaptation to mineral stress in problem soils. *Iowa St. J. Res.* 57:339–354.
Foy, C. D. 1983b. The physiology of plant adaptation to mineral stress. *Iowa St. J. Res.* 57:355–391.
Foy, C. D. 1987. Acid soil tolerance of two wheat cultivars related to soil pH, KCl-extractable aluminum and degree of aluminum saturation. *J. Plant Nutr.* 10:609–623.
Gabelman, W. H., and G. C. Gerloff. 1983. The search for and interpretation of genetic controls that enhance plant growth under deficiency levels of a macronutrient. *Plant Soil* 72:335–350.
Gabelman, W. H., G. C. Gerloff, T. Schettini, and R. Coltman. 1986. Genetic variability in root systems associated with acquisition and use. *HortScience* 21:971–973.
Gardner, W. K., D. G. Parberry, and D. A. Barker. 1982. The acquisition of phosphorus by *Lupinus albus. Plant Soil* 68:19–32.
Gerloff, G. C., and W. H. Gabelman. 1983. Genetic basis of inorganic plant nutrition. In A. Läuchli and R. L. Bieleski (eds.), Inorganic Plant Nutrition. Springer-Verlag, New York. Pp. 453–480.
Graham, R. D. 1984. Breeding for nutritional characteristics in cereals. *Adv. Plant Nutr.* 1:57–102.
Gunn, R. E., and P. R. Day. 1986. *In vitro* culture in plant breeding. In L. A. Withers and P. G. Alderson (eds.), *Proc. 41st Conf. in Agricultural Science on Plant Tissue Culture and Its Agricultural Implications, 1984*. Butterworth, Washington, D.C. Pp. 313–336.
Hall, T. C., and R. T. DeRose. 1988. Transformation of plant cells. *Applications of Plant Cell and Tissue Culture, Ciba Found. Symp.* No. 137, pp. 123–143.
Hasegawa, P. M., R. A. Bressan, S. Handa, and A. K. Handa. 1984. Cellular mechanisms of tolerance to water stress. *HortScience* 19:371–382.
Haug, A. R., and C. R. Caldwell. 1985. Aluminum toxicity in plants: The role of the root plasma membrane and calmodulin. In J. B. St. Johns, E. Berlin, and P. C. Jackson (eds.), Frontiers of Membrane Research in Agriculture. Rowman & Allanheld, Totowa, New Jersey. Pp. 359–381.
Hedley, M. J., P. H. Nye, and R. E. White. 1982. Plant-induced changes in the rhizosphere of rape (*Brassica napus* var. Emerald) seedlings. II. Origin of the pH change. *New Phytol.* 91:31–44.
Heichel, G. H., D. K. Barnes, C. P. Vance, and C. C. Sheaffer. 1989. Dinitrogen fixation technologies for alfalfa improvement. *J. Prod. Agric.* 2:24–32.
Hoener, I. R., and E. E. DeTurk. 1938. The absorption and utilization of nitrate nitrogen during vegetative growth by Illinois high protein and Illinois low protein corn. *J. Am. Soc. Agron.* 30:232–243.
Horiguchi, T. 1987. Mechanism of manganese toxicity and tolerance of plants. II. Deposition of oxidized manganese in plant tissues. *Soil Sci. Plant Nutr.* 33:595–606.
Horiguchi, T. 1988. Mechanism of manganese toxicity and tolerance of plants. IV. Effects of silicon on alleviation of manganese toxicity of rice plants. *Soil Sci. Plant Nutr.* 34:65–73.
Horst, W. J. 1983. Factors responsible for genotypic manganese tolerance in cowpea (*Vigna unguiculata*). *Plant Soil* 72:213–218.
Hovin, A. W., T. L. Tew, and R. E. Stucker. 1978. Genetic variability for mineral elements in reed canarygrass. *Crop Sci.* 18:423–427.
Howeler, R. H. 1987. Effective screening techniques for tolerance to aluminum toxicity. In L. M. Gourley and J. G. Salinas (eds.), *Sorghum for Acid Soils. Proc. Workshop on Evaluating Sorghum for Tolerance to Al-Toxic Tropical Soils in Latin America, Cali, Colombia, 1984*. Pp. 173–186.

Howeler, R. H., and L. F. Cadavid. 1976. Screening of rice cultivars for tolerance to Al-toxicity in nutrient solutions as compared to a field screening method. *Agron. J.* 68:551–555.

Jarrell, W. M., and R. B. Beverly. 1981. The dilution effect in plant nutrition studies. *Adv. Agron.* 34:197–224.

Jones, J. B., Jr. 1982. Hydroponics: Its history and use in plant nutrition studies. *J. Plant Nutr.* 5:1003–1030.

Kaddah, M. T., W. F. Lehman, B. D. Meek, and F. E. Robinson. 1975. Salinity effects on rice after the boot stage. *Agron. J.* 67:436–439.

Karr, M. C., J. Coutinho, and J. L. Ahlrichs. 1984. Determination of aluminum toxicity in Indiana soils by petri dish bioassay. *Proc. Indiana Acad. Sci.* 93:85–88.

Kohno, Y., and C. D. Foy. 1983. Differential tolerance of bush bean cultivars to excess manganese in solution and sand culture. *J. Plant Nutr.* 6:877–893.

Kramer, D. 1983. Genetically determined adaptations in roots to nutritional stress: Correlation of structure and function. *Plant Soil* 72:167–173.

Lafever, H. N., L. G. Campbell, and C. D. Foy. 1977. Differential response of wheat cultivars to Al. *Agron. J.* 69:563–568.

Larkin, P. J. 1985. *In vitro* culture and cereal breeding. In S. W. J. Bright and M. G. K. Jones (eds.), Cereal Tissue and Cell Culture. Nijhoff/Junk, The Hague. Pp. 273–296.

Läuchli, A. 1976. Genotypic variation in transport. In U. Lüttge and M. G. Pitman (eds.), Transport in Plants. Encyclopedia of Plant Physiology, Vol. 2B. Springer-Verlag, Berlin. Pp. 372–393.

Little, R. 1988. Plant soil interactions at low pH problem solving—The genetic approach. *Commun. Soil Sci. Plant Anal.* 19:1239–1257.

Loneragan, J. F. 1978. The physiology of plant tolerance to low phosphorus availability. In G. A. Jung (ed.), Crop Tolerance to Suboptimal Land Conditions. ASA Spec. Publ. No. 32. Am. Soc. Agron., Madison, Wisconsin. Pp. 329–343.

Loneragan, J. F., K. Snowball, and A. D. Robson. 1976. Remobilization of nutrients and its significance in plant nutrition. In I. F. Wardlaw and J. B. Passioura (eds.), Transport and Transfer Processes in Plants. Academic Press, New York. Pp. 463–469.

Loughman, B. C. 1987. The application of *in vivo* techniques in the study of metabolic aspects of ion absorption in crop plants. In H. W. Gabelman and B. C. Loughman (eds.), Genetic Aspects of Plant Mineral Nutrition. Nijhoff, The Hague. Pp. 269–280.

Lyness, A. S. 1936. Varietal differences in the phosphorus feeding capacity of plants. *Plant Physiol.* 11:665–688.

Mahon, J. D. 1983. Limitations to the use of physiological variability in plant breeding. *Can. J. Plant Sci.* 63:11–21.

Marschner, H. 1986. Mineral Nutrition of Higher Plants. Academic Press, London.

Meredith, C. P., A. J. Conner, and T. M. Schettini. 1988. The use of cell selection to obtain novel plant genotypes resistant to mineral stresses. *Iowa St. J. Res.* 62:523–535.

Millikan, C. R. 1961. Plant varieties and species in relation to the occurrence of deficiencies and excesses of certain nutrient elements. *J. Aust. Inst. Agric. Sci.* 27:220–233.

Moorby, J., and R. T. Besford. 1983. Mineral nutrition and growth. In A. Läuchli and R. L. Bieleski (eds.), Inorganic Plant Nutrition. Encyclopedia of Plant Physiology, New Series, Vol. 15B. Springer-Verlag, New York. Pp. 481–527.

Moore, D. P., W. E. Kronstad, and R. J. Metzger. 1976. Screening wheat for aluminum tolerance. In M. J. Wright (ed.), Plant Adaptation to Mineral Stress in Problem Soils. Cornell Univ. Press, Ithaca, New York. Pp. 287–295.

Muruli, B. I., and G. M. Paulsen. 1981. Improvement of nitrogen use efficiency and its relationship to other traits in maize. *Maydica* 26:63–73.

Nable, R. O. 1988. Resistance to boron toxicity amongst several barley and wheat cultivars: A preliminary examination of the resistance mechanism. *Plant Soil* 112:45–52.

Naismith, R. W., M. W. Johnson, and W. I. Thomas. 1974. Genetic control of relative calcium, phosphorus, and manganese accumulation on chromosome 9 in maize. *Crop Sci.* 14:845–849.

Nambiar, E. K. S. 1976. Genetic differences in the copper nutrition of cereals. II. Genotypic differences in response to copper in relation to copper, nitrogen, and other mineral contents of plants. *Aust. J. Agric. Res.* 27:465–477.

Nieman, R. H., and M. C. Shannon. 1976. Screening plants for salinity tolerance. In M. J. Wright (ed.), Plant Adaptation to Mineral Stress in Problem Soils. Cornell Univ. Press, Ithaca, New York. Pp. 359–367.

Nye, P. H. 1973. The relation between the radius of a root and its nutrient absorbing power: Some theoretical considerations. *J. Exp. Bot.* 24:783–786.

Nye, P. H., and P. B. Tinker. 1977. Solute Movement in the Soil–Root System. Univ. of California Press, Berkeley, California.

Ojima, K., and K. Ohira. 1988. Aluminum-tolerance and citric acid release from a stress-selected cell line of carrot. *Commun. Soil Sci. Plant Anal.* 19:1229–1236.

Pearson, G. A., A. D. Ayers, and D. L. Eberhard. 1966. Relative salt tolerance of rice during germination and early seedling development. *Soil Sci.* 102:151–156.

Pearson, G. A., and L. Bernstein. 1959. Salinity effects at several growth stages of rice. *Agron. J.* 51:654–657.

Pope, D. T., and H. M. Munger. 1953a. Heredity and nutrition in relation to magnesium deficiency chlorosis in celery. *Proc. Am. Soc. Hortic. Sci.* 61:472–480.

Pope, D. T., and H. M. Munger. 1953b. The inheritance of susceptibility to born deficiency in celery. *Proc. Am. Soc. Hortic. Sci.* 61:481–486.

Rauser, W. E., and N. R. Curvetto. 1980. Metallothionein occurs in roots of *Agrostis* tolerant to excess copper. *Nature (London)* 287:563–564.

Reuter, D. J., and J. B. Robinson (eds.). 1986. Plant Analysis: An Interpretation Manual. Inkata Press, Melbourne.

Ripperger, H., and K. Schreiber. 1982. Nicotianamine and analogous amino acids, endogenous iron carriers in higher plants. *Heterocycles* 17:447–461.

Ritchey, K. D., V. C. Baligar, and R. J. Wright. 1988. Wheat seedling response to soil acidity and implications for subsoil rooting. *Commun. Soil Sci. Plant Anal.* 19:1285–1293.

Romheld, V., and H. Marschner. 1981. Iron deficiency stress induced morphological and physiological changes in root tips of sunflower. *Physiol. Plant.* 53:354–360.

Rorison, I. H. (ed.). 1969. Ecological Aspects of the Mineral Nutrition of Plants. Ecol. Soc. Sheffield, 1968. Blackwell, Oxford.

Saric, M. R. 1981. Genetic specificity in relation to plant mineral nutrition. *J. Plant Nutr.* 3:743–766.

Saric, M. R. 1983. Theoretical and practical approaches to the genetic specificity of mineral nutrition of plants. *Plant Soil* 72:137–150.

Schettini, T. M., W. H. Gabelman, and G. C. Gerloff. 1987. Incorporation of phosphorus efficiency from exotic germplasm into agriculturally adapted germplasm of common bean (*Phaseolus vulgaris* L.). *Plant Soil* 99:175–184.

Scowcroft, W. R., and P. J. Larkin. 1988. Somaclonal variation. *Applications of Plant Cell and Tissue Culture, Ciba Found. Symp.* No. 137, pp. 21–35.

Seetharama, N., K. R. Krishna, T. J. Rego, and J. R. Burford. 1987. Prospects for sorghum improvement for phosphorus efficiency. In L. M. Gourley and J. G. Salinas (eds.), *Sorghum for Acid Soils. Proc. Workshop on Evaluating Sorghum for Tolerance to Al-Toxic Tropical Soils in Latin America, Cali, Colombia, 1984*. Pp. 229–249.

Shannon, M. C. 1985. Principles and strategies in breeding for higher salt tolerance. *Plant Soil* 89:227–241.

Stratton, S. D., and D. A. Sleper. 1979. Genetic variation and interrelationships of several minerals in orchardgrass herbage. *Crop Sci.* 19:477–481.

Tanaka, A. 1980. Physiological aspects of productivity in field crops. *Proc. Symp. on Potential Productivity of Field Crops under Different Environments, IRRI, Los Banos, Philippines* pp. 61–80.

Tanaka, A., and Y. Hayakawa. 1975. Comparison of tolerance to soil acidity among crop plants. II. Tolerance to high levels of aluminum and manganese. *J. Sci. Soil Manure, Jpn.* 46:19–25.

Taylor, G. J. 1988. The physiology of aluminum tolerance in higher plants. *Commun. Soil Sci. Plant Anal.* 19:1179–1194.

Thibaud, J. B., and C. Grignon. 1981. Mechanism of nitrate uptake in corn roots. *Plant Sci. Lett.* 22:279–289.

Thung, M., J. Ortega, and O. Erazo. 1987. Breeding methodology for phosphorus efficiency and tolerance to aluminum and manganese toxicities for beans (*Phaseolus vulgaris* L.). In L. M. Gourley and J. G. Salinas (eds.), *Sorghum for Acid Soils. Proc. Workshop on Evaluating Sorghum for Tolerance to Al-Toxic Tropical Soils in Latin America, Cali, Colombia, 1984*. Pp. 197–211.

Turner, R. G., and C. Marshall. 1972. The accumulation of zinc by subcellular fractions of roots of *Agrostis tenuis* Sibth. in relation to zinc tolerance. *New Phytol.* 71:671–676.

Vose, P. B. 1963. Varietal differences in plant nutrition. *Herb. Abstr.* 33:1–13.

Vose, P. B. 1982. Effects of genetic factors on nutritional requirements of plants. In P. B. Vose and S. G. Blixt (eds.), Crop Breeding: A Contemporary Basis. Pergamon, Oxford. Pp. 67–114.

Wallace, D. H., J. L. Ozbun, and H. M. Munger. 1972. Physiological genetics of crop yield. *Adv. Agron.* 24:97–146.

Wegrzyn, V. A., R. R. Hill, Jr., and D. E. Baker. 1980. Soil–fertility–crop genotype associations and interactions. *J. Plant Nutr.* 2:607–627.

Weiss, M. G. 1943. Inheritance and physiology of efficiency in iron utilization in soybeans. *Genetics* 28:253–268.

Wiersum, L. K. 1981. Problems in soil fertility characterization by means of plant nutrient requirements. *Plant Soil* 61:259–567.

Wilcox, G. E. 1982. The future of hydroponics as a research and plant production method. *J. Plant Nutr.* 5:1031–1038.

Wild, A., L. H. P. Jones, and J. H. Macduff. 1987. Uptake of mineral nutrients and crop growth: The use of flowing nutrient solutions. *Adv. Agron.* 41:171–219.

Wilke-Douglas, M., L. Perani, S. Radke, and M. Bossert. 1986. The application of recombinant DNA technology toward crop improvement. *Physiol. Plant.* 68:560–565.

Woolhouse, H. W. 1987. New plants and old problems. *Ann. Bot.* 60:189–198.

Wright, R. J., V. C. Baligar, K. D. Ritchey, and S. F. Wright. 1989. Influence of soil solution aluminum on root elongation of wheat seedlings. *Plant Soil* 113:294–298.

Young, N. D., and S. D. Tanksley. 1989. Restriction fragment length polymorphism maps and the concept of graphical genotypes. *Theor. Appl. Genet.* 77:95–101.

2

Ion Absorption and Utilization: The Cellular Level

ANTHONY D. M. GLASS

It is universally accepted that all plants require a minimum of sixteen chemical elements (C, H, O, N, S, P, K, Ca, Mg, Fe, B, Mn, Cu, Zn, Mo, and Cl) for normal growth and development. In addition to this group of "essential" elements, however, many other elements have been proposed as required for particular plant species. These include Si for diatoms (Werner and Roth, 1983) and *Equisetum arvense* (Chen and Lewin, 1969), Na^+ for *Atriplex vesicaria* and, indeed, for all C_4 species (Brownell, 1965, 1979), Se for the marine diatom *Thalassiosira pseudonana* (Price *et al.*, 1987), and Ni for all higher plants (Brown *et al.*, 1987). Clearly, some of the more recent additions may prove to be universally required, in which case the traditional group of sixteen essential elements will need to be expanded correspondingly.

The required elements serve diverse roles as structural components of the organic constituents of plants (C, H, O, N, S, and P), enzyme activators (K, Ca, Mg, and Mn), and redox reagents (Fe, Cu, and Mo), as well as functions that are, at present,

Crops as Enhancers of Nutrient Use

still poorly defined (e.g., B and Cl). Many of the elements serve several functions. For example, K is an important osmotic agent, responsible for cell enlargement (Haschke and Lüttge, 1975), stomatal opening (Humble and Hsiao, 1970), and many rapid movements of plant parts (Satter *et al.*, 1974); it is an activator of more than 50 enzymes (Evans and Wildes, 1971) and, in addition, it provides a critical ionic environment, within the cytoplasm, for protein synthesis (Leigh and Wyn Jones, 1984). Quantitative requirements for the essential elements reflect their metabolic roles. Those required in large amounts (C, H, O, N, S, P, K, Ca, and Mg) are referred to as the macronutrients. For example, K and N may each constitute between 20 and 50 g kg^{-1} of plant dry weight. By contrast Mo (one of the micronutrients) is toxic to most plant species at tissue concentrations approaching 1 g kg^{-1} by dry weight (Marschner, 1986).

A most characteristic feature of living organisms, and particularly of plants, is their capacity for selective accumulation of particular elements (and exclusion of others) to create the unique chemical milieu in which life processes can occur. This has involved the evolution of cellular mechanisms for ion accumulation, exclusion, and compartmentation as well as whole-plant adaptations and morphological/developmental strategies to ensure procurement of adequate supplies of the inorganic nutrients. Many of these are common to all plant species, particularly those mechanisms which operate at the cellular level, but some are unique to particular species and even subspecies.

Soil solution concentrations of the required elements may be so high in particular areas that they prove toxic for most plant species. Only those species or genotypes which possess specific exclusion or detoxifying mechanisms can exploit such habitats. Serpentine soils, for example, may contain as much as 360 g MgO kg^{-1} (Brooks, 1987) as well as potentially toxic levels of Ni, Cr, and Co. The flora of serpentine soils has been studied extensively and shown to contain endemic species as well as physiotypes of nonserpentine species which can grow as well on serpentine as on nonserpentine soils. More commonly, certain of the essential elements (particularly N or P) may be present in such low concentrations, or be otherwise unavailable because of soil pH (e.g., Fe) or other factors, that they severely limit plant growth and crop productivity (Dudal, 1976). Data in Table I provides estimates of phosphate concentrations for a large number of agricultural soils. Approximately 60% of the 149 soils analyzed contained phosphate concentrations below 3 μM in their soil solution (Reisenauer, 1966).

It is not surprising, therefore, that some authorities estimate that greater than 20% of the world's land mass may impose some form of mineral stress upon its vegetation (Dudal, 1976). These stresses, one presumes, must have represented the selection forces responsible for the evolution of genotypic differences in mineral nutrition. Species and subspecies differ markedly in their capacity to tolerate low nutrient stresses (Vose, 1963). These differences have been documented for over a century (see Vose, 1963, for a review of the history of this topic), yet the cellular and subcellular mechanisms responsible for the observed differences have not been defined. Intensive physiological and biochemical investigations into underlying

Table I. Phosphorus Concentrations (as PO_4) in Soil Solutions from 149 Soils[a]

Concentration (μM)	Fraction of Samples (%)
0– 0.97	25.5
1– 1.93	18.8
1.97– 3.2	16.8
3.25– 4.83	12.1
4.87– 6.45	2.7
6.48– 8.06	2.0
8.09– 9.67	4.0
9.71–12.90	6.0
12.93–16.1	4.0
>16.1	8.1

[a]Source: Reisenauer, 1966.

mechanisms of solute transport across the plasma membranes and tonoplasts of selected model systems (such as barley, maize, oat, and beet root) have tended to confirm the essential unity of processes. Comparative studies of genotypic differences in nutrient uptake and utilization efficiency have been common at the descriptive level, but lacking in details of mechanisms. As a consequence, we are unable to account for observed genotypic differences at a cellular or subcellular level.

In this chapter, cellular mechanisms involved in the acquisition and utilization of inorganic nutrients by plant roots will be reviewed. Intrinsic and extrinsic factors which influence or regulate nutrient acquisition will be discussed, and wherever possible genotypic differences and opportunities for comparative studies will be emphasized.

I. Ion Absorption by Roots

A. ION FLUXES IN SOILS

Apart from C, H, and O, which are obtained from CO_2 and H_2O, all other required elements are obtained in charged (ionic) form from soil solution. The substantial transpirational flux of water, from soil to air via the plant, is responsible for bringing significant quantities of some ions to the root surface. The extent of this mechanism of delivery is clearly a function of the bulk flow (convective flux) of soil solution (J_v) and the concentration of ions in solution. The solute flux to the root surface is given by

$$J_s = J_v C \tag{1}$$

where:

J_s = solute flux to the root (mol m^{-2} s^{-1});
J_v = convective flux to the root (m^3 m^{-2} s^{-1} = m s^{-1}; and
C = concentration of a particular ion (mol m^{-3}).

When transpiration rates are low, or when soil solution concentrations are dilute, J_v makes little contribution to the delivery of ions. Under such conditions, zones of depletion develop around the root surface and the transfer of ions across these regions becomes diffusion limited. Data in Table II (from Barber, 1984) indicate the relative importance of mass flow, diffusion, and root interception for the provision of the major inorganic nutrients to a maize root. Evidently, mass flow is capable of supplying the bulk of total nitrogen requirements (as NO_3^-) and of providing sulfur (as SO_4^{2-}), calcium, and magnesium in excess of requirements. Hence a diffusion gradient away from the root may develop for the latter nutrients. For potassium and phosphate, diffusion is the principal mechanism for supply to the root. The case study developed by Barber for this maize crop should not be viewed as immutable, however. When nutrients are in limited supply, plants adopt a variety of strategies to optimize nutrient acquisition and it is likely that the importance of interception increases under such conditions. These strategies are outlined below, and include morphological as well as cellular (biochemical) adaptations.

B. MORPHOLOGICAL FACTORS

Clearly, the quantities of different inorganic ions absorbed by a plant's root system will be a function of root biomass, specific uptake rate (uptake per mass or uptake per surface area), and morphology. Hence, when subjected to low nutrient conditions plants may respond by increasing or modifying these characteristics. Genotypic variation among species or cultivars, attributable to differences in miner-

Table II. Relative Importance of Mass Flow, Diffusion, and Root Interception in Providing Inorganic Nutrients to Maize Roots[a]

		Approximate Amount (%) Supplied by:		
Nutrient	Amount Needed for 9500 kg Grain ha^{-1}	Mass Flow	Diffusion	Root Interception
Nitrogen	190	79	20	1
Phosphorous	40	5	93	2
Potassium	195	18	80	2
Calcium	40	375	0	150
Magnesium	45	222	0	33
Sulfur	22	295	0	5

[a]Source: Barber, 1984.

al nutrition, may therefore be due to differences in one or all of these characteristics as well as to differences in internal utilization of absorbed nutrients.

Among cultivars of crop species there is abundant information to demonstrate that high rates of ion absorption are commonly associated with large root systems (Vose, 1963; Perby and Jensen, 1983). However, under conditions of high soil fertility root size becomes less critical. Indeed, root : shoot ratios have progressively declined during the development of high-yielding cereals (Evans and Dunstone, 1970). It might be anticipated that these changes would be accompanied by increased specific uptake rates, and there is evidence for sulfate and nitrate absorption by maize hybrids developed over a 45-year period (Cacco *et al.*, 1983) and for K^+ uptake by barley and wheat cultivars (Siddiqi and Glass, 1983; Woodend, 1986) that this is true.

A well-documented response to inadequacies of many inorganic nutrients, particularly N, P, and K, is the increase of root : shoot ratios (Barber, 1979; Siddiqi and Glass, 1983; Marschner, 1986). The barley cultivar 'Fergus' (Table III) increased this ratio from 0.59 to 0.87 at harvest 1 (2-week-old plants) as ambient $[K^+]$ decreased from 50 to 5 μM. Note also the ontogenetic pattern; root : shoot ratios decreased with increasing age.

Genotypic differences in root : shoot ratios are evident among species and cultivars of crop plants (Chapin, 1980; Gerloff, 1976). Data in Table III, for the efficient and inefficient barley cultivars Fergus and 'Excelsior,' show that at all level of K stress, Fergus had higher root : shoot ratios (Siddiqi and Glass, 1983). According to Chapin (1980), rapidly growing species from nutrient-rich habitats typically increase root : shoot ratios when deprived of inorganic nutrients. However, species from nutrient-poor habitats may already possess high ratios and have little flexibility to increase this further. On the basis of published data for sunflowers, Gutschick (1987) estimated that an increase of root : shoot ratio from 0.2 to 1.0, associated

Table III. Growth of Barley Cultivars at Various Ambient K^+ Levels and Harvested at 2 Weeks (H1), 4 Weeks (H2), and 6 Weeks (H3)[a]

	5 μM K^+			10 μM K^+			50 μM K^+		
Plant Parameters	H1	H2	H3	H1	H2	H3	H1	H2	H3
				Cultivar Fergus					
Plant weight (g.f.w. $plant^{-1}$)	0.56	2.26	6.35	0.71	4.13	13.52	0.73	3.92	12.33
Root : shoot ratio	0.87	0.69	0.55	0.73	0.50	0.47	0.59	0.52	0.49
				Cultivar Excelsior					
Plant weight (g.f.w. $plant^{-1}$)	0.26	0.39	1.35	0.51	2.00	3.43	0.61	3.44	9.65
Root : shoot ratio	0.63	0.56	0.45	0.55	0.45	0.39	0.45	0.50	0.34

[a]Source: Siddiqi and Glass, 1983.

with N deprivation, would require an extra carbon allocation of ~42.5 g glucose for each gram of N procured for the shoot. The acquisition of adequate amounts of nutrients through increased root : shoot ratios is therefore costly in energy terms; nevertheless, the universality of this response attests to its significance.

A satisfactory mechanistic explanation of the sequence of events responsible for increasing root : shoot ratios appears to be lacking, but root/shoot competition and hormonal interactions between root and shoot have been invoked (Troughton, 1977; Marschner, 1986).

Differences in root morphology among species are extremely common in nature (Weaver, 1919). The resulting stratification of root systems within plant communities reduces direct competition because different species tend to draw water and nutrients from different soil layers. However, within a single species, root morphology is sensitive to the physical and chemical (particularly nutritional) properties of soils. Indeed, phenotypic plasticity, particularly in the expression of physiological characteristics, may be much more extensive than genotypic differences.

Increased surface area for ion absorption under conditions of nutrient limitation is achieved by several different morphological/developmental strategies. The adaptive value of these responses is quite apparent, however, the mechanisms responsible for these effects are unknown.

Increased lateral root growth, to form dense (proteoid) clusters of roots, occurs in *Lupinus alba* under conditions of limited P supply (Römheld, 1986). Plants are also capable of selective lateral root proliferation into nutrient-rich regions of soil (Drew and Saker, 1975), thus maximizing the utilization of endogenous resources for nutrient acquisition. Preferential extension of the main axis and suppression of laterals have been observed in many plant species deprived of adequate supplies of K (Jensen, 1982), N (Bergmann, 1954), or P (Bowen *et al.*, 1973). Lengths and numbers of root hairs may also increase when N or P are in short supply (Munns, 1968; Lefebvre, 1980). The development of mycorrhizal associations, particularly in response to P deprivation, represents an important means of extending the root's absorbing power into soil remote from the root and of increasing surface area for ion absorption. These strategies are especially effective in bypassing the diffusion-limited depletion zones referred to above. However, they are relatively costly in terms of raw materials and occur on a time scale of days to weeks. Physiological adaptations occur much more rapidly, on a scale of hours to days.

C. PHYSIOLOGICAL MECHANISMS OF ION ABSORPTION

Solutes may traverse the cell membranes of epidermal cells by diffusion, facilitated transport, or active transport. The lipoidal nature of the plasma membrane provides a significant barrier to the diffusive entry of charged solutes so that only uncharged molecules such as CO_2, O_2, and NH_3 can readily enter cells by this process. In the case of NH_3, the high pK_a (9.25) means that at pH values below ~7.0 less than 1% of total (NH_3 + NH_4^+) species will be available as NH_3. Therefore, the significance of diffusive entry of NH_3 in most crops is highly questionable except at high NH_4^+ concentrations, high pH, and/or low temperatures (Glass, 1988).

Only limited information and speculation is available regarding facilitated diffusion of inorganic ions into plant root. By definition this form of transport is saturable and can only occur along an electrochemical potential difference ($\Delta\bar{\mu}$). It has consequently been invoked to explain high concentration (>1 m*M*) kinetics (Nissen, 1980) as discussed below.

The use of hydroponic facilities in which nutrient concentrations can be maintained by continuous infusion at concentrations as low as 1 μM and below (Asher and Ozanne, 1967; Loneragan and Asher, 1967; Clement *et al.*, 1978; Woodhouse *et al.*, 1978; Siddiqi and Glass, 1983) has established that many plant species (wild and cultivated) are able to achieve optimum growth rates at unexpectedly low ambient concentrations of K^+, NO_3^-, and inorganic P (Pi). Such observations provide strong evidence for active transport of these solutes (see Section D, below). For example, increasing ambient K^+ beyond 24 μM produced no further increases of yield in 8 of 14 species (Asher and Ozanne, 1967). Barley, however, continued to increase in yield up to 1000 μM K^+. At 1 μM K^+ all yields were depressed and, except for silver grass, all species showed symptoms of severe potassium deficiency. Similarly, between 14.3 μM and 14.3 m*M* NO_3^-, uptake and growth of ryegrass (cv. S23) was independent of NO_3^- concentration (Clement *et al.*, 1978). In the study by Loneragan and Asher (1967), increasing available Pi concentration beyond 1 μM caused no increase of P accumulation by silver grass, whereas, by contrast, lupins and flatweed showed no evidence of saturation even at 24 μM Pi.

Within-species variation is also considerable. Siddiqi and Glass (1983) observed that some barley cultivars (e.g., Fergus and 'Olli') reached maximum yield at 10 μM K^+ while others required 50 or 100 μM K^+. At low ambient $[K^+]$ ($\sim$5 μM), the growth of cultivars such as Excelsior was reduced more strongly than was the case for more efficient cultivars, such as Fergus (Table III).

Clearly, genotypic differences in growth responses to ambient concentrations of the required elements may represent the resultants of differences in acquisition and/or utilization of absorbed nutrients. Both processes are complex, and poorly understood, particularly the basis of genotypic differences in utilization. The next section reviews in detail the processes involved in ion absorption.

D. PHYSIOLOGICAL DETERMINANTS OF ION UPTAKE

Although the concept of metabolically mediated transport and ion "pumps" had been advanced much earlier (see, e.g., Collander, 1959, for a historical review), Epstein and co-workers were particularly involved in emphasizing the enzyme kinetic approach to ion absorption. The characteristic propensity of plant roots to maintain ions at concentrations which are orders of magnitude higher or lower than those of ambient solutions suggested a high degree of impermeability of cellular membranes (Epstein, 1976). For example, cytoplasmic K^+ concentration ($[K^+]_c$) was reported to be 1.3×10^{-1} *M* in roots of the barley cultivar 'Compana' when grown in solutions maintained at 1×10^{-5} *M* K^+ (Memon *et al.*, 1985). By contrast, $[Ca^{2+}]_c$ of most eukaryotic cells is held at or below 1×10^{-6} *M*, although external $[Ca^{2+}]$ ($[Ca^{2+}]_0$) is commonly as high as 5×10^{-3} *M* in soil solution. To traverse lipid biomembranes (against $\Delta\bar{\mu}$) demands the existence of special transport mechanisms (carriers) localized within these membranes. The high degree of selectivity (e.g., K^+ is typically absorbed in preference to Na^+, even when Na^+ is present in considerable excess of K^+ in ambient solution) as well as the competitive inhibition kinetics observed for pairs of related ions (K^+, Rb^+; Ca^{2+}, Sr^{2+}; Cl^-, Br^-; SeO_4^{2-}, SO_4^{2-}) was also consistent with the operation of specialized ion-specific carriers. Finally, when ion uptake is plotted against external ion concentration, the absorption of most ions conforms to a rectangular hyperbola, reminiscent of Michaelis–Menten enzyme kinetics, at moderately low ambient concentrations (~0.1 m*M*) (Epstein, 1976). Epstein and Hagen (1952) postulated that the carriers were enzymelike, selectively binding particular ions (substrates) prior to transport across the membrane, and proposed the use of the kinetic constants K_m (half-saturation concentration or Michaelis constant) and V_{max} (maximum velocity) to describe ion uptake characteristics (Figure 1A).

One feature of these kinetics which has not been particularly emphasized is that the uptake curve, which tends to be linear at very low concentration, may not pass through the zero intercepts as is the commonly drawn approximation. Rather, the uptake line intercepts the *X* axis, i.e., net uptake is zero at some low ion concentration. These concentrations, analogous to photosynthetic CO_2 compensation points, have been referred to as C_{min} in the literature. They correspond to ambient concentrations at which influx = efflux. Barber (1979) found values of 2 μM for K^+ and 0.2 μM for Pi in maize, while Drew *et al.* (1984) reported corresponding values of 1 and <0.1 μM for barley. Differences in nitrate compensation points were found within (Edwards and Barber, 1976; Therios *et al.*, 1979) and between (Warncke and Barber, 1974) species. As both influx and efflux values are extremely sensitive to modulation by exogenous and endogenous factors, it is not surprising to observe such differences. For example, Bieleski and Ferguson (1983) reported that C_{min} for Pi increased from <1 μM at 25°C to 1 m*M* at 5°C in *Spirodela*. This was due to the combined effects of increased efflux and decreased influx at 5°C. However, the significance of C_{min} values under field conditions remains to be documented, particularly for nutrients such as NO_3^-, whose concentrations in soil solution are so much higher than measured C_{min} values (Therios *et al.*, 1979).

A particular feature of much of the kinetic work on ion uptake by plant roots dating from the 1950s was an almost universal use of low salt (i.e., $CaSO_4$-grown plants) for uptake experiments. The use of this source of plant material established that in the low concentration range (<0.1 m*M*), the uptake of most ions appeared to be highly selective, typically energy dependent, and largely insensitive to the nature of the accompanying ion. In addition the transport systems were characterized by high affinities (low K_m values, ~ 0.01 m*M* for K^+ or Rb^+) for their substrates and by high V_{max} values (~10 μmol K^+ g^{-1}·f.w. h^{-1}) (Epstein, 1976).

It was subsequently demonstrated that at much higher concentrations (>1 m*M*), uptake curves revealed a second plateau. These biphasic kinetics, referred to as the "dual pattern of ion absorption" or the "dual isotherm of uptake," were found to

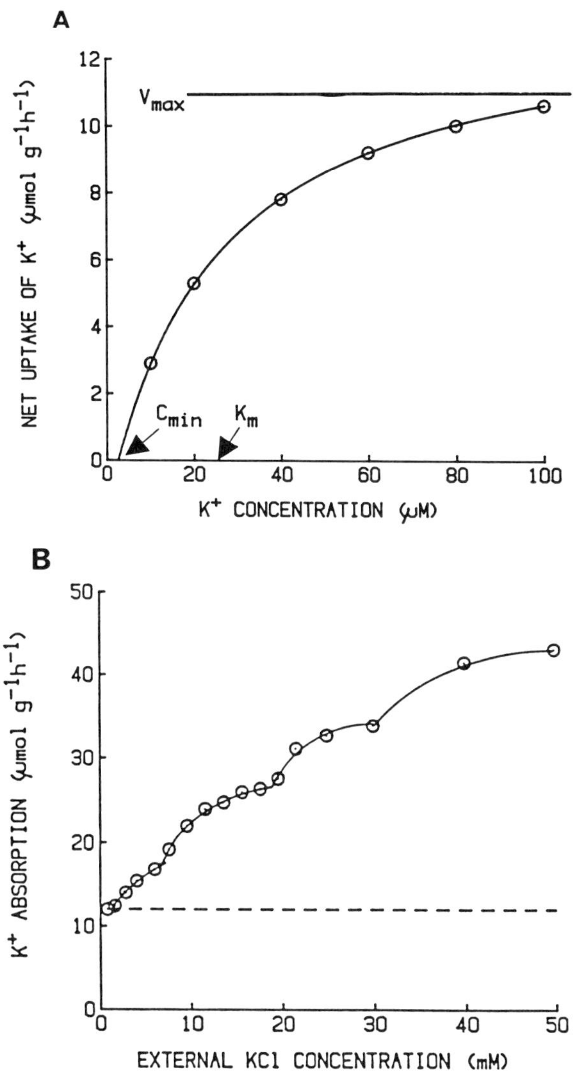

Figure 1. (A) Typical uptake curve for K^+ absorption in the low concentration range. (B) Rates of K^+ absorption by excised barley roots as a function of external KCl concentration. Dashed line shows the maximal rate (V_{max}) for the low concentration uptake mechanism (Epstein and Rains, 1965).

apply to a large number of solutes and to many plant or tissue types (Epstein, 1976). In this high concentration range ion absorption was characterized by much lower substrate affinities (high K_m values), sensitivity to the nature of the accompanying ion (e.g., K^+ uptake was strongly reduced by substituting SO_4^{2-} for Cl^-), a low Q_{10} for phosphate uptake and relative insensitivity to metabolic inhibitors (Barber, 1972), and, in the case of K^+ uptake, to inhibition by Na^+ (Epstein, 1976). This second (high concentration) curve was later demonstrated to be more complex than a single rectangular hyperbola (Figure 1B), conforming in appearance to a series of hyperbolae (Elzam *et al.*, 1964; Epstein and Rains, 1965). The situation is even further complicated in that in several other systems, for example, K^+ influx in maize (Kochian and Lucas, 1982), net NO_3^- uptake in maize (Pace and McClure, 1986), and $^{13}NO_3^-$ influx in barley (Siddiqi *et al.*, 1989), the high concentration fluxes appear to be quite linear as a function of external concentration; there is no indication of saturation even at 50 m*M*.

Despite many years of study and much discussion there is still no clear consensus regarding the mechanisms responsible for high concentration kinetics. Epstein (1966, 1976) interpreted the dual isotherm as resulting from the activities of two sets of carrier sites located in the plasmalemma. Nissen (1973, 1980) interpreted his own data and that of other workers, after linearization using Lineweaver–Burk plots, in terms of single multiphasic transport systems located in the plasmalemma. These systems are thought to undergo concentration-dependent phase changes and to mediate transport at high external concentration by facilitated transport (Nissen, 1980). This interpretation has been seriously challenged by several workers, notably Borstlap (1981, 1983) and Kochian and Lucas (1982). Based on their observed linear high concentration K^+ fluxes and the sensitivity of these fluxes to channel blockers, Kochian and Lucas (1982) interpreted these fluxes as mediated by specific K^+ channels in maize roots. Ionic channels are well known in animal systems as responsible for large fluxes of ions down their $\Delta\bar{\mu}$, through specific proteinaceous membrane pores. In recent years, ionic channels have been identified in many plant systems. Yet another explanation for the complex concentration dependence of ion uptake kinetics, developed by Sanders (1986), is based on random binding of solute and driver-ion (typically H^+, as discussed below) to a single transmembrane carrier.

Both K_m and V_{max} values for ion influx have been shown to vary among plant species and cultivars of crop species (Chapin, 1974; Glass and Perley, 1980; Siddiqi and Glass, 1983) and in selected cases (Cartwright, 1972; Glass, 1976; Lee, 1982) to vary according to nutrient pretreatment (see Table IV). It has been argued that under field conditions, where diffusion rates may determine delivery of ions to the root surface, these kinetic "constants" are of little significance in determining rates of nutrient acquisition (Barber, 1984). Nevertheless, an almost universal response of plants to nutrient deprivation is to increase V_{max} and, in some cases, to reduce K_m values. It is, therefore, difficult to believe that this response is without significance for nutrient acquisition.

E. BIOENERGETIC NATURE OF LOW CONCENTRATION AND HIGH CONCENTRATION FLUXES

Taking the minimum values of ambient ion concentrations required to optimize growth referred to above, namely, 24, 14.3, and 1 μ*M* for K^+, NO_3^-, and $H_2PO_4^{2-}$, respectively, and using estimates of cytoplasmic ion concentrations of 150 (Memon *et*

Table IV. Effect of Growth with or without SO_4^{2-}, Pi, Cl^-, or K^+ on the V_{max} and K_m Values for Absorption of Labeled SO_4^{2-}, Pi, Cl^-, or Rb^+ in a Subsequent Uptake Period

Nutrition Prior to Uptake	Ion Uptake Measured[a]	V_{max}[a] (nmol g^{-1} h^{-1})	K_m[a] (μM)
+ SO_4^{2-}	$^{35}SO_4^{2-}$	53	13.9
− SO_4^{2-}	$^{35}SO_4^{2-}$	758	17.6
+ Pi	^{32}Pi	257	6.6
− Pi	^{32}Pi	475	4.9
+ Cl^-	$^{36}Cl^-$	1010	57.4
− Cl^-	$^{36}Cl^-$	1010	23.7
+ K^+	$^{86}Rb^+$ [b]	230[b]	169[b]
− K^+	$^{86}Rb^+$	9210	50

[a]Source: Lee, 1982
[b]Source: Siddiqi and Glass, 1983.

al., 1985), 26 (Lee and Clarkson, 1986), and 18 m*M* (Lee and Ratcliffe, 1983) for K^+, NO_3^-, and $H_2PO_4^{2-}$, respectively, and a representative value of −150 mV (Thibaud and Grignon, 1981) for root transmembrane electrical potential difference, it is possible to estimate the electrochemical potential differences ($\Delta\bar{\mu}$) for K^+, NO_3^-, and $H_2PO_4^{2-}$ between the cytoplasm and external media. Data in Table V demonstrate that each of these ions must be actively transported into the cytoplasm with a

Table V. Electrochemical Potential Differences $\Delta\bar{\mu}$ (kJ mol^{-1}) for K^+, NO_3^-, and $H_2PO_4^{2-}$ between Cytoplasm and External Media at Different Values of External Ion Concentration ($[\ ]_0$)[a]

	Electrochemical Potential Differences ($\Delta\bar{\mu}$) in kJ mol^{-1}						
	$[K^+]_0$ (μmol $liter^{-1}$)						
Ion	1	10	24	100	400	1000	10,000
K^+	14.5	8.9	6.8	3.3	0	−2.3	−7.9
	$[NO_3^-]_0$ (μmol $liter^{-1}$)						
	1	10	14.3	100	1000	10,000	
NO_3^-	39.2	33.6	32.7	28	22.4	16.8	
	$[H_2PO_4^{2-}]_0$ (μmol $liter^{-1}$)						
	0.01	0.1	1	10	100	1000	10,000
$H_2PO_4^{2-}$	64	58	53	47	42	36	30

[a]Cytoplasmic K^+, NO_3^- and $H_2PO_4^{2-}$ were set at 150 m*M*, 26 m*M* and 18 m*M*, respectively, and $\Delta\Psi$ at −150 mV, from literature values (see text). Negative values of $\Delta\bar{\mu}$ indicate a downhill gradient for ion uptake.

minimum free energy requirement of 6.8, 32.7, and 52.8 kJ mol^{-1} for K^+, NO_3^-, and $H_2PO_4^{2-}$, respectively. When $[K^+]_0$ is elevated beyond 400 μM, however, uptake is no longer "uphill," but down its gradient. By contrast, NO_3^- and $H_2PO_4^{2-}$ uptake remains active even when $[ion]_0$ exceeds molar concentrations, because of the large negative value of $\Delta\Psi$ and high values of cytoplasmic NO_3^- and $H_2PO_4^{2-}$ concentrations. It should be stressed that these calculations are based on the premise that cytoplasmic [ion] is held constant at the quoted values as $[ion]_0$ is varied. There is good evidence that this is true for K^+ and $H_2PO_4^{2-}$ (Rebeille *et al.*, 1983; Lee and Ratcliffe, 1983; Memon *et al.*, 1985). However, it may be that cytoplasmic $[NO_3^-]$ is more variable (Siddiqi ***et al.***, 1989). Should cytoplasmic $[NO_3^-]$ fall to the micromolar range when external NO_3^- is removed, the situation vis-à-vis active transport would change considerably. From a purely bioenergetic viewpoint the extent of $\Delta\Psi$ and $[ion]_c$ are important determinants of $\Delta\bar{\mu}$ and hence the energy that must be expended to transport ions actively against the existing $\Delta\bar{\mu}$. However, comparative values for $\Delta\Psi$ or $[ion]_c$ are rare. Memon *et al.* (1985) found that Compana, a barley cultivar which grew well at low ambient $[K^+]$, had a lower value of cytoplasmic K^+ than did less efficient varieties. It is likely, however, that Compana's efficiency is based on a lower requirement and better utilization of K^+ than any energy saving associated with a smaller requirement for active transport.

The immediate source of free energy for these active transport steps is considered to be $\Delta\bar{\mu}_{H^+}$ or the proton motive force (p.m.f.) across the plasma membrane. H^+-translocating ATPases, located in the plasma membrane, are thought to bring about vectorial H^+ transport in an outward direction, thus generating both $\Delta\Psi$ (inside negative) and ΔpH (outside acid) (Sze, 1985). The extent of $\Delta\bar{\mu}_{H^+}$ is given by

$$\Delta\bar{\mu}_{H^+} = F\,\Delta\psi - 2.303\,RT\,\Delta pH \tag{2}$$

where:

F = the Faraday constant, 96.5 kJ mol^{-1} V^{-1}
$\Delta\Psi$ = electrical potential difference across the plasma membrane, in mV
R = the gas constant, 8.3 J mol^{-1} $degree^{-1}$
T = temperature (Celsius).

Setting $[H^+]_c$ at 10^{-7} M and $[H^+]_0$ at 10^{-5} M, with $\Delta\Psi$ at -150 mV, it is evident that the $\Delta\bar{\mu}_{H^+}$ is inwardly directed and amounts to ~26 kJ mol^{-1} H^+. Kinetic and electrophysiological evidence is consistent with a model in which H^+ and the required nutrients (K^+, NO_3^-, $H_2PO_4^{2-}$, etc.) are cotransported (symported) via specific transporters (Sze, 1985). These are probably the carriers identified kinetically by Epstein and others (Epstein, 1976). With a $H^+:K^+$ stoichiometry of 1 : 1 (Rodriguez-Navarro *et al.*, 1986) and $H^+:NO_3^-$ stoichiometry of 2 : 1 (Ullrich and Novacky, 1981), the proton gradient provides adequate free energy for the uptake of these ions under the conditions specified above and in Table V. For Pi uptake, which causes depolarization of $\Delta\Psi$ in *Lemna* (Ullrich-Eberius *et al.*, 1981), signifying a $H^+:H_2PO_4^{2-}$ stoichiometry of at least 3 : 1, it is difficult to satisfy the very large

$\Delta\bar{\mu}$ for net uptake from such dilute solutions of P as are typically found in soils (see Table I). Indeed, on the basis of their studies with *Lemna*, Ullrich-Eberius *et al.* (1981) have suggested that H^+ cotransport of divalent phosphate may be thermodynamically impossible. By contrast $\Delta\bar{\mu}_{H^+}$ was considered to be adequate to account for monovalent phosphate uptake.

Inwardly directed H^+-translocating ATPases have also been characterized at the tonoplast and, in addition, pyrophosphate-powered H^+ pumps have been identified in this membrane (Rea and Sanders, 1987). The activities of these two systems generate ΔpH values of 2–3 units (acid on the vacuolar side) and a small ΔΨ, typically 10–20 mV (positive on the vacuolar side). This $\Delta\bar{\mu}_{H^+}$ (13–18 kJ mol^{-1}, directed toward the cytoplasmic phase) provides the driving force for countertransport (antiporting) of anions such as Na^+ or K^+ into the vacuole and for symporting of anions from vacuole to cytoplasm (Figure 2).

Anion (NO_3^- or Cl^-) absorption by isolated vacuoles or by vesicles derived from tonoplast membranes is thought to occur in response to ΔΨ across the tonoplast via uniporters. Such fluxes lead to dissipation of ΔΨ, which in turn stimulates H^+ pumping into the vacuole. As a result, ΔΨ is replaced by ΔpH.

Although there has been controversy regarding the mechanism(s) of transfer of solute to the stele for transport to the shoot, the balance of opinion is in favor of a second active transport step (Bowling, 1981; Pitman, 1982). Xylem parenchyma cells, adjacent to xylem vessels, have been suggested as the sites for the second active transport steps (Läuchli, 1972), and $\Delta\bar{\mu}_{H^+}$ between the root symplasm and the vascular apoplasm has been suggested to provide the driving force for solute

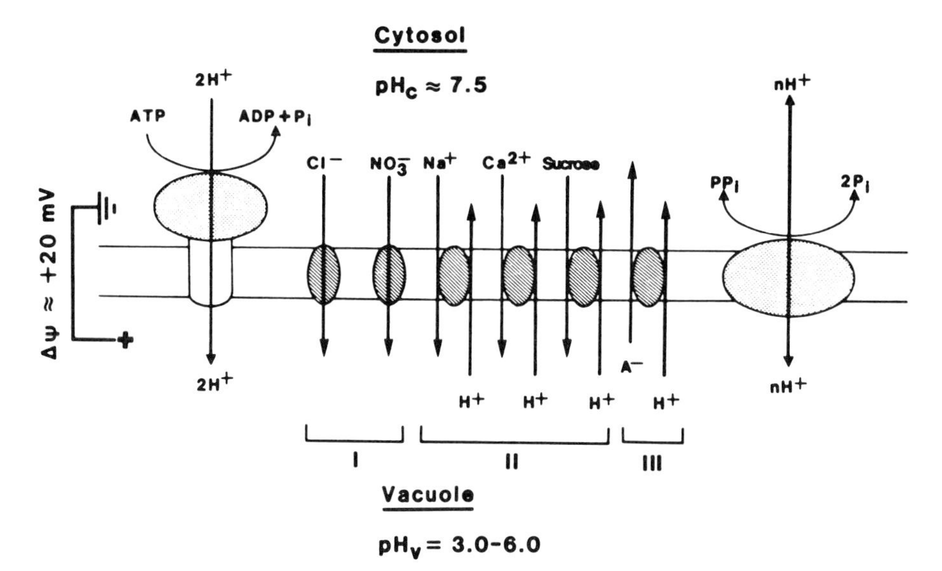

Figure 2. Electrogenic H^+ translocation by anion-sensitive tonoplast ATPase and cation-sensitive tonoplast pyrophosphatase (Rea and Sanders, 1987).

loading into the lumina of xylem tracheary elements, analogous to vacuole loading (Hanson, 1978).

Notwithstanding the importance of $\Delta\bar{\mu}_{H^+}$ as the immediate source of free energy for all of the above transport steps, it is unlikely that the magnitude of transmembrane solute fluxes is determined by the extent of the thermodynamic driving forces (Cram, 1975; Glass and Dunlop, 1979; Raven and Smith, 1978). Glass *et al.* (1981) demonstrated that K^+ influx and net uptake in 24 barley cultivars was highly correlated with H^+ efflux. However, H^+ efflux was probably stimulated as solute (K^+) uptake dissipated $\Delta\bar{\mu}_{H^+}$. Hence the quantitative correlation was probably the result, rather than the cause, of K^+ uptake. Rates of ion uptake can unquestionably be reduced by the application of metabolic inhibitors. For example, Petraglia and Poole (1980) showed that K^+ and Cl^- uptake by red beet was linearly dependent on ATP levels of the tissue, following treatment with KCN. Nevertheless, under normal conditions the magnitude of ion fluxes across the plasma membrane and the tonoplast appears to be regulated by specific feedback signals (and in the case of NO_3^- uptake, by positive feedforward signals), so that internal nutrient concentrations are maintained at prescribed levels (Glass and Siddiqi, 1984). These regulatory processes are discussed in detail in the next section.

II. Regulation of Ion Uptake

As early as 1906 it was demonstrated that wheat plants grown in solutions lacking potassium subsequently absorbed more potassium from complete solution than plants of the same age which had grown continuously in complete solutions (Breazeale, 1906). Hoagland and Broyer (1936) rediscovered this phenomenon and coined the terms "low-salt" and "high-salt" to describe salt-starved ($CaSO_4$-grown) and salt-replete roots. Roots of low-salt plants were characterized by high sugar content and a high capacity for ion uptake; by contrast, the sugar content and uptake capacity of high-salt roots were low. Hoagland and Broyer (1936) concluded that it was not sugar content per se (and hence energy supply) which was limiting ion uptake but that plants possessed some finite limits for accumulation. Clearly, the coupling between a source of free energy (now believed to be $\Delta\bar{\mu}_{H^+}$) and ion uptake represents a fundamental aspect of the mechanism(s) of ion transport. However, it is unlikely (see above) that ion fluxes are determined by energy supply. Rather, energy supply is adjusted to meet demand (Lambers *et al.*, 1983). It is evident from a large number of studies with diverse organisms that ion fluxes are regulated in response to feedback or feedforward from internal signals such as cellular ion concentrations (or concentration-dependent parameters such as volume, turgor, or osmotic potential) and relative growth rates (Pitman and Cram, 1973; Cram, 1976; Zimmerman, 1978; Glass, 1983). The end result of modulating rates of ion transport in response to such signals is to buffer the organism against variations in nutrient availability and to achieve a match between plant "demand" for nutrients and delivery (by means of transmembrane fluxes).

As discussed above, plant demand for most inorganic nutrients can be satisfied at quite low ambient concentrations, provided that nutrients are continuously replenished. Elevating $[\text{ion}]_0$ causes only small increases of $[\text{ion}]_i$ in long-term studies (Glass and Siddiqi, 1984). Under these steady-state conditions ion uptake is essentially independent of $[\text{ion}]_0$ except at values approaching C_{min}. Likewise, gradual acclimation to temperature can result in ion fluxes and $[\text{ion}]_i$ which are independent of temperature over a finite range of temperature (Raven and Smith, 1978; Siddiqi *et al.*, 1985).

Ideally these acclimated fluxes should just match the "demand" generated by the particular value of growth rate under these conditions (Deane-Drummond, 1982; Glass and Siddiqi, 1984). Beever and Burns (1980) have demonstrated that in *Neurospora crassa,* P uptake remains constant at about 3.3 μmol g^{-1}·d.w. min^{-1} in the range from 50 to 10,000 μM Pi through modulations of its low- and high-affinity transport systems. Transfer of such acclimated plants to elevated concentrations of particular ions gives evidence of increased capacity for ion uptake compared to plants grown continuously at the elevated concentration (Hoagland and Broyer, 1936; Pitman *et al.*, 1968; Cram, 1973; Cartwright, 1972). Likewise when plants grown at a particular temperature are transferred to an elevated temperature, rates of ion uptake at the new temperature are higher than for plants maintained at the higher temperature (Clarkson, 1976; Siddiqi *et al.*, 1985). Low-salt plants grown in $CaSO_4$ solutions or plants deprived of a single nutrient represent the extreme cases of this acclimation. Uptake values may be increased >10- to 20-fold by this treatment.

The mechanisms responsible for regulating ion fluxes to bring about acclimation are only poorly understood. Considering that the structures and functions of the proton-translocating ATPases are only partly resolved, and that the biochemical identities of the specific ion carriers are virtually unknown, it is perhaps not surprising that we know so little about the regulation of these transporters. Nevertheless it has been possible by means of kinetic experiments to determine which fluxes are regulated and to speculate concerning the nature of the signals responsible for regulating fluxes (Cram, 1976; Raven, 1977; Zimmerman, 1978; Glass, 1983).

The use of radioactive tracers in uptake experiments of short duration compared to the half-lives for exchange of the cytoplasmic compartment has enabled experimenters to determine that the increase or decrease of uptake which is evident following temperature or concentration changes results primarily from effects on influx (rather than efflux) in most cases. This is characteristic of K^+ (Johansen *et al.*, 1970), $H_2PO_4^{2-}$ (Lefebvre and Glass, 1982; Bieleski and Ferguson, 1983), Cl^- (Cram, 1973), SO_4^{2-} (Smith, 1975), and NO_3^- (Lee and Drew, 1986) uptake. Rapidly growing plants, as opposed to systems such as tissue slices (which may be close to flux equilibrium), will clearly have a high demand for inorganic nutrients. Under such conditions influx must necessarily be high compared to efflux. When $CaSO_4$-grown barley plants were provided with K^+ after 5 days of deprivation, the initially high influx value resulted in a rapid increase of tissue [K^+]. Net K^+ uptake and $^{42}K^+$ influx were almost identical, signifying that efflux was insignificant. As tissue [K^+] rose, net uptake and $^{42}K^+$ influx fell in parallel until steady state was reached (Johansen *et al.*, 1970). Estimates of K^+ efflux as the difference between influx and net uptake, or by direct means, using compartmental analysis, give values of ~1 μmol g^{-1} h^{-1} in intact barley roots under steady-state conditions in 10 μM K^+ solutions (Memon *et al.*, 1985). Differences among the cultivars Fergus, Compana, and 'Betzes' were insignificant. By contrast influx varied from 1.88 to 3.42 μmol g^{-1} h^{-1} so that there were significant differences in net uptake. Plants grown at steady state in 100 μM K^+ had significantly higher values of influx (~5 μmol g^{-1} h^{-1}) and efflux (~2 μmol g^{-1} h^{-1}). Hence, although during adjustment to perturbations in supply influx appears to be the target for regulation, effects on efflux are also evident.

The regulation of NO_3^- uptake has been controversial. Net uptake of NO_3^- has been shown to be inversely related to prior NO_3^- provision in barley (Smith, 1973; Deane-Drummond, 1982). Using $^{15}NO_3^-$ as a tracer for NO_3^-, Jackson *et al.* (1976) concluded that influx was reduced and efflux increased by prior loading with NO_3^-. However, using $^{36}ClO_3^-$ as a tracer for NO_3^- (Deane-Drummond and Glass, 1983) and $^{13}NO_3^-$ (Glass *et al.*, 1985) to estimate influx, it was concluded that influx was independent of NO_3^- pretreatment and that uptake was regulated by effects on efflux. Subsequently, Lee and Drew (1986) demonstrated that $^{13}NO_3^-$ influx increased significantly following NO_3^- deprivation for 3 days. Siddiqi *et al.* (1989) have demonstrated that careful time course studies are necessary to resolve this controversy. Following the provision of NO_3^- to NO_3^--starved barley, they observed rapid increases of $^{13}NO_3^-$ influx due to the "inductive" effects of added NO_3^-. During this period, induction of influx was concentration dependent and a plot of $^{13}NO_3^-$ influx against root NO_3^- gave a positive correlation. However, depending on $[NO_3^-]_0$ the peak of $^{13}NO_3^-$ influx was followed by a rapid decline of influx. This was most rapid in plants treated with 10 mM NO_3^-, but was absent altogether in plants treated in 10 μM NO_3^- (Figure 3).

Using $^{15}NO_3^-$ as a tracer for NO_3^-, Jackson *et al.* (1976) and Pearson *et al.* (1981) have demonstrated that NO_3^- efflux can represent a significant factor in determining net uptake. Although the use of long-term (0.5–3 h) studies to estimate influx and/or efflux by means of ^{15}N has been questioned because of the short half-life of NO_3^- exchange, the fact remains that short-term estimates of efflux using $^{13}NO_3^-$ (Presland and McNaughton, 1984) or $^{36}ClO_3^-$ as a tracer for NO_3^- (Deane-Drummond and Glass, 1983) confirm that efflux of NO_3^- can become substantial, particularly at elevated concentrations (see also Breteler and Nissen, 1982).

Under the somewhat artificial conditions of the laboratory, when nutrient-deprived plants are suddenly resupplied with a particular nutrient, the elevated uptake capacity can result in rapid restoration of "normal" tissue concentrations. Under natural conditions sudden flushes of nutrients are rare in soils (except perhaps for NO_3^-), although in aquatic systems this may be more common. Hence, it has been suggested that the regulation of ion fluxes associated with diminishing external concentrations is

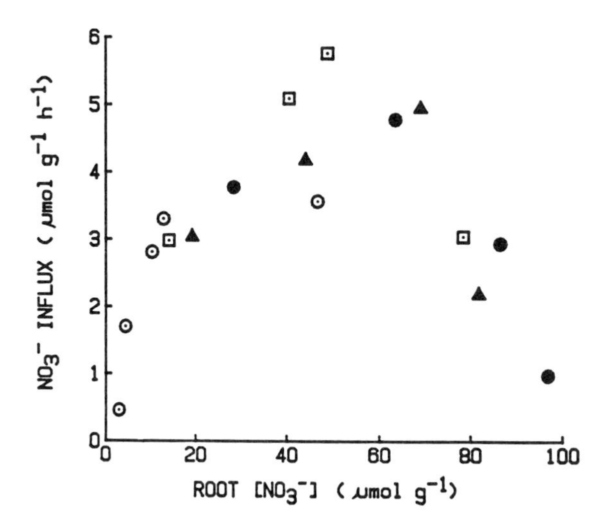

Figure 3. $^{13}NO_3^-$ influx in barley plants previously starved of NO_3^- and then treated with 10 (⊙), 100 (⊡), or 1000 (▲) μ*M* or 10 m*M* (●) NO_3^- for periods up to 4 days prior to $^{13}NO_3$- influx from 100 μ*M* NO_3 (Siddiqi *et al.*, 1989).

a means of achieving greater efficiency in acquiring the limiting resource (Clarkson, 1985).

SIGNALS FOR THE REGULATION OF FLUXES

The negative correlations between tissue concentrations of particular ions and rates of ion uptake have led to the proposal that internal ion concentrations may be critical in regulating rates of uptake in plant roots (Smith, 1973; Cram, 1973; Glass, 1975, 1976; Pitman and Cram, 1973, 1977; Raven, 1977; Rebeille *et al.*, 1983). Nitrate influx during induction is an exception since it is positively correlated with $[NO_3^-]_i$ (Siddiqi *et al.*, 1989).

On the basis of changes of K_m and V_{max} for K^+ influx associated with K^+ loading of low-salt barley roots, and the relationship between influx and root $[K^+]$, Glass (1976) proposed that influx was regulated allosterically by direct feedback from cytoplasmic K^+ on the K^+ transporter. In retrospect this model was naive because measurements of root $[K^+]$ reflect vacuolar $[K^+]$ rather than cytoplasmic $[K^+]$. Furthermore, cytoplasmic $[K^+]$ appears to be held constant (Leigh and Wyn Jones, 1984) except at extremely low root $[K^+]$. Nevertheless, the rapidity with which plasma membrane fluxes are modulated when plants are loaded with K^+ (Young and Sims, 1972; Glass, 1977) suggests that the cytoplasm is capable of exerting effects on influx.

The use of split-root experiments (Drew and Saker, 1984) showed that when only a limited portion of the entire root was provided with K^+, influx into that region (which now provided the entire K^+ supply for the plant) increased significantly. However, this increase occurred without change of root [K^+]. Drew and Saker argued that this negated the hypothesis of direct (allosteric) regulation of influx by root [K^+]. They argued, instead, for a plant "demand" factor transmitted from the shoot. Siddiqi and Glass (1987) have responded to this argument, suggesting that effects of shoot demand, which may be exerted via recycling of K^+ in phloem (LaGuardia *et al.*, 1985), may well exert effects on ion uptake by the roots but that these are expressed indirectly by changing root [K^+] or by changing kinetic responses of the uptake system to root [K^+].

These arguments apply equally to other nutrients which, in the main, have been less extensively studied. In the case of metabolized ions such as NO_3^- there is evidence that the induction of NO_3^- influx is specifically under the control of tissue [NO_3^-] (Glass, 1988; Siddiqi *et al.*, 1989). However, it has been argued that the negative feedback effects of accumulated N are exerted by a signal or signals beyond the level of NH_4 (Lee and Rudge, 1986). Exogenously applied amino acids have been found to inhibit or stimulate subsequent NO_3^- uptake in dwarf bean (Breteler and Arnozis, 1985). By contrast, Pi and SO_4^{2-} have been proposed to be the principal signals for regulating Pi and SO_4^{2-} uptake (Bieleski and Ferguson, 1983; Smith, 1975).

Clearly, we are far from understanding the complexities of ionic regulation. Considerably more study at the whole-plant level and at the biochemical level is needed. It has frequently been postulated (without direct evidence, it should be stressed) that the regulation of fluxes referred to above may be the result of changes in carrier synthesis. Recently, reports from McClure *et al.* (1987) and from Dhugga *et al.* (1988) provide evidence that NO_3^- induction is associated with changes in membrane polypeptides in maize. Likewise Fernando *et al.* (1987, 1990) have observed enhanced expression of membrane polypeptides in barley roots following K^+ deprivation. These reports are preliminary but give good indications of the direction of future work in this area.

At a more practical level, the negative correlations between measured influx and tissue ion concentrations may present serious problems for those who might wish to screen cultivars of crop species for efficient nutrient uptake. Pace and McClure (1986) noted eightfold differences in net uptake of NO_3^- on a per gram basis among 85 maize hybrids. Likewise, Woodend (1986) reported values of K^+ uptake ranging from 2.4 to 7.4 μmol g^{-1} h^{-1} among 24 wheat cultivars. Similar differences were reported for NO_3^- uptake by the same cultivars. In the Woodend study (1986) a significant negative correlation was observed between K^+ uptake rate and root [K^+] despite the fact that all plants had been grown in a common nutrient solution. Hence some part of apparent genotypic differences may be due to transient differences associated with acclimation. Also, as stated earlier, under steady-state conditions rates of uptake under nonlimiting conditions of nutrient provision may be determined by growth rates (Pitman, 1976; Clarkson, 1985). As a consequence, a screen for efficient nutrient acquisition may simply be screening for high growth rates.

III. Utilization of Absorbed Nutrients

The utilization of absorbed nutrients can be considered to commence from the moment the ion traverses the epidermal or cortical plasma membranes and enters the symplasm. Immediate utilization may involve metabolic changes, as in the case of Pi which is rapidly converted to organic phosphate, or transfer, without chemical change, to the vacuole or to the stele for onward translocation to the shoot as is the case for K^+. Subsequent redistribution of mobile elements (e.g., N, P, K, Cl) toward rapidly growing regions is characteristic even when nutrients are not in short supply. However, retranslocation is not universal and nonmobile elements such as Ca and B tend to be retained in older tissues.

Efficient utilization of absorbed nutrients may therefore involve the sum total of a large number of component processes. If, as proposed by Loneragan (1976), the whole of the plant–soil system is incorporated into a definition of plant mineral use efficiency, the complexity of the system becomes overwhelming. Nevertheless, environment × genotype interactions are well known in studies of nutrient use efficiency (Baker, 1968; Giordano *et al.*, 1982; Woodend, 1986) and Loneragan's concerns are therefore justified. However, to provide for a manageable system and provide a quantitative basis for comparisons, Gerloff (1976) and Gabelman (1976) have made use of a quotient, the efficiency ratio or utilization coefficient based on plant growth under strictly defined conditions. The efficiency ratio (E.R.) describes the plant biomass per unit of tissue nutrient and hence has units of reciprocal concentration, e.g., mg·d.w. mg^{-1} P or g.f.w. μmol^{-1} K. A criterion which Gerloff and Gabelman have emphasized in the identification of inefficient strains is that the inefficiency identified be strictly based on the utilization of the particular nutrient under examination. This eliminates genotypes possessing metabolic lesions which result in poor growth under all conditions.

Clearly, in terms of the E.R. criterion, developing strains in which this ratio is maximized would be advantageous. This goal has frequently been discussed in one form or another (Vose, 1963; Woodend, 1986). However, universal agreement on the point is not apparent. Arnon (1975) maintains that efforts in this direction have largely been unsuccessful. Fischer (1981) argues that the yields obtained from cultivars defined as efficient by this criterion would be too low to be economical. Borlaug (1976) rejects thc goal as untenable, advocating that increased productivity can only be achieved under high fertility conditions. There can be little doubt that this latter strategy has been highly effective. According to Saric (1987), 50% of the yield increases of the last few decades have been due to fertilizer application. Yet this strategy is costly in economic terms and also in terms of environmental damage. The most zealous advocate of the strategy of incorporating physiological attributes and improving nutrient use efficiency is not unaware of the Second Law of Thermodynamics. However, the improvement of characteristics involved in uptake and utilization must certainly lead to better performance under marginal conditions or lead to lower absolute yields at significantly lower cost (Woodend, 1986). According to Graham (1984), major emphasis should be directed toward nutrients such as Fe, Mn, Cu, Zn, and P, for which soil reserves are large but availability low.

With regard to the concern for yield under reduced input conditions, Siddiqi and Glass (1981) advocated the use of a modified E.R. By expressing efficiency in terms of yield per unit of tissue concentration rather than yield per unit of content, the contribution of yield is emphasized. This quotient, referred to as utilization efficiency (U.E.), has units of biomass per concentration and can sometimes provide a better selection criterion for efficient nutrient utilization. For example, Woodend (1986) reported on the utilization of K and N for 24 wheat cultivars. Data in Table VI demonstrate that two cultivars ('C306' and 'Moti'), whose shoot yields were quite different (ranked 1 and 12, respectively) under conditions of low K^+ provision, possessed E.R. values which were indistinguishable. Use of the U.E. to assess efficiency ranked C306 as the first of the 24 cultivars with Moti at position 8.

Inter- and intraspecies differences in efficiency ratios are well known (Asher and Ozanne, 1967; Loneragan and Asher, 1967; Gerloff, 1976; Woodhouse *et al.*, 1978; Nielsen and Schjørring, 1983; Pettersson and Jensen, 1983). However, the conditions under which the limiting nutrient is provided, and the potential problems arising from ionic interactions, should be cause for exercising considerable care in interpreting these differences. Some studies, e.g., those from the laboratories of Gerloff and Gabelman, involve provision of a fixed amount of a given nutrient at either a limiting or sufficient level, while other work (e.g., Asher and Ozanne, 1967; Woodend, 1986) has involved continuous maintenance of a particular limiting nutrient at a range of concentrations. Using fixed allocations E.R. estimates internal utilization alone; continuous maintenance adds the effects of differential uptake. Also, certain ionic interactions may reduce the accumulation of a particular ion and cause elevation of E.R. values. For example, Spear *et al.* (1978) showed that elevating available [K^+] in flowing cultures reduced tissue [Ca^{2+}] and [Mg^{2+}] and elevated efficiency ratios. At 0.5 μM K^+, E.R.'s for cassava, sunflower, and maize shoots were 0.43, 0.3, and 0.64, respectively, for Ca^{2+}. These values increased to 1.43, 0.72, and 2.7, respectively, as external [K^+] was raised to ~8 mM. A more serious difficulty which has been emphasized by Woodend (1986), arising from his study of K^+ utilization in

Table VI. Plant Yield, Potassium Efficiency Ratios, and Potassium Utilization Efficiencies for Two Wheat Cultivars (values in parentheses indicate rankings among 24 wheat cultivars)[a]

Cultivars	Shoot Weight (g.f.w.)	K E.R. (g mmol^{-1})	K U.E. (g^2 mmol^{-1})
C306	1.90 (1)	9.2 (3)	17.3 (1)
Moti	1.43 (12)	9.3 (2)	13.3 (8)

[a]Source: Woodend, 1986.

a diverse group of wheat cultivars, was that neither efficiency ratios nor utilization efficiencies at the vegetative stage correlated with grain yield or grain utilization efficiency. As a consequence Woodend proposed that selection for improved potassium utilization should be based on grain production rather than vegetative characteristics.

As stated above, biomass production represents the integral of many separate and interrelated processes. It has been easy to document differences in utilization but studies which have defined the mechanism(s) responsible for observed differences are rare. In their study of phosphorus nutrition in *Hordeum vulgare* (cultivar 'Kombar') and *Hordeum leporinum* (barley grass), Chapin and Bieleski (1982) concluded that the low-P-adapted barley grass had P absorption rates comparable to those of the fertilizer-demanding Kombar cultivar but at all levels of P supply *H. vulgare* produced more biomass and hence a higher E.R. than *H. leporinum*. This appears to be characteristic of many species and populations (Bradshaw *et al.*, 1960; Chapin *et al.*, 1982) from such habitats. Chapin and Bieleski concluded that efficient P utilization in the sense of a high E.R. has not been an important adaptation to nutrient stress. Rather, species occupying nutrient-poor habitats tend to have adopted a slow growth strategy, are typically unresponsive to added nutrients (Bradshaw *et al.*, 1964), and hence tend to have low E.R.'s. The advantage of this life-style (under nutrient-poor conditions) is that nutrients are accumulated when available and growth is sustained for longer than is the case for fast-growing species. The latter soon exhaust available nutrients and become susceptible to disease and other environmental stresses (Chapin, 1980). The concept of the slow growth strategy, effective as it is, does not explain the basis of the growth limitation in slow-growing plants. As Passioura (1979) has so eloquently proposed, biological systems can be organized according to hierarchical levels. At a particular level of study (e.g., tissue, cell, organelle), mechanism is resolved by proceeding to the next lower level (reductionism); significance is illuminated by moving up to the next level (synthesis). As stated above, too few studies have resolved the mechanisms responsible for differences in utilization.

One such study, by Memon *et al.* (1985), provided a satisfactory explanation of differences in K^+ utilization in barley cultivars. At 10 μM $[K^+]_o$, Compana performed significantly better than Betzes. Efficiency ratios and utilization efficiencies for roots and shoots of the two cultivars reflect the greater efficiency of Compana at low K^+. At 100 μM K^+ these differences were much less evident. (Table VII).

The use of compartmental analysis to estimate cytoplasmic and vacuolar $[K^+]$ revealed that vacuolar $[K^+]$ in Compana was significantly lower than in Betzes (Table VII). Considering that the vacuole typically occupies >85% of cell volume, the greater reduction of vacuolar $[K^+]$ in Compana made available considerable quantities of K^+ to maintain cytoplasmic $[K^+]$ at a level appropriate to sustain the higher growth rate. Increasing $[K^+]_0$ to 100 μM failed to cause significant increase of cytoplasmic $[K^+]$ in Compana, but raised cytoplasmic $[K^+]$ from 127 to 187mM in Betzes. They concluded that the optimization of growth in Betzes required a higher cytoplasmic $[K^+]$. Leigh and Wyn Jones (1984) proposed that the critical

Table VII. Distribution of K^+ in Efficient (Compana) and Inefficient (Betzes) Barley Cultivars Grown in 10 and 100 μM K^+ Solutions[a]

	A. Average Tissue Analysis (μmol g^{-1})			
	10 μM K^+		100 μM K^+	
Cultivar	Root	Shoot	Root	Shoot
Compana	25.2 ± 1.4	77.4 ± 2.1	62.0 ± 3.3	96.9 ± 1.3
Betzes	44.4 ± 1.4	93.2 ± 4.3	59.3 ± 3.9	100.2 ± 4.8
	B. Cytoplasmic and Vacuolar [K^+] (mmol liter^{-1})			
	10 μM K^+		100 μM K^+	
	Cytoplasm	Vacuole	Cytoplasm	Vacuole
Compana	133 ± 8	21 ± 0.5	140 ± 10	61 ± 1.6
Betzes	127 ± 14	33 ± 1.7	187 ± 12	56 ± 5.5

[a]Source: Memon *et al.*, 1985.

role of cytoplasmic K^+ was to provide an appropriate milieu for protein synthesis. The finding that total protein content and the rate of protein synthesis (assayed by means of [^{3}H]leucine incorporation) were significantly higher in Compana than in Betzes indicates that the differences in [K^+] in the two cultivars exert significant effects on growth via effects on protein synthesis (Memon and Glass, 1987).

IV. Summary

The application of biophysical and biochemical methods to studies of ion absorption has permitted unparalleled developments in our understanding of the mechanisms of transmembrane ion fluxes. Preliminary reports on membrane polypeptides which may be involved in the uptake of NO_3^- (McClure *et al.*, 1987; Dhugga *et al.*, 1988) and K^+ (Fernando *et al.*, 1987, 1990) have recently appeared and may be anticipated to lead to definitive characterization of the carriers, originally defined by kinetic studies. The development of gene constructs to enhance expression of particular traits (e.g., herbicide tolerance) has already been achieved in higher plants (Kay *et al.*, 1987) by using techniques which are rapidly becoming routine in biotechnology laboratories. The engineering of membrane transport to increase uptake, increase storage capacity, reduce ionic interactions, and introduce other desirable characteristics may likewise soon become possible. The complexity of the processes involved in nutrient utilization and the lack of understanding of basic mechanisms responsible for the differences which have been reported are cause for considerably less optimism that progress in improving nutrient utilization will be achieved as readily.

References

Arnon, I. 1975. Mineral Nutrition of Maize. Int. Potash Inst., Bern.

Asher, C. J., and P. G. Ozanne. 1967. Growth and potassium content of plants in solution cultures maintained at constant potassium concentrations. *Soil Sci.* 103:155–161.

Baker, L. R. 1968. Inheritance and basis for efficiency of potassium utilization in the red beet, *Beta vulgaris* L. Ph.D. Thesis, Univ. of Wisconsin, Madison.

Barber, D. A. 1972. "Dual isotherms" for the absorption of ions by plant tissues. *New Phytol.* 71:255–262.

Barber, S. A. 1979. Growth requirements for nutrients in relation to demand at the root surface. In J. L. Harley and R. Scott-Russell (eds.), The Soil–Root Interface. Blackwell, Oxford. Pp. 5–20.

Barber, S. A. 1984. Soil Nutrient Availability: A Mechanistic Approach. Wiley (Interscience), Toronto.

Beever, R. E., and D. J. W. Burns. 1980. Phosphate uptake, storage and utilization by Fungi. *Adv. Bot. Res.* 8:127–219.

Bergmann, W. 1954. Wurzelwachstum und Ernteertrag. *Z. Acker-Pflanzenbau* 97:337–368.

Bieleski, R. L., and I. B. Ferguson. 1983. Physiology and metabolism of phosphate and its compounds. In A. Läuchli and R. L. Bieleski (eds.), Inorganic Plant Nutrition. Encyclopedia of Plant Physiology, New Series, Vol. 15A. Springer-Verlag, Berlin. Pp. 422–449.

Borlaug, N. E. 1976. The green revolution: Can we make it meet expectations? *Proc. Am. Phytopathol. Soc.* 3:6–21.

Borstlap, A. C. 1981. Invalidity of the multiphasic concept of ion absorption in plants. *Plant, Cell Environ.* 4:189–195.

Borstlap, A. C. 1983. The use of model-fitting in the interpretation of "dual" uptake isotherms. *Plant, Cell Environ.* 6:407–416.

Bowen, G. D., B. Cartwright, and J. R. Mooney. 1973. Wheat root configuration under phosphate stress. In R. L. Bieleski, A. R. Ferguson, and M. M. Creswell (eds.), Mechanisms of Regulation of Plant Growth. Royal Soc. New Zealand, Wellington. Pp. 121–125.

Bowling, D. J. F. 1981. Release of ions to the xylem in roots. *Physiol. Plant.* 53:392–397.

Bradshaw, A. D., M. J. Chadwick, D. Jowett, R. W. Lodge, and R. W. Snaydon. 1960. Experimental investigations into the mineral nutrition of several grass species. III. Phosphate level. *J. Ecol.* 48:631–637.

Bradshaw, A. D., M. J. Chadwick, D. Jowett, and R. W. Snaydon. 1964. Experimental investigations into the mineral nutrition of several grass species. IV. Nitrogen level. *J. Ecol.* 52:665–676.

Breazeale, J. F. 1906. The relation of sodium to potassium in soil and solution cultures. *J. Am. Chem. Soc.* 28:1013.

Breteler, H., and P. A. Arnozis. 1985. Effect of amino compounds on nitrate utilization by roots of dwarf bean. *Phytochemistry* 2:653–657.

Breteler, H., and P. Nissen. 1982. Effect of exogenous and endogenous nitrate concentration on nitrate utilization by dwarf beans. *Plant Physiol.* 70:754–759.

Brooks, R. R. 1987. Serpentine and Its Vegetation: A Multidisciplinary Approach. Dioscorides Press, Portland, Oregon.

Brown, P. H. R. M. Welch, and E. A. Cary. 1987. Nickel: A micronutrient essential for higher plants. *Plant Physiol.* 85:801–803.

Brownell, P. F. 1965. Sodium as an essential micronutrient for a higher plant (*Atriplex vesicaria*). *Plant Physiol.* 40:460–468.

Brownell, P. F. 1979. Sodium as an essential micronutrient for plants and its possible role in metabolism. *Adv. Bot. Res.* 7:117–224.

Cacco, G., M. Saccomani, and G. Ferrari. 1983. Changes in the uptake and assimilation efficiency for sulfate and nitrate in maize hybrids selected during the period 1930 through 1975. *Physiol. Plant.* 58:171–174.

Cartwright, B. 1972. The effect of phosphate deficiency on the kinetics of phosphate absorption by sterile excised barley roots and some factors affecting the ion uptake efficiency of roots. *Soil Sci. Plant Anal.* 4:313–322.

Chapin, F. S. 1974. Morphological and physiological mechanisms of temperature compensation in phosphate absorption along a latitudinal gradient. *Ecology* 55:1180–1198.

Chapin, F. S. 1980. The mineral nutrition of wild plants. *Annu. Rev. Ecol. Syst.* 11:233–260.

Chapin, F. S., III, J. M. Follett, and K. F. O'Connor. 1982. Growth, phosphate absorption and phosphorus chemical fractions in two *Chionochloa* species from high- and low-phosphate soils. *J. Ecol.* 70:305–321.

Chapin, G. S., and R. L. Bieleski. 1982. Mild phosphorus stress in barley and a related low-phosphorus-adapted barley grass: Phosphorus fractions and phosphate absorption in relation to growth. *Physiol. Plant.* 54:309–317.

Chen, C. H., and J. Lewin. 1969. Silicone as a nutrient element for *Equisetum arvense*. *Can. J. Bot.* 47:125–131.

Clarkson, D. T. 1976. The influence of temperature on the exudation of xylem sap from detached root systems of rye (*Secale cereale*) and barley (*Hordeum vulgare*). *Planta* 132:297–304.

Clarkson, D. T. 1985. Factors affecting mineral nutrient acquisition. *Annu. Rev. Plant Physiol.* 36:77–115.

Clement, C. R., M. J. Hopper, and L. H. P. Jones. 1978. The uptake of nitrate by *Lolium perenne* from flowing nutrient solutions. *J. Exp. Bot.* 29:453–464.

Collander, R. 1959. Cell membranes: Their resistance to penetration and their capacity for transport. In F. C. Steward (ed.), Plant Physiology: A Treatise, Vol. II: Plants in Relation to Water and Solutes. Academic Press, New York. Pp. 3–102.

Cram, W. J. 1973. Internal factors regulating nitrate and chloride influx in plant cells. *J. Exp. Bot.* 24:328–341.

Cram, W. J. 1975. Relationships between chloride transport and electrical potential differences in carrot root cells. *Aust. J. Plant Physiol.* 2:301–310.

Cram, W. J. 1976. Negative feedback regulation of transport in cells. The maintenance of turgor volume and nutrient supply. In U. Lüttge and M. G. Pitman (eds.), Transport in Plants. Encyclopedia of Plant Physiology, New Series, Vol. 2A. Springer-Verlag, Berlin. Pp. 284–316.

Deane-Drummond, C. E. 1982. Mechanisms for nitrate uptake into barley (*Hordeum vulgare* L. cv. Fergus) seedlings grown at controlled nitrate concentrations in the nutrient medium. *Plant Sci. Lett.* 24:72–89.

Deane-Drummond, C. E., and A. D. M. Glass. 1983. Short-term studies of nitrate uptake into barley plants using ion-specific electrodes and $^{36}ClO_3^-$. *Plant Physiol.* 73:100–104.

Dhugga, K. S., J. G. Waives, and R. T. Leonard. 1988. Correlated induction of nitrate uptake and membrane polypeptides in corn roots. *Plant Physiol.* 87:120–125.

Drew, M. C., and L. R. Saker. 1975. Nutrient supply and the growth of the seminal root system in barley. I'. Localized compensatory increases in lateral root growth and rates of nitrate uptake when nitrate supply is restricted to only part of the root system. *J. Exp. Bot.* 26:79–90.

Drew, M. C., and L. R. Saker. 1984. Uptake and long-distance transport of phosphate, potassium and chloride in relation to internal ion concentrations in barley: Evidence of non-allosteric regulation. *Planta* 160:500–507.

Drew, M. C., L. R. Saker, S. A. Barber, and W. Jenkins. 1984. Changes in the kinetics of phosphate and potassium absorption in nutrient-deficient barley roots measured by a solution depletion technique. *Planta* 160:490–499.

Dudal, R. 1976. Inventory of the major soils of the world with special reference to mineral stress hazards. In M. J. Wright (ed.), Plant Adaptation to Mineral Stress in Problem Soils. Cornell Univ. Press, Ithaca, New York. Pp. 3–13.

Edwards, J. H., and S. A. Barber. 1976. Nitrate flux into corn roots as influenced by shoot requirements. *Agron. J.* 68:471–473.
Elzam, O. E., D. W. Rains, and E. Epstein. 1964. Ion transport kinetics in plant tissue: Complexity of the chloride absorption isotherm. *Biochem. Biophys. Res. Commun.* 15:273–276.
Epstein, E. 1966. Dual pattern of ion absorption by plant cells and by plants. *Nature (London)* 212:1324–1327.
Epstein, E. 1976. Kinetics of ion transport and the carrier concept. In U. Lüttge and M. G. Pitman (eds.), Transport in Plants. Encyclopedia of Plant Physiology, New Series, Vol. 2B. Springer-Verlag, Berlin. Pp. 70–94.
Epstein, E., and C. E. Hagen. 1952. A kinetic study of the absorption of alkali cations by barley roots. *Plant Physiol.* 27:457–474.
Epstein, E., and D. W. Rains. 1965. Carrier-mediated cation transport in barley roots: Kinetic evidence for a spectrum of active sites. *Proc. Natl. Acad. Sci. U.S.A.* 53:1320–1324.
Evans, H. J., and R. A. Wildes. 1971. Potassium and its role in enzyme activation. *Potassium in Biochemistry and Physiology. Proc. 8th Colloq. Int. Potash Inst., Bern* pp. 13–39.
Evans, L. T., and R. L. Dunstone. 1970. Some physiological aspects of the evolution of wheat. *Aust. J. Biol. Sci.* 23:725–741.
Fernando, M., J. Kulpa, M. Y. Siddiqi, and A. D. M. Glass. 1990. Potassium-dependent changes in the expression of membrane-associated proteins in barley oats. 1. Correlations with K^+ ($^{86}Rb^+$) influx and root K^+ concentration. *Plant Physiol.* 92:1128–1132.
Fernando, M., M. Y. Siddiqi, and A. D. M. Glass. 1987. Protein profiles of the roots of barley seedlings in relation to potassium and nitrogen supply. *Plant Physiol.* 88:S 164.
Fischer, R. A. 1981. Optimizing the use of water and nitrogen through breeding of crops. *Plant Soil* 58:249–278.
Gabelman, H. W. 1976. Genetic potentials in nitrogen, phosphorus and potassium efficiency. In M. J. Wright (ed.), Plant Adaptation to Mineral Stress and Problem Soils. Cornell Univ. Press, Ithaca, New York. Pp. 205–212.
Gerloff, G. C. 1976. Plant efficiencies in the use of nitrogen, phosphorus and potassium. In M. J. Wright (ed.), Plant Adaptation to Mineral Stress and Problem Soils. Cornell Univ. Press, Ithaca, New York. Pp. 161–173.
Giordano, L. de B., W. H. Gabelman, and G. C. Gerloff. 1982. Inheritance of differences in calcium utilization by tomatoes under low-calcium stress. *J. Am. Soc. Hortic. Sci.* 107:664–669.
Glass, A. D. M. 1975. The regulation of potassium absorption in barley roots. *Plant Physiol.* 56:377–380.
Glass, A. D. M. 1976. Regulation of potassium absorption in barley roots: An allosteric model. *Plant Physiol.* 58:33–37.
Glass, A. D. M. 1977. Regulation of K^+ influx in barley roots: Evidence for direct control by internal K^+. *Aust. J. Plant Physiol.* 4:313–318.
Glass, A. D. M. 1983. Regulation of ion transport. *Annu. Rev. Plant Physiol.* 34:311–326.
Glass, A. D. M. 1988. Nitrogen uptake by plant roots. *ISI Atlas Sci.* 1:151–156.
Glass, A. D. M., and J. Dunlop. 1979. The regulation of K^+ influx in excised barley roots: Relationships between K^+ influx and electrochemical potential differences. *Planta* 145:395–397.
Glass, A. D. M., and J. E. Perley. 1980. Varietal differences in potassium uptake in barley. *Plant Physiol.* 65:160–164.
Glass, A. D. M., and M. Y. Siddiqi. 1984. The control of nutrient uptake rates in relation to the inorganic composition of plants. *Adv. Plant Nutr.* 1:103–147.
Glass, A. D. M., M. Y. Siddiqi, and K. Giles. 1981. Correlations between potassium uptake and hydrogen efflux in barley varieties. A potential screening method for the isolation of nutrient efficient lines. *Plant Physiol.* 68:457–459.
Glass, A. D. M., R. G. Thompson, and L. Bordelean. 1985. Regulation of NO_3^- influx in barley. Studies using $^{13}NO_3^-$. *Plant Physiol.* 77:379–381.
Graham, R. D. 1984. Breeding for nutritional characters in cereals. *Adv. Plant Nutr.* 1:57–102.

Gutschick, V. P. 1987. A Functional Biology of Crop Plants. Timber Press, Portland, Oregon.
Hanson, J. B. 1978. Application of the chemiosmotic hypothesis to ion transport across the root. *Plant Physiol.* 62:402–405.
Haschke, K. P., and K. Lüttge. 1975. Interactions between IAA, potassium and malate accumulation and growth in *Avena* coleopite segments. *Z. Pflanzenphysiol.* 76:450–455.
Hoagland, D. R., and T. C. Broyer. 1936. General nature of the process of salt accumulation by roots with description of experimental methods. *Plant Physiol.* 11:471–507.
Humble, G. D., and T. C. Hsiao. 1970. Light-dependent influx and efflux of potassium of guard cells during stomatal opening and closing. *Plant Physiol.* 46:483–487.
Jackson, W. A., K. D. Kwik, R. J. Volk, and R. G. Butz. 1976. Nitrate influx and efflux by intact wheat seedlings: Effects of prior nitrate nutrition. *Planta* 132:149–156.
Jensen, P. 1982. Effects of interrupted K^+ supply on growth and uptake of K^+, Ca^{2+}, Mg^{2+} and Na^+ in spring wheat. *Physiol. Plant.* 56:259–265.
Johansen, C., D. G. Edwards, and J. F. Loneragan. 1970. Potassium fluxes during potassium absorption by intact barley roots of increasing potassium content. *Plant Physiol.* 45:601–603.
Kay, R., A. Chan, M. Daly, and J. McPherson. 1987. Duplication of 355 Promoter sequences creates a strong enhancer for plant genes. *Science* 236:1299–1302.
Kochian, L. V., and W. J. Lucas. 1982. Potassium transport in corn. I. Resolution of kinetics into a saturable and linear component. *Plant Physiol.* 70:1723–1731.
Läuchli, A. 1972. Translocation of inorganic solutes. *Annu. Rev. Plant Physiol.* 23:197–218.
LaGuardia, M. D., J. M. Fournier, and M. Bennloch. 1985. Effect of potassium status on K^+ (Rb^+) uptake and transport in sunflower roots. *Physiol. Plant.* 63:176–180.
Lambers, H., R. K. Szaniawski, and R. de Visser. 1983. Respiration for growth, maintenance and ion uptake. An evaluation of concepts, methods, values and their significance. *Physiol. Plant.* 58:556–563.
Lee, R. B. 1982. Selectivity and kinetics of ion uptake by barley plants following nutrient deficiency. *Ann. Bot.* 50:429–449.
Lee, R. B., and D. T. Clarkson. 1986. Nitrogen-13 studies of nitrate fluxes in barley roots. I. Compartmental analysis from measurements of ^{13}N efflux. *J. Exp. Bot.* 37:1753–1767.
Lee, R. B., and M. C. Drew. 1986. Nitrogen-13 studies of nitrate fluxes in barley roots. II. Effect of plant N-status on the kinetic parameters of nitrate influx. *J. Exp. Bot.* 37:1768–1779.
Lee, R. B., and R. G. Ratcliffe. 1983. Phosphorus nutrition and the intracellular distribution of inorganic phosphate in pea root tips: A quantitative study using ^{31}P-NMR. *J. Exp. Bot.* 34:1222–1244.
Lee, R. B., and K. A. Rudge. 1986. Effects of nitrogen deficiency on the absorption of nitrate and ammonium by barley plants. *Ann. Bot.* 57:471–486.
Lefebvre, D. D. 1980. The regulation of phosphate uptake by intact barley plants. M. S. Thesis, Univ. of British Columbia, Vancouver.
Lefebvre, D. D., and A. D. M. Glass. 1982. Regulation of phosphate influx in barley roots. Effects of phosphate deprivation and reduction of influx with provision of orthophosphate. *Physiol. Plant.* 54:199–206.
Leigh, R. A., and R. G. Wyn Jones. 1984. An hypothesis relating critical potassium concentrations for growth to the distribution and functions of this ion in the plant cell. *New Phytol.* 97:1–13.
Loneragan, J. F. 1976. Plant efficiencies in the use of B, Co, Cu, Mn and Zn. In M. J. Wright (ed.), Plant Adaptation to Mineral Stress in Problem Soils. Cornell Univ. Press, Ithaca, New York. Pp. 193–203.
Loneragan, J. F., and C. J. Asher. 1967. Response of plants to phosphate concentration in solution culture. II. Rate of phosphate absorption and its relation to growth. *Soil Sci.* 103:311–318.
Marschner, H. 1986. Mineral Nutrition of Higher Plants. Academic Press, London.
McClure, P. R., T. E. Omholt, G. M. Pace, and P.-Y. Bouthyette. 1987. Nitrate-induced changes in protein synthesis and translation of RNA in maize roots. *Plant Physiol.* 84:52–57.
Memon, A. R., and A. D. M. Glass. 1987. Genotypic differences in subcellular compartmentation of K^+: Implications for protein synthesis, growth and yield. In H. W. Gableman and B. C. Loughman (eds.), Genetic Aspects of Plant Mineral Nutrition. Nijhoff, Dordrecht, Netherlands.

Memon, A. R., M. Saccomani, and A. D. M. Glass. 1985. Efficiency of potassium utilization by barley varieties: The role of subcellular compartmentation. *J. Exp. Bot.* 36:1860–1876.
Munns, D. N. 1968. Nodulation of *Medicago sativa* in solution culture. III. Effects of nitrate on root hairs and infection. *Plant Soil* 29:33–37.
Nielsen, N. E., and J. K. Schjørring. 1983. Efficiency and kinetics of phosphorus uptake from soil by various barley genotypes. *Plant Soil* 74:225–230.
Nissen, P. 1973. Kinetics of ion uptake in higher plants. *Physiol. Plant.* 28:113–120.
Nissen, P. 1980. Multiphasic uptake of potassium by barley roots of low and high potassium content: Separate sites for uptake and transitions. *Physiol. Plant.* 48:193–200.
Pace, G. M., and P. R. McClure. 1986. Comparison of nitrate uptake kinetic parameters across maize inbred lines. *J. Plant Nutr.* 9:1095–1111.
Passioura, J. B. 1979. Accountability, philosophy and plant physiology. *Search* 10:347–350.
Pearson, C. J., R. J. Volk, and W. A. Jackson. 1981. Daily changes in nitrate influx, efflux and metabolism in maize and pearl millet. *Planta* 152:319–324.
Perby, H., and P. Jensen. 1983. Varietal differences in uptake and utilization of nitrogen and other macroelements in seedlings of barley (*Hordeum vulgare* L.). *Physiol. Plant.* 58:223–230.
Petraglia, T., and R. J. Poole. 1980. ATP levels and their effect on plasma lemma influxes of potassium chloride in red beet. *Plant Physiol.* 65:969–972.
Pettersson, S., and P. Jensen. 1983. Variation among species and varieties in uptake and utilization of potassium. *Plant Soil* 72:231–237.
Pitman, M. G. 1976. Ion uptake by plant roots. In U. Lüttge and M. G. Pitman (eds.), Transport in Plants. Encyclopedia of Plant Physiology, New Series, Vol. 2B. Springer-Verlag, Berlin. Pp. 95–128.
Pitman, M. G. 1982. Transport across plant roots. *Q. Rev. Biophys.* 15:481–554.
Pitman, M. G., A. C. Courtice, and B. Lee. 1968. Comparison of potassium and sodium uptake by barley roots on high and low salt status. *Aust. J. Biol. Sci.* 21:871–881.
Pitman, M. G., and W. J. Cram. 1973. Regulation of inorganic ion transport in plants. In W. P. Anderson (ed.), Ion Transport in Plants. Academic Press, London. Pp. 465–481.
Pitman, M. G., and W. J. Cram. 1977. Regulation of ion contact in whole plants. *Symp. Soc. Exp. Bot.* 31:391–424.
Presland, M. R., and G. S. McNaughton. 1984. Whole plant studies using radioactive 13-nitrogen. II. A compartmental model for the uptake and transport of nitrate ions by *Zea mays*. *J. Exp. Bot.* 35:1277–1288.
Price, N. M., P. A. Thompson, and P. J. Harrison. 1987. Selenium: An essential element for growth of the coastal marine diatom *Thalassiosira pseudonana* (Bacillariophyceae). *J. Phycol.* 23:1–9.
Raven, J. A. 1977. Regulation of solute transport at the cell level. *Symp. Soc. Exp. Biol.* 31:73–99.
Raven, J. A., and F. A. Smith. 1978. Effect of temperature on ion content, ion flux and energy metabolism in *Chara corallina*. *Plant, Cell Environ.* 1:231–238.
Rea, P. A., and D. Sanders. 1987. Tonoplast energization: Two H^+ pumps, one membrane. *Physiol. Plant.* 71:131–141.
Rebeille, F., R. Bligny, J. B. Martin, and R. Douce. 1983. Relationship between the cytoplasm and the vacuole phosphate pool in *Acer pseudoplatanus* cells. *Arch. Biochem. Biophys.* 225:143–148.
Reisenauer, H. M. 1966. Mineral nutrients in soil solution. In P. L. Altman and D. S. Ditmer (eds.), Environmental Biology. Fed. Am. Soc. Exp. Biol., Bethesda, Maryland. P. 507.
Rodriquez-Navarro, A., M. A. Blatt, and C. L. Slayman. 1986. A potassium–proton symport in *Neurospora crassa*. *J. Gen. Physiol.* 87:649–674.
Römheld, V. 1986. Cited in Marschner, 1986.
Sanders, D. 1986. Generalized kinetic analysis of ion-driven cotransport systems. II. Random ligand binding as a simple explanation for non-Michaelian kinetics. *J. Membr. Biol.* 90:67–87.
Saric, M. R. 1987. Progress since the first international symposium: Genetic Aspects of Plant Mineral Nutrition, Belgrade, 1982, and perspectives of future research. *Plant Soils* 99:197–209.
Satter, R. L., P. B. Applewhite, and A. W. Galston. 1974. Rhythmic potassium flux in *Albizzia*. Effect

of aminophylline, cations and inhibitors of respiration and protein synthesis. *Plant Physiol.* 54:280–285.
Siddiqi, M. Y., and A. D. M. Glass. 1981. Utilization index: A modified approach to estimation and comparison of nutrient utilization efficiency in plants. *J. Plant Nutr.* 4:289–302.
Siddiqi, M. Y., and A. D. M. Glass. 1983. Studies of the growth and mineral nutrition of barley varieties. I. Effect of potassium supply on the uptake of potassium and growth. *Can. J. Bot.* 61:671–678.
Siddiqi, M. Y., and A. D. M. Glass. 1987. Regulation of K^+ influx in barley: Evidence for a direct control of influx by K^+ concentration of root cells. *J. Exp. Bot.* 38:935–947.
Siddiqi, M. Y., A. D. M. Glass, T. J. Ruth, and M. Fernando. 1989. Studies of the regulation of nitrate influx by barley seedlings using $^{13}NO_3^-$. *Plant Physiol.* 90:806–813.
Siddiqi, M. Y., A. R. Memon, and A. D. M. Glass. 1985. The regulation of K^+ influx in barley: Effects of low temperature. *Plant Physiol.* 74:730–734.
Smith, F. A. 1973. The internal control of nitrate uptake into excised barley roots with differing salt contents. *New Phytol.* 72:769–782.
Smith, I. K. 1975. Sulphate transport in cultured tobacco cells. *Plant Physiol.* 55:303–307.
Spear, S. N., C. J. Asher, and D. G. Edwards. 1978. Response of cassava, sunflower and maize to potassium concentration in solution. III. Interactions between potassium, calcium and magnesium. *Field Crops Res.* 1:375–389.
Sze, H. 1985. H^+-translocating ATPases: Advances using membrane vesicles. *Annu. Rev. Plant Physiol.* 36:175–208.
Therios, I. N., S. A. Weinbaum, and R. M. Carlson. 1979. Nitrate compensation points of several plum clones and relationship to nitrate uptake effectiveness. *J. Am. Hortic. Sci.* 104:768–770.
Thibaud, J. B., and C. Grignon. 1981. Mechanisms of nitrate uptake in corn root. *Plant Sci. Lett.* 22:279–289.
Troughton, A. 1977. Relationship between root and shoot systems of grasses. In J. K. Marshall (ed.), The Belowground Ecosystem: A Synthesis of Plant Associated Processes. Range Sci. Ser. No. 26. Colorado State Univ., Fort Collins. Pp. 39–52.
Ullrich, W. R., and A. Novacky. 1981. Nitrate dependent membrane potential changes and their induction in *Lemna gibba*. *Plant Sci. Lett.* 22:211–217.
Ullrich-Eberius, C. I., A. Novacky, E. Fischer, and U. Lüttge. 1981. Relationship between energy dependent phosphate uptake and the electrical membrane potential in *Lemna gibba*. *Plant Physiol.* 67:797–801.
Vose, P. B. 1963. Varietal differences in plant nutrition. *Herb. Abstr.* 33:1–13.
Warncke, D. D., and S. A. Barber. 1974. Nitrate uptake effectiveness of four plant species. *J. Environ. Qual.* 3:28–30.
Weaver, J. E. 1919. The Ecological Relations of Roots. Carnegie Inst., Washington, D.C.
Werner, D., and R. Roth. 1983. Silica metabolism. In A. Läuchli and R. L. Bieleski (eds.), Inorganic Plant Nutrition. Encyclopedia of Plant Physiology, New Series, Vol. 15B. Springer-Verlag, Berlin. Pp. 682–694.
Woodend, J. J. 1986. Genetic and physiological studies on potassium and nitrogen uptake and utilization in wheat. Ph.D. Thesis, Univ. of British Columbia, Vancouver.
Woodhouse, P. J., A. Wild, and C. R. Clement. 1978. Rate of uptake of potassium by three crop species in relation to growth. *J. Exp. Bot.* 29:885–894.
Young, M., and A. P. Sims. 1972. The potassium relations of *Lemna minor* L. I. Potassium uptake and plant growth. *J. Exp. Bot.* 23:958–969.
Zimmerman, U. 1978. Physics of turgor and osmoregulation. *Annu. Rev. Plant Physiol.* 29:121–148.

3

Plant Nutrition Relationships at the Whole-Plant Level

P. B. VOSE

I. Nutritional Variation

Plant science has learned so much about the functioning of plants through studies of excised roots, leaf segments, tissue culture, cell preparation, isolated enzyme systems, and the like that it is sometimes easy to forget that plants are closely integrated systems in which root and shoot functioning are mutually affected in the determination of yield. This chapter considers some general aspects of whole-plant functioning as it applies to nutritional variation within crop species.

From the practical point of view, at least four main forms of cultivar difference in plant nutrition can be recognized: (i) differential yield response, (ii) differential nutrient uptake, (iii) differential requirement for specific elements, and (iv) differential tolerance to toxicities. These are the recognizable end-effects of many integrated

Crops as Enhancers of Nutrient Use

processes, such as absorption, translocation, assimilation, redistribution, and detoxification/internal immobilization.

We know that many mechanisms of nutritional variation exist, and in many cases certainly more than one must be operational, but in general terms they comprise the following: root morphology, root metabolism, poorly understood factors governing distribution of dry matter between root and shoot (i.e., shoot/root ratio), active nutrient uptake, translocation and/or immobilization, metabolic incorporation, enzyme activity, and redistribution. The study of genetic variation in nutrition is certainly doing much to improve our basic knowledge of the mechanisms of plant nutrition acquisition, uptake, transport, and assimilation (Läuchli, 1976) but most of the many reviews (Vose, 1963, 1984; Myers, 1960; Epstein, 1963, 1980; Gerloff, 1963; Wright, 1977; Saric, 1981; Graham, 1984) have concentrated on the practical application aspects of plant nutritional variation. From these reviews it appears that one can almost certainly obtain genetic effects for any aspect of ion transport, accumulation, and efficiency of use, or tolerance to toxic levels, for virtually any element. Much recent work has concentrated on the identification and selection of cultivars able to withstand mineral stresses in problem soils, particularly resistance to the factors concerned with acid soils, i.e., tolerance to Al and Mn toxicities and ability to withstand low available P and Ca, tolerance to heavy metals, and tolerance to salinity, especially NaCl.

It is also appreciated that cultivars which are efficient in their use of major nutrients such as N and P, and to a lesser extent K, and of certain microelements, could be at least of equal value, either at the very highest levels of soil fertility and fertilization, through the more efficient use of fertilizers, or for low-input systems. We know that some cultivars can produce more dry matter per unit of nitrogen supplied, but they are not always the highest economic yielders. Some genotypes make better use of the nitrogen they take up by using it to produce economic yield, for example, grain as opposed to straw.

Differences in plant mineral composition, although providing no evidence of basic differences in nutrient uptake, and seldom being related to cultivar differences in production, are nevertheless important in their own right. They may provide, for example, evidence on inefficient nutrient conversion, while on the other hand cation and phosphorus content may be of importance for animal nutrition. The use of foliar analysis, especially for such crops as citrus and sugarcane, makes it essential that cultivars should be recognized that have a composition varying from the mean for the species.

The examples of nutrient variation that have been reported demonstrate the range of possible improvements in crop performance or particular characteristics that may be brought about by conscious selection for nutritional characters. There are of course some problems. For example, will selection for one specific nutritional character alter the balance of the others, so that another problem, maybe a deficiency syndrome, results? Probably the answer is "no" in the case of micronutrients, as they do not greatly influence the plant's cation/anion balance. Macroelements are more problematical, but results so far indicate that quite large differences are possible. In extreme selection for salinity or Al tolerance, it is likely that if selection is pushed too far then very tolerant plants may be obtained at the sacrifice of yield and maybe other agronomic characters.

Work on a crop plant which is to have ultimate meaning in terms of yield increase, or yield maintenance, must not only be connected with the particular nutrient requirements of the crop, but be related to an overall program of selection, breeding, and agronomy.

II. Root Effects on Whole-Plant Functioning

A. ROOT VARIABILITY

Although many whole-plant factors may affect total cultivar response to nutritional situations, the root is undoubtedly preeminent in nutrient acquisition and in plant tolerance to toxicities such as Al, Mn, heavy metals, and salinity.

Root systems have many variables which may affect function: apart from rooting depth, these include deep rooting versus lateral spreading, branching or little-branched main root, many or few secondary or tertiary roots (Hackett, 1969), and many or few root hairs (Bole, 1973). There are also characteristic differences in shoot/root ratio, and in ryegrass these are certainly heritable (Vose, 1962a). In barley and wheat there are considerable cultivar differences in rooting depth and form of root. Although in cereals semidwarf types tend to have shallow root systems, a number of workers have found that no direct relationship exists between rooting capability and size of tops (Lupton *et al.*, 1974; Irvine *et al.*, 1980). It should be noted that most crop plants have a reserve of photosynthetic capacity and there is no reason why a larger root system, if desired, could not be supported. We might suppose that plants with relatively small roots in relation to tops must have especially effective root systems, and some evidence to support this is available. Overall size, and particularly lateral spread of the root system, is important in crop plants grown in rows because the larger root system can exploit a greater volume of soils [see Vose (1962a) and Nye and Tinker (1977) for discussions]. Total root size is of diminished importance when plants are grown closer together as in cereals and grassland; here rooting depth to confer drought tolerance may be more important.

Apart from the obvious weight and spatial variables of root systems there is both inter- and intraspecies variation in age distribution of secondary roots (Troughton, 1981a) and also of the extent of root life (Troughton, 1981b). Such differences are particularly important for the uptake of those elements which are primarily taken up from the younger parts of the root, such as manganese (Page *et al.*, 1962) and calcium, magnesium, and iron (Clarkson and Sanderson, 1971), which are primarily absorbed through the root apices where the endodermis is unsuberized. Age of root seems relatively unimportant for the uptake of potassium, ammonium, and phosphate ions (Clarkson and Hanson, 1980).

Root density must be particularly important for the uptake of those elements that are taken up by diffusion and contact, and especially in poor soil, a more extensively branched root system is likely to be much more effective than a smaller or less branched system. Kraus *et al.* (1987) demonstrated with ^{33}P that the P-depletion zone around the primary root of a maize plant does not exceed the zone of the root hair cylinder. This finding supports the models proposed by van Noordwijk (1983) and colleagues to determine the optimum density of roots. They suggested that for N uptake as little as 0.1 cm of roots per cm^3 of soil is required. But in the case of P nutrition, root density is effective in explaining degrees of P response of a crop within a root density range as great as 1–10 cm of roots per cm^3 of soil.

B. ROOT FUNCTIONING

When a nutrient is at a low concentration in the medium, then root surface is most likely to be the limiting factor in ion absorption (Nye, 1973). Low P is a frequent situation. Even in culture solution, Troughton (1959) found that the uptake of ^{32}P by branched roots of perennial ryegrass can be as much as 30% greater than by unbranched roots. Much later work on P-uptake prediction models (Barber, 1982) coupled with actual experiments has indicated that root size as measured by length and radius had the greatest effect on P uptake. In other species, root hairs are important for uptake, for example, in white clover (Caradus, 1982).

Lyness (1936) and Smith (1934) long ago showed a direct correlation between "P-efficiency" and the number of secondary roots in relation to primary roots, and that the root type of efficient parents was dominant in hybrids. Rabideau *et al.* (1950) studied the absorption of ^{32}P by inbred lines of maize and their hybrids and found that the hybrid with the largest root system absorbed the most P. One line and its hybrids consistently absorbed more P than did other lines. Schenk and Barber (1979) have found that P acquisition in maize can be increased by development and selection of hybrids with more fibrous root systems.

A smaller root system may be basically successful in taking up P because of inherent superior P-uptake capacity at the metabolic level. Lindgren *et al.* (1977) found that the rate of P absorption by *Phaseolus* beans from low-P concentrations was negatively correlated with the weight of the excised roots of the different lines. They ascribed the increased rate of P absorption associated with decreased root surface to superior P-uptake capacity. Similarly, Raper and Barber (1970a,b) showed that although the soybean cultivar 'Harasoy 63' had a more extensive root system than that of 'Aoda,' the capacity for nutrient absorption at high nutrient levels was twice as great in 'Aoda' than in 'Harasoy 63.' However, at low nutrient levels the capacity of each cultivar to absorb nutrients was about the same, so that 'Harasoy 63' would have an advantage because of its more extensive root system. Root morphology effects are particularly clear-cut in the case of P uptake, but they can also be found in N uptake as noted for ryegrass (Goodman, 1981).

In barley and wheat there are considerable cultivar differences in root depth and form, which can be significant for fertilizer use efficiency, for resistance to minor element deficiencies, or to drought. Hahr (1975) found that the barley cultivar 'Carlsberg II' was able to take up a larger amount of water from deeper soil layers than a mutant with a much shallower root system. It was reported by Swaminathan (1972) that in the zinc-deficient soils of the Punjab, India, when a deep-rooted wheat cultivar like 'Kalyan Sona' was grown in rotation with a deep-rooted cotton crop then zinc deficiency was aggravated. The deficiency problem was much reduced when a wheat cultivar with spreading lateral roots was used. Variation in wheat root form and depth of rooting has been widely reported, from India (Subbiah *et al.*, 1968), from the U.K. (Newbould *et al.*, 1970), and from the U.S.S.R. (Vedrov, 1974).

C. ROOT METABOLISM

Although it is probable that root morphological effects are primarily responsible for differences in P nutrition, metabolic effects also exist. Lyness (1936) had already demonstrated differential yield response and a difference in P-absorbing capacity among 21 cultivars of maize grown at different levels of P in sand culture and nutrient solution. One cultivar with a high P requirement apparently only had a limited capacity of P absorption from low-P concentrations. Lindgren *et al.* (1977) used excised roots to evaluate the differences in P absorption by 59 lines of *Phaseolus vulgaris*. P-absorption rates varied greatly between lines and relative values for efficient and inefficient lines remained constant with plant age.

Soybean cultivars can show sensitivity to high levels of P, particularly in banded fertilizers, and this appears to be related to efficiency of P uptake by the root. Howell and Foote (1964) found that some soybean cultivars showed distinct sensitivity to high P levels, e.g., 'Lincoln,' while others such as 'Chief' were tolerant. The difference in P sensitivity of these cultivars was located primarily in the roots, so that 'Lincoln' soybeans take up more phosphorus than did 'Chief' from low-N and high-P solutions. Ferreyra Harnandez (1978) was able to demonstrate significant differences in uptake of P by excised roots of two maize cultivars, and a kinetic analysis indicated differences in "carrier" concentration in the roots. Clark and Brown (1974) compared the uptake of P from nutrient solution by maize cultivars 'Pa36' and 'WH.' They found that the accumulator 'Pa36' took up higher amounts of P from low-P concentrations than did 'WH.' Phosphatase activity of 'Pa36' roots was higher than in those of 'WH,' and phosphatase activity increased with decreasing nutrient solution P. Burauel *et al.* (1990) reported that the P-efficient soybean cultivar 'Century' supplies increased carbohydrate to the roots compared to the inefficient 'Woodworth' cultivar, and also has a more active proton release.

Cultivar differences in the uptake of inorganic Fe from solution by excised rice roots had also been shown to be related to differences in carrier concentration (Shim and Vose, 1965). However, soil conditions are much more complex and iron absorption and transport are mediated by the release of hydrogen ions by the root, thus lowering the pH of the root zone to favoring Fe^{3+} solubility and reduction of Fe^{3+} to Fe^{2+}. The reduction of Fe^{3+} to Fe^{2+} is carried out by reductants released by the roots, making it possible for the Fe^{2+} to enter the root (Olsen and Brown, 1980a–c). Genotypes which respond positively to iron stress apparently have the capacity to increase or maintain hydrogen ion excretion from the roots, excrete reducing compounds from the roots, increase the rate of reduction of Fe^{3+} to Fe^{2+} at the root surface, and increase the level of citrate chelate in the root sap (Brown *et al.*, 1961; Brown and Bell, 1969; Brown and Ambler, 1973, 1974). Evidence has been presented for a specific uptake system for phytosiderophores in grasses (Römheld and Marschner, 1986) and general methods have been developed for quantifying release of H^{+} (Marschner *et al.*, 1982) and increase of reductive capacity (Marschner *et al.*, 1986).

The release of chelating compounds has also been demonstrated to influence the availability of Mn (Godo and Reisenauer, 1980) and probably a similar story comparable to the iron mechanism will ultimately be developed for most microelements; currently iron has been of major importance and has had the most work.

D. ROOT/SHOOT RELATIONS

Although it is self-evident that root and shoot functioning must be closely integrated, it remains an area which still seems poorly researched regarding nutritional effects. However, variability in translocation patterns of some elements, e.g., Mg and Na, is well documented.

Sayre (1951) found maize cultivar 'WF9' to be "Mg-efficient" and 'Oh40B' to be "Mg-inefficient," and subsequently Foy and Barber (1958) found higher Mg in the roots and stems of 'Oh40B' and attributed the difference between the inbreds for Mg accumulation to immobilization of Mg in culms and nodes. This was subsequently confirmed by Clark (1975), who found that 'Oh40B' had the ability to take up as much Mg as the efficient cultivar 'B57,' but has poor ability to translocate Mg to the tops. Probably, poor translocation of microelements is a more widespread feature than is realized—with Mg the consequences are easily recognized.

Poor translocation need not always be a negative feature, because frequently plant tolerance to salinity is, at least in part, due to Na exclusion, which can operate in two ways: discrimination against Na uptake by roots in favor of K (Epstein and Hagen, 1952; Fried and Noggle, 1958; Rains and Epstein, 1967) or alternatively against Na transport by stems. Poor stem transport of Na relative to K is well known (Wallace *et al.*, 1965) and species differences in the ability to translocate Na to plant tops are long established (Huffaker and Wallace, 1959).

In plants which are normally grafted, a considerable effect of rootstock on the response to and accumulation of many elements in a wide variety of crops, including citrus (Haas, 1948), avocado (Embleton *et al.*, 1962), and grape vine (Cook and Lider, 1964), has been found. The basis of these effects has received little research, but most stock/scion effects are due to the stock, exceptions being boron accumulation in walnut (Eaton and Blair, 1935) and sodium content of citrus (Cooper *et al.*, 1952) when the effect was found to be due to the scion. Translocation effects/defects are probably much more generally common than is appreciated, but they are usually missed in annual plants in which the only grafting carried out is experimental. Some stock/scion effects are known in annuals, e.g., sunflower (Eaton and Blair, 1935), soybeans (Kleese, 1968), and tomato (Brown *et al.*, 1971).

Reports of cultivar differences in the leaf content and accumulation of many cations are commonplace—see Vose (1984) for comprehensive references—but not much information appears to exist concerning specific mechanisms, although ion uptake and transport, and especially leaf metabolism, must be integrated processes. Leaf cells of different cultivars can vary in their capacity to take up inorganic ions, as was shown for Mn with oat leaf discs (Vose, 1962b), while the article by Glass and Siddiqui (1984) indicates clearly that uptake, transport, and mobilization respond to the demands of the shoot. Differences of leaf cation content above the normal range rarely have significance for crop yield but may be important for animal nutrition, e.g., differences in content of Ca as great as 200% have been reported.

III. Efficient Cultivars: A Whole-Plant Function

A. PRODUCTION AND HARVEST INDEX

The productivity of crops is basically concerned with the uptake and utilization of essential elements, especially nitrogen and phosphorus, coupled with photoassimilation. Nevertheless, many photosynthetic studies have not taken into account the nutritional status of the plant, and countless nutrient studies have ignored photosynthesis as the energy source for all plant productivity.

The question of general efficiency is extremely complex. The definition of efficiency is itself a matter for discussion, for we have "responding" and "nonresponding" cultivars and genotypes according to the degree of response to increases in nutrients. Cultivars exist which are "efficient" or "inefficient" converters of nutrients into dry matter, some being specifically efficient when grown at lower levels of a particular nutrient. Cultivars may be "efficient" or "inefficient" for ion uptake or translocation, or "accumulators" or "nonaccumulators" of certain elements. One thing seems clear: efficiency in plants involves whole-plant function.

I and some others have related efficiency mainly to the production of dry matter per unit of N (or other element) involved. This is valid for forage crops where all the aboveground production is utilized, but it is true in only the most general sense for cereals and pulses, where the prime interest is in the grain. Here the Harvest Index (HI),

$$\mathrm{HI} = \frac{\text{grain}}{\text{total yield}} \times 100 \qquad (1)$$

or the Nitrogen Index (NI),

$$\mathrm{NI} = \frac{\text{grain N}}{\text{total N}} \times 100 \qquad (2)$$

are more appropriate measures.

Asana *et al.* (1968) pointed out that the "most efficient" wheat, as measured by production of dry matter per unit of N, is not necessarily the most productive, because yield in cereals is partly determined by tillering, number of fertile tillers (ears), number of grains per ear, and size of grain. Brinkman and Rao (1984) found in a comparative test of oat cultivars at different N levels that the cultivar showing greatest grain yield response to nitrogen did so because of better response of spikelets per plant and weight per kernel.

Plant breeders have had their own way of coming to terms with nutrition and photoassimilation by selecting empirically for plant types which give the highest yield under high fertilization, together with disease resistance, and they have been very successful. As was pointed out by Stoy (1975), although many cases of differences in rates of photosynthesis between species and cultivars have been found (Hesketh and Moss, 1963; Stoy, 1965; Apel, 1967; Curtis *et al.*, 1969), no clear-cut relationship was found between photosynthetic activity and yield performance. However, it has been found that if one selects for plants of a certain type then photoassimilation per unit of field area, and hence productivity, can be greatly increased. For example, in the United States, increasing maize yields have come about through greater plant populations with decreased row width, and with cultivars having a reduced tendency to tiller but with angled leaves for good light interception (Duncan, 1977). In the case of wheat (Vogel *et al.*, 1963) and rice (Jennings, 1964), a canopy structure which derives from semidwarf straw, with fewer short, thick, highly angled leaves, allows the best penetration of light and the consequent high yields of modern cultivars.

This is shown by data from South Korea (Kim *et al.*, 1986), given in Table I, comparing the photosynthetic activity per unit area of rice cultivars from 1910 until 1977. At a comparatively low level of fertilization, relative photosynthetic ability has been almost doubled, and was increased by more than 50% even at a high level of fertilization. The subsequent effect on yield at different levels of fertilization is shown in Table I, where the modern cultivar 'Geumganbyeo' continues to respond even at 240 kg N ha^{-1}, whereas the yield of the older varieties is depressed. But note, too, that the modern cultivar is the highest yielder even at the lowest N level.

Thus, the dramatic yield increases in wheat and rice over the last 25 years have come about primarily through improved harvest index, i.e., dry matter has been moved from unwanted parts such as excess leaves and culm to the more desired grain component; there has not been much increase in total yield per unit of fertilizer. Nevertheless, Mengel (1982) noted that modern cultivars of wheat and rice are

Table I. Photosynthetic Ability of Rice Varieties Developed in Korea, 1910–1977[a]

Year Released	Variety	Photosynthetic Ability (mg CO_2 dm^{-2} hr^{-1}) N-P_2O_5-K_2O 60-40-40 (kg ha^{-1})	N-P_2O_5-K_2O 240-160-160 (kg ha^{-1})
1910	'Jodongji'	17.2 (78)[b]	20.0 (84)
1944	'Paldal'	18.5 (84)	23.5 (99)
1962	'Jinheung'	22.1 (100)	23.8 (100)
1972	'Tongil'	25.3 (114)	26.8 (113)
1977	'Geumganbyeo'	28.3 (128)	32.6 (137)

[a]Source: Kim *et al.*, 1986.
[b]Figures within brackets are percentages relative to Jinheung.

very efficient in their use of P and K, because the harvest index is greatly increased compared with older cultivars, thus less P and K is required as the result of fewer leaves requiring support. However, on a "per hectare" basis the requirement for N, P, and K fertilizer may increase because the planting density must be increased both to achieve maximum yield and also for weed control.

Modern cultivars tend to be faster growing than the ones they replace, thus setting up increased demand for nutrients, which they will receive inadequately under natural soil fertility, and therefore show great responsiveness to fertilizer application. This applies not only to cereals, e.g., with tomatoes Lingle and Lorenz (1969) found that cultivars bred for high production and dense planting rates can make high demands on K absorption and transport, and these cultivars are very liable to K deficiency, which is accentuated by a short growing season. The converse of this is that older crop cultivars are slower growing but can produce with a lower rate of nutrient supply. Frequently, too, landraces or undeveloped ecotypes require little fertilization, but likewise have poor inherent physiological capacity for yield response. Is it therefore not possible to improve the fundamental efficiency in utilization of nutrients by crop plants, as opposed to merely moving around dry matter from unwanted parts such as foliage, stems, or roots to desired components such as grain, fruit, or tubers? Do low-input cultivars necessarily imply low-yield cultivars? The work summarized by Gabelman and Gerloff (1982) suggests that efficiency of nutrient utilization can be improved. Growing genotypes of tomatoes and *Phaseolus* beans under nutrient stress conditions of K, P, N, and Ca revealed some to be much more efficient than others and that broad sense heritability estimates of nutrient use efficiency were found to be high and that selection for efficiency would be effective.

Ramirez (1982) compared 60 maize (*Zea mays* L.) inbreds for their capacity to produce dry matter per unit of N, P, or K and found that there was considerable variation in the efficient use of nutrients. Although dry matter production was in general well correlated with total N, P, and K accumulation, i.e., larger plants had greater total nutrient content, nevertheless efficient use of nutrients appeared independent of plant size, which suggests the possibility of developing genuinely low-input cultivars of moderate yield.

B. NITROGEN UTILIZATION

Improved nitrogen utilization continues to be a main target. Is it possible to use less N fertilizer and can we achieve better understanding of the mechanisms of nitrogen response? Can more efficient cultivars result in less nitrate contamination of groundwaters, as is being currently emphasized (Royal Society, 1983). It is known that some genotypes have better uptake capacity for nitrate or ammonium nitrogen; some cultivars produce more dry matter per unit of N applied; other genotypes have better capacity for mobilization of nitrogen from the leaves and for distribution to the grain; and some genotypes may be adapted to either high or low N nutrition.

The factors influencing grain protein are very complex, as was shown by Huffaker and Rains (1978), who found that the breakdown and retranslocation of protein from the foliage to the grain during seed filling was a critical physiological marker for grain protein content in wheat. 'Anza' and 'UC44-111' showed marked differences in nitrate content of leaves, and although 'Anza' contained considerably more nitrate, it absorbed less nitrate in short-term studies.

Cultivar differences in N utilization and dinitrogen fixation in five *Phaseolus* cultivars were demonstrated by Ruschel *et al.* (1982) in a field experiment utilizing the ^{15}N technique. Data in Table II show that there are quite big differences in both the ratio of kg grain per kg N_2 fixed and also kg grain per kg N derived from all sources (i.e., from soil, fertilizer, and N_2 fixation) and it is noteworthy that such variation can occur in highly selected cultivars.

Table II. Parameters of Dinitrogen Fixation and Utilization Efficiency of *Phaseolus* Cultivars Grown in the Field with 20 kg N ha^{-1} Given as a "Starter"[a]

	Seed	N_2 Fixed		Ratio of kg Seed per kg N		Harvest Index[b] Percent	
Cultivar	Yield (kg ha^{-1})	Total (kg ha^{-1})	As % in Crop	N_2 Fixed	All Sources	Yield	Nitrogen
'Carioca precoce'	545	46	63	12.2	7.4	19.0	26.9
'Goiano precoce'	271	25	38	13.0	4.1	11.3	22.8
'Costa Rica'	583	58	65	11.0	6.7	11.6	24.2
'Carioca'	917	65	65	15.0	9.1	18.3	31.9
'Moruna'	343	37	61	9.1	5.7	12.5	13.1
'LDS' (<0.05)	149	29	n.d.[c]	6.5	4.2	n.d.	n.d.

[a]Derived from Ruschel *et al.*, 1982.
[b]Harvest Index = seed/total × 100.
[c]n.d. = not determined.

The data suggest that efficiency of nitrogen utilization, i.e., as measured by seed production per unit of plant nitrogen, is not directly linked to efficiency of dinitrogen fixation. The cultivar 'Goiano Precoce,' for example, a "60-day" short season cultivar, has only 40% of plant nitrogen due to fixation, and while having a low yield of grain also showed a quite high harvest index for N, as well as yield of seed per kg N_2 fixed. 'Carioca,' which had the greatest seed yield, showed good nitrogen fixation and good conversion of nitrogen. These differences are encouraging as an indication of the possibility of selecting for improved N utilization. The data for 'Moruna' are instructive. This was a then recently developed cultivar selected for yield and disease resistance, yet it performed very poorly when supplied with only 20 kg N ha^{-1} of fertilizer. It was selected under high levels of fertilization (Pompeu *et al.*, 1978), which suggests the possibility of inadvertently selecting away from both highly efficient N utilization and N_2 fixation.

With herbage grasses there is not the complication of redistribution since the product is harvested primarily in the vegetative form. Early work with ryegrass, *Lolium perenne* L. (Vose and Breese, 1964), demonstrated that one could have high dry matter–low N content plants, while Antonovics *et al.* (1967) found wild populations of *Lolium perenne* L. adapted to either high or low levels of habitat N. Subsequently, Goodman (1977, 1981, 1982) has demonstrated the possibility of selection for increased nitrogen uptake and yield in ryegrass. Although Goodman (1981) found that nitrogen response is linked to nitrate reductase activity, he also noted that ion (nitrate) uptake has two components: uptake velocity per unit size of root and root size, and that these two components may vary at different stages of maturity. This suggests that if grass and cereal cultivars are to be developed which are more efficient in recovery of applied fertilizer nitrogen, with less nitrate being lost to groundwater, then probably selection for increased root size will be an essential component.

C. PHOSPHORUS UTILIZATION EFFICIENCY

Following nitrogen, phosphorus is the second most important nutrient for crop production, and many quite large differences in response have been recorded. Variation in phosphorus nutrition can be important through adaptability to low P, ability to gather P from low-P soils, response to P, tolerance to high P in band fertilization, and tolerance to P-associated Zn- and Cu-deficiency susceptibility (Sumner *et al.*, 1982). Metabolic differences in phosphorus uptake and assimilation have been described for *Phaseolus*, for maize cultivars, for tomatoes, and for soybeans. For example, Coltman *et al.* (1982) have described differences in both growth and phosphorus acquisition and utilization for tomatoes grown under P-deficiency stress.

It was previously noted that the pioneer work of Lyness (1936) showed primarily a correlation between "P-efficiency" and rooting capacity, but even so it appeared that basic differences in capacity for P utilization existed. The more recent work of

Whiteaker *et al.* (1976) involved screening over 50 lines of *Phaseolus* for efficiency of P utilization at very low levels of P. Lines were recognized as representing extremes in response to P stress and classified as efficient or inefficient based on dry weight production per unit P, and values ranged from 380 to 671 mg dry wt mg^{-1} P.

There are other whole-plant responses to P. Thus Howell and Foote (1964) reported significant soybean cultivar differences in sensitivity to high P in a comparison of 44 cultivars. At the other extreme, Jowett (1959) had found inability to respond to P, due to adaptation to low fertility, as exhibited by a Pb-tolerant population of the grass *Agrostis tenuis*. Nitrogen and phosphorus are closely related in nutrition, as was found by Hughes (1971) in a comparison of 18 soybean cultivars over 2 years at different fertilizer levels. A close connection existed between yield and leaf N, and between leaf P and N. It will be appreciated from these few examples that the "P-efficiency" is a whole-plant phenomenon of considerable complexity.

IV. Summary

Probably there are two potential end-users of more efficient cultivars: high-tech farmers who seek more efficient fertilizer use, greater economy, and less nitrogen pollution of groundwater, and farmers in developing countries who require cultivars which can produce a reasonable yield with minimum fertilizer input. From the whole-plant aspect, possibly these two objectives are not incompatible in at least one respect: selection for a larger and more branched root system. This would be one approach to selecting cultivars which might intercept N fertilizer more effectively. It would also provide for low-input systems through greater foraging ability of the plant, thus exploiting a larger volume of soil and enhancing input of P, Fe, Mn, and Zn. Present-day cultivars are able to produce photosynthates sufficient to support a larger root system and a number of techniques for selecting larger root systems have been summarized by Clarke and Townley-Smith (1984).

References

Antonovics, J., J. Lovett, and A. D. Bradshaw. 1967. The evolution of adaptation to nutritional factors in populations of herbage grasses. *Isotopes in Plant Nutrition and Physiology. Proc. Symp., IAEA, Vienna, 1966* pp. 549–567.

Apel, P. 1967. Potentielle Photosyntheseintensität von Gerstensorten des Gaterslebener Sortiments. *Kulturpflanze* 15:161–174.

Asana, R. D., P. K. Ramaiah, and M. V. K. Rao. 1968. The uptake of nitrogen, phosphorus and potassium by cultivars of wheat in relation to growth and development. *Indian J. Plant Physiol.* 9:85–107.

Barber, S. A. 1982. Soil–plant root relationships determining phosphorus uptake. In A. Scaife (ed.), *Proc. 9th Int. Plant Nutrition Colloq., Commonw. Agric. Bur., Farnham Royal, Engl.* 1:39–44.

Bole, J. B. 1973. Influence of root hairs in supply soil phosphorus to wheat. *Can. J. Soil Sci.* 53:169–175.

Brinkman, M. A., and Y. D. Rao. 1984. Response of three oat varieties to N fertilizer. *Crop Sci.* 24:973–977.

Brown, J. C., and J. E. Ambler. 1973. "Reductants" released by roots of Fe-deficient soybeans. *Agron. J.* 65:311–314.

Brown, J. C., and J. E. Ambler. 1974. Iron-stress response in tomato, *Lycopersicum esculentum* L. Sites of Fe reduction, absorption and transport. *Physiol. Plant.* 31:221–224.

Brown, J. C., and W. D. Bell. 1969. Iron uptake dependent upon genotype of corn. *Soil. Sci. Soc. Am. Proc.* 33:99–101.

Brown, J. C., R. L. Chaney, and J. E. Ambler. 1971. A new tomato mutant inefficient in the transport of iron. *Physiol. Plant.* 25:48–53.

Brown, J. C., R. S. Holmes, and L. O. Tiffin. 1961. Iron chlorosis in soybeans as related to the genotype of rootstock. 3. Chlorosis susceptibility and reductive capacity at the root. *Soil Sci.* 91:127–132.

Burauel, P., J. Wieneke, and F. Fuhr. 1990. Carbohydrate status in roots of two soybean varieties: A possible parameter to explain different efficiencies concerning phosphate uptake. *Proc. 3rd Int. Symp. on Genetic Aspects of Plant Mineral Nutrition.* Kluwer Academic, Dordrecht, The Netherlands.

Caradus, J. R. 1982. Genetic differences in the length of root hairs in white clover and their effect on phosphorus uptake. In A. Scaife (ed.), *Proc. 9th Int. Plant Nutrition Colloq., Commonw. Agric. Bur., Farnham Royal, Engl.* 1:84–88.

Clark, R. B. 1975. Differential magnesium efficiency in corn inbreds. 1. Dry-matter yields and mineral element composition. *Soil Sci. Soc. Am. Proc.* 39:488–491.

Clark, R. B., and J. C. Brown. 1974. Differential phosphorus uptake by phosphorus stressed corn inbreds. *Crop Sci.* 14:505–508.

Clarke, J. M., and T. F. Townley-Smith. 1984. Screening and selection techniques for improving drought resistance. In P. B. Vose and S. G. Blixt (eds.), Crop Breeding: A Contemporary Basis. Pergamon, Oxford. Pp. 137–162.

Clarkson, D. T., and J. B. Hanson. 1980. The mineral nutrition of higher plants. *Annu. Rev. Plant Physiol.* 31:239–298.

Clarkson, D. T., and J. Sanderson. 1971. Relationship between the anatomy of cereal roots and the absorption of nutrients and water. *Agric. Res. Counc., Letcombe Lab., Annu. Rep.* pp. 16–25.

Coltman, R., G. Gerloff, and W. Gabelman. 1982. Intraspecific variation in growth, phosphorus acquisition and phosphorus utilization in tomatoes grown under phosphorus deficiency stress. In A. Scaife (ed.), *Proc. 9th Int. Plant Nutrition Colloq., Commonw. Agric. Bur., Farnham Royal, Engl.* 1:117–122.

Cook, J. A., and L. A. Lider. 1964. Mineral composition of cloomtime grape petiole in relation to rootstock and scion variety behavior. *Proc. Am. Soc. Hortic. Sci.* 84:243–254.

Cooper, W. C., B. S. Gorton, and E. O. Olson. 1952. Ionic accumulation in citrus as influenced by rootstock and scion and concentration of salts and boron in the substrate. *Plant Physiol.* 27:191–203.

Curtis, P. E., W. L. Ogren, and R. H. Hageman. 1969. Varietal effects in soybean photosynthesis and photorespiration. *Crop Sci.* 9:323.

Duncan, W. G. 1977. Cultural manipulation for higher yields. In J. D. Eastin, F. A. Haskins, C. Y. Sullivan, and C. H. M. Van Bavel (eds.), Physiological Aspects of Crop Yields. Am. Soc. Agron., Madison, Wisconsin. Pp. 327–339.

Eaton, F. M., and G. Y. Blair. 1935. Accumulation of boron by reciprocally grafted plants. *Plant Physiol.* 10:411–424.

Embleton, T. W., M. Matsumura, W. B. Storey, and M. J. Garber. 1962. Chlorine and other elements in avocado leaves as influenced by rootstock. *Proc. Am. Soc. Hortic. Sci.* 80:230–236.

Epstein, E. 1963. Selective ion transport in plants and its genetic control. Desalination Research Conference. *N.A.S.–N.R.C. Publ.* No. 942, pp. 284–298.

Epstein, E. 1980. Impact of Applied Genetics on Agriculturally Important Plants: Mineral Metabolism. Report to the Office of Technology Assessment, U.S. Congress, Washington, D.C. Mimeo. 42 pp.

Epstein, E., and C. E. Hagen. 1952. A kinetic study of the absorption of alkali cations by barley roots. *Plant Physiol.* 27:457–474.

Ferreyra Harnandez, F. F. 1978. Absorcao de fosforo em raizes destacadas de milho (*Zea mays* L.). Ph.D. Thesis, ESALQ, Piracicaba, Univ. de Sao Paulo.

Foy, C. D., and S. A. Barber. 1958. Magnesium absorption and utilization by two inbred lines of corn. *Soil Sci. Am. Proc.* 22:57–62.

Fried, M., and J. C. Noggle. 1958. Multiple site uptake of individual cations by roots as affected by hydrogen ion. *Plant Physiol.* 33:139–144.

Gabelman, W. H., and G. C. Gerloff. 1982. The search for, and interpretation of, genetic controls that enhance plant growth under deficiency levels of a macronutrient. In M. Saric (ed.), Genetic Specificity of Mineral Nutrition of Plants. Serb. Acad. Sci. Arts, Belgrade. Pp. 301–312.

Gerloff, G. C. 1963. Comparative mineral nutrition of plants. *Annu. Rev. Plant Physiol.* 14:107–124.

Glass, A. D. M., and M. Y. Siddiqi. 1984. The control of nutrient uptake rates in relation to the inorganic composition of plants. *Adv. Plant Nutr.* 1:103–147.

Godo, G. H., and H. M. Reisenauer. 1980. Plant effects on soil manganese availability. *Soil Sci. Soc. Am. J.* 44:993–995.

Goodman, P. J. 1977. Selection for nitrogen response in *Lolium*. *Ann. Bot.* 41:243–256.

Goodman, P. J. 1981. Genetic control of nitrogen response in *Lolium* species. In C. E. Wright (ed.), Plant Physiology and Herbage Production. Occas. Symp. No. 13. Br. Grassl. Soc. Pp. 131–136.

Goodman, P. J. 1982. Genetic variation in nitrogen nutrition of grasses and cereals and possibilities of selection. In M. R. Saric (ed.), Genetic Specificity of Mineral Nutrition of Plants. Serb. Acad. Sci. Arts, Belgrade. Pp. 356–361.

Graham, R. D. 1984. Breeding for nutritional characteristics in cereals. *Adv. Plant Nutr.* 1:57–101.

Haas, A. R. 1948. Effect of the rootstock on the composition of citrus trees and fruit. *Plant Physiol.* 23:309–330.

Hackett, C. 1969. Quantitative aspects of the growth of cereal root systems. In J. Whittington (ed.), Root Growth. 15th Easter Sch. Agric. Sci., Univ. Nottingham, 1968. Butterworth, London. Pp. 57–63.

Hahr, V. 1975. Nuclear methods for detecting root activity. *Tracer Techniques for Plant Breeding. Proc. Panel, IAEA, Vienna, 1974.*

Hesketh, J. D., and D. N. Moss. 1963. Variation in response of photosynthesis to light. *Crop Sci.* 3:107–110.

Howell, R. W., and B. D. Foote. 1964. Phosphorus tolerance and sensitivity of soybeans as related to uptake and translocation. *Plant Physiol.* 39:610–613.

Huffaker, R. C., and D. W. Rains. 1978. Factors influencing nitrate acquisition by plants: Assimilation and fate of reduced nitrogen. In D. R. Nielsen and J. G. MacDonald (eds.), Nitrogen in the Environment. Vol. 2. Academic Press, New York. Pp. 1–43.

Huffaker, R. C., and A. Wallace. 1959. Sodium absorption by different plant species at different potassium levels. *Soil Sci.* 87:130–134.

Hughes, J. L. 1971. Response of soybean (*Glycine max* L. Merrill) genotypes to levels of fertility and 2,3,5-triiodobenzoic acid. Ph.D. Thesis, Univ. of Nebraska, Lincoln. (*Diss. Abstr. B* 32:3740B–3741B, 1972.)

Irvine, R. B., B. L. Harvey, and B. G. Rossnagel. 1980. Rooting capability as it relates to soil moisture extraction and osmotic potential of semidwarf and normal statured genotypes of six-rowed barley. *Can. J. Plant Sci.* 60:241–248.

Jennings, P. R. 1964. Plant type as a rice breeding objective. *Crop Sci.* 4:13–16.

Jowett, D. 1959. Adaptation of a lead-tolerant population of *Agrostis tenuis* to low soil fertility. *Nature (London)* 184:43.

Kim, H. K., R. K. Park, and S. Y. Cho. 1986. Rice Varietal Improvement in Korea. Spec. Publ. Crop Exp. Stn., RDA, Suwon, South Korea.

Kleese, R. A. 1968. Scion control of genotypic differences in Sr and Ca accumulation by soybeans under field conditions. *Crop Sci.* 8:128–219.

Kraus, M., A. Fusseder, and E. Beck. 1987. *In situ* determination of the phosphate gradient around the root by autoradiography of frozen soil sections. *Plant Soil* 97:407–418.

Läuchli, A. 1976. Genotypic variation in transport. In U. Lüttge and M. G. Pitman (eds.), Transport in

Plants. Encyclopedia of Plant Physiology, New Series, Vol. 2B. Springer-Verlag, Berlin. Pp. 372–393.

Lindgren, D. T., W. H. Gabelman, and G. C. Gerloff. 1977. Variability of phosphorus uptake in *Phaseolus vulgaris* under phosphorus stress. *J. Am. Soc. Hortic. Sci.* 102:674–683.

Lingle, J. C., and D. A. Lorenz. 1969. Potassium nutrition of tomatoes. *J. Am. Soc. Hortic. Sci.* 94:679–683.

Lupton, F. G. H., R. H. Oliver, F. B. Ellis, B. T. Barnes, K. R. Howse, P. J. Wellbank, and P. J. Taylor. 1974. Root and shoot growth of semidwarf and taller winter wheat. *Ann. Appl. Biol.* 77:129–144.

Lyness, A. S. 1936. Varietal differences in the phosphorus feeding capacity of plants. *Plant Physiol.* 11:665–688.

Marschner, H., V. Römheld, W. J. Horst, and P. Martin. 1986. Root induced changes in the rhizosphere: Importance for mineral nutrition of plants. *Z. Pflanzenernaehr. Bodenkd.* 149:441–456.

Marschner, H., V. Römheld, and H. Ossenberg-Neuhaus. 1982. Rapid method for measuring changes in pH and reducing processes along roots of intact plants. *Z. Pflanzenphysiol.* 185:407–416.

Mengel, K. 1982. Responses of various crop species and cultivars to mineral nutrition and fertilizer application. In M. Saric (ed.), Genetic Specificity of Mineral Nutrition of Plants. Serb. Acad. Sci. Arts, Belgrade. Pp. 233–245.

Myers, W. M. 1960. Genetic control of physiological processes: A consideration of differential ion uptake by plants. In R. S. Caldecott and L. A. Snyder (eds.), *Symposium of Radioisotopes in the Biosphere, Univ. Minnesota, Minneapolis* pp. 201–226.

Newbould, P., F. B. Ellis, B. T. Barnes, K. R. Howse, and F. G. H. Lupton. 1970. Intervarietal comparisons of the roots systems of winter wheat. *Agric. Res. Counc., Letcombe Lab., Annu. Rep., 1969* p. 38.

Nye, P. H. 1973. The relation between the radius of a root and its nutrient absorbing power: Some theoretical considerations. *J. Exp. Bot.* 24:783–786.

Nye, P. H., and P. B. Tinker. 1977. Solute Movement in the Soil–Root System. Blackwell, Oxford.

Olsen, R. A., and J. C. Brown. 1980a. Factors related to iron uptake by dicotyledonous and monocotyledonous plants. I. pH hydrogen ion concentration and reductant tomatoes, soybeans, oats, and maize. *J. Plant Nutr.* 2:629–645.

Olsen, R. A., and J. C. Brown. 1980b. Factors related to iron uptake by dicotyledonous and monocotyledonous plants. II. The reduction of Fe^{3+} iron as influenced by roots and inhibitors of tomatoes, soybeans, maize, and oats. *J. Plant Nutr.* 2:647–660.

Olsen, R. A., and J. C. Brown. 1980c. Factors related to iron uptake by dicotyledonous and monocotyledonous plants. III. Competition between root and external factors for Fe iron. *J. Plant Nutr.* 2:661–682.

Page, E. R., E. K. Schofield-Palmer, and A. J. MacGregor. 1962. Studies in soil and plant manganese. 1. Manganese in soil and its uptake by oats. *Plant Soil* 16:238–246.

Pompeu, A. S., L. De Almeida, L. d'Artagnan, N. C. Schmidt, and L. C. Loberto. 1978. Comportamento de linhagens e cultivars de feijoeiro (*Phaseoulus vulgaris* L.) no Vale do Paraibo, SP. *Bragantia* 37:93–101.

Rabideau, G. S., W. G. Whaley, and C. Heimsch. 1950. The absorption and distribution of radioactive phosphorus in two maize inbreds and their hybrids. *Am. J. Bot.* 37:93–99.

Rains, D. W., and E. Epstein. 1967. Preferential absorption of potassium by leaf tissue of the mangrove, *Avicenna marina,* an aspect of halophytic competence in coping with salt. *Aust. J. Biol. Sci.* 20:847–857.

Ramirez, R. 1982. Efficient use of nitrogen, phosphorus and potassium by corn (*Zea mays* L.) inbreds. In A. Scaife (ed.), *Proc. 9th Int. Plant Nutrition Colloq., Commonw. Agric. Bur., Farnham Royal, Engl.* 1:515–520.

Raper, C. D., and S. A. Barber. 1970a. Rooting systems of soybeans. 1. Differences in root morphology among varieties. *Agron. J.* 62:581–584.

Raper, C. D., and S. A. Barber. 1970b. Rooting systems of soybeans. 2. Physiological effectiveness as nutrient absorption surfaces. *Agron. J.* 62:585–588.

Römheld, V., and H. Marschner. 1986. Evidence for a specific uptake system for iron phytosiderophores in roots of grasses. *Plant Physiol.* 80:175–180.
Royal Society (U.K.). 1983. The nitrogen cycle in the United Kingdom. In W. D. Stewart (ed.), A Study Group Report. London. Pp. 1–33.
Ruschel, A. P., P. B. Vose, E. Matsui, R. L. Victoria, and S. M. Saito. 1982. Field evaluation of N_2-fixation and nitrogen utilization by *Phaseolus* bean varieties determined by ^{15}N isotope dilution. *Plant Soil* 65:397–407.
Saric, M. R. 1981. Genetic specificity in relation to plant mineral nutrition. *J. Plant Nutr.* 3:743–766.
Sayre, J. D. 1951. Magnesium needs of inbred corn lines. *Ohio Agric. Exp. Stn., Res. Bull.* No. 705.
Schenk, M. K., and S. A. Barber. 1979. Root characteristics of corn genotypes as related to P-uptake. *Agron. J.* 71:921–924.
Shim, S. C., and P. B. Vose. 1965. Varietal differences in the kinetics of iron uptake by excised rice roots. *J. Exp. Bot.* 16:216–232.
Smith, S. N. 1934. Response of inbred lines and crosses in maize to variations of N and P supplied as nutrients. *J. Am. Soc. Agron.* 26:785–804.
Stoy, V. 1965. Photosynthesis, respiration and carbohydrate accumulation in spring wheat in relation to yield. *Physiol. Plant., Suppl.* 4:1.
Stoy, V. 1975. Use of tracer techniques to study yield components in seed crops. *Tracer Techniques for Plant Breeding. Proc. Panel, IAEA, Vienna, 1974* pp. 43–53.
Subbiah, B. V., J. C. Katyal, R. L. Narasimham, and C. Dakshinamurti. 1968. Preliminary investigations on root distribution of high yielding wheat varieties. *Int. J. Appl. Radiat. Isot.* 19:385–390.
Sumner, M. E., H. R. Boerma, and R. Isaac. 1982. Differential genotypic sensitivity of soybeans to P-Zn-Cu imbalances. In A. Scaife (ed.), *Proc. 9th Int. Plant Nutrition Colloq. Commonw. Agric. Bur., Farnham Royal, Engl.* 2:652–657.
Swaminathan, M. S. 1972. The role of nuclear techniques in agricultural research in developing countries. *Proc. 4th Int. Conf. Peaceful Uses of Atomic Energy, Geneva, 1971* 12:3–32.
Troughton, A. 1959. Report on the tenure of a National Research Fellowship, New Zealand, July 1958–July 1959, W.P.B.S., Aberystwyth.
Troughton, A. 1981a. Root:shoot relationships in mature plants. *Plant Soil* 63:101–105.
Troughton, A. 1981b. Length of life of grass roots. *Grass Forage Sci.* 36:117–120.
van Noordwijk, M. 1983. Functional interpretation of root densities in the field for nutrient and water uptake. *Wurzelökologie und ihre Nutzanwendung/Root Ecology and Its Practical Application. Int. Symp. Gumpenstein, Bundesanstalt Gumpenstein, Irdning, Austria, 1982* pp. 207–226.
Vedrov, N. G. 1974. Root systems of cereal crops as an object of breeding. In E. L. Klimashevsky (ed.), Variety and Nutrition. Sib. Inst. Plant Physiol. Biochem., Irkutsk. Pp. 93–104. (In Russ.; Engl. summ.)
Vogel, O. A., R. E. Allan, and C. J. Peterson. 1963. Plant and performance characteristics of semidwarf wheats producing most efficiently in eastern Washington. *Agron. J.* 55:397–398.
Vose, P. B. 1962a. Nutritional response and shoot/root ratio as factors in the composition and yield of genotypes of perennial ryegrass, *Lolium perenne* L. *Ann. Bot.* 26:425–437.
Vose, P. B. 1962b. Manganese requirement in relation to photosynthesis in *Avena*. *Phyton* 19:133–140.
Vose, P. B. 1963. Varietal differences in plant nutrition. *Herb. Abstr.* 33:1–12.
Vose, P. B. 1984. Effects of genetic factors on nutritional requirements of plants. In P. B. Vose and S. Blixt (eds.), Crop Breeding: A Contemporary Basis. Pergamon, Oxford. Pp. 67–114.
Vose, P. B., and E. L. Breese. 1964. Genetic variation in the utilization of nitrogen by ryegrass species *Lolium perenne* and *Lolium multiflorum*. *Ann. Bot.* 28:251–270.
Wallace, A., N. Hemaidan, and S. M. Sufi. 1965. Sodium translocation in bush beans. *Soil Sci.* 100:331–334.
Whiteaker, D., G. C. Gerloff, W. H. Gabelman, and D. Lindgren. 1976. Intraspecific differences in growth of beans at stress levels of phosphorus. *J. Am. Soc. Hortic. Sci.* 101:472–475.
Wright, M. J. (ed.). 1977. Plant Adaptation to Mineral Stress in Problem Soils. Proc. Workshop, Beltsville, 1976. Cornell Univ. Press, Ithaca, New York.

4

Genetics and Breeding of Cereals for Acid Soils and Nutrient Efficiency

ENRIQUE A. POLLE and CALVIN F. KONZAK

Significant improvements in methods and materials available for adjusting soil nutritional conditions have been made in the last 50 years. New fertilizer forms that make nutrients easily available to plants have been developed. Loss of fertilizer has been reduced by the use of chemicals that retard the release of easily leachable forms. Split application, appropriate placement of fertilizers in aggregates of optimum size, and the use of slow-release fertilizers are now common. Management practices designed to fertilize a crop system rather than individual crops and to better exploit residual effects of fertilizers that react with the soil are also widely used. As a consequence, the efficiency of applied fertilizer used by irrigated crops can be as high as 70% for nitrogen and 80–90% for potassium. For P fertilizers, however, only 15–20% of the applied P is used by the first crop. The remainder may become fixed or, under some conditions, a decreasing fraction can be extracted by succeeding crops.

For the most part, plant breeders have developed plant cultivars able to produce

Crops as Enhancers of Nutrient Use

high yields with high applications of fertilizers. In an effort to maximize yields, the selection of lines is typically made under conditions employing a generous supply of nutrients. This high-input approach is the rule in advanced industrial nations because, as a result of their economic structure, farmers can afford to adjust the soils to the plants rather than vice versa. However, in recent decades, because of the urgent need for improving food production for growing populations in industrializing countries, a low-input type of agriculture has arisen (Sanchez and Buol, 1975). This type of agriculture requires economic yields with low use of fertilizers. On the other hand, pressing problems of groundwater pollution in industrial nations have produced a reaction against excessive use of agricultural chemicals as a whole and of fertilizers in particular. Thus, both the highly and lesser industrialized nations may require crops able to produce high yields with relatively low levels of fertilizer. This situation stimulated plant breeding research toward cultivars that can produce economic yields with low fertilizer inputs. This interest has been reinforced by recent oscillations in fertilizer prices that, although transitory, could be repeated. Moreover, economic limits to the soil amendment approach are increasingly being realized as more soils with mineral imbalance stresses are used for production of major crops. In this chapter some aspects of breeding crop plants for acid soils are presented. Emphasis is given to methods of screening for the desired genetically controlled features. Much of this methodology is still in the experimental stage and not previously described. Efficient screening methods are a major obstacle to the successful development of new cultivars that are better adapted to acid soils.

I. Acid Soil Stress

Breeding plants for limited and adverse nutritional conditions requires an integral approach. All of the traits required for optimum growth under a given soil condition should be included. However, knowledge of the genetics and gene sources for the various adaptive traits is still limited. Most if not all traits controlling plant stress responses are independent of each other, and thus may be recombined readily in accordance with the requirements of the stress area. Fortunately, no evident obstacle to the combination of the genetic traits responsible for adaptation to even contrasting soil mineral imbalance conditions such as Mn toxicity tolerance versus Mn deficiency tolerance has been encountered to date (Konzak and Polle, 1985).

One of the most widely distributed soil conditions is acidity. Acid soils and andosols constitute approximately 1.17 billion hectares of nonirrigated arable lands of the world (Dudal, 1976). In these soils the main nutritional problems are P deficiency and Al and Mn toxicities, in addition to the N limitation common to all soils. Iron toxicity also has been reported in acid soils (Ponnamperuma, 1976).

In this chapter, aspects of plant tolerance to excess soil Al, Mn, and Fe, and to low levels of available P, will be discussed in relation to breeding plants to overcome the various stresses. Since one of the major concerns in breeding is the efficiency of screening procedures, the following aspects for each of these elements will be treated in this order: physiological mechanisms in relation to screening procedures, screening procedures, and genetics and breeding. A final section will analyze interactions among these nutritional factors that may be encountered in the breeding of new cultivars.

A. ALUMINUM TOXICITY STRESS

The nature of soil acidity and its influence on nutritional aspects of the soil is probably one of the oldest problems investigated by soil chemists (Jenny, 1961; Thomas, 1988). However, the incorporation of Al toxicity tolerance genes in plant cultivars was achieved by breeders long before soil chemists concluded that Al^{3+} was one of the toxic elements common to acid soils.

Breeding of wheat started in southern Brazil in 1919. When wheat was cultivated there, no lime was used, and highly Al-susceptible germplasm died before producing any yield. The affected plants had reduced growth, yellowing leaves, and deformed, short roots. The causes of this symptom, called "crestamento" disease, were not known then. However, breeders in the area were able to select cultivars that were tolerant to this condition, yielding 0.8 to 1 metric tons ha^{-1}. In 1944 (cited in da Silva, 1976) Paiva was able to relate the low pH of the soil to "crestamento" and to aluminum solubility. Since 1970 better selections and improved practices have pushed yields of wheat in the area to 3 metric tons ha^{-1} (Rosa, 1988).

The nature of Al toxicity in plants has been investigated mainly by the use of nutrient solutions and generally in short-term experiments. Investigations of Al effects on plants growing under field conditions are limited.

1. Physiological Mechanisms

General physiological aspects of Al toxicity are discussed in Chapter 3. Therefore, only the physiological disturbances related to screening methods for aluminum tolerance will be reviewed here.

Two major effects of Al in nutrient solution experiments have been reported in the literature: (1) Al causes disturbances in the uptake, transport, and utilization of the mineral nutrients Ca, K, and P (Johnson and Jackson, 1964; Clark, 1977; Clarkson, 1966) and (2) when plant roots are immersed in a solution with an appropriate concentration of Al (depending on species and cultivar), root elongation may be arrested in a matter of hours. The inhibitory effect of Al varies with the order of the root, as shown by Pan *et al.* (1989) in *Glycine max*.

The growth arrest effects of Al in short-term experiments have attracted considerable attention by research workers. Up to about 1983 (Clarkson, 1965; Sampson *et al.*, 1965; Henning, 1975; Morimura *et al.*, 1978; Horst *et al.*, 1983; Matsumoto *et al.*,

1976, 1977a; Morimura and Matsumoto, 1978; Naidoo *et al.*, 1978) the conclusion of most investigators was that Al inhibits or alters DNA activities, and that this causes the cessation of mitosis in the root tip and consequently root elongation.

More recent work (Bennet *et al.*, 1985a,b; Wagatsuma *et al.*, 1987; Wallace *et al.*, 1982) suggests that the mechanism of Al injury is more complex than the arrest of the mitotic cycle, and it is probably a simultaneous effect on the mitotic activity of the meristem and on the elastic properties of the elongation zone of the roots. The latter is responsible for injury (rupture) of epidermal cells. In roots of some fast-growing species like maize, rupture of epidermal tissue of roots growing in Al solutions has been reported (Polle *et al.*, 1978a; Bennet *et al.*, 1985a). This effect is apparently due to unequal mechanical stress between the cells of the central core of the root and of the epidermal layer. As the central core enlarges longitudinally, the epidermis may not expand because of cross-linkage by Al in the pectic substance of the cell walls, as suggested for peas by Klimashevskii and Dedov (1980). These authors reported that pectins of an aluminum-tolerant pea genotype have substantially lower levels of COOH groups than a genotype with less tolerance. Thus Al tolerance may be due to the compositional nature rather than to the quantity of pectins located in the walls of epidermal cells at the elongation zone of the roots. Matsumoto *et al.* (1977b) found no relationship between total pectin content and Al tolerance in whole pea roots. It is not known if pectins in aluminum-tolerant and nontolerant cultivars are different in chemical structure. However, Vose and Randall (1962) have reported that the exchange capacity of root tips in aluminum-tolerant species is lower than that in aluminum-susceptible species.

Growing roots secrete a mucilaginous material that appears in the root tip mixed with sloughed cap cells and, in nonsterile media, with bacterial secretions (Jenny and Grossenbacher, 1963). Morre *et al.* (1967) found that maize roots grown in water secrete large quantities of this material and that the size of the droplet formed by this highly hydrated polysaccharide was not constant but fluctuated for about 3 hr. Bennet *et al.* (1985b) postulated that Al induces changes in the outer cap cells of maize roots, and these in turn influence the meristem in two ways: (1) by regulating the flow of the gel from the root cap and (2) by sending a chemical signal to the meristem. Large amounts of gel secretion as reported by Morre (1967) and by Horst *et al.* (1982) for roots growing in solutions may act as a countercurrent for the diffusion of Al from the media to the root meristem. The way in which this protective action operates may vary, due to geometrical reasons, for different species. Root tips of maize secrete the bulk of the mucilage from the root cap, but also secrete a firmer layer from the columnar epidermic cells. This layer extends to the tip of the root, forming a distinct boundary between the root cap and the meristem (Clarke *et al.*, 1979, Miki *et al.*, 1980). In monocots like corn, the cells of the root tip converge on the meristem producing a "closed" meristem arrangement. In contrast, the rows of cells of the meristem in dicots like *Pisum sativum* L. are continuous with the cells of the root cap forming an "open" type of meristem (Torrey and Feldman, 1977). The effect of the mucilage secretions on Al toxicity was investigated by Horst *et al.* (1982) in root tips of *Vigna unguiculata*. The production of gel was found to be higher in an aluminum-tolerant genotype than in a nontolerant one. Periodic elimination of the gel secretion in both cultivars showed that this material indeed protected the roots from Al arrest to the same extent.

Transient effects caused by Al in root growth also have been reported. Moore *et al.* (1976) noted that when wheat roots were exposed to nonlethal concentrations of aluminum and then allowed to regrow in an Al-free solution, a crooked portion of the root was observed when the root tip was located at the time of Al exposure. The same effect also can be observed in wheat roots grown first in a solution free of Al and then grown for a few hours in an aluminum solution: the time of the immersion in the Al solution is frequently marked by a deviation from geotropic growth by the root. Horst *et al.* (1983) found that roots of *Vigna unguiculata* in sublethal Al concentrations tended to reassume growth with time. Mitotic activity declined rapidly after the Al exposure and then part of the initial mitotic activity was restored. Aniol (1984) found that preincubation of wheat seedlings in a nutrient solution containing low doses of aluminum enabled substantial root regrowth of cultivars in a lethal aluminum concentration. All these reactions to a sudden Al exposure may be related. Levan (1945) investigated the immediate cytological effect of salt solutions in root meristems of *Allium cepa*. Levan noted that when inorganic salts were used, characteristic growth disturbances often occurred: the roots become bent into hooks or even spirals. According to Levan, this reaction may represent an incomplete "colchicine" type of tumor growth. As suggested by Horst *et al.* (1983), recovery from these injuries may be an adaptive mechanism.

As noted by Blamey *et al.* (1983), most of the studies on Al toxicity in nutrient solutions have employed Al concentrations much in excess of that found in soil solutions. Al^{3+} concentrations ranging from 0.1 to 1.1 m*M* have been used in culture solutions to which barley, wheat, triticale, rye, and rice have been exposed (Reid *et al.*, 1971; Foy, 1976; Howeler and Cadavid, 1976; Mugwira *et al.*, 1976). These Al concentrations are, in some cases, 1000 times higher than those reported for soil solutions. This difference may be due to the necessity to obtain differential responses between cultivars in a short time and/or to an overestimation of the Al present because of precipitation of Al as $Al(PO_4)$ and $Al(OH)_3$ (Blamey *et al.*, 1983).

2. Screening Methods for Aluminum Tolerance of Barley, Wheat, Triticale, and Rye

Evaluation of differential responses of cultivars of a given species to Al toxicity has been and is practiced under field conditions, in greenhouse pot experiments, and by nutrient solutions in a laboratory (Fageria *et al.*, 1988). It is obvious that for the practical use of these methods in breeding programs, the main requirement, in addition to low cost, is their sufficiency for application to large numbers of accessions. At present some of the laboratory methods are suitable for mass screening. In any case, a prerequisite for separating classes of aluminum tolerance within a species is the field identification of genotypes that can be used as reference standards with laboratory methods.

Field evaluations of cereals have been reported for wheat (Gill *et al.*, 1972, 1973) and rice (Fageria and Zimmermann, 1979; Fageria, 1982; Fageria and Barbosa Filho, 1982; Fageria *et al.*, 1988).

Studies of differential Al tolerance and the nutritional profile of sorghum in acid soils have been reported (Duncan, 1981, 1983, 1987, 1988; Duncan *et al.*, 1983; Duncan and Sutton, 1987). Registration of acid soil-tolerant sorghum germplasm has been done by Duncan (1984).

Experiments with soils in pots and also with nutrient solutions are numerous (McLean and Gilbert, 1927; MacLean and Chiasson, 1966; Kerridge *et al.*, 1971; Foy *et al.*, 1973; Moore *et al.*, 1976; Clark, 1977; Ruhe and Grogan, 1977; Aniol *et al.*, 1980; Furlani and Clark, 1981; Fageria, 1982; Kinraide *et al.*, 1985; Ben *et al.*, 1986).

In the investigation of differential Al tolerance, two approaches have been used—short-term and long-term experiments. Short-term screening procedures are based mainly on the effects of aluminum on root growth and can yield information only on features of Al tolerance related to root effects. Long-term experiments permit a more refined search for desirable features of aluminum tolerance, such as degree of response to soil amendments, interactions with drought tolerance, and other factors. For these reasons, and especially because of their high efficiency, short-term screening methods will be more useful in the first stage of selection in breeding programs.

In the following sections, four rapid screening methods as developed or used by the authors are summarized. Of these methods, specifically designed for barley, wheat, triticale, and rye, three employ a nutrient solution and one an artificial soil. They may be used in a variety of situations, depending on the laboratory facilities available.

The simplest method involves observing the pattern of root tip staining by hematoxylin after germinated seedlings are exposed to aluminum for a few hours. The method requires a laboratory with a reasonably constant temperature of about 25°C and a set of calibrating genotypes. Cultivars listed in Table I can be used as test standards. Including some of these standards in each test will allow a consistent classification of test material, even if small variations in ambient conditions occur from one test to another. The alternate methods involve measuring the reduction of root elongation when seedlings are cultured in a nutrient solution containing aluminum, in an aluminum-toxic artificial soil, or in a paper solution medium with Al. The latter requires a germinator with controlled temperature to obtain relatively uniform seedling root growth, and the others can be carried out in a growth chamber or in the laboratory. Because of the short time of growth, the nutrient solution is simplified because the plant can depend largely on minerals stored in the seed.

Table I. Classes of Al Tolerance[a]

Tolerance Class[b]	Completely Stained with mM Al	Species[c]	Name/Identification
1	0.03	Barley	Kearney-CI7580
			Hozoroi, CI8950
2	0.06	Barley	Scotbere, CI8327
			Belts 63-1424, CI11927
3	0.09	Barley	Sunrise, CI6272
			KY63-1624, CI11925
4	0.12	Barley	Dayton-CI9517
			Ohio B62-1427, CI12242
5	0.18	Wheat	Brevor-CI12385
			Waverly-CI17911
6	0.36	Wheat	Druchamp-PI193121
			Dirkwin-CI17745
		Triticale	87610TX84
7	0.72	Wheat	Chinese Spring-CI14108
			Hill 81-CI17954
8	1.40	Wheat	Atlas 66-CI12561
			Don Marco-PI344146
		Triticale	C245TX83
9	2.10	Triticale	VT083656
		63 Rye	Tschermaks Verdelte, PI290451
10	2.80	65 Rye	1-22141, PI239575
		66 Rye	Schlagler, PI254821
11	>2.80	67 Rye	Florida Black, PI323356
		68	Wrens Abruzzi, PI323368

[a]Source: E. Polle, P. A. Franks, and C. F. Konzak, unpublished.
[b]1 = least tolerant, 11 = most tolerant.
[c]Triticale and rye carry variability within the sample. Uniform populations could probably be isolated.

Maintenance labor is reduced to a minimum by short-term screening tests. However, in general, aluminum levels used in the short-term tests are substantially higher than those found in the literature on soil solution extracts in acid soils. As noted above, this is done to reduce the test time. With each of the methods, the growth time is sufficient to surpass any "shock effect" due to the aluminum treatment.

a. Hematoxylin Stain Method In the stain method (Polle *et al.*, 1978a,b), the aluminum treatment solution contains no nutrients and the Al exposure time is less than 24 hours. A shock response to aluminum occurs quickly after immersion in the aluminum solution. This response, as mentioned above, is characterized by a change in the direction of root growth. If the aluminum concentration is low enough for the cultivar to show tolerance, growth resumes gravitropically. If not, growth is arrested. In this case the outer cells of the root tip will be heavily stained by a subsequent hematoxylin treatment. Al acts as a mordant or binding agent of hematein, an oxidized component of the hematoxylin solution (Gill *et al.*, 1974). The stainable region coincides with the region of root elongation. The center of the region of root elongation is located about 2 mm from the wheat root tip (Tomos *et al.*, 1989). The older part of the root does not bind aluminum and does not stain. It appears that root cells in the stainable region elongate normally and are not stained in cultivars tolerant to the aluminum concentration in the test solution. Only a faint mark on the elongation zone present at the moment of immersion sometimes may be observed after the hematoxylin treatment. Characteristic responses of wheat genotypes to the Al treatment with the stain technique are shown in Figure 1.

After extensive testing of barley, wheat, triticale, and rye accessions with the hematoxylin method, a classification system for Al tolerance in these species was developed. The classes of Al tolerance are specified by the concentration of Al required in the test solution for completely staining the root tip of the tested genotype.

As shown in Table I, the Al concentrations chosen for the classifications range from 0.03 to 2.80 m*M* Al. The lowest and narrowest range of tolerance is shown by barley and the highest and wider range is shown by rye. Results indicated that the roots of all the barley, wheat, and triticale genotypes tested were completely stained after treatment with 0.12 m*M* Al, 1.40 m*M* Al, or 2.30 m*M* Al, respectively. Evaluation of tolerance via this method is made by visual inspection of the roots after the hematoxylin stain treatment. Data in Figure 1 show the pattern of root staining in the standard wheat testers with increasing Al concentrations. Evaluations can be made immediately after separating the roots, or a photocopier can be used to produce an acceptable reproduction for later evaluation and a permanent record.

b. Nutrient Solution Method The nutrient solution test can be carried out using the same floating trays and aerated nutrient solution as used for the hematoxylin stain method (Polle *et al.*, 1978a,b). For this test, seeds are germinated for 24 hours in water. Then, the seedlings are exposed to nutrient solutions (pH 4) containing Al. The concentrations of Al (m*M*) used are 0, 0.015, 0.03, 0.06 for barley; 0, 02,

→

Figure 1. Pattern of staining of five wheat accessions having different levels of Al tolerance. The root of each variety has been treated with Al concentrations indicated at the top of the figure. All genotypes show a stainable region that corresponds approximately to the growth of the root tip during the Al treatment. This region is only partially stained below a concentration characteristic for each genotype and occurs in the region of root elongation at the time of exposure to the aluminum treatment. Many of the root tips of the genotypes that continue to grow at the higher Al levels show a characteristic deviation of the gravitropic direction of growth. Only the region of elongation of the root is stainable, indicating that the pectins or substances that bind Al are distinct in the cell wall of this region.

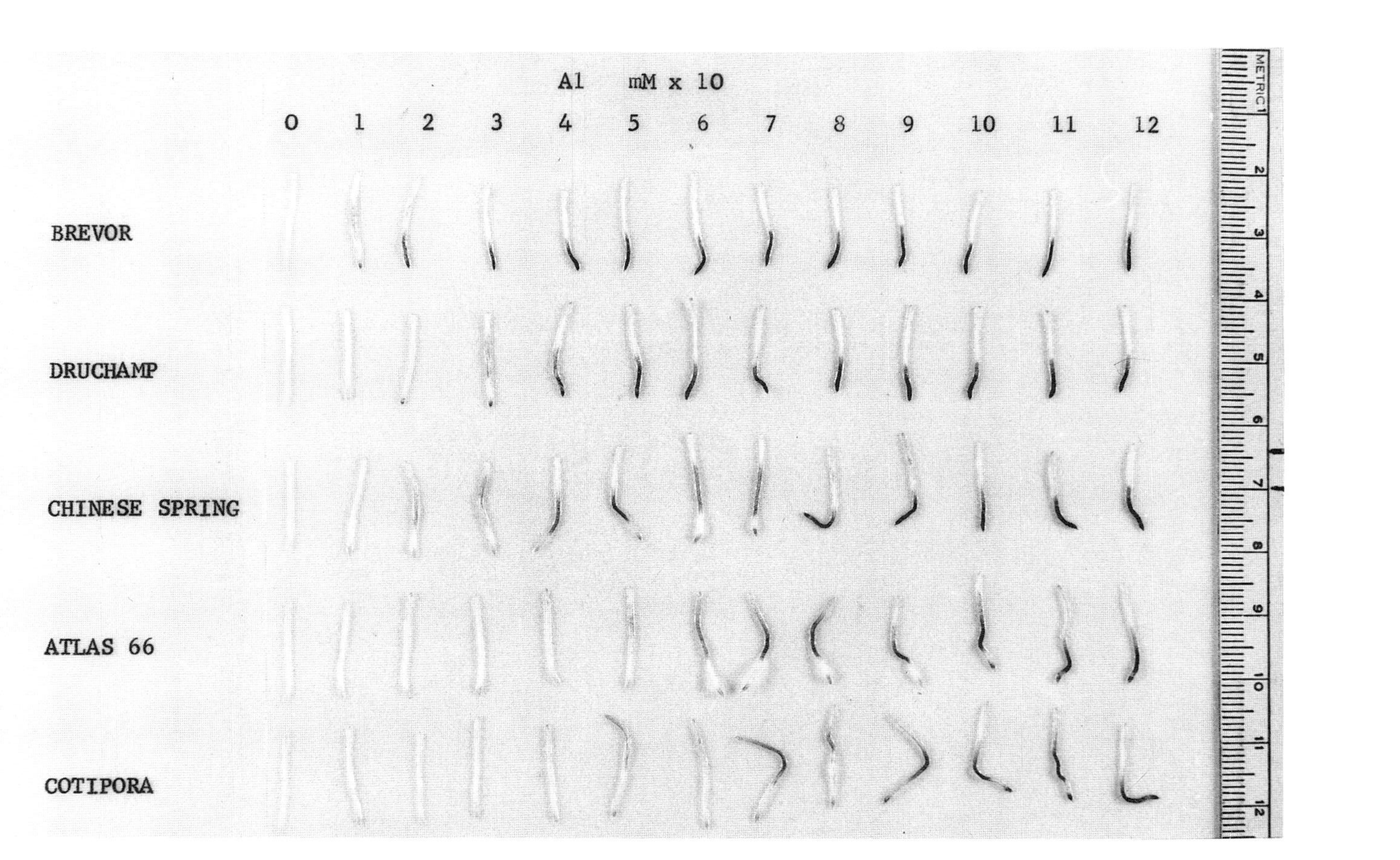
Al mM x 10
0
1
2
3
4
5
6
7
8
9
10
11
12
BREVOR
DRUCHAMP
CHINESE SPRING
ATLAS 66
COTIPORA
METRIC 1
2
3
4
5
6
7
8
9
10
11
12

0.10, 0.2 for wheat; 0, 0.3, 0.6, 1.2 for triticale; and 0, 0.3, 0.6, 1.2 for rye. During the growing period the plants are placed under light (85 μE sec^{-1} m^{-2}, 400- to 700-nm wave band in any suitable greenhouse lamp about 10 hours daily) at 20–23°C and the pH is adjusted once a day. The treatment solutions are replaced every 2 days during the test to maintain the concentrations of all minerals. The duration of the test is 7 days. At the end of the growing period the length of the roots is measured. A photographic record can be obtained using a photocopier as with the stain method. A curve for the root length versus Al concentration can be drawn for each accession.

If a single evaluation of the relative Al tolerance is desired, only two treatments are necessary, especially once the level of Al tolerated by the genotype is determined. For each gene source, however, it may be necessary to define exactly the Al concentration tolerated; and in genetic tests use of more than one Al concentration may be necessary to identify all genes and their interactions (Section I,A,3). In this case, 0 m*M* Al in addition to 0.06, 0.2, or 1.2 m*M*, respectively, can be used for barley, wheat, and triticale and rye. The relative growth, i.e., the ratio of the Al treatment to the control (without Al), can be used as a measure of Al tolerance. A set of calibrating genotypes should be run each time a test is carried out to evaluate unknown genotypes. Typical results of this test for barley are shown in Figure 2.

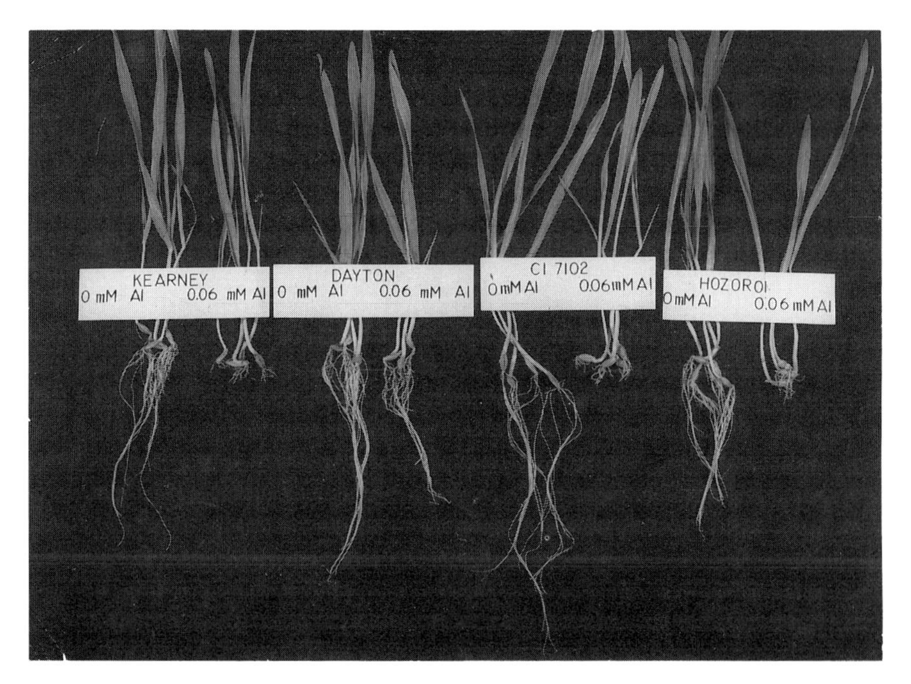

Figure 2. Four barley genotypes grown in the Al-toxic nutrient solution and in the nontoxic control solution. Dayton shows a considerable root growth with 0.06 m*M* Al, but Kearney, CI 7102, and Hozoroi do not. Courtesy of P. A. Franks.

c. Al-Toxic Artificial Soil Method The materials and percentages (dry) required for the preparation of the artificial soil are: washed quartz sand (90.2%), sphagnum air-dry peat moss (3.8%), and montmorillonite clay (6%). Industrial bentonite used for the bonding of molding sands in the foundry industry (American Colloid Co.) is an acceptable source of montmorillonite clay. To this mixture, 0.128 g of KNO_3 is added per kilogram. Four separate batches of this mixture are adjusted to pH 3.35, 3.75, 4.25, and 5, respectively, with a $Ca(OH)_2$ suspension. Plants (wheat) are grown for a week in this medium under 12 hours of light per day (85 μE sec^{-1} m^{-2}, 400- to 700-nm wave band). At the end of the growth period, root lengths are measured. Typical results with the Al-toxic artificial soil are shown in Figure 3.

d. Paper Solution Method In this method pregerminated seeds are grown against plastic plates covered with filter paper kept moistened with a test solution containing aluminum. Details of this test are in Konzak *et al.* (1976).

e. Comparative Results with the Different Test Methods The hematoxylin stain method has been found to produce data in general agreement with field results (Takagi *et al.*, 1981, 1983). A comparison between the hematoxylin stain and a solution method by Zale and Briggs (1988) showed some differences in rank among the less tolerant wheat genotypes, as referred to in the next section.

Application of the hematoxylin stain method to roots of species that secrete large quantities of mucilaginous substances and grow at a high rate may pose some

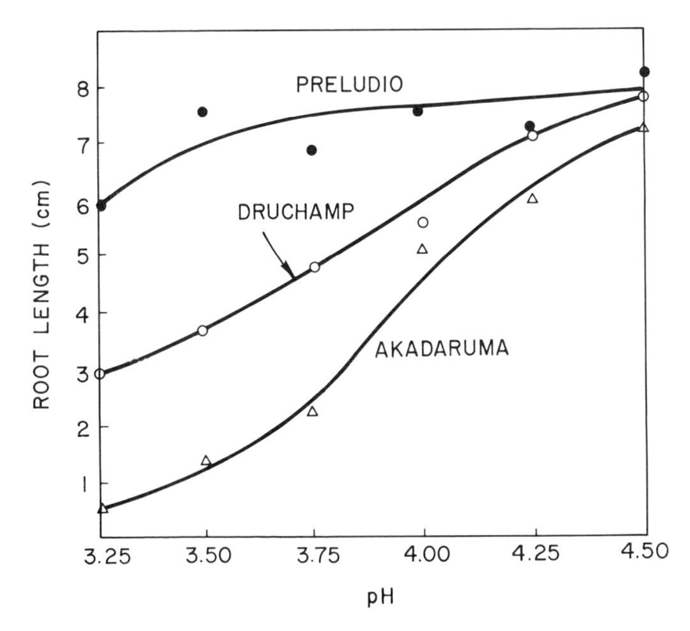

Figure 3. Root length of aluminum-tolerant Preludio compared to wheat Druchamp of intermediate Al tolerance and Al-sensitive wheat Akadaruma after being grown in the Al-toxic artificial soils.

difficulties and require adaptive modifications to the method. With maize, rupture of the cortex due to unequal elongation of the central core compared to the surface cells makes observations of the stainable region difficult. High secretions of mucigel, if periodically maximum size droplets are formed, may introduce artifacts in short-term experiments. More research needs to be done to overcome these problems. Adjustments in the Al level and shorter exposure times at the usually higher growing temperature for these species may solve the problem.

3. Genetics and Breeding

Differences in the tolerance of plant species and cultivars to mineral ion excesses or deficiencies have long been known and frequently reported (Foy, 1976; Vose, 1982). These differences undoubtedly reflect real differences in genetic controls, yet there is too little known about the mode of inheritance of such differences even for the most important crop species.

The major impediment to progress in the exploitation of the genetic differences by breeding (and greater knowledge of inheritance of these differences identified in the different plant species) has been the lack of accurate, efficient screening methods. Even so, many cultivars developed on soils with mineral imbalance problems proved to have tolerance to the major stress(es) presented. Many wheats developed in Rio Grande do Sul in south Brazil, where Al toxicity ("crestamento") is common, show unusually high tolerance to aluminum toxicity. The Al toxicity tolerance trait present in well-studied cultivars or lines such as 'Fronteira,' 'Frondoso,' 'Carazinho,' and 'BH1146' traces back to early selections made from wheats introduced from Turkey (Foy, 1976; da Silva, 1976; C. F. Konzak, unpublished). Frondoso was the source of the Al tolerance trait (as well as the high protein factor) introduced into the later-developed U.S. cultivar 'Atlas 66.'

As indicated earlier, the early Brazilian breeding work was carried out before the "crestamento" disease was known to be due to the toxic level of Al in the acid soil, demonstrating that success is achievable without detailed knowledge of inheritance of the trait and the nature of the problem. Similarly, the Al-tolerant cultivars Atlas 66, developed in North Carolina, and 'Thorne' and 'Seneca,' developed in Ohio, were selected on acidic soils (Campbell and Lafever, 1976). However, improved methods and knowledge of modes of inheritance would greatly expedite breeding progress, as well as facilitate the development of improved germplasm. Moreover, even in the acidic Brazilian soils, the mineral imbalances seem to be complex. Some soils have not only Al, but also Mn, in toxic concentrations, and phosphate may be deficient or made unavailable by the Al. For this reason, there has been rapid progress in technique development and breeding research due to the recent efforts to partition out the problems and employ laboratory methods for analyses and selection work (Polle *et al.*, 1978a,b; Polle and Konzak, 1985).

In this respect, the work of Foy (1976) and colleagues has led to the development and application of standardized laboratory methods, especially for investigating Al and Mn toxicity and Fe deficiency. Some improvements have been made to these methods and, more recently, widely applicable techniques more suitable for inheritance studies and plant breeding have been developed for Al toxicity tolerance and for Mn toxicity (Konzak and Polle, 1985).

Several studies have shown that the Al tolerance of wheats varies widely, and the tolerance levels of rye, triticale, and wheat exceed that of barley or maize (Polle and Konzak, 1985). Although none of the screening work on the cereals has been extensive, usable levels of Al tolerance have been identified in barley, hexaploid wheat, maize, sorghum, and triticale (Polle and Konzak, 1985). Of the few inheritance studies conducted in barley, a simple major dominant gene for Al tolerance has been identified in the winter barley (*H. vulgare*) cultivars 'Dayton' and 'Smooth Awn 86' (Reid, 1971). In *T. aestivum* wheat, Kerridge and Kronstad (1968) identified a single dominant gene for Al tolerance in 'Druchamp.' However, using a technique developed to differentiate levels of Al tolerance based on the Al concentration required to prevent regrowth of roots, Kerridge *et al.* (1971) and Moore *et al.* (1976) showed that Al tolerance could be considered dominant or recessive depending on the Al concentration employed to differentiate the tolerance classes. As also observed by others, Al tolerance was additively inherited—heterozygotes showed less tolerance than plants homozygous for the tolerance gene. Thus, even though they employed a solution culture technique, they obtained differentiation of tolerance classes comparable to that obtainable using the more efficient hematoxylin stain technique (Polle *et al.*, 1978a,b).

In contrast, Campbell and Lafever (1976), using solution culture root measurement methods, could readily show that Al susceptibility was recessive, but could not clearly distinguish the number of genes present in genetic combinations. This may have been because their Al concentration did not permit the identification of heterozygotes. Campbell and Lafever (1976) suggested that two to three gene loci and modifiers might be present, and that inheritance of Al tolerance was complex. Lafever and Campbell (1978) studied the inheritance of response to Al of two Al-tolerant ('Redcoat' and 'Arthur') and two Al-sensitive (Seneca and Thorne) soft red winter cultivars. Genetic differences were found between each pair for Al tolerance. Data from F_1, F_2, and backcrosses of sensitive/tolerant parents indicated that sensitivity was conditioned by a single recessive gene. Selection for Al-sensitive plants in F_2 was effective, but intermediate or tolerant plants were not as readily identified, indicating to them that the inheritance was more complex than a single gene. However, Campbell and Lafever (1981) later examined 48 F_1's representing 16 parents and noted that a large portion of the genetic variation among the F_1 apparently was additive, and the heritabilities of Al tolerance were found to be relatively high (0.57 based on parents selected as males and 0.91 for the female parents). Their apparently contradictory/confusing results may be explained in terms of those reported by Kerridge and Kronstad (1968), who could demonstrate either recessive or dominant inheritance of Al tolerance depending on the Al concentration used for the test. Their work likewise demonstrated additivity of Al tolerance because the heterozygous plants could not tolerate as high a concentration of Al as the homozygous tolerant plants. Nevertheless, the existence of genotypes with different levels of Al tolerance implies the potential for inheritance via more than a single gene, although it is not yet clear whether alleles of a single gene may be involved in some cases. Using the hematoxylin stain method (Polle *et al.*, 1978b) it should be readily possible to distinguish between different genes and different alleles for Al tolerance.

Interestingly, Prestes and Konzak (1975), using a modified liquid culture technique employing seed germination "blotter" papers [which are fixed with $AlSO_4$ (alum)], clearly identified the Chinese Spring Atlas 66 5D substitution line as carrying the Atlas 66 level of Al tolerance. However, when the same line was later tested using the hematoxylin stain method, no difference between the Atlas 66 5D substitution and Chinese Spring could be distinguished. In contrast, the Al tolerance gene of Chinese Spring could be readily identified by the greater Al susceptibility of the Chinese Spring Thatcher 4D substitution using the hematoxylin stain method (Polle *et al.*, 1978b). Atlas 66 remains one of the most Al-tolerant wheats identifiable by either solution culture or the hematoxylin stain technique. The different results are puzzling. On the other hand, Elliot (1986) has more recently confirmed the presence of an Al tolerance gene on 5D of Atlas 66.

Nodari *et al.* (1982) have shown that the Brazilian Al-tolerant wheats 'BH 1146,' 'Cotipora,' 'Lagoa Vermelha,' 'Mariga,' and 'Nobre' carry at least one dominant gene for Al tolerance. The plants, including parents, F_1, F_2, and BCF_1 and BCF_2 were grown on an Al-toxic Brazilian soil.

Regrettably, only a few of the genetics studies of the Al tolerance trait have provided definitive data, and the solution culture technique is not efficient enough for studies of F_3 progenies required to prove the genetic makeup of the F_2 plant from which the progenies were derived. With the nondestructive, highly efficient Al hematoxylin stain technique (Polle *et al.*, 1978a,b) it should be entirely feasible to classify F_1 and F_2 individuals, grow out their progeny, and test the progeny to confirm the genetic classification of the F_2 plant using tests at more than one Al concentration to identify the levels of Al controllable by the F_2 genotype. Since this stain technique has proved effective for screening Al-tolerant lines in the CIMMYT program, and its efficacy has been confirmed by Takagi *et al.* (1981, 1983), its use in genetic analyses should provide better definition of the genetics of Al tolerance. Moreover, since Al tolerance identified by the hemotoxylin stain method seems to specifically detect Al exclusion by the differentiating region behind the cap of the root, it may be possible to distinguish between tolerance mechanisms of exclusion based on membrane permeability versus other mechanisms that may relate to P extraction from Al complexes or to mucigel secretion.

Recent work by Zale and Briggs (1988) suggests, in fact, that different Al tolerance mechanisms might be detectable via differences in response to the solution culture–root length index (RLI) method and the hematoxylin score. In their test, some of the tolerant cultivars, e.g., 'Benito,' 'Chinook,' and 'Marquis,' scored only intermediate by the hematoxylin test, but had a high rating by the RLI test. In contrast, cultivar 'Bananaquit' had a comparatively lower RLI, but a high hematoxylin score. These differences are worthy of further study. The work of Zale and Briggs also helps to trace some of the lineage of Al tolerance among cultivars. This information should be useful for breeders.

In other studies, Briggs and Nyachiro (1988) obtained 9 : 7 and 15 : 1 ratios of resistant to susceptible seedlings from crosses between tolerant wheats, indicating the existence of different complementary dominant genes for Al tolerance. Their studies employed the hematoxylin stain technique, with screening at 46 ppm Al considered appropriate for differentiating resistant from susceptible plants. They obtained some spurious genetic ratios of Al-tolerant : Al-susceptible progeny from certain crosses, suggesting the existence of other mechanisms. However, there is always some risk of misclassification of single F_2 seedling plants, and while it may be appropriate for a breeding program to screen F_2 seedlings, a genetic study might better be done by analyzing F_3 progeny from F_2 plants to more precisely classify each F_2 plant as to its Al tolerance genotype. As cited earlier, Kerridge and Kronstad (1968) demonstrated that the genetic ratio could be modified by the level of Al used in the culture solution. In fact, Camargo (1981), using nutrient solution cultures, demonstrated that Atlas 66 carries two complementary dominant genes for Al tolerance, producing a 15 : 1 ratio of Al tolerant to susceptible in a cross with 'Siete Cerros.'

BH 1146, however, apparently carries only one gene, probably one of the two present in Atlas 66. One of the genes in Atlas 66 appeared to have higher Al tolerance capacity than the other, since a different genetic ratio could be demonstrated, 12 : 4 versus 15 : 1, for the same cross of BH 1146/Atlas 66 at a higher Al concentration (see also Little, 1988). As indicated before, 15 : 1 ratios would be expected if the Al concentration was too low to differentiate between the tolerance genes. Likewise, a 3 : 1 ratio may occur when use of a high Al concentration swamps the effects of a gene for lower-level tolerance. Thus, the hematoxylin stain method may offer greater precision for identifying the different genes than either culture solutions or field tests.

Based on the observed phenomena of aluminum toxic effects on the root region of elongation, which may be controlled by Al tolerance genes, the responsible mechanism appears to be aluminum exclusion and not related to pH changes in the rhizosphere as suggested by Foy *et al.* (1965). A recent study by Taylor (1988) has shown that for the genotypes investigated, i.e., Atlas 66, 'Druchamp,' 'Seneca,' 'Centurk,' and 'Scout,' Al tolerance was independent of rhizosphere pH. It is also clear that aluminum tolerance genes differ in terms of the level of aluminum tolerated without injury, and interactions of genes are indicated from the occurrence of 9 : 7 genetic ratios of tolerant to susceptible. Sapra *et al.* (1979) obtained evidence of interactions between wheat and rye chromosomes that could augment or even decrease Al tolerance in triticale.

In maize, Magnavaca (1983) and Magnavaca *et al.* (1987) studied parental lines and various hybrid and backcross generations for response to 185 μM liter^{-1} of Al in solution. Analyses of seminal root lengths indicated that additive gene effects were of primary importance, while dominance effects accounted for half as much of the genetic variation. Epistatic effects were low. Both general and specific combining ability were important.

In sorghum, Rogers (1987) studied ten genotypes in nutrient solutions and in Al-toxic soil to which seven rates of lime had been added. She observed that F_1 hybrids produced from parents with different levels of Al tolerance exhibited more root growth and had higher dry matter yields than their parents. Dominance effects were large, while additive effects were small. Heritabilities were high for both root and shoot production in a nutrient solution to which 222 μM liter^{-1} Al was added. Bastos (1983) studied the tolerance of F_1, F_2, and F_3 populations of crosses between restorer lines and F_1 hybrids to 6 and 10 ppm Al in culture solution. Al tolerance was found to be complex, involving three or more genes.

Boye-Goni (1982) and Boye-Goni and Marcarian (1985) studied a half diallel cross involving three Al-susceptible and three Al-tolerant sorghum lines identified in a screening trial. Al tolerance was predominantly additive, with three levels of dominance expressed. Heritability of Al tolerance in all plants studied to date has been high, indicating that breeding for Al tolerance should be rapid and effective.

In *Phalaris*, Culvenor *et al.* (1986b) obtained evidence that Al tolerance was due to a two-gene interaction involving at least one dominant allele at each locus. This result is not unexpected, based on results of others, since additivity of tolerance genes seems to be a rule.

In rice, Cutrim *et al.* (1982) studied F_1 and F_2 progenies of a diallel cross, testing the progeny under three concentrations of Al in solution. Tolerance was controlled by more than one gene; transgressive segregation and quantitative inheritance were evident, while specific combining ability (SCA) was more important than general combining ability. High SCA would be expected for a dominant trait controlled by a small number of genes.

Duncan (1984, 1987, 1988) has identified a number of sorghum inbred lines carrying moderate to high tolerance to acid soils with high levels of exchangeable aluminum. He is also developing Al-tolerant inbreds for sorghum hybrids. So far, however, the studies have primarily involved selecting inbreds on the basis of their response to acid soil conditions, and no genetic analyses have been done. Results of the work do, however, provide the foundation for genetic studies to better identify the genes and exploit the genetic variability for Al tolerance in sorghum.

In maize, Ruhe *et al.* (1978) found that Al tolerance was controlled by a single locus with multiple alleles. This discovery also suggests that some of the variations in Al tolerance found in wheat and other crops may be due to multiple alleles. Such differences would be worth proving in order to focus breeding applications on those alleles that confer the highest tolerance and avoid attempts to recombine allelic tolerance sources.

B. MANGANESE TOXICITY STRESS

At the turn of the nineteenth century, Bertrand (1897a,b, 1905) made two important contributions to the nutritional physiology of plants: He demonstrated that manganese was necessary to plant growth and suggested that manganese was a component of the oxidase system in plants. Manganese is not only an essential element for plants but also has an important role in the resistance to plant diseases like grey-speck disease of oats, take all, and powdery mildew (see Huber and Wilhelm, 1988; Graham, 1988). Toxic effects of manganese in soils were reported early by Funchness (1918) and Johnson (1924), and the association of manganese toxicity to acid soil conditions was firmly established later by Peech (1941), Fried and Peech (1946), and Hewitt (1952). Soil acidity conditions for Mn toxicity are not as extreme as for Al, although Mn toxicity has been reported in flax at a pH as high as 8.1 (Moraghan, 1979). In acid soil regions it usually occurs in areas of pH 5 to 6.

1. Physiological Mechanisms

The differential response to Mn toxicity by cultivars of different species has received considerably less attention than aluminum toxicity. This is probably because, in contrast to Al effects, visually distinguishable responses to Mn toxicity develop slowly in plants grown in soils and nutrient solutions (other than single-salt Mn solutions). In addition to this, manganese toxic effects in soils may be complicated with the simultaneous occurrence of aluminum toxicity. For these reasons, most of the work available on differential tolerance to manganese toxicity has been carried out in nutrient solutions or in sand culture with nutrient solution irrigation.

The degree of toxicity of manganese to plant species can be assessed by the intensity of the injury symptoms. These vary for different species. In cereals, predominant symptoms are stunted growth, chlorosis, and the development of necrotic spots. In broadleaf plants, marginal chlorosis and epinasty in young leaves are frequently observed symptoms of Mn toxicity.

In experiments with $MnCl_2$ in solution culture systems, the root elongation of wheat seedlings is more rapid than in distilled water at concentrations up to 8×10^{-5} *M*. A sharp retardation of root growth by $MnCl_2$ can be observed in about 100 hours when wheat seedlings are grown at higher $MnCl_2$ concentrations, and dark brown rings distributed at rather regular intervals along the roots are also observed (Barton and Trelease, 1927).

Morgan *et al.* (1966) found that both toxic and deficient levels of Mn markedly increased the IAA oxidase activity of cotton plants, and later Morgan *et al.* (1976) showed that growth, IAA oxidase activity, abscission, effects on internode length, and similar symptoms could be increased or decreased by raising or lowering the Mn^{2+} levels in the plant culture medium. Morgan *et al.* (1966) postulated that a causal relationship existed between IAA oxidase activity and plant responses to Mn levels. The implication was that toxic effects of Mn involved the destruction of IAA. Meudt (1971) found that Mn has a catalytic action on the autoxidation of IAA in the presence of sulfite ions. Later, Horst and Marschner (1978) were able to relate lower IAA activity to toxic levels of Mn in plant tissue. Under these conditions the IAA-dependent proton efflux was limited, causing a reduction in cell wall expansion. In summary, the reduction in growth and epinasty at toxic levels of Mn, together with the alleviating effects of Ca supply in these conditions, has been related to the effect of Mn on IAA activity.

An inspection of available plant analytical information relative to Mn tolerance between and within plant species suggests strongly that one of the major mechanisms of Mn tolerance depends on the ability of plants to grow and function with abnormally high internal concentrations of Mn rather than on their ability to exclude Mn during the absorption process by the roots (Vlamis and Williams, 1964; Foy *et al.*, 1973; Scott and Fisher, 1980; McGrath and Rorison, 1982; Nelson, 1983; Culvenor *et al.*, 1986a). In one experiment (E. Polle and C. F. Konzak, unpublished), 24 wheat cultivars were ranked according to their relative yield index (RY; see Section I,B,2,b) as Mn tolerant, intermediate, and susceptible, with RYs of 0.35–0.5, 0.5–0.7, and 0.7–1.04, respectively. The average Mn concentration in the tops of 23-day-old plants in the respective tolerance classes was 1840 mg kg^{-1} (10 cultivars), 2589 mg kg^{-1} (9 cultivars), and 2720 mg kg^{-1} (5 cultivars). However, the correlation between internal Mn concentration and Mn tolerance was not good enough to rank wheat cultivars into manganese tolerance classes. The reason for this may be related to the way in which plants adapt to grow and function with high internal manganese concentrations. This mechanism can be related to the conversion of Mn to an inactive form by chemical reactions with organic acids or proteins and/or to a physical impediment to movement of manganese to sensitive metabolic sites. Future physiological investigations of this phenomenon may open the way to improved screening methods for manganese tolerance in crop plants. Unfortunately, no rapid methods for identifying differential tolerance of plants to manganese are available at present.

One observation that may have potential for this purpose has been made by E. Polle and C. F. Konzak (unpublished). In wheat, one of the earliest responses to Mn toxicity in nutrient solutions is an increase in the dry matter concentration of Mn-susceptible genotypes. This effect is observable only a few days after planting in Mn-toxic solutions (Table II). The increase in dry matter concentration may be related to the finding of Terry *et al.* (1975) that toxic levels of Mn in sugar beets reduced the number of cells per leaf and the average leaf cell volume.

2. Screening Methods for Manganese Tolerance in Barley and Wheat

As in the aluminum screening procedures, screening for Mn tolerance requires the use of standard reference genotypes with well-established Mn tolerance responses under field conditions. Data on sensitivity to Mn toxicity in most species are still very limited. The reference wheat cultivars used in the methods outlined in this section rely mainly on observations from nutrient solution experiments. The classification of Mn tolerance groups presented here is based on the observation of over 200 wheat and 100 barley accessions.

The visual response of plants to Mn toxicity is mainly a reduction of growth and (in wheat) the appearance of interveinal chlorosis. In addition to these, barley genotypes often show necrotic spots and chlorotic areas in the older leaves. When grown with toxic levels of Mn, some wheats show a typical purplish leaf coloration as observed for P deficiency.

The method described here permits a classification in to three Mn stress tolerance groups: sensitive, medium tolerant, and tolerant. A large number of accessions have

Table II. Comparison of Mn Stress Tolerance of 18 Wheat Genotypes Evaluated by Three Methods[a]

	RY[b] 32 ppm Nutrient Solution	RY Artificial Soil	% Dry Matter (shoots, 2 weeks old)
Mn resistant			
Fielder	1.04	1.02	10.2
CM 11683	1.00	1.02	10.8
Colotana	0.97	0.91	11.0
Alondra S	0.94	0.77	10.0
Nova Prata	0.93	0.87	10.5
Tob 66	0.93	0.76	11.2
7 Cerros	0.89	0.76	11.7
Tezanos Pinto Precoz	0.87	0.85	11.3
Yding S × KaL-Bb/Hork "S"-Mo73	0.87	0.88	10.2
Mn susceptible			
Cinto 'S'	0.64	0.40	15.8
Etoile de Choisy	0.52	0.53	16.6
Zacatecas 74-Bon	0.43	0.34	20.6
Chiroca S	0.40	0.56	17.8
Glenlea	0.40	0.32	19.6
Brevor	0.36	0.36	22.5
Quetzal 75	0.36	0.42	26.2
Kafue	0.35	0.30	21.0
Trintani	—	0.43	19.3

[a]Source: E. Polle and C. F. Konzak, unpublished.
[b]RY = Relative Yield according to Eq. (1).

an intermediate type of tolerance. Extreme cases are less common. In wheat, cultivars in the highly tolerant class are 'Lerma Rojo,' 'Fielder,' 'Edwall,' 'Yecora Rojo,' and 'Nova Prata.' In the sensitive class are 'Zacatecas 74-Bon,' 'Etoile de Choissy,' 'Kafue,' 'Teal,' 'Cinto,' and 'Glenlea.' Scott and Fisher (1980) and Scott (1981) have reported high tolerance to manganese by wheat cultivars 'Cotipora' and 'Carazinho' (Brazil), with slightly less tolerance for 'Benvenuto Inca' (Argentina) and 'Egret' and 'Lance' (Australia). The most sensitive cultivars reported were 'Teal,' 'Isis,' and 'Durati' (Australia). Both 'Gatcher' and 'Spica' had severe symptoms and growth reduction. In general, Mn-tolerant cultivars had the highest levels of Mn in both roots and tops (Scott and Fisher, 1980).

In the following sections, two methods for screening Mn tolerance and a third additional tentative procedure are described. The artificial soil method has the advantage of being very compact and not laborious. Although extreme differences in visual response are noticeable (in wheat) within 2 weeks, consistent differences in plant weight can be obtained in a 3-week period. This method correlates very well with the more traditional nutrient solution method. Since the measurement of dry matter percentage is simple and the period of growth is relatively short, the method is included here as a tentative alternate.

As reported in the literature (McCool, 1935; Horiguchi, 1988), Mn effects are influenced by light intensity and temperature. The nutrient solution and the "artificial soil" methods as described here are best performed under controlled conditions in a growth chamber.

a. Mn-Toxic Artificial Soil Method Manganese toxicity in soils is the consequence of several factors, including acidity, oxidation–reduction conditions, and Mn content of the soil parent material. The organic matter content has several roles in affecting the levels of Mn that reduce plant growth: (1) as a contributor to soil acidity, (2) as a chelator of Mn in solution by decreasing free ions, and (3) as a supporter of microbial activity. Oxidizing bacteria tend to decrease Mn availability.

The artificial soil used for testing Mn tolerance in wheat and barley has a pH for the saturated paste of 5.5 and is made of commercial greenhouse vermiculite and commercial pulverized peat moss. The source of Mn is manganese carbonate. In the control treatment, this is replaced by calcium carbonate. Mineral nutrients are added from a stock nutrient solution made up with 8.12 g NH_4NO_3, 8.10 g KH_2PO_4, 0.26 g H_3BO_3, 0.38 g $CuSO_4 \cdot 5H_2O$, 1.1 g $MnCl_2 \cdot 4H_2O$, 1.6 g $ZnSO_4 \cdot 7H_2O$, and 0.01 g $(NH_4)6Mo_7O_{24} \cdot 4H_2O$ dissolved in 1 liter of final volume. One hundred milliliters of the above solution, 50 g $MnCO_3$, and 3.4 g $CaCO_3$ are added to 1 kg of a mixture by weight of one-third air-dry peat moss and two-thirds air-dry vermiculite and mixed thoroughly. The artificial soil supply should be used immediately after preparation.

For planting wheat or barley, conical plastic containers with 3.8-cm top diameter, 2° taper, and 21-cm depth are preferable. The conical containers offer economy of space (98 cones in 1860 cm^2) and allow good contact between the roots and the soil. Seeds are soaked overnight before planting. Two to four replicates, including the control and Mn soil, are run for each accession. After germination at 20–25°C (3 days), cones are transferred to a growth chamber (8 hours of dark at 15°C and 16 hours with 400 $\mu E\ sec^{-1}\ m^{-2}$ of 400- to 700-nm wave band at 20°C) and the plants thinned to one plant per cone.

The first symptoms of Mn toxicity appear in the most sensitive wheat accessions about 14 days after planting. Usually these symptoms appear earlier in barley than in wheat. The experiment can be evaluated 3 weeks after planting. The fresh weight of the tops is obtained as the measure of growth in Mn versus control. From these data the average value for each treatment and the relative yield y/y_0 are obtained (y is the yield in the Mn treatment and y_0 the yield in the control). The relative yield is the estimate of response to Mn. Relative yields between 0.5 and 0.8 correspond to the intermediate Mn tolerance class, greater than 0.8 to the tolerant class, and less than 0.5 to the Mn-sensitive class. The appearance of wheats in this test is shown in Figure 4.

Figure 4. (a) Wheat Yecora Rojo shows no effect of Mn treatment in the test with the artificial soil. (b) Wheat Zacatecas 74-Bon shows a marked decrease in growth in the Mn treatment.

b. Nutrient Solution Method Plants are grown in a nutrient solution containing 1, 1, 0.4, and 0.2 m*M* liter^{-1} of KNO_3, $Ca(NO_3)_2$, $MgSO_4$, and KH_2SO_4, and 0.07, 0.01, 0.04, 0.02, and 1 ppm of B, Cu, Zn, Mo, and Fe, respectively. The Mn concentrations are 0.13, 8, 16, and 32 ppm.

Aeration is provided during the growth period of the plants. The containers are placed in a growth chamber with an 8-hour dark period at 18°C and a 16-hour light period at 22°C. The light of the fluorescent lamp recorded at the level of the Styrofoam lid should be at least 85 μE sec^{-1} m^{-2}, in the 400- to 700-nm wave band. Low-intensity light tends to decrease symptoms of Mn toxicity. The nutrient solutions are changed on the fourth, eighth, eleventh, fourteenth, and seventeenth days after transferring plants to the Mn treatments. The plants are grown for 23 days after the initial planting in water (Figure 5). The fresh weight of the tops of the plants and that of the corresponding roots are recorded. The dry weight after drying the plants at 60°C is also determined.

The Mn effect on plant growth can be evaluated for each genotype by two estimates:

1. Mn concentration for 30% decrease in yield. This concentration can be established with a graph of fresh weight yield versus Mn concentration in the solution treatment.
2. The relative yield (RY) with 32 ppm Mn treatment.

$$RY = \frac{\text{fresh weight of the tops in the 32 ppm Mn treatment}}{\text{fresh weight of the tops in the 0.13 ppm Mn treatment}} \tag{1}$$

The relative yield is appropriate for screening a large number of accessions. In this case, treatments with 8 and 16 ppm of Mn can be omitted. Differential response of wheat cultivars in this test are shown in Figure 5.

c. Dry Weight Determination as Affected by Mn Overnight pregerminated seeds are transferred to the nutrient solution described in Section I,B,2,b, thus omitting all minor elements. On the third day the seedlings are transferred to fresh nutrient solution of the same composition but containing 32 ppm of Mn plus microelements as described in the preceding section. The plants are grown in a chamber under the conditions also described in the preceding section. The nutrient solution is changed on the seventh and the eleventh days. On the fourteenth day after the initial planting in water the plants are sampled for dry matter weight. This is done by cutting from each plant the complete first leaf and one-half of the second leaf. The fresh weight (mg) is immediately determined after cutting and after drying (overnight) at 60°C.

Mn-susceptible genotypes have dry matter percentages above 14% and Mn-tolerant genotypes below 11%. Data in Table II show the evaluation of Mn tolerance for nine wheat genotypes by the three methods described above.

Figure 5. (a) Mn-toxicity-tolerant wheat selection CM11683-A-1Y-1M-1Y-7M-OY after grown in nutrient solution with and without toxic levels of manganese. (b) Wheat cultivar Kafue, susceptible to Mn toxicity, grown under the same conditions as above.

Figure 5b. See legend on p. 103.

3. Genetics and Breeding

Although Mn toxicity is common in acidic soils, the pH range at which toxic concentrations of Mn occur is higher than that for Al. However, toxic Al and Mn concentrations apparently may occur in the same soils. Tolerance to toxic levels of Mn is found in a number of crop species, and inheritance studies as well as moderately good screening methods are available. Brown and Devine (1980) studied the F_2 of crosses involving 'Forrest' and 'Bragg' soybeans (Mn intolerant) and 'T203' and 'Lee' (Mn tolerant) soybeans using Mn-toxic soil and nutrient solutions. T203 was the most tolerant to Mn, while transgressive segregation for Mn tolerance occurred in the cross of T203 and Lee. Their results indicated that the inheritance of Mn tolerance is multigenic, and reciprocal cross differences were observed, suggesting possible maternal or cytoplasmic inheritance. The large difference in Mn tolerance between Forrest and T203, which are both Fe inefficient, was cited as an indication that the two traits are not controlled by the same allele. Heenan *et al.* (1981) investigated the genetic basis for differential tolerance of soybean cultivars to Mn toxicity. Crosses of tolerant cv. Lee × intolerant cv. Bragg gave F_1 plants with only moderate Mn toxicity symptoms, while the F_2 showed a continuous distribution skewed toward tolerance. In contrast, an F_6 progeny of 'Amredo' (tolerant) × Bragg (intolerant) showed a broader distribution.

In wheat, Camargo (1983) studied F_2 progeny of Siete Cerros × BH 1146, which carry opposite Mn and Al toxicity tolerances. Responses were determined after 15 days of seedling growth in Mn-modified nutrient solutions, or after a 72-hour exposure to toxic levels of Al. High heritabilities for the two traits were found, indicating that selection for tolerance in early generations should be effective, and that there was no obstacle to recombining the two tolerance traits. In fact, several wheat cultivars do carry combinations of Mn and Al tolerances, e.g., Yecora Rojo, Fielder, Suweon 92, and Carazinho. Foy *et al.* (1988) have determined that a small number of genes account for the major proportion of variation for Mn toxicity tolerance in wheat, making it readily feasible to transfer tolerance by backcrossing. Foy *et al.* (1988) suggested that the wide range of tolerance to Mn toxicity in wheat implied the existence of many genes of small effect, but noted that in crosses between highly tolerant and susceptible genotypes, distinct Mn tolerance classes were observed. They noted also that the "dominance" or "recessive" classification of Mn toxicity tolerance depended on the level of Mn used in differentiating the tolerance classes, much as has been shown for Al tolerance by Kerridge and Kronstad (1968). This fact implies additivity of the tolerance trait, as demonstrated by the existence of two major genes for Mn tolerance in the highly tolerant Brazilian wheat Carazinho. Foy *et al.* (1988) noted that the cv. Egret probably carries one of the Carazinho genes as compared to the Mn toxicity susceptible cv. Teal. The Mn-tolerance of Egret apparently was inherited from cv. WW15 (Syn. Anza) via Lerma Rojo, which apparently obtained the trait from the Mn-tolerant Brazilian wheat 'Supresa.' Scott and Fisher (Foy *et al.*, 1988) transferred the Egret level of Mn tolerance to two new cultivars, 'Corella' and 'Grebe,' without especially selecting for it, although field conditions for testing may have aided the selection. Scott and Fisher (1980) transferred Al tolerance from Carazinho to Egret by backcrossing using the hematoxylin technique (Polle *et al.*, 1978b), then identified six pairs of lines differing for Mn toxicity tolerance. When tested in the field, no yield differences were determined, suggesting to them that the field conditions for evaluation were too variable to show the advantage for Mn toxicity tolerance (Foy *et al.*, 1988). However, if in fact Egret carries one of the Mn tolerance genes of Carazinho, it is likely that this level of tolerance may be adequate for the test conditions, whereas the higher level of Mn toxicity tolerance of Carazinho may not be advantageous under field conditions that are not as extreme as are found in Brazil, where Carazinho was selected.

The Institute of Horticultural Plant Breeding of the Netherlands (1977; see also Eenink and Garretsen, 1977) reports the identification of one to four genes for Mn toxicity tolerance in various parental lines of lettuce. Three of the genes are linked with recombination percentages of 10, 20, and 30%.

C. IRON TOXICITY STRESS

Iron toxicity appears to be a problem limited to often water-logged, peat-type acidic soils. Toxic levels of iron may occur in lake bed soils, such as are present in the Liberty Lake and Newman Lake beds in Northeastern Washington State and in flooding-prone soils in other parts of the world. The surface 30–45 cm of the Liberty Lake soils typically dry out sufficiently for crop production, but the water table remains high and reducing conditions are common. Soil analyses may show as much as 400–500 ppm of Fe available, and while ordinary (nontolerant) wheats will not grow well, oats have been a consistently successful crop (B. Morrison, 1988 personal communication). Iron toxicity is an important yield-limiting factor for rice grown on oxisols, ultisols, and some histosols and acid sulfate soils in the tropics, causing a widespread physiological disease that has been reported in Sri Lanka, India, Malaysia, Indonesia, the Philippines, Senegal, Sierra Leone, Liberia, Nigeria, and Colombia (Ponnamperuma, 1976). Severely affected rice plants produce much less grain than unaffected plants. Iron toxicity in rice causes small brown spots on the lower leaves. These symptoms appear first on the leaf tips and later the whole leaf turns brown, purple, yellow, or orange. Lower leaves die under severe toxicity (International Rice Research Institute, 1985). They are markedly stunted, tiller poorly, head late, and produce thin, narrow panicles with many sterile florets. Roots of the plants also are poorly developed, coarse in texture, and colored dark brown. The symptoms apparently are similar to those of potassium deficiency. The disease is always associated with poor drainage and affects only lowland rice, and it appears mainly when rice fields are submerged too long. In these cases drainage is effective as a remedial measure. However, it often is not possible to apply drainage to lowland areas, and since genetic variability for tolerance to Fe toxicity has been identified, the genetic approach seems to be a practical means by which to improve crop productivity on such soils.

Genetics and Breeding

Camargo (1985) has identified Fe toxicity tolerance in the CIMMYT wheat Siete Cerros and studied the F_2 progeny of a cross between Siete Cerros and BH 1146, which is susceptible to toxic levels of Fe. Contrasted in this cross was Al tolerance possessed by BH 1146 but not by Siete Cerros, indicating the lack of relation between Al and Fe tolerance mechanisms. Using a solution culture method, Camargo (1985) obtained evidence for partial dominance of Fe toxicity tolerance, with high broad sense heritability. However, a definitive answer regarding the number of genes involved was not obtained. Because iron should stain as readily as Al with hematoxylin dye, it should be possible to use the hematoxylin stain technique to show the genetic relation of the different metal tolerance traits possessed by the two wheat genotypes.

Toxic levels of Fe also are often encountered in rice production on flooded Fe-rich acidic soils. The cultivars 'Pokkali,' 'IR 30,' 'IR 2151-190-3,' and 'IR 2153-96-1' are cited as excellent sources of tolerance to Fe toxicity (International Rice Research Institute, 1974, 1975, 1977) but no information on the inheritance of Fe toxicity tolerance was reported.

D. P DEFICIENCY TOLERANCE/P-USE EFFICIENCY

Phosphorus deficiency is a widespread problem not only in acidic ferralsols but also in neutral to acidic andosols and in alkali soils. Although andosols comprise only the equivalent of about one-tenth of the area covered by ferralsols (Dudal, 1976), they constitute important agricultural areas in the Andes of South America, in Central America, and in tropical Africa and Asia. Ferralsols with low pH and andosols with a wider range of pH exhibit high phosphorus-fixing capacities and may require heavy P fertilization to achieve economic yields. In many other areas, P deficiency can be a critical temporary condition due to cold temperatures that may affect stand establishment, disease sensitivity, and general seedling vigor.

Present world estimates indicate that there are 20,000 million metric tons of economic phosphate rock reserves and an additional 90,000 million metric tons of subeconomic deposits of rock phosphate are known to exist. These economic resources will probably last for 181 years at the present rate of world consumption of 110 million metric tons per year (Cathcart, 1980). Although fertilizer prices are cyclic, there is nonetheless a definite trend toward higher prices, causing an increasingly higher financial outlay for farmers.

It is evident that the development of crop plants more efficient in P absorption and utilization will have a significant economic impact in large areas of the world.

Breeding for P absorption efficiency does not mean exploitation of the soil P to depletion. As pointed out by Graham (1979), "deficiency of some of the less mobile nutrients is entirely due to problems of availability, not of absolute deficiency." In many areas, low P availability to developing seedlings is simply due to cold spring soil temperatures, and there is even a relation between the low P availability in cold wet soils and predisposition of wheat plants to disease (R. J. Cook, personal communication). Breeding for P efficiency may permit a reduction in the present levels of P fertilizer use and the use of less soluble and cheaper fertilizers and/or greater assurance of an adequate P nutrition at critical plant growth stages. Additionally, the residual effect of applied P fertilizers could be increased.

The scientific basis for the selection of plants for P efficiency is different in nature from that associated with Al or Mn tolerance. Al and Mn tolerance requirements can be evaluated by the ability of the plant to grow with relatively high levels of these elements, and this feature can be considered as independent of economic yields. Modern wheat cultivars selected in Brazil have the same degree of aluminum tolerance as the old tall cultivars selected in 1919, but they yield three times as much.

In the selection of P-efficient genotypes, the breeder is not interested in growth or biomass production under P deficiency stress (unless dealing with pastures or some specific crops cultivated for vegetative production). The main interest and only final measurement of P efficiency is grain yield under P deficiency stress conditions. This makes the problem of selection extremely difficult. There are, however, some measurable features in plants that can be related to P efficiency. These features may help to define an ideotype to look for in the selection procedure. Some of these features will be reviewed in the following sections.

1. Physiological Aspects of P Efficiency

There are two ways in which a plant can be more P stress tolerant than another: (1) by absorbing more P from a limited or chemically fixed soil P supply and (2) by utilizing the absorbed P more efficiently.

In a soil, nutrients can reach the plant through the mass flow of water toward the roots produced by transpiration and/or by diffusion along gradients created by root absorption. Mass flow contributions to P nutrition can be disregarded considering the extremely low concentrations of P in the soil solution (Barber *et al.*, 1963). The diffusion process is then of major interest with regard to P nutrition from the soil.

The amount of P that reaches the growing plant by diffusion can be influenced by the plant in three possible ways: (1) exploration of the soil mass by roots, enhancing the diffusion of P from the soil to the root; (2) interactions between the root and the soil, resulting in the release of P from soil particles; and (3) the ability of the root cell membranes to transport P at a higher rate from the external media, where P is usually scarce.

Mycorrhizal associations may modify this process in some species by increasing the effective root area. However, the effects of mycorrhizae tend to disappear at very low and very high soil P levels (Mosse, 1973; Malajczuk *et al.*, 1975; Abbot and Robson, 1977; Buwalda *et al.*, 1982). Higher dependence on VA mycorrhizae has been reported by Baylis (1970, 1972) for plant species lacking or having poorly developed root hairs and/or poor and coarse root branching (Baylis, 1975; St. John, 1980). Cultivar (genotype) effects on mycorrhizal associations also have been shown in wheat. Azcon and Ocampo (1981) found that mycorrhizal dependence was greatest in Siete Cerros and Cocorit, intermediate in 'Anza', and absent in 'Jupateco', 'Negrillo', and 'Champlein'. Cocorit and Siete Cerros had the smaller, and Negrillo and Champlein the larger, root systems. However, the information available on mycorrhizae does not allow an evaluation of their contribution to the absorption efficiency of P by plants.

a. Exploration of the Soil Results by Newman and Andrews (1973) showed that the absorption of K but not P per unit of root in wheat was lower the higher the root density. These results were explained by suggesting that the different diffusion coefficients of the ions caused depletion zones from 1.3 to 13 mm for K and of about 0.5 mm for P, considering that the distances between neighboring roots in high density roots was 1.5 to 4 mm. Läuchli (1967) demonstrated the accumulation of P from solutions by root hairs, and Dittmer (1937) calculated that root hairs may increase by 5–18 times the surface area in the root hair cylinder (the cylinder of soil penetrated by the root hairs from an individual root).

Results by Bhat and Nye (1974) in *Brassica napus* L. indicated that root hairs have a capacity for adapting to low levels of P in the soil. Although the level of P does not significantly affect the density of root hairs, at low P levels the length of the root hairs was four to five times longer than in the high P treatment. Studies by Bole (1973) in disomic substitution lines differing in root hair density (genotypes Chinese Spring and S-615) suggested a positive relationship between P uptake and root hair density only up to a maximum of 50 root hairs per millimeter. Bhat and Nye (1974) explain Bole's results by considering that, with higher root hair densities, zones of P depletion around root hairs rapidly overlap and a diminishing returns curve for increasing densities is to be expected. Thus, the presence of root hairs increases the P-absorbing area per unit length of root, but the benefits of this increase are limited by the geometrical factor indicated.

Studies conducted by Foehse and Jungk (1983) indicated that both the length and density of root hairs of tomato, rape, and spinach were increased by decreasing the P concentration of the rooting solution. The alteration of these two root properties increased the root surface area by a factor of 2 to 30 compared to plants grown under high P conditions. Furthermore, using split-root techniques with spinach, they concluded that the stimulation of root hair formation was dependent on the P content of the shoot, rather than the P concentration at the root surface. The nature of the signal emitted by shoot tissue low in P is presently unknown. A mathematical model depicting soil and plant parameters that influence P uptake led Silverbush and Barber (1983) to conclude that root growth rate and root radius were among the most influential factors determining P uptake. This is consistent with findings by

Steffens (1984) that P uptake by ryegrass and red clover under field conditions was positively related to root length. In addition to this, Misra *et al.* (1988) demonstrated that root hairs have a strong influence on the accessibility of P to roots in macrostructure soils.

Root area interacts with P supply in the soil. Under low P availability, plants respond with the well-known stimulation of root growth relative to leaf growth. Under these conditions and with young seedlings, after seed P reserves are exhausted, root growth increases by the retention in the roots of P absorbed from the soil (Williams, 1948).

b. Root–Soil Interactions Plant roots may modify the soil rhizosphere by several mechanisms. These include localized alteration of soil pH, soil Eh (reducing conditions), and the composition of inorganic and organic ions surrounding the roots. In addition, the release of phosphatases and other organic compounds from roots can alter P availability. Changes in these soil parameters influence the activity and form of inorganic P present in the soil solution, as well as the ability of a soil colloid to replenish P in solution during P uptake and subsequent depletion of soluble P in the root rhizosphere. Overall, the environment around the root is known to differ from the bulk soil. However, there has been little definitive work to show that these conditions increase the availability of P to the plant.

In many soils of the world where P deficiency is a problem, the prevailing mechanisms of P fixation are formation of insoluble metal phosphates and sorption of P to hydrous oxides and noncrystalline aluminosilicates. In most tropical soils and many acid or volcanic ash-derived temperate soils, the hydrous oxides of Fe, Al, and Mn and the so-called amorphous aluminosilicates (allophane and imogolite) will be involved in these reactions. The work done to date on P fixation by these soils has primarily involved equilibrium studies of precipitation and sorption of P on known minerals or various soils. Some kinetic studies have been performed to determine rates of P fixation and to infer the mechanism of fixation. Very little work has been done to determine the effect of the rhizosphere environment on P release from solid phases. The existence of genotypes with differential P response provides a unique opportunity to determine if products of the root environment are directly involved in P release. Genotypes selected for tolerance to P-deficient soils are promising candidates for utilizing root exudates for P solubilization.

Plant roots can either increase or decrease the rhizosphere pH compared to that found in the bulk soil (Nye, 1981). Changes as large as 1 to 2 pH units have been observed. Either H^+ or OH^-/HCO_3^- ions are extruded from plant roots, depending on cation/anion balance during nutrient acquisition. These processes vary with nitrogen source and differ among species (Marschner and Römheld, 1983). Little information is available on the magnitude of intraspecific differences. It is well documented that NO_3 nutrition stimulates OH^-/HCO_3^- exudation extrusion, whereas NH_4^+ uptake results in rhizosphere acidification (Riley and Barber, 1971; Smiley, 1974). P uptake by soybeans was enhanced by NH_4^+ nutrition over NO_3^-, and this effect was closely related to the relative effects on the rhizosphere pH (Riley and Barber, 1971). Root systems of corn genotypes absorbed differing proportions of NH_4^+ and NO_3^-, which implies that genotypic differences in rhizosphere acidification probably exist (Pan *et al.*, 1985).

Phosphorus adsorption to amorphous Al is an important mechanism of P fixation in volcanic ash-derived soils. Kawai (1980) demonstrated a positive relationship between amorphous Al content in 14 soils and the extent of P adsorption. Furthermore, the adsorption of P was more extensive at soil pH 4.5 in comparison with pH 7. These data imply that plants capable of decreasing rhizosphere acidity when grown on these types of soils may be capable of reducing P adsorption by soil particles in the vicinity of the root surfaces, thereby increasing P availability. Increases in rhizosphere pH of as little as 1 unit can effect a 10-fold or greater increase of P concentration in solution, depending on the solid phase present (Lindsay and Moreno, 1960; Nye, 1977).

The exudation of soluble organic compounds represents a second major mode by which plant roots can alter P availability in the vicinity of the root. The mechanisms include acidification, promotion of reducing conditions, metal ion complexing, and competitive adsorption. In wheat seedlings, Barber and Martin (1976) measured up to 20% of the total photosynthate being released into the soil as non-CO_2 carbon. Moghini *et al.* (1978) fractionated rhizosphere exudates of wheat seedlings. Approximately 38% of the rhizosphere products were carbohydrate, and about 20% were composed of 2-ketogluconic acid. The latter compound was found to be important in the dissolution of hydroxyapatite (Moghini and Tate, 1978). The importance of dissolution–complexation reactions in increasing P availability is underscored by simulation models designed to predict P uptake based on independent soil and plant parameters (Bhat *et al.*, 1976; Brewster *et al.*, 1976). They found that the concentration of P in the soil solution around the root was somehow increased above predicted values.

The redox systems most likely to react with reductants from the roots are Fe(III) = FE(II) and Mn(III, IV) = Mn(II). The initial effect on hydrous oxides or phosphates of Mn and Fe would be dissolution, which could bring sorbed or precipitated P into solution. Although the P released would be subject to equilibrium reactions at surfaces or in solution, diffusion to the root could easily occur before an equilibrium was reached. There is little knowledge at this time concerning the redox properties of the rhizosphere and no evidence of effects on P solubility.

The exudation of inorganic and organic metal-complexing agents, including SO_4^{2-}, organic acids, and amino acids, by the root is a potential mechanism for increasing solution P, at least in the short term. Complexing ligands such as citrate and oxalate are known to reduce P sorption by gibbsite and kaolinite (Nagarajah *et al.*, 1970), to participate in the dissolution of hydrous oxides (Higashi and Ikeda, 1974; Schwertman, 1964; Zutic and Stumm, 1984), and to increase the rate of Al and Mn dissolution in forest soils (Pohlman and McColl, 1984). Tinker (1975) and Tinker and Sanders (1975) have argued against such a mechanism based on the amount of exudates released by the plant. Nye (1977) was unable to discount the possibility of increased solubilization of P in the rhizosphere, however, given the large amount of exuded material in a relatively small volume about the root (Barber and Martin, 1976).

Bolan *et al.* (1984) have recently shown that mycorrhizal plants are able to obtain P from iron hydrous oxide surfaces more effectively than nonmycorrhizal plants, reopening the debate as to whether the only effect of mycorrhizae is extension of the "root" surface (Sanders and Tinker, 1971). If one considers the potential kinetic effects of organic anions on dissolution reactions as well as the equilibrium relationships, it is clear that the possibility of root exudates releasing P from solid phases cannot be discounted.

Gardner *et al.* (1982, 1983) have shown that the production of reducing and complexing agents around proteoid roots of *Lupinus albus* L. contributes to the solubilization of P. Large amounts of citrate exuded by the root and the formation of a citrate–Fe–OH–P complex were implicated in the process. The techniques of Gardner *et al.* (1982; modified from Gerretsen, 1948) need to be applied to systems other than the proteoid root-producing plants to determine their P-releasing capability.

In many soils, organic forms of P constitute 50% or more of the total soil P in the A horizon. Therefore, release of phosphatases and/or organic P solubilization by the plant root must be considered a possibility. The specificity of such a response by the plant makes it a desirable mechanism for P release. This mechanism has been postulated by Boero and Thien (1979), but experimental work thus far has produced no evidence in its support (Barber, 1984; Bolan *et al.*, 1984).

c. Membrane Transport For roots growing in a soil, the absorption rate will increase as the concentration gradient increases near the root surface. For a given concentration of P available in the soil bulk, the lower the concentration at the root surface, the higher the gradient and the absorption rate. A low P concentration can be maintained from very dilute solutions by roots when the membrane transport is high.

In experiments on ion absorption from solutions, the absorption rate J is adequately described by a rectangular hyperbola of the form

$$J = \frac{J_{\max} c}{K + C} \tag{2}$$

where K is the concentration (c) for half the maximum rate ($J_{\max}$). This equation fits the experimental absorption data well (Edwards and Barber, 1976). For a given root system, the efficiency of P uptake will increase where $J_{\max}$ increases and when K decreases. Claassen and Barber (1974) developed a method for determining $J_{\max}$ and K, based on the depletion rate of a nutrient from solution by the roots. Results with corn by Nielsen and Barber (1978) suggest that genotypes with long, fine roots absorb P (from solutions) more slowly per centimeter of root than genotypes with shorter and thicker roots. Also the ability to absorb P from dilute solutions increases in P-starved corn plants (Anghinoni and Barber, 1980).

d. Relative Contribution of Plant–Soil Properties to P Absorption In the inflow equation describing the uptake of nutrients by a root in the soil, two terms can be distinguished. One term is related to plant properties such as the absorbing power of the root, and the other, the diffusion term, is related to soil properties. Nye (1977) calculated the sensitivity of the inflow equation to each of these terms at the threshold of the nutrient concentration when the plant growth response ceases to increase further. Available experimental data were used in these calculations. Results suggested that for the major nutrients N, P, and K, the inflow is more sensitive to the diffusion terms than to the combined effects of root absorbing power and transpiration. In other words, as the ion concentration at the surface of the roots becomes lower than the bulk concentration in the soil solution, the rate-limiting step shifts from control by plant properties to become—for exceedingly low concentrations at the root surface—entirely soil dependent. Silverbush and Barber (1983) also found little effect of the P–root uptake kinetics in the P supply to the plant. Results by Nielsen and Schjorring (1982) suggest that with the soil used in their experiment, P supply at the root surface was not entirely dominated by soil diffusion. Below the threshold of plant response to P, the concentration range of our main interest, the absorption rate is most probably controlled by soil diffusion.

The results reviewed above leave little doubt that root area is related to P stress tolerance. Complex mechanisms of root–soil interaction may be operative in some species. In this case, "soil diffusion"-dominated transport may shift to "plant control" if root exudates significantly affect the concentration and/or movement of P in the root vicinity (Nye, 1977).

It is possible that when the concentration is low enough in experiments with nutrient solutions, gradients across the stagnant boundary layer adjacent to the roots may be the limiting step in ion absorption (Polle and Jenny, 1971). This type of effect may arise at concentrations of 0.01 m*M* for Rb. No information is available for P. According to Loneragan and Asher (1967), at concentrations of 0.04 μ*M* of P, very little P was absorbed by the roots of eight species. There is a rapid increase in absorption rate up to 20.0 μ*M*, a slower increase up to 1 μ*M*, and very little beyond this concentration. Low concentrations are very difficult to maintain in experiments and require sophisticated equipment.

McLachlan *et al.* (1987) observed that in P solutions being depleted by plant roots, three phases develop in tissue: a first phase where the concentration is not limiting and plant processes control the absorption rate; a second phase where solution concentration and plant roots interact to determine the absorption rate; and a third phase where the uptake is limited by the low external concentration. Thus, even in a liquid medium, if the concentration of P is low enough, the rate of uptake escapes the plant control, In a soil where the liquid phase is not in a turbulent flow as in a stirred nutrient solution, gradients outside the plants are to be expected to dominate the absorption process. For these reasons, membrane transport ability of the roots may not be a factor related to the P absorption efficiency of plants in soils.

e. Phosphorous Utilization Efficiency The efficiency with which plants use the absorbed P depends both on the efficiency with which they distribute P to functional sites and on the P requirements of cells at those sites (Loneragan, 1978). This matter is presented in detail in Chapter 5 and will not be reviewed here.

f. P-Use Efficiency and P Extraction from the Soil Very little information is available to evaluate the relative importance of extraction ability and P-use efficiency in plants. Probably the relative importance of these features varies with the species. In this and in the following sections the discussion will be centered particularly around the wheat plant. This is so because in this species the information, although still incomplete, covers more basic areas of the problem than in other species.

In Lipsett's work (1964), no genotype differences in P uptake by wheat grown under P deficiency conditions were found. However, strong genotypic differences were found for P concentration in the grain. This result suggests that the major genotype effect observed was on the internal P-use efficiency. His experiment was conducted using 5- and 7-inch-diameter pots holding four and five plants per pot. Limitations to root growth may not have allowed the distinction of genotype differences in P uptake from a restricted soil mass. According to Batten *et al.* (1986), wheat plants produce maximum grain yield when P is supplied until heading and the effect in yield due to P supply after floral initiation is higher in P-deficient plants. However, more than 50% of the yield is secured with early supply of P (before floral initiation, in the first 3 weeks). Results by Rahman and Wilson (1977) indicate that adequate P nutrition in the early stages of wheat growth increases the spikelet number per spike and the rate of spikelet initiation per day. Once the number of spikelets per spike is determined, only the number of florets per spikelet and the seed size can vary. Results by Koehler (1983) and our own (E. Polle and C. F. Konzak, unpublished) suggest that these two yield components may account for major contributions to yield in P-efficient genotypes. Thus, genotypes that have better P nutrition and distribution in the early stages of growth may have the possibility to increase both tillers and number of spikelets per spike.

Batten *et al.* (1986) found that P supplied during grain filling in normal and low-P plants raised the concentration of P in the grain, but not the grain yield. This was interpreted as evidence that even low-P plants export more P to the grain than is required for grain growth. Consequently, there is room for increasing the efficiency of P use by lowering the content of P in the grain and still maintaining high yields. The relative importance of P-use efficiency and extraction of P by plants is at present an unresolved question.

2. Genetic Variability

Genotypic variability in P deficiency stress tolerance is generally evaluated from one or more of the following measurements: (1) growth response to P supply in the support medium, (2) the P absorption rate from soils or solutions by different genotypes, or (3) the differential accumulation of P by the plants of different genotypes. However, as discussed in Section I,D,1,e, gentoype variability in growth response to P supply can be due either to differential P absorption from the media or to differential P utilization efficiency or both, while the "accumulation" of P by a plant genotype may not necessarily be due to P deficiency stress tolerance. As noted by Vose (1982), this distinction between P absorption capacity and utilization efficiency has not yet been achieved experimentally. In spite of this, results of growth response and absorption rate studies indicated the existence of wide genetic variability in both P absorption and P utilization efficiency in plants.

DeTurk *et al.* (1933), Smith (1934), and Lyness (1936) found differences in response to P fertilization and supply among maize inbreds and hybrids. Baker *et al.* (1967, 1970, 1971) obtained large differences in P accumulation by maize. Differential ability for P utilization and in absorption rates from solutions were found by Whiteaker *et al.* (1976) and Lindgren *et al.* (1977) in *Phaseolus vulgaris* L. Differences in the ability to grow on P-deficient soils and in absorption from solutions by rice was found by Koyama and Chamnek (1971), Koyama and Snitwongse (1971), and Salinas and Sanchez (1978).

Miranda and Lobato (1978) and Ben and Rosa (1983) found significant differences in response to P fertilization between wheat genotypes when grown in soils with different percentages of Al saturation, although these differences may wholly or in part reflect genotype differences in Al tolerance now known to be an independent trait (E. Polle and C. F. Konzak, unpublished). A wide range of P concentrations in the grain of Australian and U.S. wheats, grown in pots under controlled P deficiency, was reported by Lipsett (1964). The cultivar 'Bencubbin' produced higher grain yields than cultivar 'Charter' under conditions of P deficiency, and its grain had a lower P concentration. Palmer and Jessop (1977) compared the P response of a semidwarf (Israel M-68) with the standard-height cultivar 'Olympic' in a pot experiment. The results suggested an improved ability of semidwarf wheats to explore the soil rhizosphere system.

3. Screening Methods

At present, methods for screening for P efficiency in wheat and related species can only be tentative. The most suitable genotypic characteristics to look for are not known with certainty. From the previously mentioned results, one of the factors is the development of the root system under conditions of stress. This can be studied in field or pot experiments during the first 3 weeks of the wheat plant development. Methods are available for examining root development.

May *et al.* (1965, 1967) concluded from detailed studies on barley that the root growth pattern established during the first few weeks after planting has a considerable significance in the potential for later increases in root number and length. Tennant (1976) studied root growth of wheat (cv. 'Gamenya') for 30 days after planting in a quartz sand medium supplied with standard, twofold, half, and zero levels of N, P, and K. Root numbers and length followed consistent patterns of increase in the seminal and nodal root systems in all treatments. Phosphorus deficiency gave increasing delays in root component appearance with increasing order of root branching. Increasing suppression of seminal and lateral root numbers and severe suppression of nodal root growth followed. All responses to applied nutrient levels were more obvious with increasing order of lateral roots and with the nodal roots rather than the seminal root system. Root characteristics of interest can and should be observed during the first 4 weeks of growth considering, as discussed before, that P-stress-tolerant genotypes must establish themselves with vigor at an early age.

Experiments by Phillips (1979) showed strong interaction between early seminal root growth (length) and root number and wheat genotype, location and seed production, and year of seed production. However, the relative amount of root growth between genotypes was constant if the seed of all the genotypes compared was taken from the same environment. It is then advisable that seeds of genotypes for testing root characteristics should be produced at the same time and at the same site.

The conclusion that root properties are important for P absorption efficiency is based on observations in soils containing a uniform distribution of P. This condition is not present when the fertilization is made with a localized P source. In this case, the characteristics of the root system linked to P absorption efficiency needs to be established and the testing methodology developed.

The identification and evaluation of the features of root–soil interactions (secretions, surface reactions at the soil–root interface) related to P absorption efficiency are at present very elementary. Generalizations in these matters are not possible since the nature of the soil of interest may have considerable effect. However, methods of fractionation of root rhizoplane and bulk soil are available (Brown and Anwar-Ul-Haq, 1984; Cappy and Brown, 1980) and can be used in these studies. The development by Coltman *et al.* (1982) of a sand–alumina medium in which the P adsorption equilibrium can be controlled is a valuable contribution for the study of root–soil interaction under conditions of high P-fixing capacity. Field screening, if not the most rapid and economic way of selecting genotypes for P efficiency, is the most reliable.

In the authors' experience, soil diseases may interact with P deficiency—some pathogens like *Pythium* tend to appear irregularly in the field, introducing extreme experimental variability and reducing the chance to identify genotypes tolerant to P deficiency as distinct from tolerance to *Pythium*. Treatments of the soil with methyl bromide will control this disease and at the same time eliminate the mycorrhizae, which also interact to provide P. Selection of parental breeding lines for P efficiency can and should be made on the basis of grain yield. The most useful measure in this regard is the relative yield, y/y_0 (y = yield under P stress conditions; y_0 = yield without P stress). However, field results for grain yield are strongly influenced by local ambient factors other than P nutrition. In field studies, the selection of parental lines should be made on the basis of their higher relative yield, while at the same time selecting for the highest maximum yield. Data in Table III show the results from a field comparison between seven lines and 'Fielder', a spring wheat in common use in the area of the experiment. The lines in Table III are F_7 of crosses

Table III. 1987 Yields of Comparable Wheat Selections in a Volcanic Ash P-Deficient Soil Treated with Methyl Bromide (Mission variant silt loam, Idaho)[a]

Selection		Replications (metric tons ha^{-1})			
		1	2	3	Mean
39 FL × AL	P_{max}[b]	5.59	5.61	3.69	4.96
	P_1[c]	4.65	4.14	3.74	4.18
11 AL × PVN	P_{max}	4.78	4.97	5.39	5.05
	P_1	2.76	3.47	3.05	3.09
21B2 × AL	P_{max}	6.45	5.96	5.09	5.83
	P_1	3.92	4.54	4.51	4.32
Fielder	P_{max}	4.96	5.54	5.29	5.26
	P_1	3.41	3.62	3.15	3.39
4-119372 × AL	P_{max}	5.08	4.20	4.65	4.64
	P_1	3.56	3.67	3.59	3.61
9 FL × A1	P_{max}	4.07	4.62	4.96	4.55
	P_1	2.84	3.04	2.61	2.83
33 FL × Al	P_{max}	5.30	4.19	4.46	4.65
	P_1	3.80	4.13	3.02	3.65
10 AL × PVN	P_{max}	4.43	5.38	4.37	4.73
	P_1	1.82	3.29	2.09	2.40

[a]Source: E. Polle and C. F. Konzak, unpublished.
[b]P_{max} corresponds to an application of 400 kg of P_2O_5 ha^{-1} as triple superphosphate.
[c]P_1 corresponds to an application of 12.6 kg of P_2O_5 ha^{-1} as triple superphosphate.

selected in the same field under P stress conditions. The genotypes are ordered in Table III in consecutive pairs having about the same maximum yield. This permits a comparison of relative yields under limited P supply between genotypes of similar adaptation to the ambient conditions of the area.

4. Genetics and Breeding

Although there is yet little knowledge of the heritability of differences in P-use efficiency and/or P deficiency tolerance in crop plants, evidence of genotypic differences for these (or this) traits exist. E. Polle, C. F. Konzak, and T. J. Koehler (1984 unpublished) studied the response of several hundred wheat accessions to P deficiency conditions of a P-fixing volcanic ash soil in northern Idaho. Although the soil pH was high enough to avoid Al toxicity, and the Mn level was too low for Mn toxicity, marked differences among wheat genotypes were identified. These differences were expressed in terms of vegetative growth, grain yield, or sometimes both, and genotypes with a high harvest index seemed to have an advantage. The only consistent association of grain yield on the non-P-fertilized soils for some genotypes was related to their 1B/1R translocation chromosome, or by other pedigrees indicating a strong genetic control. Batten *et al.* (1984) and Batten (1986a,b) also found that a number of modern semidwarf wheats not only responded better to increased levels of P fertility compared with older tall wheats, but also were more efficient in their use of P. These wheats had higher harvest indices than the older tall wheats. Batten (1986a,b) found a positive but comparatively low correlation between grain yield and P availability and a negative relation between grain P content and grain yield. Batten's (1986a,b) results suggested that it should be possible to gain two benefits from selection and breeding for P-use efficiency: (1) better yield at moderate to low P fertility and (2) lower levels of P incorporated in the grain as phytic phosphate. Since phytic P interferes with mineral nutrition of animals and man, breeding for P-use efficiency seems to be a goal that is worth considerable effort.

P deficiency tolerance, on the other hand, may have a different genetic basis than P-use efficiency, although in many situations the two responses may be difficult to separate. P deficiency tolerance, unlike P-use efficiency, could be related to the ability of the plant root exudates to solubilize or dissolve P fixed in soil particle complexes as with Al clays. There has been some indication that the spring wheat cultivar 'Alondra' may have such an ability since it has performed unexpectedly well in Al-toxic soils of Brazil. A variant of Alondra with a moderate level of Al tolerance was identified (CIMMYT, personal communication) but its level of Al tolerance may not be adequate to explain adaptation of the Alondra to the Al-toxic soils. Alondra was fairly susceptible to Al in tests conducted at Washington State University, but proved to perform surprisingly well in the northern Idaho P-fixing soil. Interestingly, the different means by which the wheat genotypes achieve P-use efficiency may be complementary and recombinable to produce genotypes with considerably improved P-use efficiency and adaptability to a range of soil conditions. Although germplasm sources of P-use efficiency/P deficiency tolerance have recently been identified in wheat, no genetic analyses have been conducted. However, research is already under way in Brazil (Rosa, 1988) and in Washington State to develop wheat cultivars more efficient in P use (C. F. Konzak and E. Polle, 1989 unpublished). Selections have been recovered in both programs that recombine the P deficiency tolerance trait with more desirable agronomic properties using cultivars Toropi, PG1, and Alondra as sources of P deficiency tolerance.

In rice (IRRI, 1977), 8 of 314 lines were notable for tolerance to P deficiency. In each program, the differences achievable in response to P deficiency and P supplements appear to be worth attempting to exploit in breeding programs. Genetic analyses may result as secondary goals of these programs, but extensive genetic analyses and a better understanding of the biochemical genetic basis for P deficiency tolerance/P-use efficiency will probably not occur until efficient laboratory mass screening systems are developed.

It will be of considerable importance that research be pursued to better define, identify and differentiate the different mechanisms responsible for P use efficiency and P efficiency tolerance in crop plants to provide a better scientific basis for breeding plants able to yield well with low inputs of applied P fertilizers.

II. Compatibility of Genetically Controlled Tolerances to Stresses

It is appropriate to take special note of the fact that the evidence available to date indicates that the genes controlling Al and Mn toxicity tolerances and P deficiency tolerance appear to be independently inherited and recombinable. Numerous examples are now available in the wheat literature demonstrating the occurrence of multiple stress tolerances in a single wheat genotype. Thus, Suweon 92 carries both Al and Mn toxicity tolerances, while Yecora Rojo is tolerant to Al and Mn and appears to be P efficient in a volcanic ash soil. Tolerances to deficiencies or excesses of the essential minor elements Mn and Fe in plants are genetically controlled. The genetics of tolerance to manganese and iron deficiency is well documented (Graham, 1988; Fehr, 1983). The relationship, if any, between efficiency by which plants obtain these elements and tolerance to their excess availability in the soil has not been investigated. Efficiency of manganese acquisition in plants is mainly related to root modifications of the soil or root/microbial associations and to metabolic utilization within the plant. Tolerance to excess manganese seems to be related to the inactivation of manganese within the plants. It is reasonable then to conclude that these different mechanisms also are genetically independent and not in opposition. A similar situation may exist for iron, although in this case little is known of the mechanism of tolerance to excess iron. Until proven otherwise, breeders should assume that it will be possible to combine in one genotype all of the possible stress and tolerance genes known, including Mn and Fe tolerances as well as Mn, P, Zn, Fe, and Cu deficiency tolerances. Combinations of stress tolerances should greatly improve the adaptability of cultivars to a wide range of environments and reduce the potentials for crop losses due to environmental interactions.

III. Role of Genetic Engineering

Recent developments in genetic engineering offer several new avenues for improving the genetic adaptation of plants to soil mineral imbalance conditions, including increases in mineral use efficiency. Of course, it is not reasonable to suggest that N-use efficiency, for example, could reach 95 to 100%, since obviously some losses of the mineral are unrelated to plant N uptake. It is, however, reasonable to expect that higher than currently achievable efficiencies in mineral use would be attainable. Genetic engineering is unfortunately a poorly defined term that is applicable to many techniques, and it may be well not to focus on any one approach when many avenues may be worth developing.

In a number of cereals, for example, tissue culture and anther or microspore culture methods already have achieved efficiencies applicable to practical plant breeding processes. Nunez *et al.* (1985) have used anther culture in rice to produce progenies that are homozygous for Al tolerance from crosses between tolerant and susceptible strains, and experiments using anther culture methods to investigate the inheritance of Al and other tolerances in wheat suggest that it may be possible to attain unequivocal data on the existance of alleles, numbers of genes, and interactions (H. Ekiz and C. F. Konzak, 1989 unpublished).

Totipotent cells or cell groups developed by these tissue culture methods represent efficient targets for the introduction of genes responsible for useful economic traits. However, before a gene or DNA carrying the desired trait can be introduced via genetic engineering techniques, the gene must be isolated, identified, and reproduced in the laboratory. So far this part of the process has been achieved only for a few marker genes. Intensive research toward the localization, identification, isolation, and reproduction of economically important genes is now under way at several laboratories. This research includes RFLP (restriction fragment length polymorphism) mapping to better localize genes into linkage groups. Artificial synthesis of DNA and "gene" units also is becoming commonplace in modern laboratories. Moreover, an unexpected breakthrough in the development of a physical method for introducing DNA into plant cells has been achieved, including applicability to monocotyledonous plants (Klein *et al.*, 1988a,b). Research using these techniques will certainly have profound consequences on plant breeding problems.

Although it may be possible to introduce artificially constructed DNA sequences and then screen cell progenies for the desired traits, today considerable effort is being placed on the DNA sequencing of genes of potential value to breeding. Thus, a gene taken from bacteria, fungi, or other plant species could serve as the candidate for introduction by genetic engineering techniques. It is certainly within the realm of expectation that a highly effective Al-tolerance gene from rye, for example, could be introduced not only into wheats but also into barley, sorghum, maize, and rice by genetic engineering methods.

However, also critical to the exploitation of the genetic engineering techniques will be improved methods for mass screening of cells, calli, or embryoids in culture for desired traits, including Al, Mn, or salt tolerance and P- and N-use efficiencies. Too little has been done in these areas. In fact, it may be feasible to use the effective tissue culture systems to isolate mutants with improved mineral imbalance tolerance levels once efficient screening methods are developed and proved to be related to both the cellular and whole-plant levels of expression for the trait. Considerable progress in this regard has already been made by the Tissue Culture for Crops Project (TCCP) at Colorado State (Ketchum, 1989). Because mineral imbalance mechanisms often do seem to be effective at the cellular and tissue level of expression, it may be feasible as well as practical to exploit tissue culture systems for improving the Al and Mn toxicity tolerance levels of many current crop species.

The tissue culture/mutation approach has great potential for exploitation to achieve Al/Mn toxicity tolerances. For example, Conner and Meredith (1985) have shown that stable, Al-tolerant, dominant, single-gene mutants of *Nicotiana plumbaginifolia* can be obtained from tissue cultures grown with increasing levels of Al. Because even the highest levels of Al tolerance available in the gene resources of cereal crops are probably inadequate for extreme situations, or could at least be improved upon, the tissue culture approach to improved Al tolerance should receive more attention in future research.

References

Abbot, L. K., and A. D. Robson. 1977. Growth stimulation of subterranean clover with vesicular arbuscular mycorrhizas. *Aust. J. Agric. Res.* 28:639–649.

Anghinoni, I., and S. A. Barber. 1980. Phosphorus influx and growth characteristics of corn roots as influenced by phosphorus supply. *Agron. J.* 72:685–688.

Aniol, A. 1984. Induction of aluminum tolerance in wheat seedlings by low doses of aluminum in nutrient solution. *Plant Physiol.* 75:551–555.

Aniol, A., R. D. Hill, and E. N. Larter. 1980. Aluminum tolerance of rye inbred lines. *Crop Sci.* 20: 205–208.

Azcon, R., and J. A. Ocampo. 1981. Factors affecting the vesicular–arbuscular infection and mycorrhizal dependency of thirteen wheat cultivars. *New Phytol.* 87:677–685.

Baker, D. E., R. R. Bradford, and W. I. Thomas. 1967. Accumulation of Ca, Sr, Mg, P and Zn by genotypes of corn (*Zea mays* L.) under different soil fertility levels. *Isotopes in Plant Nutrition and Physiology. Proc. Symp., Vienna, 1966* pp. 465–477.

Baker, D. E., A. E. Jarrell, L. E. Marshall, and W. I. Thomas. 1970. Phosphorus uptake from soils by corn hybrids selected for high and low phosphorus accumulation. *Agron. J.* 62:103–106.

Baker, D. E., F. J. Wooding, and M. W. Johnson. 1971. Chemical element accumulation by populations of corn (*Zea mays* L.) selected for high and low accumulation of P. *Agron J.* 63:404–406.

Barber, D. A., and J. K. Martin. 1976. The release of organic substances by cereal roots into soil. *New Phytol.* 76:69–80.

Barber, S. A. 1984. Soil Nutrient Availability. Wiley, New York.

Barber, S. A., J. M. Walker, and E. H. Vasey. 1963. Mechanisms for the movement of plant nutrients from the soil and fertilizer to the plant root. *J. Agric. Food Chem.* 11:204–207.

Barton, L. V., and S. F. Trelease. 1927. Stimulation, toxicity and antagonism of calcium nitrate and manganese chloride as indicated by growth of wheat roots. *Bull. Torrey Bot. Club* 54:559–577.

Bastos, C. R. 1983. Inheritance study of aluminum tolerance in sorghum in nutrient culture. *Diss. Abstr. Int. B* 43(7): 2069B. (Abstr.)

Batten, G. D. 1986a. The uptake and utilization of phosphorus and nitrogen by diploid, tetraploid and hexaploid wheats (*Triticum* spp.). *Ann. Bot.* 58:49–59.

Batten, G. D. 1986b. Phosphorus fractions in the grain of diploid, tetraploid, and hexaploid wheat grown with contrasting phosphorus supplies. *Cereal Chem.* 63:384–387.

Batten, G. D., M. A. Khan, and B. R. Cullis. 1984. Yield responses by modern wheat genotypes to phosphate fertilizer and their implications for breeding. *Euphytica* 33:81–89.

Batten, G. D., I. F. Wardlaw, and M. J. Aston. 1986. Growth and distribution of phosphorus in wheat developed under various phosphorus and temperature regimes. *Aust. J. Agric. Res.* 37:459–469.

Baylis, G. T. S. 1970. Root hairs and phycomycetous mycorrhizas in phosphorus-deficient soil. *Plant Soil* 33:713–716.

Baylis, G. T. S. 1972. Minimum levels of available phosphorus for non-mycorrhizal plants. *Plant Soil* 36:233–234.

Baylis, G. T. S. 1975. The magnolioid mycorrhiza and mycotrophy in root systems derived from it. In F. E. Sanders, B. Mosse, and P. B. Tinker (eds.), Endomycorrhizas. Academic Press, New York. Pp. 373–389.

Ben, J. R., G. Peruzzo, and E. Minella. 1986. Comportamento de alguns genotypos de cevada em relazao a acidez do solo. *Resultados de Pesquisa do Centro Nacional de Pesquisa de Trigo a presentados na XIV Reuniao Nacional de Pesquisa de Trigo, Centro Nacional de Pesquisa de Trigo, Passo Fundo, Braz.*

Ben, J. R., and O. S. Rosa. 1983. Compartamento de algumas cultivares de trigo em relazao a forforo no solo. *Pesq. Agropecu. Bras.* 18:967–972.

Bennet, R. J., C. M. Breen, and M. V. Fey. 1985a. Aluminum uptake sites in the primary root of *Zea mays* L. *S. Afr. J. Plant Soil* 2:1–7.

Bennet, R. J., C. M. Breen, and M. V. Fey. 1985b. The primary site of aluminum injury in the root of *Zea mays* L. *S. Afr. J. Plant Soil* 2:8–17

Bertrand, G. 1897a. Sur l'intervention du manganese dans les oxydations provoquees par la laccase. *C. R. Acad. Sci.* 124:1032–1035.

Bertrand, G. 1897b. Sur l'action oxydante de sels manganeux et sur la constitution chimique des oxydases. *C. R. Acad. Sci.* 124:1355–1358.

Bertrand, G. 1905. Sur l'emploi favorable du manganese comme engrais. *C. R. Acad. Sci.* 141:1255–1257.

Bhat, K. K. S., and P. H. Nye. 1974. Diffusion of phosphate to plant roots in soil. II. Uptake along the roots at different levels of phosphorus. *Plant Soil* 41:365–382.

Bhat, K. K. S., P. H. Nye, and J. P. Baldwin. 1976. Diffusion of phosphate to plant roots in soil. IV. The concentration distance profile in the rhizosphere of roots with root hairs in a low-P soil. *Plant Soil* 44:63–72.

Blamey, F. P. C., D. G. Edwards, and C. J. Asher. 1983. Effects of aluminum, OH : Al and P : Al molar ratios, and ionic strength on soybean root elongation in solution culture. *Soil Sci.* 136:197–
207.

Boero, G., and S. Thien. 1979. Phosphatase activity and phosphorus availability in the rhizosphere of corn roots. In J. L. Harley and R. S. Russell (eds.), The Soil Root Interface. Academic Press, New York. Pp. 231–242.

Bolan, N. S., A. D. Robson, N. J. Barrow, and L. A. G. Aylmore. 1984. Specific activity of phosphorus in mycorrhizal and non mycorrhizal plants in relation to the availability of phosphorus to plants. *Soil Biol. Biochem.* 16:299–304.

Bole, J. B. 1973. Influence of root hairs in supplying soil phosphorus to wheat. *Can. J. Soil Sci.* 53:169–175.

Boye-Goni, S. R. 1982. Combining ability and inheritance of aluminum tolerance in grain sorghum (*Sorghum bicolor* (L) Moench). *Diss. Abstr. Int. B* 43(3): 582B–583B.

Boye-Goni, S. R., and V. Marcarian. 1985. Diallel analysis of aluminum tolerance in selected lines of grain sorghum. *Crop. Sci.* 25:749–754.

Brewster, J. L., D. D. S. Bhat, and P. H. Nye. 1976. The possibility of predicting solute uptake and plant growth response from independently measured soil and plant characteristics. V. The growth and phosphorus uptake of rape in soil at a range of phosphorus concentrations and a comparison of results with the predictions of a simulation model. *Plant Soil* 44:295–328.

Briggs, K. G., and J. M. Myachiro. 1988. Genetic variation for aluminum tolerance in Kenyan wheat cultivars. *Commun. Soil. Sci. Plant Anal.* 19:1273–1284.

Brown, D. A., and Anwar-Ul-Haq. 1984. A porous membrane-root culture technique for growing plants under controlled soil conditions. *Soil Sci. Soc. Am.* 48:692–695.

Brown, J. C., and T. E. Devine. 1980. Inheritance of tolerance or resistance to manganese toxicity in soybeans. *Agron. J.* 72:898–904.

Buwalda, J. G., G. J. S. Ross, D. P. Stribley, and P. B. Tinker. 1982. The development of endomycorrhizal root systems. IV. The mathematical analysis of phosphorus on the spread of vesicular–arbuscular mycorrhizal infection in root systems. *New Phytol.* 92:391–399.

Camargo, C. E. De O. 1981. Melhoramento do Trigo I. Hereditariedade da tolerancia a toxicidade do aluminio. *Bragantia* 40:33–45.

Camargo, C. E. De O. 1983. Melhoramento do trigo III. Evidencia de controle genetico na tolerancia do manganes e alumio toxico em trigo. *Bragantia* 42:91–103.

Camargo, C. E. De O. 1985. Melhoramento do trigo XI. Estudo genetico da tolerancia a toxicidade de ferro. *Bragantia* 44:87–96.

Campbell, L. G., and H. N. Lafever. 1976. Correlation of field and nutrient culture techniques of screening wheat for aluminum tolerance. In M. J. Wright (ed.), Plant Adaptation to Mineral Stress in Problem Soils. Cornell Univ. Press, Ithaca, New York. Pp. 277–286.

Campbell, L. G., and H. N. Lafever. 1981. Heritability of aluminum tolerance in wheat. *Cereal Res. Commun.* 9:281–287.

Cappy, J. J., and D. A. Brown. 1980. A method for obtaining soil-free, soil-solution grown plant root systems. *Soil Sci. Soc. Am. J.* 44:1321–1323.

Cathcart, J. B. 1980. World phosphate reserves and resources. In F. E. Kasawneh, E. C. Sample, and E. J. Kamprath (eds.), The Role of Phosphorus in Agriculture. Am. Soc. Agron., Madison, Wisconsin. Pp. 1–118.

Claassen, N., and S. A. Barber. 1974. A method for characterizing the relation between nutrient concentration and flux into roots of intact plants. *Plant Physiol.* 54:564–568.

Clark, K. J., M. E., McCully, and N. K. Miki. 1979. A developmental study of the epidermis of young roots of *Zea mays* L. *Protoplasma* 98:283–309.

Clark, R. B. 1977. Effect of aluminum on growth and mineral elements of A1-tolerant and A1-intolerant corn. *Plant Soil* 47:653–662.

Clarkson, D. T. 1965. The effect of aluminum and some other trivalent metal cations on cell division in the root apices of *Allium cepa*. *Ann. Bot.* 29:309–315.

Clarkson, D. T. 1966. Effect of aluminum on the uptake and metabolism of phosphorus by barley seedlings. *Plant Physiol.* 41:165–172.

Coltman, R. R., G. C. Gerloff, and W. H. Gabelman. 1982. A sand culture system for simulating plant responses to phosphorus in the soil. *J. Am. Soc. Hortic. Sci.* 107:938–942.

Conner, A. J., and C. P. Meredith. 1985. Large scale selection of aluminum-resistant mutants from plant cell culture: Expression and inheritance in seedlings. *Theor. Appl. Genet.* 71:159–165.

Culvenor, R. A., R. N. Oram, and D. J. David. 1986a. Genetic variability for manganese concentration in *Phalaris aquatica* growing in an acid soil.*Aust. J. Agric. Res.* 37:409–416.

Culvenor, R. A., R. N. Oram, and J. T. Wood. 1986b. Inheritance of aluminum tolerance in *Phalaris aquatica* L. *Aust. J. Agric. Res.* 37:397–408.

Cutrim, V. dos A., T. V. Nguyen, J. C. Silva, and J. D. Galvao. 1981. Inheritance of tolerance for aluminum toxicity in Brazilian rice *Oryza sativa* L. *Int. Rice Res. Newsl.* 6:9.

da Silva, A. R. 1976. Application of the genetic approach to wheat culture in Brazil. In M. J. Wright (ed.), Plant Adaptation to Mineral Stress in Problem Soils. Cornell Univ. Press, Ithaca, New York. Pp. 223–231.

DeTurk, E. E., J. R. Holbert, and B. H. Howk. 1933. Chemical transformations of phosphorus in the growing corn plant with results on two first-generation crosses. *J. Agric. Res.* 46:121–141.

Dittmer, H. J. 1937. A quantitative study of the roots and root hairs of a winter rye plant (*Secale cereale*). *Am. J. Bot.* 24:417–420.

Dudal, R. 1976. Inventory of the major soils of the world with special reference to mineral stress hazards. In M. J. Wright (ed.), Plant Adaptation to Mineral Stress in Problem Soils. Cornell Univ. Press, Ithaca, New York. Pp. 3–13.

Duncan, R. R. 1981. Variability among sorghum genotypes for uptake of elements under acid soil field conditions. *J. Plant Nutr.* 4:21–32.

Duncan, R. R. 1983. Concentration of critical nutrients in tolerant and susceptible sorghum lines for use in screening under acid soil field conditions. In M. R. Saric and B. C. Loughman (eds.), Genetic Aspects of Plant Mineral Nutrition. Nijhoff/Junk, The Hague. Pp. 101–104.

Duncan, R. R. 1984. Registration of acid soil tolerant sorghum germplasm. *Crop Sci.* 24:1006.

Duncan, R. R. 1987. Sorghum genotype comparisons under variable acid soil stress. *J. Plant Nutr.* 10:1079–1088.

Duncan, R. R. 1988. Sequential development of acid soil tolerant sorghum genotypes under field stress conditions. *Commun. Soil Sci. Plant Anal.* 19:1295–1305.

Duncan, R. R., R. B. Clark, and P. R. Furlani. 1983. Laboratory and field evaluation of sorghum for response to aluminum and acid soil. *Agron. J.* 75:1023–1026.

Duncan, R. R., and J. D. Sutton. 1987. Influence of field sampling techniques on the Al, Mn, Mg, and Ca nutritional profiles for acid soil tolerant and susceptible sorghum genotypes. In W. H. Gabelman and B. C. Loughman (eds.), Genetic Aspects of Plant Mineral Nutrition. Nijhoff, Dordrecht, Netherlands. Pp. 99–110.

Edwards, J. H., and S. A. Barber. 1976. Phosphorus uptake rate of soybean roots as influenced by plant age, root trimming and solution P concentration. *Agron. J.* 68:973–975.

Eenink, A. H., and F. Garretsen. 1977. Inheritance of insensitivity of lettuce to a surplus of exchangeable manganese in steam sterilized soils. *Euphytica* 26(1):47–53.

Elliot, M. D. 1986. Chromosomal locations of genes for aluminum tolerance, and high protein in the wheat cultivar Atlas 66 using the monosomic method. *Diss. Abstr. Int. B* 47(1):3-B.

Fageria, N. K. 1982. Differential aluminum tolerance of rice cultivars in nutrient solutions. *Pesq. Agropecu Bras.* 15:259–265.

Fageria, N. K., V. C. Baligar, and R. J. Wright. 1988. Aluminum toxicity in crop plants. *J. Plant Nutr.* 11:303–319.

Fageria, N. K., and M. P. Barbosa Filho. 1982. Upland rice varietal reactions to aluminum toxicity on an oxisol in Central Brazil. *IRRI Newsl.* 8:18–19.

Fageria, N. K., and F. J. P. Zimmermann. 1979. Screening rice varieties for resistance to aluminum toxicity. *Pesq. Agropecu. Bras.* 14:141–147.

Fehr, W. R. 1983. Modification of mineral nutrition in soybeans by plant breeding. *Iowa State J. Res.* 57:393–407.

Foehse, D., and A. Jungk. 1983. Influence of phosphate and nitrate supply on root hair formation of rape, spinach and tomato plants. *Plant Soil* 74:359–368.

Foy, C. D. 1976. General principles involved in screening plants for aluminum and manganese tolerance. In M. J. Wright (ed.), Plant Adaptation to Mineral Stress in Problem Soils. Cornell Univ. Press, Ithaca, New York. Pp. 25–267.

Foy, C. D., G. R. Burns, J. C. Brown, and A. L. Fleming. 1965. Differential aluminum tolerance of two wheat varieties associated with plant induced pH changes around their roots. *Soil Sci. Soc. Am. Proc.* 29:64–67.

Foy, C. D., A. L. Fleming, and J. W. Schartz. 1973. Opposite aluminum and manganese tolerances in two wheat varieties. *Agron J.* 65:123–126.

Foy, C. D., B. J. Scott, and J. A. Fisher. 1988. Genetic differences in plant tolerance to manganese toxicity. In R. Graham, R. J. Hannamm, and N. C. Uren (eds.), Manganese in Soils and Plants. Kluwer Academic Publishers, Dordrecht, Netherlands. Pp. 293–307.

Fried, M., and M. Peech. 1946. The comparative effects of lime and gypsum upon plants grown on acid soils. *J. Am. Soc. Agron.* 38:614–623.

Funchness, M. J. 1918. The development of soluble manganese in acid soils, as influenced by certain nitrogenous fertilizers. *Ala. Polytech. Inst. Agric. Exp. Stn., Bull.* 201:37–78.

Furlani, P. R., and R. B. Clark. 1981. Screening sorghum for aluminum tolerance in nutrient solutions. *Agron. J.* 73:587–594.

Gardner, W. K., D. A. Barber, and D. G. Parbery. 1983. The acquisition of phosphorus by *Lipinus albus* L. III. The probable mechanism by which phosphorus movement in the soil/root interface is enhanced. *Plant Soil* 70:107–124.

Gardner, W. K., D. G. Parbery, and D. A. Barber. 1982. The acquisition of phosphorus by *Lupinus albus* L. I. Some characteristics of soil/root interface. *Plant Soil* 68:19–32.

Gerretsen, F. C. 1948. The influence of microorganisms on the phosphate intake by plants. *Plant Soil* 1:51–81.

Gill, B. S., M. C. Medeiros, C. N. A. Sousa, E. P. Gomes, W. I. Linhares, and A. C. Baier. 1973. *Annu. Wheat Newsl.* 19:28–31.
Gill, B. S., M. C. Medeiros, C. N. A. Sousa, W. I. Linhares, E. P. Gomes, O. S. Rosa, A. M. Prestes, F. A. Langer, and S. R. Dotto. 1972. *Annu. Wheat Newsl.* 18:19–20.
Gill, G. W., J. K. Frost, and K. A. Miller. 1974. A new formula for half-oxidized hematoxylin solution that neither overstains nor requires differentiation. *Acta Cytol.* 18:300–311.
Graham, R. D. 1979. Nutrient efficiency objectives in cereal breeding. In A. R. Ferguson, R. I. Bieleski, and I. B. Ferguson (eds.), Plant Nutrition 1978. Proc. 8th Int. Colloq. on Plant Analysis and Fertilizer Problems, Auckland, 1978. DSIR Inf. Ser. No. 134. Gov. Printer, Wellington, New Zealand. Pp. 165–169.
Graham, R. D. 1988. Genotypic differences in tolerance to manganese deficiency. In R. Graham, R. J. Hannam, and N. C. Uren (eds.), Manganese in Soils and Plants. Kluwer Academic Publishers, Dordrecht, Netherlands. Pp. 261–276.
Heenan, D. P., L. C. Campbell, and O. G. Carter. 1981. Inheritance of tolerance to high manganese supply in soybeans glycine-max. *Crop. Sci.* 21:625–627.
Henning, S. J. 1975. Aluminum toxicity in the primary meristem of wheat roots. Ph.D. Thesis, Oregon State Univ., Corvallis. Univ. Microfilms, Ann Arbor, Michigan. (*Diss. Abstr. B* 35:5728B.)
Hewitt, E. J. 1952. A biological approach to the problems of soil acidity. *Trans. 2nd 4th Commun. Int. Soc. Soil Sci., Dublin* 1:107–118.
Higashi, T., and H. Ikeda. 1974. Dissolution of allophane by acid oxalate solution. *Clay Sci.* 4:205–211.
Horiguchi, T. 1988. Mechanism of manganese toxicity and tolerance of plants. VII. Effect of light intensity on manganese-induced chlorosis. *J. Plant Nutr.* 11:235–246.
Horst, W. J., and H. Marschner. 1978. Effect of excessive manganese supply on uptake and translocation of calcium in bean plants (*Phaseolus vulgaris* L.). *Z. Pflanzenphysiol.* 87:137–148.
Horst, W. J., A. Wagner, and H. Marschner. 1982. Mucilage protects root meristems from aluminum injury. *Z. Pflanzenphysiol.* 105:435–444.
Horst, W. J., A. Wagner, and H. Marschner. 1983. Effect of aluminum on root growth, cell division rate and mineral element contents in roots of *Vigna unguiculata* genotypes. *Z. Pflanzenphysiol.* 109:95–103.
Howeler, R. H., and L. F. Cadavid. 1976. Screening of rice cultivars for tolerance to Al-toxicity in nutrient solutions as compared with a field screening method. *Agron J.* 68:551–555.
Huber, D. M., and N. S. Wilhelm. 1988. The role of manganese in resistance to plant diseases. In R. D. Graham, R. J. Hannam, and N. C. Uren (eds.), Manganese in Soils and Plants. Kluwer Academic Publishers, Dordrecht, Netherlands. Pp. 155–173.
Institute of Horticultural Plant Breeding of the Netherlands. 1977. Annual Report for 1976. Wageningen, Netherlands.
International Rice Research Institute. 1974. Rice II. Annual Report for 1973. Los Banos, Philippines.
International Rice Research Institute. 1975. Research Highlights for 1974. Los Banos, Philippines.
International Rice Research Institute. 1977. Rice III. Annual Report for 1976. Los Banos, Philippines.
International Rice Research Institute. 1985. Field Problems of Tropical Rice. Los Banos, Philippines.
Jenny, H. 1961. Reflection on the soil acidity merry-go-round. *Soil Sci. Soc. Am. Proc.* 25:428–432.
Jenny, H., and K. Grossenbacher. 1963. Root soil boundary zones as seen in the electron microscope. *Soil Sci. Soc. Am. Proc.* 27:273–277.
Johnson, M. O. 1924. Manganese chlorosis of pineapples: Its cause and control. *Hawaii Agric. Expt. Stn., Bull.* 52:1–38.
Johnson, R. E., and W. A. Jackson. 1964. Calcium uptake and transport by wheat seedlings as affected by aluminum. *Soil Sci. Soc. Am. Proc.* 28:381–386.
Kawai, K. 1980. The relationship of phosphorus adsorption to amorphous aluminum for characterizing andosols. *Soil Sci.* 129:186–190.
Kerridge, P. C., M. D. Dawson, and D. P. Moore. 1971. Separation of degrees of aluminum tolerance in wheat. *Agron. J.* 63:586–591.
Kerridge, P. C., and W. E. Kronstad. 1968. Evidence of genetic resistance to aluminum toxicity in wheat (*Tricitcum aestivum* (Vill.) Host). *Agron. J.* 60:710–712.

Ketchum, J. F. 1989. The International Plant Biotechnology Network, Tissue Culture for Crops Project, Newsletter No. 10.

Kinraide, T. B., R. C. Arnold, and V. C. Balizar. 1985. A rapid assay for aluminum phytotoxicity at submicromolar concentrations. *Physiol. Plant.* 65:245–250.

Klein, T. M., T. Gradziel, M. E. Fromm, and J. C. Sandford. 1988a. Factors influencing gene delivery into *Zea mays* cells by high velocity microprojectiles. *Bio/Technology* 6:559–563.

Klein, T. M., E. C. Harper, Z. Svav, J. C. Sandford, M. E. Fromm, and P. Maliga. 1988b. Stable genetic transformation of intact *Nicotiana* cells by the particle bombardment process. *Proc. Natl. Acad. Sci. U.S.A.* 85:8502–8505.

Klimashevskii, E. L., and V. M. Dedov. 1980. Characteristics of an elastic cell wall of the root in relation to genotypic variance of plant resistance to aluminum ions. *Isv. Sib. Otb. Hkad. Nauk SSSR, Ser. Biol. Nauk* 1:108–112.

Koehler, T. J. 1983. Identifying efficient spring wheat accessions at low levels of phosphorus in acid soils. M.S. Thesis, Washington State Univ., Pullman.

Konzak, C. F., and E. Polle. 1985. Al toxicity tolerance, Mn toxicity, and P use efficiency (P deficiency tolerance) in common wheat, *Triticum aestivum* L. *Agron. Abstr., 77th Annu. Meet., Chicago, Ill., Am. Soc. Agron., Madison, Wis.* p. 60.

Konzak, C. F., E. Polle, and J. A. Kittrick. 1976. Screening several crops for aluminum tolerance. In M. J. Wright (ed.), Plant Adaptation to Mineral Stress in Problem Soils. Cornell Univ. Press, Ithaca, New York. Pp. 311–327.

Koyama, T., and C. H. Chamnek. 1971. Soil–plant nutrition studies on tropical rice. I. Studies on the varietal differences in absorbing phosphorus from soil low in available phosphorus (Part 1). *Soil Sci. Plant Nutr.* 17:115–126.

Koyama, T., and P. Snitwongse. 1971. Soil–plant nutrition studies on tropical rice. II. Rice varietal differences in absorbing phosphorus from soil low in available phosphorus (Part 2). *Soil Sci. Plant Nutr.* 17:186–194.

Lafever, H. M., and L. G. Campbell. 1978. Inheritance of aluminum tolerance in wheat. *Can. J. Genet. Cytol.* 20:355–364.

Läuchli, A. 1967. Unterguchungen uber verteilung und transport vou ionen in pflanzengeweben mit der rontgen-mikrosonde. *Planta* 75:185–206.

Levan, A. 1945. Cytological reactions induced by inorganic salt solutions. *Nature (London)* 156:751–752.

Lindgren, D. T., W. H. Gabelman, and G. C. Gerloft. 1977. Variability of phosphorus uptake and translocation in *Phaseolus vulgaris* L. under phosphorus stress. *J. Am. Hortic. Sci.* 102:674–677.

Lindsay, W. L., and E. C. Moreno. 1960. Phosphate phase equilibria in soils. *Soil Sci. Soc. Am. Proc.* 24:177–182.

Lipsett, J. 1964. The phosphorus content and yield of grain of different wheat varieties in relation to phosphorus deficiency. *Aust. J. Agric. Res.* 15:1–8.

Little, R. 1988. Plant soil interactions at low pH. Problem solving—The genetic approach. *Commun. Soil Sci. Plant Anal.* 19(7–12):1239–1257.

Loneragan, J. F. 1978. The physiology of plant tolerance to low phosphorus availability. In G. A. Jung (ed.), Crop Tolerance to Suboptimal Land Conditions. Am. Soc. Agron. Spec. Publ. No. 32. Madison, Wisconsin. Pp. 329–343.

Loneragan, J. F., and C. J. Asher. 1967. Response of plants to phosphate concentration in solution culture. II. Rate of phosphate absorption and its relation to growth. *Soil Sci.* 103:311–318.

Lyness, A. S. 1936. Varietal differences in the phosphorus feeding capacity of plants. *Plant Physiol.* 11:665–688.

MacLean, A. A., and T. C. Chiasson. 1966. Differential performance of two barley varieties to varying aluminum concentrations. *Can. J. Soil Sci.* 46:147–153.

Magnavaca, R. 1983. Genetic variability and the inheritance of aluminum tolerance in maize (*Zea mays* L.). *Diss. Abstr. Int. B* 43(7):2073B.

Magnavaca, R., C. O. Gardner, and R. B. Clark. 1987. Inheritance of aluminum tolerance in maize. In

W. H. Gabelman and B. C. Loughman (eds.), Genetic Aspects of Plant Mineral Nutrition. Nijhoff, Dordrecht, Netherlands. Pp. 201–212.
Malajczuk, N., A. J. McComb, and J. F. Loneragan. 1975. Phosphorus uptake and growth of mycorrhizal and unifected seedlings of *Eucalyptus calophylla* R. Br. *Aust. J. Bot.* 23:231–238.
Marschner, H., and V. Römheld. 1983. *In vivo* measurement of root induced pH changes at the soil–root interface: Effect of plant species and nitrogen source. *Z. Pflanzenphysiol.* 111:241–251.
Matsumoto, H., E. Hirasawa, H. Torikai, and E. Takahashi. 1976. Localization of absorbed aluminum in pea root and its binding to nucleic acids. *Plant Cell Physiol.* 17:127–137.
Matsumoto, H., S. Morimura, and E. Takahashi. 1977a. Binding of aluminum to DNA of DNP in pea root nuclei. *Plant Cell Physiol.* 18:987–993.
Matsumoto, H., S. Morimura, and E. Takahashi. 1977b. Less involvement of pectin in the precipitation of Al in pea root. *Plant Cell Physiol.* 18:325.
May, L. H., F. H. Chapman, and D. Aspinall. 1965. Quantitative studies of root development. I. The influence of nutrient concentration. *Aust. J. Biol. Sci.* 18:25–35.
May, L. H., F. H. Randles, D. Aspinall, and L. G. Paleg. 1967. Quantitative studies of root development. II. Growth in the early stages of development. *Aust. J. Biol. Sci.* 20:273–283.
McCool, M. M. 1935. Effect of light intensity on the manganese content of plant. *Contrib. Boyce Thompson Inst.* 7:427–437.
McGrath, S. P., and I. H. Rorison. 1982. The influence of nitrogen source on the tolerance of *Holcus lanatus* and *Bromus erectus* Huds to manganese. *New Phytol.* 91:443–452.
McLachlan, K. D., K. Yanhua, and W. J. Muller. 1987. An assessment of the depletion technique for comparative measurement of phosphorus uptake in plants. *Aust. J. Agric. Res.* 38:263–277.
McLean, F. T., and B. E. Gilbert. 1927. The relative aluminum tolerance of crop plants. *Soil Sci.* 24:163–175.
Meudt, W. J. 1971. Interactions of sulfite and manganous ion with peroxidase oxidation products of ridole 3-acetic acid. *Phytochemistry* 10:2103–2109.
Miki, N. K., K. J. Clarkee, and M. E., McCully. 1980. A histological and histochemical comparison of the mucilages on the root tips of several grasses. *Can. J. Bot.* 58: 2581–2593.
Miranda, L. N., and E. Lobato. 1978. Tolerancia de variedales de feijao e de trigo as aluminio e a baixa disponibilidade de fosforo no solo. *Rev. Bras. Ci. Solo* 2:44–50.
Misra, R. K., A. M. Alston, and A. R. Dexter. 1988. Role of root hairs in phosphorous depletion from a macrostructured soil. *Plant Soil* 107:11–18.
Moghini, A., D. G. Lewis, and J. M. Oades. 1978. Release of phosphate from calcium phosphate by rhizophere products. *Soil Biol. Biochem.* 10:277–281.
Moghini, A., and M. E. Tate. 1978. Does 2-ketogluconate chelate calcium in the pH range 2.4 to 6.4. *Soil Biol. Biochem.* 10:289–292.
Moore, D. P., W. E. Kronstad, and R. J. Metzger. 1976. Screening wheat for aluminum tolerance. In M. J. Wright (ed.), Plant Adaptation to Mineral Stress in Problem Soils. Cornell Univ. Press, Ithaca, New York. Pp. 287–295.
Moraghan, J. T. 1979. Manganese toxicity in flax growing on certain calcareous soil low in available iron. *Soil Sci. Soc. Am. J.* 43:1177–1180.
Morgan, P. W., H. E. Joham, and J. V. Amin. 1966. Effect of manganese toxicity on the indoleacetic acid oxidase system of cotton. *Plant Physiol.* 41:718–724.
Morgan, P. W., D. M. Taylor, and H. E. Joham. 1976. Manipulations of IAA-oxidase activity and auxin deficiency symptoms in intact cotton plants with manganese nutrition. *Physiol. Plant.* 37:149–156.
Morimura, S., and H. Matsumoto. 1978. Effect of aluminum on some properties and template activity of purified pea DNA. *Plant Cell Physiol.* 18:987–993.
Morimura, S., E. Takahashi, and H. Matsumoto. 1978. Association of aluminum with nuclei and inhibition of cell division in onion (*Allium cepa*) roots. *Z. Pflanzenphysiol.* 88:395–401.
Morre, D. J., D. D. Jones, and H. H. Mollenhauer. 1967. Golgi apparatus mediated polysaccharide secretion by outer root cap cells of *Zea mays*. I. Kinetics and secretory pathways. *Planta* 74:286–301.
Mosse, B. 1973. Plant growth responses to vesicular arbuscular mycorrhiza. IV. In soil given additional phosphate. *New Phytol.* 72:127–136.

Mugwira, L. M., S. M. Elgawhary, and K. I. Patel. 1976. Differential tolerances of triticale, wheat, rye and barley to aluminum in nutrient solution. *Agron. J.* 68:782–786.

Nagarajah, S., A. M. Posner, and J. P. Quirck. 1970. Competitive adsorption of phosphate with polygalacturonate and other organic anions on kaolinite and oxide surfaces. *Nature (London)* 228:83–84.

Naidoo, G., J. McD. Stewart, and R. J. Lewis. 1978. Accumulation sites of Al in snap bean and cotton roots. *Agron. J.* 70:489–492.

Nelson, L. E. 1983. Tolerance of 20 rice cultivars to excess Al and Mn. *Agron. J.* 75:134–138.

Newman, E. I., and R. E. Andrews. 1973. Uptake of phosphorus and potassium in relation to root growth and root density. *Plant Soil* 38:49–69.

Nielsen, N. E., and S. A. Barber. 1978. Differences among genotypes of corn in the kinetics of P uptake. *Agron. J.* 70:695–698.

Nielsen, N. E., and J. K. Schjorring. 1982. Efficiency and kinetics of phosphorus uptake from soil by various barley genotypes. In M. R. Saric (ed.), Genetic Specificity of Mineral Nutrition of Plants. Serb. Acad. Sci. Arts, Belgrade. Pp. 125–129.

Nodari, R. O., F. I. F. De Carvalho, and L. C. Federizzi. 1982. Genetic bases of the inheritance of aluminum toxicity tolerance in wheat triticum-aestivum genotypes. *Pesq. Agropecu. Bras.* 17:269–280.

Nunez, Z. V. M., C. P. Martinez, V. J. Naravez, and W. Roca. 1985. Using anther cultures to obtain homozygous lines of rice oryza-sativa with tolerance for aluminum toxicity. *Acta Agron. (Palmira)* 35:7–26.

Nye, P. H. 1977. The rate limiting step in plant nutrient absorption from soil. *Soil Sci.* 123:292–297.

Nye, P. H. 1981. Changes of pH across the rhizosphere induced by roots. *Plant Soil* 61:7–26.

Palmer, B., and R. S. Jessop. 1977. Some aspects of wheat cultivar response to applied phosphate. *Plant Soil* 47:63–73.

Pan, W. L., A. G. Hopkins, and W. A. Jackson. 1989. Aluminum inhibition of short lateral branches of *Glycine max* and reversal by exogenous cytokinin. *Plant Soil* 120:1–9.

Pan, W. L., W. A. Jackson, and R. H. Moll. 1985. Nitrate uptake and partitioning by corn root systems. *Plant Physiol* 77:560–566.

Peech, M. 1941. Availability of ions in light, sandy soils as affected by soil reaction. *Soil Sci.* 51:473–486.

Phillips, M. R. 1979. Environmental and genetic factors affecting wheat seedling root number and length. M.S. Thesis, Washington State Univ., Pullman.

Pohlman, A., and J. G. McColl. 1984. Mechanism of Al, Fe, Mn and Mg dissolution from two forest soils by soluble polyfunctional organic acids. *Agron. Abstr., Annu. Meet., Am. Soc. Agron.* p. 34.

Polle, E., and H. Jenny. 1971. Boundary layer effects in ion absorption by roots and storage organs of plants. *Physiol. Plant.* 25:219–224.

Polle, E., and C. F. Konzak. 1985. A single scale for Al tolerance in cereals. *Agron. Abstr., 77th Annu. Meet., Chicago, Ill. Am. Soc. Agron., Madison, Wis.* p. 67.

Polle, E., C. F. Konzak, and J. A. Kittrick. 1978a. Rapid Screening of Maize for Tolerance to Aluminum in Breeding Varieties Better Adapted to Acid Soils. Tech. Bull. No. 22. Office of Agriculture Development Support Bureau, Agency of International Development, Washington, D.C.

Polle, E., C. F. Konzak, and J. A. Kittrick. 1978b. Visual detection of aluminum tolerance levels in wheat by hematoxylin staining of seedling roots. *Crop Sci.* 18:823–827.

Ponnamperuma, F. N. 1976. Screening rice for tolerance to mineral stresses. In M. J. Wright (ed.), Plant Adaptation to Mineral Stress in Problem Soils. Cornell Univ. Press, Ithaca, New York. Pp. 341–354.

Prestes, A. M., and C. F. Konzak. 1975. Genetic association of aluminum tolerance to high protein and leaf rust resistance in Atlas 66. *Annu. Wheat Newsl.* 21:162–163.

Rahman, M. S., and J. H. Wilson. 1977. Effect of phosphorus applied as superphosphate on rate of development and spikelet number per ear in different cultivars of wheat. *Aust. J. Agric. Res.* 28:183–186.

Reid, D. A. 1971. Genetic control of reaction to aluminum in winter barley. In R. Nilan (ed.), Barley Genetics II. Proc. 2nd Int. Barley Genetics Symp. Washington State Univ. Press, Pullman. Pp. 409–413.

Reid, D. A., A. L. Fleming, and C. D. Foy. 1971. A method for determining aluminum response of barley in nutrient solution in comparison to response in Al-toxic soil. *Agron. J.* 63:600–603.
Riley, D., and S. A. Barber. 1971. Effect of ammonium and nitrate fertilization on phosphorus uptake as related to root induced pH changes at the root soil interface. *Soil Sci. Soc. Am. Proc.* 35:301–306.
Rogers, S. A. 1987. Methods of evaluation and inheritance of aluminum tolerance in sorghum. *Diss. Abstr. Int. B* 47(8):3175B.
Rosa, O. de S. 1988. Current special breeding projects at the National Research Center for Wheat (CNPT), Embrapa, Brazil. In M. M. Kohli and S. Rajaram (eds.), Wheat Breeding for Acid Soils: Review of Brazilian/CIMMYT Collaboration, 1974–1986. CIMMYT, Mexico City, Pp. 6–10.
Ruhe, R. D., and C. O. Grogan. 1977. Screening corn for Al tolerance using different Ca and Mg concentrations. *Agron. J.* 69:755–760.
Ruhe, R. D., C. O. Grogan, E. W. Stockmeyer, and H. L. Everett. 1978. Genetic control of aluminum tolerance in corn. *Crop Sci.* 18:1063–1067.
Salinas, J. G., and P. A. Sanchez. 1978. Tolerance to Al Toxicity and Low Available P. Agronomic–Economic Research on Soils of the Tropics. Annual Report for 1976–1977. Soil Science Department, North Carolina State Univ., Raleigh.
Sampson, M., D. T. Clarkson, and D. D. Davies. 1965. DNA synthesis in aluminum treated roots of barley. *Science* 148:1476–1477.
Sanchez, P. A., and S. W. Buol. 1975. Soils of the tropics and the world food crisis. *Science* 188:598–603.
Sanders, F. E., and P. B. Tinker. 1971. Mechanism of absorption of phosphate from soil by Endozone mycorrhizas. *Nature (London)* 233:278–279.
Sapra, V. T., M. A. Chowdry, and L. M. Mugwira. 1979. Genetic variability in aluminum tolerance of Triticinae. *Wheat Inf. Serv.* 50:47–50.
Schwertman, U. 1964. Differenzierung der Eisenoxide des bodens durch extraktion mit ammoniumoxalat-Losung. *Z. Pflanzenernaehr. Bodenkd.* 105:194–202.
Scott, B. J. 1981. Tolerance to acid soils. *Annu. Wheat Newsl.* 27:30.
Scott, B. J., and J. A. Fisher. 1980. Adaption of cultivars to toxic levels of aluminum and manganese. *Annu. Wheat Newsl.* 26:39.
Silverbush, M., and S. A. Barber. 1983. Sensitivity of simulated phosphorus uptake to parameters used by a mechanistic–mathematical model. *Plant Soil* 74:93–100.
Smiley, R. W. 1974. Rhizophere pH as influenced by plants, soils and nitrogen fertilizers. *Soil Sci. Soc. Am. Proc.* 38:795–799.
Smith, S. N. 1934. Response of inbred lines and crosses in maize to variations of nitrogen and phosphorus supplied as nutrients. *J. Am. Soc. Agron.* 26:785–804.
St. John, T. W. 1980. Root size, root hairs and mycorrhizal infection: A re-examination of Baylis's hypothesis with tropical trees. *New Phytol.* 84:483–487.
Steffens, D. 1984. Wurzelstudien und phosphat-aufnahme von weidelgras und rotklee unter feldbedingungen. *Z. Pflanzenernaehr. Bodenkd.* 147:85–97.
Tagaki, H., H. Namai, and K. Murakami. 1981. Evaluation of haematoxylin staining method for detecting wheat tolerance to aluminum. *Jpn. J. Breed.* 31:152–160.
Takagi, H., H. Namai, and K. Murakami. 1983. Exploration of aluminum tolerant genes in wheat. *Proc. 6th Int. Wheat Genet. Symp., Kyoto* pp. 143–146.
Taylor, G. J. 1988. The physiology of aluminum tolerance in higher plants. *Commun. Soil Sci. Plant Anal.* 19:1179–1194.
Tennant, D. 1976. Root growth of wheat. I. Early pattern of multiplication and extension of wheat roots including effects of levels of nitrogen, phosphorus and potassium. *Aust. J. Agric. Res.* 27:183–196.
Terry, N., P. S. Evans, and D. E. Thomas. 1975. Manganese toxicity effects on leaf cell multiplication and expansion and on dry matter yield of sugar beets. *Crop Sci.* 15:205–208.
Thomas, G. W. 1988. Beyond exchangeable aluminun: Another ride on the merry-go-round. *Commun. Soil Sci. Plant Anal.* 19:833–856.
Tinker, P. B. 1975. In F. E. Sanders, B. Mosse, and P. B. Tinker (eds.), Endomycorrhiza. Academic Press, New York.

Tinker, P. B. H., and F. E. Sanders. 1975. Rhizosphere microorganisms and plant nutrition. *Soil Sci.* 119:363–368.

Tomos, A. D., M. Malone, and J. Pritchard. 1989. The biophysics of differential growth. *Environ. Exp. Bot.* 29:7–23.

Torrey, J. G., and L. J. Feldman. 1977. The organization and function of the root apex. *Am. Sci.* 65:334–344.

Vlamis, J., and D. E. Williams. 1964. Iron and manganese relations in rice and barley. *Plant Soil* 20:221–231.

Vose, P. B., and D. J. Randall, 1962. Resistance to aluminum and manganese toxicities in plants related to variety and cation exchange capacity. *Nature (London)* 196:85–86.

Vose, P. B., 1982. Rationale of selection for specific nutritional characters in crop improvement with *Phaseolus vulgaris* L. as a case study. In M. R. Saric (ed.), Genetic Specificity of Mineral Nutrition of Plants. Vol. 12, No. 3. Serb. Acad. Sci. Arts, Belgrade. Pp. 313–323.

Wagatsuma, T., M. Kaneko, and Y. Hayasaka. 1987. Destruction process of plant root cells by aluminum. *Soil Sci. Plant Nutr.* 33:161–175.

Wallace, S. V., S. J. Henning, and I. C. Anderson. 1982. Elongation, Al concentration, and hematoxylin staining of aluminum-treated wheat roots. *Iowa State J. Res.* 57:97–106.

Whiteaker, D., G. C. Gerloff, W. H. Gabelman, and D. Lindgren. 1976. Intra-specific differences in growth of beans at stress levels of phosphorus. *J. Am. Soc. Hortic. Sci.* 101:472–475.

Williams, R. F. 1948. The effect of phosphorus supply on the rates of intake of phosphorus and nitrogen and upon certain aspects of phosphorus metabolism in gramineous plants. *Aust. J. Sci. Res.* 31:333–361.

Zale, J. M., and K. G. Briggs. 1988. Aluminum tolerance in Canadian spring wheats. *Commun. Soil Sci. Plant Anal.* 19:1259–1272.

Zutic, V., and W. Stumm. 1984. Effect of organic acids and fluoride on the dissolution kinetics of hydrous alumina. A model study using the rotating disk electrode. *Geochim. Cosmochim. Acta* 48:1493–1503.

5

Physiology of Cereals for Mineral Nutrient Uptake, Use, and Efficiency

RALPH B. CLARK

The growth and productivity of cereals are influenced by their genetic potentials and the environments in which they are grown. Mineral nutrition is one of the most important environmental factors affecting plant productivity. Mineral nutrients must be available in sufficient and balanced proportions if plants are to achieve their maximum growth and yield potentials. The normal practice in many production systems is to ameliorate unfavorable soil conditions so that plants will produce with limited mineral deficiency or toxicity stresses. These stresses may be continuous

Crops as Enhancers of Nutrient Use

over broad areas and may be difficult to alleviate, or some alleviations may only be temporary. Thus, having an optimum growth environment for plants at all times is not always easy to maintain. As the costs of fertilizer inputs increase and the problems of environmental pollution rise, techniques and methods to solve mineral nutrient stresses and to provide adequate supplies of mineral nutrients for plants become challenging ventures.

Genetic variability and plant ability to absorb, translocate, distribute, accumulate, and use mineral elements are important in adapting plants to specific environments. It is well documented that genotypes (cultivars, parental lines, and hybrids) within species differ extensively in ability to take up and use mineral elements. Taking advantage of these differences to select or improve plants for greater capacity to adapt to or to have greater efficiency to take up and use mineral elements is becoming more important. This is especially important in lesser-developed countries where capital inputs and resources are limited. The approach of solving plant productivity problems with more efficient use and fewer inputs to soils or growing plants effectively on soils with greater mineral nutrient stresses is becoming more practical and feasible.

Intraspecific variability in plant species has been recognized for a long time, but only recently have these differences been considered conscientiously to adapt plants to fit soil mineral nutrient stress conditions or to improve efficiency of nutrient uptake and use. Background information about genotypic differences in plants for use of mineral nutrients can be found in recent reviews (Saric, 1981; Clark, 1983; Gerloff and Gabelman, 1983; Vose, 1984, 1987). Information in this research area is expanding rapidly, and several international symposia have been held on the subject (Wright, 1976; Jung, 1978; Saric and Loughman, 1983; Gabelman and Loughman, 1987) and several books and articles have been written recently (Clark and Brown, 1980; Christiansen and Lewis, 1982; Gerloff and Gabelman, 1983; Graham, 1984; Sherrard *et al.*, 1984a,b; Vose and Blixt, 1984; Marschner, 1986; Neyra, 1986). Progress will continue to be made in this area of research, and these kinds of books, articles, and symposia proceedings will be published in the future giving additional volumes of information. Taking advantage of these differences in plants will hopefully help reduce costs and enhance plant productivity with more efficient plants.

I. Definition of Mineral Nutrient Efficiency

Mineral nutrient efficiency represents different things to different people and has little meaning unless specifically defined. The broad definition of efficiency is output divided by input (Munson, 1974). Definitions of efficiency often depend on the orientation or discipline of scientists using the word. For example, a soil scientist may define efficiency in terms of the amount of nutrient taken up by a plant relative to the amount of nutrient in or added to a soil or growth medium, while a plant physiologist may define efficiency as what a plant does with a mineral nutrient once it is inside the plant. Both of these definitions are correct, but more specific definitions are needed other than just "efficiency" or terms like "Fe-efficiency," "N-efficiency," or some other mineral nutrient efficiency.

Many terms arise in the literature relative to nutrient efficiency that makes a reader realize that everyone is not speaking the same language when he or she uses the word efficiency. Terms such as absorption efficiency, acquisition efficiency, agronomic efficiency, apparent nutrient recovery, assimilation efficiency, distribution efficiency, economic yield efficiency, efficiency quotient, efficiency ratio, instantaneous use efficiency, metabolic efficiency, mobilization efficiency, nutrient harvest index, photosynthetic efficiency, physiological efficiency, recovery efficiency, reutilization efficiency, substitution efficiency, transfer efficiency, translocation efficiency, uptake efficiency, use efficiency, utilization efficiency, utilization efficiency ratio, utilization index, and many other efficiency terms are used. Many definitions of efficiency have mathematical formulas to describe them, while others do not. Some definitions are visual assessments of traits like vigor or ratings of leaf deficiency symptoms. Regardless of the concepts used, the word efficiency needs to be defined.

The general concept for mineral efficiency is that a plant somehow performs better or does something better than another plant when given a comparable amount of the same mineral, or that a plant acquires more or does a process better with lower amounts of a specific mineral nutrient than another plant. Some have defined efficient plants as those that produce more dry matter or have a greater increase in harvested portion per unit time, area, or applied nutrient, have fewer deficiency symptoms, or have greater increment increases and higher concentrations of mineral nutrients than other plants grown under similar conditions or compared to a standard genotype (Vose and Breese, 1964; Clark and Brown, 1974a; Barber, 1976; Fox, 1978; Graham, 1984). Common definitions of efficiency are dry matter, grain, seed, or fruit yield per unit of nutrient absorbed by the plant (Loneragan and Asher, 1967; Gabelman, 1976; Chevalier and Schrader, 1977; R. H. Brown, 1978; Gabelman and Gerloff, 1978; Maranville *et al.*, 1980; Moll *et al.*, 1982a,b, 1987; Cox *et al.*, 1985; Jackson *et al.*, 1986). Some definitions give more emphasis to the economic component of the plant and not just to the dry matter yield relative to the amount of nutrient absorbed or contained within the plant or particular plant part (Enyi, 1968; Barber, 1976; Boulter and Gatehouse, 1979; Maranville *et al.*, 1980). Some definitions also give greater emphasis to the transfer of nutrients to the desired products (Hay *et al.*, 1953; Eck *et al.*, 1975; Beauchamp *et al.*, 1976; Desai and Bhatia, 1978; Halloran and Lee, 1979; Maranville *et al.*, 1980; Moll *et al.*, 1987) and to the energy required, metabolic reactions involved in particular processes, or nutrient in a plant part (Bhatia and Rabson, 1976; Boulter and Gatehouse, 1979; Field *et al.*, 1983; Fernandez, 1987; Alagarswamy *et al.*, 1988; Gardner, 1988).

Definitions of fertilizer efficiency include not only the amount of dry matter or economic yield per unit of nutrient in the plant (physiological efficiency), but also include the efficiency with which plants can absorb nutrients relative to the amount applied (Moll *et al.*, 1982a,b; Craswell and Godwin, 1984; Jackson *et al.*, 1986). Agronomic efficiency has been defined as the amount of economic yield (grain) per unit of nutrient applied. Another term (apparent nutrient recovery) is sometimes used to reflect the efficiency of a plant to obtain nutrients from the soil. Agronomic efficiency is a product of physiological efficiency and the apparent recovery that will reflect the overall efficiency with which an applied nutrient is used. That is, an increase in either physiological efficiency or apparent recovery of nutrient will increase the agronomic efficiency (Novoa and Loomis, 1981). Uptake efficiency has been defined as the total nutrient taken into the plant compared to the amount applied, utilization efficiency as the grain yield per unit of nutrient applied, and use efficiency as the grain yield per total nutrient accumulated (Moll *et al.*, 1982a,b, 1987; Jackson *et al.*, 1986).

Objections to using an efficiency ratio (amount of biomass produced per unit of nutrient in the plant) sometimes arises because this definition does not take into consideration the mode of nutrient supply and the length of growth period (Agren, 1985). Another objection has been that an efficiency ratio alone does not consider the extent of growth (yield) that a plant makes, which is the ultimate goal to be obtained, and that the ratio of dry matter to the efficiency ratio (dry matter × dry matter/unit nutrient) might be a more appropriate measure of nutrient utilization by plants (Siddiqi and Glass, 1981). Other definitions may include some morphological features of a plant, like root branching, root length, surface area, root hairs, fineness, or shoot/root ratio (Gerloff and Gabelman, 1983; Graham, 1984). Differences in nutrient uptake by roots, movement across roots and from roots to shoots, distribution within leaves, utilization or reutilization, release from cell vacuoles, metabolic processes, and potential substitution are also definitions of efficiency.

Efficiency as used with Fe nutrition is a completely different definition from that used for most of the other mineral nutrients. The first usage of the term "Fe-efficiency" tried to explain the ability or differences among genotypes to overcome or withstand an iron deficiency chlorotic condition when grown under conditions that would induce such a disorder (Weiss, 1943). Some plants were greener than others, and the greener plants were termed more "Fe-efficient," while some plants were more chlorotic than others and these were termed more "Fe-inefficient." Iron analysis showed that many of the chlorotic plants had as high or higher Fe concentrations in their leaves as the green plants, so ability to take up or accumulate Fe or the amount of production per unit of element added or absorbed had little relevancy. The usage of efficiency in this context has continued even to the present.

More recent definitions of efficiency in the Fe nutrition literature define the word relative to metabolic processes associated with plant ability to make Fe available in its growth medium or in the rhizosphere. Thus, plants that can remain green or become green under conditions that normally induce Fe deficiency chlorosis for other plants or genotypes are considered "Fe-efficient." These processes have been studied extensively and are normally associated with ability of roots to produce H ions (reduce the pH of their root environment) and reducing compounds and to reduce ferric-Fe to ferrous-Fe more effectively (Brown *et al.*, 1972; Brown, 1977, 1978a,b). Recent studies have shown that these processes are associated with most plant species, but not with the grasses. The grasses appear to have a different type of mechanism to mobilize Fe in the rhizophere; release of highly effective Fe-mobilizing compounds called phytosiderophores or mugineic acids (Takagi *et al.*, 1984; Marschner *et al.*, 1986; Römheld and Marschner, 1986a,b). Plants that have a greater ability to perform these processes or respond better to Fe deficiency stress conditions have been called more "Fe-efficient."

With the many meanings and definitions given for efficiency, the specific way it is used needs to be defined clearly.

II. Rationale for Improving Plants for Mineral Nutritional Characteristics

Large increases in productivity of cereals over the past 50 plus years have come primarily from better management, improved cultivars, and increased fertilizer applications, especially N. Little regard has been given to plant improvement for mineral nutrient characteristics because most resources have been readily available and relatively inexpensive. Thus, the concept of changing the soil to meet plant mineral nutritional needs prevailed. However, when the oil crisis came in the mid-1970s and the world economic situation became more unstable, interest in taking advantage of plant variability to help alleviate mineral nutrient constraints began to receive added emphasis. Even though considerable information was available about genotypic differences in requirements for mineral nutrients before this time, the amount of information in the past decade has expanded markedly.

To succeed in crop improvement/breeding programs for mineral nutritional characteristics, several criteria need to be met (Graham, 1984; Sherrard *et al.*, 1986). The first criterion is whether biological, ecological, and economic considerations dictate that a breeding solution would be better than an agronomic solution. A second criterion is that the site (soil) for evaluation of the desired characteristics must have the potential to permit a new genotype to express appropriate differences. A third criterion is that sufficient variability must exist for the desired characteristic so that improvement through plant breeding can be made. An understanding of the inheritance of these traits will also be needed, and hopefully the traits are highly heritable. For biochemical or physiological traits, accurate and simple methods to measure the appropriate traits are available, and the trait should show positive relationship with yield or final product in the field.

The first criterion could be met realistically for many mineral nutrients and has been of special importance in lesser-developed countries because of the lack of capital to pay for fertilizers and soil amendments. This first criterion could also feasibly be met for most nutrients in developed countries because of high costs and pollution concerns in recent years. This has been addressed particularly for N because of its high requirement in cereals, high cost, and potential pollution in groundwater. From a theoretical point of view, Graham (1984) concluded that it would be pointless to breed cereals for greater tolerance to N-deficient soils because added N either though fertilizers or rotation with N_2-fixing crops would be needed to sustain high yields. Nevertheless, wide genotypic variation for many N characteristics has been noted in crops like maize. This indicates that some N efficiency traits can be improved (Moll *et al.*, 1982a,b, 1987; Sherrard *et al.*, 1984a,b, 1986; Jackson *et al.*, 1986). Plant use of the other mineral nutrients also warrant improvement through the breeding solution (Graham, 1984). One of the greatest success stories for plant improvement relative to mineral nutrient stresses has been the improved resistance to Fe deficiency chlorosis incorporated into soybean (Clark *et al.*, 1990). Evidence is sufficient to indicate that good progress should be made with the cereal crops for resistance to Fe deficiency chlorosis; sorghum and rice are particularly susceptible to this disorder (Clark, 1982a; Vose, 1982).

The second criterion should be met for most mineral nutrients. Sites (soil) or conditions for genotypes to exhibit good expressions of variability to particular nutrients should be found somewhere or developed. Sites may not be conveniently located, and methods for screening can be developed if they are not readily available. Laboratory screening methods need to correlate well with field observations, however.

The third criterion, genotypic variability, has been reported for many mineral nutritional traits in cereals. Genotypic variation has been noted in most studies using only limited amounts of germplasm. Probably the greatest restriction in finding genotypic variability has been the relatively small germplasm bases that have been screened for specific nutrients. Large screening efforts have been conducted for mineral nutrient deficiencies and element toxicities in rice (IRRI, 1976, 1977; see also other IRRI annual reports and newsletters; Ponnamperuma, 1976) and wheat (Kohl and Rajaram, 1988).

Relatively little is known about the heritability of specific mineral nutritional traits and their positive associations with yield or final desired product. If this information was available, considerably more research would have been conducted and more progress made. This is one of the areas that needs extensive research if progress is to be made in plant improvement for mineral nutritional traits.

The criteria for improving plants for mineral nutritional traits through plant breeding can be realistically met, and improvements could be made if efforts were expended. Initial studies indicate that improvements can be made, but little attention has been given to this effort. Vose (1984) concluded that improvements have not been attained "due partly to the compartmental nature of agricultural research, and partly to the challenge that changing the soil environment represented," and partly because "the breeder of any crop plant must take so many factors into account, e.g. yield, disease resistance, growth habit, agronomic adaptability, resistance to lodging etc., and commercial factors such as earliness, quality, size, uniformity and suitability for packing and transport, that there is little inducement to consider an additional factor such as nutritional efficiency, unless forced to do so by extreme requirements."

III. Mechanisms for Genotypic Variation in Mineral Nutrients

Each mineral nutrient required for plant growth has its own unique chemistry and physiology so that mechanisms of efficiency may be different for each nutrient. Even so, certain broad processes are common for each nutrient. These processes are summarized into four major categories: (1) acquisition from the environment, (2) movement across roots and delivery to the xylem, (3) translocation and distribution into plant parts, and (4) utilization in metabolism and growth (Gerloff and Gabelman, 1983). Each of these processes can be divided into additional component parts and specific kinds of reactions or factors are associated with each part. For example, acquisition of a nutrient concerns not only uptake per se but the processes and factors that affect this process, such as plant root morphology, shoot/root ratio, extent of roots (lateral or vertical), degree of root division (root number and diameter), and factors affecting delivery to the roots. Information on the various processes for any plant species is limited, and these kinds of processes need to receive more attention. Many of these aspects will be addressed with each nutrient discussed.

A. NITROGEN

Nearly half of the protein supply for world consumption is derived from the cereals (Cooke, 1975). Since N is a major component of proteins, it is a major mineral nutrient required for growth. Cereals, which are nonnodulating and non-N_2-fixing plants, require abundant amounts of N either through added fertilizer, manure, or green crops or by growing in rotation with legumes that fix N if their production is to be maintained at high levels. Because of this, N is normally the most heavily used and most costly mineral nutrient required for cereal and other crop productivity. Worldwide, cereals (primarily wheat, maize, and rice) receive a high proportion of the N that is applied for crop production (Martinez and Diamond, 1984).

Various sources of N can be used for cereal production. Except for many organic sources of N as integral parts of plant residues, most sources of N (nitrate and ammonium) are readily soluble and readily move in soils, especially nitrate. Ammonium is usually converted to nitrate through nitrification unless nitrification inhibitors or certain environments prevail (e.g., acid or paddy soils and cool temperatures), so nitrate is considered to be the dominant form of N absorbed by most plants, including cereals (Schrader and Thomas, 1981; Barber, 1984b; Hageman,

1984). As long as adequate N is available and fairly optimum soil conditions prevail (i.e., moisture, temperature, aeration, and pH), roots of cereal crops will normally absorb sufficient N to maintain good production.

Nitrogen moves in the soil primarily by mass flow (Barber, 1984b). Root length and extensiveness may not be too important for N availability at root surfaces as long as adequate amounts of N are in the soil and adequate moisture is available to move N to the root surface.

The amount of N absorbed by cereal plants compared to the amount of N supplied to the soil or growth medium varies extensively with environmental conditions, plant genotype, and many other processes (Craswell and Godwin, 1984). It has generally been considered that only about half of the N applied to a soil in a particular season is absorbed by plants (Allison, 1955, 1966; Barber, 1976). Improvement in the amount of N taken into plants compared to the amount added (a form of N efficiency) is the subject of considerable research (Stevenson, 1982; Craswell and Godwin, 1984; Hauck, 1984; De Datta and Patrick, 1986; Haynes, 1986; Lambers *et al.*, 1986). Craswell and Godwin (1984) summarized factors affecting N uptake and use by cereal crops. These include form of fertilizer; soil water, temperature, pH, and biota; sink strength; losses from volatilization, leaching, and denitrification; competition from other plants and biota; soil properties for fixation of ammonium; C/N ratio; management and aeration; rooting depth and intensity; plant stresses; and plant genotype. Each factor is important for the overall ability of plants to absorb and use N. More specifically, some of the major processes of concern for N efficiency in cereals are uptake, reduction, translocation, and accumulation of N. The acquisition of N by plants has considerably different concerns and processes because N is one of the few mineral nutrients that is used so extensively as a structural component of plants with little used in catalytic processes so characteristic of many of the other nutrients. The processes and factors affecting N acquisition have recently been discussed and reviewed (Butz and Jackson, 1977; Haynes and Goh, 1978; Huffaker and Rains, 1978; Jackson, 1978; Schrader, 1978, 1984; Goodman, 1979; Miflin, 1980; Schrader and Thomas, 1981; Beevers and Hageman, 1983; Jackson *et al.*, 1986).

Intraspecific variability within cereal crop species for mineral nutrition has been recognized for a long time. Maize genotypes differed extensively in adaptation to soils and their productivity differed with fertilizer inputs (Mooers, 1921, 1922, 1933). Using inbred lines and single crosses of maize and different growth media, Smith (1934) noted differences among genotypes for N uptake and yield. Similar differences were noted for maize hybrids, open-pollinated cultivars, and inbred lines (Stringfield and Salter, 1934; Harvey, 1939). Harvey (1939) also noted that ammonium enhanced growth in maize more than nitrate. Low- and high-protein strains of maize had similar amounts of N per plant when grown at relatively low N conditions (25 and 50 parts per 10^6), but the high-protein strain absorbed and assimilated about twice as much N at higher (100 and 200 parts per 10^6) levels of N as did the low-protein strain (Hoener and DeTurk, 1938). Maize hybrids produced more dry matter than inbred lines when plants were grown with relatively high levels of N (Burkholder and McVeigh, 1940). Differential responses of barley genotypes to N and other mineral nutrients were also noted at an early date (Gregory and Crowther, 1928). Since these early studies, many investigations have been conducted to study cereal crops for N traits.

Since about 1950, fertilizer N usage for cereal production has risen markedly (Hageman, 1979). During this same period of time, grain yields have also risen. However, the amount of grain produced per unit of N applied has decreased dramatically (Sherrard *et al.*, 1984a). For example, the average production of maize per ton of N applied was 376 Mg during the period 1950–1955, but only 39 Mg for the 1974–1979 period (Hageman, 1979; Sherrard *et al.*, 1984a). Duvick (1977) and Welch (1979) noted that more recently developed maize hybrids (comparing releases of about 40 years) responded more to applied N than did earlier releases. Modern sorghum genotypes responded to added N to a much greater extent than did landraces (Gardner, 1988). Even though high levels of N might have reduced acquisition efficiency in recently developed maize genotypes (Rao and Rains, 1976), recently developed maize hybrids bred for improved performance with high N showed higher rates of nitrate uptake than hybrids developed many years earlier (Cacco *et al.*, 1983b). Wheat cultivars have also shown large differences in their response to N (Woodend *et al.*, 1986; McMahan, 1987), especially in regard to K_m, V_{max}, and N utilization efficiency [dry matter yield (DMY)[2] per unit N absorbed] (Woodend *et al.*, 1986).

Cereal crops need good photosynthetic productivity (source) and adequate storage capacity (sink) to assure high productivity. The source for photosynthetic productivity depends on characteristics like leaf area, leaf area duration, rates of photosynthesis, respiration, and amino acid synthesis. These are functions of the number of leaves and area per leaf (Moorby and Besford, 1983; Hageman and Below, 1984). The sink in cereals is associated with the number and site of grains per plant and their maintenance to assure proper development and function.

Vegetative yield normally increases with increased supplies of N, but relationships between grain yield and leaf or stalk N concentration and content have not always been good in cereals. However, this relationship was good in one wheat study (Papastylianou *et al.*, 1984). The amount of N acquired, assimilated, and distributed is the result of many interacting and regulating processes and reactions. All of the processes must be working together to produce the desired products. Several of these processes are considered, and in most cases genotypes vary considerably.

Genotypic differences in uptake rates of nitrate have been reported for maize (Cacco *et al.*, 1983a,b; Jackson *et al.*, 1986; Pan *et al.*, 1987), barley (Perby and Jensen, 1983), wheat (Woodend *et al.*, 1986), and sorghum (Franca, 1981). Nitrogen uptake rates appeared to accelerate with time in wheat (Rao *et al.*, 1977) and barley (Huffaker and Rains, 1978), and leveled off and even showed some decrease with time in sorghum (Franca, 1981) and barley (Perby and Jensen, 1984). Diurnal variation in nitrate uptake was also noted in maize genotypes (Pan *et al.*, 1987). Uptake rates of N decreased dramatically after 34 days of age, and total N absorbed per plant increased as plants aged with a maximum at about 44 days of age in sorghum (Franca, 1981) and at 30 to 45 days in barley (Perby and Jensen, 1984). Genotypic differences in N accumulation were also noted in sorghum (Franca, 1981) and barley (Perby and Jensen, 1986). Nitrogen accumulation was normally higher in later-maturing then in earlier-maturing barley genotypes (Perby and Jensen, 1984). Influx rates of N per root were important in barley, but not differences in influx per unit fresh weight of root (Perby, 1986). Root size (diameter and length) was also important in uptake of N by barley.

After absorption, considerable amounts of nitrate are retained in the roots for storage and metabolism, some is reduced, and large amounts are translocated to the leaves. Most of the nitrate absorbed by cereals is generally assumed to be reduced in the leaves compared to the roots of plants like *Vicia* and *Pisum* (Wallace and Pate, 1965; Beevers and Hageman, 1969; Chantarotwong *et al.*, 1976; Raven and Smith, 1976). However, considerable amounts of nitrate may be stored and reduced in roots (Jackson and Volk, 1981). Roots were reported to reduce about 30 to 40% of the nitrate absorbed in wheat (Ashley *et al.*, 1975; Cooper *et al.*, 1986), about one-third to one-half in maize (Hageman and Lambert, 1981; Keltjens *et al.*, 1986), and about 50% in barley (Gojon *et al.*, 1986). Maize and sorghum genotypes showed wide variances in the amount of N accumulated, translocated, and reduced (Franca, 1981; Jackson *et al.*, 1986). The kinetic patterns for nitrate uptake and *in vivo* reduction in barley were similar, indicating that uptake can control assimilation (Huffaker and Rains, 1978).

Although nitrate is considered the predominant form of N absorbed by cereals (Huffaker and Rains, 1978; Jackson, 1978; Schrader, 1978, 1984; Schrader and Thomas, 1981; Jackson *et al.*, 1986), ammonium can also be absorbed and may be a dominant form under some conditions. Ammonium-fed plants appear to have altered metabolism compared with nitrate-fed plants (Goyal and Huffaker, 1984; Hageman, 1984). Ammonium is also toxic to plants at lower levels than nitrate. Both ammonium and nitrate can serve as adequate sources of N for cereal production, but nitrate fertilizers are considered "safer" sources to use (Hageman, 1984). Ammonium is converted to nitrate in most soils when aeration and temperature are optimal for plant growth, so nitrate is considered to be the primary form of N for cereal production. Under most conditions, ammonium in cereal production, except rice, is relatively insignificant.

Since nitrate is the predominant form of N absorbed by cereals, this form of N must be reduced before the N can be used in metabolism. Enzymes catalyzing the reduction of nitrate to ammonium are nitrate and nitrite reductases. These two enzymes are normally considered jointly and the overall reaction is called nitrate reductase (NR). Nitrate reductase is a prevalent enzyme that occurs primarily in leaves of cereals, but has also been reported in roots (Beevers and Hageman, 1969; Oaks and Hirel, 1985).

Hageman (1979) and Sherrard *et al.* (1984a,b, 1986), indicated that the potential existed to increase maize yield through identifying and selecting genotypes for improved N traits for uptake and metabolism. Selecting for biochemical and physiological N traits was also suggested for improvement in whcat (Johnson *et al.*, 1968; Croy and Hageman, 1970; Deckard *et al.*, 1973). Metabolic processes of N considered were NR activity and nitrate contents in plants.

Selection for superior genotypes on the basis of NR activity has produced variable results (Hageman, 1979; Cacco *et al.*, 1983a). Significant correlations between NR activity and grain yield and grain N were noted for wheat (Eilrich and Hageman, 1973) and maize (Deckard *et al.*, 1973). These and other studies established that NR activity was variable in genotypes, that the enzyme activity was heritable, and that selection for NR activity would be effective in some genetic materials (Zieserl and Hageman, 1962; Hageman *et al.*, 1963; Zieserl *et al.*, 1963; Schrader *et al.*, 1966; Warner *et al.*, 1969; Croy and Hageman, 1970; Boyat and Robin, 1977; Dalling and Lyon, 1977).

A plant breeding program was initiated in Illinois to assess the direct effects of NR activity and productivity of maize (Hageman, 1979; Sherrard *et al.*, 1984a,b, 1986). After six cycles of selection, divergence in NR activity among strains of the parent was significant and continuous (Sherrard *et al.*, 1986). The NR activity in the high-NR strain increased 52% in Cycle 6 over the initial parent (Cycle O) and decreased 38% in the low-NR strain. After eight cycles of selection, NR activity in the high-NR strain was four times higher than that for the low-NR strain (Eichelberger, 1986). The data after these cycles of selection showed: (a) selection for NR activity affected enzyme activities in leaves, (b) the rate of NR decline over the grain filling season did not overlap for high- and low-NR strains, (c) the NR enzyme activity was heritable, and (d) the level of NR activity could be altered through classical plant breeding strategies. The high-NR strain had higher specific leaf weights, lower chlorophyll concentrations, and lower leaf nitrate concentrations (between anthesis and 37 days afterward) than the low-NR strain (Sherrard *et al.*, 1986). Over the period of these studies, no relationship between grain yield and leaf NR activity (seasonal average) was noted. Similar results were noted when wheat NR activity was related to grain yield and grain protein (Eilrich and Hageman, 1973; Brunetti and Hageman, 1976). In the maize studies, selecting for increased NR activity did not increase leaf reduced N concentrations (Sherrard *et al.*, 1986).

Even though these studies showed that NR activity was not closely associated with grain yield and did not appear to enhance reduced N concentrations in maize, leaf NR activity was associated with N accumulation in wheat (Dalling *et al.*, 1975; Huffaker and Rains, 1978). The maize data were interpreted to mean that factors other than NR activity influenced reduced N accumulation. Other studies noted that NR activity was influenced by the availability of nitrate at the induction and assimilation sites (Shaner and Boyer, 1976; Rao *et al.*, 1977), the availability of reductant and metabolites for regeneration of reductant (Klepper *et al.*, 1971; Brunetti and Hageman, 1976; Hageman *et al.*, 1980), and variability in the flux of nitrate to the site of reduction (Shaner and Boyer, 1976; Reed and Hageman, 1980a,b). Activity of NR also increased with late applications of N (Przemeck and Kuche, 1986).

Assuming that nitrate uptake would influence N assimilation of tissue, studies at the University of California were conducted to compare nitrate contents of tissues, leaf proteins, and remobilization of reduced N in wheat (Huffaker and Rains, 1978; Gallagher *et al.*, 1983). The high-nitrate-accumulating cultivar Anza had 30% lower N absorption rates and lower root NR activity than the low-nitrate-accumulating cultivar UC 44-111. The lower uptake rates of the high-nitrate-accumulating cultivar were unexpected, but lower tissue nitrate had been reported for genotypes with higher NR activity (Rao *et al.*, 1977). The differences in protein in these two wheat cultivars during the vegetative stage appeared to correspond with differences in translocation of reduced N from the vegetative tissues to the grain during grain fill (Huffaker and Rains, 1978; Gallagher *et al.*, 1983). Anza retained less protein in the leaves compared to the grain than did UC 44-111, and was considered to be more efficient in N translocation. Even though UC 44-111 was determined to have a greater capacity to respond to N fertilization and to reduce N in leaves than Anza, both cultivars had similar grain protein concentrations because of the ineffective translocation system in UC 44-111. Differences in N translocation among genotypes have also been reported for triticale (Lal *et al.*, 1978), maize (Hay *et al.*, 1953; Beauchamp *et al.*, 1976; Chevalier and Schrader, 1977; Rodriguez-P., 1977), sorghum (Franca, 1981; Gardner, 1988), and other wheats (Johnson *et al.*, 1967; Lal *et al.*, 1978).

Studies at the University of Illinois showed continued adverse effects of high NR activity on grain yield in high- and low-NR activity maize cultivars (Sherrard *et al.*, 1986). Higher grain yields were observed with the low-NR activity cultivars, which also had higher reduced N accumulation in the leaves. Further study with high reduced N cultivars verified that they had higher grain yields than low reduced N cultivars. It was concluded that high concentrations of reduced N and low levels of NR activity, especially during the first third of the grain filling period after anthesis, may be a useful criterion for the development of high-yielding maize hybrids. Why low-NR activity cultivars had higher grain yields was difficult to explain, except that other factors like nitrate flux into the leaf may have been involved (Shaner and Boyer, 1976; Reed and Hageman, 1980a,b). Higher concentrations of reduced N in leaves might also have been associated with maintenance of high photosynthetic activity in leaves. Other studies showed that maize leaves lost large amounts of reduced N, chlorophyll, and photosynthetic activity as the grain filling period progressed (Christensen *et al.*, 1981; Swank *et al.*, 1982). However, studies with maize selected from 1930 through 1975 showed N uptake and assimilation traits, including NR activity, to be related to grain yield (Cacco *et al.*, 1983b).

An important concept for understanding or enhancing high cereal productivity and N efficiency could be the mobilization and distribution of C and N in tissues (Cherry, 1985). High yields in cereals have been attributed in part to increases in harvest index (HI, amount of grain yield per total plant dry matter yield) by changing the amount of harvestable grain (sink) compared to the total biomass (source) (Gifford *et al.*, 1984; Day *et al.*, 1985). Limiting the amount of vegetative plant tissue compared to the reproductive parts takes advantage of maximizing the solar radiation intercepted. The source (leaves and stalk primarily) is usually not limiting in most cereals. However, the source was limiting more than the sink in sorghum

(Muchow and Wilson, 1976), and yield was related to leaf area, leaf area duration (length of time leaf area remains functional), and plant height (Sriram and Rao, 1983). Other studies with wheat showed good relationships between yield and HI as well as with yield and HI of N (amount of N in grain compared to the rest of the plant) (Dubois and Fossati, 1981; Loffler and Busch, 1982; Paccaud *et al.*, 1985). So far, plant improvement has not enhanced yields through increasing net photosynthesis per unit of photosynthetic tissue (Hageman and Below, 1984; Cherry, 1985).

Once the reproductive phase begins in cereals, mobilization of products from vegetative to the reproductive tissue also begins. If these processes are in balance or can be controlled appropriately, adequate N and C will accumulate in the grain to give high yield potentials without reducing activities of essential enzymes. Nitrogen is generally remobilized more effectively in tissue than C (Pate and Layzell, 1981). Accumulations of dry matter and reduced N in maize grain were independent of each other (Swank *et al.*, 1982). Sugar and N translocation to wheat kernels and starch synthesis in grain also appeared to be different from each other (Barlow *et al.*, 1983). Nitrogen requirements of grain can often be met from N already accumulated during the vegetative growth stage, but C requirements are not easily met from this same source (Pate and Layzell, 1981). accordingly, the proportion of N in the grain at maturity is as high as 65 to 75% of the total plant N, while the proportion of C is only about 40 to 45% of the total plant C. About 40 to 50% of the grain N in wheat was derived from the vegetative growth, and as high as 75 to 80% when N was low during the same growth stage (Spiertz and Ellen, 1978; Campbell *et al.*, 1983; Dalling, 1985). Translocation efficiency for N (total grain N to total head N) ranged from 75 to 94% in wheat cultivars (Halloran and Lee, 1979) and N translocation from shoot to grain was about 67% under irrigation and 75% under dryland conditions (McNeal *et al.*, 1968). Nitrogen absorbed from soil during grain fill of maize had little effect on grain yield (Friedrich and Schrader, 1979). Low supplies of soil N resulted in smaller sorghum leaves with lower contents of N, which lowered light conversion efficiency (Lafitte and Loomis, 1988). These low-N leaves were not able to supply as much N to panicle growth and grain as high-N leaves. Greater effects were noted for N deficiency on structural limitations of growth than were noted for reduced conversion efficiency per unit of leaf area. Establishing and maintaining the growth structure and the photosynthetic sink capacities in plants are important (Hageman and Below, 1984; Lafittc and Loomis, 1988).

Leaves and culms of wheat remobilized 80% of their N to the grain with the rest coming from roots (about 6%) and glumes (15%) (Dalling *et al.*, 1976). Other wheat studies showed that N in grain could be derived entirely from redistribution of N from vegetative organs (Simpson *et al.*, 1983). In this latter study, leaves contributed 40%, glumes 23%, clums 23%, and roots 16% of the N in the grain. Flag leaves contributed 37% of the total N to developing wheat kernels, lower leaves contributed 18%, and the peduncle and spike contributed 45% in other studies (MacKown and Van Sanford, 1986). A range of N transfer from vegetative tissue to grain varied from 58 to 87% among sorghum genotypes (ICRISAT, 1975–1976).

Other sorghum genotypes have shown large differences in N remobilization (Franca, 1981). Genotypic differences in N remobilization efficiency have also been noted in wheat (Dubois and Fossati, 1981; Loffler and Busch, 1982; Paccaud *et al.*, 1985). High-protein rice cultivars tended to translocate more leaf N to developing grains than low-protein cultivars (Perez *et al.*, 1973), and N increased more rapidly in high-protein than in low-protein wheats (Seth *et al.*, 1960). Nitrogen partitioning to tillers was also important in barley, and high external N supplies normally increased the number of tillers although genotypes differed (Perby and Jensen, 1987).

Good relationships were noted between proteolytic enzymes (acid proteinase) and the amount of N remobilized in wheat (Dalling *et al.*, 1976). Proteolytic enzyme activity increased as leaf protein decreased during grain fill in maize (Feller *et al.*, 1977). Grain N was related to the amount of N accumulated and distributed in leaves as well as with proteolytic enzymes (Reed *et al.*, 1980). Proteolytic enzyme activity was higher and increased earlier in low-NR than in high-NR activity genotypes. The low-NR activity–high-protease genotype had higher percentages of grain N and higher N harvest index (NHI) values than the high-NR activity–low-protease genotype. More rapid transfer of nutrients from leaves and culms to the grain was noted in a sorghum genotype compared to other genotypes with similar maturity and height (Murty, 1979). Selecting for biomass production and HI in sorghum was considered an effective means for selecting genotypes with greater N efficiency (ICRISAT, 1979–1980).

Differences in photosynthesis in leaves of plants containing similar amounts of N may be a mechanism for enhanced N use efficiency. This trait has been called "instantaneous N use efficiency" or "apparent photosynthesis" (Bolton and Brown, 1980; Field *et al.*, 1983; Alagarswamy *et al.*, 1988) and is measured as the carbon dioxide exchange rate per unit leaf N (CER/N). Plants with the C_4 photosynthetic pathway have been hypothesized to have greater N use efficiency (dry matter yield per unit N absorbed) than plants with the C_3 pathway (R. H. Brown, 1978; Schmitt and Edwards, 1981; Krstic and Saric, 1983).

Species of dune grass (*Ammophila arenaria*) considered to be N efficient (higher dry matter yield per unit N) had 30 to 40% higher CO_2 uptake per unit N than N-inefficient species (*Elymus mollis*) (Pavlik, 1983a,b). Using *Panicum,* which has both C_4 and C_3 photosynthetic pathways within the species, the C_4 cultivars (*P. maximum* and *P. prionitis*) had higher CER/N values than the C_3 cultivars (*P. laxum* and *P. hylaeicum*) at high levels of N, but not at low levels of N (Brown and Wilson, 1983; Wilson and Brown, 1983). Nitrogen use efficiency in these *Panicum* cultivars did not appear to be entirely related to whether the cultivars were C_4 or C_3, but appeared to be related to leaf and culm morphology. One of the *Panicum* cultivars (*P. prionitis*) had low CER/N values at the lower temperature regime used (24/19°C versus 35/30°C, day/night temperature cycles), which was attributed to its higher specific leaf weight (three to five times higher) compared to the other species (Brown and Wilson, 1983). Maize (C_4) was more N efficient than wheat or rice (C_3) on the basis of total N, soluble protein, and ribulose-1,5-diphosphate carboxylase (RuDPCase) protein at different temperatures and up to 320 parts per 10^6 CO_2 (21%

O_2) (Schmitt and Edwards, 1981). However, wheat became as N efficient as maize at saturating levels of CO_2. In other studies, C_4 plants had lower N than C_3 plants and no relationship was noted between N content and CO_2 absorption (Krstic and Saric, 1983). Maize was an ineffective user of N, however (Jocic and Saric, 1983).

Landraces and recently developed (modern) genotypes of sorghum were investigated for N efficiency traits in the field (Gardner, 1988). The landrace genotypes were more N efficient (producing 152 g dry matter per g N absorbed) than the modern genotypes (producing 124 g dry matter per g N absorbed), but the modern genotypes responded more (362% increase in dry matter) than the landraces (170% increase in dry matter) when grown at a moderate compared to a low level of N. The most N-efficient landrace genotype partitioned more dry matter to stalk tissue and developed more leaf area and a higher leaf area index than the modern genotypes. One landrace genotype had lower specific leaf areas, lower proportions of N in the soluble N fraction, and a higher net assimilation rate (g DM per m^2 ground area per day) than a comparable modern genotype (a commercial hybrid). The modern genotype had higher proportions of total N in the soluble fraction and responded more to added N than the landrace. In other studies with these genotypes, the most N-efficient genotypes (one landrace genotype and one modern genotype) remobilized N more rapidly from older to younger leaves. The genotypes differed in instantaneous N use efficiency (CO_2 gain per unit leaf N), but this trait could not be used to predict seasonlong N use efficiency of the genotypes. The organic leaf N was closely related to stomatal conductance. The genotypes also differed markedly for dark respiration rates, with the modern genotype having the lowest rate. Canopy N contributed to overall N efficiency, but did not appear to be related to the level of domestication of a particular developed genotype. The landrace genotype had thicker leaves (136 μm) and a larger phloem area (1321 μm^2) than the other genotypes (mean values of 114 μm thickness and 979 μm^2 phloem area).

Tropical grasses (normally C_4 plants) often show a greater response to applied N than C_3 grasses (Wilson and Haydock, 1971; R. H. Brown, 1978). This may result in lower N concentrations in C_4 plants than in C_3 plants (Wilson, 1975). Low leaf N concentrations were characteristic of N efficiency in the C_4 plant sorghum (Maranville *et al.*, 1980; Zweifel *et al.*, 1987). C_4 plants may also have a smaller investment of N in the photosynthetic carboxylation enzymes than C_3 plants, which might lead to their greater CER/N values or N use efficiency (R. H. Brown, 1978; Ogata *et al.*, 1983). About 40 to 50% of the soluble protein is associated with RuDPCase compared to only about 30% in C_4 plants (R. H. Brown, 1978). Phosphoenolpyruvate carboxylase (PEPCase) and pyruvate orthophosphate dikinase (PPD) account for about 14% of the soluble protein in maize (Sugiyama *et al.*, 1984), so about the same amounts of soluble protein appear to be associated with CO_2 fixation in both C_3 and C_4 plants. Even differences have been noted among C_4 species in the amount of RuDPCase (Ku *et al.*, 1979), and differences in the specific activity of the enzyme were found among wheat genotypes (Evans and Seemann, 1984).

RuDPCase, PEPCase, and PPD are proteins that may be important in coupling N metabolism to photosynthesis, which could eventually enhance grain yield

(Peterson and Huffaker, 1975; Makino *et al.*, 1984). RuDPCase appears to be an important source of N for grain when the soil does not supply N during grain development (Frith and Dalling, 1980). Proteins are commonly degraded and the N redistributed elsewhere, particularly during late grain fill and leaf senescence (Friedrich and Schrader, 1979). Since RuDPCase is a major enzyme lost during this period of time (Thomas and Stoddart, 1980), RuDPCase losses would likely accompany losses in photosynthesis, and this has been reported (Wittenbach, 1978; Friedrich and Huffaker, 1980; Evans and Seemann, 1984). PEPCase and PPD paralleled maize biomass production more closely than RuDPCase, and the proportion of soluble protein allocated to RuDPCase decreased as biomass increased (Sugiyama *et al.*, 1984). RuDPCase synthesis may have higher priority than other soluble CO_2-fixing enzymes when N is limiting. The percentage of RuDPCase increase was greater than the proportion allocated to PEPCase and PPD. RuDPCase may also be rate limiting in photosynthesis compared to other CO_2 carboxylation enzymes (Usada, 1984). These differences between C_3 and C_4 and within C_3 and C_4 plants need to be investigated further to determine the relationships between the CO_2 carboxylation enzymes and N use efficiency.

Twenty diverse pearl millet genotypes grown in the field showed few differences in total N uptake, but extensive differences in dry matter produced (Alagarswamy and Bidinger, 1982, 1987). These results indicated large differences in N efficiency (dry matter yield per unit N absorbed). When N-efficient and N-inefficient genotypes were compared for CER/N values and leaf morphology in the laboratory, differences in instantaneous N use efficiency could not explain differences between N efficiency and N inefficiency (Alagarswamy *et al.*, 1988). The N-efficient genotypes also had thicker leaves. Differences among these pearl millet genotypes for CER/N were attributed to differences in partitioning of N in the leaves to increased leaf area. Thus, leaf areas were high, which maintained low concentrations of N, but leaves still had high N contents (total N).

Sorghum varied in N uptake, N distribution, grain yield, and photosynthesis when grown at high and low N (Fernandez, 1987; Gardner, 1988). No particular pattern for N efficiency traits was noted with plant age and N level, but the genotypes considered to be more N efficient (DMY per unit N absorbed) had higher NE_1 (DM/N, dry matter per unit N absorbed), NUE (DM/N $\times$ DM, dry matter per unit N absorbed times dry matter), and PS-$NEff_2$ (μmol CO_2/N% s^{-1}, photosynthesis per percent leaf N per second) and lower $PSrate_1$ (μmol CO_2/m^2 s^{-1}, photosynthesis per square meter leaf area per second), $PSrate_2$ (μmol CO_2/leaf s^{-1}, photosynthesis per unit leaf weight per second), or PS-Eff_1 (μmol CO_2/g N s^{-1}, photosynthesis per g leaf N per second) values (Fernandez, 1987). The same genotypes had highest leaf dry matter and leaf area, which provided for greater photosynthesis and dry matter and grain yields, but lower dry matter and N harvest index values. Selection for any one particular N efficiency trait did not appear to be practical.

Selecting and breeding cereals for the many N efficiency traits and making sure each of the desired N efficiency processes is functioning appropriately would be impractical. The measurement of the best number of traits also would be almost impossible. The interrelationships of the various processes would be difficult to predict. In a mass selection study using half-sib progenies of maize, several N efficiency traits were evaluated (Muruli and Paulsen, 1981). The N-efficient progenies had higher yield than inefficient progenies at low N, but not at high N, indicating the feasibility of improving maize for N efficiency. Large numbers of traits were used to assess sorghum for N efficiency, and genotypes showed many overlaps, making interpretations relatively difficult (Maranville *et al.*, 1980; Zweifel *et al.*, 1987). However, relatively good success was achieved in assessing rice for N efficiency using nine different N efficiency traits (Broadbent *et al.*, 1987). Although cultivars overlapped in some traits, large differences were noted and consistency was obtained from season to season by evaluating the cultivars using these traits. The authors felt genetic improvement for N efficiency in rice was feasible and practical using many traits.

The assessment of only a few selected traits and their integration with yield was the approach taken by a North Carolina group to select and improve maize for N efficiency (Moll *et al.*, 1982a,b). The measured traits were aboveground total dry matter (T_w) and grain (G_w) yields, N supplied to the soil (N_s), and total N taken into the aboveground plant (N_t). Aboveground T_w and N_t were used because of the difficulty in assessing root dry matter and root N. Measuring residual soil N would also be difficult, so only N supplied was used. The traits evaluated were G_w/N_s (N use efficiency), N_t/N_s (N uptake efficiency), and G_w/N_t (N utilization efficiency). Instead of using G_w/N_s as a direct measurement of G_w and N_s, this trait was derived from $(G_w/N_t)(N_t/N_s) = G_w/N_s$. This way the contribution due to variations in N uptake could be separated from grain yield variations. These studies also considered N uptake in the plant at silking and the amount of N in the grain compared to the total plant N (Moll *et al.*, 1982a).

Eight single-cross maize hybrids differed in N efficiency traits and yield when grown in the field with low and high N (Moll *et al.*, 1982a). At low N, hybrid differences in N use efficiency were due largely to variation in utilization of acquired N. At high N, hybrid differences were attributed to variation in N uptake efficiency. Differences in N translocation and remobilization to the grain were important at the low level of N. Hybrids showed differences in the component traits that led to N efficiency.

Other studies showed that prolific maize plants (more than one ear per plant) were more N efficient than nonprolific (one ear per plant) maize plants (Bertin *et al.*, 1976; Kamprath *et al.*, 1982; Moll *et al.*, 1982a; Casnoff, 1983; Anderson *et al.*, 1984a,b). Sink capacity appeared to be important in enhancing N efficiency in the prolific genotypes (Kamprath *et al.*, 1982; Anderson *et al.*, 1984a,b). Increased levels of N applied to the soil tended to increase the number of ears per plant in prolific genotypes (Anderson *et al.*, 1984a). More prolific genotypes also accumulated more N and partitioned more N to the grain than nonprolific genotypes (Anderson *et al.*, 1984a,b). Higher grain yields at high N were due to the high number of prolific plants grown at high N (Anderson *et al.*, 1984b). Degree of prolificacy was also associated with accumulation and partitioning of both N and dry matter (Anderson *et al.*, 1985; Pan *et al.*, 1985). However, the increase in grain yield relative to N supplied did not necessarily increase N utilization efficiency (Anderson *et al.*, 1985). Remobilization of leaf N was not affected by ear number, nor did source of N have any effect on dry matter production and N uptake among prolific or nonprolific maize (Pan *et al.*, 1986). Slightly greater partitioning of N was noted for prolific plants grown with ammonium than with nitrate compared to nonprolific genotypes. In other studies, prolific genotypes absorbed more N than nonprolific genotypes, and only nitrate uptake was different (Casnoff, 1983).

Selection among hybrid progenies ($S_0 \times S_0$) for high yield at high N levels and number of ears per plant was effective in increasing number of ears per plant and N uptake (Moll *et al.*, 1987). These selection procedures did not improve yield above that of the original population hybrid, however. Selection among $S_1 \times S_1$ hybrid progenies was effective in improving yield at high N, N use efficiency, N uptake efficiency, and utilization of accumulated N efficiency. The most effective selection criterion was N use efficiency over all N levels. Ears per plant and weight per ear were highly correlated with grain yield in prolific hybrid selections. Detailed analyses of how these traits may be used in breeding strategies and how the component traits may be used to improve maize for yield and N efficiency have been described (Jackson *et al.*, 1986; Moll *et al.*, 1987).

Although numerous studies have shown wide genotypic differences among and within the cereals for N efficiency traits, the genetics of these plant responses are not well understood and appear to be complex. Most studies indicate a genetic control and genotypic differences. The heritabilities of some N efficiency traits were relatively high (Sherrard *et al.*, 1986). Genetic variation in prolificacy in maize was associated with both additive and nonadditive gene effects (Casnoff, 1983). Additive, dominance, and epistatic effects had little influence in controlling inheritance of N efficiency in other maize studies (Alvarado, 1985), but heterosis due to dominance had a major contribution to variation in most N efficiency traits evaluated. Alvarado (1985) indicated that breeding advances should best be achieved using procedures that emphasize dominant gene effects, although additive and epistatic gene effects would be beneficial. Nitrogen uptake efficiency accounted for 54% of the genetic variation in N use efficiency for yield and 72% of the genotypic variation in N use efficiency for protein (Van Sanford and MacKown, 1986).

B. PHOSPHORUS

Phosphorus is another major nutrient requiring large applications as a fertilizer to maintain high productivity of cereals. A major portion of fertilizer P used for food production worldwide is applied to cereals, and nearly 60% of the fertilizer P used in the United States is for cereal production (Hanway and Olson, 1980). Many kinds of P compounds are applied to soils (Engelstad and Terman, 1980; Sample *et al.*,

1980), but the soil availability of P from any fertilizer P compound depends on many factors, mostly those associated with P reactions after application. The phenomenon of P fixation is well known and has been the subject of extensive research, probably the most in soil–fertilizer–plant interaction studies (Sample *et al.*, 1980).

Because of high P retention by soil particles and as precipitates, P moves only limited distances from its source or location (estimated to be about 5 mm) (Barber, 1976). Since roots or crops like cereals make up a volume of less than about 1% of the total soil volume (Barber *et al.*, 1963), roots will make contact with less than 1% of the available P. Phosphorus movement by mass flow has been estimated at 1%, with 90 to 98% of P movement by diffusion (Barber, 1980). Thus, roots need to grow to the source of P if plants are to receive adequate P for plant growth. Although plants differ and conditions affecting movement and availability vary, the amount of P absorbed by plants compared to the amount applied has been estimated at about 10% (Barber, 1976).

Many soil factors affect the amount of P that will be available to plant roots for uptake. Such factors as solution P, buffering capacity, distribution in the profiles, moisture, tortuosity (thickness of water film and fineness of soil particles), and temperature have been discussed (Barber, 1980; Ozanne, 1980). Many plant factors are also important to P uptake, and some of these are discussed.

Both inter- and intraspecific variation in P nutrition have been recognized among cereal species and genotypes. Interspecific differences would be expected since plant species vary so extensively in growth habit and dimensions (Barber, 1980). Considerable evidence for intraspecific variation among genotypes has been appearing consistently since the early studies in which maize genotypes exhibited differences in response to P (DeTurk *et al.*, 1933; Smith, 1934; Lyness, 1936).

Variability in P uptake was noted among maize hybrids and differences in high- and low-P leaf concentrations were shown to be genetically controlled (Thomas *et al.*, 1960; Gorsline *et al.*, 1964a, 1965, 1968; Baker *et al.*, 1967, 1970, 1971). Similar results were noted for other cereal crops (Palmer and Jessop, 1977; Saric, 1987; Saric *et al.*, 1987). When large numbers of maize inbreds and hybrids (73) were grown under greenhouse conditions with limited soil P, seedling differences in vigor and growth were noted (Fox, 1978). Seventeen of these genotypes were grown in the field and their seedling ratings for P responses (dry matter yield in a specific time) in the greenhouse did not carry over to maturity in the field. Other studies with maize showed genotypic differences in response to P when grown in a greenhouse with a low-P soil (Clark and Brown, 1974a). Early-maturing maize lines tended to accumulate higher P than later-maturing lines (Bruetsch and Estes, 1976). The early-maturing lines also tended to have shallower rooting patterns, which were associated with their greater responsiveness to added P.

Sorghum showed wide genotypic variation in response to P (Brown and Jones, 1975, 1977; Brown *et al.*, 1977; Clark *et al.*, 1977, 1978; ICRISAT, 1979–1980; Furlani, 1981; Raju, 1985). When different sorghum genotypes were grown in soils of different pH values and from different locations, some genotypes were more tolerant to element toxicities (Al and Mn) and others were more resistant to mineral element deficiencies (P, Fe, and Cu) than other genotypes (Brown and Jones, 1975, 1977; Brown *et al.*, 1977; Clark *et al.*, 1977). Sorghum genotypes showed wide variation in many other P nutritional traits like distribution between roots and leaves and among leaves and dry matter yield per unit P absorbed (Clark *et al.*, 1978; Furlani, 1981; Raju, 1985). Many sorghum genotypes showed few differences in responses to P in greenhouse studies when grown with sufficient P (Clark *et al.*, 1977).

Large numbers of rice and wheat cultivars have been tested for tolerance and growth on P-deficient soils and wide differences have been noted in the many trials held over the years (IRRI, 1976, 1977; see also other IRRI annual reports; Kohl and Rajaram, 1988). In the case of rice, about 1000 cultivars were grown on low-P soils in Sri Lanka and in the Philippines (IRRI, 1976, 1977), and many showed greater ability to grow under these conditions than others. Rice cultivars were also tested in nutrient solutions under greenhouse conditions for ability to grow with low P, and of 314 cultivars tested, 74 were considered tolerant to the conditions used. In other large field screening tests, 28 out of 161 cultivars treated were determined to be tolerant to low P. Many rice cultivars did not respond to applied P and grew well at 1 μg P g^{-1} soil. Relative tillering ratio (number of tillers for plants grown at 1 mg P $liter^{-1}$/number of tillers for plants grown at 10 mg P $liter^{-1}$) was a better trait to assess rice cultivars for tolerance to low P than measurements of dry weights and dry weight ratios (Turner *et al.*, 1978). Wheat cultivars showed large differences in adaptation to grow with low P on many of the acid soils of Brazil (Kohl and Rajaram, 1988). Wide variability in P nutrition has also been noted during mass screening of plants at other International Agricultural Research Centers [ICRISAT (1975–1976, 1979–1980; see also other ICRISAT Annual Reports); see IITA (International Institute of Tropical Agriculture), CIAT (International Center for Tropical Agriculture), and CIMMYT (International Maize and Wheat Improvement Center) annual reports].

Mechanisms whereby cereal genotypes exhibit differential abilities to grow at low or high P are not completely understood and few have been described to any extent. Factors that need to be considered are root properties (morphology, distribution, size, length, and type); P uptake, translocation, distribution, accumulation, metabolic use, and interactions (especially with other elements) in plants; and root associations with mycorrhizae (Barber, 1980; Ozanne, 1980; Bieleski and Ferguson, 1983; Vose, 1984; Clarkson, 1985).

Since P does not move very far in soils, root extensiveness and morphology are important for root contact and P acquisition. Soils with different physical and chemical properties affected root morphology (root length, distance between roots, root radius, and root surface : shoot ratio) of maize, which also affected P uptake (Schenk and Barber, 1979). Maize root morphology was affected and growth reduced by lowered soil moisture and temperature, and P level had little effect on these root properties (Mackay and Barber, 1984, 1985a,b). Root length was decreased fivefold when maize roots were grown at 18°C compared to 25°C, and was the major factor restricting P uptake (Mackay and Barber, 1985b). Although significant, the differences in root length due to changes in soil moisture were not as great as those noted for temperature (Mackay and Barber, 1984, 1985b). Root lengths with higher amounts and density of root hairs decreased as soil moisture was reduced (Mackay and Barber, 1985a). Even though the density of root hairs decreased as soil P increased, root hair length remained unchanged. Phosphorus in the plant was not correlated with root hair properties. The results from these studies indicated that P had only secondary effects compared to soil moisture and temperature on rooting characteristics and patterns in maize plants.

Root proliferation is normally stimulated in soil zones where P is high (Ozanne, 1980), but relatively high P levels have caused decreases in root growth of maize and sorghum (Clark, 1982b) and other grasses (Ozanne, 1980). Reduced root growth relative to shoot growth may be particularly important (Asher and Loneragan, 1967; Ozanne, 1980) because a given amount of root mass or surface area will have to supply more units of shoot growth. Phosphorus-deficient maize (and other grasses) had greater decreases in shoot growth than root growth (Clark, 1970; Ozanne, 1980), and under these conditions, roots appeared to take the first opportunity to utilize P when it was initially absorbed.

Decreasing proportions of wheat roots supplied P had no effect on shoot growth (Edwards and Barber, 1976; Lu and Barber, 1985), but decreases were noted for maize (and soybean) (Jungk and Barber, 1975). Maximum P uptake in wheat occurred when P was mixed with only 20% of the soil compared to the same amount of P mixed with other proportions of the soil (Yao and Barber, 1986). Maize ear leaf P concentrations were highly correlated with root lengths in the surface (0 to 15 cm) soil even though P level had no measurable effect on root length (Kuchenbuch and Barber, 1987). This was because most roots were located in the shallow portion of the soil where most of the P was located. Root length and P concentration in plants had lower correlations as the soil depth increased. Year-to-year distribution of roots and root lengths in the soil could influence differences noted in year-to-year yields. Barley and rice genotypes showed extensive differences in root growth and ability to extract P at high and low levels of P (Hirata, 1987; Schjorring and Nielsen, 1987).

As roots grow, different regions of the root system age and change morphologically. Most regions of maize seedling roots absorbed P effectively, but translocation was different because of differential xylem development (Burley *et al.*, 1970). Progressive suberization and endodermal thickening along roots had little effect on radial movement of P in maize and wheat, and P uptake occurred over much of the root system (up to 15 to 25 cm of the root from the apex) (Bowen and Rovira, 1967; Ferguson and Clarkson, 1975). Although rates of P uptake usually decrease with age in maize and sorghum roots (Warncke and Barber, 1974; Jungk and Barber, 1975; Furlani *et al.*, 1984b), total uptake by plants is usually enhanced because of the greater number of roots.

Initial studies examining maize genotypic differences for plant accumulation of P could not be explained by P uptake kinetic properties (K_m and V_{max}) although differences were noted for leaf uptake of P (Phillips *et al.*, 1971b), and differences were noted among proportions of inorganic to organic P compounds (Phillips *et al.*,

1971a). The P levels in these studies may have been too high (above levels noted in soil solutions) to show differential P uptake among the genotypes tested. Other studies with maize showed that genotypes differed not only in P uptake (Clark and Brown, 1974b; Baligar and Barber, 1979; Nielsen and Barber, 1979; Schenk and Barber, 1979) but also in root weights, root lengths per unit root and plant weight, and maximal net P influx (I_{max}, K_m, and C_{min} values) (Nielsen and Barber, 1979). Other maize studies showed that differences among genotypes might be explained by differences in tolerance to P deficiency stress imposed by Al and root phosphatase activities (Clark and Brown, 1974b). In addition, decreased Fe uptake and lower Fe translocation from roots to shoots and P partitioning between inorganic and organic P could help explain these differences (Elliott and Läuchli, 1985). Differences among maize genotypes for P were also explained by efflux compared to influx rates, especially at low P (Elliott *et al.*, 1984). Net depletion of the total plant P absorbed was much higher for roots of plants grown at low P than at higher P.

Although rates of P uptake and parameters to measure this (I_{max}, K_m, and C_{min}) were different among maize genotypes (Nielsen and Barber, 1979), factors such as soil moisture, temperature, P status of the plant, and proportion of the root supplied P also affected these values (Jungk, 1974; Jungk and Barber, 1975; Anghinoni and Barber, 1980; Mackay and Barber, 1984, 1985b; Lu and Barber, 1985). Maize root I_{max} and K_m values increased as temperature increased (up to 25°C) (Mackay and Barber, 1984), while soil moisture stress affected these values less than temperature (Mackay and Barber, 1985b). Starving plants for P or supplying only part of maize roots with P caused I_{max} values to increase and K_m values to be unchanged (Anghinoni and Barber, 1980; Lu and Barber, 1985), and P-deficient plants had higher I_{max} values than P-sufficient plants (Jungk, 1974). Withholding P from barley plants also increased I_{max} values while K_m values remained fairly constant (Drew *et al.*, 1984).

Similar P uptake parameters and root lengths differed extensively among barley genotypes grown in the field under moderate P deficiency (Nielsen and Schjorring, 1983). Efficiencies of P uptake (net influx of P per unit DMY) were determined and shown to vary among genotypes. Plant improvement for P efficiency appeared feasible in barley, and genotypes selected should be identified for high I_{max} values and root lengths and for low C_{min} and K_m values.

Sorghum genotypes grown in nutrient solutions also showed large differences in P uptake (Clark *et al.*, 1977, 1978; Furlani, 1981; Furlani *et al.*, 1984b; Raju, 1985). The uptake rates were greater among genotypes when plants were younger (21 days of age) than after they aged (38 and 52 days of age) (Furlani *et al.*, 1984b). Genotypes that grew better at low P (Clark *et al.*, 1977, 1978) had lower P uptake rates than genotypes that grew poorly in the same solution (Furlani *et al.*, 1984b). Greater P uptake was associated with genotypes having larger root systems. Marked differences were also noted among sorghum genotypes for distribution of P between roots and shoots and among leaves, and for P efficiency (DMY per unit P absorbed) (Clark *et al.*, 1978; Furlani *et al.*, 1984b; Raju, 1985). Sorghum genotypes known for their glossy (light yellow-green color and shiny leaves) and nonglossy (normal dark green leaves) properties also showed differences in P uptake, distribution, and efficiency (Raju *et al.*, 1987b). The glossy genotypes were generally more P efficient than the nonglossy genotypes. Sorghum genotypes showed differences in intact root phosphatase activities, which was associated with the degree of P deficiency in the plants (Furlani *et al.*, 1984a). Enhanced phosphatase activities in other cereal species have been associated with the degree of P deficiency that plants were undergoing at the time (Barrett-Lennard *et al.*, 1982; McLachlan, 1984; Smyth and Chevalier, 1984; Elliott and Läuchli, 1986). Sorghum genotypes also differed in tolerance to high levels of P (Furlani *et al.*, 1986a,b) and to organic and inorganic sources of P (Furlani *et al.*, 1987a). Sorghum genotypes grown with organic sources of P produced more dry matter, had a greater degree of P toxicity ("red-speckling"), and had lower P efficiency (DMY per unit P absorbed) values (Furlani *et al.*, 1986a,b, 1987a) than did genotypes grown with inorganic P. Photosynthesis did not decrease in P-deficient sorghum leaves, but did at high P if P toxicity symptoms ('red-speckling') became too severe (Furlani *et al.*, 1986a). Sorghum genotypes from wide genetic backgrounds also showed extensive differences in uptake and efficiency (DMY per unit P absorbed) of P (and other elements) (Seetharama *et al.*, 1987).

The translocation or redistribution of P in young sorghum plants grown in nutrient solution varied extensively with genotype (Clark *et al.*, 1977, 1978; Furlani *et al.*, 1984b; Raju *et al.*, 1987b). Different sorghum genotypes grown in the field also showed differences in P distribution among leaves, stalk, and grain (R. B. Clark, unpublished). Phosphorus readily accumulates in kernels of cereal plants (Williams, 1948; Sayre, 1955; Hanway, 1962; Vanderlip, 1972; Clark, 1975b, 1988; Koehler, 1976). In studies showing the distribution of P in the various plant parts with age, especially at maturity, it was noted that over half and sometimes as much as three-fourths of the plant P was in the kernels (Williams, 1948; Sayre, 1955; Clark, 1988b). Higher proportions of the total plant P were found in oat kernels if plants had been grown with limited P compared to plants grown with sufficient P (Williams, 1948). Uptake of P from the soil during grain fill readily occurred compared to many of the other elements in maize (Sayre, 1955) and sorghum (Clark, 1988).

Sixteen wheat genotypes grown in the field showed differences in partitioning of P among plant parts, but accumulation of P in the grain was not related to HI (Sherchand and Paulsen, 1985a). When three wheat isolines with tall, semidwarf, and double dwarf characteristics were grown in hydroponics to study P nutrition differences, dwarfing increased HI, but P partitioning was not affected (Sherchand and Paulsen, 1985b). Partitioning of P was related to differences in dry matter yields. Several studies have shown that P taken into wheat plants by the time of heading would be sufficient to support maximum grain yields (Boatwright and Viets, 1966; Sutton *et al.*, 1983; Batten *et al.*, 1986). Phosphorus supplied after anthesis increased grain P, but not grain yield (Batten *et al.*, 1986; Batten and Wardlaw, 1987). A high proportion of plant P in wheat grains (58%) could be attributed to retranslocation from the leaves in plants grown with low P (0.25 m*M*) compared to only small proportions of plant P in the grain (21%) from plants grown with relatively high P (1.0 m*M*) (Batten *et al.*, 1986). Temperatures at which plants were grown before anthesis affected the amount of P retranslocated from leaves to grain (Batten *et al.*, 1986). Plants grown at lower preanthesis temperatures (15/10°C, day/night) retranslocated about 60% of the vegetative P to the grain and plants grown at higher preanthesis temperatures (30/25°C, day/night) retranslocated 80% of the P to the grain. At both temperatures, 6 to 12% of the P came from the flag leaf. Phosphorus remobilization appeared to follow losses of leaf N (and leaf senescence). Phosphorus remobilization and leaf senescence differed with genotype and appeared to be associated with leaf alkaline pyrophosphatase activities and leaf senescence in wheat (Morris *et al.*, 1985) and rice (Mondal and Chaudhuri, 1985). Leaf pyrophosphatase activity in wheat leaves was associated as closely or more closely to remobilization of N as N mobilization reactions such as proteases and hydrolysis of proteins (Morris *et al.*, 1985).

Rice cultivars exhibiting both sequential (greater and rapid senescence in the third leaf compared to the flag leaf) and nonsequential (greater and rapid senescence in the flag leaf compared to the third leaf) senescence traits showed differences in P retranslocation to the grain (Mondal and Chaudhuri, 1985). Cultivars with sequential senescence had greatest P export from the second leaf during the initial grain fill stage (0 to 7 days) while the flag leaf had the greatest P export during the later grain fill stage (7 to 14 days). The flag leaf of nonsequential senescence cultivars exported more P during these same grain fill stages than the two lower leaves. Leaves treated with a senescence-delaying compound (benzyladenine) retained their P longer and exported more P to the grain than leaves treated with a senescing compound (abscisic acid).

Export of P from main (mother) to tiller (daughter) culms was low for both rice (Mondal and Chaudhuri, 1985) and maize (Russelle *et al.*, 1984). Rice genotypes readily translocated P to the tillers, and hybrids were more effective in this than parents (Yongrui *et al.*, 1987). Tiller number was highly correlated with P concentration. Maize tillers contributed considerable P to the main culm (greater than 24% of the total ^{32}P absorbed by the plant) even though tillers made up only low proportions (less than 0.3%) of the total dry matter yield of plants at the 16-leaf stage of growth (Russelle *et al.*, 1984). Total grain P from tillers depended on total grain weight and was greater when tillers did not senesce early. Phosphorus export from leaves occurred fairly early in the life cycle of barley plants and occurred earlier for plants grown with low P than with plants grown with high P (Wieneke, 1985). Phosphorus export started from even high-P leaves before leaf dry matter increases were completed, and about 50% of the P translocated into the leaf during this same interval of time was exported. Phosphorus export from relatively young leaves was not necessarily caused by senescence and appeared to occur earlier than C and N exportation. In mature plants, 80 to 90% of the total P in the culms was retranslocated from leaves to grain in barley (Wieneke *et al.*, 1983) and 75% in maize (Waldren and Flowerday, 1979).

Even as root length is important in P uptake, cereal root associations with vesicular-arbuscular mycorrhizae (VAM) may be important in extending root length to enhance P (and other nutrient) uptake without the host having to entirely provide the extended root surface (Ozanne, 1980). Possible mechanisms for greater P uptake with VAM associations are extensions of root systems to increase absorption surface areas (Gerdemann, 1968), reduction of the diffusion distance for P (Rhodes and Gerdemann, 1975), chemical modification of P sources to make the more available for uptake and transfer (Abbott and Robson, 1984), and modification of root properties for P uptake (Smith and Gianinazzi-Pearson, 1988).

Several cereal crops had enhanced growth and absorption of P from low-P soils when roots were colonized with VAM (Krishna and Bagyaraj, 1981; Raju *et al.*, 1987a). Although sorghum genotypes varied in infection, growth, and P uptake when colonized with VAM, the differences between mycorrhizal and nonmycorrhizal associations were much greater (Pacovsky *et al.*, 1986; Raju *et al.*, 1987a). With no added P, VAM associations with roots enhanced growth equivalent to the addition of 25 kg P ha^{-1} in sorghum (Raju, 1986), to 0.12 to 0.22 m*M* of added soil P for sorghum (Pacovsky *et al.*, 1986), and to 8 kg of P ha^{-1} for pearl millet (Krishna and Dart, 1984). Additions of P to the soil reduced VAM infections and growth enhancements due to VAM (Pacovsky *et al.*, 1986; Raju, 1986). Certain VAM species were more effective in colonizing and enhancing growth and P uptake than other species (Jensen, 1984; Krishna and Dart, 1984; Raju, 1986). Mycorrhizal sorghum plants had considerably longer root lengths than nonmycorrhizal plants (Raju, 1986; Sieverding, 1986). Not only was P uptake enhanced in sorghum, but K, Cu, and Zn were also enhanced (Raju, 1986). VAM-colonized sorghum and wheat roots enhanced plant ability to withstand water stress (Ellis *et al.*, 1985; Sieverding, 1986), but enhanced drought stress was not noted in maize (Hetrick *et al.*, 1984). Enhanced P uptake was considered responsible for the enhanced drought stress in sorghum (Sieverding, 1986). Ear leaf P concentration differences in maize were closely correlated to the percentage of VAM colonization (Toth *et al.*, 1984). High leaf P maize lines had higher VAM colonizations than low leaf P lines. VAM colonization appeared to be important for genetic differences in cereals for enhanced growth and P uptake when grown on low-P soils. Phosphate-dissolving bacteria were also effective in enhancing barley adaptation to high salinity (Badr El-din and Sabar, 1983).

Inheritance studies with sorghum hybrids showed that certain male parents were more tolerant to low P than other male parents, and that male parent differences were greater than those of female parents (Furlani *et al.*,1987b). Hybrids produced from P male parents that were tolerant to low P had higher dry matter yields, higher upper/lower leaf P ratios, and higher efficiency ratios (DMY per unit P absorbed) than hybrids produced from male parents that were less tolerant to low P. Hybrids from certain female parents also showed differences in tolerance to low P. Heterosis for total dry matter yield and P efficiency (DMY per unit P absorbed) was observed in most hybrids. Greater differences were noted for sorghum plants grown at a lower (75 μmol P per plant) than at a higher (150 μmol P per plant) P level. The better growth of the male parents at low P and the transfer of the trait to their hybrids indicated that dominant genes were important, although additive genes appeared to be involved in the variability of P uptake and efficiency. In the case of triticale, heterotic efforts were localized in the shoots at higher levels of P and had no relationships to uptake and transport of P (Jensen and Jonsson, 1981). Influx and transport characters in triticale were inherited mainly from wheat. The inheritance of P efficiency has been described extensively and good progress has been made for improvements in P efficiency traits with crops like *Phaseolus vulgaris* (Whiteaker *et al.*, 1976; Fawole *et al.*, 1982a,b; Schettini *et al.*, 1987), but progress has been slower in the cereal crops.

C. POTASSIUM

Another mineral nutrient added extensively in regular fertilizer applications for cereal production is K. Most of the fertilizer K (96%) is added as KCl with other forms being of minor importance (Stewart, 1985). Four crops accounted for 66% of the K added as fertilizer in the United States, and maize and wheat accounted for 47 and 5%, respectively, of this amount (Welch and Flannery, 1985). Maize accounted for over 78% of the K used in one Corn Belt state (Illinois) (Welch and Flannery, 1985). Soil-applied K is soluble in water and is readily available to roots by mass flow and diffusion if sufficient moisture is present and other conditions are appropriate. Barber (1985) indicated that diffusion would supply more K to plant roots than mass flow. It is estimated that plants absorb about 20 to 40% of the K applied in any one season (Barber, 1976). Soil factors affecting K availability to plants are many and have been discussed (Barber, 1985; McLean and Watson, 1985).

Following the studies on genotypic differences and genetic control of K concentrations in maize leaves at the Pennsylvania State University (Thomas *et al.*, 1960; Gorsline *et al.*, 1965, 1968; Baker *et al.*, 1971), genotypic variation for K uptake in other cereals, especially barley, has been reported (Pettersson, 1978; Glass and Perley, 1980; Jensen and Pettersson, 1980; Perby and Jensen, 1983; Pettersson and Jensen, 1983; Siddiqi and Glass, 1983b). Genotypic differences in K (Rb) uptake among barley cultivars were attributed to differential influx at the plasmalemma, but net uptake differences between two genotypes with extremes in uptake patterns were small (Pettersson, 1978). The barley genotype with the high influx of Rb also had a greater efflux of Rb than the genotype with the lower influx. Barley cultivars grown with a constant K and with comparable fresh and dry weights differed in K efficiency (DMY per unit K absorbed), influx and efflux of Rb, and transport of Rb to the shoot (Jensen and Pettersson, 1980). Potassium (Rb) uptake was regulated by the K status of the plant and K in the external growth media (Pettersson and Jensen, 1978), and maximum influx was regulated over a narrow interval of K concentrations in the roots (Pettersson and Jensen, 1983). Even though K influx was lower in high-K roots than in low-K roots, net K uptake and dry matter yields were not related to K influx. Maximum K influx and K_m values were of limited use in predicting plant performance at high K levels.

Barley genotypes differed in K partitioning, which changed with plant age and root temperature (Perby and Jensen, 1984, 1986). Even though K (Rb) influx was higher in roots of plants grown at 10° versus 20°C, total K influx was lower in plants grown at 10°C because of smaller roots that could not compensate for the greater surface absorption area of plants grown at 20°C. A landrace cultivar of barley performed better at low K and low temperatures than a modern cultivar (Jensen and Perby, 1986; Perby and Jensen, 1986). The K uptake parameters V_{max} and K_m increased as relative daily increases in K increased (Pettersson, 1986). Potassium concentrations remained low in roots but increased in shoots. With continuous and stable K deficiency stress, the uptake system adjusted to provide effective K at each K stress level. Efflux of K was also small, and allosteric regulation of K influx was not apparent (Jensen and Pettersson, 1978). Maize hybrids of different maturities differed in K uptake rates and also in the ability of roots to recover control of membrane permeability (Pinton *et al.*, 1987).

Additional studies on K uptake with barley showed that K (Rb) uptake strongly correlated with H^+ secretion from roots (Glass *et al.*, 1981; Romani *et al.*, 1985), and that H^+ secretion could possibly be used as a screening tool for selection of high-K-absorbing genotypes (Glass *et al.*, 1981). Other studies showed that barley genotypes differed extensively in K uptake, accumulation, and utilization if K was sufficiently high in the growth medium (Siddiqi and Glass, 1983a,b). At low K levels (1 m*M*), all genotypes had K deficiency, but when K levels were increased to 5 m*M* or higher, wide genotypic variations were noted (Siddiqi and Glass, 1983a). Among 27 barley genotypes grown at low K (5 m*M*), those genotypes with higher absolute and relative growth rates also had higher K uptake rates, higher root weight ratios, and higher K utilization efficiencies (DMY^2 per unit K absorbed). Beyond a critical K concentration in leaves, plant growth did not respond to further increases in external K. Predicted K influxes based on kinetic constants (K_m and V_{max}) and internal K concentrations corresponded well with observed fluxes for plants grown over a fairly wide range of root K concentrations (Siddiqi and Glass, 1983b, 1986). These relationships did not hold if K levels were low, however. These uptake parameters of K increased with increased K concentrations in roots, but not in shoots (Siddiqi and Glass, 1986). Growth rates and the root/shoot status of plants affected K influx indirectly, but root K concentrations affected influx directly (Siddiqi and Glass, 1987).

Barley genotypes that varied in K_m values but not in V_{max} values were compared for pyruvate kinase activities, a K^+-requiring enzyme (Memon *et al.*, 1985a). Differences for pyruvate kinase were small, and genotypic differences for K could not be explained by the activity of this enzyme.

Differences among barley genotypes were noted in the allocation of K between the cytoplasm and the vacuole (Memon *et al.*, 1985a; Memon and Glass, 1987) and in protein synthesis (Memon and Glass, 1987). A barley cultivar that showed good responses to added K had increased cytoplasmic K compared to another cultivar that did not respond to added K and had lower cytoplasmic K (Memon *et al.*, 1985b). Another cultivar responded to added K, but also had higher root and shoot K at low levels of K. The allocation of K from the vacuole to the cytoplasm was considerably lower in the genotype that responded to low K than that noted for the genotype that did not respond to low K. Nevertheless, the cultivar that responded to added K and had higher root and shoot K had higher protein synthesis (leucine incorporation) because of its increased response to K with added K than the genotypes that did not respond to added K and had low cytoplasmic K (Memon and Glass, 1987).

Barley genotypes with known differences in K uptake and utilization were grown competitively with wild oat (*Avena fatua* L.) to determine if K uptake properties of the barley genotypes were related to their competitiveness with wild oat (Siddiqi *et al.*, 1985). Competitive differences between barley genotypes related to K uptake and utilization values reported in earlier studies (Siddiqi and Glass, 1983a,b). Wild oat lines also showed differences in K uptake and utilization even as did barley cultivars (Siddiqi *et al.*, 1987a,b).

Wheat cultivars with different growth habits (tall, semidwarf, double dwarf, and triple dwarf) showed differences in K uptake and utilization (Woodend *et al.*, 1987). The variability among the wheat groups for efficiency ratio (DMY per unit K absorbed) was small within and between groups, but considerable differences in shoot weight and utilization efficiency (DMY^2 per unit K absorbed) were noted. The tall (Indian origin especially) and triple dwarf groups were superior to the semi-dwarf and double dwarf groups for net K influx and efficiency ratio. Other studies with wheat showed that as ploidy increased, higher V_{max} values were noted, but K_m values remained unchanged (Cacco *et al.*, 1976). Studies on long-distance transport of K to grain showed that K in grain was not dependent on the K nutrition of wheat plants and was proportional to the K in treatment solutions (Haeder and Beringer, 1984). The results showed that the grain itself did not control the uptake of K.

Potassium uptake was related to grain yield and associated with hybrid vigor in maize hybrids (Frick and Bauman, 1978, 1979). The basal K uptake rate was strongly conserved by S_1 progeny of crosses between high or low augmentation potentials (K uptake by submerged and aerated roots), but the augmentation potentials of the S_1 progeny tended to be nonconserved or open (Frick and Bauman, 1979). The augmentation potentials of backcross progenies tended to be in the direction of the parental inbred. Heterosis was also noted in sorghum hybrids compared to their parent lines for K absorption and transport of excised and intact roots (Nirale *et al.*, 1982).

The results on genotypic differences among barley cultivars for K showed that cultivars could be changed and improved through genetic manipulation to enhance K uptake and use. Various authors pointed out that this is and should be a feasible approach to enhance plant competition for limited amounts of K in soils and to compete with other plant species for limited K.

D. CALCIUM AND MAGNESIUM

Both Ca and Mg are normally added with lime when such is applied to soils with some acidity. Alkaline soils normally contain inherently large quantities of Ca and

Mg, which are particularly high in calcareous soils. Probably the most common liming substance is limestone, which usually contains both Ca and Mg, and dolomitic limestone contains even higher Mg than other sources of limestone. Availabilities of Ca and Mg to plant roots depend on their levels of exchangeability with other exchange sites and their levels in soil water that moves to roots by mass flow (Barber, 1984a). Availability of these mineral nutrients is not usually a problem unless soils are highly acid. Calcium and Mg are usually readily available for uptake if moisture and other conditions are not limiting.

Root absorption of Ca and Mg are somewhat similar, but Mg is usually absorbed in lower quantities than Ca and K (Kirkby and Mengel, 1976; Mengel and Kirkby, 1982). Various mechanisms of absorption of Ca and Mg have been proposed and appear to be subject to conditions in the active root surfaces, but evidence for active transport into cells has not been conclusive (Clark, 1984). Mechanisms proposed, concepts involved, and factors affecting Ca and Mg uptake and translocation in plants have been discussed previously (Clark, 1984).

Numerous studies have shown plant genotypic differences in Ca and Mg nutrition. An early study showed that one maize line required five times more Ca in the plant for normal growth as did another line (DeTurk, 1941). Sayre (1952a,b,c, 1955) in Ohio and various scientists at Pennsylvania State University (Thomas *et al.*, 1960; Gorsline *et al.*, 1961, 1964a,b, 1965, 1968; Baker *et al.*, 1964, 1967, 1971; Bradford *et al.*, 1966; Thomas and Baker, 1966; Craig and Thomas, 1970; Naismith *et al.*, 1974) studied maize inbreds and hybrids grown under controlled conditions and in the field for differential responses in growth and accumulation of Ca and Mg (and various other mineral nutrients). These studies showed that maize genotypes differed extensively in mineral nutrients and that uptake and accumulation of nutrients were genetically controlled. Studies with barley and wheat showed differences in Ca (Sr) and other mineral nutrient uptake and accumulation were also genetically controlled (Myers, 1960; Rasmusson *et al.*, 1963, 1964, 1971; Young and Rasmusson, 1966; Fick and Rasmusson, 1967; Kleese *et al.*, 1968). The differences were sufficiently large that plant selection appeared feasible and the nutrient most amenable to selection appeared to be Ca (Rasmusson *et al.*, 1971). Other studies have reported genotypic differences in Ca and Mg uptake in barley (Jensen and Perby, 1986; Perby and Jensen, 1986).

Studies designed to understand genotypic differences in Ca and Mg nutrition have shown that low Cu in leaves was associated with Ca deficiency in maize (Brown, 1965) and sorghum (Brown *et al.*, 1977). However, if Cu became too high, Fe deficiencies appeared (Brown, 1967). Maize genotypes also showed differences in Ca and Mg concentrations with different soil pH (Lutz *et al.*, 1972a), different maturities and planting dates (Bruetsch and Estes, 1976; Gallaher and Jellum, 1976a,b), and different temperatures (Porter and Moraghan, 1975). Both Ca- and Mg-efficient maize hybrids (DMY per unit Ca or Mg absorbed) were less efficient in the use of K (Gallaher and Jellum, 1976a). Maize hybrids efficient for Mg concentrations were less efficient utilizers of Mg per unit of yield (Gallaher *et al.*, 1981). Different Ca absorption properties were also noted among maize and sorghum genotypes (Kawaski and Moritsugu, 1979). Differences were also noted among maize genotypes for Mg absorption, root cation-exchange capacities, Mg utilization, and node and culm concentrations of Mg (Foy and Barber, 1958), and differences among genotypes could not be explained by node morphology or Mg–protein complexes (Schauble and Barber, 1958).

Maize inbred lines were grown on low-Ca and low-Mg soils to determine genotypic differences in Ca and Mg nutrition (Clark and Brown, 1974a). Some genotypes grew better, had higher concentrations of Ca and Mg, and had fewer deficiency symptoms than other genotypes. Inbreds showing marked differences in these traits were grown under controlled conditions to better understand reasons for these differences (Clark, 1975a, 1978a).

Oh43 was chosen to represent maize inbreds that grew better at low levels of Ca and exhibited fewer Ca deficiencies than A251, which represented the opposite (Clark, 1978a). Oh43 produced the same amount of dry matter as A251 with one-fourth of the Ca when plants were grown separately or in the same solution to compete with each other. Oh43 grew better and had higher dry matter yields over a wider range of Ca levels than did A251. Oh43 also grew better when high levels of K and Mg were added to compete with Ca absorption. A251 had as high or higher leaf Ca concentrations as did Oh43, indicating that Ca translocation was not a problem. Oh43 also had greater efficiency ratios (DMY per unit Ca absorbed) than A251 at all Ca levels tested. A251 also had higher P and greater decreases in Mn and Mg than did Oh43 when grown with varied levels of Ca.

The maize inbred B57 responded better to low Mg than did Oh40B (Clark, 1975a). Oh40B required a solution of Mg concentration that of 10 times higher than that for B57 to produce comparable amounts of dry matter at low levels of Mg. Regardless of Mg level in solution, B57 contained higher leaf and Oh40B contained higher root Mg concentrations, indicating a difference in Mg translocation between the inbreds. B57 was more Mg efficient (DMY per unit Mg absorbed) at the lower Mg levels, but Oh40B surpassed B57 at the higher levels of Mg. Dry matter top/root ratios were not affected as dramatically in B57 at low Mg as in Oh40B. Since Mg affects root growth more extensively than shoot growth in maize (Clark, 1970), Mg requirements for Oh40B appeared to be higher than for B57. Oh40B roots also contained higher concentrations of P, Mn, Zn, Fe, Cu, Mo, and K than B57. These mineral nutrients may have interacted more extensively with Mg in Oh40B than in B57. B57 was also more tolerant to Al than Oh40B (Clark, 1977).

Sorghum, and sometimes maize, grown under controlled conditions shows extensive Ca deficiency symptoms (Murtadha *et al.*, 1988). Sorghum genotypes differed in tolerance to this disorder. For example, 'Martin' was more susceptible to Ca deficiency than 'Redlan' (R. B. Clark, personal observations). Plants grown with ammonium sources of N had more severe Ca deficiency symptoms than plants grown with nitrate and other sources of N (Murtadha *et al.*, 1988). Plants grown at 60% relative humidity and at high temperatures also showed more severe Ca deficiencies than plants grown at lower (30%) and higher (90%) relative humidities and lower temperature (Murtadha, 1986). Plants grown under high light intensities and under metal halide lamps also exhibited more severe Ca deficiency symptoms than those grown at lower light intensities and under incandescent and cool white lamps. Analysis of roots and different leaf segments for mineral nutrients indicated that this disorder was indeed Ca deficiency (Murtadha *et al.*, 1988).

Calcium and Mg nutrition in maize was reported to be controlled by two or three genes (Gorsline *et al.*, 1968) with a locus controlling the nutrition of Ca and other mineral nutrients located on chromosome 9 (Barber, 1970; Naismith *et al.*, 1974). Even though the same chromosome appeared to control accumulation of many mineral nutrients, each nutrient had separate mechanisms for genetic control. Heterosis for Ca in shoots was noted in triticale (Jensen and Jonsson, 1981).

E. SULFUR

Maize hybrids exhibited wide differences in sulfate uptake efficiency (uptake per root surface area per hour) and sulfate metabolism (Ferrari and Renosto, 1972; Cacco *et al.*, 1976). Both S uptake efficiency and metabolism (ATP-sulfurylase activity) correlated well with high levels of heterosis (Cacco *et al.*, 1978). The kinetic traits V_{max} followed closely the pattern of uptake efficiency while K_m values showed low levels of heterosis. Heterosis for sulfate uptake was also noted for triticale (Jensen and Jonsson, 1981). Genotypic variability in young plants for S uptake correlated well with grain productivity of maize hybrids, but not with translocation or ATP-sulfurylase activity (Saccomani *et al.*, 1981; Landi *et al.*, 1983). Both V_{max} and K_m values for S uptake showed significant general and specific combining abilities, and both additive and nonadditive gene effects were noted (Motto *et al.*, 1982). V_{max} values for S uptake were higher than values for K uptake (Landi *et al.*, 1983). Other studies showed that a maize genotype with higher grain-yielding ability had higher differences in S uptake and *o*-acetylserine sulfhydrylase activities under S deficiency conditions than a lower-yielding genotype (Saccomani *et al.*, 1984). Sulfur metabolism appeared better adapted than N metabolism to withstand nutrient deficiency conditions. In studying maize genotypes developed over a number of years, S uptake capacity and ATP-sulfurylase activity increased during selection (Cacco *et al.*, 1983b). Uptake of S increased more than S metabolism (ATP-sulfurylase activity) over time, and close relationships between S and N metabolism were noted in maize.

F. MANGANESE

Applications of Mn as a fertilizer have not been common for most crops, but can be important when some crops are grown on some soils. Manganese deficiencies have occurred in cereals grown on soils high in pH and organic matter (Murphy and Walsh, 1972). Manganese can become toxic to cereals when grown on low-pH soils (Foy, 1984). Cereals (and large-seeded legumes) appear to be relatively susceptible to Mn deficiency (Murphy and Walsh, 1972).

Manganese deficiency can be alleviated by soil applications or foliar sprays. Probably the most common compound used for soil or foliar applications is $MnSO_4$, although several other carriers of Mn have been used (MnO_2, $MnCO_3$, Mn frits, and Mn chelates). Treatments for reducing or preventing Mn deficiencies in plants have been discussed (Murphy and Walsh, 1972).

Movement and availability of Mn in soils depends on many factors, which have been covered extensively in earlier review articles (Ellis and Knezek, 1972; Lindsay, 1972; Wilkinson, 1972). Plant roots generally have to grow to the sites where Mn is located to obtain adequate Mn for uptake. A common alternative and effective method to solve Mn deficiencies in plants is to use foliar sprays. Root uptake of Mn appears to follow a two-phase absorption process that includes a passive initial phase and a slower metabolically sustained phase (Moore, 1972). Maize genotypes showed heterotic effects for Mn V_{max} values more extensively than for K_m values (Landi and Fagioli, 1983). Once inside the root or leaf cells, Mn is readily translocated to other leaf cells or from roots to leaves as a divalent ion or as a possible complex with other ligands (Tiffin, 1972). Redistribution of Mn in cereal leaves is relatively good (Vose, 1963).

Genotypic differences in resistance to Mn deficiency were recognized in cereal crops relatively early (Gallagher and Walsh, 1943), and differences in susceptibility to Mn deficiency have been noted among various cereal crops (Gallagher and Walsh, 1943; Neenan, 1960; Nyborg, 1970; Graham *et al.*, 1983; Graham, 1984; Marcar and Graham, 1987). Oat appeared to be considerably more susceptible to Mn deficiency than wheat or barley, which appeared to be more susceptible than rye. Triticale showed more sensitivity to Mn deficiency than wheat or barley (Graham *et al.*, 1983). Manganese concentrations in leaves appeared to be lower for most cereals (except rye) than many other plants (Gladstones and Loneragan, 1970).

Resistance to Mn deficiency in barley was large and associated in many cases with germplasm from different sources (Graham *et al.*, 1983; Graham, 1984). For example, barley cultivars from various parts of the world were grown on a Mn-deficient soil in South Australia to screen them for resistance to Mn deficiency, which is so prevalent in that area (Graham *et al.*, 1983; Graham, 1984; Longnecker *et al.*, 1988). Barley genotypes varied extensively in resistance to Mn deficiency, but those most susceptible to the disorder were derived from a line that came from Egypt (Graham *et al.*, 1983; Graham, 1984). This line had other good yield qualities, but not the ability to grow with low Mn. The most resistant genotypes were derived from lines that came from England, where they were commonly grown on limestone soils known to induce Mn deficiency problems.

Resistance or susceptibility of oat to a "grey speck" disorder was associated with Mn nutrition (Vose and Griffiths, 1961). One oat genotype had high root and low leaf Mn but was resistant to the "grey speck," while another genotype had low root and high leaf Mn but was susceptible to the "grey speck." The genotype most resistant to the "grey speck" disorder appeared to have greater mobility and redistribution of Mn than the genotype susceptible to the "grey speck."

Graham (1983) noted greater disease resistance in cereal genotypes that resisted

Mn deficiency when grown at low soil Mn. Some mineral nutrients like Mn may not be translocated effectively from one portion of the root to another portion that may lack the nutrient, making this latter root (or portion of root) more susceptible to pathogenic attack (Graham, 1984). Resistance of wheat plants to "take-all" [*Gaeumannomyces graminis* var. *tritici* (Gyt)] was associated with Mn (Graham and Rovira, 1984; Wilhelm *et al.*, 1988). Manganese-deficient plants were more susceptible to the pathogen than Mn-sufficient plants (Graham and Rovira, 1984). Factors that promoted infection of "take-all" were also factors that induced Mn deficiency. Host–pathogen balances appeared to be involved with Mn nutrition. Low Mn in wheat also reduced the amount of phenolic compounds and lignin, which were related to disease resistance (Brown *et al.*, 1984). The size and number of lesions of fungal pathogens were related to the Mn content of the wheat roots. Sterility of cereal plants was also enhanced with low Mn, and this occurred without foliar symptoms of Mn deficiency on the leaves (Graham, 1984).

Seed Mn appeared adequate to supply young wheat plants (26 days old) if seeds contained sufficient Mn (Marcar and Graham, 1986). Wheat plants grown from seeds collected from different locations to provide seeds with different Mn concentrations showed differences in susceptibility to Mn deficiency depending on the amount of seed Mn. Seeds soaked with Mn before germination provided only 15–20% of the Mn recovered in the wheat seedlings. It was concluded that seeds provided a major source of Mn, and that soil in Mn during this stage of development was relatively unimportant. Seed Mn differences could cause differential ability of genotypes to grow in soils low in Mn.

Oat genotypes differed by 30 to 50% in distribution of Mn between roots and shoots and in forms of Mn within roots (Munns *et al.*, 1963a,b,c). Genotypic rankings for Mn concentration per unit dry matter were similar to genotypic rankings for Mn per unit soluble N, Mn uptake per unit root weight, and usually with Mn content per plant (Munns *et al.*, 1963a). These genotypic differences persisted despite variations in season, pH, Ca, Fe, source of N, and Mn concentration applied to the plants. However, environmental factors like high pH and low temperature did reverse genotypic rankings because of altered Mn distribution between roots and shoots. Calcium was consistently higher in the shoots of oat genotypes exhibiting fewer Mn deficiency symptoms than in genotypes exhibiting extensive Mn deficiency symptoms when grown with low Mn (Brown and Jones, 1974). It was suggested that Ca may substitute for Mn at nonspecific sites in the plant, making Mn more available for specific chemical reactions. Wheat genotypic differences in tolerance to Mn deficiency were better correlated with Mn uptake than with Mn utilization (Marcar and Graham, 1987). Increased applications of Mn to oat genotypes also increased grain yield and protein in some cases (Murray and Benson, 1976). Improved resistance to Mn deficiency was attributed to greater uptake of Mn rather than to a higher requirement for Mn inside the plant (Nyborg, 1970). Sorghum genotypes exhibited dry matter yield differences when grown at low Mn, but did not show differences in Mn uptake or distribution (R. B. Clark, unpublished).

Graham (1984) felt that cereal improvement in resistance to Mn deficiency through breeding was a feasible alternative to extensive Mn applications or worrying about whether plants will obtain sufficient Mn to avert a deficiency during the growing season. Much of his research has centered around improving cereals for mineral nutritional traits to resist some of the serious mineral deficiency problems of Australian soils. Extensive screening of barley for ability to grow at low Mn levels has shown large genotypic differences (Graham *et al.*, 1983; Longnecker *et al.*, 1988). Manganese efficiency (ability to produce well under limited Mn conditions compared to a standard genotype) in barley appeared to be controlled by relatively simple genetics (Graham *et al.*, 1983; Sparrow *et al.*, 1983). Crosses between Mn-efficient and Mn-inefficient barley lines showed F_2 distributions (based on leaf Mn concentrations) that had a favorable recombination of Mn efficiency from resistant parents (Graham, 1984). A histogram of the F_2 progenies provided evidence that considerable overdominance occurred, which was interpreted to mean that the progenies received favorable Mn efficiency from one parent and good early seedling root vigor from the other parent. It was not clear whether or not major genes controlled these genotypic responses to Mn.

G. IRON

Iron applications are common for plants grown on many alkaline, especially calcareous, soils to correct Fe deficiency chlorosis. This disorder is especially severe for many plants grown under these conditions (Clark, 1982a; Vose, 1982). Iron can also be toxic to plants grown on many low-pH soils (Foy *et al.*, 1978). Iron deficiency chlorosis is especially severe for sorghum and rice, and to a lesser extent for maize (Tanaka and Navasero, 1966; Clark, 1982a; Vose, 1982; Singh *et al.*, 1985). The other cereal crops do not have severe Fe deficiency chlorosis problems although certain soils or conditions may induce this disorder or enhance the problem. Iron deficiency chlorosis can be enhanced by many interactions of Fe with high levels of other nutrients (P, Ca, Mn, Cu, and Zn), low temperatures, prolonged and continuous waterlogging, and many other factors (Wallace and Lunt, 1960).

Iron availability and mobility in calcareous soils in low (Lindsay and Schwab, 1982; Lindsay, 1984). Soluble Fe salts added to well-aerated soils quickly dissolve and precipitate as $Fe(OH)_3$. The activity of Fe in soils is pH dependent, and the solubility of Fe^{3+} decreases 1000-fold for each unit increase in pH. Iron deficiency chlorosis does not often occur because of low soil Fe, but rather because Fe is not available to plants. Iron is the fourth most abundant element in the earth's crust (near 5%) and is unavailable to plants because of its ability to become insoluble (Whittemore *et al.*, 1981). When Fe amendments are added to soils, these are effective for a short time (usually for only one crop or season) before the Fe becomes chemically inert within a few weeks or months (Lindsay and Schwab, 1982; Lindsay, 1984). Residual effects of Fe amendments are small for crops that follow the application, and these crops must receive their own source of Fe. When Fe is spayed on foliage, multiple sprays are needed to keep plants green because Fe is a relatively immobile nutrient inside plants. Thus, Fe must be added continually. Common Fe carries for Fe amendments include $FeSO_4$, Fe oxides, and Fe chelates (Murphy and Walsh, 1972).

For plants to obtain sufficient Fe, roots must either grow to sites of available Fe (since Fe is so insoluble and unavailable to plants) or roots must make Fe available through their own metabolism. Since roots explore or make direct or extremely close contact with less than 1% of the soil in which they grow (Barber, 1976), the movement of Fe to root surfaces can be very important (Wilkinson, 1972). Iron movement from soil to root surfaces is primarily by diffusion (Wilkinson, 1972; O'Connor *et al.*, 1975). Even so, the amount of available Fe that root surfaces contact may not be sufficient to supply plant needs. Otherwise, Fe deficiency chlorosis would not be such a problem for so many plants grown on many alkaline, calcareous soils.

Biochemical and physiological reactions of plants that make Fe available for uptake in the root rhizosphere are very important. These processes are becoming recognized as the primary means whereby plants are able to make Fe available for uptake and use. Iron is one of the few elements that shows such unique chemical properties that the normal mineral nutrient amendment concepts do not particularly apply. Availability and mobilization of Fe appear to depend on plant ability to produce compounds or to change the root rhizophere accordingly, rather than depend exclusively on soil supply. This is probably why plants differ so extensively in their ability to grow on low-Fe soils or under conditions that induce Fe deficiency chlorosis.

Many of the processes that plants use to make Fe available, often called "Fe stress response," have been reviewed and described (Brown *et al.*, 1972; Brown, 1977, 1978a,b; Clark *et al.*, 1981a; Olsen *et al.*, 1981b; Bennett *et al.*, 1982; Longnecker and Welch, 1986; Marschner *et al.*, 1986; Chaney and Bell, 1987; Kannan, 1988; Longnecker, 1988). In brief, these processes include root production of H^+ and reducing compounds like phenolics, riboflavin, and phytosiderophores; increased ability of roots to reduce Fe^{3+} to Fe^{2+}; development of transfer cells and increased number and density of hairs on roots; and increases in production of citrate in root sap. These processes enhance the availability, uptake, and transport of Fe. For the most part, genotypes with greater ability to grow with low Fe or with greater resistance to withstand Fe deficiency chlorosis (Fe efficiency) have higher activities or greater potential to perform one or more of these processes. Thus, Fe efficiency in plants has often been defined by plant ability to "turn on" the Fe stress response as described by these processes. In many genotypes within a species, the ability of plants to grow with low Fe or to have greater resistance to Fe deficiency is related to plant ability to carry out these processes. Other mechanisms suggested for improved Fe efficiency in plants include lower metabolic Fe requirements; maintenance of extracellular pools of utilizable and metabolically available Fe (e.g., phytoferritin) in the apoplast or tissues; increased availability and more effective absorption and/or translocation of Fe; and increased mobilization or redistribution of Fe from older leaves (Longnecker and Welch, 1986).

The concepts of Fe absorption by roots have rapidly changed in recent years so that older ideas of cation uptake normally ascribed to Fe (Moore, 1972) may not be completely correct. For most plants, Fe is absorbed as a divalent cation (Fe^{2+}), but Fe^{3+} normally has to be separated from its binding ligand or chelate and be reduced before uptake can occur (Chaney *et al.*, 1972). Recent evidence indicates that the Fe^{3+}-phytosiderophores can be absorbed directly by roots of some plants (grasses) and the Fe^{3+} is reduced after it is inside the cell (Marschner *et al.*, 1986; Römheld and Marschner, 1986a).

Monocotyledonous plants (including cereals) have normally been considered to be less Fe efficient than dicotyledonous plants (Christ, 1974; Brown and Olsen, 1980; Olsen and Brown, 1980a,b; Landsberg, 1981), but observations in the field have not always agreed with this concept (Chaney and Bell, 1987; R. B. Clark, personal observations). Because plant families are so different, dicotyledonous plants should not necessarily be compared to monocotyledonous plants relative to Fe nutrition. Explanations for differences noted between monocotyledonous and dicotyledonous plants need more attention. Recent research has indicated that grasses (monocotyledonous plants) have a different mechanism for Fe mobilization and uptake than that of most other plants (including the dicotyledonous plants) (Marschner *et al.*, 1986; Römheld and Marschner, 1986a,b; Römheld, 1987). The roots of grasses produce substances that mobilize Fe, and these substances have been called phytosiderophores (plant-derived iron ionophores) (Takagi *et al.*, 1984). Phytosiderophore production by grasses and its mobilization of Fe have been several of the more recent and few concepts that appear to explain some of the questions and unexplained observations about cereal Fe nutrition that have not been explained by the general models used to describe Fe mobilization (Chaney and Bell, 1987). Although research on phytosiderophores is in its initial information stage, work is progressing and needs more attention to determine whether or not this is an important mechanism for grasses to mobilize and absorb Fe.

Takagi (1976) observed that naturally occurring substances that mobilized Fe were produced by oat and rice roots. These substances could chelate Fe, and Fe in this form was readily absorbed by plants. The amounts of these substances that were produced increased if plants were grown under Fe-deficient conditions compared to plants grown with sufficient Fe. These kinds of compounds have also been isolated from oat (Fushiya *et al.*, 1980), wheat (Nomoto *et al.*, 1981), rye (Nomoto and Ofune, 1982), and barley (Takemoto *et al.*, 1978; Nomoto and Ofune, 1982) by Japanese scientists. The compounds that enhanced Fe mobilization have been identified as derivatives of mugineic acid (Nomoto and Ofune, 1982) and are termed phytosiderophores because of their chemical and physiological similarities to microbial siderophores (Neilands, 1981; Winkelman, 1982). Attempts to identify these compounds in dicotyledonous plants were unsuccessful in many plant species (Takagi *et al.*, 1984; Römheld, 1987), but one compound was isolated in soybean that was termed a phytosiderophore (Porter, 1986). Römheld (1987) found phytosiderophore production to be restricted to plants in the order of Paoles (family Paoceae, which includes the grasses) in a study that included over 100 plant species. Plants that produce phytosiderophores and use this mechanism for Fe uptake have been called "Strategy II" plants, and plants that use the previously described system (production of H^+ and reducing agents, increased root reducing capacity, increased root sap citrate, development of transfer cells, and enhanced root hair development) have been called "Strategy I" plants (Marschner *et al.*, 1986; Römheld and Marschner, 1986a). Details of the unique features of plants in each group have been described (Marschner *et al.*, 1986; Römheld and Marschner, 1986a,b; Römheld, 1987; Longnecker, 1988).

Initial studies to compare grasses (cereal crops) for phytosiderophore production indicated that those Strategy II plants producing higher amounts of the compound also showed greater resistance to Fe deficiency chlorosis (Marschner *et al.*, 1986; Römheld and Marschner, 1986b). The relative order of chlorosis resistance and phytosiderophore production for some of the cereals follows the sequence of barley $>$ oat $>$ maize $>>$ sorghum $>$ rice (Marschner *et al.*, 1986). Sorghum and rice are known for their high susceptibility to Fe deficiency chlorosis (Tanaka and Navasero, 1966; Vose, 1982; Clark, 1982a; Singh *et al.*, 1985), and Fe deficiency chlorosis is not frequently reported for barley, oat, rye, and wheat. Barley genotypes appear to produce fairly high amounts of the phytosiderophores, and this species has been used as a standard to compare other plant species and to conduct phytosiderophore studies (Takagi *et al.*, 1984, 1988; Marschner *et al.*, 1986; Römheld and Marschner, 1986a,b; Mori *et al.*, 1987, 1988; Römheld, 1987; Clark *et al.*, 1988d; Jolly *et al.*, 1988; Kawai *et al.*, 1988).

Even though barley genotypes appear to produce relatively high amounts of phytosiderophores (Takagi *et al.*, 1984; Marschner *et al.*, 1986; Jolly *et al.*, 1988), information on specific genotypic differences among barley has been limited. Barley produced phytosiderophores as plants aged so long as plants were stressed with Fe deficiency (Takagi *et al.*, 1984; Marschner *et al.*, 1986; Römheld and Marschner, 1986b; Jolly *et al.*, 1988), and the production of phytosiderophores followed daily rhythmic patterns (Takagi *et al.*, 1984; Marschner *et al.*, 1986). Phytosiderophore compounds were absorbed more effectively than other sources of Fe (Takagi *et al.*, 1984; Römheld and Marschner, 1986a). The concept is that Fe^{3+}-phytosiderophore compounds are absorbed directly by the root without an obligatory reduction of Fe^{3+} to Fe^{2+} as is required for Strategy I plants (Chaney *et al.*, 1972; Chaney and Bell, 1987).

Genotypic differences in Fe-phytosiderophore uptake and phytosiderophore production were noted in sorghum (Clark *et al.*, 1988d). As long as Fe was provided to plants as Fe-phytosiderophore, sorghum genotypes readily absorbed Fe at levels comparable to that of barley. The degree of chlorosis on plant leaves (plants of different ages also) had little effect on Fe uptake, but phytosiderophore production was reduced when plants had severe Fe deficiency chlorosis. Similar results were noted for oat, and at least a small amount of Fe was needed in the growth medium for oat plants to continue production of phytosiderophore (Jolley and Brown, 1989). Sorghum genotypes had about as much or more phytosiderophore production as barley when plants were beginning to undergo chlorosis, but phytosiderophore production in barley remained fairly high as chlorosis became more severe and sorghum phytosiderophore release decreased to essentially nothing within 4 days when chlorosis became severe (Clark *et al.*, 1988d). Essentially no phytosiderophore was produced when sorghum plants were provided sufficient Fe to keep plants green. The sorghum genotype that produced the greatest amount of phytosiderophore (higher than barley at that stage of plant development) was a forage sorghum (Sudan grass). Forage sorghums have generally shown greater resistance to Fe deficiency chlorosis than most grain sorghums grown in the field (R. B. Clark, personal observation).

It is not known whether compounds other than phytosiderophores enhance Fe absorption in grasses (cereal crops) similar to phytosiderophores. It has been suggested that cereals as well as other crop plants have enhanced Fe uptake mediated by microbial siderophores (Powell *et al.*, 1982; Cline *et al.*, 1984; Reid *et al.*, 1984; Hemming, 1986). Hydroxamate siderophores benefitted sorghum Fe absorption when soluble, nonchelated Fe sources were supplied to plants (Cline *et al.*, 1984). However, synthetic or bacterial siderophores were relatively ineffective in enhancing Fe absorption by barley (Römheld and Marschner, 1986a). It was concluded that only the natural phytosiderophores were effective in enhancing Fe uptake by barley.

Iron absorption has been greater for portions of barley and wheat roots near the apex (3–4 cm) than for portions further removed (> 5 cm) from the apex (Weavind and Hodgson, 1971; Clarkson and Sanderson, 1978). Similar results have been noted for other plant species (Ambler *et al.*, 1971; Römheld and Marschner, 1984). This phenomenon has been shown visually for plants grown in agar incorporated with a reducing compound (Marschner *et al.*, 1982, 1986). Phytosiderophore production was also greater near root tips than at other portions of the root in barley (Marschner *et al.*, 1987). Two sorghum genotypes showing differences in susceptibility to Fe deficiency chlorosis were grown at different root temperatures to determine root growth and chlorosis relationships at different temperatures (R. B. Clark, unpublished). The more resistant genotype had longer root lengths at all temperatures than the susceptible genotype even though the root dry weight were higher for the susceptible genotype. Values for the amount of root length per unit of root dry weight were higher for the resistant than for the susceptible genotype, and the amount of leaf weight per unit of root length was higher for the resistant than for the susceptible genotype. Root lengths were markedly lower at a cool temperature (12°C) than at higher temperatures (up to 28°C). These results indicated that root length (number of sites and amount of surface area) may be important for plant ability to adapt to low Fe or conditions inducing Fe deficiency. Cool temperatures are known to induce Fe deficiency in most plants (Wallace and Lunt, 1960), and cool temperature effects on chlorosis could be one reason for reduced root growth, which could restrict not only chemical reactions at low temperatures, but the number of root tips that would be functional to produce Fe-solubilizing compounds like phytosiderophores.

Even though the concept of a close and probably a very important involvement of phytosiderophores with Fe uptake by the cereals has begun to emerge, other concepts for enhanced or restricted Fe mobilization have been reported in several studies with the cereal crops. These have generally been concerned with mechanisms associated with plant ability to adapt to Fe deficiency stress and to withstand Fe deficiencies when grown at low Fe. A maize genotype susceptible to Fe deficiency had a lower ability to use Fe^{3+} than a genotype more resistant to Fe deficiency (Bell *et al.*, 1958). This Fe deficiency-susceptible genotype (ys_1) has been used as a standard to compare other maize genotypes for ability to resist Fe deficiency chlorosis and to understand the mechanisms involved (Brown and Bell, 1969; Brown and Ambler, 1970; Clark and Brown, 1974c; Römheld and Marschner, 1984). More Fe deficiency-resistant maize genotypes usually had lower solution pH values, higher leaf Fe concentrations, and lower P than the Fe deficiency susceptible standard genotype when grown with low Fe or with $CaCO_3$ to induce Fe deficiency stress (Brown and Bell, 1969; Brown and Ambler, 1970). This Fe deficiency-susceptible genotype contained higher leaf P than more resistant genotypes under treatments used to induce Fe deficiency stress. This Fe deficiency-susceptible maize genotype usually reduced the nutrient solution pH more slowly and to a lesser extent (Clark and Brown, 1974c; Römheld and Marschner, 1984) and had lower ability to reduce Fe^{3+} to Fe^{2+} at the root surface (Clark and Brown, 1974c) than more Fe deficiency-resistant genotypes. No genotypic differences in excreted compounds were found (Johnson, 1982). This standard Fe deficiency-susceptible maize genotype also had higher upper leaf concentrations of P, Mn, Zn, and Ca and lower Fe and Mg when grown at varied levels of Fe (Clark and Brown, 1974c). Another Fe deficiency-susceptible maize genotype accumulated higher P and Mn than an Fe deficiency-resistant genotype, and P/Fe or Mn/Fe ratios were unbalanced for the Fe deficiency-susceptible genotype (Odurukwe and Maynard, 1969). Iron uptake and translocation were inhibited extensively in other maize genotypes when P was added to the growth medium (Elliott and Läuchli, 1985).

Like maize, sorghum genotypes showed differences in pH lowering of nutrient solutions and ability to recover from or to resist Fe deficiency stresses (Kannan, 1980a,b, 1981a,b, 1982), and auxiliary roots had a greater effect on changing pH than seminal roots (Kannan, 1981a). More Fe-reducing compounds as noted in tomato (Olsen *et al.*, 1981a) were released by roots of a sorghum genotype considered to be more resistant to Fe deficiency chlorosis than genotypes more susceptible to the disorder (McKenzie *et al.*, 1985), although caution had to be used to distinguish effects of pH and N uptake on the release of reductant (McKenzie *et al.*, 1986). Sorghum genotypes could also be distinguished for differences in Fe deficiency resistance by measuring root homogenates for *p*-coumarate hydroxylase activity of plants grown under Fe deficiency stress (McKenzie *et al.*, 1987). Other studies showed that dibutyl phthalate was produced by sorghum genotypes, and that this compound was not found when plants were not suffering from Fe deficiency stress (Kannan *et al.*, 1984). When root exudates obtained from Fe deficiency stressed plants or synthetic dibutyl phthalate were added to chlorotic sorghum plants, the chlorotic plants became green. Both dibutyl phthalate and phthalic acid were effective in regreening sorghum genotypes (Kannan, 1986; Kannan and

Ramani, 1987). In studies with treatments of Fe-deficient sorghum leaves with serine or serine + anthranilic acid, plants regreened when leaves were treated with these compounds (Kannan and Romani, 1984). Salicylic acid also allowed sorghum to recover from Fe deficiency when applied in nutrient solutions. Phenylalanine-soaked sorghum seeds appeared to produce greener plants than plants from nontreated seeds. These compounds also enhanced root reduction of nutrient solution pH.

Rice genotypes showed differences in Fe absorption (excised roots), which were believed to be due to differences in ion carriers (Shim and Vose, 1965). Iron uptake was inhibited by Mn but not by Cu. In other studies, rice genotypes did not decrease solution pH like sorghum and maize (Kannan, 1981a), but differences in root reduction (Fe^{3+} to Fe^{2+}) capacity were noted among genotypes to show resistance to Fe deficiency chlorosis (Misal and Nerkar, 1983; Reddy and Prasad, 1986). Calcium inhibited Fe uptake in rice genotypes (Pandey and Kannan, 1982), and Ca concentrations were consistently higher in an oat cultivar that was more susceptible to Fe deficiency than in a more resistant genotype (Brown and McDaniel, 1978a). Identification of reductive compounds from barley indicated that a sugar (glucose) was associated with root reducing capacity of Fe (Hether *et al.*, 1984). The compounds typical of tomato and soybean (caffeic and/or chlorogenic acids; Olsen *et al.*, 1981a) were not found in barley (Hether *et al.*, 1984).

In most plant species, Fe is transported primarily as Fe-citrate (Tiffin, 1972). Further evidence of this was noted when citrate (or isocitrate) competed against other organic acids (malate, aconitate, and acetate) when the latter were supplied at 20-fold higher concentrations than that of citrate to complex Fe (Clark *et al.*, 1973). The other organic acids could not compete with citrate or isocitrate for Fe.

The importance of Fe in chlorophyll biosynthesis (Miller *et al.*, 1984; Pushnik *et al.*, 1984), in chloroplast metabolism (Pushnik *et al.*, 1984; Terry and Abadia, 1986), in metalo-proteins (Sandmann and Boger, 1983; Smith, 1984), and in storage molecules (Seckback, 1982; Smith, 1984) has been established. Iron deficiency chlorosis is a direct and visual symptom of the lack of chlorophyll in leaves. Iron is directly involved in chlorophyll biosynthesis and insufficient or inactive Fe within the plant will have pronounced deleterious effects on photosynthesis (Miller *et al.*, 1984; Pushnik *et al.*, 1984).

Because Fe-deficient tissue often contains higher leaf Fe concentrations than green leaves, "active" forms of Fe have been sought in plants (Leeper, 1952; Brown, 1956; Wallace and Lunt, 1960). "Active Fe" has been reported to be Fe^{2+} (DeKock *et al.*, 1979; Clarkson and Hanson, 1980; Katyal and Sharma, 1980; DeKock, 1981) and is also defined as the P/Fe ratio (DeKock *et al.*, 1979). "Active Fe" has been considered to be that portion of Fe available for or participating in metabolic reactions or incorporated into molecular structures while the rest of the Fe is probably precipitated, inactivated, or stored so that plants cannot use it. Total Fe has not usually been a reliable assay for Fe deficiency problems. Katyal and Sharma (1980) and Takkar and Kaur (1984) developed techniques to determine Fe^{2+} in leaf tissue to relate Fe deficiency chlorosis to "active Fe." Good correlations between Fe^{2+} and degree of Fe deficiency chlorosis have been reported in rice (Katyal and

Sharma, 1984; Singh *et al.*, 1985), but this fraction of Fe did not give as good results as total Fe in sorghum and maize (Pierson *et al.*, 1984b). Iron values in newly emerging leaves of sorghum and maize genotypes changed daily in response to plants attempting to grow and become green when grown with Fe deficiency stresses (Pierson *et al.*, 1984b). Similar responses were noted in dry bean and soybean genotypes (Pierson *et al.*, 1984a). Genotypes more resistant to Fe deficiency chlorosis showed fewer changes in total Fe concentrations and had a greater ability to obtain Fe for metabolism than the more susceptible genotypes. Very little Fe appeared to be remobilized in the leaves of sorghum compared to dry bean, and pH did not appear to affect Fe uptake in sorghum (Pierson *et al.*, 1986). Some remobilization of Fe was reported for rice (Kannan and Pandey, 1982).

Genotypes of the various cereal crops show large differences in Fe nutrition and some of these differences have been discussed in this section. Probably more references are available in the literature on genotypic differences to Fe nutrition than for any other mineral nutrient. A compilation of references for plant genotypic differences to Fe nutrition appeared recently (Clark and Gross, 1986).

One of the earliest examples for the genetic control of Fe deficiency in cereals was reported for maize (Beadle, 1929). A recessive chlorophyll defect was noted in a yellow stripe mutant called ys_1. Later studies showed that ys_1 could not use chelated Fe (Fe^{3+}) effectively, and that the control in Fe nutrition was in uptake and not in the plant requirement for Fe (Bell *et al.*, 1958, 1962). Many differences between Fe-deficient-susceptible ys_1 and other more resistant genotypes have been discussed (this section).

Attempts to improve sorghum, long recognized for its susceptibility to Fe deficiency chlorosis (Fisher and Reyes, 1954; Brown and Jones, 1973; Mortvedt, 1975; Clark, 1982a), for resistance to Fe deficiency chlorosis have been more difficult than for soybean (Fehr, 1983; Clark *et al.*, 1990) and dry bean (Coyne *et al.*, 1982; Zaiter *et al.*, 1987, 1988). Genetic improvement for Fe deficiency chlorosis resistance in soybean has been very successful in the relatively short time this approach has been used (Fehr, 1983). The inheritance of Fe deficiency chlorosis in dry bean was determined to be a two complementary dominant gene control (Zaiter *et al.*, 1987).

An early sorghum population was developed that showed improved resistance to Fe deficiency chlorosis, but these studies were not continued after a few cycles of selection (A. J. Cassady and D. M. Rodgers, Kansas State University, personal communication). Iron deficiency chlorosis in sorghum was not a simply inherited trait (Mikesell *et al.*, 1973; Esty *et al.*, 1980). It was suggested that improved resistance to this disorder in sorghum could probably be best attained by concentrating genes to decrease the accumulation of P and other antagonizing mineral elements (Mikesell *et al.*, 1973). Other studies showed that sorghums more susceptible to Fe deficiency chlorosis accumulated higher Ca (Mushi, 1965) and P (Esty *et al.*, 1980). The inheritance of resistance to Fe deficiency chlorosis in sorghum appeared to be dominant or overdominant (Esty *et al.*, 1980).

Broad-sense heritability estimates for chlorosis readings and agronomic traits were high (0.86 ± 0.15 for chlorosis readings), as is often the case with these estimates, over two years for field-grown sorghum (Williams *et al.*, 1986; Clark *et al.*, 1988e). The studies considered 100 S_1 families from a sorghum population grown on a calcareous Fe-deficient soil with and without added Fe. Heritability estimates for many traits were relatively high (0.78 to 0.85), but some yield traits were relatively low (0.32 to 0.52). Gains in agronomic traits between the base population and the first cycle of selection were predicted to be fairly large for plants grown both with and without Fe. The relatively high heritability estimates, even though inflated, indicated that improvement for resistance to Fe deficiency chlorosis should be feasible and worthwhile. Assuming linear changes each year (which may not be the actual case), the predicted mean average and maximum chlorosis ratings could be reduced about 0.6 to 0.7 units per cycle of selection (1 = green to 5 = severe chlorosis) to be below 1.0 within four to five cycles of selection (Williams *et al.*, 1987). Assuming that the practical expectation of realized gain would only be half the theoretical predicted gain, significant improvement could be made to reduce Fe deficiency chlorosis by S_1 testing and recurrent selection. Increased grain yield and improved agronomic traits should be directly associated with Fe deficiency chlorosis because the disorder ultimately affects yield (Williams *et al.*, 1986; Clark *et al.*, 1988e). Heterosis for resistance to Fe deficiency chlorosis in sorghum was also noted (Kannan, 1981c).

H. BORON

Cereals and monocotyledonous plants show few B deficiencies and have lower requirements for B than dicotyledonous plants (Murphy and Walsh, 1972; Dugger, 1983; Pilbeam and Kirkby, 1983). Monocotyledonous plants generally require about one-fourth the amount of B inside the plant as dicotyledonous plants (Murphy and Walsh, 1972). Monocotyledonous plants can be separated into at least two groups of susceptibility to B deficiency: (1) those that show early vegetative growth reduction and final death (maize and barley) and (2) those that show deficiency in the reproductive stage (wheat, oat, and rye) (Dugger, 1983). Boron toxicities tend to be more of a problem than deficiencies in cereals (Gupta *et al.*, 1985; Riley, 1987a). Because of the low B requirement of cereals, B applications are usually not needed. However, when B is applied, soil applications of borax or sodium tetraborate are commonly used to alleviate B deficiency problems (Murphy and Walsh, 1972). Soil applications usually last longer than foliar applications. Deficiency and toxicity symptoms of B on cereals have been described (Gupta, 1979).

The uptake of B as an active or passive process by plants has still not been fully resolved (Dugger, 1983). Boron appears to be important for the function of plant membranes like P, Ca, and Rb uptake and ATPase activity (Pollard *et al.*, 1977; Pilbeam and Kirkby, 1983), and for many other metabolic processes like pollen germination, cell wall biosynthesis, auxin relationships, sugar metabolism and translocation, and nucleic acid metabolism (Dugger, 1983; Pilbeam and Kirkby,

1983). Boron appears to be readily absorbed by cereals and distributed throughout the plant with older leaves accumulating higher B then newer leaves (Dugger, 1983; Riley, 1987a). Boron appears to be relatively immobile in the phloem and deficiencies normally arise in newly developing tissue (Dugger, 1983).

Cereal crop genotypic differences have been noted for B accumulation in maize (Gorsline *et al.*, 1965; Clark, 1975a; Mozafar, 1987) and for B toxicity in barley (Christensen, 1934; Riley, 1987a). Barley appeared to be semitolerant to and more sensitive to B toxicity than wheat and oat (Riley, 1987a).

Maize appears to have a fairly high B requirement (Berger, 1949). Boron deficiency interferes with the differentiation of staminate and pistillate spikes in maize and causes barren stalks or malformed (nubbin) ears (Berger *et al.*, 1957; Mozafar, 1987). Even though soil tests showed B to be adequate in soils, maize responded to B applications to overcome a barren ear malady (Mozafar, 1987).

The effects of B toxicity in barley were primarily during vegetative growth, and older leaves showed more severe symptoms than younger leaves (Riley, 1987a). The greater accumulation of B in the older leaves enhanced senescence. Boron toxicity compensation losses in photosynthetic leaf area were by increases in height of primary tillers and area of the flag leaf. High B also enhanced lodging. Root growth was affected more than shoot growth when B was toxic. Boron toxicity symptoms occurred before decreases were noted in dry matter yield. The harvest index and grain fill were affected only at high levels of B.

I. COPPER

Copper deficiency has been associated with soils high in organic matter and with highly weathered sandy mineral soils (Kubota and Allaway, 1972; Murphy and Walsh, 1972). Copper deficiency is not as common as other micronutrient deficiencies in the United States (Kubota and Allaway, 1972), but is a common problem for wheat grown on many soils in Australia (Graham, 1981, 1984). The sensitivity of small grain cereals to Cu deficiency follows the sequence of wheat > oat > barley > rye (Piper, 1942; Smilde and Henkens, 1967; Rahimi and Bussler, 1973), and rye appears to be very tolerant to low-Cu conditions compared to wheat (Graham, 1987b; Graham *et al.*, 1981; Graham and Nambiar, 1981; Harry and Graham, 1981).

Copper applications are usually incorporated with other major fertilizers and added in many forms, but $CuSO_4$ is a common carrier (Murphy and Walsh, 1972). Copper is not usually lacking in soils, but is unavailable (tied up in mineral lattices) to plants (Graham, 1981, 1984; Loneragan, 1981). Some movement of Cu occurs in soils, but root exploration has a large influence on the amount of Cu that can move to the root surface or be taken into the plant (Wilkinson, 1972; Loneragan, 1981). The uptake of Cu by roots is generally considered an active process (Moore, 1972). In maize, Cu uptake (V_{max}) showed heterotic effects in hybrids compared to inbred parents, but K_m values were only slightly higher in hybrids compared to the parents

(Landi and Fagioli, 1983). Rice genotypes showed marked differences in uptake and affinities (K_m) for Cu (Bowen, 1987). Many other elements interfere and interact with Cu uptake (Olsen, 1972).

Genotypic differences have been noted for Cu nutrition in wheat (Smilde and Henkens, 1967; Nambiar, 1976a; Graham, 1984), oats (Rademacher, 1940; Brown and McDaniel, 1978b), maize (Brown, 1967), and sorghum (Brown *et al.*, 1977). Copper deficiency delayed maturity and reduced straw and grain yields in wheat (Smilde and Henkens, 1967; Nambiar, 1976a). Grain yields were reduced because of fewer fully developed grains (Smilde and Henkens, 1967). Even though oat and barley had reduced grain yields, they had increased straw yields. An oat genotype sensitive to Cu deficiency had higher P concentrations than a genotype tolerant to Cu deficiency (Brown and McDaniel, 1978b). The high P appeared to inactivate Ca to cause a Ca deficiency whose symptoms were typical of plants grown at low Cu. Similar results were noted for sorghum genotypes that showed differential sensitivity to Cu deficiency (Brown *et al.*, 1977). Depressed Ca has also been noted in Cu-deficient wheat (Brown, 1965), barley (Brown, 1965), and maize (Brown, 1967).

Although wheat genotypes showed large differences in leaf Cu concentrations at the mid-tillering stage, these differences were not apparent at maturity (Nambiar, 1976b). Copper concentrations in the flag leaf at maturity did not reflect the Cu status of plants, and the distribution of Cu in plants could not explain genotypic differences in resistance to Cu deficiency. However, grain accumulations of N and P were related to genotypic differences to Cu nutrition. Those genotypes with high grain protein were potentially more susceptible to Cu deficiency than genotypes with low grain protein.

Copper deficiency may cause male sterility in wheat (Graham, 1975; Graham and Pearce, 1979). In addition, when organic ligands capable of binding Cu were added to wheat, male sterility occurred and visual symptoms appearing on pollen, anthers, ears, grain, and leaves grown with these ligands were typical of Cu deficiency (Graham, 1986). Sterility in wheat plants may have been related to low carbohydrates noted in Cu-deficient plants (Graham, 1975). Without adequate Cu, wheat plants did not develop reproductive organs (Brown *et al.*, 1958; Graham, 1975). Wheat plants grown with adequate Cu accumulated reducing sugars and had greater reducing capacity in culms compared to Cu-deficient plants (Brown and Clark, 1977). Copper-deficient wheat plants appeared to have insufficient energy to perform needed functions. Copper deficiency also caused concentrations of aspartic acid, alanine, and serine to be higher and aminobutyric acid to be lower in wheat leaves. Nitrate-N, P, and K concentrations were higher and Ca and Cu lower in Cu-deficient wheat leaves compared to Cu-sufficient leaves. Other physiological and biochemical functions of Cu have been reviewed (Bussler, 1981; Walker and Webb, 1981).

Rye was considerably more resistant to Cu deficiency when grown on low-Cu soil conditions than wheat and triticale, and triticale showed an intermediate response (Graham, 1978b, 1984; Graham *et al.*, 1981; Harry and Graham, 1981).

Rye had a more extensive root system (longer root length) than wheat even though root dry matter yields of wheat were high (Graham, 1978b; Graham *et al.*, 1981). Rye roots were smaller in diameter and branching was more extensive than that noted for wheat (Graham *et al.*, 1981). Rye also had higher Cu concentrations in the leaves than wheat when grown under similar conditions. Rye had higher absorption rates of Cu and translocated more Cu from roots to shoots than wheat. Triticale had intermediate responses to Cu relative to rye and wheat. Copper concentrations in leaves of rye, wheat, and triticale depended on soil pH (Harry and Graham, 1981). Wheat, which is sensitive to Cu deficiency, had lower Cu concentrations in leaves than rye, which is tolerant to Cu deficiency, when grown in soil at varied pH values. Triticale was intermediate to wheat and rye in leaf Cu concentrations, but had low leaf concentrations at high pH like that of wheat and high concentrations at low pH like that of rye.

The beneficial Cu nutrition properties of rye (Cu efficiency) could be transferred to triticale, which were also transferred to wheat (Graham, 1978a,b). Even though rye and hexaploid triticale had good tolerance to Cu deficiency, octoploid triticale had lower tolerance to Cu deficiency than hexaploid triticale (Graham, 1978b). In other studies on the genetic transfer of Cu efficiency, it was noted that the Cu efficiency trait in rye was carried on the long arm of chromosome 5 (5RL) and could be transferred to wheat (Graham, 1984; Graham *et al.*, 1987). The presence of this chromosome in four wheat genotypes increased yield by an average of 100% when grown on a Cu-deficient soil (Graham *et al.*, 1987). These wheat plants also had greater culm extension, slightly higher leaf Cu concentrations, and greater Cu uptake in proportion to yield. This was the first time that a micronutrient trait had been introduced from alien chromatin.

These results showed that the concept of breeding cereals for greater ability to grow on low-Cu (and other micronutrient) soils can be accomplished using genetic approaches. Superior Cu efficiency characters from rye were transferred to Cu-inefficient wheat to make wheat more tolerant to Cu deficiency. The Cu efficiency in triticale was inherited from rye. Copper efficiency was a dominant trait and probably controlled by a single gene (Graham, 1984). The Cu efficiency in rye was not associated with Mn or Zn efficiency (ability to grow well in Mn- and Zn-deficient soils). The Cu efficiency in rye was not because of a lower shoot requirement for Cu, but because of a greater uptake and translocation of Cu from roots to shoots.

J. ZINC

Zinc deficiency in cereals has been recognized for many years. Maize and sorghum have been reported to have relatively low resistance and oat, rye, and wheat to have high resistance to Zn deficiency (Lucas and Knezek, 1972). Barley and rice have been reported to be intermediate to these other cereals in resistance to Zn deficiency. Zinc deficiency has been a common mineral nutrient problem in rice and genotypes have been screened extensively for this disorder at the International

Rice Research Institute (IRRI, 1976, 1977). Maize has normally shown greater susceptibility to Zn deficiency than sorghum (R. B. Clark, personal observations; Shukla and Raj, 1987) and pearl millet (Shukla and Raj, 1987). Conditions associated with Zn deficiency have varied, but usually are found on calcareous, alkaline, low organic matter, cool, disturbed (leveled for irrigation), compacted, and high-P soils (Murphy and Walsh, 1972).

Various sources of Zn are available, but $ZnSO_4$ is probably the most common carrier for applications either to soils or as a foliar spray (Murphy and Walsh, 1972). Soil applications are more common than foliar sprays for cereal crops and have a fairly long residual effect on subsequent crops. Root extension throughout the soil, especially in the surface soil, greatly influences the amount of Zn taken into the plant. Uptake of Zn is considered an active process (Moore, 1972). Zinc interacts with many other mineral nutrients and the Zn–P interaction is among the most well-known mineral interaction (Olsen, 1972). Ratios of P/Zn or P-induced Zn deficiency problems are commonly reported. Zinc is translocated in the xylem as a cation and unlike Cu and Fe does not bind extensively with organic ligands (Tiffin, 1972).

Genotypic differences in tolerance to Zn deficiency have been reported for maize (Massey and Loeffel, 1967; Halim *et al.*, 1968; Giordano and Mortvedt, 1969; Clark and Brown, 1974a; Terman *et al.*, 1975; Randhawa and Takkar, 1976; Shukla and Raj, 1976; Clark, 1978b; Peaslee *et al.*, 1981; Kannan, 1983; Ramani and Kannan, 1985a,b), sorghum (Shukla *et al.*, 1973; Randhawa and Takkar, 1976; Brown and Jones, 1977; Kannan and Ramani, 1982), rice (IRRI, 1976, 1977; Ponnamperuma, 1976; Randhawa and Takkar, 1976; Mahadevappa *et al.*, 1981; Bowen, 1986, 1987), wheat (Sharma *et al.*, 1971; Shukla and Raj, 1974; Randhawa and Takkar, 1976), barley (Randhawa and Takkar, 1976), and oat (Brown and McDaniel, 1978b). Reasons for such differences are not fully understood, but most reports show that they are related to root ability to extract, to make greater contact (root extensiveness) with Zn, or to increase Zn availability in the soil (Shukla *et al.*, 1973; Shukla and Raj, 1974, 1976), but not necessarily to decreases in concentrations of Zn within the plants (Halim *et al.*, 1968; Peaslee *et al.*, 1981).

Rice genotypes showed differences in Zn absorption that were not attributed to differences in root surface area (Bowen, 1986, 1987). The rice genotypes studied showed marked differences in V_{max} and K_m values. The genotypes varied by two-fold for Zn affinity. Leaf absorption of Zn by rice genotypes with resistance and susceptibility to Zn deficiency were not different. Genotypes more resistant to Zn deficiency had higher Zn uptake rates and lower affinity for Zn than genotypes more susceptible to Zn deficiency. Genotypic differences in Zn uptake rates have been reported for other cereal crops (Shulka *et al.*, 1973; Ramani and Kannan, 1985a,b). However, Zn uptake could not explain genotypic differences in some studies (Massey and Loeffel, 1967; Peaslee *et al.*, 1981). Some factors like high P affected Zn concentrations in shoots, but not Zn uptake by roots (Halim *et al.*, 1968; Warnock, 1970). Imbalances of Zn with other elements (high Cu, P, Mn, and Fe) decreased the concentration of Zn in plants and often induced Zn deficiency (Halim *et al.*, 1968; Warnock, 1970; Olsen, 1972; Shukla and Raj, 1976, 1987; Brown and McDaniel, 1978b; Clark, 1978b).

From a screening study of maize inbreds grown in soil known to induce Zn deficiency (Clark and Brown, 1974a), two genotypes that showed resistance and susceptibility to Zn deficiency were chosen to study genotypic differences in Zn nutrition (Clark, 1978b). The genotype more susceptible to Zn deficiency had higher concentrations of P, Ca, Mg, Mn, Fe, and Cu, showed greater changes in the mineral nutrients as Zn in nutrient solution varied, and had more imbalances of the minerals relative to Zn than did the genotype more resistant to Zn deficiency. The more Zn deficiency-resistant genotype had higher dry matter yields, higher shoot/root ratios of dry matter, higher Zn concentrations and contents (total Zn), and higher dry matter yield per unit Zn absorbed (more Zn efficient), translocated more Zn from roots to shoots, and developed fewer Zn deficiency symptoms than the more Zn deficiency-susceptible genotype. The genotype more susceptible to Zn deficiency responded more to added Zn than the genotype more resistant to Zn deficiency. The genotype more resistant to Zn deficiency appeared to have a lower Zn requirement inside the plant than the genotype more susceptible to Zn deficiency. Other studies showed that this more resistant Zn deficiency genotype had a higher short-term (excised root) Zn absorption rate than the more susceptible genotype, but total Zn absorption could not explain differences between the genotypes (Kimball, 1978).

Maize hybrids differing in ability to grow with low Zn showed no differences in Zn absorption by roots or differences in translocation from roots to shoots (Peaslee *et al.*, 1981). The hybrid that produced the most dry matter at low Zn also produced more dry matter per unit Zn absorbed (more Zn efficient) than the other hybrid. When the hybrids were grown at low Zn, less ^{65}Zn was found in tissues than when Zn was supplied in adequate amounts. Zinc translocation was greater in the younger leaves than in the older leaves, and the more Zn-inefficient hybrid translocated more Zn than the Zn-efficient hybrid.

Understanding mechanisms for differential responses among cereal genotypes has been an objective of research in the All-India Coordinated Micronutrient Program (Randhawa and Takkar, 1976). Studies were conducted to understand Zn uptake rates, absorption differences at high and low levels of Zn, K_m and V_{max} values for genotypes, responses of genotypes to Zn levels, mineral nutrient imbalances with Zn (P/Zn, Mn/Zn, and Fe/Zn ratios), and Zn-requiring enzyme (carbonic anhydrase) activities with Zn concentration or Zn differences in genotypes. High correlations were noted between Zn concentration in leaves and Zn applications to soil ($r = 0.93$) and between nutrient/Zn ratios and plant Zn concentrations ($r = 0.96$). Differences among genotypes appeared to be due mainly to variations in Zn concentrations in plants.

Zinc concentrations in leaves of field-grown maize were related to Zn in the seed, but not in greenhouse-grown maize (Massey and Loeffel, 1967). Maize inbreds also showed large differences in total kernel Zn and in Zn associated with various components of the kernel. Considerable amounts of the total plant Zn have been found in maize kernels (Massey and Loeffel, 1967; Clark, 1975b). As plants aged, about one-third of the plant Zn was found in kernels and one-fourth of the kernel Zn came from the stalk and leaves during the time of tasseling to maturity (Massey and

Loeffel, 1967). Early resistance of maize genotypes to Zn deficiency was attributed to higher seed Zn (Halim *et al.*, 1968). Zinc decreased rapidly in maize kernel endosperms, but did not increase in the scutellum, root, or shoot of maize seedlings (Ramani and Kannan, 1985b). Maize genotypic differences in tolerance to Zn deficiency were not attributed to seed Zn. Zinc mobility in young maize plants appeared to be low. Genotypes more resistant to Zn deficiency had healthier roots than genotypes more susceptible (decayed root tips) to Zn deficiency (Halim *et al.*, 1968).

Resistance to Zn deficiency appeared to be controlled polygenetically in rice and seemed to be a dominant trait (Mahadevappa *et al.*, 1981). Heritability of resistance to Zn deficiency varied. Heterosis for resistance to Zn deficiency was noted in hybrids compared to the parents in sorghum (Kannan and Ramani, 1982) and maize (Kannan, 1983). Zinc deficiency was not related to plant ability to produce H^+ (lower solution pH). Transport of Zn was also higher in maize hybrids than in their parents (Ramani and Kannan, 1985a). Heterosis for Zn absorption and translocation appeared to be low. Absorption and translocation of Zn appeared to be inherited more from the female parent than from the male parent.

K. MOLYBDENUM

Deficiencies of Mo are extensive in legumes and *Brassica* sp. plants (Lucas and Knezek, 1972; Clark, 1984), but have also been reported in some cereals, especially maize (Clark, 1984). Molybdenum deficiencies have been reported throughout the world, particularly for cereals grown in Australia (Kubota and Allaway, 1972; Murphy and Walsh, 1972; Riley, 1987b). Molybdenum deficiencies can occur extensively on plants grown on acid soils and on soils high in Fe and Al oxides (Lucas and Knezek, 1972). Cereals have low Mo requirements compared to legumes (Lucas and Knezek, 1972; Murphy and Walsh, 1972). The residual effect of Mo is usually fairly long in subsequent crops. Several sources of Mo are available, but sodium molybdate is a common source added to soils (Murphy and Walsh, 1972). Seed treatments with Mo have been effective means for preventing Mo deficiencies. Increasing the soil pH is another effective method to increase Mo availability, especially if the soil is acid. For each unit increase in pH, Mo availability increases 100-fold (Lindsay, 1972).

Like most micronutrients, Mo supplies to roots are greater if roots grow to the site where the mineral is located in the soil, since movement of Mo is relatively low (Wilkinson, 1972). Evidence for active uptake of Mo by roots has been limited (Moore, 1972). Movement of Mo inside plants through the xylem is most likely as an anion, but Mo complexation with other organic molecules cannot be ruled out (Tiffin, 1972). The function of Mo in cereal plants is primarily in the enzyme nitrate reductase (Clark, 1984).

Genotypic differences of cereals to Mo deficiency have been reported for maize (Noonan, 1953; Dios and Broyer, 1965; Wier and Hudson, 1966; Lutz *et al.*, 1972b; Brown and Clark, 1974) and would likely be found in other cereals like wheat, since wheat responded extensively to added Mo when grown on Mo-deficient soils in western Australia (Riley, 1987b). Maize genotypes of various origins showed differences in resistance to Mo deficiency, but genotypes of American origin appeared to be more susceptible to Mo deficiency than those of Spanish origin (Dios and Broyer, 1965). These differences could not be attributed to differences in seed Mo. However, seed Mo did determine whether maize was susceptible or resistant to Mo deficiency in other studies (Wier and Hudson, 1966). Some maize genotypes had consistently lower leaf Mo when grown on a specific soil at varied pH levels (Lutz *et al.*, 1972b).

A study was conducted where numerous maize inbreds were grown on various soils to induce mineral nutrient deficiencies (Clark and Brown, 1974a), and genotype differences for Mo deficiency were noted when the inbreds were grown on an acid soil (Brown and Clark, 1974). One inbred would not grow on this acid soil even though it was P efficient (Clark and Brown, 1974b) and tolerant to Al (Clark, 1977). Chemical analyses of leaf material consistently showed that the genotype that had "whiptail" symptoms was lower in Mo than the other genotype. Once Mo was added to the soil, the disorder was alleviated and diagnosed as Mo deficiency (Brown and Clark, 1974). Other chemical analysis of the leaves showed that the Mo deficiency-susceptible genotype had higher nitrate than the genotype that grew normally. Leaf nitrate concentrations decreased when Mo was added. High leaf nitrate has been noted for plants that show Mo deficiency (Mulder, 1948).

Added nitrate, but not increased light intensity, enhanced the accumulation/development of a Mo cofactor (Mo-bound enzyme as a molybdopterin moiety) to its highest level in maize leaves (Campbell *et al.*, 1987). On the other hand, nitrate reductase activity was enhanced by the addition of both nitrate and increased light intensity.

The absorption of Mo by roots and its transport to the shoots showed heterotic effects in sorghum hybrids compared to their parents grown at varied levels of Mo (Ramani and Kannan, 1986). Hybrid vigor was noted when the effects of Mo level on nitrate reductase activity were studied. The heterosis in the hybrids followed inheritance patterns more closely related to the male parents than to the female parents.

L. OTHER BENEFICIAL MINERAL NUTRIENTS

Several mineral nutrients have shown beneficial effects on the growth of cereals, but have not yet been proven to be essential to their growth and development. These include mineral nutrients like Si, Al, Ni, and possibly others. Information on the essentiality and physiology of many of these nutrients in higher plants has been reviewed (Bollard, 1983; Werner and Roth, 1983).

1. Silicon

Rice growth benefits from Si, and Si may be essential to the growth of this cereal (Werner and Roth, 1983). Deficiency symptoms of Si have also been described in rice. The absence of Si in the growth medium reduced rice growth rate, but did not cause growth to cease completely (Okuda and Takahashi, 1965). Transpiration rates were lower in Si-deficient compared to Si-sufficient rice plants (Yoshida *et al.*, 1959; Mitsui and Takatoh, 1963). Enhanced growth of rice and barley with Si was attributed to detoxification of high Fe in plants (Okuda and Takahashi, 1965). Silicon reportedly detoxified excess Mn by causing a more uniform distribution of Mn within bean (Horst and Marschner, 1978) and barley (Horiguchi and Morita, 1987) plants.

Sorghum grown on an acid ultisol in Colombia, South America, at 60% Al saturation had higher Si in genotypes more tolerant than in genotypes more susceptible to the acid conditions (Clark *et al.*, 1988b). Whenever lime was added to alleviate or reduce acid soil stresses (i.e., change from 60% to 40% Al saturation), sorghum leaves consistently showed greater increases in Si than any other mineral nutrient (Clark and Gourley, 1987, 1988; Clark *et al.*, 1988b). Even though genotypes increased in Ca as lime was added, greater increases in Si than Ca were noted (Clark and Gourley, 1987, 1988). Pearl millet, which is highly tolerant to acid soil stresses and more tolerant than sorghum (Flores *et al.*, 1988), had higher concentrations of Si than sorghum when grown on acid soils (Clark *et al.*, 1988c). All pearl millet genotypes had higher Si concentrations than acid soil-tolerant sorghum. Widely prevailing acid soil stresses include toxicities of Al, Mn, and Fe (Foy *et al.*, 1978; Foy, 1984). Silicon may benefit plant growth by detoxifying excess Al, Mn, and Fe in soils or solutions and/or by complexing and detoxifying these elements within plants.

Laboratory results showed that Si could alleviate Mn and Al toxicity effects in sorghum, and genotypes showed differences in their responses to these toxic nutrients and to Si (Galvez *et al.*, 1987, 1989; Galvez and Clark, 1988). Even without toxic levels of Mn, Si benefited dry matter yields, but the Si effects were greater if Mn and Al were at toxic levels (Galvez *et al.*, 1987; Galvez and Clark, 1988). Silicon generally had greater effects on shoots than roots. If Mn or Al was at extremely high levels, Si was unable to overcome the toxicity effects. If Si was added at sufficiently high levels, it too became toxic (L. Galvez and R. B. Clark, personal observations).

2. Aluminum

Although Al has been considered to be a toxic element for plant growth, under conditions where Al is low, beneficial effects have been noted (Foy *et al.*, 1978; Foy, 1984). Beneficial effects of Al have been noted in maize (Clark, 1977), rice (Howeler and Cadavid, 1976), wheat (Foy and Fleming, 1978), and sorghum (R. B. Clark, unpublished). Where beneficial effects of Al have been noted, genotypes more tolerant to Al have usually shown greater effects compared to genotypes more susceptible to Al toxicity (Clark, 1977; Foy and Fleming, 1978). Mechanisms for beneficial effects of Al have been described (Foy, 1984) and include increasing solubility and availability of Fe in calcareous or high pH conditions, correcting and preventing Fe deficiency by displacing bound Fe, blocking negatively charged sites in cell walls to promote P uptake, correcting and preventing excess P uptake (P toxicity), delaying root deterioration of plants growing with low Ca by slowing growth to prevent depletion of Ca, altering the distribution of growth regulators, preventing Cu and Mn toxicities, acting as a fungicide, and reducing undesirable shoot growth in N-rich plants. It has been suggested that Al prevents excess P uptake and helps alleviate potential P toxicity problems in maize (Clark, 1977) and sorghum. Phosphate levels as low as 1 to 3 μM can induce "red-speckling" in sorghum, which is a symptom of P toxicity (Furlani *et al.*, 1986a), and sorghum genotypes differ in the induction of red-speckling due to P (Furlani *et al.*, 1986b). Aluminum toxicity effects have been associated with restricted P uptake and distribution (Bollard, 1983).

3. Nickel

Beneficial effects of Ni on plant growth have been reported (Mishra and Kar, 1974; Bollard, 1983), and the essentiality of Ni for some higher plants is beginning to emerge (Eskew *et al.*, 1983, 1984; Brown *et al.*, 1987). Germination of rice seeds was enhanced with Ni and activities of some enzymes were higher in rice seedlings grown with Ni (Das *et al.*, 1978). Nickel was shown to be essential to soybean growth and possibly for other higher plants (Eskew *et al.*, 1983, 1984), and marked beneficial effects were noted in wheat, barley, and oat grown with Ni (Brown *et al.*, 1987). Proton extrusion was also enhanced by maize roots treated or grown with Ni (Morgutti *et al.*, 1984; Cocucci and Morgutti, 1986).

Wheat, barley, and oat grown without Ni showed symptoms resembling a complex of Fe, Cu, Zn, and Mn deficiencies (Brown *et al.*, 1987). These symptoms appeared in the youngest leaves, indicating that Ni is immobile in the phloem. Iron was positively correlated with Ni in plant tissues. Nickel toxicities have also been described as Fe deficiencies in plants (Crooke *et al.*, 1954; Crooke, 1955), and these symptoms are typical in leaves of cereals (Agarwala *et al.*, 1977; Clark *et al.*, 1981b). Reduced shoot and root dry matter yields occurred in barley plants grown with low seed Ni compared to plants grown with higher seed Ni (Brown *et al.*, 1987). Characteristic deficiency symptoms on oat plants grown without Ni could be alleviated by adding Ni. Oat plants matured earlier and had earlier leaf senescence when grown without than when grown with Ni.

A common trait of Ni-deficient plants has been the accumulation of urea in tissues (Walker *et al.*, 1985). Urea at concentrations 15- to 20-fold higher were found in wheat, barley, and oat leaves and roots of plants grown without Ni compared to plants grown with Ni (Brown *et al.*, 1987). The only identified role of Ni in higher plants has been in the enzyme urease (Welch, 1981), and Ni-deficient plants might be expected to accumulate urea. These results and other associated information have led to the conclusion that Ni may be essential to higher plants, including the cereals.

4. Other Elements

Beneficial effects of elements like V, Se, and Co have also been reported for higher plants, but information is relatively limited for their effects on cereals. Vanadium enhanced plant height and leaf area in maize (Singh, 1971), but essentiality for V has not been established for higher plants (Bollard, 1983). Vanadium has been shown to substitute for Mo to some extent in many higher plant reactions. Selenium accumulates in many grasses, but reports of Se in the cereals are lacking except as a toxic element (Bollard, 1983). Selenium can substitute for S in many reactions in plants. Cobalt, which is required by N_2-fixing symbionts with legumes, stimulates coleoptile elongation of oat (Thimann, 1956), but the essentiality of Co in cereals has not been established (Bollard, 1983).

IV. Summary

The term "mineral nutrient efficiency" has been used and defined in many ways. Hence, any use of the term needs to be accompanied by an appropriate definition. Cereal genotypes vary extensively in responses to mineral nutrients, and taking advantage of desirable mineral nutrition traits to improve plant growth and production under mineral stress conditions has not been fully utilized. Considerably more could be done to take advantage of genotypic differences to help reduce inputs, help plants utilize marginal or mineral stressed land, or utilize inputs added to soils. Many of the micronutrients in soils are present in sufficient quantitites to supply plants for many years without amendments, but present genotypes used do not have appropriate ability to use or adapt to these apparent deficient conditions. Taking advantage of the plant instead of relying exclusively on soil amendments has great potential. Over the past years, scientists have changed soils to meet plant needs, and this does not always need to be done nor is it economically advisable. Resources, costs, and environmental concerns may dictate that alternate approaches be considered. Taking advantage of genotypic differences in tolerance/resistance to mineral nutrient deficiency stresses and improving plants for desired traits appears to be feasible and practical for many mineral nutrients. Plant genotypes can be important in determining efficient use of mineral nutrients and in adapting to mineral stress or marginal nutrient conditions.

References

Abbott, L. K., and A. D. Robson. 1984. The effect of VA mycorrhizae on plant growth. In C. L. Powell and D. J. Bagyaraj (eds.), VA Mycorrhizae. CRC Press, Boca Raton, Florida. Pp. 113–130.

Agarwala, S. C., S. S. Bisht, and C. P. Sharma. 1977. Relative effectiveness of certain heavy metals in producing toxicity and symptoms of iron deficiency in barley. *Can. J. Bot.* 55:1299–1307.

Agren, G. I. 1985. Theory for growth of plants derived from the nitrogen productivity concept. *Physiol. Plant.* 64:17–28.

Alagarswamy, G., and F. R. Bidinger. 1982. Nitrogen uptake and utilization by pearl millet [*Pennisetum americanum* (L.) Leeke]. In A. Scaife (ed.), Plant Nutrition 1982. Proc. 9th Int. Plant Nutrition Colloq., Warwick, England. Commonw. Agric. Bur., Slough, England. Pp. 12–16.

Alagarswamy, G., and F.R. Bidinger. 1987. Genotypic variation in biomass production and nitrogen use efficiency in pearl millet [*Pennisetum americanum* (L.) Leeke]. In W. H. Gabelman and B. C. Loughman (eds.), Genetic Aspects of Plant Mineral Nutrition. Nijhoff, Dordrecht, Netherlands. Pp. 281–286.

Alagarswamy, G., J. C. Gardner, J. W. Maranville, and R. B. Clark. 1988. Measurement of instantaneous nitrogen use efficiency among pearl millet genotypes. *Crop Sci.* 28:681–685.

Allison, F. E. 1955. The ensigma of soil nitrogen balance sheets. *Adv. Agron.* 7:213–250.

Allison, F. E. 1966. The fate of nitrogen applied to soils. *Adv. Agron.* 18:219–258.

Alvarado, L. R. 1985. Gene effects controlling grain yield and nitrogen use efficiency traits in maize. Ph.D. Thesis, Univ. of Nebraska, Lincoln. (*Diss. Abstr. B* 46:3268B–3269B, 1986.)

Ambler, J. E., J. C. Brown, and H. G. Gauch. 1971. Sites of iron reduction in soybean plants. *Agron. J.* 63:95–97.

Anderson, E. L., E. J. Kamprath, and R. H. Moll. 1984a. Nitrogen fertility effects on accumulation, remobilization and partitioning of N and dry matter in corn genotypes differing in prolificacy. *Agron. J.* 76:397–404.

Anderson, E. L., E. J. Kamprath, R. H. Moll, and W. A. Jackson. 1984b. Effect of N fertilization on silk synchrony, ear number, and growth of semiprolific maize genotypes. *Crop Sci.* 24:663–666.

Anderson, E. L., E. J. Kamprath, and R. H. Moll. 1985. Prolificacy and N fertilizer effects on yield and N utilization in maize. *Crop Sci.* 25:598–602.

Anghinoni, I., and S. A. Barber. 1980. Phosphorus influx and growth characteristics of corn roots as influenced by phosphorus supply. *Agron. J.* 72:685–688.

Asher, C. J., and J. F. Loneragan. 1967. Response of plants to phosphate concentration in solution culture. I. Growth and phosphorus content. *Soil Sci.* 103:225–233.

Ashley, D. A., W. A. Jackson, and R. J. Volk. 1975. Nitrate uptake and assimilation by wheat seedlings during initial exposure to nitrate. *Plant Physiol.* 55:1102–1106.

Badr El-din, S. M. S., and M. S. M. Saber. 1983. Effect of phosphate dissolving bacteria on P-uptake by barley plants grown in a salt affected calcareous soil. *Z. Pflanzenernaehr. Bodenkd.* 146:545–550.

Baker, D. E., R. R. Bradford, and W. I. Thomas. 1967. Accumulation of Ca, Sr, Mg, P, and Zn by genotypes of corn (*Zea mays* L.) under different soil fertility levels. *Isotopes in Plant Nutrition and Physiology. Proc. Symp. IAEA, Vienna, 1966* pp. 465–477.

Baker, D. E., A. E. Jarrell, L. E. Marshall, and W. I. Thomas. 1970. Phosphorus uptake from soils by corn hybrids selected for high and low phosphorus accumulation. *Agron. J.* 62:103–106.

Baker, D. E., W. I. Thomas, and G. W. Gorsline. 1964. Differential accumulations of strontium, calcium and other elements by corn (*Zea mays* L.) under greenhouse and field conditions. *Agron. J.* 56:352–355.

Baker, D. E., F. J. Wooding, and M. W. Johnson. 1971. Chemical element accumulation by populations of corn (*Zea mays* L.) selected for high and low accumulation of P. *Agron. J.* 63:404–406.

Baligar, V. C., and S. A. Barber. 1979. Genotypic differences of corn for ion uptake. *Agron. J.* 71:870–883.

Barber, S. A. 1976. Efficient fertilizer use. In F. L. Patterson (ed.), Agronomic Research for Food. Spec. Publ. No. 26. ASA, Madison, Wisconsin. Pp. 13–29.

Barber, S. A. 1980. Soil–plant interactions in the phosphorus nutrition of plants. In F. E. Khasawneh, E. C. Sample, and E. J. Kamprath (eds.), The Role of Phosphorus in Agriculture. ASA, CSSA, and SSSA, Madison, Wisconsin. Pp. 591–615.

Barber, S. A. 1984a. Liming materials and practices. In F. Adams (ed.), Soil Acidity and Liming. ASA, CSSA, and SSSA, Madison, Wisconsin. Pp. 171–209.

Barber, S. A. 1984b. Nutrient balance and nitrogen use. In R. D. Hauck (ed.), Nitrogen in Crop Production. ASA, CSSA, and SSSA, Madison, Wisconsin. Pp. 87–95.

Barber, S. A. 1985. Pottassium availability at the soil–root interface and factors influencing potassium uptake. In R. D. Munson (ed.), Potassium in Agriculture. ASA, CSSA, and SSSA, Madison, Wisconsin. Pp. 309–324.

Barber, S. A., J. M. Walker, and E. H. Vasey. 1963. Mechanisms for the movement of plant nutrients from the soil and fertilizer to the plant root. *J. Agric. Food Chem.* 11:204–207.

Barber, W. D. 1970. An investigation of the genetic control of accumulation of phosphorus and other elements by corn leaves using chromosomal translocations. Ph.D. Thesis, Pennsylvania State Univ., College Park. (*Diss. Abstr. B* 30:3033B–3034B.)

Barlow, E. W. R., G. R. Donovan, and J. W. Lee. 1983. Water relations and composition of wheat ears grown in liquid culture: Effect of carbon and nitrogen. *Aust. J. Plant Physiol.* 10:99–108.

Barrett-Lennard, E. G., A. D. Robson, and H. Greenway. 1982. Effect of phosphorus deficiency and water deficit on phosphatase activities from wheat leaves. *J. Exp. Bot.* 33:682–693.

Batten, G. D., and I. F. Wardlaw. 1987. Senescence of the flag leaf and grain yield following late foliar and root applications of phosphate on plants of differing phosphorus status. *J. Plant Nutr.* 10:735–748.

Batten, G. D., I. F. Wardlaw, and M. J. Aston. 1986. Growth and the distribution of phosphorus in wheat developed under various phosphorus and temperature regimes. *Aust. J. Agric. Res.* 37:459–469.

Beadle, G. W. 1929. Yellow stripe: A factor for chlorophyll deficiency in maize located in the Pr pr chromosome. *Am. Nat.* 63:189–192.

Beauchamp, E. G., L. W. Kannenberg, and R. B. Hunter. 1976. Nitrogen accumulation and translocation in corn genotypes following silking. *Agron. J.* 68:418–422.

Beevers, L., and R. H. Hageman. 1969. Nitrate reduction in higher plants. *Annu. Rev. Plant Physiol.* 20:495–522.

Beevers, L., and R. H. Hageman. 1983. Uptake and reduction of nitrate: Bacteria and higher plants. In A. Läuchli and R. L. Bieleski (eds.), Inorganic Plant Nutrition. Encyclopedia of Plant Physiology, New Series, Vol 15A. Springer-Verlag, New York. Pp. 351–375.

Bell, W. D., L. Bogorad, and W. J. McIlrath. 1958. Response of the yellow-stripe maize mutant (ys_1) to ferrous and ferric iron. *Bot. Gaz.* 120:36–39.

Bell, W. D., L. Bogorad, and W. J. McIlrath. 1962. Yellow-stripe phenotype in maize. I. Effects of ys_1 locus on uptake and utilization of iron. *Bot. Gaz.* 124:1–8.

Bennett, J. H., R. A. Olsen, and R. B. Clark. 1982. Modification of soil fertility by plant roots: Iron stress–response mechanism. *What's New Plant Physiol.* 13(1):1–4.

Berger, K. C. 1949. Boron in soils and crops. *Adv. Agron.* 1:321–351.

Berger, K. C., T. Heikkinen, and E. Zube. 1957. Boron deficiency, a cause of blank stalks and barren ears in corn. *Soil Sci. Soc. Am. Proc.* 21:629–632.

Bertin, C., A. Panouille, and S. Rautou. 1976. Obtention de varietes de mais *prolifiques en epis* productives in grain et a large adaptation ecologique. *Ann. Amelior. Plant.* 26:387–418. (In Fr.)

Bhatia, C. R., and R. Rabson. 1976. Bioenergetic considerations in cereal breeding for protein improvement. *Science* 194:1418–1421.

Bieleski, R. L., and I. B. Ferguson. 1983. Physiology and metabolism of phosphate and its compounds. In A. Läuchli and R. L. Bieleski (eds.), Inorganic Plant Nutrition. Encyclopedia of Plant Physiology, New Series, Vol. 15A. Springer-Verlag, New York. Pp. 422–449.

Boatwright, G. O., and F. G. Viets. 1966. Phosphorus absorption during various growth stages of spring wheat and intermediate wheat grass. *Agron. J.* 58:185–188.

Bollard, E. G. 1983. Involvement of unusual elements in plant growth and nutrition. In A. Läuchli and R. L. Bieleski (eds.), Inorganic Plant Nutrition. Encyclopedia of Plant Physiology, New Series, Vol. 15B. Springer-Verlag, New York. Pp. 695–744.

Bolton, J. K., and R. H. Brown. 1980. Photosynthesis of grass species differing in carbon dioxide fixation pathways. V. Response of *Panicum maximum, Panicum milioides,* and tall fescue (*Festuca arundinacea*) to nitrogen nutrition. *Plant Physiol.* 66:97–100.

Boulter, D., and J. A. Gatehouse. 1979. Some biochemical aspects of the efficiency of plants. *Qual. Plant.—Plant Foods Hum. Nutr.* 29:187–195.

Bowen, G. D., and A. D. Rovira. 1967. Phosphorus uptake along attached and excised roots measured by an automatic scanning method. *Aust. J. Biol. Sci.* 20:369–378.

Bowen, J. E. 1986. Kinetics of zinc uptake by two rice cultivars. *Plant Soil* 94:99–107.

Bowen, J. E. 1987. Physiology of genotypic differences in zinc and copper uptake in rice and tomato. In W. H. Gabelman and B. C. Loughman (eds.), Genetic Aspects of Plant Mineral Nutrition. Nijhoff, Dordrecht, Netherlands. Pp. 413–423.

Boyat, A., and P. Robin. 1977. Relations entre productivite qualite du grain et activite nitrate reductase chez les cereales. *Ann. Amelior. Plant.* 27:389–410. (In Fr.)

Bradford, R. R., D. E. Baker, and W. I. Thomas. 1966. Effect of soil treatments on chemical element accumulation of four corn hybrids. *Agron. J.* 58:614–617.

Broadbent, F. E., S. K. De Datta, and E. V. Laureles. 1987. Measurement of nitrogen utilization efficiency in rice genotypes. *Agron. J.* 79:786–791.

Brown, J. C. 1956. Iron chlorosis. *Annu. Rev. Plant Physiol.* 7:171–190.

Brown, J. C. 1965. Calcium movement in barley and wheat as affected by copper and phosphorus. *Agron. J.* 57:617–621.

Brown, J. C. 1967. Differential uptake of Fe and Ca by two corn genotypes. *Soil Sci.* 103:331–338.

Brown, J. C. 1977. Genetically controlled factors involved in absorption and transport of iron by plants. In K. N. Raymond (ed.), Bioinorganic Chemistry—II. Adv. Chem. Ser. No. 162. Am. Chem. Soc., Washington, D.C. Pp. 93–103.

Brown, J. C. 1978a. Mechanism of iron uptake by plants. *Plant Cell Environ.* 1:249–257.

Brown, J. C. 1978b. Physiology of plant tolerance to alkaline soils. In G. A. Jung (ed.), Crop Tolerance to Suboptimal Land Conditions. Spec. Publ. No. 32. ASA, Madison, Wisconsin. Pp. 257–276.

Brown, J. C., and J. E. Ambler. 1970. Further characterization of iron uptake in two genotypes of corn. *Soil Sci. Soc. Am. Proc.* 34:249–252.

Brown, J. C., J. E. Ambler, R. L. Chaney, and C. D. Foy. 1972. Differential responses of plant genotypes to micronutrients. In J. J. Mortvedt, P. M. Giordano, and W. L. Lindsay (eds.), Micronutrients in Agriculture. SSSA, Madison, Wisconsin. Pp. 389–418.

Brown, J. C., and W. D. Bell. 1969. Iron uptake dependent upon genotype of corn. *Soil Sci. Soc. Am. Proc.* 33:99–101.

Brown, J. C., and R. B. Clark. 1974. Differential response of two maize inbreds to molybdenum stress. *Soil Sci. Soc. Am. Proc.* 38:331–333.

Brown, J. C., and R. B. Clark. 1977. Copper as essential to wheat reproduction. *Plant Soil* 48:509–523.

Brown, J. C., R. B. Clark, and W. E. Jones. 1977. Efficient and inefficient use of phosphorus by sorghum. *Soil Sci. Soc. Am. J.* 41:747–750.

Brown, J. C., and W. E. Jones. 1973. Needed: A sorghum for iron-poor soils. *Crops Soils* 26(1):10–11.

Brown, J. C., and W. E. Jones. 1974. Differential response of oats to manganese stress. *Agron. J.* 66:624–626.

Brown, J. C., and W. E. Jones. 1975. Phosphorus efficieny as related to iron inefficiency in sorghum. *Agron. J.* 67:468–472.

Brown, J. C., and W. E. Jones. 1977. Fitting plants nutritionally to soils. III. Sorghum. *Agron. J.* 69:410–414.

Brown, J. C., and M. E. McDaniel. 1978a. Factors associated with differential response of oat cultivars to iron stress. *Crop Sci.* 18:551–556.

Brown, J. C., and M. E. McDaniel. 1978b. Factors associated with differential response of two oat cultivars to zinc and copper stress. *Crop Sci.* 18:817–820.

Brown, J. C., and R. A. Olsen. 1980. Factors related to iron uptake by dicotyledonous and monocotyledonous plants. III. Competition between root and external factors for Fe. *J. Plant Nutr.* 2:661–682.

Brown, J. C., L. O. Tiffin, and R. S. Holmes. 1958. Carbohydrate and organic acid metabolism with C^{14} distribution affected by copper in Thatcher wheat. *Plant Physiol.* 33:38–42.

Brown, P. H., R. D. Graham, and D. J. D. Nicholas. 1984. The effect of manganese and nitrate supply on the levels of phenolics and lignin in young wheat plants. *Plant Soil* 81:437–440.

Brown, P. H., R. M. Welch, E. E. Cary, and R. T. Checkai. 1987. Beneficial effects of nickel on plant growth. *J. Plant Nutr.* 10:2125–2135.

Brown, R. H. 1978. A difference in N use efficiency in C_3 and C_4 plants and its implications in adaptation and evolution. *Crop Sci.* 18:93–98.

Brown, R. H., and J. R. Wilson. 1983. Nitrogen response of *Panicum* species differing in CO_2 fixation pathways. II. CO_2 exchange characteristics. *Crop Sci.* 23:1154–1159.

Bruetsch, T. F., and G. O. Estes. 1976. Genotype variation in nutrient uptake efficiency in corn. *Agron. J.* 68:521–523.

Brunetti, N., and R. H. Hageman. 1976. Comparison of *in vivo* and *in vitro* assays of nitrate reductase in wheat (*Triticum aestivum* L.) seedlings. *Plant Physiol.* 58:583–587.

Burkholder, P. R., and I. McVeigh. 1940. Growth and differentiation of maize in relation to nitrogen supply. *Am. J. Bot.* 27:414–424.

Burley, J. W. A., F. I. O. Nwoke, G. L. Leister, and R. A. Popham. 1970. The relationship of xylem maturation to the absorption and translocation of ^{32}P. *Am. J. Bot.* 57:504–511.

Bussler, W. 1981. Physiological functions and utilization of copper. In J. F. Loneragan, A. D. Robson, and R. D. Graham (eds.), Copper in Soils and Plants. Academic Press, New York. Pp. 213–234.

Butz, R. G., and W. A. Jackson. 1977. A mechanism for nitrate transport and reduction. *Phytochemistry* 16:409–417.

Cacco, G., G. Ferrari, and G. C. Lucci. 1976. Uptake efficiency of roots in plants at different ploidy levels. *J. Agric. Sci.* 87:585–589.

Cacco, G., G. Ferrari, and M. Saccomani. 1978. Variability and inheritance of sulfate uptake efficiency and ATP-sulfurylase activity in maize. *Crop Sci.* 18:503–505.

Cacco, G., G. Ferrari, and M. Saccomani. 1983a. Genetic variability of the efficiency of nutrient utilization by maize (*Zea mays* L.) In M. R. Sarić and B. C. Loughman (eds.), Genetic Aspects of Plant Nutrition. Nijhoff, The Hague, Netherlands. Pp. 435–439.

Cacco, G., M. Saccomani, and G. Ferrari. 1983b. Changes in the uptake and assimilation efficiency for sulfate and nitrate in maize hybrids selected during the period 1930 to 1975. *Physiol. Plant.* 58:171–174.

Campbell, C. A., H. R. Davidson, and T. N. McCaig. 1983. Disposition of nitrogen and soluble sugars in manitou spring wheat as influenced by N fertilizer, temperature and duration and stage of moisture stress. *Can. J. Plant Sci.* 63:73–90.

Campbell, W. H., D. L. DeGracia, and E. R. Campbell. 1987. Regulation of molybdenum cofactor of maize leaf. *Phytochemistry* 26:2149–2150.

Casnoff, D. M. 1983. Nitrogen: Its effect on the expression of prolificacy and its utilization by non-prolific and prolific genotypes of maize (*Zea mays* L.). Ph.D. Thesis, Univ. of Nebraska, Lincoln. (*Diss. Abstr. B* 44:2621B–2622B, 1984.)

Chaney, R. L., and P. F. Bell. 1987. Complexity of iron nutrition: Lessons for plant–soil interaction research. *J. Plant Nutr.* 10:963–994.

Chaney, R. L., J. C. Brown, and L. O. Tiffin. 1972. Obligatory reduction of ferric chelates in iron uptake by soybeans. *Plant Physiol.* 50:208–213.

Chantarotwong, W., R. C. Huffaker, B. L. Miller, and R. C. Granstadt. 1976. *In vivo* nitrate reduction in relation to nitrate uptake, nitrate content, and *in vitro* nitrate reductase activity in intact barley seedlings. *Plant Physiol.* 57:519–522.

Cherry, J. H. 1985. Approaches to yield enhancement: Biochemical processes. In J. E. Harper, L. E. Schrader, and R. W. Howell (eds.), Exploitation of Physiological and Genetic Variability to Enhance Crop Productivity. Am. Soc. Plant Physiol., Rockville, Maryland. Pp. 72–78.

Chevalier, P., and L. E. Schrader. 1977. Genotypic differences in nitrate absorption and partitioning in N among plant parts in maize. *Crop Sci.* 17:897–901.

Christ, R. A. 1974. Iron requirement and iron uptake from various iron compounds by different plant species. *Plant Physiol.* 54:582–585.

Christensen, J. J. 1934. Nonparasitic leaf spots of barley. *Phytopathology* 24:726–742.

Christensen, L. E., F. E. Below, and R. H. Hageman. 1981. The effects of ear removal on senescence and metabolism of maize. *Plant Physiol.* 68:1180–1185.

Christiansen, M. N., and C. F. Lewis (eds.). 1982. Breeding Plants for Less Favorable Environments. Wiley, New York.

Clark, R. B. 1970. Mineral deficiencies of corn. Their effects on growth and interactions between nutrients. *Agrichem. Age .MDNM*/13:4, 8.

Clark, R. B. 1975a. Differential magnesium efficiency in corn inbreds. I. Dry-matter yields and mineral element composition. *Soil Sci. Soc. Am. Proc.* 39:488–491.

Clark, R. B. 1975b. Mineral element concentration of corn plant parts with age. *Commun. Soil Sci. Plant Anal.* 6:451–464.

Clark, R. B. 1977. Effect of aluminum on growth and mineral elements of Al-tolerant and Al-intolerant corn. *Plant Soil* 47:653–662.

Clark, R. B. 1978a. Differential response of corn inbreds to calcium. *Commun. Soil Sci. Plant Anal.* 9:729–744.

Clark, R. B. 1978b. Differential response of maize inbreds to Zn. *Agron. J.* 70:1057–1060.

Clark, R. B. 1982a. Iron deficiency in plants grown in the Great Plains of the U.S. *J. Plant Nutr.* 5:251–268.

Clark, R. B. 1982b. Nutrient solution growth of sorghum and corn in mineral nutrition studies. *J. Plant Nutr.* 5:1039–1057.

Clark, R. B. 1983. Plant genotype differences in the uptake, translocation, accumulation, and use of mineral elements required for plant growth. In M. R. Sarić and B. C. Loughman (eds.), Genetic Aspects of Plant Nutrition. Nijhoff, The Hague, Netherlands. Pp. 49–70.

Clark, R. B. 1984. Physiological aspects of calcium, magnesium, and molybdenum deficiencies in plants. In F. Adams (ed.), Soil Acidity and Liming. 2nd Ed. ASA, CSSA, and SSSA, Madison, Wisconsin. Pp. 99–170.

Clark, R. B. 1988. Mineral nutrient requirements and deficiency/excess disorders of sorghum. *Crop Res.* 1:16–35.

Clark, R. B., and J. C. Brown. 1974a. Differential mineral uptake by maize inbreds. *Commun. Soil Sci. Plant Anal.* 5:213–227.

Clark, R. B., and J. C. Brown. 1974b. Differential phosphorus uptake by phosphorus-stressed corn inbreds. *Crop Sci.* 14:505–508.

Clark, R. B., and J. C. Brown. 1974c. Internal root control of iron uptake and utilization in maize genotypes. *Plant Soil* 40:669–677.

Clark, R. B., and J. C. Brown. 1980. Role of the plant in mineral nutrition as related to breeding and genetics. In L. S. Murphy, E. C. Doll, and L. F. Welch (eds.), Moving up the Yield Curve: Advances and Obstacles. Spec. Publ. No. 39 ASA, Madison, Wisconsin. Pp. 45–70.

Clark, R. B., J. C. Brown, R. A. Olsen, and J. H. Bennett. 1981a. Biological aspects of iron in plants. In Multimedia Criteria for Iron and Its Compounds. EPA, Cincinnati, Ohio. Pp. 272–338.

Clark, R. B., D. P. Coyne, W. M. Ross, and B. E. Johnson. 1990. Genetic aspects of plant resistant to iron deficiency. *Proc. Int. Congr. Plant Physiology, Soc. Plant Physiol. Biochem., New Delhi* (in press).

Clark, R. B., C. I. Flores, and L. M. Gourley. 1988b. Mineral element comparisons of pearl millet and sorghum grown on acid soil. *Agron. Abstr., ASA, Madison, Wisconsin* p. 232.

Clark, R. B., C. I. Flores, and L. M. Gourley. 1988c. Mineral element concentrations of acid soil tolerant and susceptible sorghum genotypes. *Commun. Soil Sci. Plant Anal.* 19:1003–1017.

Clark, R. B., and L. M. Gourley. 1987. Evaluation of mineral elements of sorghum grown on acid tropical soils. In L. M. Gourley and J. S. Salinas (eds.), *Sorghum for Acid Soils. Proc. Int.*

Sorghum/Millet Collaborative Research Support Program (INTSORMIL), ICRISAT, CIAT Workshop, CIAT, Cali, Colombia pp. 251–270.

Clark, R. B., and L. M. Gourley. 1988. Mineral element concentrations of sorghum genotypes grown on tropical acid soil. *Commun. Soil Sci. Plant Anal.* 19:1019–1029.

Clark, R. B., and R. D. Gross. 1986. Plant genotype differences to iron. *J. Plant Nutr.* 9:471–491.

Clark, R. B., J. W. Maranville, and H. J. Gorz. 1978. Phosphorus efficiency of sorghum grown with limited phosphorus. In A. R. Ferguson, R. L. Bieleski, and I. B. Ferguson (eds.), Plant Nutrition 1978. Proc. 8th Int. Colloq. Plant Analysis and Fertilizer Problems, Auckland. *Inf. Ser.—N.Z. Dep. Sci. Ind. Res.* No. 134, pp. 93–99.

Clark, R. B., J. W. Maranville, and W. M. Ross. 1977. Differential phosphorus efficiency in sorghum. *Proc. 10th Grain Sorghum Research Utilization Conf., Wichita, Kans.* Sorghum Improv. Conf. North Am., Lubbock, Texas. Pp. 1–2.

Clark, R. B., P. A. Pier, D. Knudsen, and J. W. Maranville. 1981b. Effect of trace element deficiencies and excesses on mineral nutrients in sorghum. *J. Plant Nutr.* 3:357–374.

Clark, R. B., V. Römheld, and H. Marschner. 1988d. Iron uptake and phytosiderophore release by roots of sorghum genotypes. *J. Plant Nutr.* 11:663–676.

Clark, R. B., L. O. Tiffin, and J. C. Brown. 1973. Organic acids and iron translocation in maize genotypes. *Plant Physiol.* 52:147–150.

Clark, R. B., E. P. Williams, W. M. Ross, G. M. Herron, and M. D. Witt. 1988e. Effect of iron deficiency chlorosis on growth and yield traits of sorghum. *J. Plant Nutr.* 11:747–754.

Clarkson, D. T. 1985. Factors affecting mineral nutrient acquisition by plants. *Annu. Rev. Plant Physiol.* 36:77–115.

Clarkson, D. T., and J. B. Hanson. 1980. The mineral nutrition of higher plants. *Annu. Rev. Plant Physiol.* 31:239–298.

Clarkson, D. T., and J. Sanderson. 1978. Sites of absorption and translocation of iron in barley roots: Tracer and microautographic studies. *Plant Physiol.* 61:731–736.

Cline, G. R., C. P. P. Reid, P. E. Powell, and P. J. Szaniszlo. 1984. Effects of a hydroxamate siderophore on iron absorption by sunflower and sorghum. *Plant Physiol.* 76:36–39.

Cocucci, S. M., and S. Morgutti. 1986. Stimulation of proton extrusion by K^+ and divalent cations (Ni^{2+}, Co^{2+}, Zn^{2+}) in maize root segments. *Physiol. Plant.* 68:497–501.

Cooke, G. W. 1975. Sources of protein for people and livestock; The amounts now available and future prospects. In Fertilizer Use and Protein Production. Int. Potash Inst., Bern. Pp. 29–51.

Cooper, H. D., D. T. Clarkson, H. E. Ponting, and B. C. Loughman. 1986. Nitrogen assimilation in field-grown winter wheat: Direct measurements of nitrate reduction in roots using ^{15}N. *Plant Soil* 91:397–400.

Cox, M. C., C. O. Qualset, and D. W. Rains. 1985. Genetic variation for nitrogen assimilation and translocation in wheat. I. Dry matter and nitrogen accumulation. *Crop Sci.* 25:430–435.

Coyne, D. P., S. S. Korban, D. Knudsen, and R. B. Clark. 1982. Inheritance of iron deficiency in crosses of dry beans (*Phaseolus vulgaris* L.). *J. Plant Nutr.* 5:573–585.

Craig, W. F., and W. I. Thomas. 1970. Prediction of chemical accumulation in maize double-cross hybrids from single-cross data. *Crop Sci.* 10:609–610.

Craswell, E. T., and D. C. Godwin. 1984. The efficiency of nitrogen fertilizers applied to cereals in different climates. *Adv. Plant Nutr.* 1:1–55.

Crooke, W. M. 1955. Further aspects of the relationship between nickel toxicity and iron supply. *Ann. Appl. Biol.* 43:465–476.

Crooke, W. M., J. G. Hunter, and O. Vergnano. 1954. The relationship between nickel toxicity and iron supply. *Ann. Appl. Biol.* 41:311–324.

Croy, L. I., and R. H. Hageman. 1970. Relationship of nitrate reductase activity to grain protein production in wheat. *Crop Sci.* 10:280–285.

Dalling, M. J. 1985. The physiological basis of nitrogen redistribution during grain filling in cereals. In J. E. Harper, L. E. Schrader, and H. W. Howell (eds.), Exploitation of Physiological and Genetic Variability to Enhance Crop Productivity. Am. Soc. Plant Physiol., Rockville, Maryland. Pp. 55–71.

Dalling, M. J., G. Boland, and J. H. Wilson. 1976. Relation between acid proteinase activity and redistribution of nitrogen during grain development in wheat. *Aust. J. Plant Physiol.* 3:721–730.

Dalling, M. J., G. M. Halloran, and J. H. Wilson. 1975. The relation between nitrate reductase activity and grain nitrogen productivity in wheat. *Aust. J. Agric. Res.* 26:1–10.

Dalling, M. J., and R. H. Lyon. 1977. Level of activity of nitrate reductase at the seedling stage as a predictor of grain nitrogen yield in wheat (*Triticum aestivum* L.). *Aust. J. Agric. Res.* 28:1–4.

Das, P. K., M. Kar, and D. Mishra. 1978. Nickel nutrition of plants. I. Effect of nickel on some oxidase activities during rice (*Oryza sativa* L.) seed germination. *Z. Pflanzenphysiol.* 90:225–234.

Day, G. E., G. M. Paulsen, and R. G. Sears. 1985. Relationships among important traits in the nitrogen economy of winter wheat. *J. Plant Nutr.* 8:357–368.

Deckard, E. L., R. J. Lambert, and R. H. Hageman. 1973. Nitrate reductase activity in corn leaves as related to yields of grain and grain protein. *Crop Sci.* 13:343–350.

De Datta, S. K., and W. H. Patrick, Jr. (eds.). 1986. Nitrogen Economy of Flooded Rice Soils. Nijhoff, Dordrecht, Netherlands.

DeKock, P. C. 1981. Iron nutrition under conditions of stress. *J. Plant Nutr.* 3:513–521.

DeKock, P. C., A. Hall, and R. H. E. Inkson. 1979. Active iron in plant leaves. *Ann. Bot.* 43:737–740.

Desai, R. M., and C. R. Bhatia. 1978. Nitrogen uptake and nitrogen harvest index in durum wheat cultivars varying in their grain protein concentration. *Euphytica* 27:561–566.

DeTurk, E. E. 1941. Plant nutrient deficiency symptoms. Physiological basis. *Ind. Eng. Chem.* 33:648–653.

DeTurk, E. E., J. R. Holbert, and B. W. Howk. 1933. Chemical transformations of phosphorus in the growing corn plant, with results on two first generation crosses. *J. Agric. Res.* 46:121–141.

Dios, R. V., and T. V. Broyer. 1965. Deficiency symptoms and essentiality of molybdenum in corn hybrids. *Agrochimica* 9:273–284.

Drew, M. C., L. R. Saker, S. A. Barber, and W. Jenkins. 1984. Changes in the kinetics of phosphate and potassium absorption in nutrient-deficient barley roots measured by a solution-depletion technique. *Planta* 160:490–499.

Dubois, J.-B., and A Fossati. 1981. Influence of nitrogen uptake and nitrogen partitioning efficiency on grain yield and grain protein concentration of twelve winter wheat genotypes (*Triticum aestivum* L.). *Z. Pflanzenzuecht.* 86:41–49.

Dugger, W. M. 1983. Boron in plant metabolism. In A. Läuchli and R. L. Bieleski (eds.), Inorganic Plant Nutrition. Encyclopedia of Plant Physiology, New Series, Vol. 15B. Springer-Verlag, New York. Pp. 626–650.

Duvick, D. V. 1977. Genetic rates of gain in hybrid maize yields during the past 40 years. *Maydica* 22:187–196.

Eck, H. V., G. C. Wilson, and T. Martinez. 1975. Nitrate reductase activity of grain sorghum leaves as related to yeilds of grain, dry matter, and nitrogen. *Crop Sci.* 15:557–561.

Edwards, J. H., and S. A. Barber. 1976. Phosphorus uptake rate of soybean roots as influenced by plant age, root trimming, and solution P concentration. *Agron. J.* 68:973–975.

Eichelberger, K. D. 1986. Response to eight cycles of divergent phenotypic recurrent selection for nitrate reductase activity and correlated responses in maize (*Zea mays* L.). M.S. Thesis, Univ. of Illinois, Urbana-Champaign.

Eilrich, G. L., and R. H. Hageman. 1973. Nitrate reductase activity and its relationship to accumulation of vegetative and grain nitrogen in wheat (*Triticum aestivum* L.). *Crop Sci.* 13;59–66.

Elliott, G. C., and A. Läuchli. 1985. Phosphorus efficiency and phosphate–iron interaction in maize. *Agron. J.* 77:399–403.

Elliott, G. C., and A. Läuchli. 1986. Evaluation of an acid phosphatase assay for detection of phosphorus deficiency in leaves of maize (*Zea mays* L.). *J. Plant Nutr.* 9:1469–1477.

Elliott, G. C., J. Lynch, and A. Läuchli. 1984. Influx and efflux of P in roots of intact maize plants. *Plant Physiol.* 76:336–341.

Ellis, B. G., and B. D. Knezek. 1972. Adsorption reactions of micronutrients in soils. In J. J. Mortvedt, P. M. Giordano, and W. L. Lindsay (eds.), Micronutrients in Agriculture. SSSA, Madison, Wisconsin. Pp. 59–78.

Ellis, J. R., H. Larsen, and M. Boosalis. 1985. Drought resistance of wheat plants inoculated with vesicular-arbuscular mycorrhizae. *Plant Soil* 86:369–378.
Engelstad, O. P., and G. L. Terman. 1980. Agronomic effectiveness of phosphate fertilizers. In F. E. Khasawneh, E. C. Sample, and E. J. Kamprath (eds.), The Role of Phosphorus in Agriculture. ASA, CSSA, and SSSA, Madison, Wisconsin. Pp. 311–332.
Enyi, B. A. C. 1968. Comparative studies of upland and swamp rice varieties (*Oryza sativa* L.). I. Effect of soil moisture on growth and nutrient uptake. *J. Agric. Sci.* 71:1–13.
Eskew, D. L., R. M. Welch, and E. E. Cary. 1983. Nickel: An essential micronutrient for legumes and possibly all higher plants. *Science* 222:621–623.
Eskew, D. L., R. M. Welch, and W. A. Norvell. 1984. Nickel in higher plants. Further evidence for an essential role. *Plant Physiol.* 76:691–693.
Esty, J. C., A. B. Onken, L. R. Hossner, and R. Matheson. 1980. Iron use efficiency in grain sorghum hybrids and parental lines. *Agron. J.* 71:589–592.
Evans, J. R., and J. R. Seemann. 1984. Differences between wheat genotypes in specific activity of ribulose-1,5-biphosphate carboxylase and the relationship to photosynthesis. *Plant Physiol.* 74:759–765.
Fawole, I., W. H. Gabelman, and G. C. Gerloff. 1982a. Genetic control of root development in beans (*Phaseolus vulgaris* L.) grown under phosphorus stress. *J. Am. Soc. Hortic. Sci.* 107:98–100.
Fawole, I., W. H. Gabelman, G. C. Gerloff, and E. V. Nordheim. 1982b. Heritability of efficiency in phosphorus utilization in beans (*Phaseolus vulgaris* L.) grown under phosphorus stress. *J. Am. Soc. Hortic. Sci.* 107:94–97.
Fehr, W. R. 1983. Modification of mineral nutrition in soybeans by plant breeding. *Iowa State J. Res.* 57:393–407.
Feller, U. K., T. T. Soong, and R. H. Hageman. 1977. Leaf proteolytic activities and senescence during grain development of field-grown corn (*Zea mays* L.). *Plant Physiol.* 59:290–294.
Ferguson, I. B., and D. T. Clarkson. 1975. Ion transport and endodermal suberization in the roots of *Zea mays*. *New Phytol.* 75:69–80.
Fernandez, P. G. 1987. Studies on physiological mechanisms of nitrogen use efficiency in sorghum. Ph.D. Thesis, Univ. of Nebraska, Lincoln. (*Diss. Abstr. B* 48:1248B, 1988.)
Ferrari, G., and F. Renosto. 1972. Comparative studies on the active transport by excised roots of inbred and hybrid maize. *J. Agric. Sci.* 79:105–108.
Fick, G. N., and D. C. Rasmusson. 1967. Heritability of Sr-89 and Ca-45 accumulation in barley seedlings. *Crop Sci.* 7:315–317.
Field, C., J. Merino, and H. A. Mooney. 1983. Compromises between water-use efficiency and nitrogen-use efficiency in five species of California evergreens. *Oecologia* 60:384–389.
Fisher, F. L., and L. Reyes. 1954. Chlorosis in sorghum in the Rio Grande Plains of Texas. *Tex. Agric. Exp. Stn., Prog. Rep.* No. 1737.
Flores, C. I., R. B. Clark, and L. M. Gourley. 1988. Agronomic traits of pearl millet grown on infertile acid soil. *Agron. Abstr., ASA, Madison, Wisconsin.* p. 108.
Fox, R. H. 1978. Selection for phosphorus efficiency in corn. *Commun. Soil Sci. Plant Anal.* 9:13–37.
Foy, C. D. 1984. Physiological effects of hydrogen, aluminum, and manganese toxicities in acid soil. In F. Adams (ed.), Soil Acidity and Liming. 2nd Ed. ASA, CSSA, and SSSA, Madison, Wisconsin. Pp. 57–97.
Foy, C. D., and S. A. Barber. 1958. Magnesium absorption and utilization by two inbred lines of corn. *Soil Sci. Soc. Am. Proc.* 22:57–62.
Foy, C. D., R. L. Chaney, and M. C. White. 1978. The physiology of metal toxicity in plants. *Annu. Rev. Plant Physiol.* 29:511–566.
Foy, C. D., and A. L. Fleming. 1978. The physiology of plant tolerance to excess available aluminum and manganese in acid soils. In G. A. Jung (ed.), Crop Tolerance to Suboptimal Land Conditions. ASA, CSSA, and SSSA, Madison, Wisconsin. Pp. 301–328.
Franca, G. E. de. 1981. Differences in dry-matter yield and the uptake, distribution, and use of nitrogen by sorghum genotypes. Ph.D. Thesis, Univ. of Nebraska, Lincoln. (*Diss. Abstr. B* 41:5018B.)

Frick, H., and L. F. Bauman. 1978. Heterosis in maize as measured by K uptake properties of seedling roots. *Crop Sci.* 18:99–103.

Frick, H., and L. F. Bauman. 1979. Heterosis in maize as measured by K uptake properties of seedling roots: Pedigree analyses of inbreds with high and low augmentation potential. *Crop Sci.* 19:707–710.

Friedrich, J. W., and R. C. Huffaker. 1980. Photosynthesis, leaf resitances, and ribulose-1,5-biphosphate carboxylase degradation in senescing barley leaves. *Plant Physiol.* 65:1103–1107.

Friedrich, J. W., and L. E. Schrader. 1979. N deprivation in maize during grain-filling. II. Remobilization of ^{15}N and ^{35}S and the relationship between N and S accumulation. *Agron. J.* 71:466–472.

Frith, G. J. T., and M. J. Dalling. 1980. The role of peptide hydrolases in leaf senescence. In K. V. Thimann (ed.), Senescence in Plants. CRC Press, Boca Raton, Florida. Pp. 117–130.

Furlani, A. M. C. 1981. Differences in phosphorus uptake, distribution, and use by sorghum genotypes grown with low phosphorus. Ph.D. Thesis, Univ. of Nebraska, Lincoln. (*Diss. Abstr. B* 42:846B.)

Furlani, A. M. C., R. B. Clark, J. W. Maranville, and W. M. Ross. 1984a. Root phosphatase activity of sorghum genotypes grown with organic and inorganic sources of phosphorus. *J. Plant Nutr.* 7:1583–1595.

Furlani, A. M. C., R. B. Clark, J. W. Maranville, and W. M. Ross. 1984b. Sorghum genotype differences in phosphorus uptake rate and distribution in plant parts. *J. Plant Nutr.* 7:1113–1126.

Furlani, A. M. C., R. B. Clark, J. W. Maranville, and W. M. Ross. 1987a. Organic and inorganic sources of phosphorus on growth and phosphorus uptake in sorghum genotypes. *J. Plant Nutr.* 10:163–186.

Furlani, A. M. C., R. B. Clark, W. M. Ross, and J. W. Maranville. 1987b. Differential phosphorus uptake, distribution, and efficiency by sorghum inbred parents and their hybrids. In W. H. Gabelman and B. C. Loughman (eds.), Genetic Aspects of Plant Mineral Nutrition. Nijhoff, Dordrecht, Netherlands. Pp. 287–298.

Furlani, A. M. C., R. B. Clark, C. Y. Sullivan, and J. W. Maranville. 1986a. Induction of leaf red-speckling by phosphorus on sorghum grown under controlled conditions. *Crop Sci.* 26:551–557.

Furlani, A. M. C., R. B. Clark, C. Y. Sullivan, and J. W. Maranville. 1986b. Sorghum genotype differences to leaf "red-speckling" induced by phosphorus. *J. Plant Nutr.* 9:1435–1451.

Fushiya, S., Y. Sato, S. Nozoe, K. Nomoto, T. Takemoto, and S. Takagi. 1980. Avenic acid A, a new amino acid possessing an iron-chelating activity. *Tetrahedron Lett.* 21:3071–3072.

Gabelman, W. H. 1976. Genetic potentials in nitrogen, phosphorus, and potassium efficiency. In M. J. Wright (ed.), Plant Adaptation to Mineral Stress in Problem Soils. Cornell Univ. Press, Ithaca, New York. Pp. 205–212.

Gabelman, W. H., and G. C. Gerloff. 1978. Isolating plant germplasm with altered efficiencies in mineral nutrition. *HortScience* 13:682–684.

Gabelman, W. H., and B. C. Loughman (eds.). 1987. Genetic Aspects of Plant Mineral Nutrition. Nijhoff, Dordrecht, Netherlands.

Gallagher, L. W., K. M. Soliman, D. W. Rains, C. O Qualset, and R. C. Huffaker. 1983. Nitrogen assimilation in common wheats differing in potential nitrate reductase activity and tissue nitrate concentrations. *Crop Sci.* 23:913–919.

Gallagher, P. H., and T. Walsh 1943. The susceptibility of cereal varieties to manganese deficiency. *J. Agric. Sci.* 33:197–203.

Gallaher, R. N., and M. D. Jellum. 1976a. Elemental and/or action ratio efficiency of corn hybrids grown on an infertile soil inadequate in magnesium. *Commun. Soil Sci. Plant Anal.* 7:653–664.

Gallaher, R. N., and M. D. Jellum. 1976b. Influence of soils and planting dates on mineral element efficiency of corn hybrids. *Commun. Soil Sci. Plant Anal.* 7:665–676.

Gallaher, R. N., M. D. Jellum, and J. B. Jones, Jr. 1981. Leaf magnesium concentration efficiency versus yield efficiency of corn hybrids. Commun. *Soil Sci. Plant Anal.* 12:345–354.

Galvez, L., and R. B. Clark. 1988. Effects of silicon on mineral element composition of sorghum grown with toxic aluminum. *Agron. Abstr., ASA, Madison, Wisconsin.* p. 235.

Galvez, L., R. B. Clark, L. M. Gourley, and J. W. Maranville. 1987. Silicon interactions with manganese and aluminum toxicity in sorghum. *J. Plant Nutr.* 10:1139–1147.

Galvez, L., R. B. Clark, L. M. Gourley, and J. W. Maranville. 1989. Effects of silicon on mineral element composition of sorghum grown with excess manganese. *J. Plant Nutr.* 12:547–561.

Gardner, J. C. 1988. Strategies of nitrogen efficiency among landrace and domesticated sorghum cultivars. Ph.D. Thesis, Univ. of Nebraska, Lincoln.

Gerdemann, J. W. 1968. Vesicular–arbuscular mycorrhiza and plant growth. *Annu. Rev. Phytopathol.* 6:397–418.

Gerloff, G. C., and W. H. Gabelman. 1983. Genetic basis of inorganic plant nutrition. In A. Läuchli and R. L. Bieleski (eds.), Inorganic Plant Nutrition. Encyclopedia of Plant Physiology, New Series, Vol. 15B. Springer-Verlag, New York. Pp. 453–480.

Gifford, R. M., J. H. Thorne, W. D. Hitz, and R. T. Giaquinta. 1984. Crop productivity and photoassimilate partitioning. *Science* 225:801–808.

Giordano, P. M., and J. J. Mortvedt. 1969. Response of several corn hybrids to level of water-soluble zinc in fertilizers. *Soil Sci. Soc. Am. Proc.* 33:145–148.

Gladstones, J. S., and J. F. Loneragan. 1970. Nutrient elements in herbage plants, in relation to soil adaptation and animal nutrition. *Proc. 11th Int. Grassland Congr., Brisbane, Australia* pp. 350–354.

Glass, A. D. M., and J. E. Perley. 1980. Varietal differences in potassium uptake by barley. *Plant Physiol.* 65:160–164.

Glass, A. D. M., M. Y. Siddiqi, and K. I. Giles. 1981. Correlations between potassium uptake and hydrogen efflux in barley varieties. *Plant Physiol.* 68:457–459.

Gojon, A., L. Passama, and P. Robin. 1986. Root contribution to nitrate reduction in barley seedlings (*Hordeum vulgare* L.). *Plant Soil* 91:339–342.

Goodman, P. J. 1979. Genetic control of inorganic nitrogen assimilation of crop plants. In E. J. Hewitt and C. V. Cutting (eds.), Nitrogen Assimilation of Plants. Academic Press, New York. Pp. 165–176.

Gorsline, G. W., D. E. Baker, and W. I. Thomas. 1965. Accumulation of eleven elements by field corn (*Zea mays* L.). *Pa. State Agric. Exp. Stn., Bull.* No. 725.

Gorsline, G. W., J. L. Ragland, and W. I. Thomas. 1961. Evidence for inheritance of differential accumulation of calcium, magnesium, and potassium by maize. *Crop Sci.* 1:155–156.

Gorsline, G. W., W. I. Thomas, and D. E. Baker. 1964a. Inheritance of P, K, Mg, Cu, B, Zn, Mn, Al, and Fe concentrations in corn (*Zea mays* L.) leaves and grain. *Crop Sci.* 4:207–210.

Gorsline, G. W., W. I. Thomas, and D. E. Baker. 1968. Major gene inheritance of Sr–Ca, Mg, K, P, Zn, Cu, B, Al–Fe, and Mn concentrations in corn (*Zea mays* L.). *Pa. State Agric. Exp. Stn., Bull.* No. 746.

Gorsline, G. W., W. I. Thomas, D. E. Baker, and J. L. Ragland. 1964b. Relationship of strontium–calcium within corn. *Crop Sci.* 4:154–156.

Goyal, S. S., and R. C. Huffaker. 1984. Nitrogen toxicity in plants. In R. D. Hauck (ed.), Nitrogen in Crop Production. ASA, CSSA, and SSSA, Madison, Wisconsin. Pp. 97–118.

Graham, R. D. 1975. Male sterility in wheat plants deficient in copper. *Nature (London)* 254:514–515.

Graham, R. D. 1978a. Nutrient efficiency objectives in cereal breeding. In A. R. Ferguson, R. L. Bieleski, and I. B. Ferguson (eds.), Plant Nutrition 1978. Proc. 8th Int. Colloq. Plant Analysis and Fertilizer Problems, Auckland. *Inf. Ser.—N.Z. Dep. Sci. Ind. Res.* No. 134, pp. 165–170.

Graham, R. D. 1978b. Tolerance of *Triticale,* wheat and rye to copper deficiency. *Nature (London)* 271:542–543.

Graham, R. D. 1981. Absorption of copper by plant roots. In J. F. Loneragan, A. D. Robson, and R. D. Graham (eds.), Copper in Soils and Plants. Academic Press, New York. Pp. 141–163.

Graham, R. D. 1983. Effects of nutrient stress on susceptibility to disease with particular reference to trace elements. *Adv. Bot. Res.* 10:221–276.

Graham, R. D. 1984. Breeding for nutritional characteristics in cereals. *Adv. Plant Nutr.* 1:57–102.

Graham, R. D. 1986. Induction of male sterility in wheat using organic ligands with high specificity for binding copper. *Euphytica* 35:621–629.

Graham, R. D., G. D. Anderson, and J. S. Ascher. 1981. Absorption of copper by wheat, rye and some hybrid genotypes. *J. Plant Nutr.* 3:679–686.

Graham, R. D., J. S. Ascher, P. A. E. Ellis, and K. W. Shepherd. 1987. Transfer to wheat of the copper efficiency factor carried on rye chromosome arm 5RL. *Plant Soil* 99:107–114.
Graham, R. D., J. W. Davies, O. H. B. Sparrow, and J. S. Ascher. 1983. Tolerance of barley and other cereals to manganese-deficient calcareous soils of South Australia. In M. R. Sarić and B. C. Loughman (eds.), Genetic Aspects of Plant Nutrition. Nijhoff, The Hague, Netherlands. Pp. 339–345.
Graham, R. D., and E. K. S. Nambiar. 1981. Advances in research on copper deficiency in cereals. *Aust. J. Agric. Res.* 32:1009–1037.
Graham, R. D., and D. T. Pearce. 1979. The sensitivity of hexaploid and octoploid triticales and their parent species to copper deficiency. *Aust. J. Agric. Res.* 30:791–799.
Graham, R. D., and A. D. Rovira. 1984. A role for manganese in the resistance of wheat plants to take-all. *Plant Soil* 78:441–444.
Gregory, F. G., and F. Crowther. 1928. A physiological study of varietal differences in plants. Part I. A study of the comparative yields of barley varieties with different manurings. *Ann. Bot.* 42:757–770.
Gupta, U. C. 1979. Boron nutrition of crops. *Adv. Agron.* 31:273–307.
Gupta, U. C., Y. W. Jame, C. A. Campbell, A. J. Leyshon, and W. Nicholaichuk. 1985. Boron toxicity and deficiency: A review. *Can. J. Soil Sci.* 65:381–404.
Haeder, H. E., and H. Beringer. 1984. Long distance transport of potassium in cereals during grain filling in detached ears. *Physiol. Plant.* 62:433–438.
Hageman, R. H. 1979. Integration of nitrogen assimilation in relation to yield. In E. J. Hewitt and C. V. Cutting (eds.), Nitrogen Assimilation of Plants. Academic Press, New York. Pp. 591–611.
Hageman, R. H. 1984. Ammonium versus nitrate nutrition of higher plants. In R. D. Hauck (ed.), Nitrogen in Crop Production. ASA, CSSA, and SSSA, Madison, Wisconsin. Pp. 67–85.
Hageman, R. H., and F. E. Below. 1984. The role of nitrogen in the productivity of corn. *Proc. 39th Annu. Corn Sorghum Res. Conf., Am. Seed Trade Assoc., Washington, D.C.* pp. 145–156.
Hageman, R. H., and R. J. Lambert. 1981. Recurrent divergent and mass selections in maize with physiological and biochemical traits: Preliminary and projected application. In J. M. Lyons, R. C. Valentine, D. A. Phillips, D. W. Rains, and R. C. Huffaker (eds.), Genetic Engineering of Symbiotic Nitrogen Fixation and Conservation of Fixed Nitrogen. Plenum, New York. Pp. 581–598.
Hageman, R. H., A. J. Reed, R. A. Femmer, R. A. Sherrard, and M. J. Dalling. 1980. Some new aspects of the *in vivo* assay for nitrate reductase in wheat (*Triticum aestivum* L.) leaves. I. Reevaluation of nitrate pool sizes. *Plant Physiol.* 65:27–32.
Hageman, R. H., J. F. Zieserl, and E. R. Leng. 1963. Levels of nitrate reductase activity in inbred lines and F_1 hybrids in maize. *Nature (London)* 197:263–265.
Halim, A. H., C. E. Wassom, and R. Ellis, Jr. 1968. Zinc deficiency symptoms and zinc and phosphorus interactions in several strains of corn (*Zea mays* L.). *Agron. J.* 60:267–271.
Halloran, G. M., and J. W. Lee. 1979. Plant nitrogen distribution in wheat cultivars. *Aust. J. Agric. Res.* 30:779–789.
Hanway, J. J. 1962. Corn growth and composition in relation to soil fertility. II. Uptake of N, P, and K and their distribution in different plant parts during the growing season. *Agron. J.* 54:217–222.
Hanway, J. J., and R. A. Olson. 1980. Phosphate nutrition of corn, sorghum, soybeans, and small grains. In F. E. Khasawneh, E. C. Sample, and E. J. Kamprath (eds.), The Role of Phosphorus in Agriculture. ASA, CSSA, and SSSA, Madison, Wisconsin. Pp. 681–692.
Harry. S. P., and R. D. Graham. 1981. Tolerance of triticale, wheat and rye to copper deficiency and low and high soil pH. *J. Plant Nutr.* 3:721–730.
Harvey, P. H. 1939. Hereditary variation in plant nutrition. *Genetics* 24:437–461.
Hauck, R. D. (ed.). 1984. Nitrogen in Crop Production. ASA, CSSA, and SSSA, Madison, Wisconsin.
Hay, R. E., E. B. Earley, and E. E. DeTurk. 1953. Concentration and translocation of nitrogen compounds in the corn plant (*Zea mays*) during grain development. *Plant Physiol.* 28:606–621.
Haynes, R. J. (ed.). 1986. Mineral Nitrogen in the Plant–Soil System. Academic Press, New York.
Haynes, R. J., and K. M. Goh. 1978. Ammonium and nitrate nutrition of plants. *Biol. Rev. Cambridge Philos. Soc.* 53:465–510.

Hemming, B. C. 1986. Microbial–iron interactions in the plant rhizosphere. A review. *J. Plant Nutr.* 9:505–521.

Hether, N. H., R. A. Olsen, and L. L. Jackson. 1984. Chemical identification of iron reductants exuded by plant roots. *J. Plant Nutr.* 7:667–676.

Hetrick, B. A. D., J. A. Hetrick, and J. Bloom. 1984. Interaction of mycorrhizal infection, phosphorus level, and moisture stress in growth of field corn. *Can. J. Bot.* 62:2267–2271.

Hirata, H. 1987. Varietal differences of rice in phosphorus absorption from phosphorus compounds in soil. *J. Plant Nutr.* 10:1997–2005.

Hoener, I. R., and E. E. DeTurk. 1938. The absorption and utilization of nitrate nitrogen during vegetative growth by Illinois high protein and Illinois low protein corn. *J. Am. Soc. Agron.* 30:232–243.

Horiguchi, T., and S. Morita. 1987. Mechanism of manganese toxicity and tolerance of plants. VI. Effect of silicon on alleviation of manganese toxicity of barley. *J. Plant Nutr.* 10:2299–2310.

Horst, W. J., and H. Marschner. 1978. Effect of silicon on manganese tolerance of bean plants. (*Phaseolus vulgaris* L.). *Plant Soil* 50:287–303.

Howeler, R. H., and L. F. Cadavid. 1976. Screening of rice cultivars for tolerance to Al-toxicity in nutrient solutions as compared with a field screening method. *Agron. J.* 68:551–555.

Huffaker, R. C., and D. W. Rains. 1978. Factors influencing nitrate acquisition by plant: Assimilation and fate of reduced nitrogen. In D. R. Nielson and J. G. MacDonald (eds.), Nitrogen in the Environment. Vol. 2. Academic Press, New York. Pp. 1–43.

International Crops Research Institute for the Semi-Arid Tropics (ICRISAT). 1975–1976. Sorghum: Physiology. ICRISAT Annual Report. ICRISAT, Patancheru, India. Pp. 41–43.

International Crops Research Institute for the Semi-Arid Tropics (ICRISAT). 1979–1980. Sorghum: Physiology, Selection for Nitrogen and Phosphorus Efficiency. ICRISAT Annual Report. ICRISAT, Patancheru, India. Pp. 28–29.

International Rice Research Institute (IRRI). 1976. IRRI Annual Report. IRRI, Los Banos, Philippines.

International Rice Research Institute (IRRI). 1977. IRRI Annual Report. IRRI, Los Banos, Philippines.

Jackson, W. A. 1978. Nitrate acquisition and assimilation by higher plants: Processes in the root system. In D. R. Nielson and J. G. MacDonald (eds.), Nitrogen in the Environment. Vol. 2. Academic Press, New York. Pp. 45–88.

Jackson, W. A., W. L. Pan, R. H. Moll, and E. J. Kamprath. 1986. Uptake, translocation, and reduction of nitrate. In C. A. Neyra (ed.), Biochemical Basis of Plant Breeding. Vol. 2: Nitrogen Metabolism. CRC Press, Boca Raton, Florida. Pp. 73–108.

Jackson, W. A., and R. J. Volk. 1981. Nitrate transport processes and compartmentation in root systems. In J. M. Lyons, R. C. Valentine, D. A. Phillips, D. W. Rains, and R. C. Huffaker (eds.), Genetic Engineering of Symbiotic Nitrogen Fixation and Conservation of Fixed Nitrogen. Plenum, New York. Pp. 517–532.

Jensen, A. 1984. Responses of barley, pea, and maize to inoculation with different vesicular-arbuscular mycorrhizal fungi in irradiated soil. *Plant Soil* 78:315–323.

Jensen, P., and A.-S. Jonsson. 1981. Heterosis and ion transport in hexaploid varieties of rye–wheat (triticale) compared to the parental species. *Physiol. Plant.* 53:342–346.

Jensen, P., and H. Perby. 1986. Growth and accumulation of N, K^+, Ca^{2+}, and Mg^{2+} in barley exposed to various nutrient regimes and root/shoot temperatures. *Physiol. Plant.* 67:159–165.

Jensen, P., and S. Pettersson. 1978. Allosteric regulation of potassium uptake in plant roots. *Physiol. Plant.* 42:207–213.

Jensen, P., and S. Pettersson. 1980. Varietal variation in uptake and utilization of potassium (rubidium) in high-salt seedlings of barley. *Physiol. Plant.* 48:411–415.

Jocic, B., and M. R. Saric. 1983. Efficiency of nitrogen, phosphorus, and potassium use by corn, sunflower, and sugarbeet for the synthesis of organic matter. In M. R. Saric and B. C. Loughman (eds.), Genetic Aspects of Plant Nutrition. Nijhoff, The Hague, Netherlands. Pp. 123–127.

Johnson, G. V. 1982. Application of high performance liquid chromatography in the characterization of iron stress response. *J. Plant Nutr.* 5:499–514.

Johnson, V. A., P. J. Mattern, and J. W. Schmidt. 1967. Nitrogen relations during spring growth in varieties of *Triticum aestivum* L. differing in grain protein content. *Crop Sci.* 7:664–667.

Johnson, V. A., J. W. Schmidt, and P. J. Mattern. 1968. Cereal breeding for better protein impact. *Econ. Bot.* 22:16–25.

Jolly, V. D., and J. C. Brown. 1989. Iron efficient and inefficient oats. I. Differences in phytosiderophore release. *J. Plant Nutr.* 12:423–435.

Jolly, V. D., J. C. Brown, and M. J. Blaylock. 1988. An iron chelating compound released by barley roots in response to Fe-deficiency stress. *J. Plant Nutr.* 11:77–91.

Jung, G. A. (ed.). 1978. Crop Tolerance to Suboptimal Land Conditions. Spec. Publ. No. 32. ASA, Madison, Wisconsin.

Jungk, A. 1974. Phosphate uptake characteristics of intact root systems in nutrient solution as affected by plant species, age and P supply. In J. Wehrmann (ed.), Plant Analysis and Fertilizer Problems. Proc. 7th Int. Colloq. Vol. 1. Ger. Soc. Plant Nutr., Hannover, F.R.G. Pp. 185–196.

Jungk, A., and S. A. Barber. 1975. Plant age and the phosphorus uptake characteristics of trimmed and untrimmed corn root systems. *Plant Soil* 42:227–239.

Kamprath, E. J., R. H. Moll, and N. Rodriguez. 1982. Effects of nitrogen fertilization and recurrent selection on performance of hybrid populations of corn. *Agron. J.* 74:955–958.

Kannan, S. 1980a. Correlative influence of pH reduction on recovery from iron chlorosis in sorghum varieties. *J. Plant Nutr.* 2:507–516.

Kannan, S. 1980b. Differences in iron stress response and iron uptake in some sorghum varieties. *J. Plant Nutr.* 2:347–358.

Kannan, S. 1981a. Regulation of Fe-stress response in some crop varieties: Anomaly of a mechanism for recovery through non-redemptive pH reduction. *J. Plant Nutr.* 4:1–19.

Kannan, S. 1981b. The reduction of pH and recovery from chlorosis in Fe-stressed sorghum seedlings: The principal role of adventitious roots. *J. Plant Nutr.* 4:73–78.

Kannan, S. 1981c. Differences in Fe-stress response in sorghum hybrids and their parental lines: Evidence of heterosis. *Z. Pflanzenphysiol.* 103:285–290.

Kannan, S. 1982. Genotypic differences in iron uptake and utilization in some crop cultivars. *J. Plant Nutr.* 5:531–542.

Kannan, S. 1983. Cultivar differences for tolerance to Fe and Zn deficiency: A comparison of two maize hybrids and their parents. *J. Plant Nutr.* 6:323–337.

Kannan, S. 1986. Effects of dibutyl phthalate and phthalic acid on chlorosis recovery in iron-deficiency stressed sorghum cultivars. *J. Plant Nutr.* 9:1543–1551.

Kannan, S. 1988. Physiological responses associated with Fe-deficiency stress in different plant species. *J. Plant Nutr.* 11:1185–1192.

Kannan, S., and D. P. Pandey. 1982. Absorption and transport of iron in some crop cultivars. *J. Plant Nutr.* 5:395–403.

Kannan, S., and S. Ramani. 1982. Zinc-stress response in some sorghum hybrids and parent cultivars: Significance of pH reduction and recovery from chlorosis. *J. Plant Nutr.* 5:219–227.

Kannan, S., and S. Ramani. 1984. Effects of some chemical treatments on the recovery from chlorosis in Fe deficiency stressed sorghum cultivars. *J. Plant Nutr.* 7:631–639.

Kannan, S., and S. Ramani. 1987. Mechanism of Fe-deficiency tolerance in crop cultivars: Effects of dibutyl phthalate and caffeic acid on Fe-chlorosis recovery. *J. Plant Nutr.* 10:1051–1058.

Kannan, S., S. Ramani, and A.V. Patankar. 1984. Excretion of dibutyl phthalate by sorghum roots under Fe-stress: Evidence of its action on chlorosis recovery. *J. Plant Nutr.* 7:1717–1729.

Katyal, J. C., and B. D. Sharma. 1980. A new technique of plant analysis to resolve iron chlorosis. *Plant Soil* 55:105–119.

Katyal, J. C., and B. D. Sharma. 1984. Association of soil properties and soil and plant iron deficiency response in rice (*Oryza sativa* L.). *Commun. Soil Sci. Plant Anal.* 15:1065–1081.

Kawai, S., S. Takagi, and Y. Sato. 1988. Mugineic acid—family phytosiderophores in root-secretions of barley, corn and sorghum varieties. *J. Plant Nutr.* 11:633–642.

Kawaski, T., and M. Moritsugu. 1979. A characteristic symptom of calcium deficiency in maize and sorghum. *Commun. Soil Sci. Plant Anal.* 10:41–56.

Keltjens, W. G., J. W. Nieuwenhuis, and J. A. Nelemans. 1986. Nitrogen retranslocation in plants of maize, lupin and cocklebur. *Plant Soil* 91:323–327.
Kimball, J. G. 1978. Zinc accumulation by four *Zea mays* cultivars. M.S. Thesis, Univ. of Kentucky, Lexington.
Kirkby, E. A., and K. Mengel. 1976. The role of magnesium in plant nutrition. *Z. Pflanzenernaehr. Bodenkd.* 139:209–222.
Kleese, R. A., D. C. Rasmusson, and L. H. Smith. 1968. Genetic and environmental variation in mineral element accumulation in barley, wheat, and soybeans. *Crop Sci.* 8:591–593.
Klepper, L. A., D. Flesher, and R. H. Hageman. 1971. Generation of reduced nicotinamide adenine dinucleotide for nitrate reduction in green leaves. *Plant Physiol.* 48:580–590.
Koehler, F. E. 1976. Plant food taken up by 108 Bu/A wheat while it grows. *Better Crops Plant Food* 60(1):16–18.
Kohl, M. M., and S. Rajaram (eds.). 1988. Wheat Breeding for Acid Soils: Review of Brazilian/CIMMYT Collaboration, 1974–1986. CIMMYT, El Baton, Mexico.
Krishna, K. R., and D. J. Bagyaraj. 1981. Note on the effect of VA mycorrhiza and soluble phosphate fertilizer on sorghum. *Indian J. Agric. Sci.* 51:688–690.
Krishna, K. R., and P. J. Dart. 1984. Effect of mycorrhizal inoculation and soluble phosphorus fertilizer on growth and phosphorus uptake of pearl millet. *Plant Soil* 81:247–256.
Krstic, B., and M. R. Sarić. 1983. Efficiency of nitrogen utilization and photosynthetic rate in C_3 and C_4 plants. In M. R. Sarić and B. C. Loughman (eds.), Genetic Aspects of Plant Nutrition. Nijhoff, The Hague, Netherlands. Pp. 255–260.
Ku, M. S. B., M. R. Schmitt, and G. C. Edwards. 1979. Quantitative determination of RuBP carboxylase-oxygenase protein in leaves of several C_3 and C_4 plants. *J. Exp. Bot.* 30:89–98.
Kubota, J., and W. H. Allaway. 1972. Geographic distribution of trace element problems. In J. J. Mortvedt, P. M. Giordano, and W. L. Lindsay (eds.), Micronutrients in Agriculture. SSSA, Madison, Wisconsin. Pp. 525–554.
Kuchenbuch, R. O., and S. A. Barber. 1987. Yearly variation of root distribution with depth in relation to nutrient uptake and corn yield. *Commun. Soil Sci. Plant Anal.* 18:255–263.
Lafitte, H. R., and R. S. Loomis. 1988. Growth and composition of grain sorghum with limited nitrogen. *Agron. J.* 80:492–498.
Lal, P., G. G. Reddy, and M. S. Modi. 1978. Accumulation and redistribution of dry matter and N in triticale and wheat varieties under water stress condition. *Agron. J.* 70:623–626.
Lambers, H., J. J. Neeteson, and I. Stulen (eds.). 1986. Fundamental, Ecological and Agricultural Aspects of Nitrogen Metabolism in Higher Plants. Nijhoff, Dordrecht, Netherlands.
Landi, S., and F. Fagioli. 1983. Efficiency of manganese and copper uptake by excised roots of maize genotypes. *J. Plant Nutr.* 6:957–970.
Landi, S., M. Saccomani, and G. Cacco. 1983. Grain yield and efficiency of K^+ and SO_4^{2-} uptake in maize. *Agrochimica* 27:73–78.
Landsberg, E.-C. 1981. Organic acid synthesis and release of hydrogen ions in response to Fe deficiency stress of mono- and dicotyledonous plant species. *J. Plant Nutr.* 3:579–591.
Leeper, G. W. 1952. Factors affecting availability of inorganic nutrients in soils with special reference to micronutrient metals. *Annu. Rev. Plant Physiol.* 3:1–16.
Lindsay, W. L. 1972. Inorganic phase equilibria of micronutrients in soils. In J. J. Mortvedt, P. M. Giordano, and W. L. Lindsay (eds.), Micronutrients in Agriculture. SSSA, Madison, Wisconsin. Pp. 41–57.
Lindsay, W. L. 1984. Soil and plant relationships associated with iron deficiency with emphasis on nutrient interactions. *J. Plant Nutr.* 7:489–500.
Lindsay, W. L., and A. P. Schwab. 1982. The chemistry of iron in soils and its availability to plants. *J. Plant Nutr.* 5:821–840.
Loffler, C. M., and R. H. Busch. 1982. Selection for grain protein, grain yield, and nitrogen partitioning efficiency in hard red spring wheat. *Crop Sci.* 22:591–595.
Loneragan, J. F. 1981. Distribution and movement of copper in plants. In J. F. Loneragan, A. D.

Robson, and R. D. Graham (eds.), Copper in Soils and Plants. Academic Press, New York. Pp. 165–188.
Loneragan, J. F., and C. J. Asher. 1967. Response of plants to phosphate concentration in solution culture. II. Rate of phosphate absorption and its relation to growth. *Soil Sci.* 103:311–318.
Longnecker, N. 1988. Iron nutrition of plants. In ISI Atlas of Science: Animal and Plant Science. Inst. Sci. Inf. (ISI), Philadelphia, Pennsylvania. Pp. 143–150.
Longnecker, N. E., R. D. Graham, K. W. McCarthy, O. H. B. Sparrow, and J. Egan. 1988. Screening for manganese efficiency in barley. *Abstr. 3rd Int. Symp. Genetic Aspects of Plant Mineral Nutrition, Inst. Crop Sci. Plant Breed., Fed. Res. Cent., Braunschweig, F.R.G.*
Longnecker, N., and R. Welch. 1986. The relationship among iron-stress response, iron-efficiency and iron uptake of plants. *J. Plant Nutr.* 9:715–727.
Lu, N., and S. A. Barber. 1985. Effect of the fraction of the root system supplied with P on P uptake and growth characteristics of wheat roots. *J. Plant Nutr.* 8:799–809.
Lucas, R. E., and B. D. Knezek. 1972. Climatic and soil conditions promoting micronutrient deficiencies in plants. In J. J. Mordvedt, P. M. Giordano, and W. L. Lindsay (eds.), Micronutrients in Agriculture. SSSA, Madison, Wisconsin. Pp. 265–288.
Lutz, J. A., Jr., C. F. Genter, and G. W. Hawkins. 1972a. Effect of soil pH on element concentration and uptake by maize. I. P, K, Ca, Mg, and Na. *Agron. J.* 64:581–583.
Lutz, J. A., Jr., C. F. Genter, and G. W. Hawkins. 1972b. Effect of soil pH on element concentration and uptake by maize. II. Cu, B, Zn, Mn, Mo, Al, and Fe. *Agron. J.* 64:583–585.
Lyness, A. S. 1936. Varietal differences in the phosphorus feeding capacity of roots. *Plant Physiol.* 11:665–688.
Mackay, A. D., and S. A. Barber. 1984. Soil temperature effects on root growth and phosphorus uptake by corn. *Soil Sci. Soc. Am. J.* 48:818–823.
Mackay, A. D., and S. A. Barber. 1985a. Effect of soil moisture and phosphate level on root hair growth of corn roots. *Plant Soil* 86:321–331.
Mackay, A. D., and S. A. Barber. 1985b. Soil moisture effects on root growth and phosphorus uptake by corn. *Agron. J.* 77:519–523.
MacKown, C. T., and D. A. Van Sanford. 1986. Postanthesis nitrate assimilation in winter wheat. *Plant Physiol.* 81:17–21.
Mahadevappa, M., H. Ikehashi, and P. Aurin. 1981. Screening rice genotypes for tolerance to alkalinity and zinc deficiency. *Euphytica* 30:253–257.
Makino, A., T. Mae, and K. Ohira. 1984. Relation between nitrogen and ribulose-1,5-biphosphate carboxylase in rice leaves from emergence through senescence. *Plant Cell Physiol.* 25:429–437.
Maranville, J. W., R. B. Clark, and W. M. Ross. 1980. Nitrogen efficiency in grain sorghum. *J. Plant Nutr.* 2:577–589.
Marcar, N. E., and R. D. Graham. 1986. Effect of seed manganese content on the growth of wheat (*Triticum aestivum*) under manganese deficiency. *Plant Soil* 96:165–173.
Marcar, N. E., and R. D. Graham. 1987. Genotypic variation for manganese efficiency in wheat. *J. Plant Nutr.* 10:2049–2055.
Marschner, H. 1986. Mineral Nutrition of Higher Plants. Academic Press, New York.
Marschner, H., V. Römheld, and M. Kissel. 1986. Different strategies in higher plants in mobilization and uptake of iron. *J. Plant Nutr.* 9:695–713.
Marschner, H., V. Römheld, and M. Kissel. 1987. Localization of phytosiderophore release and of iron uptake along intact barley roots. *Physiol. Plant.* 71:157–162.
Marschner, H., V. Römheld, and H. Ossenberg-Neuhaus. 1982. Rapid method for obtaining changes in pH and reducing processes along roots of intact plants. *Z. Pflanzenphysiol.* 105:407–416.
Martinez, A., and R. B. Diamond. 1984. Nitrogen use in world crop production. In R. D. Hauck (ed.), Nitrogen in Crop Production. ASA, CSSA, and SSSA, Madison, Wisconsin. Pp. 3–21.
Massey, H. F., and F. A. Loeffel. 1967. Factors in interstrain variation in zinc content of maize (*Zea mays* L.) kernels. *Agron. J.* 59:214–217.

McKenzie, D. B., L. R. Hossner, and R. J. Newton. 1985. Root reductant release as a measure of sorghum cultivar iron efficiency. *J. Plant Nutr.* 8:847–857.

McKenzie, D. B., L. R. Hossner, and R. J. Newton. 1986. The influence of NH_4^+-N vs. NO_3–N nutrition on root reductant release by Fe-stressed sorghum. *J. Plant Nutr.* 9:1289–1301.

McKenzie, D. B., L. R. Hossner, and R. J. Newton. 1987. Measurement of p-coumarate hydroxylase activity as a qualitative test for sorghum cultivar Fe-efficiency. *J. Plant Nutr.* 10:15–24.

McLachlan, K. D. 1984. Effect of drought, aging and phosphorus status on leaf acid phosphatase activity in wheat. *Aust. J.Agric. Res.* 35:777–787.

McLean, E. O., and M. E. Watson. 1985. Soil measurements of plant-available potassium. In R. D. Munson (ed.), Potassium in Agriculture. ASA, CSSA, and SSSA, Madison, Wisconsin. Pp. 277–308.

McMahan, M. A. 1987. Fertilizer use and high yields are compatible with quality environment. *Proc. Int. Mineral Corp. (IMC) World Food Production Conf., Madrid (Better Crops with Plant Food, 1988, Potash Phosphate Inst., Atlanta, Georgia)* pp. 3–8.

McNeal, F. H., G. O. Boatright, M. A. Berg, and C. A. Watson. 1968. Nitrogen in plant parts of seven spring wheat varieties at successive stages of development. *Crop Sci.* 8:535–537.

Memon, A. R., and A. D. M. Glass. 1987. Genotypic differences in subcellular compartmentation of K^+: Implications for protein synthesis, growth and yield. In W. H. Gabelman and B. C. Loughman (eds.), Genetic Aspects of Plant Mineral Nutrition. Nijhoff, Dordrecht, Netherlands. Pp. 323–329.

Memon, A. R., M. Saccomani, and A. D. M. Glass. 1985a. Efficiency of potassium utilization by barley varieties: The role of subcellular compartmentation. *J. Exp. Bot.* 36:1860–1876.

Memon, A. R., M. Y. Siddiqi, and A. D. M. Glass. 1985b. Efficiency of K^+ utilization by barley varieties: Activation of pyruvate kinase. *J. Exp. Bot.* 36:79–90.

Mengel, K., and E. A. Kirkby. 1982. Principles of Plant Nutrition. 3rd Ed. Int. Potash Inst., Bern.

Miflin, B. J. 1980. Nitrogen metabolism and amino acid biosynthesis in crop plants. In P. S. Carlson (ed.), The Biology of Crop Productivity. Academic Press, New York. Pp. 255–296.

Mikesell, M. E., G. M. Paulsen, R. Ellis, Jr., and A. J. Casady. 1973. Iron utilization by efficient and inefficient sorghum lines. *Agron. J.* 65:77–80.

Miller, G. W., J. C. Pushnik, and G. W. Welkie. 1984. Iron chlorosis, a world wide problem: The relation of chlorophyll biosynthesis to iron. *J. Plant Nutr.* 7:1–22.

Misal, M. B., and Y. S. Nerkar. 1983. Genotypic variation for iron reductive capacity of rice roots. *Cereal Res. Commun.* 11:291–292.

Mishra, D., and M. Kar. 1974. Nickel in plant growth and metabolism. *Bot. Rev.* 40:395–452.

Mitsui, S., and H. Takatoh. 1963. Nutritional study of silicon in graminaceous crops, Part I. *Soil Sci. Plant Nutr.* 9:49–53.

Moll, R. H., E. J. Kamprath, and W. A. Jackson. 1982a. Analysis and interpretation of factors which contribute to efficiency of nitrogen utilization. *Agron. J.* 74:562–564.

Moll, R. H., E. J. Kamprath, and W. A. Jackson. 1982b. The potential for genetic improvement in nitrogen use efficiency in maize. *27th Annu. Corn Sorghum Research Conf., Am. Seed Trade Assoc., Washington, D.C.* Publ. No. 37, pp. 163–175.

Moll, R. H., E. J. Kamprath, and W. A. Jackson. 1987. Development of nitrogen-efficient prolific hybrids of maize. *Crop Sci.* 27:181–186.

Mondal, W. A., and M. A. Chaudhuri. 1985. Comparison of phosphorus mobilization during monocarpic senescence in rice cultivars with sequential and non-sequential leaf senescence. *Physiol. Plant.* 65:221–227.

Mooers, C. A. 1921. The agronomic placement of varieties. *J. Am Soc. Agron.* 13:337–352.

Mooers, C. A. 1922. Varieties of corn and their adaptability to different soil. *Tenn. Agric. Exp. Stn., Bull.* No. 126.

Mooers, C. A. 1933. The influence of soil productivity on the order of yield in a varietal trial in corn. *J. Am. Soc. Agron.* 25:796–800.

Moorby, J., and R. T. Besford. 1983. Mineral nutrition and growth. In A. Läuchli and R. L. Bieleski (eds.), Inorganic Plant Nutrition. Encyclopedia of Plant Physiology, New Series, Vol. 15B. Springer-Verlag, New York. Pp. 481–527.

Moore, D. P. 1972. Mechanisms of micronutrient uptake by plants. In J. J. Mortvedt, P. M. Giordano, and W. L. Lindsay (eds.), Micronutients in Agriculture. SSSA, Madison, Wisconsin. Pp. 171–198.

Morgutti, S., G. A. Sacchi, and S. M. Cocucci. 1984. Effects of Ni^{2+} on proton extrusion, dark CO_2 fixation and malate synthesis in maize roots. *Physiol. Plant.* 60:70–74.

Mori, S., M. Hachisuka, S. Kawai, S. Takagi, and N. Kishi-Nishizawa. 1988. Peptides related to phytosiderophore secretion in Fe-deficient barley roots. *J. Plant Nutr.* 11:653–662.

Mori, S., N. Nishizawa, S. Kawai, Y. Sato, and S. Takagi. 1987. Dynamic state of mugineic acid and analogous phytosiderophores in Fe-deficient barley. *J. Plant Nutr.* 10:1003–1011.

Morris, F., V. J. Flynn, and M. L. Reilly. 1985. Nitrogen and phosphorus mobilisation in maturing/senescing wheat flag leaves. *Field Crops Res.* 12:71–80.

Mortvedt, J. J. 1975. Iron chlorosis. *Crops Soils* 27(9):10–12.

Motto, M., M. Saccomani, and G. Cacco. 1982. Combining ability estimates of sulfate uptake efficiency in maize. *Theor. Appl. Genet.* 64:41–46.

Mozafar, A. 1987. Nubbins (partially barren ears of maize [*Zea mays* L.]): A review of major causes including mineral nutrient deficiency. *J. Plant Nutr.* 10:1509–1521.

Muchow, R. C., and G. L. Wilson. 1976. Photosynthetic and storage limitations to yield in *Sorghum bicolor* L. Moench. *Aust. J. Agric. Res.* 27:489–500.

Mulder, E. C. 1948. Importance of molybdenum in the nitrogen metabolism of microorganisms and higher plants. *Plant Soil* 1:94–119.

Munns, D. N., L. Jacobson, and C. M. Johnson. 1963a. Uptake and distribution of manganese in oat plants. II. A kinetic model. *Plant Soil* 19:193–204.

Munns, D. N., C. M. Johnson, and L. Jacobson. 1963b. Uptake and distribution of manganese in oat plants. I. Varietal variation. *Plant Soil* 19:115–126.

Munns, D. N., C. M. Johnson, and L. Jacobson. 1963c. Uptake and distribution of manganese in oat plants. III. An analysis of biotic and environmental effects. *Plant Soil* 19:285–295.

Munson, R. D. 1974. Plant breeding and nutrient concentration or uptake: A perspective. Mimeo of talk presented at ASA meetings, Chicago, Illinois.

Murphy, L. S., and L. M. Walsh. 1972. Correction of micronutrient deficiencies with fertilizers. In J. J. Mortvedt, P. M. Giordano, and W. L. Lindsay (eds.), Micronutrients in Agriculture. SSSA, Madison, Wisconsin. Pp. 347–387.

Murray, G. A., and J. A. Benson. 1976. Oat response to manganese and zinc. *Agron. J.* 68:615–616.

Murtadha, H. M. 1986. Effects of nitrate/ammonium ratio, nitrogen source, temperature, relative humidity, and light intensity on growth and calcium uptake, translocation, and accumulation in sorghum [*Sorghum bicolor* (L.) Moench]. Ph.D. Thesis, Univ. of Nebraska, Lincoln. (*Diss. Abstr.B* 47:861B.)

Murtadha, H. M., J. W. Maranville, and R. B. Clark. 1988. Calcium deficiency in sorghum grown in controlled environments in relation to nitrate/ammonium ratio and nitrogen source. *Agron. J.* 80:125–130.

Murty, B. R. 1979. Selection of parental material, breeding methods and evaluation procedures in developing improved crop varieties. *Indian J. Genet. Plant Breed.* 39:305–314.

Muruli, B. I., and G. M. Paulsen. 1981. Improvement of nitrogen use efficiency and its relationship to other traits in maize. *Maydica* 26:63–75.

Mushi, A. A. A. 1965. Studies on the uptake and metabolism of Fe^{59} in chlorosis susceptible and resistant grain sorghum. Ph.D. Thesis, Texas A&M Univ., College Station. (*Diss. Abstr. B* 27:2568B, 1967.)

Myers, W. M. 1960. Genetic control of physiological processes: Consideration of different ion uptake by plants. In R. S. Caldecott and L. A. Snyder (eds.), Radioisotopes in the Biosphere. Univ. of Minnesota Press, Minneapolis. Pp. 201–226.

Naismith, R. W., M. W. Johnson, and W. I. Thomas. 1974. Genetic control of relative calcium, phosphorus, and manganese accumulation on chromosome 9 in maize. *Crop Sci.* 14:845–849.

Nambiar, E. K. S. 1976a. Genetic differences in the copper nutrition of cereals. I. Differential responses of genotypes to copper. *Aust. J. Agric. Res.* 27:453–463.

Nambiar, E. K. S. 1976b. Genetic differences in the copper nutrition of cereals. II. Genotypic differences in response to copper in relation to copper, nitrogen and other mineral contents of plants. *Aust. J. Agric. Res.* 27:465–477.

Neenan, M. 1960. The effects of soil acidity on the growth of cereals, with particular reference to the differential reaction of varieties thereto. *Plant Soil* 12:324–328.

Neilands, J. B. 1981. Microbial iron compounds. *Annu. Rev. Biochem.* 50:715–731.

Neyra, C. A. (ed.). 1986. Biochemical Basis of Plant Breeding. Vol. 2: Nitrogen Metabolism. CRC Press, Boca Raton, Florida.

Nielsen, N. E., and S. A. Barber. 1979. Differences among genotypes of corn in the kinetics of P uptake. *Agron. J.* 70:695–698.

Nielsen, N. E., and J. K. Schjorring. 1983. Efficiency and kinetics of phosphorus uptake from soil by various barley genotypes. *Plant Soil* 72:225–230.

Nirale, A. S., S. Kannan, and S. Ramani. 1982. Heterosis in ion uptake patterns in some sorghum hybrids and parents: A study with ^{86}Rb absorption and transport. *J. Plant Nutr.* 5:15–26.

Nomoto, K., and Y. Ofune. 1982. Studies on structure, synthesis and metal complexes of mugineic acid. *J. Synth. Org. Chem., Jpn.* 40:401–414. (In Jpn.; Engl. summ.)

Nomoto, K., H. Yoshioka, M. Arima, S. Fushiya, S. Takagi, and T. Takemoto. 1981. Structure of 2'-deoxymugineic acid, a novel amino acid possessing an iron-chelating activity. *Chimia* 35:249–250.

Noonan, J. B. 1953. Molybdenum deficiency in maize and other crops in the Taree District. *Agric. Gaz. N.S.W.* 64:422–424.

Novoa, R., and R. S. Loomis. 1981. Nitrogen and plant production. *Plant Soil* 58:177–204.

Nyborg, M. 1970. Sensitivity to manganese deficiency of different cultivars of wheat, oats, and barley. *Can. J. Plant Sci.* 50:198–200.

Oaks, A., and B. Hirel. 1985. Nitrogen metabolism in roots. *Annu. Rev. Plant Physiol.* 36:345–365.

O'Connor, G. A., W. L. Lindsay, and S. R. Olsen. 1975. Iron diffusion to plant roots. *Soil Sci.* 119:285–289.

Odurukwe, S. O., and D. N. Maynard. 1969. Mechanism of the differential reponse of WF9 and Oh40B corn seedlings to iron nutrition. *Agron. J.* 61:694–697.

Ogata, S., T. Kubo, K. Fujita, and K. Kouno. 1983. Studies on interaction between carbon and nitrogen metabolism in C_3 and C_4 plants. 1. Effects of nitrogen nutrition on photosynthetic rates, and ribulose-1,5-biphosphate carboxylase and phosphoenolpyruvate carboxylase activities. *J. Jpn. Soc. Grassl. Sci.* 29:1–8. (In Jpn.; Engl. summ.)

Okuda, A., and E. Takahashi. 1965. The role of silicon. In The Mineral Nutrition of the Rice Plant. Johns Hopkins Press, Baltimore, Maryland. Pp. 123–146.

Olsen, R. A., J. H. Bennett, D. Blume, and J. C. Brown. 1981a. Chemical aspects of the Fe stress response mechanism in tomatoes. *J. Plant Nutr.* 3:905–921.

Olsen, R. A., and J. C. Brown. 1980a. Factors related to iron uptake by dicotyledonous and monocotyledonous plants. I. pH and reductant. *J. Plant Nutr.* 2:629–645.

Olsen, R. A., and J. C. Brown. 1980b. Factors related to iron uptake by dicotyledonous and monocotyledonous plants. II. The reduction of Fe^{3+} as influenced by roots and inhibitors. *J. Plant Nutr.* 2:647–660.

Olsen, R. A., R. B. Clark, and J. H. Bennett. 1981b. The enhancement of soil fertility by plant roots. *Am. Sci.* 69:378–384.

Olsen, S. R. 1972. Micronutrient interactions. In J. J. Mortvedt, P. M. Giordano, and W. L. Lindsay (eds.), Micronutrients in Agriculture. SSSA, Madison, Wisconsin. Pp. 243–264.

Ozanne, P. G. 1980. Phosphate nutrition of plants—A general treatise. In F. E. Khasawneh, E. C. Sample, and E. J. Kamprath (eds.), The Role of Phosphorus in Agriculture. ASA, CSSA, and SSSA, Madison, Wisconsin. Pp. 559–589.

Paccaud, F. X., A. Fossati, and H. S. Cao. 1985. Breeding for yield and quality in winter wheat: Consequences for nitrogen uptake and partitioning efficiency. *Z. Pflanzenzuecht.* 94:89–100.

Pacovsky, R. S., G. J. Bethlenfalvay, and E. A. Paul. 1986. Comparisons between P-fertilized and mycorrhizal plants. *Crop Sci.* 26:151–156.

Palmer, B., and R. S. Jessop. 1977. Some aspects of wheat cultivar response to applied phosphate. *Plant Soil* 47:63–73.

Pan, W. L., J. J. Camberato, W. A. Jackson, and R. H. Moll. 1986. Utilization of previously accumulated and concurrently absorbed nitrogen during reproductive growth of maize. *Plant Physiol.* 82:247–253.

Pan, W. L., W. A. Jackson, and R. H. Moll. 1985. Nitrate uptake and partitioning by corn (*Zea mays* L.) root systems and associated morphological differences among genotypes and stages of root development. *J. Exp. Bot.* 36:1341–1351.

Pan, W. L., R. H. Teyker, W. A. Jackson, and R. H. Moll. 1987. Diurnal variation in nitrate, potassium, and phosphate uptake in maize seedlings: Considerations in screening genotypes for uptake efficiency. *J. Plant Nutr.* 10:1819–1833.

Pandey, D. P., and S. Kannan. 1982. Absorption and transport of Fe and Rb in rice cultivars differing in their Fe-stress response: An analysis of the patterns of uptake in relation to the tolerance. *J. Plant Nutr.* 5:27–43.

Papastylianou, I., R. D. Graham, and D. W. Puckridge. 1984. Diagnosis of the nitrogen status of wheat at tillering and prognosis for maximum grain yield. *Commun. Soil Sci. Plant Anal.* 15:1423–1436.

Pate, J. S., and D. B. Layzell. 1981. Carbon and nitrogen partitioning in the whole plant—A thesis based on empirical modeling. In J. D. Bewley (ed.), Nitrogen and Carbon Metabolism. Nijhoff, The Hague, Netherlands. Pp. 94–134.

Pavlik, B. M. 1983a. Nutritient and productivity relations of dune grasses *Ammophila arenaria* and *Elymus mollis*. I. Blade photosynthesis and nitrogen use efficiency in the laboratory and field. *Oecologia* 57:227–232.

Pavlik, B. M. 1983b. Nutrient and productivity relations of the dune grasses *Ammophila arenaria* and *Elymus mollis*. II. Growth and patterns of dry matter and nitrogen allocation as influenced by nitrogen supply. *Oecologia* 57:233–238.

Peaslee, D. E., R. Isarangkura, and J. E. Leggett. 1981. Accumulation and translocation of zinc by two corn cultivars. *Agron. J.* 73:729–732.

Perby, H. 1986. Cultivar differences in macro-element nutrition of barley (*Hordeum vulgare* L.) as influenced by mineral supply, temperature, and plant age. Ph.D. Thesis, Univ. of Lund, Lund.

Perby, H., and P. Jensen. 1983. Varietal differences in uptake and utilization of nitrogen and other macro-elements in seedlings of barley, *Hordeum vulgare*. *Physiol. Plant.* 58:223–230.

Perby, H., and P. Jensen. 1984. Net uptake and partitioning of nitrogen and potassium in cultivars of barley during ageing. *Physiol. Plant.* 61:559–565.

Perby, H., and P. Jensen. 1986. Variation in growth and accumulation of N., K^+, Ca^{2+}, and Mg^{2+} among barley cultivars exposed to various nutrient regimes and root/shoot temperatures. *Physiol. Plant.* 67:166–172.

Perby, H., and P. Jensen. 1987. Vegetative adaptation to N stress regimes in two barley cultivars with different N requirement. In W. H. Gabelman and B. C. Loughman (eds.), Genetic Aspects of Plant Mineral Nutrition. Nijhoff, Dordrecht, Netherlands. Pp. 361–367.

Perez, C. M., G. B. Cagamapang, B. V. Esmama, R. U. Monservati, and B. O. Juiliano. 1973. Protein metabolism in leaves of developing grains of rices differing in grain protein content. *Plant Physiol.* 51:537–542.

Peterson, L. W., and R. C. Huffaker. 1975. Loss of ribulose-1,5-diphosphate carboxylase and increase in proteolytic activity during senescence of detached primary barley leaves. *Plant Physiol.* 55:1009–1015.

Pettersson, S. 1978. Varietal differences in rubidium uptake efficiency of barley roots. *Physiol. Plant.* 44:1–6.

Pettersson, S. 1986. Growth, contents of K^+ and kinetics of K^+ (^{85}Rb) uptake in barley cultured at different low supply rates of potassium. *Physiol. Plant.* 66:122–128.

Pettersson, S., and P. Jensen. 1978. Allosteric and non-allosteric regulation of rubidium influx in barley roots. *Physiol. Plant.* 44:110–114.

Pettersson, S., and P. Jensen. 1983. Variation among species and varieties in uptake and utilization of potassium. In M. R. Saric and B. C. Loughman (eds.), Genetic Aspects of Plant Nutrition. Nijhoff, The Hague, Netherlands. Pp. 151–157.

Phillips, J. W., D. E. Baker, and C. O. Clagett. 1971a. Identification of compounds which account for variation in P concentration in corn hybrids. *Agron. J.* 63:541–543.
Phillips, J. W., D. E. Baker, and C. O. Clagett. 1971b. Kinetics of P absorption by excised roots and leaves of corn hybrids. *Agron. J.* 63:517–520.
Pierson, E. E., R. B. Clark, D. P. Coyne, and J. W. Maranville. 1984a. Plant genotype differences to ferrous and total Fe in emerging leaves. II. Dry beans and soybeans. *J. Plant Nutr.* 7:355–369.
Pierson, E. E., R. B. Clark, D. P. Coyne, and J. W. Maranville. 1986. Iron deficiency stress effects on total iron in various leaves and nutrient solution pH in sorghum and beans. *J. Plant Nutr.* 9:893–907.
Pierson, E. E., R. B. Clark, J. W. Maranville, and D. P. Coyne. 1984b. Plant genotype differences to ferrous and total Fe in emerging leaves. I. Sorghum and maize. *J. Plant Nutr.* 7:371–387.
Pilbeam, D. J., and E. A. Kirkby. 1983. The physiological role of boron in plants. *J. Plant Nutr.* 6:563–582.
Pinton, R., Z. Varanini, A. Maggioni, and H. Frick. 1987. Potassium flux in corn roots during augmentation of ion uptake. *J. Plant Nutr.* 10:1975–1982.
Piper, C. S. 1942. Investigations on copper deficiency in plants. *J. Agric. Sci.* 32:143–178.
Pollard, A. S., A. J. Parr, and B. C. Loughman. 1977. Boron in relation to membrane function in higher plants. *J. Exp. Bot.* 28:831–841.
Ponnamperuma, F. N. 1976. Screening rice for tolerance to mineral stresses. In M. J. Wright (ed.), Plant Adaptation to Mineral Stress in Problem Soils. Cornell Univ. Press, Ithaca, New York. Pp. 341–353.
Porter, J. R. 1986. Production of a putative phytosiderophore by soybeans in response to iron deficiency stress. *J. Plant Nutr.* 9:1113–1121.
Porter, O. A., and J. T. Moraghan. 1975. Differential response of two corn inbreds to varying root temperature. *Agron. J.* 67:515–518.
Powell, P. E., P. J. Szaniszlo, G. R. Cline, and C. P. P. Reid. 1982. Hydroxamate siderophores in the iron nutrition of plants. *J. Plant Nutr.* 5:653–673.
Przemeck, E., and M. Kuche. 1986. Accumulation and reduction of nitrate in cereal plants dependent on N supply. *Plant Soil* 91:405–410.
Pushnik, J. C., G. W. Miller, and J. H. Manwaring. 1984. The role of iron in higher plant chlorophyll biosynthesis, maintenance and chloroplast biogensis. *J. Plant Nutr.* 7:733–758.
Rademacher, B. 1940. Uber die Veranderungen des Kupfergehaltes, den Verlauf der Kupferaufnahme und den Kupferentzug deim Hafer. *Z. Pflanzenernaehr. Bodenkd.* 19:80–108.
Rahimi, A., and W. Bussler. 1973. The effect of copper deficiency on the tissue structure of higher plants. *Z. Pflanzenernaehr. Bodenkd.* 135:183–195. (In Ger.; Engl. summ.)
Raju, P. S. 1985. Differential phosphorus nutrition in sorghum genotypes. M.S. Thesis, Univ. of Nebraska, Lincoln.
Raju, P. S. 1986. Vesicular–arbuscular mycorrhizal infection effects on growth and uptake of phosphorus and mineral elements by sorghum. Ph.D. Thesis, Univ. of Nebraska, Lincoln. (*Diss. Abstr. B* 47:4353B, 1987.)
Raju, P. S., R. B. Clark, J. R. Ellis, and J. W. Maranville. 1987a. Vesicular–arbuscular mycorrhizal infection effects on sorghum growth, phosphorus efficiency, and mineral element uptake. *J. Plant Nutr.* 10:1331–1339.
Raju, P. S., R. B. Clark, R. K. Maiti, and J. W. Maranville. 1987b. Phosphorus uptake, distribution, and use by glossy and nonglossy sorghum. *J. Plant Nutr.* 10:2017–2024.
Ramani, S., and S. Kannan. 1985a. An examination of zinc uptake patterns by cultivars of sorghum and maize: Differences amongst hybrids and their parents. *J. Plant Nutr.* 8:1199–1210.
Ramani, S., and S. Kannan. 1985b. Studies on Zn uptake and influence of Zn and Fe on chlorophyll development in young maize cultivars. *J. Plant Nutr.* 8:1183–1189.
Ramani, S., and S. Kannan. 1986. Molybdenum uptake and nitrate reductase activity in sorghum (*Sorghum bicolor* L. Moench). Evidence for heterosis. *Z. Pflanzensuecht.* 97:334–339.
Randhawa, N. S., and P. N. Takkar. 1976. Screening of crop varieties with respect to micronutrient stresses in India. In M. J. Wright (ed.), Plant Adaptation to Mineral Stress in Problem Soils. Cornell Univ. Press, Ithaca, New York. Pp. 393–400.

Rao, K. P., and D. W. Rains. 1976. Nitrate absorption by barley. I. Kinetics and energetics. *Plant Physiol.* 57:55–58.
Rao, K. P., D. W. Rains, C. O. Qualset, and R. C. Huffaker. 1977. Nitrogen nutrition and grain protein in two spring wheat genotypes differing in nitrate reductase activity. *Crop Sci.* 17:283–286.
Rasmusson, D. C., A. J. Hester, G. N. Fick, and I. Byrne. 1971. Breeding for mineral content in wheat and barley. *Crop Sci.* 11:623–626.
Rasmusson, D. C., L. H. Smith, and R. A. Kleese. 1964. Inheritance of Sr-89 accumulation in wheat and barley. *Crop Sci.* 4:586–589.
Rasmusson, D. C., L. H. Smith, and W. M. Myers. 1963. Effect of genotype on accumulation of strontium-89 in barley and wheat. *Crop Sci.* 3:34–37.
Raven, J. A., and F. A. Smith. 1976. Nitrogen assimilation and tranport in vascular land plants in relation to intracellular pH regulation. *New Phytol.* 76:415–431.
Reddy, C. K., and G. V. S. S. Prasad. 1986. Varietal response to iron chlorosis in upland rice. *Plant Soil* 94:289–292.
Reed, A. J., F. E. Below, and R. H. Hageman. 1980. Grain protein accumulation and the relationship between leaf nitrate reductase and protease activities during grain development in maize (*Zea mays* L.). I. Variation between genotypes. *Plant Physiol.* 66:164–170.
Reed, A. J., and R. H. Hageman. 1980a. Relationship between nitrate uptake, flux, and reduction and the accumulation of reduced nitrogen in maize (*Zea Mays* L.). I. Genotypic variation. *Plant Physiol.* 66:1179–1183.
Reed, A. J., and R. H. Hageman. 1980b. Relationship between nitrate uptake, flux, and reduction and the accumulation of reduced nitrogen in maize (*Zea mays* L.). II. Effect of nutrient nitrate concentration. *Plant Physiol.* 66:1184–1189.
Reid, C. P. P., D. E. Crowley, H. J. Kim, P. E. Powell, and P. J. Szaniszlo. 1984. Utilization of iron by oat when supplied as ferrated synthetic chelate or as ferrated hydroxamate siderophore. *J. Plant Nutr.* 7:437–447.
Rhodes, L. H., and J. W. Gerdemann. 1975. Phosphate uptake zones of mycorrhizal and nonmycorrhizal onions. *New Phytol.* 75:555–561.
Riley, M. M. 1987a. Boron toxicity in barley. *J. Plant Nutr.* 10:2109–2115.
Riley, M. M. 1987b. Molybdenum deficiency in wheat in Western Australia. *J. Plant Nutr.* 10:2117–2123.
Rodriguez-P., M. S. 1977. Varietal differences in maize in the uptake of nitrogen and its translocation to the grain. Ph.D. Thesis, Cornell Univ., Ithaca, New York. (*Diss. Abstr. B* 38:5690B–5691B, 1978.)
Römheld, V. 1987. Existence of two different strategies for the acquisition of iron in higher plants. In D. van der Helm, J. B. Neilands, and G. Winkelmann (eds.), Iron Transport in Microbes, Plants, and Animals. Verlag Chemie, Weinheim. Pp. 353–374.
Römheld, V., and H. Marschner. 1984. Plant-induced pH changes in the rhizosphere of "Fe-efficient" and "Fe-inefficient" soybean and corn cultivars. *J. Plant Nutr.* 7:623–630.
Römheld, V., and H. Marschner. 1986a. Evidence for a specific uptake system for iron phytosiderophores in roots of grasses. *Plant Physiol.* 80:175–180.
Römheld, V., and H. Marschner. 1986b. Mobilization of iron in the rhizosphere of different plant species. *Adv. Plant Nutr.* 2:155–204.
Romani, G., M. T. Marre, M. Bellando, G. Alloatti, and E. Marre. 1985. H^+ extrusion and potassium uptake associated with potential hyperpolarization in maize and wheat root segments treated with permanent weak acids. *Plant Physiol.* 79:734–739.
Russelle, M. P., J. A. Schild, and R. A. Olson. 1984. Phosphorus translocation between small, non-reproductive tillers and the main plant of maize. *Agron. J.* 76:1–4.
Saccomani, M., G. Cacco, and G. Ferrari. 1981. Efficiency of the first steps of sulfate utilization by maize hybrids in relation to their productivity. *Physiol. Plant.* 53:101–104.
Saccomani, M., G. Cacco, and G. Ferrari. 1984. Effect of nitrogen and/or sulfur deprivation on the uptake and assimilation steps of nitrate and sulfate in maize seedlings. *J. Plant Nutr.* 7:1043–1057.
Sample, E. C., R. J. Soper, and G. J. Racz. 1980. Reactions of phosphate fertilizers in soils. In F. E.

Khasawneh, E. C. Sample, and E. J. Kamprath (eds.), The Role of Phosphorus in Agriculture. ASA, CSSA, and SSSA, Madison, Wisconsin. Pp. 263–310.

Sandmann, G., and P. Boger. 1983. The enzymological function of heavy metals and their role in electron transfer processes of plants. In A. Läuchli and R. L. Bieleski (eds.), Inorganic Plant Nutrition. Encyclopedia of Plant Physiology, New Series, Vol. 15B. Springer-Verlag, New York. Pp. 563–596.

Saric, M. R. 1981. Genetic specificity in relation to plant mineral nutrition. *J. Plant Nutr.* 3:743–766.

Saric, M. R. 1987. Progress since the first international symposium: "Genetic Aspects of Plant Mineral Nutrition," Beograd, 1982, and perspectives of future research. In W. H. Gabelman and B. C. Loughman (eds.), Genetic Aspects of Plant Mineral Nutrition. Nijhoff, Dordrecht, Netherlands. Pp. 617–629.

Saric, M. R., B. Krstic, and Z. Stankovic. 1987. Genetic aspects of mineral nutrition of wheat. I. Concentrations of N, P, K, Ca, and Mg in leaves. *J. Plant Nutr.* 10:1539–1545.

Saric, M. R., and B. C. Loughman (eds.). 1983. Genetic Aspects of Plant Nutrition. Nijhoff, The Hague, Netherlands.

Sayre, J. D. 1952a. Accumulation of radioisotopes in corn leaves. *Ohio Agric. Exp. Stn., Res. Bull.* No. 723.

Sayre, J. D. 1952b. Mineral accumulation in corn leaves. *Proc. 28th Annu. Meet. National Joint Committee on Fertilizer Application, Natl. Fert. Assoc., Washington, D.C.* pp. 16–26.

Sayre, J. D. 1952c. Magnesium. . . important element in corn nutrition. *Crops Soils* 5:16–17.

Sayre, J. D. 1955. Mineral nutrition of corn. In G. F. Sprague (ed.), Corn and Corn Improvement. Academic Press, New York. Pp. 293–314.

Schauble, C. E., and S. A. Barber. 1958. Magnesium immobility in the nodes of certain corn inbreds. *Agron. J.* 50:651–653.

Schenk, M. K., and S. A. Barber. 1979. Phosphate uptake by corn as affected by soil characteristics and root morphology. *Soil Sci. Soc. Am. J.* 43:880–881.

Schettini, T. M., W. H. Gabelman, and G. C. Gerloff. 1987. Incorporation of phosphorus efficiency from exotic germplasm into agriculturally adapted germplasm of common bean (*Phaseolus vulgaris* L.). In W. H. Gabelman and B. C. Loughman (eds.), Genetic Aspects of Plant Mineral Nutrition. Nijhoff, Dordrecht, Netherlands. Pp. 559–568.

Schjorring, J. K., and N. E. Nielsen. 1987. Root length and phosphorus uptake by four barley cultivars grown under moderate deficiency of phosphorus in field experiments. *J. Plant Nutr.* 10:1289–1295.

Schmitt, M. R., and G. E. Edwards. 1981. Photosynthetic capacity and nitrogen use efficiency of maize, wheat, and rice: A comparison between C_3 and C_4 photosynthesis. *J. Exp. Bot.* 32:459–466.

Schrader, L. E. 1978. Uptake, accumulation, assimilation, and transport of nitrogen in higher plants. In D. R. Nielsen and J. G. MacDonald (eds.), Nitrogen in the Environment. Vol. 2. Academic Press, New York. Pp. 104–141.

Schrader, L. E. 1984. Functions and transformations of nitrogen in higher plants. In R. D. Hauck (ed.), Nitrogen in Crop Production. ASA, CSSA, and SSA, Madison, Wisconsin. Pp. 55–65.

Schrader, L. E., D. M. Peterson, E. R. Leng, and R. H. Hageman. 1966. Nitrate reductase activity of maize hybrids and their parental inbreds. *Crop Sci.* 6:169–173.

Schrader, L. E., and R. J. Thomas. 1981. Nitrate uptake, reduction and transport in the whole plant. In J. D. Bewley (ed.), Nitrogen and Carbon Metabolism. Nijhoff, The Hague, Netherlands. Pp. 49–93.

Seckback, J. 1982. Ferreting out the secrets of plant ferritin—A review. *J. Plant Nutr.* 5:369–394.

Seetharama, N., R. B. Clark, and J. W. Maranville. 1987. Sorghum genotype differences in uptake and use-efficiency of mineral elements. In W. H. Gabelman and B. C. Loughman (eds.), Genetic Aspects of Plant Mineral Nutrition. Nijhoff, Dordrecht, Netherlands. Pp. 437–443.

Seth, J., T. T. Hebert, and G. K. Middleton. 1960. Nitrogen utilization in high and low protein wheat varieties. *Agron. J.* 52:207–209.

Shaner, D. L., and J. S. Boyer. 1976. Nitrate reductase activity in maize (*Zea mays* L.) leaves. I. Regulation by nitrate flux. *Plant Physiol.* 58:499–504.

Sharma, C. P., S. C. Agarwala, P. N. Sharma, and S. Ahmad. 1971. Performance of eight high yielding wheat varieties in some zinc deficient soils of Uttar Pradesh and their response to zinc amendments in pot culture. *J. Indian Soc. Soil Sci.* 19:93–100.

Sherchand, K., and G. M. Paulsen. 1985a. Genotypic variation in partitioning of phosphorus in relation to nitrogen and dry matter during wheat grain development. *J. Plant Nutr.* 8:1161–1170.

Sherchand, K., and G. M. Paulsen. 1985b. Partitioning of phosphorus in different height isolines of winter wheat. *J. Plant Nutr.* 8:1147–1160.

Sherrard, J. H., R. J. Lambert, F. E. Below, R. T. Dunand, M. J. Messmer, M. R. Willman, C. S. Winkels, and R. H. Hageman. 1986. Use of physiological traits, especially those of nitrogen metabolism for selection in maize. In C. A. Neyra (ed.), Biochemical Basis of Plant Breeding. Vol. 2: Nitrogen Metabolism. CRC Press, Boca Raton, Florida. Pp. 109–130.

Sherrard, J. H., R. J. Lambert, M. J. Messmer, F. E. Below, and R. H. Hageman. 1984a. Plant breeding for efficient plant use of nitrogen. In R. D. Hauck (ed.), Nitrogen in Crop Production. ASA, CSSA, and SSSA, Madison, Wisconsin. Pp. 363–378.

Sherrard, J. H., R. J. Lambert, M. J. Messmer, F. E. Below, and R. H. Hageman. 1984b. Search for useful physiological and biochemical traits in maize. In P. B. Vose and S. G. Blixt (eds.), Crop Breeding, a Contemporary Basis. Pergamon, New York. Pp. 51–66.

Shim, S. C., and P. B. Vose. 1965. Varietal differences in the kinetics of iron uptake by excised rice roots. *J. Exp. Bot.* 16:216–232.

Shukla, U. C., S. K. Arora, Z. Singh, K. G. Prasad, and N. M. Safaya. 1973. Differential susceptibility in some sorghum (*Sorghum vulgare*) genotypes to zinc deficiency in soil. *Plant Soil* 39:423–427.

Shukla, U. C., and H. Raj. 1974. Influence of genetic variability on zinc response in wheat. *Soil Sci. Soc. Am. Proc.* 38:477–479.

Shukla, U. C., and H. Raj. 1976. Zinc response in corn as influenced by genetic variability. *Agron. J.* 68:20–22.

Shukla, U. C., and H. Raj. 1987. Relative response of corn (*Zea mays* L.), pearl-millet [*Pennisetum typhoides*. (Burm F.) Stapf and C. E. Hubb], sorghum (*Sorghum vulgare*), and cowpea [*Vigna unguiculata* (L.) Walp] to zinc deficiency in soil. *J. Plant Nutr.* 10:2057–2067.

Siddiqi, M. Y., and A. D. M. Glass. 1981. Utilization index: A modified approach to the estimation and comparison of nutrient utilization efficiency in plants. *J. Plant Nutr.* 4:289–302.

Siddiqi, M. Y., and A. D. M. Glass. 1983a. Studies of the growth and mineral nutrition of barley varieties. I. Effect of potassium supply on the uptake of potassium and growth. *Can. J. Bot.* 61:671–678.

Siddiqi, M. Y., and A. D. M. Glass. 1983b. Studies of the growth and mineral nutrition of barley varieties. II. Potassium uptake and its regulation. *Can. J. Bot.* 61:1551–1558.

Siddiqi, M. Y., and A. D. M. Glass. 1986. A model for the regulation of K^+ influx, and tissue potassium concentrations by negative feedback effect upon plasmalemma influx. *Plant Physiol.* 81:1–7.

Siddiqi, M. Y., and A. D. M. Glass. 1987. Regulation of K^+ influx in barley: Evidence for a direct control on influx by K^+ concentration of root cells. *J. Exp. Bot.* 38:935–947.

Siddiqi, M. Y., A. D. M. Glass, A. I Hsiao, and A. N. Minjas. 1985. Wild oat/barley interactions: Varietal differences in competitiveness in relation to K^+ supply. *Ann. Bot.* 56:1–7.

Siddiqi, M. Y., A. D. M. Glass, A. I. Hsiao, and A. N. Minjas. 1987a. Genetic differences among wild oat lines in potassium uptake and growth in relation to potassium supply. *Plant Soil* 99:93–105.

Siddiqi, M. Y., A. D. M. Glass, A. I. Hsiao, and A. N. Minjas. 1987b. Genetic differences among wild oat lines in potassium uptake and growth in relation to potassium supply. In W. H. Gabelman and B. C. Loughman (eds.), Genetic Aspects of Plant Mineral Nutrition. Nijhoff, Dordrecht, Netherlands. Pp. 369–381.

Sieverding, E. 1986. Influence of soil water regimes on VA mycorrhiza. IV. Effect of root growth and water relations of *Sorghum bicolor*. *Z. Acker-Pflanzenbau* 157:36–42.

Simpson, R. J., H. Lambers, and M. J. Dalling. 1983. Nitrogen redistribution during grain growth in wheat (*Triticum aestivum* L.). IV. Development of a quantitative model of the translocation of nitrogen to the grain. *Plant Physiol.* 71:7–14.

Singh, B. B. 1971. Effect of vanadium on the growth, yield and chemical composition of maize (*Zea mays* L.). *Plant Soil* 34:209–213.

Singh, B. P., R. A. Singh, M. K, Sinha, and B. N. Singh. 1985. Evaluation of technique for screening Fe-efficient genotypes of rice in calcareous soil. *J. Agric. Sci.* 105:193–197.

Smilde, K. W., and C. H. Henkens. 1967. Sensitivity to copper deficiency of different cereals and strains of cereals. *Neth. J. Agric. Sci.* 15:249–258.

Smith, B. N. 1984. Iron in higher plants: Storage and metabolic role. *J. Plant Nutr.* 7:759–766.

Smith, S. E., and V. Gianinazzi-Pearson. 1988. Physiological interactions between symbionts in vesicular–arbuscular mycorrhizal plants. *Annu. Rev. Plant Physiol.* 39:221–244.

Smith, S. N. 1934. Response of inbred lines and crosses in maize to variations of nitrogen and phosphorus supplied as nutrients. *J. Am. Soc. Agron.* 26:785–804.

Smyth, D. A., and P. Chevalier. 1984. Increases in phosphatase and B-glucosidase activities in wheat seedlings in response to phosphorus-deficient growth. *J. Plant Nutr.* 7:1221–1231.

Sparrow, O. H. B., R. D. Graham, W. J. Davies, and J. S. Ascher. 1983. Genetics of tolerance of barleys to manganese deficiency. *Proc. Aust. Plant Breed. Conf., Adelaide, Australia.*

Spiertz, J. H. J., and J. Ellen. 1978. Effects of nitrogen on crop development and grain growth of winter wheat in relation to assimilation and utilization of assimilates and nutrients. *Neth. J. Agric. Sci.* 26:210–231.

Sriram, N., and J. S. Rao. 1983. Physiological parameters influencing sorghum yield. *Indian J. Agric. Sci.* 53:641–649.

Stevenson, F. J. (ed.). 1982. Nitrogen in Agricultural Soils. Agron. Monogr. No. 22. ASA, CSSA, and SSSA, Madison, Wisconsin.

Stewart, J. A. 1985. Potassium sources, use, and potential. In R. D. Munson (ed.), Potassium in Agriculture. ASA, CSSA, and SSSA, Madison, Wisconsin. Pp. 83–98.

Stringfield, G. H., and R. M. Salter. 1934. Differential response of corn varieties to fertility levels and to seasons. *J. Agric. Res.* 49:991–1000.

Sugiyama, T., M. Mizuno, and M. Hayashi. 1984. Partitioning of nitrogen among ribulose-1,5-biphosphate carboxylase/oxygenase, phosphoenolpyruvate carboxylase, and pyruvate orthophosphate dikinase as related to biomass productivity in maize seedlings. *Plant Physiol.* 75:665–669.

Sutton, P. J., G. A. Peterson, and D. H. Sander. 1983. Dry matter production in tops and roots of winter wheat as affected by phosphorus availability during various growth stages. *Agron. J.* 75:657–663.

Swank, J. C., F. E. Below, R. J. Lambert, and R. H. Hageman. 1982. Interaction of carbon and nitrogen metabolism in the productivity of maize. *Plant Physiol.* 70:1185–1190.

Takagi, S. 1976. Naturally occurring iron-chelating compounds in oat- and rice-root washings. I. Activity measurement and preliminary characterization. *Soil Sci. Plant Nutr.* 22:423–433.

Takagi, S., S. Kamei, and M.-H. Yu. 1988. Efficiency of iron extraction from soil by mugineic acid family phytosiderophores. *J. Plant Nutr.* 11:643–651.

Takagi, S., K. Nomoto, and T. Takemoto. 1984. Physiological aspect of mugineic acid, a possible phytosiderophore of graminaceous plants. *J. Plant Nutr.* 7:469–477.

Takemoto, T., K. Nomoto, S. Fushiya, R. Ouchi, G. Kusano, H. Hikino, S. Takagi, Y. Matsuura, and M. Kondo. 1978. Structure of mugineic acid, a new amino acid possessing an iron-chelating activity from root washings of water-cultured *Hordeum vulgare* L. *Proc. Jpn. Acad., Ser. B* 54:469–473.

Takkar, P. N., and N. P. Kaur. 1984. HCl method for Fe^{2+} estimation to resolve iron chlorosis in plants. *J. Plant Nutr.* 7:81–90.

Tanaka, A., and S. A. Navasero. 1966. Chlorosis of the rice plant induced by high pH of culture solution. *Soil Sci. Plant Nutr.* 12:213–219.

Terman, G. L., P. M. Giordano, and N. W. Christensen. 1975. Corn hybrid yield effects on phosphorus, manganese, and zinc absorption. *Agron. J.* 67:182–184.

Terry, N., and J. Abadia. 1986. Function of iron in chloroplasts. *J. Plant Nutr.* 9:609–646.

Thimann, K. V. 1956. Studies on the growth and inhibition of isolated plant parts. V. The effects of cobalt and other metals. *Am. J. Bot.* 43:241–250.

Thomas, H., and J. L. Stoddart. 1980. Leaf senescence. *Annu. Rev. Plant Physiol.* 31:83–130.

Thomas, W. I., and D. E. Baker. 1966. Application of biochemical genetics in quality improvements and plant nutrition. II. Studies in the inheritance of chemical element accumulation in corn (*Zea mays* L.). *Qual. Plant. Mater. Veg.* 13:98–104.

Thomas, W. I., G. W. Gorsline, C. W. Korman, J. L. Ragland, and C. C. Wernham. 1960. Corn performance studies 1959. *Pa. Agric. Exp. Stn., Prog. Rep.* No. 220.

Tiffin, L. O. 1972. Translocation of micronutrients in plants. In J. J. Mortvedt, P. M. Giordana, and W. L. Lindsay (eds.), Micronutrients in Agriculture. SSSA, Madison, Wisconsin. Pp. 199–229.

Toth, R., T. Page, and R. Castleberry. 1984. Differences in mycorrhizal colonization of maize selections for high and low ear leaf phosphorus. *Crop Sci.* 24:994–996.

Turner, F. T., C. N. Bollich, and J. E. Scott. 1978. Phosphorus-efficient varieties. *IRRI Newsl.* 3(3):13–14.

Usada, H. 1984. Variations in the photosynthesis rate and activity of photosynthetic enzymes in maize leaf tissues of different ages. *Plant Cell Physiol.* 25:1297–1301.

Vanderlip, R. L. 1972. How a sorghum plant develops. *Kans. Agric. Exp. Stn., Coop Ext. Circ.* No. C-447.

Van Sanford, D. A., and C. T. MacKown. 1986. Variation in nitrogen use efficiency among soft red winter wheat genotypes. *Theor. Appl. Genet.* 72:158–163.

Vose, P. B. 1963. The translocation and redistribution of manganese in *Avena*. *J. Exp Bot.* 14:448–457.

Vose, P. B. 1982. Iron nutrition in plants: A world overview. *J. Plant Nutr.* 5:233–249.

Vose, P. B. 1984. Effects of genetic factors on nutritional requirements of plants. In P. B. Vose and S. G. Blixt (eds.), Crop Breeding, a Contemporary Basis. Pergamon, New York. Pp. 67–114.

Vose, P. B. 1987. Genetical aspects of mineral nutrition—Progress to date. In W. H. Gabelman and B. C. Loughman (eds.), Genetic Aspects of Plant Mineral Nutrition. Nijhoff, Dordrecht, Netherlands. Pp. 3–13.

Vose, P. B., and S. G. Blixt (eds.). 1984. Crop Breeding, a Contemporary Basis. Pergamon, New York.

Vose, P. B., and E. L. Breese. 1964. Genetic variation in the utilization of nitrogen by ryegrass species *Lolium perenne* and *L. multiflorum*. *Ann. Bot.* 28:251–270.

Vose, P. B., and D. J. Griffiths. 1961. Manganese and magnesium in the grey speck syndrome of oats. *Nature (London)* 191:299–300.

Waldren, R. P., and A. D. Flowerday. 1979. Growth stages and distribution of dry matter, N, P, and K in winter wheat. *Agron. J.* 71:391–397.

Walker, C. D., R. D. Graham, J. T. Madison, E. E. Cary, and R. M. Welch. 1985. Effects of Ni deficiency on some nitrogen metabolites in cowpea (*Vigna unguiculata* L. Walp.). *Plant Physiol.* 79:474–479.

Walker, C. D., and J. Webb. 1981. Copper in plants: Forms and behavior. In J. F. Loneragan, A. D. Robson, and R. D. Graham (eds.), Copper in Soils and Plants. Academic Press, New York. Pp. 189–212.

Wallace, A., and O. R. Lunt. 1960. Iron chlorosis in horticultural plants. A review. *Proc. Am. Soc. Hortic. Sci.* 75:819–841.

Wallace, W., and J. S. Pate. 1965. Nitrate reductase in the field pea (*Pisum arvense* L.). *Ann. Bot.* 29:655–671.

Warncke, D. D., and S. A. Barber. 1974. Root development and nutrient uptake by corn grown in solution culture. *Agron. J.* 66:514–516.

Warner, R. L., R. H. Hageman, J. W. Dudley, and R. J. Lambert. 1969. Inheritance of nitrate reductase activity in *Zea mays* L. *Proc. Natl. Acad. Sci. U.S.A.* 62:785–792.

Warnock, R. E. 1970. Micronutrient uptake and mobility within corn plants. (*Zea mays* L.) in relation to phosphorus-induced zinc deficiency. *Soil Sci. Soc. Am. Proc.* 34:765–769.

Weavind, T. E. F., and J. F. Hodgson. 1971. Iron absorption by wheat roots as a function of distance from the root tip. *Plant Soil* 34:697–705.

Weiss, M. G. 1943. Inheritance and physiology of efficiency in iron utilization in soybeans. *Genetics* 28:253–268.

Welch, L. F. 1979. Nitrogen use and behaviour in crop production. *Ill. Agric. Exp. Stn., Bull.* No. 761.

Welch, L. F., and R. L. Flannery. 1985. Potassium nutrition of corn. In R. D. Munson (ed.), Potassium in Agriculture. ASA, CSSA, and SSSA, Madison, Wisconsin. Pp. 647–664.

Welch, R. M. 1981. The biological significance of nickel. *J. Plant Nutr.* 3:345–356.
Werner, D., and R. Roth. 1983. Silica metabolism. In A. Läuchli and R. L. Bieleski (eds.), Inorganic Plant Nutrition. Encyclopedia of Plant Physiology, New Series, Vol. 15B. Springer-Verlag, New York. Pp. 682–694.
Whiteaker, G., G. C. Gerloff, W. H. Gabelman, and D. Lindgren. 1976. Intraspecific differences in growth of beans at stress levels of phosphorus. *J. Am. Soc. Hortic. Sci.* 101:472–475.
Whittemore, D., D. K. Nordstrom, and S. S. Que Hee. 1981. Physical and chemical properties. In Multimedia Criteria for Iron and Compounds. EPA, Cincinnati, Ohio. Pp. 2–29.
Wieneke, J. 1985. $^{32}P/^{33}P$-uptake and remobilization before and during tillering of spring barley. *Angew. Bot.* 59:393–408. (In Ger.)
Wieneke, J., C. Schimansky, and K. Isermann. 1983. Distribution of fertilizer and soil-derived phosphate in spring barley as a function of the P-availability in the soil as well as the level and placement of $^{32}P/^{33}P$-labeled fertilizer. *Landwirtsch. Forsch.* 36:348–361. (In Ger.)
Wier, R. G., and A. Hudson. 1966. Molybdenum deficiency in maize in relation to seed reserves. *Aust. J. Exp. Agric. Anim. Husb.* 6:35–41.
Wilhelm, N. S., R. D. Graham, and A. D. Rovira. 1988. The genotypes of both the wheat host and its pathogen take-all control the status of manganese in the environment and the rate of infection. *Abstr. 3rd Int. Symp. Genetic Aspects of Plant Mineral Nutrition, Inst. Crop Sci. Plant Breed., Fed. Res. Cent. Agric. Braunschweig, F.R.G.*
Wilkinson, H. F. 1972. Movement of micronutrients to plant roots. In J. J. Mortvedt, P. M. Giordano, and W. L. Lindsay (eds.), Micronutrients in Agriculture. SSSA, Madison, Wisconsin. Pp. 139–169.
Williams, E. P., R. B. Clark, W. M. Ross, G. M. Herron, and M. D. Witt. 1987. Variability and correlation of iron-deficiency symptoms in a sorghum population evaluated in the field and growth chamber. *Plant Soil* 99:127–137.
Williams, E. P., W. M. Ross, R. B. Clark, G. M. Herron, and M. D. Witt. 1986. Iron-deficiency chlorosis: Its heritability and effects on agronomic traits in a sorghum population. *J. Plant Nutr.* 9:423–433.
Williams, R. F. 1948. The effects of phosphorus supply on the rates of intake of phosphorus and nitrogen and upon certain aspects of phosphorus metabolism in gramineous plants. *Aust. J. Sci. Res., Ser. B* 1:333–361.
Wilson, J. R. 1975. Comparative response to nitrogen deficiency of a tropical and temperate grass in the interrelation between photosynthesis, growth, and the accumulation of non-structural carbohydrate. *Neth. J. Agric. Sci.* 23:104–112.
Wilson, J. R., and R. H. Brown. 1983. Nitrogen response of *Panicum* species differing in CO_2 fixation pathways. I. Growth analysis and carbohydrate accumulation. *Crop Sci.* 23:1148–1153.
Wilson, J. R., and K. P. Haydock. 1971. The comparative response of tropical and temperate grasses to varying levels of nitrogen and phosphorus nutrition. *Aust. J. Agric. Res.* 22:573–587.
Winkelmen, G. 1982. Fungal siderophores–membrane reactive ionophores. *J. Plant Nutr.* 5:645–651.
Wittenbach, V. A. 1978. Breakdown of ribulose biphosphate carboxylase and change in proteolytic activity during dark-induced senescence of wheat seedlings. *Plant Physiol.* 62:604–608.
Woodend, J. J., A. D. M. Glass, and C. O Person. 1986. Intraspecific variation for nitrate uptake and nitrogen utilization in wheat (*T. aestivum* L.) grown under nitrogen stress. *J. Plant Nutr.* 9:1213–1225.
Woodend, J. J., A. D. M. Glass, and C. O. Person. 1987. Genetic variation in the uptake and utilization of potassium in wheat (*Triticum aestivum* L.) varieties grown under potassium stress. In W. H. Gabelman and B. C. Loughman (eds.), Genetic Aspects of Plant Mineral Nutrition. Nijhoff, Dordrecht, Netherlands. Pp. 383–391.
Wright, M. J. (ed.). 1976. Plant Adaptation to Mineral Stress in Problem Soils. Cornell Univ. Press, Ithaca, New York.
Yao, J., and S. A. Barber. 1986. Effect of one phosphorus rate placed in different soil volumes on P uptake and growth of wheat. *Commun. Soil Sci. Plant Anal.* 17:819–827.
Yongrui, W., L. Zhenshang, and C. Kunchao. 1987. Translocation and distribution of ^{32}P, ^{35}S and ^{14}C-glucose in hybrid rices and their parents. *J. Plant Nutr.* 10:1623–1630.

Yoshida, S., Y. Onishi, and K. Kitagishi. 1959. Role of silicon in rice nutrition. *Soil Sci. Plant Nutr.* 5:127–133.

Young, W. I., and D. C. Rasmusson. 1966. Variety differences in strontium and calcium accumulation in seedlings of barley. *Agron. J.* 58:481–483.

Zaiter, H. Z., D. P. Coyne, and R. B. Clark. 1987. Genetic variation and inheritance of resistance of leaf iron-deficiency chlorosis in dry beans. *J. Am. Soc. Hortic. Sci.* 112:1019–1022.

Zaiter, H. Z., D. P. Coyne, and R. B. Clark. 1988. Genetic variation, heritability, and selection response to iron deficiency chlorosis in dry beans. *J. Plant Nutr.* 11:739–746.

Zieserl, J. F., and R. H. Hageman. 1962. Effect of genetic composition on nitrate reductase activity in maize. *Crop Sci.* 2:512–515.

Zieserl, J. F., W. L. Rivenbark, and R. H Hageman. 1963. Nitrate reductase activity, protein content, and yield of four maize hybrids at varying plant populations. *Crop Sci.* 3:27–32.

Zweifel, T. R., J. W. Maranville, W. M. Ross, and R. B. Clark. 1987. Nitrogen fertility and irrigation influence on grain sorghum nitrogen efficiency. *Agron. J.* 79:419–422.

6

Legume Genetics and Breeding for Stress Tolerance and Nutrient Efficiency[1]

THOMAS E. DEVINE, JOE H. BOUTON, and TADESSE MABRAHTU

[1]Portions of this paper are reprinted by permission of the publisher from "Genetic Fitting of Crops to Problem Soils" by Thomas E. Devine in *Breeding Plants for Less Favorable Environments,* edited by Meryl N. Christiansen and Charles F. Lewis. Copyright © 1982 by John Wiley & Sons.

Crops as Enhancers of Nutrient Use

The goal of improving plant species for use by humans may take many forms. Specific objectives will be a function of breeders' creative visions of the new forms of the species that can be synthesized and the forms that current and future technology can utilize for human welfare. Overcoming the factors restricting crop productivity is often a feasible and fruitful objective. When cultivation of a crop species is to be extended beyond the region where it was domesticated from indigenous wild forms, genetic alteration of the crop to cope with new edaphic conditions is often needed. The introduction of fertilizer use, particularly since the advent of the widespread use of chemically synthesized fertilizer, often presents the crop species with an enriched edaphic milieu. Unfortunately, little research has been recorded on genetic modification of crop species for enhanced efficiency in the utilization of nutrients. More extensive research is available on the genetic control of plant tolerance to mineral deficiencies and toxicities in the soil. These tolerances may profoundly affect the capacity of the crop to efficiently utilize nutrients in the edaphic milieu.

I. Formulation of Goals

Accurate analysis of the problem to be solved and the goal to be achieved are critical to the effective use of genetics in the construction of useful new genotypes. It would be a serious error to dismiss the task of determining the factors restricting crop production in any particular soil as merely perfunctory. The collaboration of soil scientists and plant breeders is needed to assure the accurate definition of the problems, both current and potential, and the assessment of genetic resources available for alleviating or ameliorating these problems.

Effective and efficient use of soil resources in a particular area for crop production requires an implicit or explicit assessment of both the potential to adapt the soil to fit the plant and the potential to adapt the plant to fit the soil. For example, large areas of acid soils are found on every continent (Wambeke, 1976). Although liming and fertilization may alleviate these conditions, for many developing countries these practices can be a prohibitive expense. Even if it were affordable, subsoils are rarely changed by liming and fertilization.

A wide range of acid soil tolerance has been documented among the different genera of forage legumes. Alfalfa was found to be extremely susceptible to low soil pH and less tolerant than other legumes, especially white clover (Tanaka *et al.*, 1984). *Sericea lespedeza* was extremely tolerant and alfalfa very susceptible to acid soil conditions in a study by Joost and Hoveland (1986). In their study, alfalfa root growth rate was reduced by 69% in acid soil, but some sericea lines showed no reduction. Big trefoil gave higher yields than white and red clover and birdsfoot trefoil on undeveloped, acid soils (Scott and Lowther, 1981). Therefore, selection of a tolerant legume species to grow in acid soils may provide an effective solution. However, if the desired forage species is a sensitive species, then cultivars tolerant of the acid soil conditions need to be developed.

In Georgia, a state in the southeastern United States with predominantly acid soils, subsoil liming increased alfalfa yield by almost 50% over conventional liming of surface soils, indicating that deeper rooting achieved through an ability to overcome subsoil toxicity has a significant effect on both stand life and productivity (Bouton *et al.*, 1986; Sumner *et al.*, 1986). If, through selection and breeding, one could develop acid soil-tolerant germplasms of legumes, similar enhancement in performance may be realized. Acid soil tolerance could then be incorporated, along with other desirable agronomic traits, in new cultivars to provide benefits to farmers and consumers.

II. Assessment of Stress Factors in the Edaphic Environment and Their Amendability to Resolution through Plant Breeding

Before starting a program to synthesize new crop cultivars specifically adapted to particular edaphic conditions, several criteria should be considered. First, the specific soil problems, both present and potential, in the region for which the breeder is attempting to develop cultivars must be accurately identified. Second, the problem must be sufficiently acute and widespread to merit the investment of resources required to develop adapted cultivars. Third, the nature of the problem must be such that it is not amenable to alleviation more economically by other means such as liming, fertilization, or selection of another crop species adapted to the stress condition. Fourth, the problems must be assessed as at least partially resolvable through plant breeding. The latter criterion is a composite of several assumptions, namely that: (a) techniques exist, or can be devised, to assay plant response to the pertinent edaphic stresses; (b) useful genetic variation for the characteristics needed can either be synthesized through mutation or is already available, either in agronomically suitable cultivars or in noncultivated forms of the crop species or in related species that can be introgressed into the crop cultivars; (c) the character is heritable; and (d) the estimated improvement in adaptation is sufficient to be of applied use. The expertise of both soil scientists and plant breeders is needed from the very inception of a program for problem definition and goal development.

III. Breeding Principles

The opportunity to capitalize on crop adaptation to defined soil conditions may arise at any stage in a breeding program. Ideally, however, such an opportunity would be most valuable at the initiation of the program. We will consider the latter case.

After the nature of the soil stress condition (e.g., nutrient deficiency, Al or Mn toxicity, saline toxicity) is defined, the breeder will test a diverse range of germplasm for response to the stress. If adequate tolerance or nutrient uptake efficiency can be found in otherwise agronomically adapted cultivars, the problem is resolved by choosing proper cultivars for planting and by using such germplasm to breed cultivars for the future.

If the level of tolerance in adapted cultivars is not deemed adequate, the germplasm collection of the crop species would be screened for sources of the desired variability. If the intraspecific variation is not adequate, the more arduous task of locating sources of the desired genes in related species may be undertaken. Such genes will then have to be transferred by interspecific hybridization to crop cultivars. Usually, undesirable alleles must be eliminated in this process. Mutagenesis may be employed in an effort to produce the desired variability in cases where such variation is not readily available.

In searching for desired variants in natural populations, it would be helpful to have information on the geographical distribution of problem soils. Such information would be used to identify ecotypes that may have evolved a level of tolerance through natural selection. Unfortunately, detailed information on specific soil characteristics is often lacking, and indeed, information on the origin of plant introductions maintained in germplasm collections is sometimes so general that little can be surmised as to the edaphic adaptation of individual plant introductions. Hopefully, future germplasm collection programs will provide more detailed information.

The choice of breeding methods to be used for the characteristics in question will depend on several considerations: (a) the method of reproduction (self-pollinated vs. cross-pollinated, vegetative vs. sexual, tolerance to inbreeding, seed increase methods); (b) mode of gene action (multigenic vs. monogenic, dominant vs. recessive, heterosis, epistasis); (c) sources of tolerance available (crop cultivars, noncultivated forms of the crop species, related species); and (d) priority assigned in relation to other agronomic traits (disease, insect, and nematode resistance, quality factors).

Fundamental to the development of crop cultivars for tolerance to edaphic stresses is the development of techniques and procedures for assaying plant response to the pertinent stress factor. Ideally the assay procedures should have several characteristics.

The assay should correctly simulate the appropriate stress. The area in which the crop is to be grown should be accurately characterized for the significant stress factor or factors. For example, is the problem one of pH per se, or Al toxicity or Mn toxicity, or Ca and Mg unavailability or a combination of these? Because of the very complex interaction of pH, mineral solubility, and the complexing of mineral constituents in nutrient media, care must be taken to assure that the intended stress is actually being imposed.

The assay procedure should provide values that are both accurate and precise. Fluctuations in values due to environmental effects should be minimized. The assay should characterize the genotype of the zygote being tested rather than the residual maternal effects due to the genotype of seed-bearing parent or the seed production environment. This will be particularly important in early seedling tests.

Devine (1976) tested the effect of seed source on the expression of soybean seedling tolerance to Al toxicity in an acid Al-toxic Tatum soil. The comparison of seedlings from seed of the cultivars 'Kent' and 'Dare' from sources in Maryland, Virginia, Oklahoma, and Arkansas indicated that environmental effects of seed production did not mask genetic expression of tolerance by the seedling. However, in analyzing plant introductions where variation in seed size is much greater than that encountered in commercial lots of U.S. cultivars, Devine *et al.* (1979) found that seed weight was significantly correlated with seedling tolerance. Furthermore, Hanson and Kamprath (1979) reported that one cycle of selection for Al tolerance resulted in a 21% increase in seed weight and volume.

The test environment should simulate the stress environment in which the crop will be grown. Prolonged selection under artificial conditions in the greenhouse or laboratory may result in loss of fitness for field adaptation. This may result from relaxation of selection for tolerance to disease, insect, or nematode damage, or a change in the selection pressure for physiological responses such as winter-hardiness, date of flowering, or rate or regrowth after cutting.

It is desirable to use an assay technique that favors maximum expression of genetic variation without producing undesirable genotype by environment interactions. For example, the level of Ca in a nutrient solution designed to induce Al toxicity stress will markedly affect the severity of root stunting symptoms. It is important to know whether the genotypes selected for Al tolerance at the level of Ca that permits maximum expression of genetic variation for tolerance would be essentially the same genotypes selected at other Ca levels, and if there is an important difference, which set of genotypes would best express the desired tolerance in the crop production environment.

The assay procedures should be rapid and inexpensive to permit testing of a large number of genotypes. This would be especially important in the case of the cross-pollinated polyploid crop species, such as the perennial forage species, which are especially subject to inbreeding depression. In these cases, if there are low frequencies for the desired genes in the initial populations, several thousand or tens of thousands of individuals may have to be screened to obtain sufficient resistant plants for intercrossing in a phenotypic recurrent mass selection program. The cost of various alternative techniques may vary considerably from program to program. Complex laboratory equipment may be difficult to obtain and operate in some locations, and research assistants to perform detailed plant measurements may not be available.

Usually, for many characteristics, the seedling stage is most efficient for screening very large numbers of plants. Care must be taken to assure that seedling response predicts mature plant response for the character in question.

The development of screening techniques for determining plant response to soil stresses is an active area of research. Techniques vary with the species under study and with the particular stress or complex of stresses being imposed.

Sartain and Kamprath (1978) studied Al tolerance of several soybean cultivars in short-term (48-hour exposure to Al) solution culture and long-term (growth to full bloom) soil culture in the greenhouse. In solution culture studies the cultivars 'Lee,' 'Lee 68,' 'York,' 'Dare,' and 'Ogden' were Al tolerant. The cultivars Lee, 'Bragg,' 'Pickett 71,' and York were more tolerant in the soil studies. These investigators concluded that short-term solution culture studies accounted only for the effects of Al on cell elongation and cell division, while soil growth studies reflected the continued effect of Al on top growth, root growth, and nutrient uptake.

IV. Genetic Variability

Genetic variability occurs both within and between species. Variation within a species is readily capitalized upon by breeders in developing new cultivars. Variability between species requires more intricate and complex operations for utility, particularly in the legume family, where barriers to sexual hybridization between species are formidable. Embryo rescue techniques for interspecific hybrids and advances in biotechnology offer avenues for circumventing these barriers. In breeding for any trait, it is important to assess the genetic variability available for use. If sufficient genetic variability is not available in agronomic cultivars, a search for the desired genetic variability in new germplasm sources such as plant introductions, undomesticated forms, and other species may be undertaken. Induced mutation of critical loci may prove to be an effective means to construct the desired genotype in an otherwise agronomically desirable background.

If the desired characteristic is conditioned by a polygenic system with minor incremental effects contributed by numerous genes, systems of recurrent selection could be employed to produce a more intense expression of the trait.

Variability for tolerance to conditions associated with acid soils has been documented both among and within different germplasms of species such as alfalfa

(Bouton *et al.*, 1986; Brooks *et al.*, 1982; Buss *et al.*, 1975a,b; Dessureaux, 1960, 1969; Devine *et al.*, 1976; Simpson *et al.*, 1977), white clover (Snaydon, 1962a,b), subterranean clover (Evans *et al.*, 1987; Osborne *et al.*, 1980), birdsfoot trefoil (Allison, 1987), and Siratro (Hutton *et al.*, 1978).

V. Inheritance and Heritability

Determining the genetic mechanism (monogenic vs. multigenic, major gene vs. minor genes, qualitative vs. quantitative, and dominant vs. recessive) that controls a particular edaphic response is important in formulating the breeding strategy. Traits controlled principally by a single gene with a discrete qualitative effect may often be readily transferred to the pertinent cultivars by backcrossing. For traits controlled by many genes, resulting in qualitative inheritance, other breeding systems may prove more efficient. With quantitatively inherited traits, the degree of heritability will be important in predicting success in breeding and in developing the suitable breeding procedures. Heritability in a broad sense refers to that portion of the phenotypically expressed variation that is due to genetic rather than environmental variation. In a more narrow sense, it is a measure of the degree to which a trait can be modified by selection.

Dessureaux (1959) concluded that manganese tolerance in alfalfa was heritable and controlled by additive genes with little or no dominance. He later (Dessureaux, 1960) reported that reciprocal differences associated with seed weight occurred in crosses of alfalfa plants, and concluded that maternal influences were unimportant in assessing manganese toxicity of alfalfa seedlings, but that seed size may affect characters used to measure tolerance.

Croughan *et al.* (1978) selected a salt-tolerant line of alfalfa cells by repeated subculturing of cells on 1% NaCl media. The selected line grew better than the unselected control culture in media with high levels of NaCl. The selected line performed poorly in the absence of NaCl and appeared to require a substantial amount of NaCl for its growth. Salt tolerance appeared to be stable under maintained selection pressure in culture. In this, as in many other programs, regeneration of plants will be necessary to determine the stability and heritability of tolerance through seed propagation. Cell and tissue culture techniques are potentially powerful selection techniques that may prove valuable when integrated into long-term crop improvement programs for nutrient use efficiency.

Weiss (1943) identified a gene locus with major effect on the efficiency of Fe uptake in soybean. However, Cianzio and Fehr (1980, 1982) found evidence for additional genes modifying the expression of Fe deficiency chlorosis in soybean, and Fehr's program (Fehr, 1982, 1984; Hintz *et al.*, 1987) of recurrent selection based on a multigenic model with quantitative inheritance has proven effective in developing lines with superior resistance to Fe deficiency chlorosis. This demonstrates an important, but frequently neglected, principle: The identification of specific genetic loci with major effect on any trait should not lead those reading the literature to the unwarranted and naive assumption that there is only one genetic locus affecting expression of the trait.

VI. Acid Soil Stress

Aluminum and/or manganese toxicities often pose problems in acid soils, and many research reports deal specifically with these problems while others deal with acid soils without defining specific toxicities and soil conditions. Acid soil conditions are found in tropical and subtropical countries and in cultivated soils in the temperate and subarctic areas. Acid soils pose an important problem throughout the southeastern United States (Melton *et al.*, 1988). High acidity is a prevailing feature of old, highly weathered and leached soils of Brazil (Malavolta *et al.*, 1976). The "Cerrado" soils with pH values in the range of 4.8 to 5.2 (Lopes and Cox, 1977) occupy between 20 and 25% of the land area of Brazil. In these soils, Al saturation of the cation-exchange complex on the order of 60% or higher is almost the rule, and Ca deficiency is common (Malavolta *et al.*, 1979). Genetic variability in acid soil tolerance has been documented both among and within different germplasms of species such as alfalfa (Bouton *et al.*, 1986; Brooks *et al.*, 1982; Buss *et al.*, 1975a,b; Dessureaux, 1960, 1969; Devine *et al.*, 1976, Simpson *et al.*, 1977), white clover (Snaydon, 1962a,b), subterranean clover (Evans *et al.*, 1987; Osborne *et al.*, 1980), birdsfoot trefoil (Allison, 1987), and Siratro (Hutton *et al.*, 1978).

Methods used to determine variation in acid soil tolerance both among and within germplasms vary among scientists (Table I). Similar methods are used to screen germplasms in selection programs. These range from documenting top or root growth in acid soil, to calculating relative tolerance by comparing top or root growth in acid soil to that in limed, fertile soil (relative growth), to documenting the uptake of the toxic elements, Al and Mn, or deficiencies in Ca and P, usually associated with acid soils.

It is not certain that increasing acid soil tolerance in a forage legume will result in better agronomic performance. In Georgia, a conventional yield trial conducted in typical agricultural soils (i.e., limed and fertilized topsoils but acid subsoils) showed that an acid-tolerant alfalfa germplasm labeled 'Georgia Exp. A-2' gave 6% higher forage yield than the check cultivar 'Apollo' (Bouton and Monson, 1985). However, in a subsequent yield trial, another related acid soil-tolerant germplasm labeled 'Georgia Exp. A-3' was found to give the same yield as Apollo (Bouton *et al.*, 1988). Although an increased acid soil tolerance of up to 30% was estimated for A-2 (Bouton *et al.*, 1986), the above data indicate that a very high level of acid soil tolerance may be needed to affect a crop's agronomic performance in soils with limed and fertilized topsoils but acid subsoils. This is still the envisioned farming scenario because trying to grow even acid-tolerant alfalfa in soils with complete acid

Table I. Methods Used to Screen and/or Select Different Forage Legume Species for Acid Soil Tolerance

Species	Screening Procedure	Selection Criteria	Reference
Alfalfa (*Medicago sativa* L.)	Greenhouse pots containing field soil of differential pH	Seedling vigor	Brooks *et al.* (1982)
	Field plots of differential soil pH	Forage yield	Bouton and Sumner (1983)
	Greenhouse pots containing field soil of differential pH	Top and root growth; mineral uptake	Buss *et al.* (1975a,b)
	Solution culture with varying levels of Al or Mn	Seedling dry weight; size of unifoliate leaf; uptake of Mn	Dessureaux (1960, 1969)
	Greenhouse pots containing field soil of differential pH	Top growth; root vigor	Devine *et al.* (1976)
	Solution culture with varying levels of Al	Seedling vigor	Elgin (1979)
	Field plots with acid subsoils	Plant vigor	Simpson *et al.* (1977)
Birdsfoot trefoil (*Lotus corniculatus*)	Greenhouse root boxes containing soil profiles of differential pH	Relative root growth and herbage yield	Allison (1987)
Leucaena (*Leucanena leucocephala*)	Greenhouse pots containing field soil of differential pH	Relative root yield	Oakes and Foy (1984)
Sericea lespedeza (*Lespedeza cuneata*)	Greenhouse root boxes containing soil profiles of differential pH	Root growth rate	Joost and Hoveland (1986)
	Petri plates containing high Al	Relative root growth	Joost *et al.* (1986)
Siratro (*Macroptilium atropurpureum*)	Sand culture in glass house with varying Mn levels	Dry matter yield of tops; ratio of root to total dry matter	Hutton *et al.* (1978)
Subterranean clover (*Trifolium subterraneum*)	Solution culture of varying Al and Mn levels	Shoot weight; root weight and length; uptake of P and Ca	Osborne *et al.* (1980)
	Solution culture of varying Mn levels	Plant dry weight; nodulation and nitrogenase activity; Mn content and distribution	Evans *et al.* (1987)
White clover (*Trifolium repens*)	Greenhouse pots containing field soil of differential pH and fertility	Plant dry weight; visual symptoms of lime-induced chlorosis	Snaydon (1962a,b)

profiles was not successful (Bouton and Sumner, 1983), even though performance was recently greatly enhanced in these conditions when alfalfa was properly inoculated with *Rhizobium meliloti* (Bouton and Hartel, 1987).

A. ALUMINUM TOXICITY

Under acid soil conditions (pH 4.5 and lower), Al in the soil solution can be toxic to soybean roots. Even though Al toxicity in surface soils can be alleviated by liming, Al toxicity in subsoils may still limit root penetration by Al-sensitive plants. Liming of subsoils is usually prohibitively expensive. Aluminum toxicity in acid subsoils is thought to limit soybean yields in the southeastern United States and in large areas of Brazil.

Devine *et al.* (1976) established that recurrent phenotypic selection was effective in altering the tolerance of alfalfa to an acid, Al-toxic soil. They subjected a single heterogeneous germplasm pool to two successive cycles of selection for seedling tolerance or susceptibility to Al toxicity during growth in a Bladen soil in growth chambers. The population derived by selection for tolerance was significantly more acid soil tolerant in subsequent growth chamber tests on Bladen soil than the population derived by selection for susceptibility. These results indicated that Al tolerance is a heritable trait in alfalfa. Brooks *et al.* (1982), using divergent recurrent selection from the same parental base population, developed in the second cycle two alfalfa germplasms, one productive in an unlimed, Al-toxic Cecil soil and one productive in the same soil that had been limed and fertilized. In subsequent experiments, the acid soil-selected germplasm, when compared to the germplasm selected in limed, fertilized soil, rooted more deeply in acid subsoils (Bouton *et al.*, 1982) and produced higher yields at several soil pH levels, especially where P was adequate (Bouton and Sumner, 1983). The Al tolerance of this acid soil-selected germplasm, identified as 'Cycle 2 Acid,' was further confirmed when it was tested in acidic soil profiles (Bouton *et al.*, 1986).

Differential tolerance to excess Al and a varying ability to utilize P have been observed in different cultivars of dry beans (CIAT, 1977). In tropical and subtropical regions, beans are grown extensively in acid soils, but this species is generally considered sensitive to acidic conditions. Among nine tropical and seven temperate legumes tested in a nitrogen- (N) deficient oxisol by Munns and Fox (1977), bean was the third most responsive to lime, and even more responsive than alfalfa, a well-known acidic-sensitive legume (Spain *et al.*, 1975). These investigators consistently found that black-seeded bean cultivars were less sensitive to acid soils with high Al concentrations than cultivars with nonblack seeds.

Franco and Munns (1982) reported similar results with shoot growth at low Al levels. However, the light-brown-seeded cultivar, 'Carioca,' had more total root length and tended to be less sensitive to higher levels of Al than the black-seeded cultivar, 'Venezuela 350.' Previous studies by Long and Foy (1970) and Foy *et al.* (1967, 1972) showed that 'Dade' and 'Romano' snap beans differ significantly in their tolerance to Al when grown in acid Bladen soil and in nutrient solutions. However, the specific physiological or biochemical causes of such differences are still unknown.

Jones (1961) suggested a possible explanation for differential tolerance among plant species. He stated that plant organic acids (OA) may act as chelating agents that prevent precipitation of Al at physiological pH values. Lee and Foy (1986) reported that aluminum stress reduced total organic acid concentrations in the roots of both Al-tolerant and Al-sensitive bean cultivars; however, the roots of the Al-tolerant Dade contained higher concentrations than the Al-sensitive Romano cultivar in both the presence and absence of Al stress. Aluminum stress reduced OA levels in the Al-sensitive cultivar Romano to a greater extent than in the Al-tolerant cultivar Dade. Changes in the malic and citric acids contents of snap bean root are the most marked biochemical indicators of Al-induced injury measured to date. This evidence indicates that the Al-tolerant cultivar Dade has a higher potential for chelation and detoxification of Al than does the Al-sensitive cultivar Romano under Al stress conditions.

Klimashevskii and Chernysheva (1980) reported that Al-tolerant cultivars of pea, corn, and barley contained substantially higher concentrations of citric acid than Al-sensitive cultivars of the same species. Recent evidence suggests that OA treatments can prevent Al-induced conformational changes in calmodulin, a Ca protein that regulates ATPase activity in membranes (Haug and Caldwell, 1985; Suhayda and Haug, 1984, 1985). It has been suggested that in Al injury, the element is present in the nuclei and at the surface of the roots (Aimi and Murakami, 1964; Clarkson, 1969), as revealed by various staining procedures. It has been reported that in pea tissue, Al is localized in the meristematic zone (cell walls and cell nuclei) as well as in the epidermis (Matsumoto and Morimura, 1980). Aluminum injury may prevent cell division, as suggested by other investigations in which the binding of Al to DNA has been demonstrated both *in vitro* and *in vivo* (Matsumoto *et al.*, 1976a,b, 1977a,b).

Armiger *et al.* (1968) measured tolerance to Al toxicity for several U.S.-bred soybean cultivars and plant introductions by assaying root and top growth after 43 days on an acid Al-toxic Bladen soil as a percentage of growth on the same soil limed to correct Al toxicity. Devine (1976) tested seed of two U.S.-bred soybean cultivars produced at four different locations for growth in pots on an acid Al-toxic Tatum soil to determine whether the differential environmental effects during seed production affected cultivar response to Al toxicity. He concluded that the major factor affecting response was the genotype of the zygote rather than environmental effects during seed production.

Sapra *et al.* (1975) tested a large number of soybean plant introductions for Al toxicity in nutrient culture. Six-day-old seedlings were transferred to Al+ or Al− solution culture and grown for 12 days, then the length of the longest root was determined for each plant. The ratios of root lengths at 8 ppm Al/0 ppm Al were used as indicators of tolerance to Al toxicity.

Mason (1982) evaluated 3105 soybean plant introductions in maturity groups IV and V from the U.S. germplasm collection for Al tolerance on acid Tatum soil in the field. Ten tolerant and ten susceptible lines were selected for testing in nutrient solution and soil pot tests. He concluded that the solution culture tests were more labor-intensive and had more potential sources for error than the soil pot test. Visual evaluation of root morphology and top and root fresh weight after 14 days growth in Al-toxic soil were used as criteria for tolerance.

Sartain and Kamprath (1978) compared a solution culture technique with a soil pot procedure for assaying Al tolerance in soybean. In the solution culture procedure they calculated a relative extension rate (RER) of the primary root. Roots of seedlings were measured 72 hours after germination, then subjected to solutions with 0, 9.0, 18.0, or 37.0 μM Al for 48 hours and measured again. Al tolerance was measured by determining the Al concentration at which no further significant reduction in RER occurred. For example, cultivars with a minimum RER at 9.0 μM Al were judged less tolerant than those with an RER at 18.0 μM. In the pot culture procedure, plants were grown to full bloom in a Lynchburg sandy loam with an initial Al saturation of 81%. Plants grown in the same soil limed to give an Al saturation of 4% were used as a comparison. Dry weights of tops and roots and root lengths were determined. Tops and roots were analyzed for Ca, Al, and P. They found no significant correlation between root length and top growth or between Ca and P concentrations in the tops and top growth. The Al concentration in plants also did not correlate with relative Al tolerance. They concluded that studies with soil give an integrated effect of soil acidity on plant growth in contrast to short-term root elongation studies, which provide information about the specific effects of Al on cell division and elongation.

Devine *et al.* (1979) examined several parameters for assaying Al tolerance in hydroponic systems in tests of soybean germplasm lines from Korea. Three-day-old seedlings were transferred to Al+ or Al− solutions for a 72-hour growth period. Plants were measured for primary root length (PRL), distance from the root tip to the most recently emerged secondary root (RPS), and length of the three longest lateral roots (LRL). A visual damage score was also assigned. The cultivars 'Perry' and 'Chief,' tolerant and susceptible, respectively (Foy *et al.*, 1969), served as checks. For comparisons across tests in series, values for Perry and Chief were averaged to yield a test standard value. Other lines were evaluated by computing their value as a percentage of the test standard value. Significant differences among lines were found for the Al+/Al− PRS and LRL with the greatest range expressed for LRL.

To determine the influence of seed reserves on the expression of aluminum tolerance, seed weight was tested for correlation with other measurements (Table II). Seed weight was positively correlated ($p < .01$) with PRL as percentage of standard and with the PRS as percentage of standard, but this was not significantly correlated with the LRL as percentage of standard. This suggests that seed reserves strongly influence the aluminum response of primary root growth. Correlations of seed weight with the PRL, PRS, and LRL at 0 ppm Al were not significant. But, under Al stress at 222 μM Al, seed weight was significantly correlated with PRL and LRL and approached significance with PRS. This suggests that under aluminum

Table II. Correlation (r) Values for Root Parameters with Seed Weight and Visual Scores[a]

Observations	Seed Weight	Visual Score
Visual score	−.50 NS[b]	—
PRL as % of standard	.85**	−.28 NS
PRL at 0 μ*M* Al	.35 NS	−.35 NS
PRL at 222 μ*M* Al	.64*	−.51 NS
PRS at % of standard	.93**	−.51 NS
PRS at 0 μ*M* Al	−.44 NS	.14 NS
PRS at 222 μ*M* Al	.57 NS	−.52 NS
LRL as % of standard	.54 NS	−.75*
LRL at 0 μ*M* Al	.44 NS	−.51 NS
LRL at 222 μ*M* Al	.64*	−.78*
LRL ratio Al+/Al−	.69*	−.73*

[a]Source: Devine *et al.*, 1979.
[b]NS, not significant.
*Significant at the 5% level.
**Significant at the 1% level.

stress the influence of seed reserves on growth is greater than in the absence of Al stress. Additional correlations were made to determine the influence of the factors measured on the assignment of visual damage scores. The visual damage score was not correlated with seed weight. Nor was visual score correlated with any of the measurements at 0 μ*M* Al. The visual score was not correlated with PRL or PRS either at 222 μ*M* Al or as percentage of standard, suggesting that these measurements had little or no influence on the assignment of visual ratings. However, the LRL at 222 μ*M* Al and as percentage of standard were negatively correlated with the visual score, indicating the LRL was an important factor influencing this score. A previous study (Devine, 1976) reported that variation in seed lots of the same soybean cultivar produced at different locations had little effect on Al tolerance in comparison with effect of the genotype of the zygote. In that study, conducted with adapted U.S. cultivars, seed weight within a cultivar did not vary appreciably. In the study of Korean germplasm, however, seed weight varied from 10.9 to 28.8 g/100 seed, a factor of 2.6. Similar variation would be expected in screening the Soybean Germplasm Collection. Results indicated that seed weight exerted an influence on early seedling expression of Al tolerance. Caution should be used in imputing long-term physiological tolerance to lines expressing tolerance at this stage.

Hanson and Kamprath (1979) selected for Al tolerance based on root growth rate of 5-day-old soybean seedlings under Al stress in solution. After three cycles of divergent selection from a broad-based population, tolerance was found to be heritable (.67). Selection at the seedling stage identified genetic differences for tolerance in established plants as well as tolerance apparently unique to the seedling stage.

Aluminum tolerance screening in solution culture is labor-intensive and requires meticulous care to maintain the hydroponic solutions. In addition, conditions in the solution culture may not mimic those of the soil. For example, roots may be unable to form a normal rhizosphere in the flux of an aerated solution, root hair growth may be modified, and the absence of physical resistance to root penetration may have physiological and morphological effects. Important mechanisms of tolerance or resistance that are operational in soil may fail to be expressed in the solution culture. Campbell *et al.* (1988) reported that selection of alfalfa for Al tolerance in nutrient solution resulted in more vigorous plants, but also concluded that Al tolerance in nutrient solution and soil involve different mechanisms. Pot culture of seedlings in soil prepared specifically for testing for Al tolerance should provide conditions closer to the production environment. However, pot tests are usually conducted on seedlings, and initial seed weight, seed quality, and juvenile characteristics may produce results uncharacteristic of the mature plant. Mature plants may be evaluated in field planting under stress of Al toxicity. However, uniformity of soil conditions is difficult to achieve in the field.

B. MANGANESE TOXICITY

Differential tolerance to excess Mn occurs among cultivars of many plant species (Blatt and Van Diest, 1981; Foy, 1973, 1983; Foy *et al.*, 1981; Heenan and Campbell, 1981; Heenan *et al.*, 1981; Kohno and Foy, 1983b; Mugwira *et al.*, 1981; Ohki *et al.*, 1980). Such differences have been associated with differences in (a) oxidizing power of plant roots; (b) absorption and translocation rate of Mn; (c) Mn entrapment in nonmetabolic centers; (d) internal tolerance to excess Mn; and (e) uptake and distribution of Si and Fe (Foy, 1983; Foy *et al.*, 1978). Ohki (1981) and Temple-Smith and Koen (1982) considered a 10% reduction in dry matter production to be a critical threshold level for Mn toxicity in soybean and bush bean.

Dessureaux (1959) studied manganese tolerance in alfalfa and concluded that tolerance was heritable and controlled by additive genes with little or no dominance. He reported (Dessureaux, 1960) that reciprocal differences for manganese tolerance occurred in crosses of alfalfa plants. These differences were evident in the number of trifoliate leaves developed, dry matter production, and unifoliate leaf area. Differences were generally associated with seed weight. However, reciprocal differences were not detected in visual symptom ratings. He concluded that maternal influences were unimportant in assessing the reaction of alfalfa seedlings to manganese toxicity, but that seed size may sometimes affect characters used to measure tolerance.

Hutton *et al.* (1978) predicted from their data that improving Mn tolerance in Siratro should be possible by selection and/or breeding. Mn-sensitive bush bean cultivars contained lower Mn concentration in the leaves than the Mn-tolerant ones, when Mn toxicity symptoms developed (Kohno and Foy, 1983b). It was also concluded that the differential tolerance of bush bean cultivars to excess Mn is due to a difference in the ability to tolerate a high level of Mn accumulation within the leaves.

Brown and Jones (1977) used grafts of Mn-tolerant and -susceptible tops and rootstocks to determine if tolerance to Mn toxicity is controlled by roots or tops in soybean. They found that regardless of the rootstock the Mn-sensitive tops of the cultivars Bragg and Forrest developed Mn toxicity symptoms and the grafted tops of Lee and 'T203' were more tolerant. They concluded that manganese tolerance is controlled by the tops of the plants rather than the roots. Mn concentration in the tops did not differ regardless of rootstock or the manifestation of symptoms.

Brown and Devine (1980) reported continuous variation in plant growth under Mn toxicity stress in F_2 progeny of crosses of soybean lines differing in tolerance to excess Mn. This suggested that tolerance was quantitatively inherited. Transgressive segregation was also observed. Reciprocal differences in the crosses suggested cytoplasmic involvement in the expression of tolerance. Heenan *et al.* (1981) also studied the inheritance of Mn tolerance in a segregating F_2 soybean population using visual scores and reported that the F_2 progeny had a continuous frequency distribution characteristic of multigenic inheritance. The moderate tolerance of the F_1 and the skewedness of the F_2 generation suggested that tolerance was partially dominant. Ohki *et al.* (1980) reported that 'Lee 74' was more sensitive to Mn toxicity than its progenitor soybean cultivar Lee. Lee 74 was bred by incorporating resistance to phytophthora root rot (*Phytophthora megasperma* Drechs var. *sojae*) and root-knot nematode (*Meloidogyne incognita*) into Lee. They suggested the Mn-efficient character of Lee may have been partially lost during this process. This demonstrates the complexity involved in breeding cultivars that combine a multiplicity of desirable agronomic features.

Brazilian scientists (Mascarenhas *et al.* 1982, 1984) report that neither length of the primary root nor dry weight of the root was a good indicator of Mn toxicity response in soybean. Plant height and dry weight of the leaves or top growth was a good indicator. Miranda *et al.* (1982) found a good relationship between scores based on visual symptoms, dry matter of top growth, and the index of Mn concentration/dry matter production, and therefore suggested that scores for visual symptoms of Mn toxicity could be used efficiently in breeding soybean for Mn tolerance.

Environmental conditions can affect the expression of Mn toxicity. Heenan and Carter (1977) and Mascarenhas *et al.* (1985) demonstrated that high temperature (28°C) reduced expression of Mn toxicity relative to cooler temperatures. The Mn concentration in the plant can also be affected by the interaction of soil temperature and Fe deficiency in the soil (Moraghan *et al.*, 1986). Zn (Abd-Elgawad and Knezek, 1983) Ca (Heenan and Carter, 1975), and Fe (Blatt and Van Diest, 1981) also affect Mn concentration in plant tissue. These environmental variables should be considered in designing screening methods.

C. MANGANESE TOXICITY AND IRON INTERRELATIONS

Kohno and Foy (1983a) showed that increasing Fe concentration in the nutrient solution reduced Mn concentration in the tissues of bush bean cultivar 'Wonder Crop 2' and alleviated Mn toxicity symptoms on the leaves. However, Mn treatment that produced toxicity symptoms did not decrease the Fe concentration in the plant tops. It was concluded that interveinal chlorosis, a typical symptom of Mn toxicity observed in bush bean leaves, was due to Mn accumulation and not to Mn-induced Fe deficiency. These symptoms are quite diverse among different plant species and include marginal chlorosis of leaves, leaf puckering, chlorosis of young leaves resembling Fe deficiency, and necrotic spots, particularly on old leaves (Foy, 1973; Foy *et al.*, 1978). Brown necrotic spots on leaves have also been attributed to Mn accumulation in bean (Horst and Marschner, 1978). Manganese-sensitive bush bean cultivars contained lower Mn concentration in the leaves than Mn-tolerant cultivars, when Mn toxicity symptoms developed. However, Fe concentration in the leaves showed no relationship to Mn tolerance. Therefore, it was concluded that the differential tolerance of bush bean cultivars to excess Mn is due to a difference in the ability to tolerate a high level of Mn accumulation within the leaves (Kohno and Foy, 1983b).

The distribution pattern of Mn, in the study by Kohno *et al.* (1984), indicated that Mn-sensitive Wonder Crop 2 tended to accumulate relatively more Mn in tops, particularly the youngest trifoliate leaves, while Mn-tolerant 'Greenlord' tended to accumulate relatively more Mn in the roots and older leaves and less in younger leaves. Hence, the superior Mn tolerance of Greenlord may be due, at least in part, to reduced translocation of Mn to actively growing tissues. In contrast to the Mn distribution pattern, Fe distribution data indicated that Mn-sensitive Wonder Crop 2 accumulated more Fe in the roots and less in the tops than Greenlord (Kohno *et al.*, 1984). On the other hand, for a unit of root mass, Mn-tolerant Greenlord translocated more Fe from roots to tops than did Wonder Crop 2. Kohno (1986) also reported that manganese concentrations in the primary leaves of beans increased under the higher vanadium (V) treatment in Wonder Crop 2, but not in Greenlord. In contrast, Fe concentration in the leaves of Wonder Crop 2 decreased markedly with increasing V concentration in the solution. Enhanced Mn uptake and greater reduction of Fe uptake to Wonder Crop 2 than Greenlord resulted from the interference and/or competition of V with Fe before entering the root tissue. Vanadium treatment resulted in Fe–Mn imbalance within the tissue. The development of Mn toxicity symptoms in Mn-sensitive Wonder Crop 2 caused by this treatment may be due to reduction and enhancement of Fe and Mn uptake, respectively.

Threshold internal concentrations of Mn required for toxicity have not been clearly defined for peanut, but values ranging from 550 to 1600 ppm have been suggested for many tropical legumes (Andrew and Hegarty, 1969). Morris and Pierre (1949) evaluated the minimum concentrations of Mn necessary for injury to various legumes in culture solutions and concluded that, for peanuts, Mn concentration in tissue above 1000 ppm caused significant reduction in plant growth.

D. TOLERANCE TO ZINC TOXICITY

Earley (1943) reported cultivar differences in soybean to zinc toxicity, with 'Hudson Manchu' and 'Peking' being tolerant. White *et al.* (1979a), using reciprocal grafts of the Zn-tolerant soybean cultivar 'Wye' and Zn-tolerant cultivar York, determined that the rootstock genotype controlled Zn absorption and translocation, but that the scion genotype controlled relative Zn tolerance. White *et al.* (1979b) studied the response of 20 soybean cultivars to phytotoxic levels of soil Zn. Some cultivars were stimulated by the added soil Zn while others were slightly retarded. Plant Mn was also increased by the Zn additions to the soil. These authors concluded that excessive Mn may react with Zn synergistically, antagonistically, or not at all, and that there is little information on which to base a judgment.

VII. Alkaline Soils

A. ZINC DEFICIENCY

Tests in Michigan indicated that heavy application of P may induce Zn deficiency in pea beans (Ellis, 1965). Zn deficiency occurs on lake-deposited soils in a region of east-central Michigan. These soils are usually calcareous. Fertilization with P at 0, 97, 195, 390, and 780 kg ha^{-1} produced yields of 1, 952, 1624, 1333, 1120, and 594 kg ha^{-1}, respectively. However, with inclusion of zinc sulfate at 4 lbs of Zn per acre, yields averaged 2509 kg ha^{-1}. The cultivar Sanilac was more susceptible to Zn deficiency than 'Saginaw.' Sanilac yielded 504 kg ha^{-1} without Zn fertilization and 2621 kg ha^{-1} with Zn fertilization, but Saginaw produced 1210 kg ha^{-1} without added Zn and 2352 kg ha^{-1} with Zn fertilization. Paulsen and Rotimi (1968) used the P-tolerant soybean cultivar 'Chief' and the P-sensitive cultivar 'Lincoln' to study the P–Zn interaction. High P decreased growth of Lincoln more than Chief, but decreased Zn concentration in both cultivars equally. Added Zn overcame the effect of P on Chief but not on Lincoln. The effect of P on Zn appeared to originate in the roots and to occur on translocation of Zn to the upper plant parts.

B. MANGANESE DEFICIENCY

The critical Mn concentration of plant tissue for deficiency in peanut has been estimated at 10–15 ppm (Gheesling and Perkins, 1970; Ohki, 1974). Lombin (1983) suggested that peanuts may extract soil Mn more effectively under conditions of low Mn level than cotton, which has a greater ability to take up Mn under conditions of high Mn levels. Both crops are evidently capable of withstanding low levels of soil Mn, probably, because of their intrinsically low requirements.

Lombin and Bates (1982) studied the relative responses of alfalfa, peanuts, and soybeans to Mn additions on two calcareous soils from Ontario, Canada, in a greenhouse pot test. All three crops failed to respond to Mn additions on the Pontypool soil, which apparently had adequate Mn. With the Plainfield soil, however, addition of Mn increased alfalfa yield by 75%, but did not significantly affect peanut or soybean yields. They suggested that the absence of Mn deficiency with soybean might be associated with a relatively high soil Mn absorption efficiency in soybean. Ohki *et al.* (1980) studied the sensitivities of soybean cultivars to Mn deficiency in nutrient solution. The ranking of cultivars for sensitivity was partially dependent on levels of deficiency stress. Davis, Lee 74, Bragg, and Pickett 71 were the most sensitive cultivars under severe stress.

C. IRON DEFICIENCY

Forage legumes, especially the *Trifolium* genus, develop iron deficiency chlorosis when grown in high-pH soils. When Snaydon (1962b) tested white clover populations selected from either calcareous or acid soils, he found that each grew better on its native soil type. However, plants from acid soil-selected populations were found to be susceptible to lime-induced iron chlorosis when grown on calcareous soil, while plants from calcareous soil populations showed no chlorosis. Gildersleeve and Ocumpaugh (1988) tested cultivars of seven *Trifolium* species in calcareous soils prone to produce iron chlorosis and found cultivar differences for dry matter production and expression of the chlorosis symptoms. These researchers also documented a screening procedure and isolated genotypes of both arrowleaf and crimson clovers that were resistant to iron chlorosis (R. R. Gildersleeve and W. R. Ocumpaugh, 1988, personal communication).

Weiss (1943) identified a gene locus with a major effect on the efficiency of iron utilization and designated the recessive allele *fe* conditioning low efficiency and the dominant allelle *Fe* conditioning high efficiency. Brown *et al.* (1961) attributed the differences in susceptibility of 'Hawkeye' (non susceptible) and PI54619-5-1 (chlorosis susceptible) to the reductive capacity of their roots and their root capacity to absorb iron. Brown *et al.* (1967) associated the differential iron efficiency of Hawkeye and PI54619-5-1 with their ability to reduce Fe^{3+} to Fe^{2+} at the root.

Iron deficiency chlorosis often occurs on soybeans grown on calcareous soils in the midwestern United States (Randall, 1976; Fehr and Trimble, 1982). Chaney (1984) has identified bicarbonate in soil with low Fe availability as the fundamental cause of soybean chlorosis. Bicarbonate interacts with Fe availability in the soil and interferes with the regulatory control of Fe stress response, the adaptive increase in the ability of roots to absorb and translocate Fe to the shoots.

An outstanding example of what can be achieved by modifying the plant genotype for adaptation to a specified edaphic condition is found in the soybean breeding program at Iowa State University. Tests of plant introductions and cultivars failed to identify genotypes free of chlorosis symptoms on some of the calcareous soils in Iowa. Although a qualitatively inherited gene was identified that had a major effect on Fe chlorosis (Weiss, 1943), Fehr (1982) found that iron chlorosis was governed by a multigenic system and treated it as a quantitatively inherited character in his breeding program (Cianzio and Fehr, 1980). Chlorosis was evaluated at field sites in Iowa selected for the propensity to induce chlorosis. The main culms of the young plants were cut off and the new leaves showed a more intense expression of chlorosis than the original foliage. Recurrent selection resulted in the development of more resistant germplasm and the release of cultivars such as 'Weber' and 'Lakota' with superior tolerance to iron chlorosis.

The Iowa program also addressed the question of yield for chlorosis-resistant versus nonresistant lines on calcareous versus noncalcareous soils. Since some of the highest-yielding cultivars in maturity groups I to III were susceptible to chlorosis, the possibility that chlorosis resistance, by its very nature, restricted yield was posed. In a study of several soybean populations, Hintz *et al.* (1987) found that phenotypic correlations between yield of lines on noncalcareous soil and chlorosis resistance either were not significant or were negatively associated. They concluded either that there was no physiological association between chlorosis resistance and yield on noncalcareous soil or that the association favored selection of high-yielding lines with chlorosis resistance. The interpretation may then be drawn that the highest-yielding lines in the maturity groups studied were chlorosis susceptible because the most intensive effort had been invested in developing high-yielding lines for soils not presenting the iron chlorosis problem. With time and effort, it should be possible to develop lines highly resistant to iron deficiency chlorosis with high yield potential.

In the meantime, however, Fehr (1982) analyzed the use of blends of chlorosis-susceptible high-yielding lines and chlorosis-resistant lines with lower yield potential. Usually, only a small portion of a field presents the potential for iron deficiency. For the use of such a field a producer has several alternatives. The field may be planted (a) to a high-yielding chlorosis-susceptible line only, (b) to a lower-yielding chlorosis-resistant line, (c) the chlorosis-inducing portion of the field to a resistant line and the rest of the field to a susceptible line, or (d) to a blend of resistant and susceptible lines. By Fehr's calculation the best yield would be obtained with alternative c; however, this would require additional time and effort at planting. The use of alternative d was found to be next best. Because the calcareous spots often are a relatively low percentage of the field, the best yield for a field generally would be achieved when the high-yielding susceptible cultivar is at the highest percentage possible in the blend. In Fehr's tests a relatively small percentage of the resistant cultivar is needed to approach the yield of a pure stand of the cultivar under iron deficiency stress. This reflects the ability of soybean plants to adjust their growth and yield to the amount of space available. In a resistant–susceptible blend, the resistant plants on calcareous soil take advantage of the space provided by the weak susceptible ones when chlorosis symptoms are expressed.

VIII. Salinity and Other Problems

Salt tolerance was examined among 11 tropical legumes, 10 temperate legumes, and 11 tropical grasses using pots containing soil with increasing levels of NaCl

(Russell, 1976). The grasses showed greater persistence at the high NaCl levels than the legumes. Alfalfa was the most salt tolerant and silverleaf desmodium and safari white clover were the least salt tolerant of the forage legumes. Using percentage of total leaves exhibiting partial necrosis as the screening criterion, variation for salt tolerance was documented among different genotypes from the alfalfa cultivar 'CUF 101' (Noble *et al.*, 1984). Successful selection for salt tolerance among different alfalfa genotypes was speculated. Similarly, significant differences were found among cultivars of subterranean clover for both percentage germination and growth in saline conditions (West and Taylor, 1981). A NaCl-tolerant alfalfa cell line grew better than an unselected line at high levels of NaCl in cell culture (Croughan *et al.*, 1978). This suggested that the line was NaCl tolerant, but a concurrent finding of poor growth under normal NaCl levels complicates an adequate interpretation for the usefulness of the trait.

Abel (1969) established the existence of a major gene pair in soybeans, *Ncl ncl*, controlling chloride exclusion from plant tops. Chloride exclusion was completely dominant. Chloride-"including" genotypes contained over 7000 ppm chloride whereas chloride-excluding genotypes usually contained less than 1000 ppm chloride. Gauch and Magistad (1943) found differences among strains of strawberry clover in tolerance to salt (NaCl). The absence of breeding programs to develop salt-tolerant cultivars of soybean suggests that this has not been considered a priority objective.

IX. Micronutrients

A. CALCIUM

Vose and Koontz (1960) reported variation in Ca and strontium (Sr) content among legume species used for forage across three soil types. Birdsfoot trefoil was lowest in Ca and Sr content across all three soils; however, in general, the legumes tested were not ranked in the same order of Sr content for each soil.

Snaydon and Bradshaw (1969) found that natural populations of white clover from contrasting soil types differed in response to Ca, Mg, and K in sand culture. Populations from acid soils low in Ca were less responsive to Ca but more responsive to Mg compared to populations from calcareous soils. Populations from calcareous soils had depressed yields at high K levels. Shoot Ca concentrations were higher in populations from acid soils than those from calcareous soils. Ca uptake per unit of root weight was also higher. They concluded that root cation-exchange capacities of populations from acid soils were significantly higher than those from calcareous soils, but that the observed differences in Ca, Mg, and K were not the result of differences in root cation-exchange capacity.

Bradshaw and Snaydon (1959) compared the Ca response in sand culture of a population of white clover from a chalk downland site with one from a base-poor upland site. They found that at low calcium (4–8 ppm Ca) the upland population outyielded the chalk downland population, with the latter suffering severe calcium deficiency symptoms while the former was quite healthy. But at high Ca, the chalk downland population outyielded the upland population.

B. CALCIUM AND MANGANESE

Vose and Koontz (1960) and Vose and Jones (1963) found major differences among white clovers in Ca content. For the Ca–Mn interaction, three genotypic types were observed: (a) resistance to Mn toxicity if sufficient Ca was present, (b) sensitivity to Mn independent of Ca level, and (c) little effect of Ca level on Mn response.

Vose and Jones (1963) studied the interaction of manganese and calcium with cultivars of white clover and the effects of these interactions on nodulation. In general, Ca depressed Mn uptake and reduced the severity of manganese toxicity symptoms. The cultivar 'Szolnok' resisted manganese toxicity if sufficient calcium was present, while the cultivar 'Dutch' was very sensitive to Mn regardless of Ca level. The cultivar 'S100 Nomark' had the least response to difference in Ca level. Nodulation was normal with low-Mn treatment but with high Mn the nodules were small, collapsed, greenish, and appeared abortive. Under low-Ca conditions the cultivars reacted in a similar way to increasing manganese level, but under high-Ca conditions Szolnok was anomalous in having greater nodule numbers and nodule volume as Mn increased from 0.5 to 20 ppm.

C. MOLYBDENUM

Andrew and Milligan (1954) reported species differences between subterranean clover and *Medicago hispida reticulata* for the molybdenum (Mo) requirement for the ability to fix nitrogen in a red-brown earth soil with a pH of 5.7 and a lack of free lime in the profile. Similar variation among several species of medics was also reported.

Harris *et al.* (1965) tested seed of the soybean cultivar 'Hill' produced at nine different research sites in six states for response to Mo in a field test. The Mo content of the seed ranged from 0.6 to 2.5 ppm for all sites except one, which had been treated with 22.4 ppm Mo. Mo treatment resulted in an average bean yield increase of 504 lbs per acre from all seed sources. With the exception of the seed with high Mo, the plants from the other seed sources showed a response to Mo treatment as a darker green foliage appearing 2 months after planting and remaining throughout the season. A highly significant correlation (r = .66) between leaf N content and bean yield indicated that the yield response to Mo treatment was a response to additional N made available from increased N fixation. The lack of significant response from the high-Mo seed source illustrates the importance of selection of seed source in studies designed to measure the effects of applied Mo. From experiments conducted at 14 locations in Indiana, no response was obtained from applied Mo if untreated beans harvested from the soil contained more than 1.6 ppm Mo (Lavy and Barber, 1963).

Before Mo fertilization came into common usage, the scald disease (Mo deficiency) of beans in Australia was often overcome by planting seeds that had been grown in nondeficient areas (Anderson, 1956).

According to Harris *et al.*, (1965):

> Two types of Mo deficiency symptoms have been described. For the sake of convenience they will be referred to as type I and type II symptoms. Peterson and Purvis (1961) described the symptoms resulting when the Mo requirement for metabolism of the soybean plant was not satisfied as ' . . . leaves twisted on stem, and necrotic areas adjacent to midrib, between the veins, and along the margins.' They had to deplete the Mo reserves in the seed by planting one generation on a Mo-deficient medium before these type I symptoms were manifested. Type II symptoms occur when the Mo requirement for metabolism of the soybean plant is satisfied, but the reserves in the seed and soil are inadequate to supply the Mo needs for symbiotic N fixation. The symptoms of this deficiency are typically those of N deficiency—a general yellowing of the foliage.

D. COPPER

Andrew and Bryan (1955) found that in the coastal lowlands of southern Queensland, the addition of copper sulfate to a low-humic gley soil improved growth of *Trifolium repens* and *Trifolium pratense* L. Later, the same investigators (Andrew and Bryan, 1958) found that *Desmodium uncinatum* did not produce a growth response to copper (Cu) addition on a lateritic podzolic soil, although white clover did respond markedly.

Using the definition of Cu requirement as the Cu content in the plant material coinciding with maximum growth, Andrew and Thorne (1962) reported a lack of significant difference in Cu requirements among 10 pasture legume species. However, in both soil and water culture experiments, the extraction efficiencies (total Cu uptake per unit of soil or water) of *Stylosanthes* and berseem clover were low while those of *Desmodium,* white clover, and strawberry clover were relatively high, particularly at low-Cu applications. At high applications, all species were equally efficient.

In Cu uptake per unit weight of root tissue, *Stylosanthes* was more efficient than *Phaseolus,* white clover, and barrel medic, particularly at low-Cu applications. However, alfalfa (lucerne) and Phaseolus have a narrow top/root ratio while others, such as *Stylosanthes indigofera* and *Centrosema,* have a wide ratio. They suggested that the deeper and larger root system of *Desmodium* compared with white clover gives it an advantage in nutrient uptake, and that it is a less responsive species to additions of copper in the field than white clover.

E. COPPER, ZINC, AND IRON

Polson and Adams (1970) reported that the navy bean cultivar Saginaw grew normally in a solution containing 5 ppm Zn, while the cultivar Sanilac was extremely reduced at this concentration. A differential Fe–Cu interaction induced by high Cu levels was observed. Differences were not attributed to differential uptake or distribution of nutrients since elemental composition was not significantly different.

F. ZINC, IRON, AND PHOSPHORUS

Ambler and Brown (1969) reported that the cultivar Sanilac (more susceptible to zinc deficiency) takes up more Fe and P than the cultivar Saginaw (less susceptible to zinc deficiency) where Zn is low and Fe and P are relatively high in the growth medium. They postulated that the uptake of Fe and P appears to accentuate Zn deficiency symptoms in Sanilac.

G. MANGANESE

Walsh and Cullinan (1945) identified "Marsh Spot" disease as occurring on peas grown on manganese-deficient soils in County Galway in Ireland. Symptoms of Marsh Spot include dark brown lesions on the flat surface of the cotyledons. They reported well-defined differences in susceptibility among cultivars, with 'Blues' and 'Laxton's Superb' being tolerant and 'Onward' and 'Giant Stride' being susceptible.

Kang and Fox (1980) reported genetic variation among lines of cowpea for tolerance to excess manganese in field tests. Horst (1983) obtained similar ranking of 29 cowpea genotypes in both sand and water culture evaluations for manganese tolerance. He suggested a quick and nondestructive method of assaying for tolerance by applying Mn to the petioles of intact plants and rating the degree of induced Mn toxicity symptoms.

H. SULFUR

Sulfur (S) can be an important factor affecting yield, oil, and protein in oil seeds (Saggar and Dev, 1974). Application of S increased the amino acids cystine, cysteine, and methionine in soybean seeds (Kumar *et al.*, 1981). Chandel *et al.* (1989) demonstrated cultivar differences in soybean for S content.

I. MAGNESIUM

Kleese *et al.* (1968) reported that P, Ca, Sr, Mg, and Mn accumulations in soybean seed were highly and positively correlated, suggesting that accumulation of these elements may be controlled by a coordinated genetic system. Two possibilities were suggested: (a) the genes controlling accumulation of the different elements may be linked genetically or (b) a single genetic system governs the accumulation of all these elements.

X. Phosphorus

Seay and Henson (1958) reported a fivefold range among 30 clones of red clover in yield of P at the first cut.

Chisholm and Blair (1988) compared the phosphorus efficiency of two pasture legume species, 'Ladino' white clover and 'Verano' *Stylosanthes hamata*. They found that the amount of dry matter accumulated per unit of P indicated that white clover was more P efficient than Verano in the early stages of growth.

The ability of a genotype to extract and accumulate particular nutrients in low supply in the soil does not necessarily provide the entire resolution of crop productivity. The efficiency of use of the nutrients in producing economic yield must be considered.

About one-third of the total farmland of New Zealand is classed as "hill country." Hill country soils are typically moist and infertile with low phosphorous. Pastures in these areas often have a low legume content and are usually dominated by grasses of low productivity. With fertilization and improved management, stocking capacity can be increased up to threefold. To improve production in these areas, research efforts have been directed toward development of white clover cultivars adapted to productivity and survival under grazing by sheep in this low-P environment (Caradus and Williams, 1981). Caradus (1983) found significant differences among white clover populations for P uptake per unit of root length in solution culture following growth at high P but not following growth at low P. Populations originating from low-P soils had lower rates of P uptake per unit of root length than those from high-P soils. Caradus and Snaydon (1986a) reported that white clover populations collected from low-P soils had a higher percentage survival and higher yields than populations collected from high-P soils. Populations from low-P soils had finer root systems than populations from high-P soils when grown in pure stands.

Caradus and Snaydon (1986b) compared seven methods of evaluating white clover populations for tolerance to low P. Methods ranged from short-term solution culture in growth cabinets to heavily grazed sward on low-P acid soil. They concluded that field tolerance to soils with low P can be accurately identified only by screening in realistic field conditions, i.e., in grazed, grass-dominated swards on low-P soils. Caradus and Williams (1981) postulated a model white clover ideotype for survival and production on low-P infertile hill country sites that was designed to conserve P by partitioning it into storage tissues. The model plant would have small leaves, prostrate habit, and very dense stolon branching. Field results indicated that plants selected on the basis of this model performed better than plants with medium-sized leaves in low fertility hill country soils, but were inferior on high fertility lowland soils. In a glasshouse study, the plants adapted to the low fertility soils had lower total P and lower growth rates than large-leaved types. Genotypes differed greatly in the portion of P partitioned to leaves, stolons, and roots. The hill country selections had a higher proportion of P in stolons and roots. The range of variation suggested opportunity for selection of desired types.

Hart and Colville (1988) studied shoot dry weight and several characteristics of the youngest mature leaf of white clover populations from high-P soils and low-P soils in glasshouse soil pot tests at a range of P levels. They reported a significant population group × P interaction for leaf total P concentration, inorganic P concentration, and soluble protein concentration. They concluded that field performance tests are an absolute necessity in determining the utility of pasture plants differing in P responses and nutritional characters in controlled environment experiments.

In evaluating the usefulness of pasture and forage species it is important to bear in mind that the final marketable product is a livestock product, i.e., meat, milk, fiber, hides, etc. Although plant productivity and survival are important, the contribution of the crop to livestock nutrition and performance needs to be evaluated. The crop's contribution to the mineral nutrition of the livestock may be supplemented by minerals supplied in salt licks or additives in dietary supplements.

The ranking of common bean cultivars for shoot and root growth rate and nutrient absorption and translocation in a flowing nutrient solution culture system and the ranking for seed yield on a soil low in available P were found to be the same in a study by Salinas (1978). Whiteaker *et al.* (1976) identified bean accessions differing in the efficiency of P utilization and root development in a low-P nutrient solution culture. Fawole *et al.* (1982a,b) found that these traits were governed by quantitative inheritance. Schettini *et al.* (1987) reported that the inbred backcross breeding method (Bliss, 1981) had been used successfully to transfer P efficiency from exotic germplasm into the common bean cultivar Sanilac.

Chick-pea usually grows well without P application in vertisols with low soil solution levels of P (Itoh, 1987). Phosphorous depletion curves for roots of chick-pea, pigeon pea, soybean, and maize indicated that chick-pea and pigeon pea absorbed P less rapidly than maize or soybean in solution culture. Other factors such as extensive root hair development, mycorrhizae, and the ability to solubilize soil P may enhance the uptake of P from low-P concentrations in the soil in chick-pea and pigeon pea.

Wide variation is found in soybean cultivars for tolerance to excess P (Howell and Bernard, 1961). A dominant allele, designated *Np*, has a major effect on phosphorous tolerance (Bernard and Howell, 1964). The use of some commercial soil mixes with the high P levels conducive to culture of African violets may be toxic to soybean growth (T. E. Devine, 1989 personal communication).

XI. Potassium

Coelho and Blue (1978) tested five species of *Stylosanthes* (*S. guianensis, S. vicosa, S. fruticosa, S. hamata,* and *S. scabra*) for differential response to levels of K application. They reported that oven-dry herbage and mineral concentrations were generally different among species.

Shea *et al.* (1967) identified a gene locus in snap beans controlling the efficiency of K utilization. The symbol *ke* was proposed to designate a recessive allele conditioning efficient utilization. Inefficient utilization was completely dominant at this locus. The differential response was attributed to K utilization rather than K uptake.

Evidence suggests the soybean cyst nematode injury is more severe in infertile soils, particularly soils low in K (Wrather *et al.*, 1984). Fertilizer containing KCl as the source of K is usually of lower cost. Abel (1969) identified a dominant allele designated *NCl* that conditioned chloride exclusion in crosses of Lee, a chloride excluder, with 'Jackson,' a chloride accumulator. Hanson and Charles (1989) described differential soybean cultivar response to inoculation with soybean cyst nematode and K and Cl nutrition.

Raper and Barber (1970a) reported that the soybean cultivar 'Harosoy 63' had a more extensive root system than the cultivar 'Aoda.' Root surface area differed by greater than twofold. Raper and Barber (1970b) compared the accumulation per unit of root area for K, Ca, Mg, and Zn for Aoda and Harosoy 63 and found the Aoda roots absorbed all nutrients better than Harsoy 63, with the greatest difference in the case of K. The efficiency of K absorption for Aoda roots was almost double that for Harosoy 63 at high K levels but nearly equal at low levels. The larger Harosoy 63 root system has an advantage when competition with surrounding root systems occurs. At low population density, greater K absorption occurs with the smaller Aoda root system. In the field, the larger root system of Harosoy 63 would be an advantage if the availability of nutrients in the soil were low. The ability of Aoda roots to absorb K at a more rapid rate than Harosoy 63 roots as the K concentration increases is a distinct advantage when K is not uniformly distributed throughout the soil. When K is banded, Aoda would have an advantage per unit root surface. The greater root surface area and volume would enable Harosoy 63 plants to compete with Aoda plants at high levels of available nutrients.

XII. Symbiotic Nitrogen Fixation

Legume crops are distinguished from virtually all other crops by their ability to access the supply of atmospheric nitrogen through a symbiotic association with rhizobial bacteria. Successful symbiotic nitrogen fixation permits the culture of legume crops without application of synthetic nitrogen fertilizer. Selection and use of rhizobial strains adapted to particular edaphic conditions and compatible with the host crop genotype are important in achieving good nitrogen fixation.

A. RHIZOBIA

The legume–bacterial symbiosis is affected by acid soil conditions. Liming with P fertilization increased nodulation and N content as well as plant weight in alfalfa inoculated with *Rhizobium meliloti* (Bouton *et al.*, 1981b). Munns (1965) found differences among *R. meliloti* strains for ability to nodulate in acid soils and suggested that a practical advantage may be gained by selecting acid soil-tolerant strains. Bouton *et al.* (1981a) also found differences among strains of *R. meliloti* for ability to nodulate and fix N_2 with acid-tolerant alfalfa. Lowendorf and Alexander (1983) reported that *R. meliloti* strains previously chosen for ability to increase alfalfa performance in acid soil showed greater numerical increase in acid soil than other strains when in the presence of the growing alfalfa plant. The need for adapted *Rhizobium* strains is therefore compelling.

B. BEANS

Beans are widely cultivated in temperate and semitropical regions and provide an important source of protein for the human diet. Cultivated forms range from the shorter determinate bush bean type to the taller indeterminate climbing type. The bush bean types have insufficient fixation for adequate crop yields and require application of synthetic nitrogen fertilizer. Bush beans have a relatively short time interval from germination to flowering. With the onset of flowering and seed development, the demand on photosynthate supplies for seed development are thought to compete with the demand for energy involved in nitrogen fixation. Graham and Rosas (1977) found that cultivars of climbing beans were superior to bush bean cultivars in acetylene reduction. The difference was not due to nodule development, but was associated with greater specific nodule activity.

Graham (1981) suggested that at least three factors could be contributing to the variability in N_2 fixation observed in *Phaseolus vulgaris:* (a) supply of carbohydrates to the nodule, (b) relative rates of N uptake from soil, and (c) time to flowering. He reported that climbing cultivars appeared to transfer a greater proportion of their nonstructural carbohydrate to the nodules than did bush cultivars. Graham reported that seedlings of bush bean cultivars absorb soil N more rapidly than climbing cultivars (Graham and Rosas, 1977) and suggested that they might decrease carbohydrate supply to nodules and lower fixation. Developing pods compete for photosynthetic products and thus reduce nodule growth and limit N_2 fixation (Lawn and Brun, 1974). Graham (1981) reported that photoperiod sensitivity had been used to delay flowering and improve N_2 fixation, but that few naturally late-flowering bush bean genotypes had been identified.

Bliss (1986) reported substantial variability in common bean germplasm for N_2 fixation and traits supporting fixation. Superior lines selected from inbred backcross populations of crosses of a low-fixing cultivar and a high-fixing donor had higher total N_2 fixed per plant and a higher percentage of total N from fixation than the recurrent parent. He suggested that total nitrogen concentration of plants growing on a low-N soil would be a useful selection criterion for breeding. Graham (1981) described a CIAT breeding program for improvement of N_2 fixation in bush type beans utilizing recurrent selection and introgression of desirable germplasm from CIAT nurseries.

C. SOYBEAN

Rhizobial strains that increase soybean yields have been identified (Abel and Erdman, 1964; Caldwell and Vest, 1970) and efforts are under way to construct new highly superior strains of *Rhizobium* (Williams and Phillips, 1983) Much of the world's soybean area contains an abundant, heterogeneous and persistent population of rhizobia ranging from poor to good in their N_2 fixation efficiency with current soybean cultivars. At present there is no economically feasible method of establishing introduced strains against the competition of indigenous strains (Ham, 1976; Vest *et al.*, 1973). Therefore, two approaches to the problem of indigenous rhizobia may be taken: (a) breed the host plant for adaptation to the indigenous rhizobia or (b) attempt to control the genotype of the rhizobia admitted to symbiosis.

1. Breeding for Compatibility with Indigenous Rhizobia

Selecting soybean germplasm for adaptation to indigenous rhizobia may have been implicitly practiced by breeders selecting for performance on low-nitrogen soils since nitrogen fixation would have been integrated with the other components of agronomic performance used as selection criteria. New techniques for measuring N_2 fixation directly may improve the efficiency of selection. A striking example of specific selection of plant germplasm for adaptation to indigenous rhizobia has been conducted by IITA in Africa. Soybean is not an indigenous crop in Africa and U.S.-bred soybean cultivars do not adequately nodulate and fix nitrogen with indigenous African bradyrhizobia. The introduction of compatible bradyrhizobia results in markedly improved fixation of the U.S.-bred lines in Africa. However, production and distribution of inoculum may be restricted in developing countries. Nangju (1980) reported that soybeans from southeast Asia nodulated successfully with indigenous African rhizobia resulting in significant N_2 fixation. United States-bred cultivars, however, have the advantage of higher grain yield potential and better resistance to lodging and pod shattering. Nangju suggested breeding to combine the desirable agronomic characteristics of U.S.-bred cultivars with the symbiotic potential for N_2 fixation and the indigenous bradyrhizobia characteristic of the southeast Asian germplasm. Such cultivars have been developed.

Pal (1989) evaluated the performance of the promiscuous cultivars and check cultivars in field tests in the Nigerian savanna under applied nitrogen, inoculation with cultured strains of *R. japonicum,* and untreated control treatments. In 1985, under the control treatment the highest-yielding lines were 'Samsoy 2' (1812 kg ha^{-1}) and 'M90' (1972 kg ha^{-1}), while in the treatment inoculated with a mixture of two strains of *R. japonicum,* these lines yielded 1930 and 2170 kg ha^{-1}, respectively. Within these cultivars, these yields were not significantly different between these treatments. In 1986, the highest-yielding lines in the control treatment were 'Samsoy 1' (3350 kg ha^{-1}) and 'JGX297-6F' (3630 kg ha^{-1}). However, Samsoy 1 yielded 3110 kg ha^{-1} and JGX297-6F yielded 3940 kg ha^{-1} when inoculated with *R. japonicum* strain IRj2114. These later yields, however, were not significantly greater than those in the control treatment. In other lines, the inoculated strains of *R. japonicum* contributed significantly greater amounts of symbiotically fixed N to total N uptake than the native rhizobia. In such lines, further yield increases in the promiscuous cultivars were deemed possible and Pal recommended the isolation and testing of effective strains of *R. japonicum* adapted to the savanna environment and subsequent production of commercial inoculants in order to increase yields of currently available promiscuous cultivars in the Nigerian savanna.

2. Controlling Rhizobial Specificity

a. Concepts Devine and Weber (1977) proposed a genetic system for host specificity of the rhizobial strains admitted to symbiosis. The system requires (a) development of host cultivars that substantially exclude infection by the indigenous *Rhizobium* strains, (b) identification or development of *Rhizobium* strains having the genetic potential to infect these specific host cultivars, and (c) development of the technology to manipulate the genetic system of *Rhizobium* in order to couple this specific nodulating ability with high nitrogen fixation. Achievement of these three elements would provide farmers with a production package of cultivars that would not nodulate with indigenous strains and with inoculum of several superior strains that would nodulate these cultivars.

b. The rj_1 Gene By backcrossing the rj_1 gene, the nonnodulating gene, into agronomic lines of soybean, one can achieve the first requirement—exclusion of the indigenous *Rhizobium* strains from symbiosis. To achieve the second requirement, genetic information is needed in *Rhizobium* that will "break down the resistance" of the rj_1 genotype and permit nodulation. Although bradyrhizobia that nodulate the rj_1 phenotype have been found, a strain that gives adequate nodulation is still under construction.

c. Multigeneic Incompatibility Systems A variation of the exclusion scheme utilizing incompatibility alleles other than rj_1 has been suggested (Devine and Breithaupt, 1980). This would involve the accumulation at multiple loci of alleles excluding various groups of bradyrhizobia from effective nodulation. In comparing this approach with the use of the rj_1 allele, it should be borne in mind that the single allele rj_1 may be readily transferred by backcrossing, whereas accumulation of several incompatible alleles, such as Rj_2, Rj_4, etc., is more difficult in a backcrossing program and may be complicated by the need to break linkages with agronomically undesirable alleles.

For many years, soybean was thought to be nodulated only by slow-growing rhizobia (bradyrhizobia). In the early 1980s, Asiatic fast-growing rhizobia were found to nodulate and fix nitrogen with many soybean lines from Asia (Keyser *et al.*, 1982; Devine, 1985; Dowdle and Bohlool, 1985). If a host plant can be developed that excludes nodulation by the slow-growing rhizobial strains found in the United States but nodulates and fixes nitrogen very efficiently with the fast-growing rhizobial strains, a system of host specificity of the bacteria admitted to symbiosis could be constructed. If the host genotype has a strong preference for nodulation with the fast-growing rhizobia, this might be sufficient to achieve host specificity (Devine, 1985).

To employ the system of host–plant exclusion of indigenous bradyrhizobia with selective receptivity to desirable strains, researchers at the University of Minnesota identified lines in the plant introduction collection that did not nodulate normally with the bradyrhizobia indigenous to the Minnesota test site (Kiven *et al.*, 1981). These lines did nodulate with desirable strains of bradyrhizobia. Unfortunately, the ability to exclude the indigenous strains from nodulation was not consistent across test sites in Minnesota. Although most of the strains of *Bradyrhizobium japonicum* in the northern Midwest of the United States are classified by serological response into a single serogroup, these organisms vary quite markedly in other characteristics, including N_2 fixation efficiency. They also vary in their ability to nodulate host plants carrying genes that may prevent nodulation with some members of the serogroup, thus frustrating the efforts to exclude the whole serogroup (Keyser and Cregan, 1987).

d. Naturally Occurring Systems of Host Specificity Some strains of bradyrhizobia that nodulate soybean also produce, both *in vitro* and during symbiosis, a substance called rhizobitoxine that interferes with chlorophyll synthesis. Afflicted plants show a chlorotic appearance in the upper leaves. Devine *et al.* (1988) established that the bradyrhizobia producing these symptoms during symbiosis were found only in a particular DNA homology group (group 2) of bradyrhizobia that was sufficiently different from other bradyrhizobia to warrant the suggestion that they should receive separate taxonomic classification (Hollis *et al.*, 1981). Research by Kuykendall *et al.* (1988) on fatty acid composition and multiple antibiotic resistance patterns affirms this distinction.

When these bradyrhizobia that induce chlorosis on soybean during symbiosis are inoculated onto another host species, such as cowpea, they can nodulate very well and produce very vigorous growth in nitrogen-limited medium, which indicates very successful nitrogen fixation, and they do not induce chlorosis symptoms (Devine, 1988). Apparently, these bradyrhizobia are physiologically more compatible with cowpea than with soybean.

Information on the relative frequency of bradyrhizobial strains inducing chlorosis during symbiosis in North American soils is limited. Rhizobitoxine-induced chlorosis symptoms have not been reported in the northern Midwest. However, during the initial years of the development of soybean production for grain in the United States, chlorosis symptoms were encountered in sizable areas of production fields in the southeastern United States. In more recent years such symptoms are rare (E. Hartwig, 1988 personal communication). It appears the chlorosis-inducing strains are better adapted to sandy or well-drained soils and sites subject to droughty conditions and high temperatures. The reduced incidence of symptoms in the Southeast suggests that the long-term effects of inoculation of fields with non-chlorosis-inducing strains may have reduced the relative frequency of chlorosis-inducing bacteria (Dunigan *et al.*, 1984).

A recent survey by J. J. Fuhrman (1988 poster session) of 18 farms in Delaware indicated that 37% of the nodules sampled contained bradyrhizobia with the colony morphology characteristic of the chlorosis-inducing strains. Nitrogen content of plant top growth was lower for strains with the colony morphology characteristic of chlorosis-inducing strains. In general, rhizobitoxine production by the isolates was more strongly related to the N content of soybean top growth than was uptake hydrogenase. It appears, then, that the chlorosis-inducing strains belong to a DNA homology group distinct from those containing the strains most efficient in fixation with soybean, that the chlorosis-inducing strains have a physiology more compatible with other legumes than with soybean, and that they are less efficient in nitrogen fixation with soybean than other strains of *Bradyrhizobium japonicum*.

The Rj_4 allele in soybean conditions an ineffective nodulation response with specific strains of bradyrhizobia while the rj_4 allele permits normal nodulation with these strains. Two near isogenic soybean lines (99.95% genetically identical) differing specifically in the Rj_4 versus rj_4 alleles have been synthesized (Devine and O'Neill, 1986) and used to test an array of chlorosis-inducing and non-chlorosis-inducing bradyrhizobia for response with the Rj_4 allele. The Rj_4 allele was found to interdict nodulation with 33% of the chlorosis-inducing rhizobia and only 3% of the non-chlorosis-inducing rhizobia (Devine, 1988). Seventy-seven percent of the strains interdicted by the Rj_4 allele are chlorosis-inducing strains and 90% have the antibiotic resistance profile characteristic of DNA homology group 2. Thus, the Rj_4 allele may have a positive value to the host plant in areas with significant populations of chlorosis-inducing rhizobia by protecting the plant from nodulation with some of these rhizobia. The frequency of ineffective nodulation with the chlorosis-inducing strain USDA 61 shows a progressive loss with domestication from the wild species *Glycine soja* (60%), to the Asiatic plant introductions of the domesticated species, *G. max* (30%), and with selection for agronomic type in North America as demonstrated by the preliminary test lines (14%) and the more highly selected uniform test lines (9%) (Devine, 1987). The apparent attrition of the frequency of the Rj_4 allele in the breeding of North American cultivars should be of concern to soybean breeders in the southeastern United States.

D. ALFALFA

Alfalfa is highly valued for its high-quality forage. The USDA research team located at the University of Minnesota has conducted a multidisciplinary research program directed toward improving symbiotic nitrogen fixation concurrent with breeding for yield and quality. The method used is a recurrent selection procedure in which a sequential culling is practiced during each cycle. Selection is imposed independently in populations previously selected for multiple pest resistance and agronomic performance; then the independent germplasm pools are intercrossed to produce heterosis for forage yield in addition to improved nitrogen fixation (Barnes *et al.*, 1984, 1987).

The selection program is initiated by planting a large number of seed of a population in nil-N sand in benches in the glasshouse. One week after seeding, seedlings are inoculated with a mixture of strains of *Rhizobium meliloti* characterized by high or low competence for nitrogen fixation with alfalfa. Subsequent selection of the most vigorous plants is based on the premise that plants that nodulate primarily with the rhizobia with low competence for fixation will be discarded. Subsequently, surviving plants are sequentially evaluated and culled for lower vigor of shoot growth, nodule mass and root mass, low nitrogenase activity, and lower nodule enzyme (GOGAT, PEC, and total soluble protein) activities. Surviving plants are then intercrossed to initiate another cycle of selection.

Two cycles of selection were imposed in two nondormant germplasm pools for increased amounts of forage, crowns, and roots during autumn and for increased concentrations of N stored in the taproot. Strain crosses between the two germplasm pools resulted in the new cultivar 'Nitro' (Barnes *et al.*, 1988). This cultivar reportedly provides more N (140 kg N ha^{-1} vs. 106 kg N ha^{-1}) in the seedling year for plow-down than dormant check cultivars and a 7% improvement in nitrogen fixation ability over the parental germplasm pools (Barnes, *et al*, 1987).

Researchers in California (Teuber and Phillips, 1988) demonstrated in glasshouse tests and preliminary field trials that a multiple trait selection method can increase forage yield and quality in 'Hairy Peruvian,' an alfalfa germplasm unadapted to California conditions. With similar selection in the agronomically adapted cultivar 'Moapa 69,' glasshouse tests indicated the same method of selection was successful. These researchers concluded that selection for dry weight and N concentration must be done with dependence on both symbiotically fixed and supplied nitrogen to develop alfalfa populations that express increased forage yield and higher crude protein concentration in both N dependency environments under glasshouse conditions.

XIII. Mycorrhizae and Legumes in Problem Soils

Mycorrhizal fungi are widespread and their mutualistic associations with plants are found in both natural habitats and cultivated fields. However, the role of mycorrhizae in enhancing nutrient absorption, especially in conditions where essential nutrients are limiting, is so important that no discussion of plant nutrition can be made without consideration of these fungi (Harley, 1986).

Interactions of all components in a legume cropping system (i.e., plant host–*Rhizobium*–Mycorrhizae–soil fertility) are complex, making the study of these interactions very difficult. However, in soil, inoculation of legumes with mycorrhizae has been found to increase plant growth. Increased growth at low levels of P was reported for alfalfa (Karow and Lindsey, 1984; Satterlee *et al.*, 1983), soybean (Skipper and Struble, 1984), and white clover (Hall *et al.*, 1977). Increased growth after mycorrhizal inoculation was also found in unlimed soils in both alfalfa (Kucey, 1984) and soybeans (Skipper and Smith, 1979). Similarly, better rhizobial nodulation was reported for soybeans (Skipper and Smith, 1979) and dual inoculation with both rhizobia and mycorrhizae doubled alfalfa yield over uninoculated controls (Azcon-Aguilar and Barea, 1981).

The specificity of mycorrhizae and their plant hosts is low with nonspecificity generally felt to be the norm in most mutualistic associations (Harley, 1986). However, there does seem to be specificity for some mycorrhizae and their legume hosts as reported by Lioi *et al.* (1986). Similarly, Skipper and Smith (1979) found certain soybean cultivar—mycorrhizal species associations to be more responsive in unlimed soils than other associations. Heckman and Angle (1987) found that soybean cultivars were differentially colonized by vesicular-arbuscular mycorrhizal (VAM) fungi, and Carling and Brown (1980) reported differential cultivar responses in growth and yield to various fungal species. Furthermore, van Nuffelen and Schenck (1984) found that the fungi may develop differently under the host plant's influence. This differential response of specific cultivars or genotypes has also been documented for alfalfa (Lambert *et al.*, 1980) and white clover (Hall *et al.*, 1977). These findings, therefore, indicate that plant or mycorrhizal genetic factors may influence their mutualism, thus providing the potential for the use of genetic selection to enhance the association. At the very least, these reports indicate that screening legumes for use in problem soils should be conducted under natural conditions in the presence of mycorrhizae.

XIV. Summary

Genetic modification of crop plants for enhanced efficiency and utilization of nutrients or tolerance to nutrient toxicities is a worthwhile objective for both grain and forage legumes. Two approaches are possible: (a) replacement of agriculturally acceptable intolerant or inefficient species with new tolerant or efficient species or (b) selection for tolerance or efficiency within the agriculturally acceptable but intolerant or inefficient species. In either case, one must assess the factors in the edaphic environment that limit production and then use this information to clearly define the research objectives necessary to improve production.

For genetic selection and plant breeding to be successful, one must choose an appropriate breeding method that will vary depending on the species' mode of reproduction, the degree and type of genetic variability, gene action, heritability of the trait itself, the development and use of an appropriate screening procedure, and testing of selected material to ensure genetic advance. The major topics that have been successfully researched in one or more of these areas and presented in this chapter include the following: acid soil stress; Al, Mn, or Zn toxicity; Fe, Mn, and Zn deficiency; salinity; and efficient use of Ca, Cu, Fe, K, Mg, Mn, Mo, P, S, and Zn.

The interaction of *Rhizobia* with legumes for symbiotic nitrogen fixation is also strongly influenced by the soil environment, therefore selection and use of rhizobial strains adapted to the particular edaphic conditions and compatible with the host legume is important in achieving good nitrogen fixation. Finally, the mutualistic association of legumes with mycorrhizal fungi is important in the efficient use of nutrients and in overcoming toxic soil conditions and should also be addressed during assessment and selection.

References

Abd-Elgawad, M., and B. D. Knezek. 1983. Studies on manganese toxicity in two soybean cultivars. *Ann. Agric. Sci., Moshotor* 19:603–615.

Abel, G. H. 1969. Inheritance of the capacity for chloride inclusion and chloride exclusion by soybeans. *Crop Sci.* 9:697–698.

Abel, G. H., and L. H. Erdman. 1964. Response of Lee soybeans to different strains of *Rhizobium japonicum*. *Agron. J.* 56:423–424.

Aimi, R., and T. Murakami. 1964. Cell-physiological studies on the growth of crop plants. *Bull. Natl. Inst. Agric. Sci., Ser. D* 11:331–393.

Alison, M. W. 1987. Response of birdsfoot trefoil cultivars to defoliation and soil acidity. Ph.D. Thesis, Univ. of Georgia, Athens. (*Diss. Abstr.* 88-06772.)

Ambler, J. E., and J. C. Brown. 1969. Cause of differential susceptibility to zinc deficiency in two varieties of navy beans (*Phaseolus vulgaris* L.). *Agron. J.* 61:41–43.

Anderson, A. J. 1956. Molybdenum as a fertilizer. *Adv. Agron.* 8:163–202.

Andrew, C. S., and W. W. Bryan. 1955. Pasture studies on the coastal lowlands of subtropical Queensland. I. Introduction and initial plant studies. *Aust. J. Agric. Res.* 6:265–290.

Andrew, C. S., and W. W. Bryan. 1958. Pasture studies on the coastal lowlands of subtropical Queensland. III. The nutrient requirements and potentialities of *Desmodium uncinatum* and white clover on a lateritic podzolic soil. *Aust. J. Agric. Res.* 9:267–285.

Andrew, C. S., and M. P. Hegarty. 1969. Comparative response to manganese excess in eight tropical and four temperate pasture legume species. *Aust. J. Agric. Res.* 20:687–686.

Andrew, C. S., and P. M. Thorne. 1962. Comparative responses to copper of some tropical and temperate pasture legumes. *Aust. J. Agric. Res.* 13:821–835.

Andrew, W. D., and R. T. Milligan. 1954. Different molybdenum requirements of medics and subterranean clover on a red brown soil at Wagga, New South Wales. *J. Aust. Inst. Agric. Sci.* 20:123–124.

Armiger, W. H., C. D. Foy, A. L. Fleming, and B. E. Caldwell. 1968. Differential tolerance of soybean varieties to an acid soil high in exchangeable aluminum. *Agron. J.* 60:67–70.

Azcon-Aguilar, C., and J. M. Barea. 1981. Field inoculation of *Medicago* with V-A mycorrhiza and *Rhizobium* in phosphate-fixing agricultural soil. *Soil Biol. Biochem.* 13:19–22.

Barnes, D. K., G. H. Heichel, C. P. Vance, and W. R. Ellis. 1984. A multiple-trait breeding program for improving the symbiosis for N_2 fixation between *Medicago sativa* L. and *Rhizobium meliloti*. *Plant Soil* 82:303–314.

Barnes, D. K., G. H. Heichel, C. P. Vance, and C. C. Schaeffer. 1987. Breeding for N_2 fixation and N nutrition in alfalfa. In T. Huguet and C. P. Vance (eds.), Plant Genes Involved in Nitrogen Fixation and Productivity of Alfalfa. Proc. Workshop USDA–INRA, Toulouse, France. p. 89.

Barnes, D. K., C. C. Schaeffer, G. H. Heichel, D. M. Smith, and R. N. Peaden. 1988. Registration of 'Nitro' alfalfa. *Crop Sci.* 28:718.

Bernard, R. L., and R. W. Howell. 1964. Inheritance of phosphorus sensitivity in soybeans. *Crop Sci.* 4:298–299.

Blatt, C. R., and A. Van Diest. 1981. Evaluation of a screening technique of manganese toxicity in relation to leaf manganese distribution and interaction with silicon. *Neth. J. Agric. Sci.* 29:297–304.

Bliss, F. A. 1981. Utilization of vegetable germplasm. *HortScience* 16:129–132.

Bliss, F. A. 1986. Breeding progress for increased nitrogen fixation in common bean (*Phaseolus vulgaris*). *Agron. Abstr.* pp. 57–58.

Bouton, J., H. Calvert, J. Dobson, D. Fisher, F. Newsome, E. Worley, and P. Worley. 1988. Performance of alfalfa varieties in northern Georgia, 1984–1986. *Univ. Ga. Agric. Exp. Stn., Res. Rep.* No. 533.

Bouton, J. H., J. E. Hammel, and M. E. Sumner. 1982. Alfalfa, *Medicago sativa* L., in highly weathered acid soil. IV. Root growth into acid subsoil of plants selected for acid tolerance. *Plant Soil* 65:187–192.

Bouton, J. H., and P. G. Hartel. 1987. Recurrent selection of alfalfa for acid soil tolerance. *Agron. Abstr.* p. 56.

Bouton, J. H., and W. G. Monson. 1985. Performance of alfalfa varieties in Georgia, 1982–84. *Univ. Ga. Agric. Exp. Stn., Res. Rep.* No. 476.

Bouton, J. H., and M. E. Sumner, 1983. Alfalfa, *Medicago sativa* L., in highly weathered, acid soils. V. Field performance of alfalfa selected for acid tolerance. *Plant Soil* 74:431–436.

Bouton, J. H., M. E. Sumner, and J. E. Giddens. 1981a. Alfalfa, *Medicago sativa* L., in highly weathered, acid soils. II. Yield and acetylene reduction of a plant germplasm and *Rhizobium meliloti* inoculum selected for tolerance to acid soil. *Plant Soil* 60:205–211.

Bouton, J. H., M. E. Sumner, J. E. Hammel, and H. Shahandeh. 1986. Yield of an alfalfa germplasm selected for acid soil tolerance when grown in soil with modified subsoils. *Crop Sci.* 26:334–336.

Bouton, J. H., J. K. Syers, and M. E. Sumner. 1981b. Alfalfa, *Medicago sativa* L., in highly weathered, acid soils. I. Effect of lime and P application on yield and acetylene reduction of young plants. *Plant Soil* 59:455–463.

Bradshaw, A. D., and R. W. Snaydon. 1959. Population differentiation within plant species in response to soil factors. *Nature (London)* 183:129–130.

Brooks, C. O., J. H. Bouton, and M. E. Sumner. 1982. Alfalfa, *Medicago sativa* L., in highly weathered, acid soils. III. The effects of seedling selection in an acid soil on alfalfa growth at varying levels of phosphorus and lime. *Plant Soil* 65:27–33.

Brown, J. C., and T. E. Devine. 1980. Inheritance of tolerance or resistance to manganese toxicity in soybeans. *Agron. J.* 72:898–904.

Brown, J. C., R. S. Holmes, and L. O. Tiffin. 1961. Iron chlorosis in soybeans as related to the genotype of rootstalk. 3. Chlorosis susceptibility and reductive capacity at the root. *Soil Sci.* 91:127–132.

Brown, J. C., and W. E. Jones. 1977. Manganese and iron toxicities dependent on soybean variety. *Commun. Soil Sci. Plant Anal.* 8:1–15.

Brown, J. C., C. R. Weber, and B. E. Caldwell. 1967. Efficient and inefficient use of iron by two soybean genotypes and their isolines. *Agron. J.* 59:459–462.

Buss, G. R., J. A. Lutz, Jr., and G. W. Hawkins. 1975a. Effect of soil pH and plant genotype on element concentration and uptake by alfalfa. *Crop Sci.* 15:614–617.

Buss, G. R., J. A. Lutz, Jr., and G. W. Hawkins. 1975b. Yield response of alfalfa cultivars and clones to several pH levels in Tatum subsoil. *Agron. J.* 67:331–334.

Caldwell, B. E., and G. Vest. 1970. Effects of *Rhizobium japonicum* strains on soybean yields. *Crop Sci.* 10:19–21.

Campbell, T. A., J. H. Elgin, Jr., C. D. Foy, and J. E. McMurtrey, III. 1988. Selection in alfalfa for tolerance to toxic levels of aluminum. *Can. J. Plant Sci.* 68:743–753.

Caradus, J. R. 1983. Genetic differences in phosphorus absorption among white clover populations. *Plant Soil* 72:379–383.

Caradus, J. R., and R. W. Snaydon. 1986a. Response to phosphorus of populations of white clover. 1. Field studies. *N.Z. J. Agric. Res.* 29:155–162.

Caradus, J. R., and R. W. Snaydon. 1986b. Response to phosphorus of populations of white clover. 3. Comparison of experimental techniques. *N.Z. J. Agric. Res.* 9:169–178.

Caradus, J. R., and W. M. Williams. 1981. Breeding for improved white clover production in New Zealand hill country. *Plant Physiology and Herbage Production, Proc. British Grassland Society Symp., Nottingham, England* pp. 163–167.

Carling, D. E., and M. F. Brown. 1980. Relative effects of vesicular-arbuscular mycorrhizal fungi on the growth and yield of soybeans. *Soil Sci. Soc. Am. J.* 44:528–532.

Chandel, A. S., G. P. Rao, and S. C. Saxena. 1989. Effect of sulfur nutrition on soybean (*Glycine max* L. Merrill). In A. J. Pascale (ed.), *Proc. World Soybean Research Conf. IV, Buenos Aires.* Orientacion Grafica Editora, Buenos Aires. Pp. 343–368.

Chaney, R. L. 1984. Diagnostic practices to identify iron deficiency in higher plants. *J. Plant Nutr.* 5:251–268.

Chisholm, R. H., and G. J. Blair. 1988. Phosphorus efficiency in pasture species. I. Measures based on total dry weight and P content. *Aust. J. Agric. Res.* 39:807–816.

Cianzio, S. R., and W. R. Fehr. 1980. Genetic control of iron-deficiency chlorosis in soybean. *Iowa State J. Res.* 54:367–375.

Cianzio, S. R., and W. R. Fehr. 1982. Variation in inheritance of resistance to iron-deficiency chlorosis in soybeans. *Crop Sci.* 22:433–434.

CIAT. 1977. Ann. Report. Cali, Columbia.

Clarkson, D. T. 1969. Metabolic aspects of aluminum toxicity and some possible mechanisms for resistance. In I. N. Rorison (ed.), Ecological Aspect of the Mineral Nutrition of Plants. *Symp. Br. Ecol. Soc.* 9:381–387.

Coelho, R. W., and W. G. Blue. 1978. Potassium nutrition of five species of the tropical legume stylosanthes in an aeric haplaquod. *Proc.—Soil Crop Sci. Soc. Fla.* 38:90–93.

Croughan, T. P., S. J. Stavarek, and D. W. Rains. 1978. Selection of a NaCl tolerant line of cultured alfalfa cells. *Crop Sci.* 18:959–963.

Dessureaux, L. 1959. Heritability of tolerance to manganese toxicity in lucerne. *Euphytica* 8:260–265.

Dessureaux, L. 1960. The reaction of lucerne seedlings to high concentrations of manganese. *Plant Soil* 13:114–122.

Dessureaux, L. 1969. Effect of aluminum of alfalfa seedlings. *Plant Soil* 30:93–98.

Devine, T. E. 1976. Genetic potentials for solving problems of soil mineral stress: Aluminum and manganese toxicities in legumes. In M. J. Wright (ed.), Plant Adaption to Mineral Stress in Problem Soils. Cornell Univ. Press, Ithaca, New York. Pp. 65–72.

Devine, T. E. 1985. Nodulation of soybean plant introduction lines with the fast-growing rhizobial strain USDA 205. *Crop Sci.* 25:354–356.

Devine, T. E. 1987. A comparison of rhizobial strain compatibilities of *Glycine max* and its progenitor species *Glycine soja*. *Crop Sci.* 27:635–639.

Devine, T. E. 1988. Role of the nodulation restrictive allele Rj_4 in soybean evolution. *J. Plant Physiol.* 132:453–455.

Devine, T. E., and B. H. Breithaupt. 1980. Significance of incompatibility reactions of *Rhizobium japonicum* strains with soybean host genotypes. *Crop Sci.* 20:269–271.

Devine, T. E., C. D. Foy, A. L. Fleming, C. H. Hanson, T. A. Campbell, J. E. McMurtrey, III, and J. W. Schwartz. 1976. Development of alfalfa strains with differential tolerance to aluminum toxicity. *Plant Soil* 44:73–79.

Devine, T. E., C. D. Foy, D. L. Mason, and A. L. Fleming. 1979. Aluminum tolerance in soybean germplasm. *Soybean Genet. Newsl.* 6:24–27.

Devine, T. E., L. D. Kuykendall, and J. J. O'Neill. 1988. DNA homology group and the identity of

bradyrhizobial strains producing rhizoitoxine-induced foliar chlorosis on soybeans. *Crop Sci.* 28:938–941.
Devine, T. E., and J. J. O'Neill. 1986. Registration of BARC-2 (Rj_4) and BARC-3 (Rj_4) soybean germplasm. *Crop Sci.* 26:1263–1264.
Devine, T. E., and D. F. Weber. 1977. Genetic specificity of nodulation. *Euphytica* 26:527–535.
Dowdle, S. F., and B. B. Bohlool. 1985. Predominance of fast-growing *Rhizobium japonicum* in a soybean field in the People's Republic of China. *Appl. Environ. Microbiol.* 50:1171–1176.
Dunigan, E. P., P. K. Bollich, R. L. Hutchinson, P. M. Hicks, F. C. Zaunbrecher, S. G. Scott, and R. P. Mowers. 1984. Introduction and survival of an inoculant strain of *Rhizobium japonicum* in soil. *Agron. J.* 76:463–466.
Earley, E. B. 1943. Minor element studies with soybeans. I. Varietal reaction to concentrations of zinc in excess of the nutritional requirement. *J. Am. Soc. Agron.* 35:1012–1023.
Elgin, J. H. 1979. Alfalfa breeding research at the Beltsville Agricultural Research Center. *Proc. 36th Southern Pasture and Forage Crop Improvement Conf., Beltsville, Maryland.* USDA–ARS, New Orleans, Louisiana. Pp. 30–31.
Ellis, B. G. 1965. Response and susceptibility zinc deficiency—A symposium. *Crop Soils* 18:10–13.
Evans, J., B. J. Scott, and W. J. Lill. 1987. Manganese tolerance in subterranean clover (*Trifolium subterraneum* L.) genotypes grown with ammonium nitrate or symbiotic nitrogen. *Plant Soil* 97:207–215.
Fawole, I., W. H. Labelman, and G. C. Gerloff. 1982a. Genetic control of root development in beans (*Phaseolus vulgaris* L.) grown under phosphorus stress. *J. Am. Soc. Hortic. Sci.* 107:98–100.
Fawole, I., W. H. Labelman, G. C. Gerloff, and E. V. Nordheim. 1982b. Heritability of efficiency in phosphorus utilization in beans (*Phaseolus vulgaris* L.) grown under phosphorus stress *J. Am. Soc. Hortic. Sci.* 107:94–97.
Fehr, W. R. 1982. Control of iron-deficiency chlorosis in soybean by plant breeding. *J. Plant Nutr.* 5:611–621.
Fehr, W. R. 1984. Current practices for correcting iron deficiency in plant with emphasis on genetics. *J. Plant Nutr.* 7:347–354.
Fehr, W. R., and M. W. Trimble. 1982. Minimizing soybean yield loss from iron-deficiency chlorosis. *Coop. Ext. Serv., Iowa State Univ.* No. Pm-1059.
Foy, C. D. 1973. Manganese and plants. In Manganese. *N. A. S.–N. R. C., Publ.* pp. 51–76.
Foy, C. D. 1983. Plant adaptation to mineral stress in problem soils. *Iowa State J. Res.* 57:339–354.
Foy, C. D., W. H. Armiger, A. L. Fleming, and W. J. Zaumeyer. 1967. Differential tolerance of dry bean, snapbean and lima bean varieties to an acid soil high in exchangeable aluminum. *Agron. J.* 59:561–563.
Foy, C. D., R. L. Chaney, and M. C. White. 1978. The physiology of metal toxicity in plants. *Annu. Rev. Plant Physiol.* 29:511–566.
Foy, C. D., A. L. Fleming, and W. H. Amiger. 1969. Differential aluminum tolerance of soybean varieties in relation to calcium nutrition. *Agron. J.* 61:505–511.
Foy, C. D., A. L. Fleming, and G. C. Gerloff. 1972. Differential aluminum tolerance to two snapbean varieties. *Agron. J.* 64:815–818.
Foy, C. D., H. W. Webb, and J. E. Jones. 1981. Adaptation of cotton genotypes to an acid, manganese toxic soil. *Agron. J.* 73:107–111.
Franco, A. A., and D. N. Munns. 1982. Acidity and aluminum restraints on nodulation, nitrogen fixation, and growth of *Phaseolus vulgaris* in solution culture. *Soil Sci. Soc. Am. J.* 46:296–301.
Gauch, H. G., and O. C. Magistad. 1943. Growth of strawberry clover varieties and of alfalfa and ladino clover as affected by salt. *J. Am. Soc. Agron.* 35:871–880.
Gheesling, R. H., and H. F. Perkins. 1970. Critical levels of manganese and magnesium in cotton at different stages of growth. *Agron. J.* 62:29–32.
Gildersleeve, R. R., and W. R. Ocumpaugh. 1988. Variation among *Trifolium* species for resistance to iron-deficiency chlorosis. *J. Plant Nutr.* 11:727–737.
Graham, P. H. 1981. Some problems of nodulation and symbiotic nitrogen fixation in *Phaseolus vulgaris* L.: A review. *Fields Crops Res.* 4:93–112.

Graham, P. H., and J. C. Rosas. 1977. Growth and development of indeterminate bush and climbing cultivars of *Phaseolus vulgaris* L. inoculated with *Bradyrhizobium*. *J. Agric. Sci.* 88:503–508.
Hall, I. R., R. S. Scott, and P. D. Johnstone. 1977. Effect of vesicular-arbuscular mycorrhizas on response of 'Grasslands Huia' and 'Tamar' white clovers to phosphorus. *N.Z. J. Agric. Res.* 20:349–355.
Ham, G. E. 1976. Competition among strains of rhizobia. In L. D. Hill (ed.), World Soybean Research. Interstate Printers & Publishers, Danville, Illinois. Pp. 144–150.
Hanson, R. G., and K. E. Charles. 1989. Potassium and CL nutrition of soybean cultivar with genetic resistance to soybean cyst nematode. In A. J. Pascale (ed.), *Proc. World Soybean Research Conf. IV, Buenos Aires*. Orientacion Grafica Editora, Buenos Aires. Pp. 352–362.
Hanson, W. D., and E. J. Kamprath. 1979. Selection for aluminum tolerance in soybean based on seedling root growth. *Agron. J.* 71:581–586.
Harley, J. L. 1986. Mycorrhizal studies: Past and future. In V. Gianinazzi-Pearson and S. Gianinazzi (eds.), Physiological and Genetical Aspects of Mycorrhizae. Proc. 1st European Symp. Mycorrhizae, Dijon, France, 1985. INRA, Paris. Pp. 25–33.
Harris, H. B., M. B. Parker, and B. J. Johnson. 1965. Influence of molybdenum content of soybean seed and other factors associated with seed source on progeny response to applied molybdenum. *Agron. J.* 57:397–399.
Hart, A. L., and C. Colville. 1988. Differences among attributes of white clover genotypes at various levels of phosphorus supply. *J. Plant Nutr.* 11:189–207.
Haug, A., and C. R. Caldwell. 1985. Aluminum toxicity in plants: The role of the root plasma membrane and calmodulin. In J. B. St. John, E. Berlin, and P. C. Jackson (eds.), Frontiers of Membrane Research in Agriculture—Beltsville Symp. No. 9. Rowman & Allanheld, Totowa, New Jersey.
Heckman, J. R., and J. S. Angle. 1987. Variation between soybean cultivars in vesicular-arbuscular mycorrhiza fungi colonization. *Agron. J.* 79:428–430.
Heenan, D. P., and L. C. Campbell. 1981. Influence of potassium and manganese on growth and uptake of magnesium by soybeans [*Glycine max* (L.) Merr. cv. Bragg]. *Plant Soil* 61:447–456.
Heenan, D. P., L. C. Campbell, and O. G. Carter. 1981. Inheritance of tolerance to high manganese supply in soybeans. *Crop Sci.* 21:625–627.
Heenan, D. P., and O. G. Carter. 1975. Response of two soybean cultivars to manganese toxicity as affected by pH and calcium levels. *Aust. J. Agric. Res.* 26:967–974.
Heenan, D. P., and O. G. Carter. 1977. Influence of temperature on the expression of manganese toxicity by two soybean varieties. *Plant Soil* 47:219–227.
Hintz, R. W., W. R. Fehr, and S. R. Cianzio. 1987. Population developments for the selection of high-yielding soybean cultivars with resistance to iron-deficiency chlorosis. *Crop Sci.* 27:707–710.
Hollis, A. B., W. E. Kloos, and G. H. Elkan. 1981. DNA: DNA hybridization studies of *Rhizobium japonicum* and related Rhizoiaceae. *J. Gen. Microbiol.* 123:215–222.
Horst, W. J. 1983. Factors responsible for genotypic manganese tolerance in cowpea (*Vigna unguiculata*). *Plant Soil* 72:213–218.
Horst, W. J., and H. Marschner. 1978. Symptoms of manganese toxicity in beans (*Phaseolus vulgaris* L.). *Z. Pflanzenernaehr. Bodenkd.* 141:129–142.
Howell, R. W., and R. L. Bernard. 1961. Phosphorus response of soybean varieties. *Crop Sci.* 1:311–313.
Hutton, E. M., W. T. Williams, and C. S. Andrew. 1978. Differential tolerance to manganese in introduced and bred lines of *Macroptilium atropurpureum*. *Aust. J. Agric. Res.* 29:67–79.
Itoh, S. 1987. Characteristics of phosphorus uptake of chickpea in comparison with pigeonpea, soybean and maize. *Soil Sci. Plant Nutr.* 33:417–422.
Jones, L. H. 1961. Aluminum uptake and toxicity in plants. *Plant Soil* 13:292–310.
Joost, R. E., and C. S. Hoveland. 1986. Root development of *Sericea lespedeza* and alfalfa in acid soils. *Agron. J.* 78:711–714.
Joost, R. E., C. S. Hoveland, E. D. Donnelly, and S. L. Fales. 1986. Screening *Sericea lespedeza* for aluminum tolerance. *Crop Sci.* 26:1250–1251.

Kang, B. T., and R. L. Fox. 1980. A methodology for evaluating the manganese tolerance of cowpea (*Vigna uniguiculata*) and some preliminary results of field trials. *Field Crops Res.* 3:199–210.

Karow, J., and D. Lindsey. 1984. N, P, K, VAM infection, and growth response of alfalfa. In R. Molina (ed.), *Proc. 6th North American Conf. Mycorrhizae, Bend, Oregon.* For. Res. Lab., Corvallis, Oregon. P. 390.

Keyser, H. H., B. B. Bohlool, T. S. Hu, and D. F. Weber. 1982. Fast growing rhizobia isolated from root nodules of soybean. *Science* 215:1631–1632.

Keyser, H. H., and P. B. Cregan. 1987. Nodulation and competition for nodulation of selected genotypes among *Bradyrhizobium japonicum* serogroup 123 isolates. *Appl. Environ. Microbiol.* 53:2631–2635.

Kiven, C. S., G. E. Ham, and J. W. Lambert. 1981. Recovery of introduced *Rhizobium japonicum* strains by soybean genotypes. *Agron. J.* 73:900–905.

Kleese, R. A., D. C. Rusmusson, and L. H. Smith. 1968. Genetic and environmental variation in mineral element accumulation in barley, wheat, and soybeans. *Crop Sci.* 8:591–593.

Klimashevskii, E. L., and N. F. Chernysheva. 1980. Genetical variability of plant resistance to H^+ (hydrogen) and Al^{3+} (aluminum) toxicity in root zone: Theory and practical aspects (cereal, peas). *Skh. Biol.* 15:270–277.

Kohno, Y. 1986. Vanadium induced manganese in bush bean plants grown in solution culture. *J. Plant Nutr.* 9:1261–1272.

Kohno, Y., and C. D. Foy. 1983a. Manganese toxicity in bush bean as affected by concentration of manganese and iron in the nutrient solution. *J. Plant Nutr.* 6:363–386.

Kohno, Y., and C. D. Foy. 1983b. Differential tolerance of bush bean cultivars to excess manganese in solution and sand culture. *J. Plant Nutr.* 6:873–877.

Kohno, Y., C. D. Foy, A. L. Fleming, and D. T. Krizek. 1984. Effect of Mn concentration on the growth and distribution of Mn and Fe in two bush bean cultivars grown in solution culture. *J. Plant Nutr.* 7:547–566.

Kucey, M. N. 1984. VAM effects on alfalfa grown in limed and sterilized soil. In R. Molina (ed.), *Proc. 6th North American Conf. Mycorrhizae, Bend, Oregon.* For. Res. Lab., Corvallis, Oregon. P. 407.

Kumar, V., M. Singh, and N. Singh. 1981. Effect of sulphate, phosphate, and molybdate application on quality of soybean grain. *Plant Soil* 59:3–8.

Kuykendall, L. D., M. A. Roy, J. J. O'Neill, and T. E. Devine. 1988. Fatty acids, antibiotic resistance, and deoxyribonucleic acid homology groups of *Bradyrhizobium japonicum. Int. J. Syst. Bacteriol.* 38:358–361.

Lambert, D. H., H. Cole, Jr., and D. E. Baker. 1980. Variation in the response of alfalfa clones and cultivars to mycorrhizae and phosphorus. *Crop Sci.* 20:615–681.

Lavy, T. L., and S. A. Barber. 1963. A relationship between the yield response of soybeans to molybdenum applications and the molybdenum content of the seed produced. *Agron. J.* 55:154–155.

Lawn, R. J., and W. A. Brun. 1974. Symbiotic nitrogen fixation in soybean. I. Effect of photosynthetic source sink manipulations. *Crop Sci.* 14:11–16.

Lee, E. H., and C. D. Foy. 1986. Aluminum tolerances of two snapbean cultivars related to organic acid contents by high-performance liquid chromatography. *J. Plant Nutr.* 12:1481–1498.

Lioi, L., M. Giovannetti, and C. M. Hepper. 1986. Host–endophyte specificity in mycorrhizal infection of *Hedysarum coronarium* L. In V. Gianinazzi-Pearson and S. Gianinazzi (eds.), Physiological and Genetical Aspects of Mycorrhizae. Proc. 1st European Symp. Mycorrhizae, Dijon, France, 1985. INRA, Paris. Pp. 551–554.

Lombin, G. 1983. Comparative tolerance and susceptibility of cotton and peanut to high rates of Mn and B application under greenhouse conditions. *Soil Sci. Plant Nutr.* 29:363–368.

Lombin, G. L., and T. E. Bates. 1982. Comparative responses of peanuts, alfalfa, and soybeans to varying rates of boron and manganese on two calcareous Ontario soils. *Can. J. Soil Sci.* 62:1–9.

Long, F. L., and C. D. Foy. 1970. Plant varieties as indicators of aluminum toxicity in the A_2 horizon of a Norfolk soil. *Agron. J.* 62:679–681.

Lopes, A. S., and F. R. Cox. 1977. A survey of the fertility status of surface soils under 'cerrado' vegetation in Brazil. *Soil Sci. Soc. Am. J.* 41:742–746.

Lowendorf, H. S., and M. Alexander. 1983. Selecting *Rhizobium meliloti* for inoculation of alfalfa planted in acid soils. *Soil Sci. Soc. Am. J.* 47:935–938.
Malavolta, E., J. P. Dantas, R. S. Morais, and F. D. Nogueira. 1979. Calcium problems in Latin America. *Commun. Soil Sci. Plant Anal.* 10:29–40.
Malavolta, E., J. R. Sarruge, and V. C. Bittencourt. 1976. Toxidez do aluminio e manganase. In M. G. Terri (ed.), *IV simp. sobre o cerrado*. Coord. Livaria Italia Editora and Editora da Universidade de Sao Paulo. Pp. 275–301.
Mascarenhas, H. A. A., C. E. O. Camargo, and S. M. P. Falivene. 1984. Compartamento de cultivares de soja em solucao nutritiva contendo differentes niveis de manganes. *Bragantia* 43:201–209.
Mascarenhas, H. A. A., C. E. O. Camargo, S. M. P. Favilene, and E. A. Bulisani. 1985. Tolerencia da soja ao manganes en solucao nutritiva am tres temperaturas. *Bragantia* 44:531–539.
Mascarenhas, H. A. A., M. A. C. Miranda, L. C. S. Ramos, P. R. Furlani, and O. C. Bataglia. 1982. Comportamento de tres cultivares de soja em diversos niveis de manganes em solucao nutritivia. *Bragantia* 41:229–230.
Mason, D. L. 1982. Aluminum tolerance soybean germplasm. M.S. Thesis, Univ. of Maryland, College Park.
Matsumoto, H., E. Hirasawa, S. Morimura, and E. Takahashi. 1976a. Localization of aluminum in tea leaves. *Plant Cell Physiol.* 17:627–631.
Matsumoto, H., E. Hirasawa, H. Torikai, and E. Takahashi. 1976b. Localization of absorbed aluminum in pea root and its binding to nucleic acids. *Plant Cell Physiol.* 17:127–137.
Matsumoto, H., and S. Morimura. 1980. Repressed template activity of chromatin of pea roots treated by aluminum (toxicity). *Plant Cell Physiol.* 21:951–959.
Matsumoto, H., S. Morimura, and E. Takahashi. 1977a. Less involvement of pectin in the precipitation of aluminum in pea root. *Plant Cell Physiol.* 18:325–335.
Matsumoto, H., S. Morimura, and E. Takahashi. 1977b. Binding of aluminum to DNA of DNP in pea root nuclei. *Plant Cell Physiol.* 18:987–993.
Melton, B., J. B. Moutray, and J. H. Bouton. 1988. Geographic adaptation and cultivar selection. In A. A. Hanson, D. K. Barnes, and R. R. Hill, Jr. (eds.), Alfalfa and Alfalfa Improvement. Am. Soc. Agron., Madison, Wisconsin. Pp. 595–620.
Miranda, M. A. C., H. A. A. Mascarenhas, E. A. Bulisani, and J. M. A. S. Valadares. 1982. Comportamento de dois cultivares de soja em funcao do maganes do solo. *Bragantia* 41:141–143.
Moraghan, J. J., T. P. Freeman, and D. Whited. 1986. Influence of FeEDDHA and soil temperature on the growth of two soybean varieties. *Plant Soil* 95:57–67.
Morris, H. D., and W. H. Pierre. 1949. Minimum concentrations of manganese necessary for injury to various legumes in culture-solutions. *Agron. J.* 41:107–112.
Mugwira, L. M., M. Floyd, and S. U. Patel. 1981. Tolerance of triticale lines to manganese in soil and nutrient solution. *Agron. J.* 73:319–322.
Munns, D. N. 1965. Soil acidity and growth of a legume. *Aust. J. Agric. Res.* 16:733–741.
Munns, D. N., and R. L. Fox. 1977. Comparative lime requirements of tropical and temperate legumes. *Plant Soil* 46:533–548.
Nangju, D. 1980. Soybean response to indigenous rhizobia as influenced by cultivar orgin. *Agron. J.* 72:403–406.
Noble, C. L., G. M. Halloran, and D. W. West. 1984. Identification and selection for salt tolerance in lucerne (*Medicago sativa* L.). *Aust. J. Agric. Res.* 35:239–252.
Oakes, A. J., and C. D. Foy. 1984. Acid soil tolerance of leuceana species in greenhouse trials. *J. Plant Nutr.* 7:1759–1774.
Ohki, K. 1974. Manganese nutrition of cotton under two boron levels. II. Critical Mn levels. *Agron. J.* 66:572–575.
Ohki, K. 1981. Manganese critical levels for soybean growth and physiological processes. *J. Plant Nutr.* 3:271–284.
Ohki, K., D. O. Wilson, and O. E. Anderson. 1980. Manganese deficiency and toxicity sensitivities of soybean cultivars. *Agron. J.* 72:713–716.
Osborne, G. J., J. E. Pratley, and W. P. Stewart. 1980. The tolerance of subterranean clover (*Trifolium subterraneum* L.) to aluminum and manganese. *Field Crops Res.* 3:347–358.

Pal, U. R. 1989. Comparative contributions of native rhizobia vs. strains of *Rhizobium japonicum* to nitrogen uptake of promiscuous soybeans in Nigerian savanna. In A. J. Pascale (ed.), *Proc. World Soybean Research Conf. IV, Buenos Aires*. Orientacion Grafica Editora, Buenos Aires. Pp. 363–368.

Paulsen, G. M., and O. A. Rotimi. 1968. Phosphorous–zinc interaction in two soybean varieties differing in sensitivity to phosphorus nutrition. *Soil Sci. Soc. Am. Proc.* 32:73–76.

Peterson, N. K., and E. R. Purvis. 1961. Development of molybdenum deficiency symptoms in certain crop plants. *Soil Sci. Soc. Am. Proc.* 25:111–117.

Polson, D. E., and M. W. Adams. 1970. Differential response of navy beans (*Phaseolus vulgaris* L.) to zinc. I. Differential growth and elemental composition at excessive Zn levels. *Agron. J.* 62:557–560.

Randall, G. W. 1976. Correcting iron chlorosis in soybeans. *Minn. Agric. Ext. Serv., Soils Fact Sheet* No. 27.

Raper, C. D., Jr., and S. A. Barber. 1970a. Rooting systems of soybeans. I. Differences in root morphology among varieties. *Agron. J.* 62:581–584.

Raper, C. D., Jr., and S. A. Barber. 1970b. Rooting systems of soybeans. II. Physiological effectiveness as nutrient absorption surfaces. *Agron. J.* 62:585–587.

Russell, J. S. 1976. Comparative salt tolerance of some tropical and temperate legumes and tropical grasses. *Aust. J. Exp. Agric. Anim. Husb.* 16:103–109.

Saggar, S., and G. Dev. 1974. Uptake of sulphur by different varieties of soybean. *Indian J. Agric. Sci.* 44:345–349.

Salinas, J. V. 1978. Differential response of some cereal and bean cultivars to aluminum and phosphorus stress in an Oxisol in Central Brazil. Ph.D. Thesis, North Carolina State Univ., Raleigh. (*Diss. Abstr.* 79-05509.)

Sapra, V. T., T. Mebrahtu, and L. Mugwira. 1975. Evaluation of Soybean Germplasm Maturity Groups V, VI, and VII for Agronomic Characters and Aluminum Tolerance. Spec. Publ. Alabama A & M Univ., Normal.

Sartain, J. B., and E. J. Kamprath. 1978. Aluminum tolerance of soybean cultivars based on root elongation in solution culture compared with growth in acid soil. *Agron. J.* 70:17–20.

Satterlee, L., B. Melton, B. McCastin, and D. Miller. 1983. Mycorrhizae effects on plant growth, phosphorus uptake, and N_2 (C_2H_4) fixation in two alfalfa populations. *Agron. J.* 75:715–716.

Schettini, T. M., W. H. Lableman, and G. C. Gerloff. 1987. Incorporation of phosphorus efficiency from exotic germplasm into agriculturally adapted germplasm of common bean (*Phaseolus vulgaris* L.). *Plant Soil* 99:175–184.

Scott, R. S., and W. L. Lowther. 1981. Production of big trefoil on acid, low fertility soils in New Zealand. In J. A. Smith and V. W. Hays (eds.), *Proc. 14th Int. Grassland Congr., Lexington, Kentucky*. Westview Press, Boulder, Colorado. Pp. 550–552.

Seay, W. A., and L. Henson. 1958. Variability in nutrient uptake and yield of donally propagated kenland red clover. *Agron. J.* 50:165–168.

Shea, P. F., W. H. Gabelman, and G. C. Gerloff. 1967. The inheritance of efficiency in potassium utilization in snapbeans (*Phaseolus vulgaris* L.). *Proc. Am. Soc. Hortic. Sci.* 91:286–293.

Simpson, J. R., A. Pinkerson, and J. Luzdovskis. 1977. Effects of subsoil calcium on the root growth of some lucerne genotypes (*Medicago sativa* L.) in acidic soil profiles. *Aust. J. Agric. Res.* 28:629–638.

Skipper, H. D., and G. W. Smith. 1979. Influence of soil pH on the soybean–endomycorrhiza symbiosis. *Plant Soil* 53:559–563.

Skipper, H. D., and J. E. Struble. 1984. Influence of *Glomus claroideum* (VAM fungus) and phosphorus levels on soybean growth in fumigated microplots. In R. Molina (ed.), *Proc. 6th North American Conf. Mycorrhizae, Bend, Oregon*. For. Res. Lab., Corvallis, Oregon. P. 253.

Snaydon, R. W. 1962a. Micro-distribution of *Trifolium repens* L. and its relation to soil factors. *J. Ecol.* 50:133–143.

Snaydon, R. W. 1962b. The growth and competitive ability of contrasting natural populations of *Trifolium repens* L. on calcareous and acid soils. *J. Ecol.* 50:439–447.

Snaydon, R. W., and A. D. Bradshaw. 1969. Differences between natural populations of *Trifolium repens* L. in response to mineral nutrients. II. Calcium, magnesium and potassium. *J. Appl. Ecol.* 68:185–202.

Spain, J. M., C. A. Francis, R. H. Howeler, and F. Calvo. 1975. Differential species and varietal tolerance to soil acidity in tropical crops and pastures. In E. Bornemisze and A. Alvarado (eds.), Soil Management in Tropical America. Soil Sci. Dep. North Carolina State Univ., Raleigh. Pp. 308–329.
Suhayda, C. G., and A. Haug. 1984. Organic acids prevent aluminum induced conformational changes in calmodulin. *Biochem. Biophys. Res. Commun.* 119:376–381.
Suhayda, C. G., and A. Huag. 1985. Citrate chelation as a potential mechanism against aluminum toxicity in cells: The role of calmodulin. *Can. J. Biochem. Cell Biol.* 63:1167–1175.
Sumner, M. E., H. Shahandeh, J. K. Bouton, and J. H. Hammel. 1986. Amelioration of acid soil profile through deep liming and surface application of gypsum. *Soil Sci. Soc. Am. J.* 50:1254–1258.
Tanaka, A., K. Hitsuda, and Y. Tsuchihashi. 1984. Tolerance to low pH and low available phosphorous of various field and forage crops. *Soil Sci. Plant Nutr.* 30:39–49.
Temple-Smith, M. G., and T. B. Koen. 1982. Comparative response of poppy *Papaver semniferum* L. and eight crop and vegetable species top manganese excess in solution cultures. *J. Plant Nutr.* 5:1153–1169.
Teuber, L. R., and D. A. Phillips. 1988. Influences of selection method and nitrogen environment on breeding alfalfa for increased forage yield and quality. *Crop Sci.* 28:599–604.
van Nuffelen, M., and N. C. Schenck. 1984. Spore germination, penetration, and root colonization of six species of vesicular-arbuscular mycorrhizal fungi on soybean. *Can. J. Bot.* 62:624–628.
Vest, G., D. F. Weber, and C. Sloger. 1973. Nodulation and nitrogen fixation. In B. E. Caldwell (ed.), Soybeans: Improvement, Production and Uses. Am. Soc. Agron., Madison, Wisconsin. Pp. 353–390.
Vose, P. B., and D. G. Jones. 1963. The interaction of manganese and calcium on nodulation and growth in varieties of *Enfolium repens. Plant Soil* 18:372–385.
Vose, P. B., and H. V. Koontz. 1960. The uptake of strontium and calcium from soils by grasses and legumes and the possible significance in relation to Sr-90 fallout. *Hilgardia* 29:579–585.
Walsh, T., and S. T. Cullinan. 1945. Investigations on Marsh Spot disease in peas. *Proc. R. Ir. Acad., Sect. B* 50:279–285.
Wambeke, A. Van. 1976. Formation, distribution and consequences of acid soils in agricultural development. In M. J. Wright (ed.), Plant Adaptation to Mineral Stress in Problem Soils. Cornell Univ. Press, Ithaca, New York. Pp. 15–24.
Weiss, M. G. 1943. Inheritance and physiology of efficiency in iron utilization in soybeans. *Genetics* 28:253–268.
West, D. W., and J. A. Taylor. 1981. Germination and growth of cultivars of *Trifolium subterraneum* L. in the presence of sodium chloride salinity. *Plant Soil* 62:221–230.
White, M. C., R. L. Chaney, and A. M. Decker. 1979a. Role of roots and shoots of soybean in tolerance to excess soil zinc. *Crop Sci.* 19:126–128.
White, M. C., A. M. Decker, and R. L. Chaney. 1979b. Differential cultivar tolerance in soybean to phytotoxic levels of soil Zn. I. Range of cultivar response. *Agron. J.* 71:121–126.
Whiteaker, G., G. C. Gerloff, W. H. Gabelman, and D. Lindgren. 1976. Intraspecific differences in growth of beans at stress levels of phosphorus. *J. Am. Soc. Hortic. Sci.* 101:472–475.
Williams, L. F., and D. A. Phillips. 1983. Increased soybean productivity with a *Rhizobium japonicum* mutant. *Crop Sci.* 23:246–250.
Wrather, J. A., S. C. Anand, and V. H. Drapkin. 1984. Soybean cyst nematode control. *Phytopathology* 68:829–833.

7

Mechanisms Improving Nutrient Use by Crop and Herbage Legumes

J. R. CARADUS

In the last two decades there has been a great deal of interest in selecting plant material to suit the nutrient status of soils, first, as an alternative to applying large amounts of fertilizer and, second, in an attempt to improve the efficiency of fertilizer use by plants, in which fertilizer is applied in adequate amounts. Leaving aside work aimed at breeding plants that are tolerant of toxic soils, whether it be low pH, salinity, or heavy metals, most interest has been directed toward selecting plants with improved N, P, or Fe nutrition. The macronutrients K, Mg, Ca, and S and micronutrients such as B, Mn, Zn, Cu, and Mo have received less attention.

The present chapter examines mechanisms that either allow variants within legume crop and herbage species to use nutrients more efficiently when supplied in adequate amounts or maintain production when nutrient levels are deficient.

Crops as Enhancers of Nutrient Use

Interspecific comparisons of mineral nutrition of legumes have been previously reviewed (Hallsworth, 1958; Griffith, 1974; Andrew and Jones, 1978; Munns and Mosse, 1980; Goodman and Edwards, 1983; Robson, 1983).

Definitions of nutrient efficiency can be grouped into three broad categories, as in the case of definitions for P efficiency: (a) greater yield than others in low-P soil (Clark, 1975; Haag *et al.*, 1978; Sarić, 1982), (b) greater yield per amount of P absorbed (usually termed efficiency of P utilization or the P utilization quotient) (Loneragan and Asher, 1967; Godwin and Wilson, 1976; Blair and Cordero, 1978; Ramirez, 1982; Camargo, 1984; Coltman *et al.*, 1985; Elliott and Läuchli, 1985), and (c) greater yield per unit of applied (or available) P (Vose, 1963; Barber, 1976; Blair and Cordero, 1978; Caradus and Dunlop, 1978).

Selection for greater yields in low-P soils, while a valid aim, cannot be equated with increasing P efficiency. The criterion of increasing yield per amount of P absorbed is simply the inverse of %P and cannot strictly be regarded as a measure of P efficiency. This measure is greatly influenced by, and interacts with, so many factors other than plant genotype (Bates, 1971; Lambert and Toussaint, 1978) that its usefulness is very limited. The only criterion that truly measures P efficiency is yield per unit of applied or available P. The present paper will not be restricted to examining plant mechanisms that only improve nutrient efficiency but will also examine mechanisms that enable production to be at least maintained when nutrient supply is restricted, i.e., low-nutrient tolerance.

I. Mechanisms Improving Nutrient Uptake

Nutrient uptake per plant, at a given soil nutrient level, might be improved by (a) a root system that provides greater contact with the nutrient in question, particularly in the case of immobile nutrients such as P; (b) a greater uptake per unit of root, due to enhanced uptake mechanisms; and/or (c) an ability to utilize nutrient forms that are relatively unavailable or poorly available to plants, such as in the case of insoluble inorganic P. Each of these attributes may be significantly affected by association with mycorrhizal fungi.

In the case of legumes, N assimilation is related to the N-fixing ability of the symbiosis between *Rhizobium* and the host plant. The interaction between host plant and *Rhizobium* isolate may be of importance in improving the amount of N fixed and assimilated.

A. ROOT STRUCTURE

1. Root Size

Since mineral nutrients are absorbed through roots, it has been demonstrated that for relatively immobile nutrients such as P and to a lesser extent K, increased root system size would result in increased nutrient uptake (Barber, 1976; Loneragan,

1978; Silberbush and Barber, 1983). This would become increasingly important in soils of low nutrient status, as a large root comes into contact with a greater soil volume, and as long as the concentration of nutrients at the root surface is greater than C_{min} (the minimum concentration at which a nutrient can be absorbed).

Cultivar differences in root size are quite common (Mitchell and Russell, 1971; Adepetu and Akapa, 1977; Schettini *et al.*, 1987) and in a few instances have been related to differences in nutrient uptake. Bean genotypes with improved growth at deficient levels of P can have nearly twice as much root as nonproductive genotypes (Gabelman, 1976). Differences between white clover populations and cultivars for P uptake per plant at low levels of P have been related to differences in root size and absolute growth rate (Caradus and Snaydon, 1986a).

However, the apparent importance of root size in determining P uptake may simply be a result of root size being a component of total plant size. In two separate experiments comparing 10 white clover populations, P uptake per plant, at deficient levels of P, was correlated not only with root size ($r = +0.97$ and $+0.91$) but also shoot size ($r = +0.73$ and $+0.89$) (Caradus and Snaydon, 1986a). The problem here is knowing whether increased root size of plants absorbing larger amounts of P is the result or cause of larger shoot size. This will be dealt with more extensively in Section I, C, 1.

At adequate levels of nutrient supply, large root systems may not be advantageous. Two studies with alfalfa have shown that clones and cultivars with large root systems have low concentrations of Ca and Na in shoots (Chloupek, 1980) and low rates of Rb uptake (Simpson *et al.*, 1979). While root size differences of white clover populations were highly correlated with differences in P uptake per plant at high levels of P supply, multiple regression analyses showed that leaf area and absolute growth rate were better determinants of P uptake per plant at these high P levels (Caradus and Snaydon, 1986a). In a comparison of 14 lines of bean, using excised roots pretreated at moderately deficient levels of P, lines with low root dry weights compensated by having high P uptake rates per unit root weight (Lindgren *et al.*, 1977).

2. Root Morphology

Variation for root system morphology has been shown to exist within many legume species: for example, taprooted and non-taprooted types of white clover (Caradus, 1977, 1981b), differences in degree of root branching among alfalfa cultivars (Smith, 1951; Avendano and Davis, 1966; McIntosh and Miller, 1980), and differences in lateral root density among bean cultivars (Geisler and Krützfeldt, 1983). Other genetic differences have been observed for length of longest root in white clover (Ennos, 1985), number of lateral roots in alfalfa (Pederson *et al.*, 1984), root surface area in soybean (Raper and Barber, 1970a), and taproot development in bean (Zobel, 1975). It is generally assumed that a frequently branched, fine root system is more effective in absorbing nutrients (Vose, 1963; Godwin and Wilson, 1976; Clarkson and Hanson, 1980; Barber, 1984). Comparisons between species tend to support this (Pavlychenko and Harrington, 1934; Andrew, 1966;

Jeffery, 1967; McLachlan, 1976; Strong and Soper, 1973; Caradus, 1980), but there is less evidence within legume species.

White clover populations collected from low-P soils had finer root systems than populations collected from high-P soils, when grown in pure swards in a low-P soil (Caradus and Snaydon, 1986d). The populations from low-P soils had lower root elongation rates, shorter average root lengths, and their root production rates were more responsive to P than populations from high-P soils (Caradus and Snaydon, 1988a). An increase in root surface area due to furrowed cell walls of root-tip cells of a synthetic tetraploid red clover increased nutrient absorption compared with other populations (Tyutyunnikov *et al.*, 1978).

It has been proposed, from both interspecific comparisons (Itoh and Barber, 1983a,b) and theoretical comparisons (Bouldin, 1961; Nye, 1966; Läuchli, 1967; Drew and Nye, 1969), that greater root hair length would increase uptake of nutrients such as P, simply because genotypes with longer root hairs contact more soil than those with short root hairs. Selection for increased root hair length has been achieved in white clover (Caradus, 1979). However, increasing root hair length only had a significant effect on plant dry weight, because of increased P absorption, when roots were nonmycorrhizal (Caradus, 1981c). Genotypes with shorter root hairs tended to come more mycorrhizal (Crush and Caradus, 1980; Caradus, 1982) and therefore to reduce the importance of root hair length.

Although the surface area of the root systems of two soybean cultivars, compared by Raper and Barber (1970a) differed by more than 80%, the difference in shoot dry matter production was not significant. However, at high population densities or in nutrient-deficient soils, a more extensive root system was an advantage in uptake of K (Raper and Barber, 1970b).

3. Root Distribution

In many soils, nutrients such as P when applied as fertilizer do not move more than a few centimeters into the soil profile (Sharpley, 1986). In these conditions distribution of roots close to the soil surface may be advantageous. Differences in root distribution in a soil profile have been shown between soybean cultivars (Silberbush and Barber, 1984) and between white clover lines, observed using both glass-fronted cases (Caradus, 1981a) and ^{32}P in swards (Goodman and Collison, 1982). Small-leaved populations of white clover had a greater percentage of root tips in top layers than did large-leaved populations (Caradus, 1981a, 1984). A comparison of eight legume species found that relative P accumulation rates appeared to be positively associated with the extent of surface rooting (Davis, 1981).

B. INTERACTIONS WITH MICROORGANISMS

1. Rhizobium–Higher Plant Symbiosis

The economic importance of legumes can be attributed to their ability to form a symbiotic relationship with *Rhizobium* species and so fix atmospheric N. However, biological N-fixation can limit production (Wynne *et al.*, 1979; Hart *et al.*, 1981; Trimble *et al.*, 1984; Sorwli and Mytton, 1986). A considerable number of investigations have endeavoured to improve the efficiency of the symbiosis by either selecting for (a) more efficient *Rhizobium* strains, (b) high N-fixing host plants, (c) an improved *Rhizobium* strain–host plant combination, (d) increased tolerance of the symbiotic association to NO_3^-, or (e) reduced N loss to the soil due either to root and nodule decomposition or to direct excretion of nitrogenous compounds from the roots.

A number of studies have compared either (a) several *Rhizobium* strains against one host cultivar or genotype to identify more efficient *Rhizobium* strains, (b) several host cultivars or genotypes with a single *Rhizobium* strain to identify high N-fixing host material, or (c) several *Rhizobium* strains against several host cultivars or genotypes to identify improved *Rhizobium* strain–host plant combinations. Studies identifying more efficient *Rhizobium* strains have been adequately reviewed in the past (Burton, 1972; Date, 1976; Frederick, 1978) and would suggest that within any *Rhizobium* species a large variation of efficiency among strains exists. However, the premise that a strain of *Rhizobium* introduced into a site with an existing *Rhizobium* population will be unlikely to have any long-term benefits (Roughley *et al.*, 1976) is becoming accepted. Host cultivar and soil type can have large effects on the competitive ability of strains. Cultivars can differ in their selection of *Rhizobium* strains when presented with a mixture. This has been shown in comparisons of cultivars of white clover grown in pots of soil (Jones and Hardarson, 1979) and on agar (Russell and Jones, 1975), subterranean clover grown on agar (Mytton and De Felice, 1977), in pots of soil (Dughri and Bottomley, 1984), and in the field (Roughley *et al.*, 1976), and red clover grown on agar (Russell and Jones, 1975). However, Harrison *et al.* (1987) found no significant difference between four white clover cultivars in their ability to select *Rhizobium* isolates of different electrophoretic type (i.e., discrete genetic class) from an unamended soil solution.

The success of most studies in identifying significant variation between host cultivars for nodulation and N-fixation parameters (Table I) has led to the selection of high and low N-fixing types (Nutman, 1981) within alfalfa (Seetin and Barnes, 1977; Duhigg *et al.*, 1978; Viands *et al.*, 1981), cowpea (Zary and Miller, 1980), red clover (Nutman *et al.*, 1971), and white clover (Mytton and Hughes, 1985). Bidirectional selection for acetylene reduction in alfalfa produced plants with significant differences in all three studies (Table II), with realized heritabilities for high acetylene reduction selection ranging from 0.01 to 0.78. However, Satterlee *et al.* (1983) were unable to repeat the work of Duhigg *et al.* (1978) and found no difference between selection groups for acetylene reduction when grown in low-P soil. Graham and Rosas (1979) noted that at low soil P, N-fixation is restricted, and this may explain this disparity.

Selection for high and low levels of acetylene reduction, within alfalfa, caused a concomitant change in root mass, fibrous root score, and nodule number score, although for the latter character the difference was not always significant (Table II). Viands *et al.* (1981) found that 31–42% of the variation for acetylene reduction

Table I. Summary of Studies Identifying Variation between Host Cultivars or Genotypes for N-Fixation

Species	No. of Lines (l), Cultivars (c), or Genotypes (g)	Characters Measured	Significance[a]	Reference
Arachis hypogaea	ng[b]	Dry matter	*	Wynne *et al.* (1979)
		N-fixation	*	
Cicer arietinum	5(c)	Nodule number	*	Patil and Moniz (1974)
		N-fixation	*	
Glycine max	116(l)	Chlorosis	*	Johnson and Means (1960)
Glycine max[c]	16(c)	Specific N-fixation	*	McGinnity *et al.* (1978)
Lupinus sp.	ng	Dry weight	*	Rulinskaya (1980)
		N content	*	
Medicago sativa	40(c)	Dry weight	*	Leach (1968)
Medicago sativa	4(c)	Nodule number	*	Duke and Doehlert (1981)
		Nodule mass	*	
		Acetylene reduction	*	
Medicago sativa	10(c)	Nodule number	ns	Hoffman and Melton (1981)
		Dry weight	ns	
		N content	ns	
		Acetylene reduction	ns	
Phaseolus vulgaris	11(c)	Dry weight	*	Rennie and Kemp (1981a)
		Plant N content	*	
		Nodule number	*	
		Nodule mass	*	
Trifolium incarnatum	5(c)	Acetylene reduction	*	Smith *et al.* (1979)
		Shoot dry weight	*	
		Root vigor	*	
Trifolium repens	5(c)	Yield	*	Bonish (1980)
Vigna unguiculata	75(c)	Nodulation	*	Maherchandani and Rana (1977)
Vigna unguiculata	>100(g)	N-fixation	*	Zary and Miller (1977)
Vigna unguiculata[c]	38(c)	Acetylene reduction	*	Smittle and Brantley (1978)

[a]*, Significant effect; ns, no significant effect.
[b]ng, Not given.
[c]Field study.

Table II. Effect of One Cycle of Selection for High Acetylene Reduction Rates in Alfalfa[u]

Selection Group	Parents	Progeny			Realized Heritability for High Acetylene Reduction
	Acetylene Reduction	Acetylene Reduction	Root Score[z]	Nodule Number Score[z]	
(i) High	17.6[b]	2.4 a	2.7 a[d]	2.8 a	0.55
Low	10.6	1.1 b	1.6 b	1.7 b	
(ii) High	49.2[c]	15.4 a[c]	2.7 a[e]	2.5 a	0.78
Low	2.4	7.7 b	2.3 b	2.3 a	
(iii) High } gene pool A		1.5 a[c]	3.0 a[d]	2.7 a	0.25
Low } gene pool A		0.8 b	2.5 b	2.5 a	
High } gene pool B		0.6 a[c]	3.3 a[d]	2.9 a	0.01
Low } gene pool B		0.4 b	2.9 b	2.6 b	

[a]Source: (i) Seetin and Barnes, 1977; (ii) Duhigg *et al.*, 1978; (iii) Viands *et al.*, 1981.
[b]μmol ethylene pot^{-1} hr^{-1}.
[c]μmol ethylene plant^{-1} hr^{-1}.
[d]Score of fibrous roots.
[e]Score of root system size.
[f]1–5 scale, with quantity increasing with score. Means within a column not followed by the same letter are significantly different ($p = 0.05$); these comparisons do not apply between studies.

could be attributed to nodule mass score, while nodule number score only attributed for 0–3% of the variation, depending on gene pool.

Differences between selections for high and low N-fixing genotypes of cowpea were shown in a field study by Zary and Miller (1980). Maximum nitrogenase activity generally coincided with peak flowering and high N-fixing genotypes had higher activity throughout the entire growing season.

Nutman *et al.* (1971) were able to select for increased nitrogen fixation in red clover, simply by selecting for shoot yield, when plants were inoculated with a single strain of rhizobia. The effect of selection was less when the material was tested with other strains of rhizobia.

Total nodule volume has been increased by selection in white clover but after the first generation of selection this was not accompanied by an increase in plant growth (Mytton and Jones, 1971). However, selection for high N-fixing lines of white clover has resulted in a 31% increase in total N accumulation and a 20% increase in dry matter yield over unselected material (Mytton and Hughes, 1985). When supplied with nonlimiting supply of nitrate the high N-fixing lines accumulated 17% more N than unselected material.

Percentage plant N derived from the atmosphere (Ndfa) has been shown to differ between two cultivars of bean in one study (Rennie and Kemp, 1983a) but not in another (Rennie and Kemp, 1984).

The energy cost of N-fixation to the plant in comparison to utilization of nitrate-N can be considerable (Phillips, 1980). Ryle *et al.* (1979b) have calculated that plants fixing their own N require 11–13% more of their fixed carbon per day than equivalent plants lacking nodules and utilizing nitrate-N. Comparison of soybean, cowpea, and white clover has shown that all three legumes exhibited similar respiratory losses from nodulated roots per unit of N fixed (Ryle *et al.*, 1979a), suggesting that genetic variation for this character may be limited. However, Mahon (1979) found that respiration related to N-fixation activity in peas varied significantly among four cultivars and was relatively unaffected by *Rhizobium* isolate.

Significant correlations between plant vigor or dry matter production and nitrogen fixation or acetylene reduction have been observed in several species, e.g., alfalfa (Seetin and Barnes, 1977; Duhigg *et al.*, 1978; Major *et al.*, 1979; Sheehy *et al.*, 1980; Viands *et al.*, 1981), cicer milkvetch (Major *et al.*, 1979), *Medicago* sp. (Ivory, 1976), bean (Westermann and Kolar, 1976; Rennie and Kemp, 1981a,b), pea (Hobbs and Mahon, 1981, 1982), sainfoin (Major *et al.*, 1979), soybean (Ronis *et al.*, 1985), and crimson clover (Smith *et al.*, 1979). Viands and Barnes (1979) recommended that selection for high-yielding plants with large nodule mass should precede selection for rate of acetylene reduction. Genotypes of white clover selected for dry matter yield under nonlimiting fertilizer N have given a 21% yield and N-fixation increase over unselected material (Mytton and Hughes, 1985).

The driving mechanism behind the relationship between dry matter yield and rate of nitrogen fixation appears to be photosynthetic rate (Hardy and Havelka, 1976; Bethlenfalvay and Phillips, 1977; Hardy, 1977). Tan *et al.* (1977) argued that high rates of photosynthesis stimulated high rates of nitrogen fixation, which resulted in high dry matter production. Hobbs and Mahon (1981) obtained a negative correlation between the energy cost of nitrogen fixation and harvest weight of peas, again suggesting that photosynthate supply was limiting fixation. Although nitrogen deficiency of plants induced by ineffective nodulation has been shown to greatly reduce leaf area and photosynthetic rate per ground area, photosynthetic rate per unit leaf area were unaffected (Boller and Heichel, 1984). Genotypes of soybean with delayed leaf senescence have higher nitrogen fixation rates than plants that senesce earlier (Abu-Shakra *et al.*, 1978). The use of grafting techniques in soybean allowed doubling of shoot/root ratio by simply cutting off one root system (Streeter, 1974). When the shoot/root ratio was doubled during flowering or seed formation stages, no significant effect on total N or dry matter of shoots compared to ungrafted controls at maturity was found. The rate of acetylene reduction per nodule weight was 75% greater with roots having two shoots than with grafted plants with only one shoot per root, 2 days after doubling the shoot/root ratio. Two explanations were given by Streeter (1974) for this effect, the latter explanation being favored: (a) rapid removal of N-fixation products from nodules due to the demand of two shoots, assuming that rate of fixation is limited by the buildup of fixation products; and (b) a proportional increase in supply of carbohydrate to nodules when number of shoots was doubled, assuming that rate of fixation is limited by carbohydrate supply. The use of reciprocal intercultivar grafts of soybean genotypes has shown that differences among scion genotypes in total nodule activity per plant were due mainly to differences in nodule fresh weight per plant, which in turn was strongly related to the capacity of the shoot genotype to provide photosynthetic assimilates (Lawn *et al.*, 1974). Rootstock genotype effects on total nodule activity per plant (Lawn and Brun, 1974b) were due to inherent differences in specific nodule activity (Lawn *et al.*, 1974).

Other studies, however, have produced conflicting results. N-fixation rate per unit leaf area was similar between groups of pea plants selected for either high or low carbon dioxide exchange rate per unit leaf area (Mahon, 1982). Sheehy *et al.* (1980) found few differences in yield of six alfalfa plants representing the range of photosynthetic efficiency and symbiotic N-fixation. Westermann and Kolar (1976) observed a two- to threefold difference in nitrogenase activity among cultivars of bean that had similar total dry matter production.

Studies attempting to identify improved *Rhizobium* strain–host plant combinations are numerous (Table III), and most demonstrate large interactions between *Rhizobium* strain and host genotype. However, the majority of these studies were undertaken in artificial conditions with no verification in the field. This work is complicated by the ability of introduced and "native" *Rhizobium* strains to compete successfully for nodulation sites on soil-grown plants.

Mineral N is known to depress nitrogen fixation (Raggio and Raggio, 1962; Dixon, 1969) and when legumes are grown in mixtures with other species, as occurs in pastures, application of mineral N can reduce legume content (Denehy and Morrison, 1979). However, mineral N is often applied because rate of N fixation can limit plant growth (Green and Cowling, 1961; Trimble *et al.*, 1984; Wynne *et al.*, 1979). This practice has stimulated attempts to identify genotypes and cultivars that can fix N at moderately high levels of N application or in the case of pasture legumes are tolerant of and able to survive N application.

Interactions between *Rhizobium* strain and nitrogen level have been observed in alfalfa for nodulation (Heichel and Vance, 1979), and between host cultivar and nitrogen in soybean for N-fixation (Danso *et al.*, 1987). Variation in the ability of white clover cultivars and genotypes to nodulate in the presence of ammonium nitrate has also been shown (Rys, 1986). A heritable component that did not appear to be related to nodulation in the absence of N was found. However, inhibition of nodulation and nitrogenase activity by the presence of nitrate was so great that it was concluded that to have any agricultural significance, greater genetic improvements would be required than could be demonstrated in a study of 130 genotypes (Mytton and Rys, 1985).

Many studies have examined the ability of white clover cultivars to tolerate mineral N application when grown in grass swards (Table IV). Of the 21 published experiments, 7 showed no significant interaction between cultivar and N level, 4 had inconsistently significant interactions (e.g., significant for 2 out of 3 years), 4 had uncertain interactions based on the information given, although examination of the data suggests that the interactions may have been significant, while only 6 had

Table III. Summary of Studies Identifying Improved Combinations of *Rhizobium* Strain and Host Plant Cultivar or Genotype

Species	No. of Lines (l), Cultivars (c), or Genotypes (g)	Number of *Rhizobium* Strains	Characters Measured	Significance[a] Host Effect	*Rhizobium* Effect	Interaction	Reference
Arachis hypogaea	3(c)	2	Dry weight	*	ns	ng[b]	Ayala Briceno (1975)
			Plant % N	*	ns	ng	
			Plant N content	*	*	ng	
			Nitrogenase activity	*	*	ng	
Arachis hypogaea	11(c)	3	Shoot dry weight	*	*	*	Singleton *et al.* (1978)
			Shoot N content	*	*	*	
Arachis hypogaea	2(c)	9	Nodule number	*	*	*	Elkan *et al.* (1981)
			N-fixation	*	*	*	
			Nodule mass	*	*	*	
			Plant dry weight	*	*	ns	
Cicer arietinum	10(c)	ng	Nodulation	ng	ng	*	Maherchandani and Rana (1977)
Glycine max	11(c)	8	Chlorosis	*	*	*	Johnson and Means (1960)
Glycine max[c]	4(c)	30	Yield	*	*	ns	Caldwell and Vest (1970)
Glycine max	4(c)	3	Nodule mass	ng	ng	*	McGinnity *et al.* (1980)
			Specific N-fixation	ng	ng	*	
Glycine max	3(c)	6	Seed yield	*	*	*	Ham (1980)
Glycine max	2(c)	2	N-fixation	*	*	ns	Israel (1981)
Glycine max	10(c)	5	Nodule number	*	*	*	Sarić and Fawzia (1982)
			N-fixation	*	*	*	
Medicago sativa	4(c)	21	Shoot yield	*	*	*	Erdmann and Means (1953)
Medicago sativa	14(c)	6	Shoot N content	*	*	*	Gibson (1962)
	7(c)	4	Shoot dry weight	*	*	*	
Medicago sativa	8(c)	13	Plant N content	*	*	*	Burton (1972)

Medicago sativa[c]	4(c)	12	Shoot weight	*	*	ns	Kehr *et al.* (1979)
			Shoot N content	*	*	ns	
Medicago sativa[c]	4(c)	5	Nitrogenase activity	*	*	*	Miller and Sirois (1981)
Medicago sativa	14(l)	5	Total dry weight	*	*	*	Tan (1981)
			Plant N content	*	*	*	
			Plant % N	*	*	*	
			Acetylene reduction	*	*	*	
			Nodule score	*	*	*	
Phaseolus vulgaris	2(c)	20	N-fixation	*	*	ns	Rennie and Kemp (1983a)
			Acetylene reduction	ns	*	*	
Phaseolus vulgaris	8(c)	5	Shoot N content	*	*	*	Rennie and Kemp (1983b)
	24(c)	3	Shoot N content	*	*	*	
			Acetylene reduction	*	*	*	
Phaseolus vulgaris	3(c)	10	Plant dry weight	*	*	ns	Pacovsky *et al.* (1984)
			Nodule weight	*	*	ns	
			N fixation	*	*	ns	
Phaseolus vulgaris	2(c)	2	Total N content	*	*	*	Hungaria and Neves (1987a)
			% N senesced leaves	*	*	*	
			% N seeds	*	*	*	
			N harvest index	*	*	*	
			Plant dry weight	*	*	*	
			Seed dry weight	*	*	*	
	5(c)	6	Nodule weight	*	*	ng	
			Shoot N content	*	*	ng	
			Nodule efficiency	*	*	ng	
			Nitrogenase activity	*	*	ng	
			N concentration in xylem sap	*	*	ng	
			Rate of N transport	*	*	ng	
			Ureide-N	*	*	ng	

(continued)

Table III. *(continued)*

Species	No. of Lines (l), Cultivars (c), or Genotypes (g)	Number of *Rhizobium* Strains	Characters Measured	Significance[a]			Reference
				Host Effect	*Rhizobium* Effect	Interaction	
Pisum sativum	4(c)	4	Acetylene reduction	*	*	ns	Mahon (1979)
Pisum sativum	45(c)	250	Nodule number	*	ng	ng	Tikhonovich (1981)
			Nitrogenase activity	*	ng	ng	
Stylosanthes sp.	72(l)	22	Nodulation	*	*	*	Date and Norris (1979)
Stylosanthes sp.	56(l)	13	Nodulation	ng	ng	*	Date *et al.* (1979)
Trifolium pratense	ng	ng	Yield	ng	ng	*	Mavrichev and Smirnova (1977)
			N content	ng	ng	*	
Vicia faba	6(l)	6	Shoot dry weight	*	*	*	Mytton *et al.* (1977)

[a]*, Significant effect; ns, no significant effect.
[b]ng, Not given in reference.
[c]Field study.

Table IV. Summary of Trials Investigating White Clover Cultivar by Nitrogen Interactions

Number of Cultivars	Nitrogen Levels (kg N ha^{-1})	Significance of Interaction[a]	Reference
3	0,12,30	* (in 2 out of 3 years)	Reid (1961a,b)
9	0,126	* (at 1 out of 2 sites)	Aldrich (1970)
3	0,240	*	Hoen (1970)
3	0,120,240	(*)[b]	Connolly (1971)
2	100,200,300	(*)	Rais and Královec (1973, 1975)
2	0,45 (field)	* (in 1 out of 3 harvests)	Brock and Hoglund (1974)
2	0,22.5,45 (glasshouse)	ns	
2	0,50 ppm (growth cabinet)	* (root dry weight)	Hoglund and Brock (1974)
4	0,134,268	*	Murphy and Connolly (1975)
12	Fixed N or NH_4NO_3 (growth cabinet)	* (in pot but not tube expt.)	Mytton (1976)
4	0,30,60,90	ns	Department of Agriculture (1977), Laidlaw (1980)
3	0,120,240	(*)	Kasper (1977)
17	0,160–200	*	Ingram and Aldrich (1981)
6	0,224	(*)	Wilman and Asiegbu (1982a,b)
2	0,400	ns	Eltilib (1982)
4	0,360	ns	Boyd and Frame (1983)
4	20,120	ns	Harris *et al.* (1983)
4	0,1.8,5.6,17.6,56 (glasshouse)	*	Hoglund and Brock (1983)
2	0,100	*	Morrison *et al.* (1983)
4	50,200 (growth cabinet)	ns	Caradus (1984)
5	10,200 (growth cabinet)	ns	Caradus (1984)

[a]*, Significant effect; ns, no significant effect.
[b]Based on the information given it is possible that the interaction was significant.

indisputably significant interactions. Interestingly, none of the cultivars that showed tolerance to high N were specifically bred for that purpose and no single cultivar was consistently high N tolerant.

Ingram and Aldrich (1981) found that large-leaved white clovers were more tolerant to high N than small-leaved types; mean high-N tolerance ratios (yield at high N/yield at low N) were 0.63 ± 0.03 and 0.48 ± 0.02, respectively. Wilman and Asiegbu (1982a,b) also found that stolon length was less affected by N level for those cultivars more tolerant of high N ($r = +0.94$, $p < 0.01$). However, they gave data that showed no relationship between high-N tolerance ratio and mean cultivar stolon length ($r = -0.39$ n.s.), mean stolon diameter ($r = +0.51$ n.s.), mean petiole length before harvest ($r = +0.57$ n.s.), mean leaf weight ($r = +0.48$ n.s.), mean proportion of leaf in clover dry matter harvested ($r = +0.28$ n.s.), or the mean height difference between ryegrass and white clover *in situ* before harvest ($r = -0.36$ n.s.).

Intercultivar differences for nitrogen loss to the soil from nodulated plants through decomposition of root and nodular tissue and/or through the direct excretion of nitrogenous substances from the roots have been shown for soybean (Burton *et al.*, 1983).

As a result of these studies, yield improvements are most likely to succeed by selection for high N-fixing genotypes, although, even in this case, selection for high-yielding genotypes would possibly be as successful as selections for high N-fixation per se. The uncertainty of long-term benefits from "elite" *Rhizobium* strains, and combinations of "elite" strains and host genotypes, makes these options less attractive.

2. Mycorrhizae

The symbiotic relationship formed with vesicular-arbuscular (VA) mycorrhizal fungi improves the mineral nutrition of the host plant, particularly in soils that are deficient in P (Sanders and Tinker, 1971; Bowen *et al.*, 1975) and to a lesser extent K (Powell, 1975). Bieleski (1973) provides arguments for the importance of mycorrhizal hyphae in presenting a larger surface area for absorption and exploration of a greater volume of soil than root alone, even if well endowed with root hairs. Mosse (1973) has reviewed reports showing that plants inoculated with VA mycorrhizal fungi often take up more P and grow better than nonmycorrhizal plants. Enhanced P absorption by mycorrhizal roots of soybean is due more to an increase in the number of uptake sites per unit area of root rather than increased affinity of roots for P (Karunaratne *et al.*, 1986). However, the extent and significance of intraspecific variation in effectiveness of host plant–mycorrhizal strain combinations have been poorly researched and results to date are inconclusive.

Some studies have examined the interaction between genotypes or populations within species and presence or absence of mycorrhizal inoculation to identify host material that can form more effective symbioses. Research of this type has been mostly restricted to white clover and alfalfa. At low levels of available P the more responsive white clover cultivar 'Tamar' had more of its roots infected by mycorrhizae than 'Huia' (Hall *et al.*, 1977). However, differences between cultivars in response to mycorrhizal inoculation can be affected by soil type. On a highly P-retentive soil no differences between two white clover selections and two *Lotus pedunculatus* selections in the growth responses to mycorrhizal inoculation were found, while on a less P-retentive soil inoculation stimulated the growth of only one lotus and one clover selection (Powell, 1982). More complex interactions between soil type, endomycorrhizal fungal populations, and host plant selection have been observed (Crush, 1978). In another study the lack of clover type × mycorrhizal interactions for growth parameters suggested that selection for more effective mycorrhizal relationships may not be worth pursuing in a plant breeding program aimed at developing a white clover with improved P nutrition (Crush and Caradus, 1980).

Significant alfalfa cultivar × mycorrhiza inoculation level interactions have been observed for shoot dry weight, shoot % P (Satterlee *et al.*, 1983), and shoot % Cu and % Zn (Lambert *et al.*, 1980). However, the interaction between clones within cultivars and mycorrhizal inoculation level was significant for shoot dry weight and concentration of mineral nutrients such as P, Cu, Zn, Fe, Mn, K, Ca, and Mg (Lambert *et al.*, 1980). They concluded that screening of alfalfa lines or genotypes for percentage content of P, Cu, and Zn should be conducted under mycorrhizal conditions.

Few studies have compared the effect of different VA mycorrhizal fungi on different cultivars. Significant fungal strain × cultivar interactions for shoot dry weight (Skipper and Smith, 1979) and seed yield (Schenck *et al.*, 1975) have been recorded for soybean. Comparison of three leafless pea cultivars with the same phenotype infected by three *Glomus* species showed a significant interaction between cultivar and mycorrhizal species for shoot dry weight. While the infectivity of the three *Glomus* species remained the same, irrespective of the host plant cultivar, mycorrhizal effectiveness at improving plant growth varied with host plant genotype (Estaún *et al.*, 1987). Two cultivars of cowpea behaved differently to infection by three *Glomus* species (Ollivier *et al.*, 1983). For one cultivar, growth was stimulated by all three *Glomus* species, whereas for the other only two *Glomus* species had a stimulating effect.

3. Acidifying Bacteria and Fungi

Soil microorganisms are known to dissolve insoluble inorganic phosphates (Gerretsen, 1948; Sen and Paul, 1957; Hayman, 1975) by releasing organic acids such as citric, oxalic, lactic, and tartaric acids (Bajpai and Rao, 1971; Banik and Dey, 1982). The importance of such soil microorganisms in improving the nutrition of plants is open to debate. Khalafallah *et al.* (1982) obtained a significant enhancement of P uptake and yield by faba bean inoculated with phosphate-dissolving bacteria. Brown (1974), in reviewing the effect and significance of seed and root treatment with cultures of bacteria thought to release P either from insoluble inorganic phosphates or from organic P compounds, suggested that stimulated plant growth may be due to gibberellin release from the bacteria. Barea *et al.* (1984) have reported evidence for alfalfa that phosphate-solubilizing bacteria may be synergistic with mycorrhizal function. However, again it is uncertain if this effect was due to the bacteria producing growth regulators or due to their ability to release P from insoluble phosphates (Barea *et al.*, 1976, 1984; Azcón-Aguilar and Barea, 1978; Azcón *et al.*, 1978).

On the assumption that acidifying soil bacteria and fungi are important in the nutrition of plants, no evidence of intraspecific variation for effectiveness of these microorganisms has emerged.

C. PHYSIOLOGICAL CONSIDERATIONS

1. Growth Rate, Yield, and Nutrient Demand

Comparison of ecologically contrasting species has generally shown that species from low fertility habitats tend to have lower relative growth rates than those from fertile habitats (Parsons, 1968; Rorison, 1968; Grime and Hunt, 1975; Chapin, 1980, 1982). Slow-growing species make less demand on a limited supply, function closer to their maximum growth rate and metabolic rate under low-P conditions, and may therefore be "fitter" in a low-P environment (Bradshaw, 1969). However, from intraspecific comparisons one can conclude that low relative growth rate is not itself a major adaptive feature of populations within species. For example, no apparent difference in relative growth rate between populations of white clover collected from soils of high-P and low-P status when grown in either sand or solution culture (Caradus and Snaydon, 1986a) was found.

Many studies of mineral nutrition have concentrated on the size of the root system to explain differences in P uptake (Barber, 1982, 1984; Steffens, 1984). However, it is also recognized that the shoot system is likely to have a large effect on nutrient uptake, if only by the demand for nutrients that it imposes. For example, several studies have shown that the rate of P absorption, from a given concentration, is strongly influenced by demand within the plant (Edwards, 1968; Wild and Breeze, 1981; White, 1973), created by growth rate and nutrient concentration in the plant tissues (Rorison, 1968; Nye and Tinker, 1977, p. 214; Caradus, 1983, 1986). The correlation between shoot yield and nutrient content has been observed within many species, e.g., groundnuts (Achuta Rao *et al.*, 1974) and peas (Jaiswal *et al.*, 1975). One of the difficulties in determining the relative importance of shoot and root factors is that root size and shoot size can be closely correlated ($r = +0.84$; Caradus, 1977; Troughton, 1960) and that nutrient uptake per plant can be strongly correlated with both root and shoot weight (Caradus and Snaydon, 1986a). Because of this close relationship between shoot and root size, it is difficult to determine whether a large root system per se is responsible for a larger shoot as a result of greater nutrient uptake or whether a larger root system is produced as a result of greater photosynthetic production by the larger shoot (Dambroth and El Bassam, 1982).

Attempts to determine the effects of root and shoot systems independently have employed the use of root and shoot pruning and split-root and reciprocal grafting techniques. A recent set of experiments has employed these techniques to determine the effect of shoot and root system on P uptake by white clover populations (Caradus and Snaydon, 1986b,c). Root pruning reduced P uptake rate per plant but did not affect P uptake rate per unit root until "stored" inorganic P within the plant was reduced, either over a period of time or by removal of part of the shoot. Similarly, splitting root systems between plus and minus P solutions reduced P uptake rate per plant but had no effect on P uptake per unit of root supplied with P. Shoot pruning reduced both P uptake rate per plant and P uptake rate per unit root (Table V). The P uptake rates of all white clover populations studied were equally affected by the root pruning treatment; however, the populations differed in response of P uptake per unit root weight to shoot pruning (Table VI).

In soybean, root pruning has increased maximum P influx (I_{max}) by 50% for plants up to 46 days old, but had no effect on plants older than 46 days (Edwards and Barber, 1976). Unfortunately in this study, no indication of the significance of the difference between root pruning treatments was given. In another study (Borkert and Barber, 1983), also with soybean, although I_{max} increased it was unable to compensate for the decrease in root surface area, as manipulated by split-root technique, contacting the P solution. P uptake per plant decreased as roots in P solution decreased.

Reciprocal grafting of two morphologically different white clover lines confirmed the importance of the shoot in determining P uptake (Caradus and Snaydon, 1986c). The genotype of the scion determined the size of the shoot and root system, growth rate, and P uptake per plant. Neither scion nor rootstock had an overriding effect on P uptake per unit root. In soybeans the accumulation of a number of nutrients, including P, N, Na, B, and Ca, has been shown to be controlled by the scion when root–scion grafts were made between high and low accumulators (Kleese, 1967, 1968; Kleese and Smith, 1970). Polson and Smith (1972) further determined that scion control of ion accumulation was accomplished by a self-contained mechanism within the shoot itself. Foote and Howell (1964), however, showed that the rootstock genotype of soybeans regulated the accumulation of P to toxic levels in the shoots.

2. Uptake Mechanisms

Under conditions in which the rate-determining step in nutrient uptake is located in the root, nutrient uptake will increase if root length per unit plant weight and maximal net influx per centimeter root (I_{max}) increase and the Michaelis–Menten constant (K_m) and minimum concentration (C_{min}) decrease (Nielsen, 1979). However, documentation of cultivar differences within legume species of I_{max}, K_m, and C_{min} is rare. Differences in C_{min} have been shown between cultivars of soybean (Silberbush and Barber, 1983) and white clover (Mouat, 1983). Silberbush and Barber (1983, 1984) also observed differences between cultivars of soybeans for I_{max}, but not K_m.

Reciprocal grafts of Fe deficiency-tolerant and -intolerant soybeans have shown that the genotype of the rootstock was a controlling factor in the uptake of iron (Brown *et al.*, 1958, 1961). The Fe deficiency-tolerant cultivar had greater uptake capacity for Fe from relatively low Fe supply. Sain and Johnson (1986), using cell

Table V. Effect of Shoot and Root Pruning on P Uptake per Plant and P Uptake per Unit Root Weight of White Clover[a]

Uptake Rate		Expt.[b]	Shoot			Root			$MSR_{0.05}$[c] or $MSD_{0.05}$[d]
			Unpruned	Pruned	*p*	Unpruned	Pruned	*p*	
P uptake per plant	(μg P day^{-1})	1	nm[e]	nm	—	120	84	**	×1.23
	(μg P hr^{-1})	2	39	10	***	27	14	***	×1.29
P uptake per unit root	(μg P mg^{-1} day^{-1})	1	nm	nm	—	31	34	ns	—
	(μg P mg^{-1} DW hr^{-1})	2	0.39	0.12	***	0.26	0.26	ns	0.04

[a]Source: Caradus and Snaydon, 1986b; Caradus, 1984.
[b]Experiment 1 is the mean of six populations and experiment 2 is the mean of eight populations.
[c]MSR, minimum significant ratio.
[d]MSD, minimum significant difference.
[e]nm, Not measured.

Table VI. Effect of Shoot Pruning on P Uptake Rate per Unit Root Weight (μg P mg^{-1} DW hr^{-1}) of Four White Clover Cultivars[a]

Cultivar	Unpruned	Pruned
Kent Wild White	0.39	0.16
Huia	0.37	0.10
Blanca	0.36	0.26
Regal	0.34	0.06
p	*	
$MSD_{0.05}$[b]	0.17	

[a]Source: Caradus, 1984.
[b]MSD, minimum significant difference.

suspension culture, found that cells of an Fe deficiency-tolerant soybean cultivar had greater K_m and I_{max} values for Fe uptake than cells of a susceptible cultivar. They suggest that the tolerant cultivar may increase the number of Fe-reducing and/or Fe-transporting sites, thus increasing their ability for Fe uptake.

3. Root Exocellular Acid Phosphatase Activity

Although a number of suggestions have been put forward on the role of external root phosphatase activity, its role is still obscure. Possible roles include (a) the hydrolysis of phosphate esters releasing inorganic P (Bieleski and Johnson, 1972; Boero and Thien, 1979; Dracup *et al.*, 1984), (b) assimilation of P as may occur via phosphatases of VA mycorrhizae (Bartlett and Lewis, 1973; Ho and Zak, 1979), or (c) phosphate transport agents (Woolhouse, 1969; Williamson, 1973). However, because plants continue to grow poorly in P-deficient soil, despite increases in root exocellular acid phosphatase activity, Alexander and Hardy (1981) concluded that high phosphatase activity cannot compensate for an inadequate supply of inorganic P.

Intraspecific variation for exocellular acid phosphatase activity has been observed in white clover (Caradus and Snaydon, 1987b). It has been suggested (McLachlan, 1976) on the basis of species comparisons that populations adapted to low-P soils would have a lower acid phosphatase activity than populations not adapted to low-P soil. However, Caradus and Snaydon (1987b) found no clear correlation between exocellular acid phosphatase activity of populations and the P status of soil from which they were originally collected.

Negative correlations between acid phosphatase activity and plant productivity have been shown in species comparisons (McLachlan, 1976) and in population comparisons within legume species, when supplied with adequate levels of P (Caradus and Snaydon, 1987b). Environmental effects such as salinity (El-Fouly and Jung, 1972), drought (Barrett-Lennard *et al.*, 1982), and N deficiency (Horovitz, 1971) that may reduce yield also cause an increase in acid phosphatase activity. No satisfactory explanations for this effect have been proposed, but it would appear that acid phosphatase production is a response to P deficiency, brought about by a number of factors.

4. Root Cation-Exchange Capacity

The role of root cation-exchange capacity (CEC) in ion absorption processes remains uncertain. An extensive review has been given by Haynes (1980). Plant species with high root CEC have high divalent/monovalent cation ratios in their shoots (Drake *et al.*, 1951; Dunham *et al.*, 1956; Huffaker and Wallace, 1958; Mouat, 1960; Heintze, 1961). In general, dicotyledonous species have root CECs approximately double that of monocotyledonous species (Drake *et al.*, 1951; Heintze, 1961; Asher and Ozanne, 1961; Crooke and Knight, 1971a) and also have a greater uptake of divalent cations relative to monovalent cations (Elgabaly and Wiklander, 1949; Loneragan *et al.*, 1968; Loneragan and Snowball, 1969). In some studies a strong positive correlation has been demonstrated between root CEC and shoot cation content (Beeson, 1941; Asher and Ozanne, 1961; Crooke and Knight, 1962; Vose, 1963), though others have obtained poor correlations (Bear, 1950; Cunningham and Nielsen, 1963; Crooke *et al.*, 1964). Root CEC seems to bear little relation to uptake of monovalent cations (Mengel, 1961; Cunningham and Nielsen, 1963).

Although Crooke and Knight (1962) stated that root CEC could not influence plant anion content, several studies have successfully related CEC and various aspects of P nutrition. For example, the ability of species to utilize Al-, Fe-, and Ca-P in soil and phosphate rock has been correlated with root CEC (Drake and Steckel, 1955; Mitsui *et al.*, 1956; Asher and Ozanne, 1961; Fox and Kacar, 1964). It has been hypothesized that P release from largely insoluble inorganic sources occurs because of the binding of calcium and/or aluminum by plant root colloids, so promoting the dissolution of the phosphate rock in the immediate vicinity of the root (Drake and Steckel, 1955; Fox and Kacar, 1964). Johnston and Olsen (1972) have provided some evidence for this. More indirect evidence comes from studies that have shown positive correlations between plant % P and root CEC (Asher and Ozanne, 1961; Crooke and Knight, 1971b), although such correlations may be coincidental with the concentration of cations, such as calcium (Asher and Ozanne, 1961; Crooke and Knight, 1971b). Others have obtained either no significant correlations or inconsistent correlations between root CEC and P concentration (Drake and Steckel, 1955; Crooke and Knight, 1962; Kansal *et al.*, 1974). Several workers (Elgabaly, 1962; Franklin, 1970) have suggested that root CEC regulates anion uptake because negative charges tend to repel anions such as phosphate and chlorine and so reduce uptake. This was probably the basis for speculation by Mouat and Walker (1959) that differences in root CEC may provide the reason why white clover, with high root CEC, is a poor competitor with *Argrostis tenuis* for P.

Several researchers have suggested that CEC may be a useful character distinguishing genotypes that are tolerant of low-P soils. Heintze (1961) speculated that species making most demands on soil fertility tend to have a higher CEC than those species that are less exacting in their requirements. However, no relationship between CEC and ability to tolerate low-P soils in a study (Caradus and Snaydon, 1987c) of white clover populations collected from high-P and low-P soils was found. Also, in only one out of three experiments was there a significant correlation between root CEC of populations and their response to P.

5. Root Exudates and Root-Induced pH Changes

Changes in soil pH around roots have often been demonstrated; differences of as much as one unit have been recorded (Kirkby and Mengel, 1967; Riley and Barber, 1969, 1971; Jarvis and Hatch, 1985). Longer-term changes in the pH of the soil as a whole have been found beneath subterranean clover-dominated pastures (Williams, 1980; Bromfield *et al.*, 1983); this appears to be the result of the entry of nitrogen into the system by dinitrogen fixation and its loss as nitrate (Helyar, 1976; Haynes, 1983). The difference between predicted (by a simulation model) and observed P uptake by soybean may be due to acidification of the rhizosphere by root exudates increasing P concentration in soil solution and hence P uptake (Silberbush and Barber, 1984).

Changes in rhizosphere pH affect the availability of nutrients such as P (Bagshaw *et al.*, 1972; Grinsted *et al.*, 1982) and iron (Jolley *et al.*, 1986b). Several species, including buckwheat (van Ray and van Diest, 1979), rape (Grinsted *et al.*, 1982; Hedley *et al.*, 1982a, 1983), and lupin (Gardner *et al.*, 1982b), can acidify soil and, as a result, these species can utilize otherwise insoluble phosphates. The acidifying effects of N-fixing legumes may have similar results (Nyatsanga and Pierre, 1973; Israel and Jackson, 1978; Aguilar and van Diest, 1981). However, other studies have suggested that no difference exists between species in their ability to extract more insoluble P. Experiments using ^{32}P-labeled soil have shown that a number of different plant species take up P from the same source and that differences in total P uptake relate to the ability of species to extract P from the labile pool rather than by dissolving more insoluble P (Nye and Foster, 1958; Probert, 1972). The $^{32}P/^{31}P$ ratio was the same for the different species examined. It would have been lower in species capable of utilizing some of the differently labeled insoluble P. Hedley *et al.* (1982b) have proposed that for some species, decreasing pH in the rhizosphere may dissolve nonexchangeable soil inorganic P.

Various explanations of root-induced pH changes have been proposed, including (a) the release of carbon dioxide as a result of root respiration (Metzger, 1928; Overstreet and Jenny, 1939), although Nye (1968) and Johnston and Olsen (1972) consider this unlikely because, under aerobic conditions, carbon dioxide quickly diffuses away; (b) secretion of organic acids (Kirkby, 1981; Gardner *et al.*, 1981, 1982a,b, 1983); (c) enhanced hydrogen ion efflux resulting from Fe deficiency (Brown, 1978; Kannan, 1981; Marschner *et al.*, 1982) or P deficiency (Hedley *et al.*, 1982a); (d) excretion of hydrogen ions by microorganisms associated with roots (Subba Rao, 1974); and (e) imbalance of cation/anion uptake (Ledin and Wilklander, 1974; McLachlan, 1976; Raven and Smith, 1976; Israel and Jackson, 1978;

Hedley *et al.*, 1982a; Aguilar and van Diest, 1981; Jarvis and Robson, 1983c), which is particularly affected by source of nitrogen (Mulder, 1948; Kirkby and Mengel, 1967; Riley and Barber, 1971; Smiley, 1974; Nye, 1981; Jarvis and Robson, 1983a,b).

Few studies have attempted to identify intraspecific differences within legume species for root-induced pH changes. Jarvis and Robson (1983c) compared 10 cultivars of subterranean clover reliant on N from N-fixation and found that differences between cultivars in their effects on soil acidity were largely related to differences in growth rather than to marked differences in the concentration of total cations or inorganic anions. The ability of some soybean cultivars to tolerate iron deficiency, often referred to as Fe efficiency, has been partly attributed to their ability to release more hydrogen ions and "reductant" from their roots (Brown *et al.*, 1967; Ambler *et al.*, 1971; van Egmond and Aktas, 1977; Wallace, 1986; Jolley *et al.*, 1986a). Olsen and Brown (1980), however, only found a release of "reductant" from the roots of the Fe-efficient cultivar of soybean, supplied with nitrate-N, in their 6-day study and this had no effect on solution pH. While van Egmond and Aktas (1977) measured hydrogen ion release from roots of the same Fe-efficient cultivar of soybean also supplied with nitrate-N, this did not occur until 12 days after removal of Fe supply. With plants supplied nitrate-N, reducing solution pH has also been shown as a mechanism allowing Fe-efficient peanut and chick-pea cultivars to tolerate Fe deficiency [see Kannan (1982) and (1981), respectively]. However, Fe-efficient cultivars of pigeon pea and lentil did not affect solution pH more than inefficient cultivars [see Kannan (1981) and (1983), respectively], and it is therefore assumed that they release "reductant."

6. Nutrient Efflux

On the basis that measured efflux rates of P are similar to uptake rates at external P concentrations close to C_{min}, Bieleski (1973) argued that reducing efflux could be more important to the plant than increasing the affinity of the uptake mechanism. Studies examining nutrient efflux (Lüttge and Higinbotham, 1979) have been conducted but no record of intraspecific variation for nutrient efflux was found.

D. ONTOGENY

Stage of growth and time of flowering can have a large effect on plant nutrition, particularly N nutrition of legumes. N-fixation appears to be suppressed soon after flowering, declining during pod filling, and this effect has been considered a major limitation to crop yield. Higher N-fixation rates were associated with late compared with early cultivars of bean (Graham, 1979; Rennie and Kemp, 1981a, 1984; McFerson, 1983), soybean (Hardy *et al.*, 1971; Lawn and Brun, 1974a; Patterson and LaRue, 1983), and pigeon pea (Rao *et al.*, 1981). Early cultivars of soybean responded more to added N than mid-late and late types (Scherer, 1975; Kuz'min, 1976), as a result of nitrogenase activity peaking before the cessation of flowering in earlier cultivars and after flowering ceased in later cultivars (Ham *et al.*, 1976). The decline in nitrogenase activity was associated with the development of the pods as a competing assimilate sink since the decline in nitrogenase activity in each cultivar coincided with the time when pod growth rate first exceeded total top growth rate. However, Riggle *et al.* (1984) concluded that factors other than carbohydrate availability may be responsible for the decline in specific nodule activity during late seed filling, and grafting experiments (Malik, 1983) suggest that the decline is more related to the physiological state of the shoot than the presence of reproductive structures. Gomes and Sodek (1987) have suggested that under favorable growth conditions the diversion of assimilates by reproductive structures (fruits) in soybean is not the primary cause of the decline in nodule activity, but competition by fruits may be important when the production of photoassimilates is limited by low light or temperature.

In other recent studies with bean (Hungaria and Neves, 1987a,b), the presence of pods has been shown to stimulate N-fixation because of the high rates of nitrogen metabolism in these organs, i.e., a sink effect. The importance of this observation is related to reports (Weber, 1966) that nitrogen derived from fixation can be more efficiently used for seed nutrition than mineral-N. N applied to a cultivar that nodulates well stimulates vegetative rather than seed yield, but this effect was not apparent for a cultivar that nodulates poorly and requires N fertilizer for high seed yield production (Hungaria and Neves, 1987b).

Differences in yield within maturity groups of soybean have been attributed to either the partitioning of more photosynthate to the seed or a longer filling period associated with high carbon dioxide uptake and acetylene reduction rates late in the filling period (Gay *et al.*, 1980). Determinate and indeterminate cultivars of similar maturity date have similar characteristics, in terms of initiation, rate of increase, and termination of N-fixation, but determinate cultivars may fix up to 25% of total N prior to flowering compared with 10% for indeterminate cultivars (Hardy *et al.*, 1971).

Israel (1981) found that two cultivars of soybean of differing maturity class, nodulated by a single strain of *Rhizobium*, achieved essentially the same yields of seed dry matter and seed N by expression of N-fixation capability at different times in the growth cycle. High N-fixation capability before reproductive development coupled with remobilization of vegetative tissue N was associated with the yield potential of 'Davis' (earlier flowering) soybeans, while high N-fixation capability during reproductive growth was associated with the yield potential of 'Ransom' (later flowering) soybeans.

Redistribution of leaf N during seed fill was delayed in later cultivars of soybean (Boote *et al.*, 1978) so that the harvest index for N in the seed was lower than in early cultivars (Harper, 1979; Kalaidzhieva, 1983). However, Cornet *et al.* (1980) found that late cultivars of lima bean had higher N content in their seed than early cultivars. Maturity of all but one winter pea cultivar was delayed by application of N, whereas the maturity of all spring cultivars was unaffected (Trevino and Murray, 1975).

Studies examining the effect of maturity on accumulation of nutrients other than N are few. Saggar and Dev (1974) found that the amount of S removed from soil by plants was greater after flowering than before in the soybean cultivars 'Hampton' and 'Dare,' whereas in 10 other cultivars more S was removed before flowering. Early-maturing 'Amsoy' soybean accumulated the most Mg in its seeds, while the late-maturing 'Calland' accumulated the most K, Ca, Fe, and B, but relatively little P and Mo. The midseason cultivar 'Corsoy' contained the most Mo (Dhillon *et al.*, 1975).

II. Mechanisms Improving Nutrient Use

A continual problem faced by plant breeders in their attempts to improve the use of nutrients within plants has been the difficulty of determining and defining what to measure so that differences can be quantified. The most commonly used index is the nutrient utilization quotient (i.e., dry weight per P absorbed), which, as mentioned in the introduction, has serious limitations. Barrow (1978) points out that because of this definition the most efficient use of a nutrient occurs when the internal supply is lowest, a plant may appear to be an efficient user of nutrient merely because it is inefficient at obtaining the nutrient. Barrow (1978) suggests that a more useful measure is the rate of growth per unit concentration of nutrient within the plant. This would be a better alternative than the utilization quotient when small numbers of genotypes or cultivars are being compared but becomes a logistic problem when large numbers are being screened in breeding experiments.

Mechanisms other than high dry matter yield per unit of nutrient absorbed that may improve the use of nutrients within plants include harvest index and plant type, partitioning of nutrient between different pools within the plant, translocation and partitioning of nutrients within the plant, redistribution of previously assimilated nutrients, and leaf death rate.

A. HARVEST INDEX

In crop species mobilization of nutrients to the harvested part of the plant can be important in determining yield and quality. In soybean especially the high N demand by seeds, relative to other crops such as cereals and pulses, causes a fast depletion of vegetative N. The achievement of high soybean yields may be prevented by rapid vegetative N decline shortening the seed-filling period (Sinclair and de Wit, 1975, 1976). This was confirmed by Salado-Navarro *et al.* (1985), who found that high seed protein genotypes of soybean exhibited faster N partitioning and dry matter allocation into seeds, but had a shorter seed-filling duration and lower yield.

Jeppson *et al.* (1978) found that differences in harvest N index of soybean genotypes at maturity were generally due to greater, rather than more rapid, movement of N out of the stems, leaflets, and petioles. This was unaffected by N fertilization and it was suggested that harvest N index may be a useful selection parameter to improve N use efficiency in soybeans. The strong positive correlation between harvest index and harvest N index (Jeppson *et al.*, 1978; Harper, 1979; Schweitzer and Harper, 1985) would suggest that selection for harvest index alone may be sufficient. In this area Schweitzer and Harper (1985) suggest that efforts to optimize partitioning of dry matter in soybeans should be directed toward reproductive timing in late-maturing genotypes where vegetative duration is excessive.

In herbage species it may be important in some low-input systems that the removal of nutrients by the grazing animal is kept to a level that, while satisfying the animal's requirements, does not further deplete the already deficient environment in which the plant is growing. Nutrients might be conserved if they were partitioned into tissues below grazing height (Caradus and Williams, 1981). Genotypes of white clover that allocate a high proportion of total P to shoots and have low shoot % P were large leaved and had high herbage yields and harvest indices. Partitioning relatively large amounts of P below grazing height into stolon and root was associated with a low harvest index (Caradus, 1986). However, in a glasshouse study Caradus and Snaydon (1988b) were unable to show differences between populations of white clover collected from high-P and low-P soils for partitioning of dry matter between leaf, stolon, and root, suggesting that natural selection has not favored this mechanism for improving low-P tolerance.

B. PHYSIOLOGY OF NUTRITION

1. Nutrient Concentration

Whereas total nutrient content (%) is almost always positively correlated with total yield, negative correlations between nutrient concentration, especially of N, P, and K, and yield have been found in a number of species (Table VII). In several legume species it has been shown that lines or genotypes high in one mineral nutrient are high in most others (Kleese *et al.*, 1968; Hill and Jung, 1975; Dickinson *et al.*, 1983; Miller *et al.*, 1984; Raboy *et al.*, 1984).

Many studies have used the reciprocal of nutrient concentration as a measure of P efficiency. However, where selections have been made within species for high and low nutrient concentration, e.g., % P, this has not necessarily always meant an associated change in shoot yield. Selections for high % P have also been made in forage legumes as P, an important animal nutrient, can often be at deficient levels in herbage (McCaslin and Gledhill, 1980; Smith and Cornforth, 1982).

Kendall and Hill (1980), Hill (1981), Hill and Lanyon (1983), and Miller *et al.*

Table VII. Correlations, Using Intraspecific Comparisons, between Nutrient Concentration (%) and Yield

Species	No. of Cultivars (c) or Genotypes (g)	Nutrient	Correlation	Reference
Cicer arietinum	16(c)	S	−0.29[a]	Rang *et al.* (1980)
			−0.66*[b]	
Glycine max	175(c)	N	−0.75*	Musorina and Kovalevich (1985)
		P	−0.41*	
		K	−0.54*	
		Ca	−0.68*	
	38(c)	P	−0.27	Raboy *et al.* (1984)
		Zn	−0.20	
		Ca	+0.07	
		Mg	+0.01	
	49(g)	P	−0.66*	Sumner *et al.* (1982)
Glycine soja	20(c)	P	+0.08	Raboy *et al.* (1984)
		Zn	+0.18	
		Ca	−0.07	
		Mg	−0.38	
Medicago sativa	326(g)	N	+0.10	Heinrichs *et al.* (1969)
		P	−0.31*	
		K	−0.62*	
		Ca	+0.27*	
		S	+0.04	
		Mg	+0.10	
	4(c)	P	−0.32[c]	Hill and Barnes (1977)
		Ca	−0.73[c]	
		K	+0.02[c]	
		Mg	−0.40[c]	
	120(g)	N	−0.03 to −0.46* (depending on harvest and germplasm)	Phillips *et al.* (1982)
Phaseolus vulgaris	104(c)	N	−0.76*	Malavolta and Amaral (1978)
		P	−0.76*	
		K	−0.64*	
	90(c)	N	−0.67*	Malavolta *et al.* (1980)
		P	−0.76*	
		K	−0.64*	
Trifolium repens	8(g)	P	−0.92*	Robinson (1942)
		K	−0.72*	
		Ca	−0.13	
	98(g)	P	−0.61*	Caradus (1986)

[a]For seed.
[b]Correlated with harvest index.
[c]Genotypic correlations.
*Significance ($p < 0.05$).

(1987) describe the results of bidirectional selection programs for % P in herbage of alfalfa, and Miller *et al.* (1984) compared a selection containing high P concentration with unselected material also in alfalfa. These programs sought to identify P-efficient genotypes and also determine whether the nutrient quality of herbage could be improved by selection. The high % P selections had lower germination (Kendall and Hill, 1980), lower Ca : P ratio (Kendall and Hill, 1980; Hill and Lanyon, 1983; Miller *et al.*, 1984, 1987), higher Zn and B concentrations (Miller *et al.*, 1987), higher protein and lower fiber content (Hill, 1981), and higher levels of other nutrient elements (Miller *et al.*, 1984) than the low % P selections. In some studies low % P selections were higher yielding (Kendall and Hill, 1980; Hill, 1981; Miller *et al.*, 1987) than selections for high % P. However, Hill and Lanyon (1983) found no difference in yield between high and low % P selections at two sites and Miller *et al.* (1984, 1987) showed that selections for high % P did not adversely affect yield compared with an unselected control. Miller *et al.* (1984) found that the selection for high % P was more responsive to added P than the control population, whereas Miller *et al.* (1987) were unable to detect this difference.

Selection for either high or low shoot N concentration in alfalfa grown on both N from N-fixation and ammonium nitrate has been successful (Phillips *et al.*, 1982; Teuber *et al.*, 1984; Daday *et al.*, 1987). Selection against the normal inverse correlation between forage dry weight and tissue N concentration was possible. Selection for both high dry weight and high N concentration produced 17 and 52% increases in dry weight depending on germplasm source and about a 5% increase in N concentration (Teuber *et al.*, 1984).

These studies have shown that P and N concentration of legume species can be genetically altered by selection. Such selections may also affect a number of other plant characters, but do not always affect yield. However, a general relationship exists between yield and nutrient concentration such that high-yielding genotypes tend to have low levels of N, P, and K (Table VII). Whether this is a causal relationship or not is difficult to determine. However, the above studies highlight that at least for P, selection for reduced concentration may not have any agronomic advantage.

Although selection for increased P concentration was successful, Hill and Lanyon (1983) conclude that in terms of increasing P concentration to provide an adequate level for animal requirements for alfalfa, the levels were still too low to meet animal requirements. From estimates of genetic correlation, Hill and Jung (1975) have concluded that selection for greater P and lower Ca concentrations in alfalfa would simultaneously increase K concentration and reduce Mg concentration. Therefore attempts to change, by breeding, the concentration of one or two elements in forage so that cattle diets might be improved may not be successful without consideration of other elements. Effects of P fertilizer on herbage P concentration will probably be greater than effects of selection.

Comparison of four white clover cultivars used commercially on soils of relatively high fertility with four ecotypes surviving on soils of low fertility showed that the latter group had the highest % P levels (Spencer *et al.*, 1980), suggesting that natural selection acts to increase P concentration in tissues as a mechanism allowing plants to tolerate P-deficient soils. Species adapted to high and low nutrient sites have exhibited similar relationships (Chapin, 1980).

Phosphorus is partitioned in plants in various fractions (Bieleski, 1973), and these can be broadly classified into inorganic P and organic P. Some plants adapted to low-P soil may accumulate larger amounts of inorganic P in their tissues during periods of P availability, and these "reserves" can then be used in times of P stress. This has been demonstrated in species comparisons by Hart and Jessop (1982) for *Stylosanthes* compared with white clover, but not for *Lotus pedunculatus* and white clover (Hart and Jessop, 1983). Only one intraspecific study (Caradus and Snaydon, 1987a) has considered this possible adaptive feature that may allow plants to tolerate low-P soils, when P fertilizer is applied only once or twice annually at maintenance levels. Populations of white clover collected from low-P soils were shown to accumulate more inorganic P in their leaf tissue, especially when grown at high-P, and were then able to reduce these inorganic P levels to lower concentrations when the P supply was reduced than were populations collected from, and presumably adapted to, high-P soils (Table VIII). The ratio of inorganic P (P_i) to total P (P_T) in leaf tissue had a better relationship with yield than % P_T itself. In two studies comparing white clover populations, total plant dry weight was more closely correlated with P_i/P_T ratio than % P_T (study 1 with 4 populations, $r = -0.98$* and -0.78 ns, respectively; study 2 with 10 populations grown at an adequate P level, $r = -0.48$ ns and -0.18 ns, respectively, and a deficient P level, $r = -0.83$** and -0.48 ns, respectively) (Caradus, 1984; Caradus and Snaydon, 1987a).

At low P availability, nodules of white clover (A. L. Hart, personal communication), peas (Jakobsen, 1985), and soybeans (Israel, 1987) are the predominant sink

Table VIII. Concentration of Organic P (% P_o), Inorganic P (% P_i), and the Ratio of Inorganic P to Total P (P_i/P_T) in Green Leaves, and Concentration of Total P (% P_T) in Dead Leaves of White Clover Populations Adapted to High-P and Low-P Soils when Grown at High- and Low-P Levels in Solution Culture[a]

	P Level	Population Group		
	(m*M*)	Low P	High P	$MSR_{0.05}$[b]
% P_o, green leaves	0	0.16	0.18	
	0.3	0.71	0.59	×1.27
% P_i, green leaves	0	0.002	0.006	
	0.3	0.071	0.038	×1.95
P_i/P_T, green leaves	0	0.012	0.030	
	0.3	0.091	0.060	×1.75
% P_T, dead leaves	0	0.13	0.14	
	0.3	0.32	0.22	×1.36

[a]Source: Caradus and Snaydon, 1987a, by courtesy of Marcel Dekker, Inc.
[b]MSR, minimum significant ratio.

for P compared with shoots and roots, although this is not the case at high levels of P supply. Comparison of three bean cultivars has shown variation for P concentration in nodule tissue (Pereira and Bliss, 1987). The cultivar with the lowest P concentration in nodules also had low P concentrations in the shoot and root and greatest nodule mass and nitrogen accumulation in the shoot.

Tolerance or suceptibility to Fe chlorosis is not correlated with total Fe concentration of leaves in dry bean (Coyne *et al.*, 1982; Zaiter *et al.*, 1986; Hurley *et al.*, 1986) and soybean (Al-Shawk *et al.*, 1986; Jolley *et al.*, 1986b). However, rates of accumulation and depletion of Fe may be associated with tolerance to Fe chlorosis. Pierson *et al.* (1984) found that Fe deficiency-tolerant genotypes of dry bean increased total Fe more rapidly in response to Fe stress and did not decrease total Fe as low as Fe-susceptible genotypes. It has also been consistently shown that cultivars of chick-pea (Agarwala *et al.*, 1979; Singh *et al.*, 1986) and soybean (Al-Shawk *et al.*, 1986) susceptible to Fe deficiency accumulate more Ca and Mn and increase ratios of these to Na and Fe than cultivars tolerant of Fe chlorosis. Because of the important part played by Fe in photosynthesis, it is not surprising that the most Fe deficiency-tolerant cultivars of soybean maintain relatively high net photosynthetic rates at very low levels of Fe supply (Blaylock *et al.*, 1986; Davis *et al.*, 1986).

In many legume crops and herbage species, quality of product is just as important as quantity produced. The concentration of some nutrients in plants may be associated with a number of quality factors such as seed and leaf protein, dry matter digestibility, concentration of compounds such as alkaloids, and cotyledon cracking. Improved quality of product may be achieved by improved nutrient use. For instance, low S concentrations and high P concentrations have been associated with high seed protein levels in a number of species (Table IX). Similarly, leaf protein levels are correlated with % P levels; alkaloid content has been related to levels of Ca and K; cotyledon cracking is associated with low levels of Mg, Ca, and K; and "pops" disorder is similarly associated with low levels of Ca (Table IX).

2. Translocation

Genetic variation for translocation of nutrients has been demonstrated in a number of legume species; some of these differences have been related to improvements in nutrient use. For example, the incidence of cotyledon cracking in beans due to low concentrations of Ca and Mg (Table IX) in cotyledons and cell walls has been shown by grafting studies to be due to difficulties in translocation from roots to shoot and possible immobilization of these nutrients in the shoot (Duczmal, 1976). Lines of chick-pea showing symptoms of iron deficiency in a field study were shown to have more ^{59}Fe in roots and less in stems than a line not exhibiting Fe chlorosis possibly due to some restriction in translocation of ^{59}Fe (Malewar *et al.*, 1982). Rai *et al.* (1984) also considered that differences in ability to translocate Fe may have explained differences among lentil genotypes in their sensitivity to Fe deficiency.

Sweet *Lupinus augustifolius* phenotypes have been found to be susceptible to

Table IX. Association between Quality Characters and Concentrations of Nutrient Elements in Plants, Using Intraspecific Comparisons

Quality Character	Nutrient (%)	Correlation	Species	Reference
Seed protein	S	−0.47	*Cicer arietinum*	Rang *et al.* (1980)
	S	−0.62*	*Vicia faba*	Sjödin *et al.* (1981)
	S	Negative	*Vigna unguiculata*	Boulter and Evans (1976)
	P	+0.74*	*Glycine max*	Raboy *et al.* (1984)
	Zn	+0.44*		
	Ca	+0.12		
	Mg	+0.13		
	P	+0.37	*Glycine soja*	
	Zn	−0.04		
	Ca	+0.08		
	Mg	−0.14		
Leaf protein	P	+0.46*	*Medicago sativa*	Hill and Barnes (1977)
	Ca	+0.10		
	Mg	+0.28		
	K	+0.06		
Dry matter digestibility	N	Positive*	*Medicago sativa*	Davies (1979)
	P	−0.05		Hill and Barnes (1977)
	Ca	+0.11		
	K	−0.06		
	Mg	+0.12		
Alkaloid content	Ca	+0.88*	*Lupinus mutabilis*	Blasco Lamenca *et al.* (1981)
	K	−0.88*		
Cotyledon cracking	Mg	−0.52*	*Phaseolus vulgaris*	Aqil and Boe (1975)
	Ca	+0.02		
	N	+0.09		
	Mg	−0.96**[a]		Duczmal (1976)
	Ca	−0.90*[a]		
	K	−0.80*[a]		
	Na	−0.21[a]		
"Pops" disorder	Ca	Negative	*Arachis hypogea*	Burkhart and Collins (1942); Slack and Morrill (1972); Beringer and Taha (1976)

[a]Correlated with nutrient (%) in cell walls of cotyledons.
*$p < 0.05$
**$p < 0.01$

seed splitting due to low concentrations of Mn in the seed; they also show significant yield responses to Mn. These phenotypes possess the *incundus* gene, which Walton (1978) suggests may reduce Mn translocation to the seed as a secondary function of its alkaloid synthesis function.

Premature abscission of flowers and young fruits of cowpea can be a major factor limiting yield, in association with distribution of P and possibly other nutrients between racemes. Adedipe and Ormrod (1975) found that a cultivar exhibited a relatively low degree of premature abscission of fruits at upper racemes because the fruits at the lowest raceme did not act as a predominant sink for mineral nutrients as occurred in a cultivar with relatively high degree of premature abscission of upper raceme fruits.

Manganese chlorosis in navy bean has been attributed to a reduced Mn transport, in terms of concentration and total amount of Mn in the phloem sap, from root to shoot as a result of retention of Mn in the root (Singh, 1974a,b, 1975, 1976). Similarly, iron deficiency chlorosis has been associated with reduced translocation of iron from roots to shoots (Longnecker and Welch, 1986). They found that soybeans resistant to iron deficiency chlorosis had a greater proportion of total plant iron in their shoots than susceptible cultivars irrespective of iron supply.

Treatment with gibberellic acid has been shown to increase N translocation from cotyledons to the axis of dwarf pea seedlings but to have no enhancement effect on translocation in a tall pea cultivar (Garcia-Luis and Guardiola, 1975). This effect in dwarf pea cultivars may be due to the direct effect of gibberellic acid as a growth regulator and/or caused by the removal of metabolites by the axis.

3. Redistribution of Nutrients in the Plant

The ability to redistribute nutrients from old or senescing parts to growing points in herbage species or from vegetative to reproductive tissue in the case of crop species may be an important mechanism allowing plants to improve the utilization of absorbed nutrients. It has been suggested that in legume crop species such as soybean, where seed growth makes an extraordinary demand on N that cannot be satisfied by N-fixation alone, there would be gains in seed yield from increasing the amount of N available for mobilization (Sinclair and de Wit, 1976).

Variation between cultivars for ability to mobilize N from leaves to seed has been reported for soybean (Borst and Thatcher, 1931; Israel, 1981; Zeiher *et al.*, 1982; Spaeth and Sinclair, 1983; Loberg *et al.*, 1984; Salado-Navarro *et al.*, 1985) and bean (Dickson and Hackler, 1975). However, variation in mobilized leaf N accounted for only 30% of variation in a seed yield in a comparison of 18 soybean lines (Loberg *et al.*, 1984) and was unrelated to yield in another study of 8 cultivars (Zeiher *et al.*, 1982). While cultivar differences in N redistribution were also not closely related to duration of seed fill (Egli *et al.*, 1987), Zeiher *et al.* (1982) did find that the proportion of seed N that came from N redistribution was related to the amount of N available in the vegetative tissue and pod walls at the beginning of seed growth. The amount of potentially redistributable N at this stage was directly related to the weight of vegetative parts, which is closely associated with cultivar maturity. As the ratio of potentially redistributable N at beginning of seed growth to amount of seed N at maturity increased, the seed obtained more of its N from redistribution and less directly from N-fixation and/or uptake of mineral N.

In both soybean (Salado-Navarro *et al.*, 1985) and bean (Dickson and Hackler, 1975), high-protein lines redistribute a higher proportion of N from their vegetative parts to their seed, but have lower yields than low-protein lines.

Israel (1981) found that the N-fixation capacity of *Rhizobium* strains can regulate the remobilization of vegetative tissue N for reproductive tissue. N stress resulting from ineffective *Rhizobium* strains resulted in only small amounts of N being restored in vegetative tissue for subsequent remobilization.

Chapin (1980) concluded from reviewing the literature that there was limited evidence that species adapted to infertile soils were more effective in retranslocating nutrients before leaf abscission than species from fertile habitats. Caradus and Snaydon (1986e) found no evidence that white clover populations from low-P soils were more effective in retranslocating P from senescing leaves than populations adapted to high-P soils. The % P of dead leaves did not differ significantly between the population groups when plants were grown at low P (Table VIII). This would suggest that differences in retranslocation of P from leaves, prior to senescence or abscission, were not a significant adaptive feature of white clover populations growing on low-P soils (Caradus and Snaydon, 1987a).

Gabelman and Gerloff (1983) observed that bean lines with low % P and % K tended to retain more of the element in roots and older leaves when nutrient stressed than lines containing high concentrations of these nutrients. Gabelman (1976) suggested that the high level of mobility of P and K in lines with high concentrations of these elements may have reached a point where normal metabolic function was restricted. Comparison of two white clover cultivars has shown that the more P-responsive cultivar was better able to remobilize P from senescing tissue to meristems than the less P-responsive cultivar (Scott, 1977).

Cultivars of groundnuts have been shown to differ in their susceptibility to "pops," a condition where pods show empty cavities with shriveled or aborted nuts, or healthy-looking nuts with depressions at the center of the cotyledons and/or with blackening of the embryo, with small-seeded types being less affected than large-seeded types (Slack and Morrill, 1972). This disorder has been attributed to an insufficient supply of Ca to the fruits (Burkhart and Collins, 1942). Groundnuts can absorb Ca through the nut shell and it has been shown that small-seeded cultivars translocate more Ca from the shells to the nuts than large-seeded cultivars (Beringer and Taha, 1976).

4. Leaf Death Rate

Species adapted to low-P soils appear to have a lower proportion of dead leaf to total leaf material when P stressed than species from high-P soils (Beadle, 1954, 1966; Specht and Groves, 1966; Grime and Hunt, 1975). Population of white clover collected from low-P soils had a lower proportion of dead leaf to total leaf than populations from high-P soils when grown in solution culture (Caradus and Snaydon, 1986e).

Selection for delayed leaf senescence has been successful in soybean (Phillips *et al.*, 1978, 1984) in an attempt to maintain N-fixation and carbon assimilation during pod filling and increase seed production. However, seed yield was inversely related to the delayed leaf senescent phenotype, suggesting that maximum soybean seed yield can only be achieved with plants whose leaves senesce during seed development. Seed yield of delayed leaf senescent plants may have been limited either by the availability of an adequate number of ovules or by the absence of an appropriate senescence signal required to mobilize materials out of the leaves. It would appear that the longer pod filling period characteristic of some soybean genotypes is associated more with a slower rate of leaf senescence than with a later initiation of senescence (Egli *et al.*, 1987).

5. Ontogeny and Plant Type

Cultivars of white clover able to persist and tolerate mineral N application when grown in a grass sward tend to be large leaved and have longer petioles (Ahloowalia, 1977; Davies, 1978; Wilman and Asiegbu, 1982a). On soils low in mineral N, spreading types of white clover may generally have better performance (Hoglund and Williams, 1984).

P concentration of seeds was higher in tall cultivars of pea than in short cultivars (Paprocki *et al.*, 1980), whereas it was lower in shoots of erect *Stylosanthes humilis* lines than in prostrate lines (Ive and Fisher, 1974). P concentration was also lower in the later-maturing lines than in early lines of *Stylosanthes*.

N content and protein N content showed two peaks during development of peanut cultivars (Velu and Gopalakrishnan, 1985). These peaks occurred earlier in bunch cultivars than in spreading and semispreading cultivars. In bunch cultivars the second peak was more pronounced than the first, but it was only slight in the other two cultivar groups. Comparison of adzuki bean cultivars has shown that in early stages of growth, N concentration in plant tissues was highest in the early cultivar, but after flowering the midseason cultivars were superior in this respect (Sawaguchi and Nomura, 1980).

Nutrient concentration in peanuts grown under nutrient-sufficient conditions has been related to market type, i.e., whether large- or small-leaved Virginia type, or Spanish or Valencia types (Hallock *et al.*, 1971). Lines of the large-seeded Virginia type were generally highest in K, Ca, and B, intermediate in Mg, and lowest in P of the four types. The small-seeded Virginia type lines were highest in Mg, equivalent in Cu and Mn to the large-seeded lines, and intermediate in P and K. The Spanish and Valencia lines were highest in P, lowest in Cu, but equivalent to the small-seeded Virginia lines in K, Ca, B, and Mn.

Some studies (Graham and Halliday, 1977; Graham and Rosas, 1977) have shown that growth habit of bean cultivars is associated with differences in N-fixation. Climbing cultivars had higher absolute and specific acetylene reduction rates compared with bush cultivars. Other studies (Rennie and Kemp, 1981a,b) have not been able to demonstrate this relationship, although they did find that larger-seeded beans produced large plants with greater leaf weight and area and that this controlled the absolute amount of N-fixation.

III. Effect of Genotype × Environment Interactions on Improved Nutrient Uptake and Use

Demonstration of intraspecific differences in nutrient uptake and use are often confounded by environmental effects, both abiotic and biotic. Abiotic factors that may affect nutrient uptake and use include other nutrients, temperature, water, and soil disturbance; biotic factors include plant pathogens and pests, grazing animals (defoliation by cutting will be included in this category), and competition from other plants. In this section emphasis will be on how these environmental effects can interact with cultivars that show differences in their mineral nutrition.

A. ABIOTIC EFFECTS

1. Other Nutrients

Heavy metal toxicity appears to inactivate part, but not all, of the mechanisms allowing some soybean cultivars to tolerate Fe deficiency (Smith *et al.*, 1985; Wallace *et al.*, 1986). Although elements such as cadmium reduce Fe uptake by cultivars tolerant of Fe deficiency, they have a much greater effect on cultivars intolerant of Fe deficiency. Smith *et al.* (1985) suggest that response to cadmium by soybean cultivars is dependent, directly or indirectly, on their relative ability to utilize Fe. Similarly, trace element stress can induce Fe deficiency more readily in cultivars intolerant of Fe deficiency than in tolerant cultivars (Wallace *et al.*, 1986). Nitrogen, added as nitrate, has been shown to increase Fe deficiency chlorosis in intolerant cultivars of soybean, whereas yields of Fe deficiency-tolerant cultivars were enhanced by nitrate (Aktas and van Egmond, 1979). Excess P is probably a major contributing factor to Fe deficiency under field conditions (Inskeep and Bloom, 1984) because of the effect of P on the solubility and translocation of Fe in plants. Susceptibility to Fe deficiency is accentuated more in Fe deficiency-sensitive cultivars of soybean than in tolerant cultivars by high levels of P fertilizer (Tiffin, 1966; Moraghan, 1987).

P uptake by some cowpea cultivars is decreased by added Zn whereas for other cultivars adequate amounts of Zn have been shown to be essential for maintaining high rates of P uptake (Safaya, 1978). Differences between genotypes of soybean (Ambler and Brown, 1969; Sumner *et al.*, 1982), cowpea (Safaya and Singh, 1977), and pidgeon pea (Shukla and Raj, 1980) to deficiency of Zn can be related to differences in the P/Zn ratio of shoots. Cultivars that maintain a narrower P/Zn ratio respond less to applied Zn. Soybean cultivars differing in their sensitivity to toxic levels of P also differed in the effect that added Zn had on the response to P (Paulsen and Rotimi, 1968). Added Zn overcame the effect of P toxicity on the insensitive cultivar but had little effect on the sensitive cultivar. However, as added P increased, Zn concentration in both cultivars decreased equally.

Sensitivity to toxic levels of P is also thought to be closely linked to N metabolism and ammonium toxicity (Park and Stutte, 1973). Growth of P-sensitive soybean cultivars was inhibited more by ammonium-N and urea-N than that of less sensitive cultivars. However, application of nitrate-N eliminated the effects of P toxicity in two sensitive cultivars (Howell, 1964).

Iron has been shown to induce Mn chlorosis in some lines of navy bean (Singh, 1974a). Nonchlorotic lines adjust to high Fe supply by reducing total Fe transport and increasing Mn transport, in the phloem, from root to shoot (Singh, 1976).

2. Temperature

Reduced root temperatures are known to adversely affect N-fixation rates (Roughley, 1970), however, cultivars of both alfalfa and pea bean have been shown to differ in their ability to fix N at low temperatures. Duke and Dochlert (1981) found that nonhardy cultivars of alfalfa had a greater reduction in nodulation and acetylene reduction that low-temperature-hardy cultivars. They suggest that under hardening conditions the aerial portions of nonhardy cultivars were strong sinks for root carbohydrates, since stems and leaves were actively growing, while nodules may be relatively greater sinks in hardy cultivars. Therefore if nonhardy cultivars nodulate less than hardy cultivars at low temperatures, they could become N deficient when temperature falls below zero.

Rennie and Kemp (1981a) were unable to find a relationship between cold temperature tolerance and N-fixation of 11 pea bean cultivars. However, a further study (Rennie and Kemp, 1982) showed that low temperature tolerance during early root growth, rather than at germination and seedling emergence, can be associated with early nodule formation and acetylene reduction at 10°C. This tolerance could be a competitive advantage since earlier nodulation and N_2-fixation would effectively prolong the N_2-fixing life of such a cultivar.

The rate of leaf Fe chlorosis in dry beans is more rapid and the severity of symptoms is greater under lower than higher temperatures. Zaiter *et al.* (1986) speculated that many Fe ligand binding sites in the plant that are involved in the translocation of Fe throughout the plant may become inactive under low-temperature regimes. They were, however, able to demonstrate that some cultivars are more tolerant of Fe chlorosis at low temperatures than others. In soybean, Fe deficiency was more severe in both Fe-deficient sensitive and tolerant cultivars at higher soil temperatures (24°C) than at lower temperatures (16°C). This may have been due to greater P concentration in shoots at 24°C, with P having an antagonistic effect on Fe solubility and transport (Biddulph and Woodbridge, 1952), and/or an increased rate of plant growth and greater dilution of Fe in young tissue at 24°C.

Comparison of two alfalfa cultivars (Levesque and Ketcheson, 1963) showed that at 10°C and at a low level of applied P, 50% of total P found in one cultivar came from fertilizer applied whereas in the other only 33% came from this source. At 18° and 26°C the fraction was less than 30% and was similar for each cultivar.

Differences between cultivars at an adequate level of applied P were less affected by temperature.

Romero *et al.* (1981) found differences between cultivars of alfalfa for response to added K but cultivar differences were not related to winterhardiness level.

3. Moisture

Few interactions between cultivars, nutrient level, and moisture level have been demonstrated. Subjection to drought at the stage of flower bud formation and anthesis caused a reduction in N concentration of all leaves of drought-susceptible pea cultivars but only in the lower leaves of the drought-tolerant cultivar (Dolgopolova, 1976; Dolgopolova *et al.*, 1979).

4. Soil Disturbance

A cultivar of pea with a more vigorous root system was more responsive to deep ploughing and N-P-K fertilizer treatment than one with a less vigorous root system (Gritsenko and Yashin, 1978).

5. Light Intensity and Quality

Interactions between cultivar and P nutrition and the effect of light intensity have been observed in subterranean clover (Millikan, 1957) and cowpea (Adedipe and Ormrod, 1974). The distribution of P in leaf, stem, and root was unaffected by light intensity in one cultivar of cowpea but in another, higher light intensities reduced the amount of P in the leaf and increased it in the root.

B. BIOTIC EFFECTS

1. Plant Pathogens and Pests

Two areas of research will be addressed here: (a) the effect of nutritional variants on disease and pest resistance and (b) the effect of diseases and pests on nutritional variants.

Nutrient content of plants may be related to resistance or susceptibility to pests and pathogens. The two nutrient elements most commonly mentioned in this regard are N and Ca. Cultivars resistant to pathogens and pests may have low N contents and low N-fixation rates. Cultivars of dry bean that are susceptible to insect damage (Tester, 1977) and *Macrophominia phaseolina* (Gangopadhyay and Wyllie, 1973) had higher % N in leaves than did resistant cultivars. Lines of alfalfa selected for high values of N-fixation, as a result of increased nodule mass and shoot yield, were usually more susceptible to *Corynebacterium insidiosum* than those selected for low values (Viands *et al.*, 1980). Because the greater concentration of N in susceptible cultivars was related to greater potential yields and higher protein and amino acid contents (Tester, 1977), resistant cultivars were simply nutritionally deficient.

Pathogenic infections have been shown to have a direct effect on the N metabolism of plants, but the effect varies depending on the susceptibility of the cultivar. Cowpea mosaic virus infection increased total % N, but the increase was greater in the susceptible than in the resistant cultivar (Khatri and Chenulu, 1973). Nitrate reductase activity was accelerated more in the susceptible than in the resistant cultivar, although accumulation of certain free amino acids was observed in the resistant cultivar, while there was a decrease in concentration of the same amino acids in the susceptible cultivar. On the other hand, Blazhev *et al.* (1984) found that % N, % P, and % K contents of a clone of alfalfa resistant to *Fusarium oxysporum* f. sp. *medicaginis* decreased with infection but increased in the resistant clone.

Nematode infection of soybean reduces nodulation (McGinnity *et al.*, 1980) and N-fixation (Lehman *et al.*, 1971), particularly in susceptible cultivars. However, McGinnity *et al.* (1980) observed that some susceptible cultivars despite high cyst formation can retain high nodule numbers and nodule mass, which in turn result in high N-fixation rates. Baldwin *et al.* (1979) showed that the effect of nematode infection on N-fixation varied with time. After 50 days, nematode infection suppressed N-fixation of a soybean cultivar that was moderately resistant to *Meloidogyne incognita,* but only after 75 days following inoculation with nematodes. N-fixation exceeded that of nematode-free controls. For a susceptible cultivar, nematode infection by day 50 stimulated N-fixation but at 75 days depressed it.

High Ca contents have been correlated with high (Rawat and Shaw, 1983) and low (Johnson and Campbell, 1982) damage due to pests, although often the effect has been due to associated effects of other nutrients such as K (Rawat and Shaw, 1983) and Mg (Sethi and Sharma, 1976). In a field trial of 63 groundnut lines, Johnson and Campbell (1982) found significant negative correlation ($r = -0.69$) between damage caused by the two-spotted spider (*Tetranychus urticae*) and contents of carbohydrate and Ca. Rawat and Shaw (1983) compared 13 pea cultivars and showed that leaf contents of wax, amino acids, and calcium were positively correlated with oviposition by *Caliothrips indicus* (Bagnall). Amino acid, Ca, K, and ascorbic acid contents were positively, and carbohydrate content negatively, correlated with degree of infestation. In cowpea, Sethi and Sharma (1976) found higher P, K. Ca, and Fe concentrations in the nematode-susceptible cultivar but higher Mg levels in the resistant cultivar. The total concentration of Ca and Mg was similar in both cultivars when infested with *Heterodera cajoni,* but with *Meloidogyne* infestation the Ca content of the susceptible cultivar was reduced, resulting in an increase in Mg content, while the reverse trend occurred in the resistant cultivar.

Infection by *Corynebacterium insidiosum* of alfalfa has a large effect on plant nutrition, but again the magnitude and direction of the effect can vary with cultivar. Mn uptake following infection was reduced by 90% in a susceptible cultivar and only by 11% in a resistant cultivar. However, Zn uptake was 17% higher in the susceptible cultivar and 54% lower in the resistant cultivar following infection.

When infected, the susceptible cultivar reduced Mn uptake and translocation to the shoot but the roots showed excessive Zn accumulation (Kudelová *et al.*, 1978). Uptake of P (Hanker and Kudelová, 1979a) and S (Taimr *et al.*, 1975) was also reduced more in susceptible than in resistant cultivars of alfalfa by *C. insidiosum* inoculation. However, longer-term studies showed that, although P uptake by resistant plants was greater than that by susceptible plants, irrespective of inoculation, inoculation did reduce P uptake of resistant and increase P uptake by susceptible plants compared with uninoculated controls (Hanker and Kudelová, 1983). This was possibly due to increased leakage of P from roots of infected resistant plants. Whereas inoculation by *C. insidiosum* had very little effect on the partitioning of P to different metabolic infractions in resistant alfalfa cultivars, in inoculated susceptible plants inorganic P increased and organic P decreased compared with uninoculated controls (Hanker and Kudelová, 1979b).

Differences in interaction of nematodes and mycorrhizae in nematode-susceptible and -resistant cultivars of soybean have been shown (Schenck *et al.*, 1975). Seed yield of the normally nematode-susceptible cultivar was increased by inoculation with the mycorrhizae *Endogone macrocarpa* as a result of a reduction in the number of *Meloidogyne incognita* juveniles associated with the roots. Yield of the normally nematode-resistant cultivar was decreased by inoculation with *E. calospora* because of a slight increase in nematode numbers. However, nematode numbers were greatest when this cultivar was inoculated with *E. macrocarpa,* although seed yield was unaffected.

2. Grazing and Defoliation

Defoliation increases requirements for nutrients. Ozanne and Howes (1971) found that the P requirement of subterranean clover increased with frequency of defoliation. Hoglund and Brock (1983) found that some cultivars of white clover responded to added P and N differently depending on frequency of cutting. Comparison of alfalfa clones tolerant of frequent defoliation showed that the most tolerant clones had higher concentrations of nonstructural carbohydrates, starch, organic and inorganic P, and amide-N in roots and crowns than did harvest tolerant clones (Chatterton *et al.*, 1977). These differences were not evident in leaf material, which led to the conclusion that while the major function of the leaf component was converting solar energy to chemical energy, the crown and roots regulated the distribution and utilization of photoassimilates.

3. Competition from Other Plants

Competition between plants can be interspecific, as occurs in mixed species swards, or intraspecific, as occurs in crops with high plant densities.

Climbing bean cultivars have shown a reduction in N-fixation per plant with increases in plant density while prostrate cultivars were unaffected (Graham and Rosas, 1978). A cultivar × N level × plant spacing interaction for red kidney bean was due to a cultivar with smaller plants that had upright leaves and responded to N by producing higher yields per unit area than a nonresponsive cultivar with horizontal leaves (Bravo and Wallace, 1974). The K content of tall pea cultivars, but not short cultivars, was affected by sowing density; P content was unaffected by sowing density (Paprocki *et al.*, 1980). However, the N-fixation of soybean plants was higher at higher rather than lower densities irrespective of cultivar (Bello *et al.*, 1980).

The differential response to soils varying in Ca, Mg, P, and Fe levels of white clover populations was more marked under competitive conditions, especially interspecific competition (Snaydon, 1961).

IV. Summary

Intraspecific variation in the mineral nutrition of legume crop and herbage species is extensive and suggests that it may be possible to develop cultivars that are tolerant of low-nutrient levels or are capable of using nutrients more efficiently when applied as fertilizer. Plant characters that may have the greatest impact in this regard are root system morphology, root-induced pH changes, leaf death rate, the ratio of inorganic to total P, and the relationship between flowering and nitrogen fixation. The importance of the involvement of microorganisms, such as *Rhizobium,* mycorrhizae, and acidifying bacteria, in the nutrition of legume species is well documented but the possibilities of selections for improvements in their effectiveness are less certain.

References

Abu-Shakra, S. S., D. A. Phillips, and R. C. Huffacker. 1978. Nitrogen fixation and delayed leaf senescence in soybeans. *Science* 199:973–975.

Achuta Rao, K., S. S. Kandlikar, V. Jaya Mohan Rao, P. V. Anantha Raman, and N. G. P. Rao. 1974. Physiological characterisation of groundnut types. *Proc. 2nd Gen. Congr. SABRAO, New Delhi* pp. 1034–1040. (*Plant Breed. Abstr.* 47:7122.)

Adedipe, N. O., and D. P. Ormrod. 1974. Effects of light intensity on growth and carbohydrate and phosphorus distributions in the cowpea (*Vigna unguiculata*). *HortScience* 9:281. (Abstr.)

Adedipe, N. O., and D. P. Ormrod. 1975. Absorption of foliar-applied ^{32}p by successive leaves and distribution patterns in relation to early fruiting and abscission in the cowpea (*Vigna unguiculata*), *Ann. Bot.* 39:639–646.

Adepetu, J. A., and L. K. Akapa. 1977. Root growth and nutrient uptake characteristics of some cowpea varieties. *Agron. J.* 69:940–943.

Agarwala, S. C., S. C. Mehrotra, S. S. Bight, and C. P. Sharma. 1979. Mineral nutrient element composition of three varieties of chickpea grown at normal and deficient levels of iron supply. *J. Indian Bot. Soc.* 58:153–162 (*Plant Breed. Abstr.* 50:4616.)

Aguilar, A. S., and A. van Diest. 1981. Rock phosphate mobilization induced by the alkaline uptake pattern of legumes utilizing symbiotically fixed nitrogen. *Plant Soil* 61:27–42.

Ahloowalia, B. S. 1977. Strategy of forage crop improvement. *Farm Food Res.* 8:120–122.
Aktas, M., and F. van Egmond. 1979. Effect of nitrate nutrition on iron utilization by an Fe-efficient and an Fe-inefficient soybean cultivar. *Plant Soil* 51:253–274.
Aldrich, D. T. A. 1970. Interaction between nitrogen fertiliser use and the productivity of white clover. In J. Lowe (ed.), White Clover Research. *Occas. Symp. Br. Grassl. Soc.* No. 6, pp. 227–232.
Alexander, I. J., and K. Hardy. 1981. Surface phosphatase activity of Sitka spruce mycorrhizas from a serpentine soil. *Soil Biol. Biochem.* 13:301–305.
Al-Shawk, A. M., R. L. Westerman, and D. L. Week. 1986. Iron chlorosis in soybeans. *J. Plant Nutr.* 9:355–371.
Ambler, J. E., and J. C. Brown. 1969. Cause of differential susceptibility to zinc deficiency in two varieties of navy beans (*Phaseolus vulgaris* L.). *Agron. J.* 61:41–43.
Ambler, J. E., J. C. Brown, and H. G. Gauch. 1971. Sites of iron reduction in soybean plants. *Agron. J.* 63:95–97.
Andrew, C. S. 1966. A kinetic study of phosphate absorption by excised roots of *Stylosanthes humilis, Phaseolus lathyroides, Desmodium uncinatum, Medicago sativa* and *Hordeum vulgare. Aust. J. Agric. Res.* 5:611–624.
Andrew, C. S., and R. K. Jones. 1978. The phosphorus nutrition of tropical forage legumes. In C. S. Andrew and E. J. Kamprath (eds.), Mineral Nutrition of Legumes in Tropical and Subtropical Soils. CSIRO, Australia. Pp. 295–311.
Aqil, B. A., and A. A. Boe. 1975. Occurrence of cotyledonal cracking in snapbeans and its relation to nutritional status in the seed. *HortScience* 10:509–510.
Asher, C. J., and P. G. Ozanne. 1961. The cation exchange capacity of plant roots and its relationship to uptake of insoluble nutrients. *Aust. J. Agric. Res.* 12:755–766.
Avendano, R. E., and R. L. Davis. 1966. Lateral root development in progenies of creeping and noncreeping-rooted *Medicago sativa* L. *Crop Sci.* 6:198–201.
Ayala Briceno, L. B. 1975. Biochemical evaluation of nitrogen fixation by some peanut rhizobia. *Diss. Abstr. Int. B* 36:1997B.
Azcón, R., C. Azcón-Aguilar, and J. M. Barea. 1978. Effects of plant hormones present in bacterial cultures on the formation and responses of vesicular arbuscular endomycorrhiza. *New Phytol.* 80:359–364.
Azcón-Aguilar, C., and J. M. Barea. 1978. Effects of interactions between different culture fractions of 'phosphobacteria' and *Rhizobium* on mycorrhizal infection, growth and nodulation of *Medicago sativa. Can. J. Microbiol.* 24:520–524.
Bagshaw, R., L. V. Vaidyanathan, and P. H. Nye. 1972. The supply of nutrient ions by diffusion to plant roots in soil. VI. Effects of onion plant roots on pH and P desorption characteristics in a sandy soil. *Plant Soil* 37:627–639.
Bajpai, P. D., and M. V. B. S. Rao. 1971. Phosphate-solubilizing bacteria. II. Extracellular production of organic acids by selected bacteria solubilizing insoluble phosphate. *Soil Sci. Plant Nutr.* 17:44–45.
Baldwin, J. G., K. R. Barker, and L. A. Nelson. 1979. Effects of *Meloidogyne incognita* on nitrogen fixation in soybean. *J. Nematol.* 11:156–161.
Banik, S., and B. K. Dey. 1982. Available phosphate content of an alluvial soil is influenced by inoculation of some isolated phosphate solubilizing microorganisms. *Plant Soil* 69:353–364.
Barber, S. A. 1976. Efficient fertilizer use. Agronomic research for food. *ASA Spec. Publ.* No. 26, pp. 13–29.
Barber, S. A. 1982. Soil–plant root relationships determining phosphorus uptake. In A. Scaife (ed.), Plant Nutrition 1982. Proc. 9th Int. Plant Nutrition Coll., Warwick University. Pp. 39–44.
Barber, S. A. 1984. Mechanisms of phosphorus uptake by plants and their application to plant selection for improved phosphorus uptake efficiency. *Phosphorus: Indispensable Element for Improved Agricultural Production. 3rd Int. Congr. Phosphorus Compounds, Brussels, 1983.* Imphos., Casablanca. Pp. 429–443.
Barea. J. M., R. Azcón, and C. Azcón-Aguilar. 1984. Interaction between phosphate solubilizing bacteria and vesicular-arbuscular mycorrhiza to improve plant utilization of rock phosphate in non-

acidic soils. *Phosphorus: Indispensable Element for Improved Agricultural Production. 3rd Int. Congr. Phosphorus Compounds, Brussels, 1983.* Imphos., Casablanca. Pp. 127–144.

Barea, J. M., E. Navarro, and E. Montoya. 1976. Production of plant growth regulators by rhizosphere phosphate-solubilizing bacteria. *J. Appl. Bacteriol.* 40:129–134.

Barrett-Lennard, E. G., A. D. Robson, and H. Greenway. 1982. Effect of phosphorus deficiency and water deficit on phosphatase activities from wheat leaves. *J. Exp. Bot.* 33:682–93.

Barrow, N. J. 1978. Problems of efficient fertiliser use. In A. R. Ferguson, R. L. Bieleski, and I. B. Ferguson (eds.), Plant Nutrition 1978. Proc. 8th Int. Colloq. Plant Analysis and Fertilizer Problems, Aucklund. *Inf. Ser.—N.Z. Dep. Sci. Ind. Res.* No. 134, pp. 37–52.

Bartlett, E. M., and D. H. Lewis. 1973. Surface phosphatase activity of mycorrhizal roots of beech. *Soil Biol. Biochem.* 5:249–257.

Bates, T. E. 1971. Factors affecting critical concentrations in plants and their evaluation: A review. *Soil Sci.* 112:116–131.

Beadle, N. C. W. 1954. Soil phosphate and the delimitation of plant communities in Eastern Australia. *Ecology* 35:370–375.

Beadle, N. C. W. 1966. Soil phosphate and its role in moulding segments of the Australian flora and vegetation, with special reference to xeromorphy and sclerophylly. *Ecology* 47:992–1007.

Bear, F. E. 1950. Cation and anion relationships in plants and their bearing on crop quality. *Agron. J.* 42:176–178.

Beeson, K. C. 1941. The mineral composition of crops with particular reference to the soils in which they are grown. *Misc. Publ—U.S. Dep. Agric.* No. 369.

Bello, A. B., W. A. Ceron-Diaz, C. D. Nickell, F. E. El-Sherif, and L. C. Davis. 1980. Influence of cultivar, between-row spacing and plant population on fixation by soybeans. *Crop Sci.* 20:751–755.

Beringer, H., and M. A. Taha. 1976. Calcium absorption by two cultivars of groundnut (*Arachis hypogea*). *Expt. Agric.* 12:1–7.

Bethlenfalvay, G. J., and D. A. Phillips. 1977. Photosynthesis and symbiotic nitrogen-fixation in *Phaseolus vulgaris* L. In A. Hollaender (ed.), Genetic Engineering for Nitrogen Fixation. *Basic Life Sci.* 9:401–409.

Biddulph, O., and C. G. Woodbridge. 1952. The uptake of phosphorus by bean plants with particular reference to the effect of iron. *Plant Physiol.* 27:431–444.

Bieleski, R. L. 1973. Phosphate pools, phosphate transport and phosphate availability. *Annu. Rev. Plant Physiol.* 24:225–252.

Bieleski, R. L., and P. N. Johnson. 1972. The external location of phosphatase activity in phosphorus deficient *Spirodela oligorrhiza. Aust. J. Biol. Sci.* 25:707–720.

Blair, G. J., and S. Cordero. 1978. The phosphorus efficiency of three annual legumes. *Plant Soil* 50:387–398.

Blasco Lamenca, M., D. R. Horqque, and J. de Cabanyes. 1981. Chemical characteristics of the seeds of *Lupinus mutabilis* harvested in Cuzco, Peru. *Turrialba* 31:258–260. (*Plant Breed. Abstr.* 52:6769.)

Blaylock, A. D., T. D. Davis, V. D. Jolley, and R. H. Walser. 1986. Influence of cobalt and iron on photosynthesis, chlorophyll and nutrient content in regreening chlorotic tomatoes and soybeans. *J. Plant Nutr.* 9:823–838.

Blazhev. V., A. Kristov, N. Blazheva, and A. Topchieva. 1984. Effect of *Fusarium* wilt on content of nitrogen, phosphorus and potassium in lucerne plants. *Fiziol. Rast.* 10:71–76. (*Plant Breed. Abstr.* 57:389.)

Boero, G., and S. Thien. 1979. Phosphatase activity and phosphorus availability in the rhizosphere of corn roots. In J. L. Harley and R. S. Russell (eds.), The Soil–Root Interface. Academic Press, London. Pp. 231–242.

Boller, B. C., and G. H. Heichel. 1984. Canopy structure and photosynthesis of alfalfa genotypes differing in nodule effectiveness. *Crop. Sci.* 24:91–96.

Bonish, P. M. 1980. Nodulation of white clover: Plant influences on the effectiveness of *Rhizobium trifolii. N.Z. J. Agric. Res.* 23:239–242.

Boote, K. J., T. R. Zweifel, W. K. Robertson, and R. N. Gallaher. 1978. Relationship of photosynthesis to leaf nitrogen during seed fill of eight soybean genotypes. *Agron. Abstr.* pp. 70–71.

Borkert, C. M., and S. A. Barber. 1983. Effect of supplying phosphorus to a portion of the soybean root system on root growth and phosphorus uptake kinetics. *J. Plant Nutr.* 6:895–910.

Borst, H. L., and L. E. Thatcher. 1931. Life history and composition of the soybean plant. *Ohio Agric. Exp. Stn., Bull.* No. 494.

Bouldin, D. R. 1961. Mathematical description of diffusion processes in the soil–plant system. *Soil Sci. Soc. Am. Proc.* 25:476–480.

Boulter, D., and I. M. Evans. 1976. Characterization of storage protein with special reference to cowpea. In R. A. Luse and K. O. Rachie (eds.), *Proc. IITA Collaborators Meet. Grain Legume Improvement. Seed Quality/Biochemistry, Int. Inst. Trop. Agric., Ibadan, Nigeria.* Pp. 115–117. (*Plant Breed. Abstr.* 47:7956.)

Bowen, G. D., D. I. Bevege, and B. Mosse. 1975. Phosphate physiology of vesicular-arbuscular mycorrhizas. In F. E. Sanders, B. Mosse, and P. B. Tinker (eds.), Endomycorrhizas. Academic Press, London. Pp. 241–260.

Boyd, A. G., and J. Frame. 1983. Response of white clover to various management factors. In A. J. Corrall (ed.), Efficient Grassland Farming: Proc. 9th Gen. Meet. European Grassland Federation, Reading, England, 1982. *Occas. Symp., Br. Grassl. Soc.* No. 14, pp. 213–216.

Bradshaw, A. D. 1969. An ecologist's viewpoint. In I. H. Rorison (ed.), Ecological Aspects of the Mineral Nutrition of Plants. Symp. Br. Ecol. Soc. No. 9, pp. 415–427.

Bravo, A., and D. H. Wallace. 1974. Differential performance of two red kidney bean varieties to spacing and nitrogen fertilization. *HortScience* 9:280. (Abstr.)

Brock, J. L., and J. H. Hoglund. 1974. Growth of Grasslands Huia and Grasslands 4700 white clovers. II. Effect of nitrogen and phosphorus. *N.Z. J. Agric. Res.* 17:47–54.

Bromfield, S. M., R. W. Cumming, D. T. David, and C. H. Williams. 1983. Change in soil pH, manganese and aluminum under subterranean clover pastures. *Aust. J. Exp. Agric. Anim. Husb.* 23:181–191.

Brown, J. C. 1978. Mechanism of ion uptake by plants. *Plant Cell Environ.* 1:249–257.

Brown, J. C., R. S. Holmes, and L. O. Tiffin. 1958. Iron chlorosis in soybeans as related to the genotype of rootstock. *Soil Sci.* 86:75–82.

Brown, J. C., R. S. Holmes, and L. O. Tiffin. 1961. Iron chlorosis in soybeans as related to the genotype of rootstock. 3. Chlorosis susceptibility and reductive capacity at the root. *Soil Sci.* 91:127–132.

Brown, J. C., C. R. Weber, and B. E. Caldwell. 1967. Efficient and inefficient use of iron by two soybean genotypes and their isolines. *Agron. J.* 59:459–462.

Brown, M. E. 1974. Seed and root bacterization. *Annu. Rev. Plant Phytopathol.* 12:181–197.

Burkhart, L., and E. R. Collins. 1942. Mineral nutrients in peanut plant growth. *Soil Sci. Soc. Am. Proc.* 6:272–280.

Burton, J. C. 1972. Nodulation and symbiotic nitrogen fixation. In C. H. Hanson (ed.), Alfalfa Science and Technology. *ASA Agron. Series* No. 15, pp. 229–246.

Burton, J. C., C. A. Brim, and J. O. Rankings. 1983. Performance of nonnodulating and nodulating soybean isolines in mixed culture with nodulating cultivars. *Crop Sci.* 23:469–473.

Caldwell, B. E., and G. Vest. 1970. Effects of *Rhizobium japonicum* strains on soybean yields. *Crop. Sci.* 10:19–21.

Camargo, C. E. de O. 1984. Effect of various concentrations of phosphorus in the nutrient solution and in the soil on the behaviour of wheat cultivars. *Bragartia* 43:63–86. (*Plant Breed. Abstr.* 56:1649.)

Caradus, J. R. 1977. Structural variation of white clover root systems. *N.Z. J. Agric. Res.* 20:213–219.

Caradus, J. R. 1979. Selection for root hair length in white clover (*Trifolium repens* L.). *Euphytica* 28:489–494.

Caradus, J. R. 1980. Distinguishing between grass and legume species for efficiency of phosphorus use. *N.Z. J. Agric. Res.* 23:75–81.

Caradus, J. R. 1981a. Root growth of white clover (*Trifolium repens* L.) lines in glass-fronted containers. *N.Z. J. Agric. Res.* 24:43–54.

Caradus, J. R. 1981b. Root morphology of some white clovers from New Zealand hill country. *N.Z. J. Agric. Res.* 24:349–351.

Caradus, J. R. 1981c. Effect of root hair length on white clover growth over a range of soil phosphorus levels. *N.Z. J. Agric. Res.* 24:353–358.
Caradus, J. R. 1982. Genetic differences in the length of root hairs in white clover and their effect on phosphorus uptake. In A. Scaife (ed.), Plant Nutrition 1982. Proc. 9th Int. Plant Nutrition Colloq., Warwick University. Pp. 84–88.
Caradus, J. R. 1983. Genetic differences in phosphorus absorption among white clover populations. *Plant Soil* 72:379–383.
Caradus, J. R. 1984. The phosphorus nutrition of populations of white clover (*Trifolium repens* L.). Ph.D. Thesis, Univ. of Reading.
Caradus, J. R. 1986. Variation in partitioning and percent nitrogen and phosphorus content of the leaf, stolon and root of white clover genotypes. *N.Z. J. Agric. Res.* 29:367–379.
Caradus, J. R., and J. Dunlop. 1978. Screening white clover plants for efficient phosphorus use. In A. R. Ferguson, R. L. Bieleski, and I. B. Ferguson (eds.), Plant Nutrition 1978. Proc. 8th Int. Colloq. Plant Analysis and Fertilizer Problems, Auckland. *Inf. Ser.—N.Z. Dep. Sci. Ind. Res.* No. 134, pp. 75–82.
Caradus, J. R., and R. W. Snaydon. 1986a. Plant factors influencing phosphorus uptake by white clover from solution culture. I. Population differences. *Plant Soil* 93:153–163.
Caradus, J. R., and R. W. Snaydon. 1986b. Plant factors influencing phosphorus uptake by white clover from solution culture. II. Root and shoot pruning and split-root studies. *Plant Soil* 93:165–174.
Caradus, J. R., and R. W. Snaydon. 1986c. Plant factors influencing phosphorus uptake by white clover from solution culture. III. Reciprocal grafting. *Plant Soil* 93:175–181.
Caradus, J. R., and R. W. Snaydon. 1986d. Response to phosphorus of populations of white clover. 1. Field studies. *N.Z. J. Agric. Res.* 29:155–162.
Caradus, J. R., and R. W. Snaydon. 1986e. Response to phosphorus of populations of white clover. 2. Glasshouse and growth cabinet studies. *N.Z. J. Agric. Res.* 29:163–168.
Caradus, J. R., and R. W. Snaydon. 1987a. Aspects of the phosphorus nutrition of white clover populations. I. Inorganic phosphorus content of leaf tissue. *J. Plant Nutri.* 10:273–286.
Caradus, J. R., and R. W. Snaydon, 1987b. Aspects of the phosphorus nutrition of white clover populations. II. Root exocellular acid phosphate activity. *J. Plant Nutr.* 10:287–302.
Caradus, J. R., and R. W. Snaydon. 1987c. Aspects of the phosphorus nutrition of white clover populations. III. Cation exchange capacity. *J. Plant Nutr.* 10:303–318.
Caradus, J. R., and R. W. Snaydon. 1988a. Aspects of the phosphorus nutrition of white clover populations. IV. Root growth and morphology. *J. Plant Nutr.* 11:277–287.
Caradus, J. R., and R. W. Snaydon 1988b. Aspects of the phosphorus nutrition of white clover populations. V. Partitioning of dry matter. *J. Plant Nutr.* 11:289–302.
Caradus, J. R., and W. M. Williams. 1981. Breeding for improved white clover production in New Zealand hill country. In C. E. Wright (ed.), Plant Physiology and Herbage Production. *Occas. Symp., Br. Grassl S.* No. 13, pp. 163–168. University of Nottingham.
Chapin, F. S. 1980. The mineral nutrition of wild plants. *Annu. Rev. Ecol. Syst.* 11:233–260.
Chapin, F. S. 1982. Patterns of phosphorus absorption and chemistry as adaptations to infertile soils. In A. Scaife (ed.), Plant Nutrition 1982. Proc. 9th Int. Plant Nutrition Colloq., University of Warwick, Warwick, England. Pp. 95–100.
Chatterton, N. J., S. Akao, G. E. Carlson, and W. E. Hungerford. 1977. Physiological components of yield and tolerance to frequent harvests in alfalfa. *Crop Sci.* 17:918–923.
Chloupek, O. 1980. Size of the root system and productivity of lucerne. *Z. Acker- Pflanzenbau* 149:107–116.
Clark, R. B. 1976. Plant efficiencies in the use of Ca, Mg and Mo. In M. J. Wright (ed.), Plant Adaptation to Mineral Stress in Problem Soils. Cornell University Press, Ithaca, New York. Pp. 175–191.
Clarkson, D. T., and J. B. Hanson. 1980. The mineral nutrition of higher plants. *Annu. Rev. Plant Physiol.* 31:239–298.
Coltman, R. R., G. C. Gerloff, and W. H. Gabelman. 1985. Differential tolerance of tomato strains to maintained and deficient levels of phosphorus. *J. Am. Soc. Hortic. Sci.* 110:140–144.

Connolly, V. 1971. Breeding better white clover varieties. *Farm Food Res.* 2:68–69.
Cornet, D., M. Meulemans, and E. Otoul. 1980. A promising nitrogen-fixing association: The symbiosis between lime bean (*phaseolus lunatus* L.) and *Rhizobium*. *Bull. Rech. Agron Gembloux* 15:17–29. (*Plant Breed. Abstr.* 51:8582.)
Coyne, D. P., S. S. Korban, D. Knudson, and R. B. Clark. 1982. Inheritance of iron deficiency in crosses of dry beans (*Phaseolus vulgaris* L.). *J. Plant Nutr.* 5:575–585.
Crooke, W. M., and A. H. Knight. 1962. An evaluation of published data on the mineral composition of plants in the light of the cation exchange capacities of their roots. *Soil Sci.* 93:365–373.
Crooke, W. M., and A. H. Knight. 1971a. Crop composition in relation to soil pH and root cation exchange capacity. *J. Sci. Food Agric.* 22:235–241.
Crooke, W. M., and A. H. Knight. 1971b. Root cation exchange capacity and organic acid content of tops as indices of varietal yield. *J. Sci. Food Agric.* 22:389–392.
Crooke, W. M., A. H. Knight, and J. Keay. 1964. Mineral composition, cation exchange properties and uronic acid content of various tissues of conifers. *For. Sci.* 10:415–427.
Crush, J. 1978. Changes in effectiveness of soil endomycorrhizal fungal populations during pasture development. *N.Z. J. Agric. Res.* 21:683–686.
Crush, J. R., and J. R. Caradus. 1980. Effect of mycorrhizas on growth of some white clovers. *N.Z. J. Agric. Res.* 23:233–237.
Cunningham, R. K., and K. F. Nielsen. 1963. Evidence against relationships between root cation exchange capacity and cation uptake by plants. *Nature (London)* 200:1344–1345.
Daday, H. V., M. Lawrence, R. I. Forrester, M. I. Whitecross, and J. V. Possingham. 1987. Nuclear DNA regulates the level of ribulose 1,5-bisphosphate carboxylase oxygenase in *Medicago sativa* L. *Theor. Appl. Genet.* 73:856–862.
Dambroth, M., and N. El Bassam. 1982. Low input varieties: Definition, ecological requirements and selection. In M. R. Sarić (ed.), *Proc. 1st Int. Symp. Genetic Specificity of Mineral Nutrition of Plants, Belgrade*. Pp. 325–336.
Danso, S. K. A., C. Hera, and C. Douka. 1987. Nitrogen fixation in soybean as influenced by cultivar and *Rhizobium* strain. *Plant Soil* 99:163–174.
Date, R. A. 1976. Principles of *Rhizobium* strain selection. In P. S. Nutman (ed.), Symbiotic Nitrogen Fixation in Plants. Cambridge Univ. Press, London. Pp 137–150.
Date, R. A., R. L. Burt, and W. T. Williams, 1979. Affinities between various *Stylosanthes* species as shown by rhizobial, soil pH and geographic relationships. *Agro-Ecosystems* 5:57–67.
Date, R. A., and D. O. Norris. 1979. *Rhizobium* screening of *Stylosanthes* species for effectiveness in nitrogen fixation. *Aust. J. Agric. Res.* 30:85–104.
Davies, H. 1978. Variety and nitrogen interaction in white clover. The way ahead with white clover. *NIAB Fellows Conf. Rep.* No. 4, pp. 4–6.
Davies, W. E. 1979. Investigations of *in vitro* digestibility in lucerne. *Biul. Inst. Hodowli Aklim. Rosl.* No. 135, pp. 138–150. (*Plant Breed. Abstr.* 51:7280.)
Davis, M. R. 1981. Growth and nutrition of legumes on a high-country yellow-brown earth subsoil. I. Phosphate response of *Lotus*, *Trifolium*, *Lupinus*, *Astralagus* and *Coronilla* species and cultivars. *N.Z. J. Agric. Res.* 24:321–332.
Davis, T. D., V. D. Jolley, R. H. Walser, J. C. Brown, and A. D. Blaylock. 1986. Net photosynthesis of Fe-efficient and Fe-inefficient soybean cultivars grown under varying iron levels. *J. Plant Nutr.* 9:671–681.
Denehy, H. L., and J. Morrison. 1979. The seasonal production of ryegrass/white clover swards given fertilizer nitrogen in spring. *Grass Forage Sci.* 34:66–67.
Department of Agriculture, Northern Ireland. 1977. Agronomy Section Annual Report on Research and Technical Work, pp. 91–95.
Dhillon, G. S., P. E. Smith, and P. R. Henderlong. 1975. Effect of dates of sowing, varieties and seed size of soybean on the chemical composition of the seed produced. *Indian J. Ecol.* 2:172–179. (*Plant Breed. Abstr.* 48:1816.)
Dickinson, D. B., V. Raboy, and F. E. Below. 1983. Variation in seed phosphorus, phytic acid, Zn, Ca, Mg and protein among lines of *Glycine max* (soybean) and *Glycine soja*. *Agron Abstr.* p. 61.

Dickson, M. H., and L. R. Hackler. 1975. Protein quantity and quality in high yielding beans. In M. Milner (ed.), Nutritional Improvement in Food Legumes by Breeding. III: Nutritionally Related Factors in Legumes Requiring Genetic Improvement. Wiley, New York. Pp. 185–192.
Dixon, R. O. D. 1969. Rhizobia (with particular reference to relationships with host plant). *Annu. Rev. Microbiol.* 23:137–158.
Dolgopolova, L. N. 1976. Nitrogen exchange in fodder peas in relation to drought resistance. *Nauchn. Tr. Vses. Nauchno-Issled. Inst. Zernobobovykh Kul't.* 5:124–134. (*Plant Breed. Abstr.* 49:842.)
Dolgopolova, L. N., A. P. Lakhanov, and R. F. Chernen'kaya. 1979. Amino acid composition of the protein in pea leaves in relation to resistance to soil drought. *Fiziol. Biokhim. Kul't. Rast.* 11:158–162. (*Plant Breed. Abstr.* 49:8599.)
Dracup, M. N. H., E. G. Barrett-Lennard, H. Greenway, and A. D. Robson. 1984. Effect of phosphorus deficiency on phosphatase activity of cell walls from roots of subterranean clover. *J. Expt. Bot.* 35:466–480.
Drake, M., and J. E. Steckel. 1955. Solubilization of soil and rock phosphate as related to root cation exchange capacity. *Am. Soc. Soil Sci. Proc.* 19:449–450.
Drake, M., J. Vengris, and W. G. Colby. 1951. Cation exchange capacity of plant roots. *Soil Sci.* 72:139–147.
Drew, M. C., and P. H. Nye. 1969. The supply of nutrient ions by diffusion to plant roots in soil. II. The effect of root hairs on the uptake of potassium by roots of ryegrass (*Lolium multiflorum*). *Plant Soil* 31:407–424.
Duczmal, K. W. 1976. Uptake and translocation of nutrients and transverse cracking of bean cotyledons. *Acta Soc. Bot. Pol.* 45:401–410.
Dughri, M. H., and P. J. Bottomley. 1984. Soil acidity and the composition of an indigenous population of *Rhizobium trifolii* in nodules of different cultivars of *Trifolium subterraneum* L. *Soil Biol. Biochem.* 16:405–411.
Duhigg, P. M., B. A. Melton, and A. A. Baltensperger. 1978. Selection for acetylene reduction rates in Mesilla alfalfa. *Crop Sci.* 18:813–816.
Duke, S. H., and D. C. Dochlert. 1981. Root respiration, nodulation and enzyme activities in alfalfa during cold acclimation. *Crop Sci.* 21:489–495.
Dunham, C. W., C. L. Hamner, and S. Asen. 1956. Cation exchange properties of the roots of some ornamental plant species. *Am. Soc. Hortic. Sci. Proc.* 68:556–563.
Edwards, D. G. 1968. Cation effects on phosphate absorption from solution by *Trifolium subterraneum*. *Aust. J. Biol. Sci.* 21:1–11.
Edwards, J. H., and S. A. Barber. 1976. Phosphorus uptake rate of soybean roots as influenced by plant age, root trimming and solution phosphorus concentration. *Agron. J.* 68:973–975.
Egli, D. B., J. C. Swank, and T. W. Pfeiffer. 1987. Mobilization of leaf N in soybean genotypes within varying durations of seedfill. *Field Crops Res.* 15:251–258.
El-Fouly, M. M., and J. Jung. 1972. Activity of acid phosphatase (E. C. 3.1.3.2) in wheat seedlings associated with soil stress. *Plant Soil* 36:497–500.
Elgabaly, M. M. 1962. On the mechanism of anion uptake by plant roots. 2. Effect of the cation exchange capacity of plant roots on chloride uptake. *Soil Sci.* 93:350–352.
Elgabaly, M. M., and L. Wiklander. 1949. Effect of exchange capacity of clay minerals and acidoid content of plant on uptake of calcium by excised barley and pea roots. *Soil Sci.* 67:419–424.
Elkan, G. H., J. C. Wynne, and T. J. Schneeweis, 1981. Isolation and evaluation of strains of *Rhizobium* collected from centres of diversity in South America. *Trop. Agric.* 58:297–305.
Elliott, G. C., and A. Läuchli. 1985. Phosphorus efficiency and phosphate—iron interaction in maize. *Agron. J.* 77:399–403.
Eltilib, A. M. A. 1982. Effects of nitrogen fertilizer on the yield and composition of grass—white clover mixtures. Thesis. Univ. of Newcastle-upon-Tyne, U.K. (*Herb. Abstr.* 54:2542.)
Ennos, R. A. 1985. The significance of genetic variation for root growth within a natural population of white clover (*Trifolium repens* L.). *J. Ecol.* 73:615–624.
Erdmann, L. W., and U. M. Means. 1953. Strain variation of *Rhizobium meliloti* on three varieties of *Medicago sativa*. *Agron. J.* 45:625–629.

Estaún, V., C. Calvet, and D. S. Hayman. 1987. Influence of plant genotype on mycorrhizal infection: Response of three pea cultivars. *Plant Soil* 103:295–298.

Foote, B. D., and R. W. Howell. 1964. Phosphorus tolerance and sensitivity in soybean as related to uptake and translocation. *Plant Physiol.* 39:610–613.

Fox, R. L., and B. Kacar. 1964. Phosphorus mobilization in a calcareous soil in relation to surface properties of roots and cation uptake. *Plant Soil* 20:319–330.

Franklin, R. E. 1970. Effect of adsorbed cations on phosphorus absorption by various plant species. *Agron. J.* 62:214–216.

Frederick, L. R. 1978. Effectiveness of *Rhizobia*–legume associations. In C. S. Andrew and E. J. Kamprath (eds.), Mineral Nutrition of Legumes in Tropical and Subtropical Soils. CSIRO, Melbourne. Pp. 265–276.

Gabelman, W. H. 1976. Genetic potentials in nitrogen, phosphorus, and potassium efficiency. In M. J. Wright (ed.), Plant Adaptation to Mineral Stress in Problem Soils. Cornell Univ. Press. Ithaca, New York. Pp. 205–212.

Gabelman, W. H., and G. C. Gerloff. 1983. The search for an interpretation of genetic controls that enhance plant growth under deficiency levels of a macronutrient. *Plant Soil* 72:335–350.

Gangopadhyay, S., and T. D. Wyllie. 1973. Biochemical comparison of nine soybean varieties in relation to their susceptibility to *Macrophomina phaseolina*. *Indian J. Mycol. Plant Pathol.* 3:131–140. (*Plant Breed. Abstr.* 46:985.)

Garcia-Luis, A., and J. L. Guardiola. 1975. Effects of gibberellic acid on the transport of nitrogen from the cotyledons of young pea seedlings. *Ann. Bot.* 39:325–330.

Gardner, W. K., D. A. Barber, and D. G. Parbery. 1983. The acquisition of phosphorus by *Lupinus albus* L. III. The probable mechanism by which phosphorus movement in the soil/root interface is enhanced. *Plant Soil* 70:107–124.

Gardner, W. K., D. G. Parbery, and D. A. Barber. 1981. Proteoid root morphology and function in *Lupinus albus*. *Plant Soil* 60:143–147.

Gardner, W. K., D. G. Parbery, and D. A. Barber. 1982a. The acquisition of phosphorus by *Lupinus albus* L. I. Some characteristics of the soil/root interface. *Plant Soil* 68:19–32.

Gardner, W. K., D. G. Parbery, and D. A. Barber. 1982b. The acquisition of phosphorus by *Lupinus albus* L. II. The effect of varying P supply and soil type on some characteristics of the soil/root interface. *Plant Soil* 68:33–41.

Gay, S., D. B. Egli, and D. A. Reicosky. 1980. Physiological aspects of yield improvement. *Agron. J.* 72:387–391.

Geisler, G., and B. Krützfeldt. 1983. Investigations into the effect of 'nitrogen' on the morphology, dry matter formation and nutrient uptake efficiency of the root systems of maize, spring barley and field bean varieties, having regard to temperature conditions. I. Root morphology. *Z. Acker- Pflanzenbau* 152:336–353.

Gerretsen, F. C. 1948. The influence of microorganisms on the phosphate intake by the plant. *Plant Soil* 1:51–81.

Gibson, A. H. 1962. Genetic variation in the effectiveness of nodulation of lucerne varieties. *Aust. J. Agric. Res.* 13:388–399.

Godwin, D. C., and E. J. Wilson. 1976. Prospects of selecting plants with increased phosphorus efficiency. In G. Blair (ed.), Prospects for improving efficiency of phosphorus utilization. *Rev. Rural Sci.* 3:131–139.

Gomes, M. A. F., and L. Sodek. 1987. Reproduction development and nitrogen fixation in soybean (*Glycine max* (L.) Merril). *J. Exp. Bot.* 38:1982–1987.

Goodman, P. J., and M. Collison. 1982. Varietal differences in uptake and ^{32}P labelled phosphate in clover plus ryegrass swards and monocultures. *Ann. Appl. Biol.* 100:559–565.

Goodman, P. J., and J. Edwards. 1983. Mineral nutrition of leafy legumes. In D. G. Jones and D. R. Davies (eds.), Temperate Legumes: Physiology, Genetics and Nodulation. Series in Applied Biology. Pitman, London. Pp 103–118.

Graham, P. H. 1979. Influence of temperature on growth and nitrogen fixation in cultivars of *Phaseolus vulgaris* L; inoculated with *Rhizobium*. *J. Agric. Sci.* 93:365–370.

Graham, P. H., and J. Halliday. 1977. Inoculation and nitrogen fixation in the genus *Phaseolus*. In J. M.

Vincent, A. S. Whitney, and J. Bose (eds.), Exploiting the Legume–*Rhizobium* Symbiosis in Tropical Agriculture. *Coll. Trop. Agric., Misc. Publ. (Univ. Hawaii)* No. 145.

Graham, P. H., and J. C. Rosas. 1977. Growth and development of indeterminate bush and climbing cultivars of *Phaseolus vulgaris* L. inoculated with *Rhizobium*. *J. Agric. Sci.* 88:503–508.

Graham, P. H., and J. C. Rosas. 1978. Nodule development and nitrogen fixation in cultivars of *Phaseolus vulgaris* L. as influenced by planting density. *J. Agric. Sci.* 90:19–29.

Graham, P. H., and J. C. Rosas. 1979. Phosphorus fertilization and symbiotic nitrogen fixation in common bean. *Agron. J.* 71:925–926.

Green, S. O., and D. W. Cowling. 1961. The nitrogen nutrition of grassland. In C. L. Skidmore (ed.), *Proc. 8th Int. Grassland Congr., University of Reading, 1960*. Pp. 126–129.

Griffith, W. K. 1974. Satisfying the nutritional requirements of established legumes. In D. A. Mays (ed.), Forage Fertilization. ASA, CSSA, and SSSA, Madison, Wisconsin. Pp. 147–169.

Grime, J. P., and R. Hunt. 1975. Relative growth rate: Its range and adaptive significance in a local flora. *J. Ecol.* 63:393–422.

Grindsted, M. J., M. J. Hedley, R. E. White, and P. H. Nye. 1982. Plant induced changes in the rhizosphere of rape (*Brassica napus* var. Emerald) seedlings. I. pH change and the increase in phosphorus concentration in the soil solution. *New Phytol.* 91:19–29.

Gritsenko, V. V., and I. S. Yashin. 1978. Reaction of pea varieties to three-layer ploughing and fertilizer treatment. *Izv. Timiryazevsk. Skh. Akad.* 6:26–37. (*Plant Breed. Abstr.* 49:7726.)

Haag, W. L., M. W. Adams, and J. V. Wiersma. 1978. Differential responses of dry bean genotypes to nitrogen and phosphorus fertilization of a central American soil. *Agron. J.* 70:565–568.

Hall, I. R., R. S. Scott, and P. D. Johnstone. 1977. Effect of vesicular arbuscular mycorrhizas on the response of 'Grasslands Huia' and 'Tamar' white clovers to phosphorus. *N.Z. J. Agric. Res.* 20:349–355.

Hallock, D. L., D. C. Martens, and M. W. Alexander. 1971. Distribution of P, K, Ca, Mg, B, Cu, Mn and Zn in peanut lines near maturity. *Agron. J.* 63:251–256.

Hallsworth, E. G. (ed.). 1958. Nutrition of the Legumes. Proc. University of Nottingham 5th Eastern School in Agricultural Science. Butterworth, London.

Ham, G. E. 1980. Interactions of *Glycine max* and *Rhizobium japonicum*. In R. J. Summerfield and A. H. Bunting (eds.), Advances in Legume Science. Royal Botanic Gardens, Kew. Pp. 289–296.

Ham, G. E., R. J. Lawn, and W. A. Brun. 1976. Influence of inoculation, nitrogen fertilizers and photosynthetic source-sink manipulations on field-grown soybeans. In P. S. Nutman (ed.), Symbiotic Nitrogen-Fixation in Plants. Cambridge Univ. Press. London. Pp. 239–253.

Hanker, I., and A. Kudelová. 1979a. Changes in growth and in uptake, distribution and translocation of phosphorus in susceptible and resistant alfalfa plants induced by *Corynebacterium insidiosum*. *Biol. Plant.* 21:136–143.

Hanker, I., and A. Kudelová. 1979b. Changes in phosphorus metabolism in alfalfa plants induced by bacterial wilt. *Biol. Plant.* 21:144–148.

Hanker, I., and A. Kudelová. 1983. The uptake, distribution, leakage, and incorporation of ^{32}P into organic compounds in alfalfa plants susceptible and resistant to the bacterial wilt and the effect of *Corynebacterium insidiosum* upon these processes. *Biol. Plant.* 25:279–287.

Hardy, R. W. F. 1977. Rate-limiting steps in biological photoproductivity. In A. Hollaender (ed.), Genetic Engineering for Nitrogen Fixation. *Basic Life Sci.* 9:369–399.

Hardy, R. W. F., R. C. Burns, R. R. Herbert, R. D. Holsten, and E. K. Jackson. 1971. Biological nitrogen fixation: A key to world protein. In T. A. Lie and E. G. Mulder (eds.), Biological Nitrogen Fixation in Natural and Agricultural Habitats. *Plant Soil, Spec. Vol.* pp. 561–590.

Hardy, R. W. F., and U. D. Havelka. 1976. Photosynthate as a major factor limiting nitrogen fixation by field grown legumes with emphasis on soybeans. In P. S. Nutman (ed.), Symbiotic Nitrogen Fixation in Plants. Cambridge Univ. Press. London. Pp. 421–439.

Harper, J. E. 1979. Nitrogen and dry matter partitioning between seed and storer of soybeans. *Agron. Abstr.* p. 102.

Harris, W., I. Rhodes, and S. S. Mee. 1983. Observations on environment and genotypic influences on the overwintering of white clover. *J. Appl. Ecol.* 20:609–624.

Harrison, S. P., J. P. W. Young, and D. G. Jones. 1987. *Rhizobium* population genetics: Effect of clover variety and inoculum dilution on the genetic diversity sampled from natural populations. *Plant Soil* 103:147–150.

Hart, A. L., and D. Jessop. 1982. Concentration of total, inorganic and lipid phosphorus in leaves of white clover and stylosanthes. *N.Z. J. Agric. Res.* 25:69–76.

Hart, A. L., and D. Jessop. 1983. Phosphorus fractions in trifoliate leaves of white clover and lotus at various levels of phosphorus supply. *N.Z. J. Agric. Res.* 26:357–361.

Hart, A. L., D. Jessop, and J. Galpin. 1981. The response to phosphorus of white clover and lotus inoculated with rhizobia or given potassium nitrate. *N.Z. J. Agric. Res.* 24:27–32.

Hayman, D. S. 1975. Phosphorus cycling by soil micro-organisms and plant roots. In N. Walker (ed.), Soil Microbiology. Butterworth, London. Pp. 67–92.

Haynes, R. J. 1980. Ion exchange properties of roots and ionic interactions within the root apoplasm: Their role in ion accumulation by plants. *Bot. Rev.* 46:75–99.

Haynes, R. J. 1983. Soil acidification induced by leguminous crops. *Grass Forage Sci.* 38:1–11.

Hedley, M. J., R. E. White, and P. H. Nye. 1982b. Plant induced changes in the rhizosphere of rape (*Brassica napus* var. Emerald) seedlings. III. Changes in L value, soil phosphate fractions and phosphate activity. *New Phytol.* 91:45–56.

Hedley, M. J., P. H. Nye, and R. E. White 1983. Plant induced changes in the rhizosphere of rape (*Brassica napus* var. Emerald) seedlings. IV. The effect of rhizosphere phosphorus status on the pH, phosphatase activity and depletion of soil-phosphorus positions in the rhizosphere and on the cation-anion tolerance in the plants. *New Phytol.* 95:69–82.

Hedley, M. J., P. H. Nye, and R. E. White. 1982a. Plant induced changes in the rhizosphere of rape (*Brassica napus* var. Emerald) seedlings. II. Origin of the pH change. *New Phytol.* 91:31–44.

Heichel, C. H., and C. P. Vance. 1979. Nitrate-nitrogen and *Rhizobium* strain roles in alfalfa seedling nodulation and growth. *Crop Sci.* 19:512–518.

Heinrichs, D. H., J. E. Torelsen, and F. G. Warder. 1969. Variation of chemical constituents and morphological characters within and between alfalfa population. *Can. J. Plant Sci.* 49:293–305.

Heintze, S. G. 1961. Studies on cation exchange capacity of roots. *Plant Soil* 13:365–383.

Helyar, K. R. 1976. Nitrogen cycling and soil acidification. *J. Aust. Inst. Agric. Sci.* 42:217–212.

Hill, R. R. 1981. Selection for phosphorus and lignin content in alfalfa. *Agric. Rev. Man. (U.S. Dep. Agric., Sci. Educ. Adm.)* ARM-NC-19:56. (*Plant Breed. Abstr.* 52:4007.)

Hill, R. R., and R. F. Barnes. 1977. Genetic variability for chemical composition of alfalfa. II. Yield and traits associated with digestibility. *Crop Sci.* 17:948–952.

Hill, R. R., and G. A. Jung. 1975. Genetic variability for chemical composition of alfalfa. I. Mineral elements. *Crop Sci.* 15:652–657.

Hill, R. R., and L. E. Lanyon. 1983. Phosphorus fertilizer response in experimental alfalfas selected for different phosphorus concentrations. *Crop Sci.* 23:973–976.

Ho, I., and B. Zak. 1979. Acid phosphate activity of six ectomycorrhizal fungi. *Can. J. Bot.* 57:1203–1205.

Hobbs, S. L. A., and J. D. Mahon. 1981. Genetic variability in physiological characteristics of *Pisum sativum* L. *Plant Physiol.* 67:113. (Abstr.)

Hobbs, S. L. A., and J. D. Mahon. 1982. Heritability of N_2 (C_2H_2) fixation rates and related characters in peas (*Pisum sativum* L.). *Can. J. Plant Sci.* 62:265–276.

Hoen, K. 1970. Performance in mixed swards of three perennial ryegrass and three white clover varieties at two nitrogen levels. *Ir. J. Agric. Res.* 9:215–223.

Hoffman, D., and B. Melton. 1981. Variation among alfalfa cultivars for indices of nitrogen fixation. *Crop Sci.* 21:8–10.

Hoglund, J. H., and J. L. Brock. 1974. Growth of 'Grasslands Huia' and 'Grasslands 4700' white clovers. I. Effects of temperature and nitrogen. *N.Z. J. Agric. Res.* 17:41–45.

Hoglund, J. H., and J. L. Brock. 1983. Effects of defoliation frequency and nitrogen and phosphorus nutrition on performance of four white clover cultivars. *N.Z. J. Agric. Res.* 26:109–114.

Hoglund, J. H., and W. M. Williams. 1984. Genotypic variation in white clover growth and branching in response to temperature and nitrogen. *N.Z. J. Agric. Res.* 27:19–24.

Horovitz, C. T. 1971. Exocellular acid phosphatase of wheat roots. *Isr. J. Bot.* 20:41–43.

Howell, R. W. 1964. Influence of nutrient balance on response of sensitive soybean varieties to high phosphorus. *Agron. J.* 56:233–234.

Huffaker, R. C., and A. Wallace. 1958. Possible relationships of cation exchange capacity of plant roots to cation uptake. *Soil Sci. Soc. Am. Proc.* 22:392–394.

Hungaria, M., and M. C. P. Neves. 1987a. Cultivar and *Rhizobium* strain effect on nitrogen fixation and transport in *Phaseolus vulgaris* L. *Plant Soil* 103:111–121.

Hungaria, M., and M. C. P. Neves. 1987b. Partitioning of nitrogen from biological fixation and fertilizer in *Phaseolus vulgaris*. *Physiol. Plant.* 69:55–63.

Hurley, A. K., V. D. Jolley, J. C. Brown, and R. H. Walser. 1986. Response of dry beans to iron deficiency stress. *J. Plant Nutr.* 9:805–814.

Ingram, J., and D. T. A. Aldrich. 1981. Herbage production from white clover varieties grown in mixtures with grass at two nitrogen levels. In C. E. Wright (ed.), Plant Physiology and Herbage Production. University of Nottingham. *Occas. Symp., Br. Grassl. Soc.* No. 13, pp. 173–177.

Inskeep, W. P., and P. R. Bloom. 1984. A comparative study of soil solution chemistry associated with chlorotic and nonchlorotic soybeans in western Minnesota. *J. Plant Nutr.* 7:513–531.

Israel, D. W. 1981. Cultivar and *Rhizobium* strain effects on nitrogen fixation and remobilisation by soybeans. *Agron. J.* 73:509–516.

Israel, D. W. 1987. Investigation of the role of phosphorus in symbiotic dinitrogen fixation. *Plant Physiol.* 84:835–840.

Israel, D. W., and W. A. Jackson. 1978. The influence of nitrogen nutrition on ion uptake and translocation by leguminous plants. In C. S. Andrew and E. G. Kamprath (eds.), Mineral Nutrition of Legumes in Tropical and Subtropical Soils. CSIRO, Melbourne. Pp. 113–130.

Itoh, S., and S. A. Barber. 1983a. Phosphorus uptake by six plant species as related to root hairs. *Agron. J.* 75:457–461.

Itoh, S., and S. A. Barber. 1983b. A numerical solution of whole plant nutrient uptake for soil–root systems with root hairs. *Plant Soil* 70:403–414.

Ive, J. R., and M. J. Fisher. 1974. Performance of Townsville stylo (*Stylosanthes humilis*) lines in pure swards and with the annual grass (*Digitaria ciliasis*) under various defoliation treatments at Katherine, *N.Z. Aust. J. Exp. Agric. Anim. Husb.* 14:495–500.

Ivory, D. A. 1976. Agronomic variation in *Medicago scutellata* and *M. orbicularis* in south-eastern Queensland. *Trop. Grassl.* 10:165–173.

Jaiswal, S. P., G. Kaur, J. C. Kumar, K. S. Nandpuri, and J. C. Thackur. 1975. Chemical constituents of green pea and their relationship with some plant characters. *Indian J. Agric. Sci.* 45:47–52. (*Plant Breed. Abstr.* 47:10036.)

Jakobsen, I. 1985. The role of phosphorus in nitrogen fixation by young pea plants (*Pisum sativum*). *Physiol. Plant.* 64:190–196.

Jarvis, S. C., and D. J. Hatch. 1985. Rates of hydrogen ion efflux by nodulating legumes grown in flowing solution culture with continuous pH monitoring and adjustment. *Ann. Bot.* 55:41–51.

Jarvis, S. C., and A. D. Robson. 1983a. The effects of nitrogen nutrition of plants on the development of acidity in Western Australian soils. I. Effects of subterranean clover grown under leaching conditions. *Aust. J. Agric. Res.* 34:341–354.

Jarvis, S. C., and A. D. Robson. 1983b. The effects of nitrogen nutrition of plants on the development of acidity in Western Australia. II. Effects of differences in cation/anion balance between plant species grown under non-leaching conditions. *Aust. J. Agric. Res.* 34:355–366.

Jarvis, S. C., and A. D. Robson. 1983c. A comparison of the cation/anion balance of the ten cultivars of *Trifolium subterraneum* L. and their effects on soil acidity. *Plant Soil* 75:235–243.

Jeffery, D. W. 1967. Phosphate nutrition of Australian heath plants. I. The importance of proteoid roots in *Banksia* (Proteaceae). *Aust. J. Bot.* 15:403–411.

Jeppson, R. G., R. R. Jackson, and H. H. Hadley. 1978. Variation in mobilisation of plant nitrogen to the grain in nodulating and non-nodulating soybean genotypes. *Crop. Sci.* 18:1058–1062.

Johnson, D. R., and W. V. Campbell. 1982. Variation in the foliage nutrients of several peanut lines and their association with damage received by the two spotted spider. *J. Ga. Entomol. Soc.* 17:69–72. (*Plant Breed. Abstr.* 53:803.)

Johnson, H. W., and U. M. Means. 1960. Interactions between genotypes of soybeans and genotypes of nodulating bacteria. *Agron. J.* 52:651–654.

Johnston, W. B., and R. A. Olsen. 1972. Dissolution of fluorapatite by plant roots. *Soil Sci.* 114:29–36.

Jolley, V. D., J. C. Brown, T. D. Davis, and R. H. Walser. 1986a. Increased Fe-efficiency in soybeans through plant breeding related to increased response to Fe-deficiency stress. I. Iron stress response. *J. Plant Nutr.* 9:373–386.

Jolley, V. D., J. C. Brown, T. D. Davis, and R. H. Walser. 1986b. Increased Fe-efficiency in soybeans through plant breeding related to increased response to Fe-deficiency stress. II. Mineral nutrition. *J. Plant Nutr.* 9:387–396.

Jones, D. G., and G. Hardarson. 1979. Variation within and between white clover varieties in their preference for strains of *Rhizobium trifolii*. *Ann. Appl. Biol.* 92:221–228.

Kalaidzhieva, S. 1983. Effect of biological and mineral nitrogen on yield and protein and oil content of the seed in soybean. *Nauchni Tr. Vissh Selskostop. Inst. "Vasil Kolarov"* 28:91–104. (*Plant Breed. Abstr.* 56:11232.)

Kannan, S. 1981. Regulation of iron stress response in some crop varieties: Anomaly of a mechanism for recovering through non-redumptive pH reaction. *J. Plant Nutr.* 4:1–19.

Kannan, S. 1982. Genotypic differences in iron uptake and utilization in some crop plants. *J. Plant Nutr.* 5:531–542.

Kannan, S. 1983. Iron deficiency stress tolerance and pH reduction: Some contrasts in the cultivars of sesame and lentil. *J. Plant Nutr.* 6:1025–1031.

Kansal, B. D., D. R. Bhumdla, and J. S. Kanwar. 1974. Variations in fertilizer response of different varieties of wheat and rice. *Indian J. Agric. Sci.* 44:55–59.

Karunaratne, R. S., J. H. Baker, and A. V. Barker, 1986. Phosphorus uptake by mycorrhizal and non-mycorrhizal roots of soybean. *J. Plant Nutr.* 9:1303–1313.

Kasper, J. 1977. Investigation of the development and productivity of meadow fescue in pure stand and in mixture with different white clover cultivars under increasing rates of nitrogen. *Ved. Pr. Vysk. Ustavu Luk Pasienkov Banskej Bystrici* 12:95706. (*Herb. Abstr.* 50:3956.)

Kehr, W. R., R. C. Sorensen, and J. C. Burton. 1979. Field performance of 12 strains of *Rhizobium* on 4 alfalfa varieties. *Rep. 26th Alfalfa Improvement Conf., South Dakota State University, Brookings, 1978* p. 25.

Kendall, W. A., and R. R. Hill. 1980. Greenhouse tests for evaluation of phosphorus in breeding lines of alfalfa. *Agron. Abstr.* p. 59.

Khalafallah, M. A., M. S. M. Saber, and H. K. Abd-El-Maksoud. 1982. Influence of phosphate dissolving bacteria on the efficiency of superphosphate in a calcereous soil cultivated with *Vicia faba*. *Z. Pflanzenernaehr. Bodenkd.* 145:455–459.

Khatri, H. L., and V. V. Chenulu. 1973. Metabolism of resistant and susceptible cowpea varieties infected with cowpea mosaic virus. III. Changes in some aspects of nitrogen metabolism. *Indian Phytopathol.* 26:708–712. (*Plant Breed. Abstr.* 45:5083.)

Kirkby, E. A. 1981. Plant growth in relation to nitrogen supply. In F. E. Clarke and T. Rosswall (eds.), Terrestrial Nitrogen Cycles, Processes, Ecosystem Strategies and Management Impacts. *Ecol. Bull.* No. 33, pp. 249–267.

Kirkby, E. A., and K. Mengel. 1967. Ionic balance in different tissues of the tomata plant in relation to nitrate, urea and ammonium nutrition. *Plant Physiol.* 42:6–14.

Kleese, R. A. 1967. Relative importance of stem and root in determining genotypic differences in Sr-89 and Ca-45 accumulation in soybeans (*Glycine max* L.). *Crop. Sci.* 7:53–55.

Kleese, R. A. 1968. Scion control of genotypic differences in Sr and Ca accumulation in soybeans under field conditions. *Crop Sci.* 8:128–129.

Kleese, R. A., D. C. Rasmusson, and L. J. Smith. 1968. Genetic and environmental variation in mineral element accumulation in barley, wheat and soybeans. *Crop Sci.* 8:591–593.

Kleese, R. A., and L. J. Smith. 1970. Scion control of genotypic differences in mineral salts accumulation in soybean (*Glycine max* L. Merr.) seeds. *Ann. Bot.* 34:183–188.

Kudelová, A., A. Bergmannová, V. Kudela, and L. Taimr. 1978. The effect of bacterial wilt on the uptake of manganese and zinc in alfalfa. *Acta Phytopathol. Acad. Sci. Hung.* 13:121–132. (*Plant Breed. Abstr.* 49:9237.)

Kuz'min, M. S. 1976. The effect of fertilizers on the growth, development and yield of some soybean varieties. *Nauchno-tekh. Byull. Vses. Nauchno-Issled. Inst.* No. 2, pp. 13–25. (*Plant Breed. Abstr.* 49:834.)

Läuchli, A. 1967. Investigations on the distribution and transport of ions in plant tissues with the X-ray microanalyzer. I. Experiments on vegetative organs of *Zea mays*. *Planta* 75:185–206.

Laidlaw, A. S. 1980. The effects of nitrogen fertilizer applied in spring on swards of ryegrass sown with four cultivars of white clover. *Grass Forage Sci.* 35:295–299.

Lambert, D. H., H. Cole, and D. E. Baker. 1980. Variation in the response of alfalfa clones and cultivars to mycorrhizae and phosphorus. *Crop Sci.* 20:615–618.

Lambert. J., and B. Toussaint. 1978. An investigation of the factors influencing the phosphorus content of herbage. *Phosphorus Agric.* No. 73, pp. 1–12.

Lawn, R. J., and W. A. Brun. 1974a. Symbiotic nitrogen fixation in soybeans. I. Effect of photosynthetic source-sink manipulations. *Crop. Sci.* 14:11–16.

Lawn, R. J., and W. A. Brun. 1974b. Symbiotic nitrogen fixation in soybeans. III. Effect of supplemental nitrogen and intervarietal grafting. *Crop Sci.* 14:22–25.

Lawn, R. J., K. S. Fischer, and W. A. Brun. 1974. Symbiotic nitrogen fixation in soybeans. II. Interrelationships between carbon and nitrogen assimilation. *Crop Sci.* 14:17–22.

Leach, G. J. 1968. The effectiveness of nodulation of a wide range of lucerne cultivars. *Aust. J. Exp. Agric. Anim. Husb.* 8:323–326.

Ledin, S., and L. Wiklander. 1974. Exchange acidity of wheat and pea roots in salt solutions. *Plant Soil* 41:403–413.

Lehman, P. S., D. Huisingh, and K. R. Barker. 1971. The influence of races of *Heterodera glycines* on nodulation and nitrogen-fixing capacity of soybean. *Phytopathology* 61:1239–1244.

Levesque, M., and J. W. Ketcheson. 1963. The influence of variety, soil temperature and phosphorus fertilisers on yield and phosphorus uptake by alfalfa. *Can J. Plant Sci.* 43:355–360.

Lindgren, D. T., W. H. Gabelman, and G. C. Gerloff. 1977. Variability in phosphorus uptake and translocation in *Phaseolus vulgaris* L. under phosphorus stress. *J. Am. Soc. Hortic. Sci.* 102:674–677.

Loberg, G. L., R. Shibles, D. E. Green, and J. J. Hanway. 1984. Nutrient mobilization and yield of soybean genotypes. *J. Plant Nutr.* 7:1311–1327.

Longeragan, J. F. 1978. The physiology of plant tolerance to low phosphorus availability. In G. A. Jung (ed.), Crop Tolerance to Suboptimal Land Conditions. *ASA Spec. Publ.* No. 32, pp. 329–343.

Longeragan, J. F., and C. J. Asher. 1967. Response of plants to phosphate concentration in solution culture. II. Rate of phosphate absorption and its relation to growth. *Soil Sci.* 103:311–318.

Loneragan, J. F., J. S. Gladstones, and W. J. Simmons. 1968. Mineral elements in temperate crop and pasture plants. II. Calcium. *Aust. J. Agric. Res.* 19:353–364.

Loneragan, J. F., and K. Snowball. 1969. Calcium requirements of plants. *Aust. J. Agric. Res.* 20:465–478.

Longnecker, N., and R. Welch. 1986. The relationship among ion-stress response, iron efficiency and iron uptake of plants. *J. Plant Nutr.* 9:715–727.

Lüttge, U., and N. Higinbotham. 1979. Transport in Plants. Springer-Verlag, New York.

Maherchandani, N., and O. P. S. Rana. 1977. Gamma radiation induced natural variability for nodulation in legumes. *J. Nucl. Agric. Biol.* 6:75–77. (*Plant Breed. Abstr.* 50:6880.)

Mahon, J. D. 1979. Environmental and genotypic effects on the respiration associated with symbiotic nitrogen fixation in peas. *Plant Physiol.* 63:892–897.

Mahon, J. D. 1982. Field evaluation of growth and nitrogen fixation in peas selected for high and low photosynthetic CO_2 exchange. *Can. J. Plant Sci.* 62:5–17.

Major, D. J., M. R. Hanna, S. Smoliak, and R. Grant. 1979. Estimating nodule activity of sainfoin, alfalfa and cicer milkvetch seedlings. *Agron. J.* 71:983–985.

Malavolta, E., and F. A. L. Amaral. 1978. Nutritional efficiency of 104 bean varieties (*Phaseolus vulgaris* L.). In A. R. Ferguson, R. L. Bieleski, and I. B. Ferguson (eds.), Plant Nutrition 1978. Proc. 8th Int. Colloq. Plant Analysis and Fertilizer Problems, Auckland. *Inf. Ser—N.Z. Dep. Sci. Ind. Res.* No. 134, pp. 316–318.

Malavolta, E., M. O. C. Brasil Sobrinho, H. E. C. Rezende, G. C. Vitti, and F. A. L. Amaral. 1980. On the relationship between efficiency of utilization of nitrogen, phosphorus and potassium by several bean varieties (*Phaseolus vulgaris* L.) and yield potential. *Rev. Agric.* (*Piracicaba, Braz.*) 55:275–278. (*Plant Breed. Abstr.* 52:1717.)

Malewar, G. U., D. K. Jadhav, and C. P. Ghonsikar. 1982. Absorption and transport of ^{59}Fe in genotypes of chickpea. *Int. Chickpea Newsl.* 6:13–14. (*Plant Breed. Abstr.* 52:9932.)

Malik, N. S. A. 1983. Grafting experiments on the nature of the decline in N_2-fixation during fruit development in soybean. *Physiol. Plant.* 57:561–564.

Marschner, H., V. Römheld, and H. Ossenberg-Neuhans. 1982. Rapid method for measuring changes in pH and reducing processes along roots of intact plants. *Z. Pflanzenphysiol.* 105:407–416.

Mavrichev. P. I., and T. V. Smirnova. 1977. Efficacy of symbiosis of *Rhizobium trifolii* with different varieties of red clover. *Byull. Vses. Nauchno-Issled. Inst. Skh. Mikrobiol.* No. 19, pp. 15–21. (*Plant Breed. Abstr.* 49:5973.)

McCaslin, B. D., and R. J. Gledhill. 1980. Alfalfa fertilization in New Mexico. *N. M. Agric. Exp. Stn., Bull.* 675.

McFerson, J. R. 1983. Genetic and breeding studies of dinitrogen fixation in common bean (*Phaseolus vulgaris* L.). *Diss. Abstr. Int. B* 44:701B. (*Plant Breed. Abstr.* 54:1960.)

McGinnity, P. J., G. Kapusta, and O. Myers. 1978. The interaction effects of the soybean cyst nematode and *Rhizobium japonicum* on nodulation and N_2-fixation. *Agron. Abstr.* p. 143.

McGinnity, P. J., G. Kapusta, and O. Myers. 1980. Soybean cyst nematode and *Rhizobium* strain influences on soybean nodulation and N_2-fixation. *Agron. J.* 72:785–789.

McIntosh, M. S., and D. A. Miller. 1980. Development of root-branching in three alfalfa cultivars. *Crop Sci.* 20:807–809.

McLachlan, K. D. 1976. Comparative phosphorus responses in plants to a range of available phosphorus situations. *Aust. J. Agric. Res.* 27:323–341.

Mengel, K. 1961. The Donnan distribution of cations in the free space of plant roots and its significance for the active cation uptake. *Z. Pflanzenernaehr. Dueng Bodenkd.* 95:240–259.

Metzger, W. H. 1928. The effect of growing plants on solubility of soil nutrients. *Soil Sci.* 25:273–280.

Miller, D. J., B. A. Melton, B. D. McCaslin, N. Waissman, and E. Olivares. 1984. Breeding alfalfa for phosphorus concentration in southern New Mexico. *Rep. 29th Alfalfa Improvement Conf.* p. 64.

Miller, D. J., N. Waissman, B. Melton, C. Currier, and B. McCaslin. 1987. Selection for increased phosphorus in alfalfa and effects on other characteristics. *Crop Sci.* 27:22–26.

Miller, R. W., and J. C. Sirois. 1981. Symbiotic effectiveness of alfalfa cultivar–*R. meliloti* strain combinations in controlled environments. *Can. J. Plant Pathol. Abstr.* 3:117.

Millikan, C. R. 1957. Effects of environmental factors on the growth of two varieties of subterranean clover (*Trifolium subterraneum* L.). *Aust. J. Agric. Res.* 8:225–245.

Mitchell, R. L., and W. J. Russell. 1971. Root development and rooting patterns of soybean (*Glycine max* (L.) Merrill) evaluated under field conditions. *Agron. J.* 63:313–316.

Mitsui, S., M. Nakagawa, A. Baba, K. Tensho, and K. Kumazawa. 1956. Dynamic studies on nutrient uptake by sap plants. X. Contact solutional uptake of fixed magnesium phosphate (P^{32}) by acidoidal plant root and unsaturated soil colloid. *J. Sci. Soil, Tokyo* 26:497–501.

Moraghan, J. T. 1987. Effects of phosphorus and iron fertilizers on the growth of two soybean varieties at two soil temperatures. *Plant Soil* 104:121–127.

Morrison, J., H. Denehy, and P. F. Chapman. 1983. Possibilities for the strategic use of fertilizer nitrogen on white clover/grass swards. In A. J. Corrall (ed.), Efficient Grassland Farming. Proc. 9th Gen. Meet. European Grassland Federation, University of Reading. *Occas. Symp., Br. Grassl Soc.* No. 14, pp. 227–231.

Mosse, B. 1973. Advances in the study of vesicular arbuscular mycorrhiza. *Annu. Rev. Phytopathol.* 11:171–196.

Mouat, M. C. H. 1960. Interspecific difference in strontium uptake by pasture plants as a function of root cation exchange capacity. *Nature (London)* 188:513–514.

Mouat, M. C. H. 1983. Phosphate uptake fro extended soil solutions by pasture plants. *N.Z. J. Agric. Res.* 26:483–487.

Mouat, M. C. H., and T. W. Walker. 1959. Competition for nutrients between grasses and white clover. II. Effect of root cation exchange capacity and rate of emergence of associated species. *Plant Soil* 11:41–52.

Mulder, E. G. 1948. Investigations on the nitrogen nutrition of pea plants. *Plant Soil* 1:179–212.

Munns, D. N., and B. Mosse. 1980. Mineral nutrition of legume crops. In R. J. Summerfield and A. H. Bunting (eds.), Advances in Legume Science. Royal Botanic Gardens, Kew. Pp. 115–126.

Murphy, W. E., and V. Connolly. 1975. The interactions of new cultivars of grass and legumes with fertilizers. *Fertilizer Use and Protein Production. Proc. 11th Colloq. Int. Potash Institute, Rønne-Bornholm* pp. 129–136.

Musorina, L. I., and M. D. Kovalevich. 1985. Relationships between fresh herbage chemical composition and morphobiological traits of soybean plants. *Skh. Biol.* No. 9, pp. 36–38.

Mytton, L. R. 1976. The relative performance of white clover genotypes with rhizobial and mineral nitrogen in agar culture and in soil. *Ann. Appl. Biol.* 82:577–587.

Mytton, L. R., and J. De Felice. 1977. The effect of mixtures of *Rhizobium* strains on the dry matter production of white clover grown in agar. *Ann. Appl. Biol.* 87:83–93.

Mytton, L. R., M. H. El-Sherbeeny, and D. A. Lawes. 1977. Symbiotic variability in *Vicia faba*. 3. Genetic effects of host plant, *Rhizobium* strain and of host × strain interaction. *Euphytica* 26:785–791.

Mytton, L. R., and D. M. Hughes. 1985. Breeding for increased N_2-fixation in white clover. *Annu. Rep. Welsh Plant Breed. Stn. 1984* p. 44.

Mytton, L. R., and D. G. Jones. 1971. The response to selection for increased nodule tissue in white clover (*Trifolium repens* L.). In T. A. Lie and E. G. Mulder (eds.), Biological Nitrogen Fixation in Natural and Agricultural Habitats. *Plant Soil, Spec. Vol.* pp. 17–25.

Mytton, L. R., and G. J. Rys. 1985. The potential for breeding white clover (*Trifolium repens* L.) with improved nodulation and nitrogen fixation when grown with combined nitrogen (ii). Assessment of genetic variation in *Trifolium repens*. *Plant Soil* 88:197–211.

Nielsen, N. E. 1979. Is the efficiency of nutrient uptake genetically controlled? In J. L. Harley and R. S. Russell (eds.), The Soil–Root Interface. Academic Press, New York. P. 429. Abstr.

Nutman, P. S. 1981. Hereditary host factors affecting nodulation and nitrogen fixation. In A. H. Gibson and W. E. Newton (eds.), Current Perspectives in Nitrogen Fixation. Australian Acad. Sci., Canberra. Pp. 194–204.

Nutman, P. S., H. Mareckova, and L. Raichewa. 1971. Selection for increased nitrogen fixation in red clover. In T. A. Lie and E. G. Mulder (eds.), Biological Nitrogen Fixation in Natural and Agricultural Habitats. *Plant Soil, Spec. Vol.* pp. 27–31.

Nyatsanga, T., and W. H. Pierre. 1973. Effect of nitrogen fixation by legumes on soil acidity. *Agron. J.* 65:936–940.

Nye, P. H. 1966. The effect of the nutrient intensity and buffering power of a soil, and the absorbing power, size and root hairs of a root, on nutrient absorption by diffusion. *Plant Soil* 25:81–105.

Nye, P. H. 1968. Processes in the root environment. *J. Soil Sci.* 19:205–215.

Nye, P. H. 1981. Changes of pH across the rhizosphere induced by roots. *Plant Soil* 61:7–26.

Nye, P. H., and W. N. M. Foster. 1958. A study of the mechanism of soil-phosphate uptake in relation to plant species. *Plant Soil* 9:338–352.

Nye, P. H., and P. B. Tinker. 1977. Solute Movement in the Soil–Root System. Studies in Ecology, Vol. 4. Blackwell, Oxford.

Ollivier, B., Y. Berthean, H. G. Diem, and V. Gianinazzi-Pearson. 1983. Influence de la variété de *Vigna unguiculata* dans l'expression de trois associations endomycorhiziennes à vésicules et arbuscules. *Can. J. Bot.* 61:354–358.

Olsen, R. A., and J. C. Brown. 1980. Factors related to iron uptake by dicotyledonous and monocotyledonous plants. I. pH and reductant. *J. Plant Nutr.* 2:629–645.

Overstreet, R., and H. Jenny. 1939. Studies pertaining to the cation absorption of plant in soil. *Soil Sci. Soc. Am. Proc.* 4:125–130.

Ozanne, P. G., and K. M. W. Howes. 1971. The effects of grazing on the phosphorus requirement of an annual pasture. *Aust. J. Agric. Res.* 22:81–92.

Pacovsky, R. S., H. G. Bayne, and G. J. Bethlenfalvay. 1984. Symbiotic interactions between strains of *Rhizobium phaseoli* and cultivars of *Phaseolus vulgaris* L. *Crop Sci.* 24:101–105.

Paprocki, S., G. Fordonski, and S. Glowacka. 1980. Effect of sowing density on yield and nutritional value in new pea varieties. *Zesz. Nauk. Akad. Roln.-Tech. Olsztynie. Roln.* No. 30, pp. 151–159. (*Plant Breed. Abstr.* 52:4383.)

Park, H., and C. A. Stutte. 1973. Effects of NH_4-N, NO_3-N and urea-N on the growth of soybean plants different in phosphorus sensitivity. *J. Korean Agric. Chem. Soc.* 16:118–127. (*Plant Breed. Abstr.* 45:10692.)

Parsons, R. F. 1968. Ecological aspects of the growth and mineral nutrition of three mallee species of *Eucalyptus*. *Oecologia Plant.* 3:121–136.

Patil, B. D., and L. Moniz. 1974. Differential response of gram varieties to an efficient isolate of *Rhizobia* from gram (*Cicer arietinum* L.). *Res. J. Mahatma Phule Agric. Univ.* 5:42–46. (*Plant Breed. Abstr.* 45:3247.)

Patterson, T. G., and T. A. LaRue. 1983. Nitrogen fixation by soybeans: Seasonal and cultivar effects and comparison of estimates. *Crop. Sci.* 23:488–492.

Paulsen, G. M., and O. A. Rotimi. 1968. P–Zn interaction in two soybean varieties differing in sensitivity to phosphorus nutrition. *Soil Sci. Soc. Am. Proc.* 32:73–76.

Pavlychenko, T. K., and J. B. Harrington. 1934. Competitive efficiency of weeds and cereal crops. *Can. J. Res.* 10:77–94.

Pederson, G. A., R. R. Hill, and W. A. Kendall. 1984. Genetic variability for root characters in alfalfa populations differing in winterhardiness. *Crop Sci.* 24:465–468.

Pereira, P. A. A., and F. A. Bliss. 1987. Nitrogen fixation and plant growth of common bean (*Phaseolus vulgaris* L.) at different levels of phosphorus availability. *Plant Soil* 104:79–84.

Phillips, D. A. 1980. Efficiency of symbiotic nitrogen fixation in legumes. *Annu. Rev. Plant Physiol.* 31:29–49.

Phillips, D. A., R. O. Pierce, S. A. Edie, K. W. Foster, and P. F. Knowles. 1984. Delayed leaf senescence in soybeans. *Crop. Sci.* 24:518–522.

Phillips, D. A., D. W. Rains, R. C. Valentine, and R. C. Huffaker. 1978. Enhancing biological nitrogen fixation. *Cereal Foods World* 23:26–29. (*Plant Breed. Abstr.* 48:10317.)

Phillips, D. A., L. R. Teuber, and S. S. Jue. 1982. Variation among alfalfa genotypes for reduced nitrogen concentration. *Crop Sci.* 22:606–610.

Pierson, E. E., R. B. Clark, D. P. Coyne, and J. W. Maranville. 1984. Plant genotype differences to ferrous and total iron in emerging leaves. II. Dry bean and soybeans. *J. Plant Nutr.* 7:355–369.

Polson, D. E., and L. J. Smith. 1972. Nature of scion control of mineral accumulation in soybeans. *Agron. J.* 64:381–384.

Powell, C. L. 1975. Potassium uptake by endotrophic mycorrhizas. In F. E. Sanders, B. Mosse, and P. B. Tinker (eds.), Endomycorrhizas. Academic Press, London. Pp. 461–468.

Powell, C. L. 1982. Phosphate response curves of mycorrhizal and nonmycorrhizal plants. III. Cultivar effects in *Lotus pedunculatus* Cav. and *Trifolium repens* L. *N.Z. J. Agric. Res.* 25:217–222.

Probert, M. E. 1972. The dependence of isotopically exchangeable phosphate (L-value) on phosphate uptake. *Plant Soil* 36:141–148.

Raboy, V., D. B. Dickinson, and F. E. Below. 1984. Variation in seed total phosphorus, phytic acid, zinc, calcium, magnesium and protein among lines of *Glycine max* and *G. soja*. *Crop Sci.* 24:431–434.

Raggio, M., and N. Raggio. 1962. Root nodules. *Annu. Rev. Plant Physiol.* 13:109–128.

Rai, R., V. Prasad, S. K. Choudhury, and N. P. Sinha. 1984. Iron nutrition and symbiotic N_2-fixation of lentil (*Lens culinaris*) genotypes in calcareous soil. *J. Plant Nutr.* 7:399–405.

Rais, I., and J. Královec. 1973. The evaluation of Czechoslovak and introduced varieties of cocksfoot in simple mixtures with white clover. I. Forage yields and the distribution of production in the growing period. *Rostl. Vyroba* 19:1287–1295.

Rais, I., and J. Královec. 1975. The evaluation of Czechoslovak and introduced varieties of cocksfoot in simple mixtures with white clover. II. Botanical composition of the stand and nutrient content of the forage. *Rostl. Vyroba* 21:317–324.

Ramirez, R. 1982. Efficient use of nitrogen, phosphorus, and potassium by corn (*Zea mays* L.) inbreds. In A. Scaife (ed.), Plant Nutrition 1982. Proc. 9th Int. Plant Nutrition Colloq., Warwick University. Pp. 515–520.

Rang, A., T. S. Sandhu, and B. S. Bhullar. 1980. Protein and amino-acid association with yield and its components in gram. *Indian J. Genet. Plant Breed.* 40:423–426.

Rao, J. V. D. K. K., P. J. Dart, T. Matsumoto, and J. M. Day. 1981. Nitrogen fixation by pigeon pea. *Proc. Int. Workshop Pigeon Peas, ICRISAT* 1:15–19. (*Plant Breed. Abstr.* 52:5451.)

Raper, C. D., and S. A. Barber. 1970a. Rooting systems of soybeans. I. Differences in root morphology among varieties. *Agron. J.* 62:581–584.

Raper, C. D., and S. A. Barber. 1970b. Rooting systems of soybeans. II. Physiological effectiveness as nutrient absorption surfaces. *Agron. J.* 62:585–558.

Raven, J. A., and F. A. Smith. 1976. Nitrogen assimilation and transport in vascular land plants in relation to intracellular pH regulation. *New Phytol.* 76:415–431.

Rawat, R. R., and S. S. Shaw. 1983. Relative susceptibility/resistance of different pea varieties against *Caliothrips indicus* (Bagnall). *Natl. Semin. Breeding Crop Plants for Resistance to Pests and Diseases, Coimbatore, India* pp. 37–38. (*Plant Breed. Abstr.* 54:7870.)

Reid, D. 1961a. Factors influencing the role of clover in grass–clover leys fertilised with nitrogen at different rates. 1. The effects of the variety of companion grass on the yields of total herbage and of clover. *J. Agric. Sci.* 56:143–153.

Reid, D. 1961b. Factors influencing the role of clover in grass–clover leys fertilised with nitrogen at different rates. 2. The effects of variety of white clover on the yields of total herbage and of clover. *J. Agric. Sci.* 52:155–160.

Rennie, R. J., and G. A. Kemp. 1981a. Selection for dinitrogen-fixing ability in *Phaseolus vulgaris* L. at two low-temperature regimes. *Euphytica* 30:87–95.

Rennie, R. J., and G. A. Kemp. 1981b. Dinitrogen fixation in pea beans (*Phaseolus vulgaris*) as affected by growth stage and temperature regime. *Can. J. Bot.* 59: 1181–1188.

Rennie, R. J., and G. A. Kemp. 1982. Dinitrogen fixation in *Phaseolus vulgaris* at low temperatures: Interaction of temperature, growth stage and time of inoculation. *Can. J. Bot.* 60:1423–1427.

Rennie, R. J., and G. A. Kemp. 1983a. N_2 fixation in field beans quantified by ^{15}N isotope dilution. I. Effects of strains of *Rhizobium phaseoli*. *Agron. J.* 75:640–644.

Rennie, R. J., and G. A. Kemp. 1983b. N_2 fixation in field beans quantified by ^{15}N isotope dilution. II. Effect of cultivar of beans. *Agron. J.* 75:645–649.

Rennie, R. J., and G. A. Kemp. 1984. ^{15}N-determined time course for N_2 fixation in two cultivars of field bean. *Agron. J.* 76:146–154.

Riggle, B. D., W. J. Wiebold, and W. J. Kenworthy. 1984. Effect of photosynthate source-sink manipulation on dinitrogen fixation of male-fertile and male-sterile soybean isolines. *Crop Sci.* 24:5–8.

Riley, D., and S. A. Barber. 1969. Bicarbonate accumulation and pH changes at the soybean (*Glycine max* L. Merr.) root–soil interface. *Soil Sci. Soc. Am. Proc.* 33:905–908.

Riley, D., and S. A. Barber. 1971. Effect of ammonium and nitrate fertilization on phosphorus uptake as related to root-induced pH changes at the root–soil interface. *Soil Sci. Soc. Am. Proc.* 35:301–306.

Robinson, R. R. 1942. The mineral content of various clones of white clover when grown on different soils. *J. Am. Soc. Agron.* 34:933–939.

Robson, A. D. 1983. Mineral nutrition. In W. J. Broughton (ed.), Nitrogen Fixation 3. Legumes. Clarendon, Oxford. Pp. 36–55.

Romero, N. A., C. C. Sheaffer, and G. L. Malzer. 1981. Potassium response of alfalfa in solution sand and soil culture. *Agron. J.* 73:25–28.

Ronis, D. H., D. J. Sammons, W. J. Kenworthy, and J. J. Meisinger. 1985. Heritability of total and fixed nitrogen content of the seed in two soybean populations. *Crop Sci.* 25:1–4.

Rorison, I. H. 1968. The response to phosphorus of some ecologically distinct plant species. *New Phytol.* 67:913–923.

Roughley, R. J. 1970. The influence of root temperature, *Rhizobium* strain and host selection on the structure and nitrogen fixing efficiency of the root nodules of *Trifolium subterraneum*. *Ann. Bot.* 34:631–646.

Roughley, R. J., W. M. Blowes, and D. F. Herridge. 1976. Nodulation of *Trifolium subterraneum* by introduced *Rhizobia* in competition with naturalized strains. *Soil Biol. Biochem.* 8:403–407.

Rulinskaya, N. S. 1980. Effectiveness of symbiosis between some lupin species and varieties and *Rhizobium lupini*. *Sb. Nauchn. Tr. Beloruss. Skh. Akad.* No. 65, pp. 76–81. (*Plant Breed. Abstr.* 51:3296.)

Russell, P. E., and D. G. Jones. 1975. Variation in the selection of *Rhizobium trifolii* by varieties of red and white clover. *Soil Biol. Biochem.* 7:15–18.

Ryle, G. J. A., C. E. Powell, and A. J. Gordon. 1979a. The respiratory costs of nitrogen fixation in soybean, cowpea and white clover. I. Nitrogen fixation and the respiration of the nodulated root. *J. Expt. Bot.* 30:135–144.

Ryle, G. J. A., C. E. Powell, and A. J. Gordon. 1979b. The respiratory costs of nitrogen fixation in soybean, cowpea and white clover. II. Comparisons of the cost of nitrogen fixation and the utilization of combined nitrogen. *J. Exp. Bot.* 30:145–153.

Rys, G. J. 1986. Cultivar variation in the nodulation of white clover seedlings provided with nitrogen. In T. A. Williams and G. S. Wratt (eds.), *DSIR Plant Breeding Symp., N.Z. Agronomy Society* Spec. Publ. No. 5, pp. 326–329.

Safaya, N. M. 1978. Growth and nutrient fluxes in cowpea as affected by phosphate and zinc fertilization. *Agron. Abstr.* p. 161.

Safaya, N. M., and B. Singh. 1977. Differential susceptibility of two varieties of cowpea (*Vigna unguiculata* (L.) Walp) to P-induced Zn deficiency. *Plant Soil* 48:279–290.

Saggar, S., and G. Dev. 1974. Uptake of sulphur by different varieties of soybean. *Indian J. Agric. Sci.* 44:345–349.

Sain, S. L., and G. V. Johnson. 1986. Characterization of iron uptake by iron-efficient and iron-inefficient soybeans in cell suspension culture. *J. Plant Nutr.* 9:729–750.

Salado-Navarro, L. R., K. Hinson, and T. R. Sinclair. 1985. Nitrogen partitioning and dry matter allocation in soybeans with different seed protein concentration. *Crop Sci.* 25:451–455.

Sanders, F. E., and P. B. Tinker. 1971. Mechanisms of absorption of phosphate from soil by *Endogone* mycorrhizas. *Nature (London)* 233:278–279.

Sarić, M. R. 1982. Theoretical and practical approaches to the genetic specificity of mineral nutrition of plants. In M. R. Sarić (ed.), *Proc. 1st Int. Symp. Genetic Specificity of Mineral Nutrition of Plants, Belgrade.* Pp. 9–20.

Sarić, Z., and A. H. Fawzia. 1982. Nitrogen fixation in soybean depending on variety of *Rhizobium japonicum* strain. In M. R. Sarić (ed.), *Proc. 1st Int. Colloq. Genetic Specificity of Mineral Nutrition of Plants, Belgrade.* Pp. 289–293.

Satterlee, L., B. Melton, B. McCaslin, and D. Miller. 1983. Mycorrhizae effects on plant growth, phosphorus uptake and N_2 (C_2H_2) fixation in two alfalfa populations. *Agron. J.* 75:715–716.

Sawaguchi, M., and K. Nomura. 1980. Growth and nitrogen uptake by adzuki bean varieties. *Bull. Hokkaido Prefect. Agric. Exp. Stn.* No. 43, pp. 1–11. (*Plant Breed. Abstr.* 51:8595.)

Schenck, N. C., R. A. Kinloch, and D. W. Dickson. 1975. Interaction of endomycorrhizal fungi and root knot nematode on soybean. In F. E. Sanders, B. Mosse, and P. B. Tinker (eds.), Endomycorrhizas. Academic Press, London. Pp 607–617.

Scherer, C. H. 1975. Performance of varieties and lines of soybean in varying soil fertility conditions. *Anais, V. reunião geral de cultura do arroz, Cachoeirinha, Brazil, EMBRAPA-IRGA* pp. 133–135. (*Plant Breed. Abstr.* 46:4874.)

Schettini, T. M., W. H. Gabelman, and G. C. Gerloff. 1987. Incorporation of phosphorus efficiency from exotic germplasm into agriculturally adapted germplasm of common bean (*Phaseolus vulgaris* L.). *Plant Soil* 99:175–184.

Schweitzer, C. E., and J. E. Harper. 1985. Effect of hastened flowering on seed yield and dry matter partitioning in diverse soybean genotypes. *Crop Sci.* 25:995–998.

Scott, R. S. 1977. The phosphate nutrition of white clover. *Proc. 38th Grassland Conf., N.Z. Grassland Association* pp. 151–159.

Seetin, M. W., and D. K. Barnes. 1977. Variation among alfalfa genotypes for rate of acetylene reduction. *Crop Sci.* 17:783–787.

Sen, A., and N. B. Paul. 1957. Solubilization of phosphate by some common soil bacterias. *Curr. Sci.* 26:222.

Sethi, C. L., and N. K. Sharma. 1976. Nature of resistance of cowpea to the root-know nematode, *Meloidogyne incognita* and the pea cyst nematode, *Heterodera cajani*. *Indian J. Nematol.* 6:81–85.

Sharpley, A. N. 1986. Disposition of fertilizer phosphorus applied to winter wheat. *Soil Sci. Soc. Am. J.* 50:953–958.

Sheehy, J. E., K. A. Fishbeck, and D. A. Phillips. 1980. Relationships between apparent nitrogen fixation and carbon exchange rate in alfalfa. *Crop Sci.* 20:491–495.

Shukla, U. C., and H. Raj. 1980. Zinc response in pigeon pea as influenced by genotypic variability. *Plant Soil* 57:323–333.

Silberbush, M., and S. A. Barber. 1983. Prediction of phosphorus and potassium uptake by soybeans with a mechanistic mathematical model. *Soil Sci. Soc. Am. J.* 47:262–265.

Silberbush, M., and S. A. Barber. 1984. Phosphorus and potassium uptake of field-grown soybean cultivars predicted by a simulation model. *Soil Sci. Soc. Am. J.* 48:592–596.

Simpson, J. R., A. Pinkerton, and J. Lazdovskis. 1979. Interacting effects of subsoil acidity and water on the root behaviour and shoot growth of some genotypes of lucerne (*Medicago sativa* L.). *Aust. J. Agric. Res.* 30:609–619.

Sinclair, T. R., and C. T. de Wit. 1975. Comparative analysis of photosynthate and nitrogen requirements in the production of seeds by various crops. *Science* 189:565–567.

Sinclair, T. R., and C. T. de Wit. 1976. Analysis of the carbon and nitrogen limitations to soybean yield. *Agron. J.* 68:319–324.

Singh, K. K. 1974a. Iron-induced manganese chlorosis in navy bean lines (*Phaseolus vulgaris* L.) and variation in responses. *Plant Sci.* 6:41–45.

Singh, K. K. 1974b. Exudate delivery, and iron and manganese transport determining differential responses of navy bean lines (*Phaseolus vulgaris* L.) to iron-induced manganese chlorosis. *Plant Sci.* 6:54–58.

Singh, K. K. 1975. Genotypes of rootstock and scion determining differential responses of navy bean lines (*Phaseolus vulgaris* L.) to iron-induced manganese chlorosis. *Plant Sci.* 7:13–18.

Singh, K. K. 1976. Exudate delivery and iron and manganese transport determining differential responses of navy bean (*Phaseolus vulgaris* L.) to iron-induced manganese chlorosis. *Recent Advances in Plant Science. Session 5. Plant Physiology and Biochemistry. B.C. Agricultural University, Kalyani, India, 1975* p. 48. (*Field Crop Abstr.* 30:647.)

Singh, R. A., N. P. Sinha, B. P. Singh, and S. G. Sharma. 1986. Reaction of chickpea genotypes to iron deficiency in a calcareous soil. *J. Plant Nutr.* 9:417–422.

Singleton, P. W., W. G. Sanford, and K. R. Stockinger. 1978. Greenhouse studies on the symbiotic effectiveness of 11 tropical peanut cultivars. *Agron. Abstr.* p. 85.

Sjödin, J., P. Martensson, and T. Magyarosi, 1981. Selection for improved protein quality in field bean (*Vicia faba* L.). *Z. Pflanzenzuecht.* 86:221–230.

Skipper, H. D., and G. W. Smith. 1979. The influence of soil pH on the soybean–endomycorrhizal symbiosis. *Plant Soil* 53:559–563.

Slack, T. E., and L. G. Morrill. 1972. A comparison of a large-seeded (NC2) and a small-seeded (Starr) peanut (*Arachis hypogaea* L.) cultivar as affected by levels of calcium added to the fruit zone. *Soil Sci. Soc. Am. Proc.* 36:87–90.

Smiley, R. W. 1974. Rhizosphere pH as influenced by plants, soils and nitrogen fertilizers. *Soil Sci. Soc. Am. Proc.* 38:795–799.

Smith, D. 1951. Root branching of alfalfa varieties and strains. *Agron. J.* 43:573–575.

Smith, G. C., E. G. Brennan, and D. J. Greenhalgh. 1985. Cadmium sensitivity of soybean related to efficiency in iron nitrification. *Environ. Exp. Bot.* 25:99–106.

Smith, G. R., H. L. Peterson, and W. E. Knight. 1979. Screening crimson clover (*Trifolium incarnatum* L.) for dinitrogen (C_2H_2) fixation efficiency. *Agron. Abstr.* p. 78.

Smith, G. S., and I. S. Cornforth. 1982. Concentrations of N,P,S, Mg and Ca in North Island pastures in relation to plant and animal nutrition. *N.Z. J. Agric. Res.* 25:373–387.

Smittle, D. A., and B. B. Brantley. 1978. Relationships of nitrogen fixation to growth and yield of southernpea. *HortScience* 13:26. (Abstr.)

Snaydon, R. W. 1961. Competitive ability of natural populations of *Trifolium repens* and its relation to differential response to soil factors. *Heredity* 16:522. (Abstr.)

Sorwli, F. K., and L. R. Mytton. 1986. Nitrogen limitations to field bean productivity: A comparison of combined nitrogen applications with *Rhizobium* inoculation. *Plant Soil* 94:267–275.

Spaeth, S. C., and Sinclair, T. R. 1983. Variation in nitrogen accumulation and distribution among soybean cultivars. *Field Crops Res.* 7:1–12.

Specht, R. L., and R. H. Groves. 1966. A comparison of the phosphorus nutrition of Australian heath plants and introduced economic plants. *Aust. J. Bot.* 14:201–221.

Spencer, K., A. G. Govaars, and F. W. Hely, 1980. Early phosphorus nutrition of eight forms of two clover species. *Trifolium ambiguum* and *T. repens. N.Z. J. Agric. Res.* 23:457–475.

Steffens, D. 1984. Root studies and phosphorus uptake of ryegrass and red clover under field conditions. *Z. Pflanzenernaehr. Bodenkd.* 147:85–97.

Streeter. J. G. 1974. Growth of two soybean shoots on a single root. Effect on nitrogen and dry matter accumulation by shoots and on the rate of nitrogen fixation by nodulated roots. *J. Exp. Bot.* 25:189–198.

Strong, W. M., and R. J. Soper. 1973. Utilisation of pelleted phosphorus by flax, wheat, rape and buckwheat from a calcareous soil. *Agron. J.* 65:18–21.

Subba Rao, N. S. 1974. Prospects of bacterial fertilisation in India. *Fert. News* 19:32–36.

Sumner, M. E., H. R. Boerma, and R. Isaac. 1982. Differential sensitivity of soybeans to P–Zn–Cu imbalances. In A. Scaife (ed.), Plant Nutrition 1982. Proc. 9th Int. Plant Nutrition Colloq., Warwick University. Pp. 652–657.

Taimr, L., A. Kudelová, V. Kudela, and E. Bergmannová. 1975. Effect of bacterial wilt on uptake and translocation of phosphorus, sulphur, calcium and manganese in alfalfa plants. *Zentralbl. Bakteriol., Parasitenkd., Infektionskr. Hyg., Abt. 2* 130:367–386. (*Plant Breed. Abstr.* 46:10210.)

Tan, G.-Y. 1981. Genetic variation for acetylene reduction rate and other characters in alfalfa. *Crop Sci.* 21:485–488.

Tan, G.-Y., W. K. Tan, and P. D. Walton. 1977. Next assimilation rate and relative nitrogen assimilation rate in relation to the dry matter prodution of alfalfa cultivars. *Agron. Abstr.* p. 73.

Tester, C. F. 1977. Constituents of soybean cultivars differing in insect resistance. *Phytochemistry* 16:1899–1901.

Teuber, L. R., R. P. Levin, T. C. Sweeney, and D. A. Phillips. 1984. Selection for nitrogen concentration and forage yield in alfalfa. *Crop Sci.* 24:553–558.

Tiffin, L. O. 1966. Iron translocation. *Plant Physiol.* 41:510–514.

Tikhonovich, I. A. 1981. Study of the interaction of genotypes of leguminous plants and nodule bacteria in the course of symbiotic nitrogen fixation. *Ekol. Genet. Rast. Zhivotnykh. Tez. Dokl. Vses. Knof., 2 Kishinev, Mold. SSR* pp. 148–149. (*Plant Breed. Abstr.* 53:1091.)

Trevino, I. C., and G. A. Murray. 1975. Nitrogen effects on growth, seed yield and protein of seven pea cultivars. *Crop Sci.* 15:500–502.

Trimble, M. W., D. K. Barnes, G. H. Heichel, and C. C. Sheaffer. 1984. Use of soil nitrogen levels and cutting frequency to evaluate the performance of alfalfa genotypes in a breeding programme. In D. R. Viands and J. H. Elgin (eds.), *Rep. 29th Alfalfa Improvement Conf.* p. 60.

Troughton, A. 1960. Further studies on the relationship between shoot and root systems of grasses. *J. Br. Grassl. Soc.* 15:41–47.

Tyutyunnikov, A. I., A. A. Yakovlev, and A. A. Sukhorukov. 1978. Interrelationships between the structural organisation of the external surface of the cell walls in the apical part of the root and plant yield. *Dokl. Vses. Akad. Skh. Nauk im. V. I. Lenina* No. 5, pp. 11–13. (*Plant Breed. Abstr.* 48:9785.)

van Egmond, F., and M. Aktas. 1977. Iron nutrition aspects of the ionic balance of plants. *Plant Soil* 48:685–703.

van Ray, B., and A. van Diest. 1979. Utilization of phosphate from different sources by six plant species. *Plant Soil* 51:577–589.

Velu, G., and S. Gopalakrishnan. 1985. Varietal and habitual variability in carbohydrate–nitrogen relationships of *Arachis hypagaea* L. *Madras Agric. J.* 72:330–335. (*Plant Breed. Abstr.* 57:2245.)

Viands, D. R., and D. K. Barnes. 1979. Response from selection in alfalfa for factors associated with nitrogen fixation. In D. K. Barnes (ed.), *Rep. 26th Alfalfa Improvement Conf.* p. 27.

Viands, D. R., D. K. Barnes, and F. I. Frosheiser. 1980. An association between resistance to bacterial wilt and nitrogen fixation in alfalfa. *Crop Sci.* 20:699–703.

Viands, D. R., D. K. Barnes, and G. H. Heichel. 1981. Nitrogen fixation in alfalfa-responses to bidirectional selection for associated characters. *U.S. Dep. Agric. Tech. Bull.* No. 1643.

Vose, P. B. 1963. Varietal differences in plant nutrition. *Herb. Abstr.* 33:1–13.

Wallace, A. 1986. A multiplicity of mechanisms in the plant kingdom which can contribute to iron deficiency. *J. Plant Nutr.* 9:781–786.

Wallace, A., A. M. Abou-zamzam, and J. W. Cha. 1986. Influence of iron efficiency in soybeans on concentration of many trace elements in plant parts and implications on iron-efficiency mechanisms. *J. Plant Nutri.* 9:787–803.

Walton, G. H. 1978. The effect of manganese on seed yield and the split seed disorder of sweet and bitter phenotypes of *Lupinus angustifolius* and *L. cosentinii*. *Aust. J. Agric. Res.* 29:1177–1189.

Weber, C. R. 1966. Nodulating and non-nodulating soybean isolines. II. Response to applied nitrogen and modified soil conditions. *Agron. J.* 58:46–49.

Westermann, D. T., and J. J. Kolar. 1976. Symbiotic N_2 (C_2H_2) fixation by *Phaseolus vulgaris* L. *Agron. Abstr.* p. 79.

White, R. E. 1973. Studies on mineral ion absorption by plants. II. The interaction between metabolic activity and the rate of phosphorus uptake. *Plant Soil* 38:509–523.

Wild, A., and V. G. Breeze. 1981. Nutrient uptake in relation to growth. In C. B. Johnson (ed.), Physiological Processes Limiting Plant Productivity. *Proc. Nottingham University 30th Easter School in Agricultural Science, Butterworth, London.* Pp. 331–344.

Williams, C. H. 1980. Soil acidification under clover pastures. *Aust. J. Exp. Agric. Anim. Husb.* 20:561–567.

Williamson, B. 1973. Acid phosphatase and esterase activity in orchid mycorrhiza. *Planta* 112:149–158.

Wilman, D., and J. E. Asiegbu. 1982a. The effects of clover variety, cutting interval and nitrogen application on herbage yields, proportions and heights of perennial ryegrass and white clover swards. *Grass Forage Sci.* 37:1–14.

Wilman, D., and J. E. Asiegbu. 1982b. The effects of variety, cutting interval and nitrogen application on the morphology and development of stolons and leaves of white clover. *Grass Forage Sci.* 37:15–28.

Woolhouse, H. W. 1969. Differences in the properties of the acid phosphatases of plant roots and their significance in the evaluation of edaphic ecotypes. In I. H. Rorison (ed.), Ecological Aspects of the Mineral Nutrition of Plants. *Symp. Br. Ecol. Soc.* No. 9, pp. 357–380.

Wynne, J. C., S. T. Ball, G. H. Elkan, and T. J. Schreeweis. 1979. Cultivar, inoculum and nitrogen effects on nitrogen fixation of peanuts. *Agron. Abstr.* p. 82.

Zaiter, H. Z., D. P. Coyne, R. B. Clark, and D. S. Nuland. 1986. Field, nutrient solution and temperature effect upon iron leaf chlorosis of dry beans (*Phaseolus vulgaris* L.). *J. Plant Nutr.* 9:397–415.

Zary, K. W., and J. C. Miller. 1977. Intraspecific variability for dinitrogen fixation in cowpea (*Vigna unguiculata* (L.) Walp.). *HortScience* 12:402. (Abstr.)

Zary, K. W., and J. C. Miller, 1980. The influence of genotype on diurnal and seasonal patterns of nitrogen fixation in southern pea (*Vigna unguiculata* (L.) Walp.). *J. Am. Soc. Hortic. Sci.* 105:699–701.

Zeiher, C., D. B. Egli, J. E. Leggett, and D. A. Reicosky. 1982. Cultivar differences in nitrogen redistribution in soybeans. *Agron. J.* 74:375–379.

Zobel, R. W. 1975. The genetics of root development. In J. G. Toreey and D. T. Clarkson (eds.), The Development and Function of Roots. Academic Press, New York. Pp. 261–278.

8

Role of Foliar Fertilization on Plant Nutrition

SESHADRI KANNAN

I. Overview

The aquatic plants which account for nearly two-thirds of the plant kingdom (Wittwer and Bukovac, 1969) grow in a one-phase environment—in a medium which contains all the factors necessary for growth, namely, water, carbon dioxide, inorganic salts, and light. All the components of these plants are capable of absorbing

Crops as Enhancers of Nutrient Use

inorganic solutes and also fixing carbon dioxide. The terrestrial plants, on the other hand, have evolved into a system to survive and grow in a two-phase environment. The roots provide necessary anchorage and also abstract nutrients and water from the soil while the leaf serves as the major seat of photosynthesis. In the course of evolution, the roots have nearly lost their ability to carry on photosynthesis, but the leaves have retained their ancestral trait to absorb the solutes—a feature which provides the very foundation for foliar feeding and similar agronomical practices.

Foliar fertilization is probably as old as any crop improvement measures practiced in agriculture. Scientists are now aware that foliage of trees could absorb and translocate iron and sulfur compounds released into the atmosphere during ore-smelting activity (Baes and McLaughlin, 1984). Increased yields of wheat and cotton are associated with foliar uptake of sulfur dioxide (Olsen, 1957; Fowler and Unsworth, 1979). Plants acquire several elements like S, N, Mg, Fe, and Cu directly from clouds and fog through a process called "occult precipitation" (Dollard *et al.*, 1983). The absorption of atmospheric ammonia by plant leaves is well established (Hutchinson *et al.*, 1972), and environmentalists are interested in measuring the "dry deposition" of NH_3 and SO_2 on the vegetative surfaces with a view to refine the global budgets of the N and S cycles (Aneja *et al.*, 1986).

The early recorded evidence on foliar absorption relates to the uptake of Fe (Gris, 1844), NH_3 (Mayer, 1874), and K (Bohm, 1877). However, beneficial effects from foliar sprays of inorganic nutrients were demonstrated several years later. The evidence by Hamilton *et al.* (1943) that apple leaves could absorb N in appreciable quantities led to an extensive usage of urea sprays on fruit trees, sugarcane, pineapple, and vegetable crops in many countries (Dadykin, 1951; Mayberry and Wittwer, 1952). It was also shown that, apart from the leaves, trunk and bark of fruit trees also could absorb and translocate nutrients from aqueous sprays (Tukey *et al.*, 1952). Now trunk injection of $FeSO_4$ is regularly practiced for correcting Fe chlorosis of fruit trees (Raese and Staiff, 1988; Yoshikawa, 1988). Foliar application proved very effective for correcting micronutrient deficiencies, and spraying of Fe, Mn, Zn, Cu, or B salts is practiced for several crops (Parker and Southwick, 1941).

During the last three decades, studies on foliar nutrition have received considerable attention. The advent of radioisotopes and the development of techniques for plant analysis have greatly stimulated research on both the basic and applied aspects of foliar absorption, translocation, and utilization. Recent interest in foliar nutrition is also due to the greater awareness of soil–water pollution resulting from indiscriminate or excessive soil fertilization and adverse soil conditions which favor soil fixation of nutrients. For some elements, root absorption is slow and translocation to shoot is poor. Foliar supply of nutrients can increase photosynthetic efficiency by delaying the onset of leaf senescence. Several aspects of the mechanisms of foliar uptake, cuticular penetration, and cell-to-cell translocation are now well documented.

II. Leaf Structure and Components

The leaf consists of a number of structures. The cuticle is the structure which covers the entire surface of the leaf, including the stomatal pores and the epidermal hairs or trichomes, and thus constitutes the first barrier to absorption. It is noncellular and consists of two or more layers, the outermost being generally made up of cutin and wax, and the one beneath having in addition a layer of cellulose (Sitte and Rennier, 1963). The cuticle of *Agave americana* L. has six layers: epicuticular wax, cuticle proper, exterior and interior cuticular layers, and exterior and interior cellin walls (Wattendorff and Holloway, 1980). However, the number is only an approximation, since the boundaries of the layers merge with one another. Cuticle thickness varies depending on the influence of environmental factors during the growth of the leaf. It is likely that the cuticular membrane covering the stomatal cavity is thinner than in other places, since gas exchange through the stomates is rapid.

The shape and structure of the cuticle undergo changes along with the growth of the organs (Miller, 1984). There is a continuous deposition of epicuticular wax during the growth of the leaf, but it decreases and completely stops with cessation of growth (Hallam, 1970). The cuticle remains hydrophilic in the early stages and becomes hydrophobic with maturity (Miller, 1982). Thus the permeability of the cuticle depends on the stage of development, a feature which determines the efficiency of foliar absorption. The epicuticular wax is the most resistant to the entry of water and offers the first barrier to solutes as well. The waxes appear as tubes, platelets, or conelike projections (Hardin and Phillips, 1985) and vary in their wax content and chemical composition depending on the plant species (Freeman and Turner, 1985). The cutin layer beneath the epicuticular wax layer is generally more hydrophilic (Hall and Donaldson, 1963) and may thus serve as polar pathways to solutes. However, the lipids in the cuticles of some species like citrus do not affect water entry even though these are hydrophobic (Haas and Schönherr, 1979).

The surface of the cuticle is not smooth and structures like trichomes, which are uni- or multicellular projections of different shapes, are present. Trichomes are classified into glandular and nonglandular forms. The physical profile and density of both types vary between taxa and even organs of the same plant (Uphof, 1962) and trichome morphogenesis shows contrasting developmental patterns (Turner *et al.*, 1980). For example, in *Cyphomandra*, trichome initials arise for a limited period of ontogeny and decrease in density during the enlargement of the organs. However, in the leaf of *Cannabis*, new initials arise regularly, thus maintaining a nearly constant density. The long-stiped glandular trichome is of evolutionary significance because it is derived from the short-stiped form and possesses a dehiscence mechanism whereby the glandular head abscises from the stipe at a specialized site of weakness between the neck cell and the gland head (Mahlberg, 1985).

Trichomes are also covered by cuticle. Several electron micrograph pictures reveal the presence of long, pointed, singular and multiple hairs on the upper and lower surfaces, especially on the midribs and veins of pecan leaves (Meyer and Meola, 1978). Differences in the leaf surface features between the scab-susceptible and scab-resistant pecan cultivars have been observed (Wetzstein and Sparks, 1983). There are also differences in the density of nonglandular hairs between the immature and mature leaves and also between the abaxial and adaxial surfaces of pecan (Grauke *et al.*, 1987). Stomalike structures which are larger than stomates are distributed on the midrib regions. The bulbous base of the trichome extends over the entire thickness of the epidermal cells, with the basal accessory cells bulging around the trichome (Figure 1). Since cuticle development at the base of the trichomes is less than in the rest of the trichomes (Sifton, 1963) there may be greater absorption of solutes through these basal regions (Schönherr and Bukovac, 1970). Furthermore, the presence of a large number of trichomes especially on the abaxial surface would facilitate greater foliar uptake because of the increased surface area provided by them (Leon and Bukovac, 1978).

The surface of the cuticles is covered by a layer of wax, formed from the wax precursors secreted from within or below the cuticles and deposited on the surface. Long ago DeBary (1871) postulated the existence of wax-exuding pores and channels in the cuticle, but so far there has been no proof for the excretion of wax precursors through them, although the presence of "micro-channels" (Scott *et al.*, 1948) and anastomosing microchannels for the purpose of wax extrusion has been indicated (Hallam, 1964). The presence of pores and canals in the dewaxed leaf, stem, and fruit cuticles could be observed through photomicrographs (Miller, 1984, 1985, 1986). The transcuticular canals are oriented perpendicular to the outer and inner membrane surfaces and terminate as discrete pores in the dewaxed cuticular membranes (Figure 2) and in the dewaxed isolated cuticular segment of the adaxial leaf surface of *Hoya carmosa* (Figure 3). According to Schönherr and Huber (1977) the isoelectric point of the cuticle is about pH 3, indicating the possibility of transcuticular transport of cations and anions through the pores and canals, depending on the charges and the pH of the solute. However, the epicuticular layer is important for the entry of solutes through the pores, since the permeability increased by either dewaxing or abrading the leaf surface (Schönherr and Bukovac, 1970).

Certain threadlike structures were detected in the cell wall by Schümacher (1942) and Lambertz (1954) considered them to be present in all angiosperms. Franke (1971) proposed that these structures serve as pathways for solutes across the cell wall, and designated them as "ectodesmata" or "ectoteichodes" (Franke, 1975). Additional evidence has now been provided for the appearance of ectoteichodes in wheat leaves (Panić and Franke, 1979; Werner, 1981). An interesting feature is that when the leaves are infected by *Erysiphe graminis tritici*, the number of ectoteichodes within the periclinal and along the anticlinal epidermal walls decreases in the diseased leaves in contrast to the healthy ones (Figure 4). These bodies disappeared at the time when conidia production was high. According to Panić and Franke (1979) their appearance is associated with the cell metabolism. Werner (1981) has studied this in detail and has provided strong microscopic and

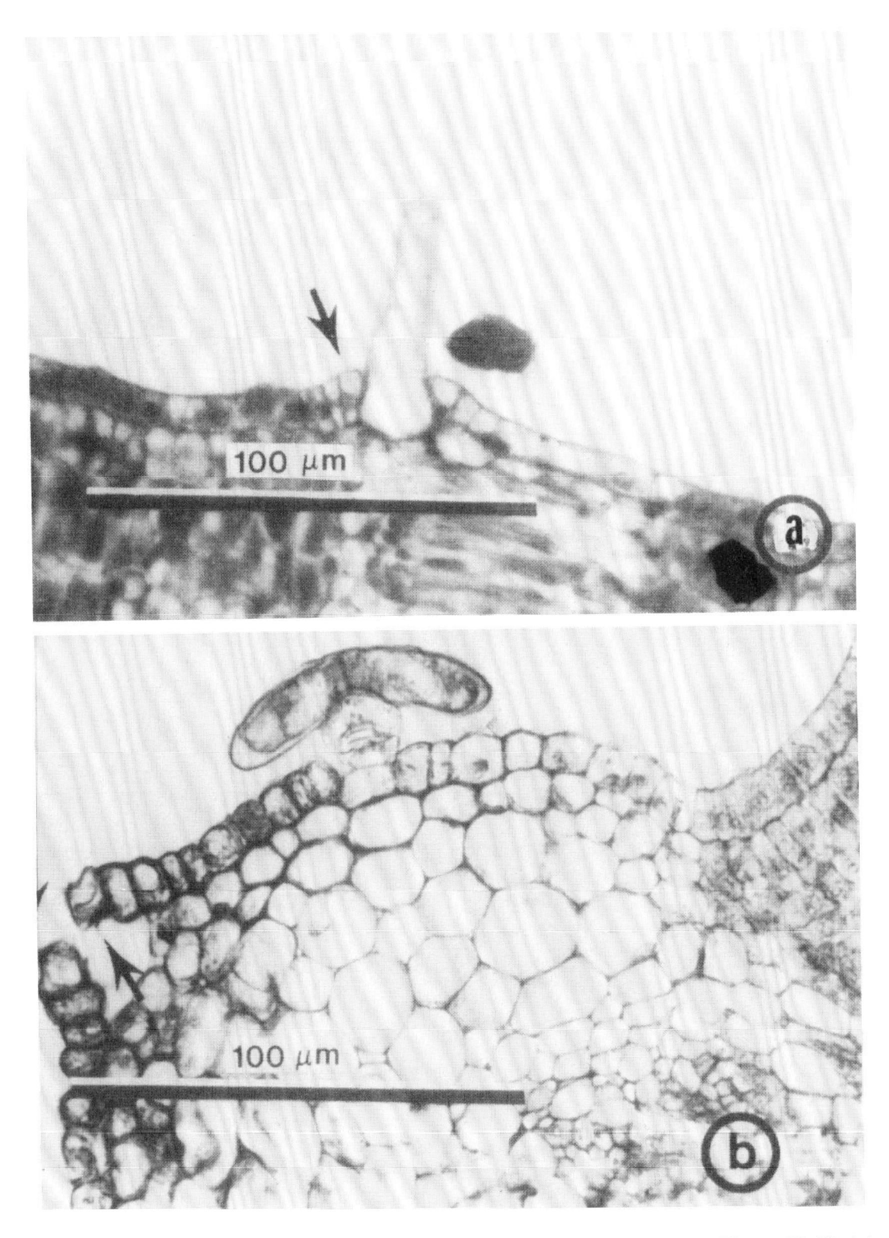

Figure 1. Light microscopic view of immature leaf of pecan (*Carya illinoepsis,* Wang., K. Koch.) 'Delicious,' showing (a) the acicular hair and the accessory cells bulging around the base and (b) a veinal stoma with large aperture. Note the cuticular lips and ledges in the substomatal chamber (arrows). Source: Grauke *et al.,* 1987. Reproduced with permission from copyright owners, the authors.

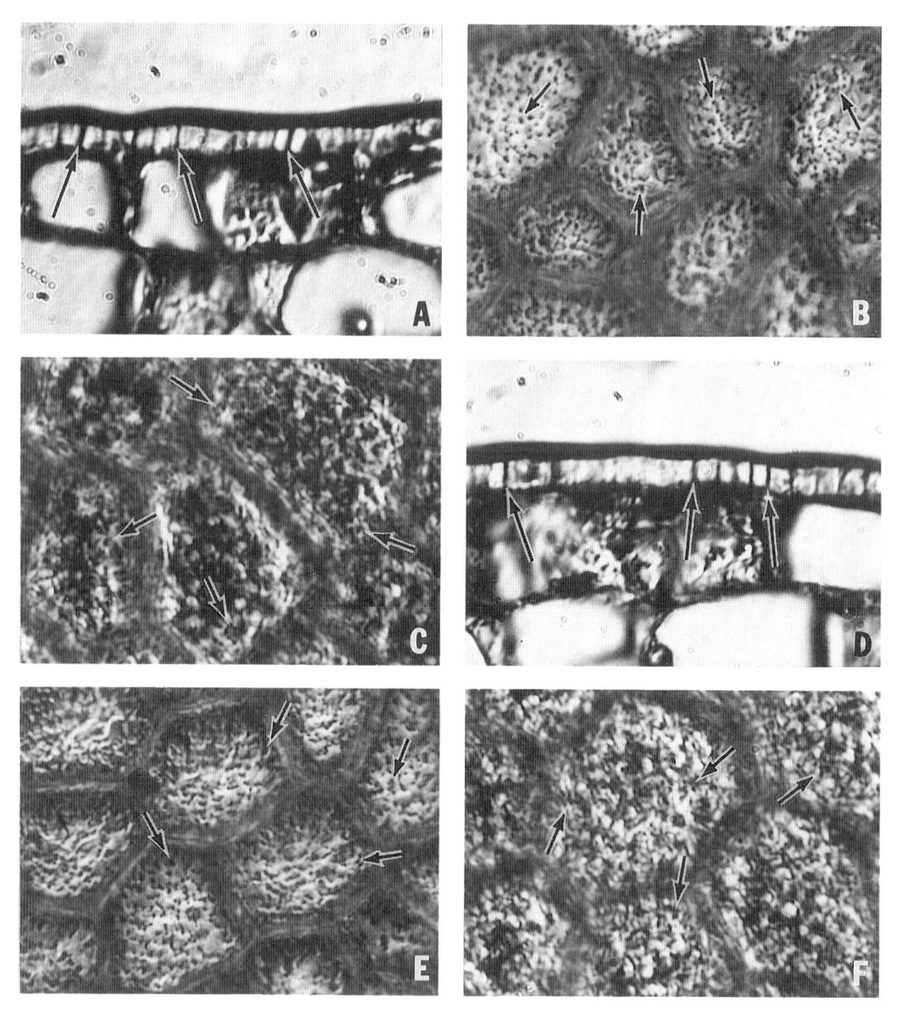

Figure 2. (A–F) *Hoya carnosa*. (A) Adaxial leaf surface transection; (B) outer surface of an adaxial isolated cuticular membrane (ICM); (C) inner surface of an adaxial ICM; (D) abaxial surface transection; (E) outer surface of an abaxial ICM; (F) inner surface of an abaxial ICM. Note the topographical "granularity" of the inner surfaces. Arrows indicate either several of the distinct anticlinally oriented transcuticular canals in the transections or several of the ubiquitous, minute cuticular pores in the isolated membranes. The thick dark line appearing on the surface of the lighter, striated-appearing cuticular membrane is an artifact due to inadvertent tilting of the tissue during sectioning. Magnification, ×425. Source: Miller, 1986. Reproduced with permission from the Annals of Botany Company.

autoradiographic evidence for the presence of ectoteichodes in several plant species.

The nutrient elements passing through the cuticle reach the water free space from where the leaf cells absorb them. The transport of ions across the plasmalemma of the cells is an active process which is mediated through the energy from both respiration and photosynthesis.

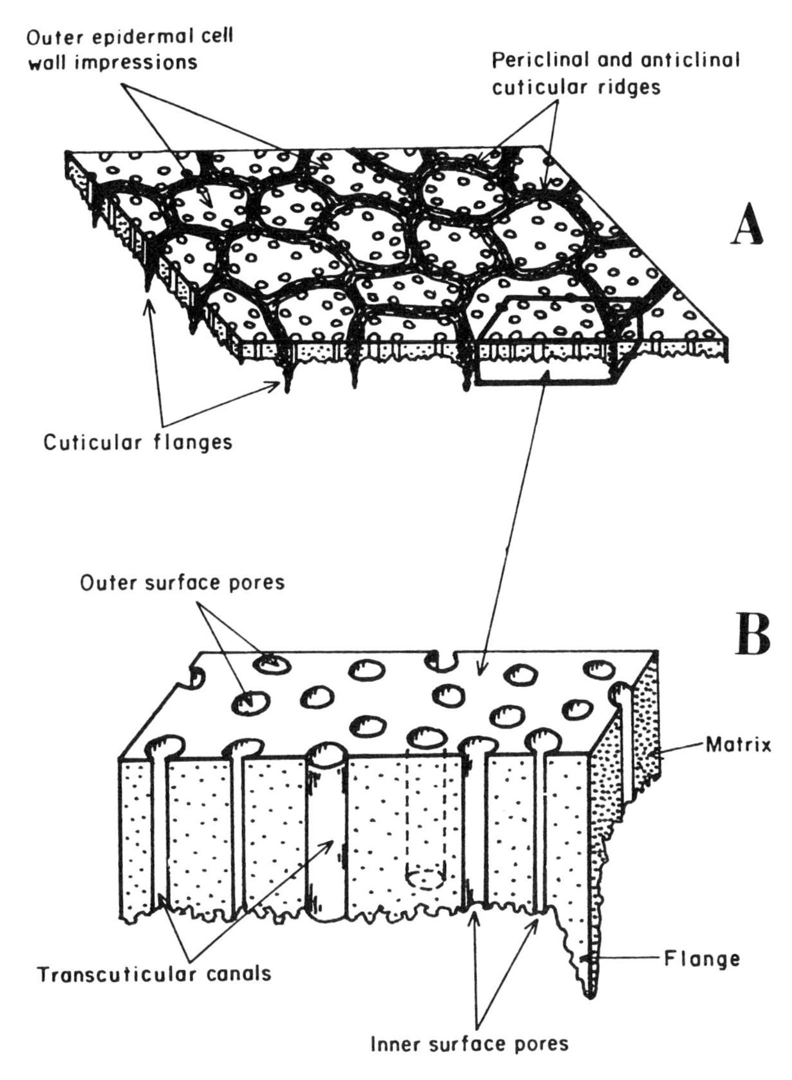

Figure 3. Diagrammatical interpretation of a dewaxed isolated cuticular membrane segment (A) of the adaxial leaf surface of *Hoya carnosa*, depicting the general morphology of its cuticular pores and related transcuticular canals (enlarged in B). The number of pores and canals is underrepresented and their respective diameters are exaggerated. Source: Miller, 1986. Reproduced with permission from the Annals of Botany Company.

III. Efficiency of Foliar Uptake

Inorganic nutrients supplied to the leaves follow certain pathways before reaching the leaf cells. As mentioned earlier, the epicuticular wax layer is essentially hydrophobic and the cutin layer beneath consists of polyesterified hydroxy fatty acids

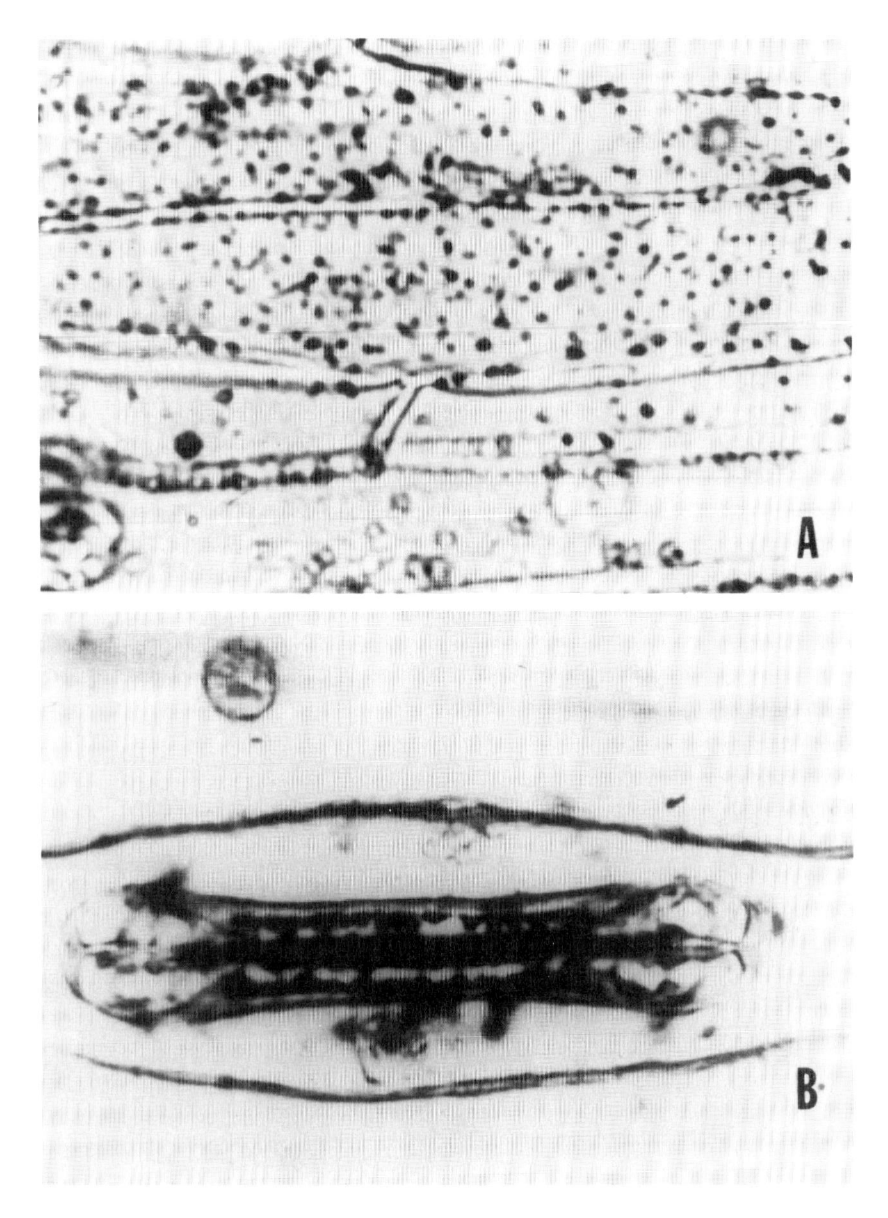

Figure 4. Normal distribution of ectoteichodes in a surface view (A) (×533) and in the stomatal region (B) (×832) of a healthy leaf of *Triticum aestivum*. The typical lining of ectoteichodes along the anticlinal walls and their scattered distribution in the periclinal walls are seen (A). The ectoteichodes are seen lined along the stomatal pore and crowded in guard cells of a stoma (B). Source: Panić and Franke, 1979. Reproduced with permission from Verlag Eugen Ulmer.

(Martin and Juniper, 1970) and is more permeable to water (Hall and Donaldson, 1963) and therefore to the dissolved inorganic ions. Other components of cuticle which consist of proteins and pectinaceous substances are also more permeable to water and solutes.

A. PERMEABILITY OF CUTICLES

Between the cuticular membrane and the epidermal cell wall is a layer of pectin (Martin and Juniper, 1970) which can be dissolved by suitable enzymes so that the cuticle can be separated intact. Cuticle can also be separated by chemical treatment. With the enzymatic method, the leaf segments are vacuum infiltrated and incubated in an acetate buffer at pH 3.8 containing 5% pectinase and 0.2% each of cellulase and hemicellulase (Yamada *et al.*, 1964). The chemical method involves the incubation of the leaf slices with 1.6% ammonium oxalate and 0.4% oxalic acid at pH 4.7 and 35°C (Mazliak, 1963). Some workers dewax the cuticle and treat with a $ZnCl_2/HCl$ mixture and this treatment is suitable for cuticles with a density of 0.12 mg/cm^{-2} and higher (Holloway and Baker, 1968).

A number of methods have been developed for studying the cuticular permeability. The apparatus used (Yamada *et al.*, 1964; Kannan, 1969) consists of a large test tube containing deionized water into which a smaller tube with the cuticle fixed is kept immersed. A known amount of radioisotope solution is placed inside the smaller tube and the penetration through the cuticle is measured by assaying the samples drawn from the large tube. A double-chamber technique is employed by McFarlane and Berry (1974), who estimated the permeability coefficients for the inorganic elements and found them to obey Fick's law of diffusion. Studies with isolated cuticles show that the cuticular penetration is largely a diffusion process although the greater rates of penetration of ions like Rb, Ca, SO_4, and Cl have been attributed to a "facilitated diffusion" in the presence of urea (Yamada *et al.*, 1965). A few studies have shown a negative correlation between ion size and cuticular penetration. The order of penetration through the *Euonymus* cuticles is CS = Rb > K > Na/Ba > Sr > Ca (Haile-Mariam, 1965). Permeability also differed with the size of the molecules. Larger molecules like FeEDDHA (ferric ethylene diamine di-orthohydroxy acetate) penetrated at a slower rate than $FeSO_4$ (Kannan, 1969). Differences in the rates of penetration between the stomatous and astomatous cuticles were also obtained. The inorganic ions penetrated at a greater rate through the stomatous than the astomatous cuticles derived from the same plant species in *Euonymus* (Kannan, 1969) and in pear leaves (Chamel, 1980). While cuticular membrane permits the penetration of ions, it also retains some of them by a process of sorption (Chamel and Neumann, 1987).

B. UPTAKE BY LEAF CELLS

After passing through the cuticles, the nutrient elements accumulate in the "free space," a region which remains outside the plasma membrane of the leaf cells.

Studies with root uptake have shown that the cell walls offer very little resistance and the space between the cell wall and the plasmalemma is only an extension of the aqueous medium from which ions are transported across the plasmalemma of the cells (Epstein, 1973). The absorption of ions by the leaf cells is also analogous to that by root cells.

Ion uptake by leaf cells has been investigated using a number of tissue systems, such as narrow (1 mm wide) leaf slices, leaf disks, and leaf cells enzymatically separated from the leaf. The studies reveal that the uptake of K, Na, Rb, Fe, Mn, phosphate, and chloride is mediated by active processes (Smith and Epstein, 1964; Jyung *et al.*, 1965; Smith and Robinson, 1971; Kannan and Ramani, 1974; MacDonald and Macklon, 1975; Bowen, 1981). The absorption is concentration dependent, affected by metabolic inhibitors, and enhanced by light (Kannan, 1970). In general, ion uptake by leaf cells follows the same patterns as obtained with the roots. Dual mechanisms were recorded for Rb and chloride (Kannan and Ramani, 1974) and dual as well as single multiphasic mechanisms for several ions have been reported in many leaf systems (Epstein, 1973; Nissen, 1973; Bowen, 1969). Bowen (1981) obtained a single multiphasic pattern in the absorption of Zn, Cu, Mo, and B by the callus from the sugarcane leaf. The use of enzymatically separated leaf cell protoplasts helps in removing the cell wall barrier in the ion uptake process. According to Goldstein and Hunziker (1985), phosphate uptake by wheat leaf cell protoplasts was metabolic and sensitive to inhibitors, in the same way as the root cell protoplasts. Rb uptake by pea mesophyll protoplasts was also inhibited by CCCP (carbonyl cyanide *m*-chlorophenylhydrazone), but differentially in the light and the dark (Rahat and Reinhold, 1983). Jacoby (1975) found that light enhanced Rb accumulation in the stomatal regions of the bean leaf, while larger amounts of Na, K, or Rb were retained in the vascular tissues in the light.

An understanding of the mechanisms of ion absorption by leaf cells is important since it helps explain the differences in response to foliar sprays by different plants. It has been shown that Fe is absorbed more effectively from $FeSO_4$ than from FeEDDHA; however, Fe from FeEDDHA is available to plants for a longer time. Urea increased Fe absorption by leaf cells (Kannan and Wittwer, 1965) and perhaps this explains the better response of Fe sprays with urea (Reed, 1988). Factors such as light and temperature which increase cellular uptake will also enhance foliage uptake.

C. TRANSPORT IN THE LEAF

The pathways of nutrients within the leaf follow two routes to reach the vascular tissues before transport out of the leaf. The free space between the cells forms a continuum providing the apoplastic route. The free space in a leaf accounts for nearly 3–5% of the total volume and varies among plant species (Crowdy and Tanton, 1970). The second route is the symplastic transport, which takes place between the cells through the cytoplasmic continuum, i.e., the plasmadesmata, and is dependent on the cytoplasmic activities and the metabolic energy. The mechanisms of apoplastic and symplastic transport of inorganic ions in the leaf are not understood. Our knowledge of the pathways of nutrient transport between the epidermal, mesophyll, and spongy parenchyma cells to the vascular tissues is largely derived from that of the photoassimilates. It is based on the consideration that the downward transport from the leaf takes place through the phloem, similar to the transport of the photosynthates. The transport of solutes from sites of photosynthesis in the mesophyll to the phloem is accomplished through the symplastic continuum. The solutes enter the apoplast at some point close to the sieve element or companion cell complex followed by active uptake by the phloem cells (Franceschi and Giaquinta, 1983a). Harris and Chaffey (1985) have identified structures known as "plasmatubules" in the transfer cells of leaf minor veins of *Pisum sativum* L. and found them to be associated with sites where high solute flux occurs between apoplast and symplast. These structures are tubular invaginations of the plasmalemma, providing the connecting links for transfer of solutes from leaf cells to vascular systems.

There are some interesting observations on the symplastic transport of substances in leaves. Erwee *et al.* (1985) employed fluorescent probes, i.e., 6-carboxy-fluorescein and lucifer yellow CH, which have the unique property of not passing through plasmalemma and are thus ideal for revealing the nonplasmatic connections. The dye 6-carboxy-fluorescein moved freely in the epidermal cells of *Commelina cyanea,* showing that these are symplastically linked, similar to those between the mesophyll cells (Figure 5). Injection of the dye into the epidermal cells of the leaves of *Vicia faba* and *Antheophora pubescens* did not show symplastic connection between the guard cells and the epidermal cells. The dye injected into the mesophylls moved into the vascular cells and then into the minor veins. The results of the dye movements suggest that solutes would move symplastically to the vascular tissues before moving out of the leaf. The establishment of functional symplastic connections from the mesophyll to the minor veins has been further shown using a novel dye-tracer method (Madore *et al.*, 1986). The fluorescent dye lucifer yellow CH is encapsulated in phospholipid vesicles (liposomes) and injected into the vacuoles of leaf cells of *Beta vulgaris* or *Ipomea tricolor*. The release and subsequent cell-to-cell movement of the dye were followed by fluorescence microscopy. The presence of plasmadesmata between all cells from the mesophyll to the minor veins was demonstrated.

Data from the double-labeling with $^{14}CO_2$ and $^{32}PO_4$ show that the pattern of P movement to the growing regions resembles that of sugars, i.e., via the phloem, and are largely in the form of inorganic phosphate (Marshall and Wardlaw, 1973). Franceschi and Giaquinta (1983a,b) identified a highly specialized layer called the paraveinal mesophyll (PVM) in the leaves of some legumes. The PVM is one layer thick and forms a cellular network. These have fewer chloroplasts and are nonphotosynthetic tissues interspersed between the palisade and spongy mesophyll. In soybean, the position of PVM is such that the assimilates would pass through this network before loading the phloem. If the nutrients are transported the same way, PVM may facilitate their movement from the cells to the phloem.

Erwee *et al.* (1985) made the important observation that guard cells in the leaf

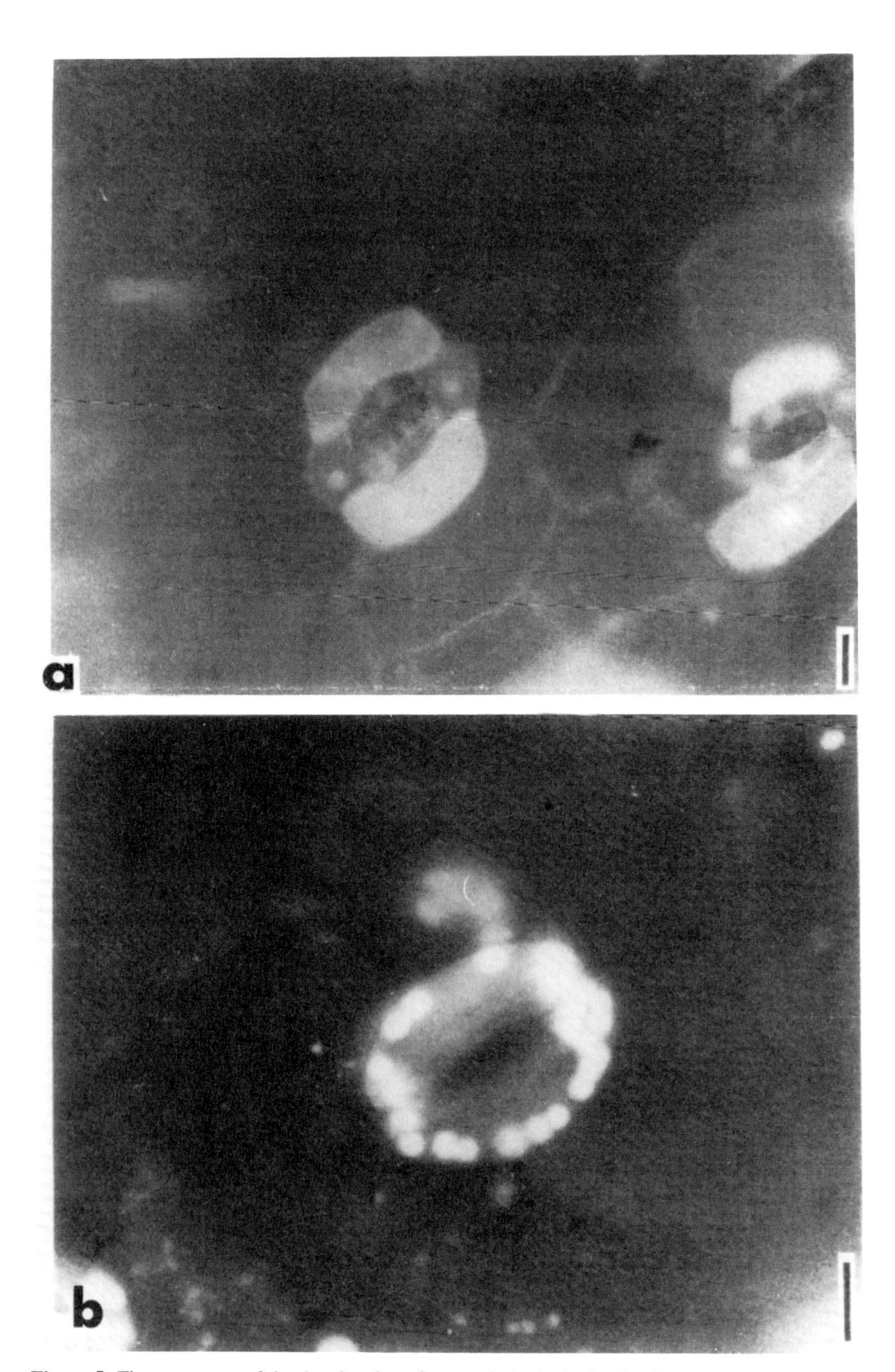

Figure 5. The movement of the dye 6-carboxyfluorescein in the leaf cells of *Vicia faba*. (a) The dye is seen to move into the epidermal and subsidiary cells. Cells containing the dye appear white as opposed to the guard cells which remain dark and do not contain the dye. Bar: 25μm. (b) Injection of the dye into the guard cells results in the dye remaining in the guard cells. No dye is seen in the surrounding epidermal cells. Bar: 20 μm. Source: Erwee *et al.*, 1985. Reproduced with permission from Blackwell Scientific Publications, Ltd.

are isolated from cell-to-cell communication (Figure 6). Thus ions such as K absorbed by guard cells would move between the guard cells, but not into the adjoining epidermal cells. Their finding is also supported by the fact that ABA (abscisic acid) caused stomatal closure by decreasing the osmotic pressure of guard cells, but adjoining cells remained unaffected (Mansfield and Jones, 1971). This effect was paralleled by a reduction in uptake and concentration of K(^{86}Rb) in the guard cells only. An interesting study on the ion transport was conducted with protoplasts isolated from the epidermal tissue of *Commelina*. The presence of K brought about volume changes and swelling of the guard cell protoplasts and the addition of $CaCl_2$ inhibited the swelling. The flux of K occurring in guard cell protoplasts was sufficient to explain the turgor changes and the stomatal opening and closing (Fitzsimons and Weyers, 1986).

Intercellular transport of inorganic solutes within the leaf is very rapid. For example, the movement of ^{45}Ca in oat leaf was traced by histoautoradiography and high-resolution tract-autoradiography (Ringoet *et al.*, 1971) and it was found to be distributed through the mesophylls of the whole leaf in 30 minutes. Furthermore, 50% of ^{45}Ca was located in the chloroplasts, and its initial entry was through the guard cells.

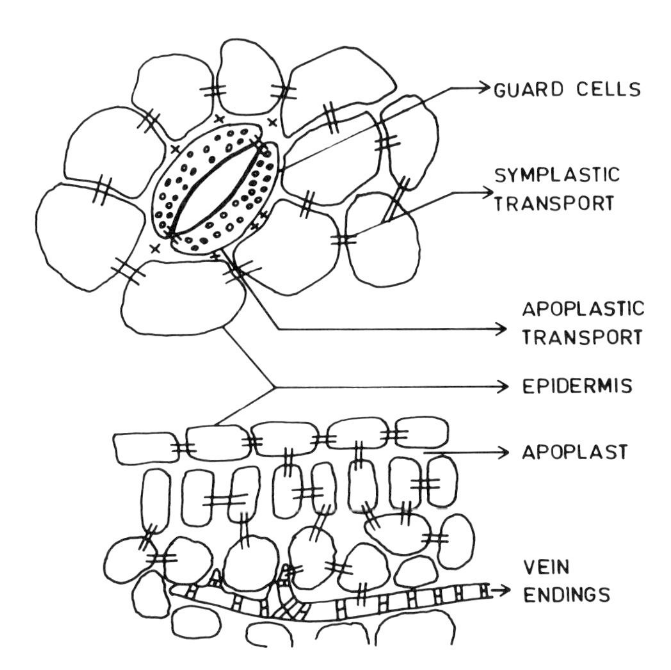

Figure 6. Diagrammatical representation of cell-to-cell communication among leaf cells vis-à-vis guard cells. Note that the epidermis, spongy and pallisade mesophyll cells, and vascular tissue (bottom) are all linked in a continuous symplast (=), but guard cells are symplastically isolated from the surrounding cells (top). The transport between guard cells and other cells is mostly apoplastic, while the two guard cells of a stomate are probably symplastically connected. Source: Kannan, 1986. Reproduced with permission from CRC Press Inc., Boca Raton, Florida.

Nutrient transport between the leaf cells and then into the conducting vessels is important for outward transport. Radioisotopes have been employed to trace the direction of movement of the elements in the leaf. The credit for such experiments goes to Müller and Leopold (1966), who were interested in the influence of growth substances like kinetin on the direction of transport of ^{86}Rb, ^{22}Na, ^{32}P, and ^{36}Cl. Using a chromatogram scanner they found that ^{32}P moved preferentially to the base, indicating a natural mobilization center at the base. Since the transport did not follow the water flow in the xylem and was blocked by steam killing, it was considered to take place through the phloem. Ringoet *et al.*, (1968) developed a technique to measure the rate of transport of ^{45}Ca supplied to oat leaves by *in vivo* assay with a beta-sensitive semiconductor device, and found that ^{45}Ca supplied above the midrib of the leaf was absorbed and transported in the vascular bundles; however, larger amounts were located in the epidermal and mesophyll cells. Migration of Ca applied below and above 0.02 *M* concentrations was in the acropetal and basipetal directions, respectively. According to them, the poor mobility of Ca is due to a greater absorption and accumulation capacity of the leaf tissues rather than its inability to move through the phloem. Millikan and Hanger (1969) found that when ^{45}Ca was supplied on the surface of the leaf, it was immobile. But when it was injected into the midrib, it was rapidly translocated and found in all the vascular tissues.

The transport patterns of ^{59}Fe, ^{54}Mn, and ^{65}Zn in corn leaves were examined using a modified chromatogram scanner fitted with a gamma-ray detector. When supplied to the tip, middle, or base of the leaf, all these elements moved toward the base, suggestive of an attractive force near the base. Furthermore, when the isotopes were placed on one side of the leaf, these elements did not migrate to the other side across the midrib (Figure 7) (Kannan, 1978). Chamel and Eloy (1983) developed a technique using spark-source mass spectrometry and laser probe spectrographs for tracing the movement of stable isotopes in leaves. They found that B entered the leaf, but became bound to some polysaccharides within the leaf and was only partially mobile.

IV. Foliar Uptake—Long-Distance Transport

A. RELATIVE MOBILITY

In general, the nutrients absorbed by the leaves are transported to other parts of the plant within a short period. However, the rate of translocation depends on the nature of the elements and on the plant species. Bukovac and Wittwer (1957) and Biddulph *et al.* (1958) developed "leaf drop and washing" and "leaf dipping" methods to measure the rates of transport of various elements. Jyung and Wittwer (1964) used a "leaf immersion" technique and introduced parameters like "specific absorption," V_{max} and K_m by which the mobility of different elements can be

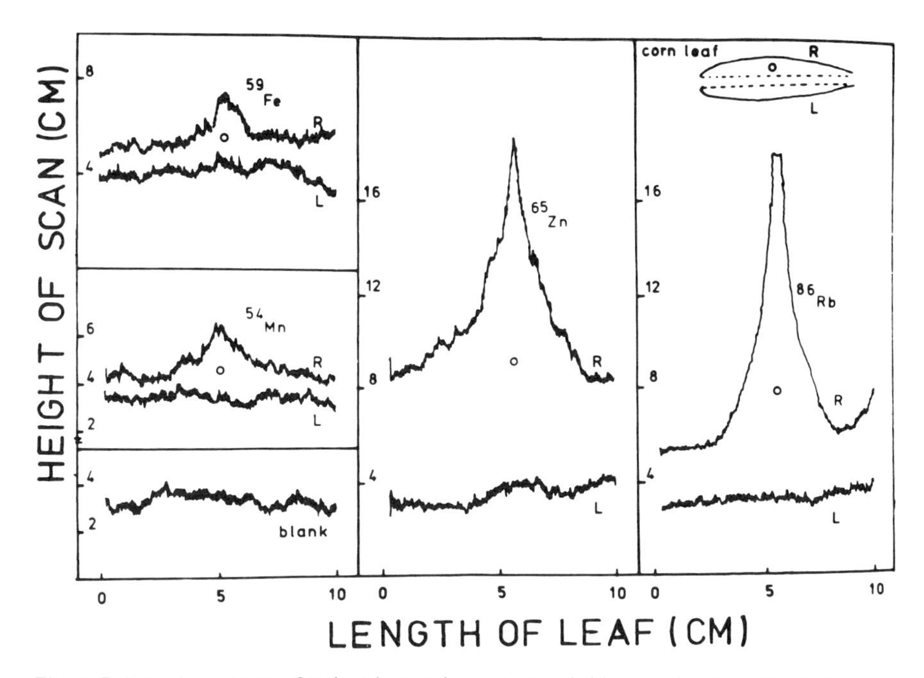

Figure 7. Lateral movement of cations in corn leaves as revealed by scanning the radioactivity in the right (R) and left (L) halves of the leaf 24 hours after isotope application. The isotopes were placed in the middle of the leaf (circle) and removed before scanning. The two halves of the leaf were separated and scanned in a modified chromatogram scanner. Source: Kannan, 1978. Reproduced with permission from the American Society of Plant Physiologists.

compared. Foliar absorption and translocation have been examined by "leaf injection," "leaf dipping," and "leaf droplet" methods (Jacoby and Plessner, 1971; Kannan and Keppel, 1976; Penot and Gallou, 1977). Bukovac and Wittwer (1957) classified the nutrient elements into freely mobile [K(Rb), Na, P, Cl, and S], partially mobile (Fe, Mn, Zn, Cu, and Mo), and immobile (Ca and Mg) elements. Penot (1972) obtained the order of mobility in *Tradescantia viridis* as $^{32}P > ^{86}Rb > ^{99}Mo > ^{35}S > ^{36}Cl > ^{24}Na$.

The transport of micronutrients from the leaf is generally slower than that of the major elements like P and K. However, the rate of transport can be increased when these are in the chelated form. In a comparative study on the mobility of Mo and Rb in bean plants, it was found that a larger percentage of Mo remained in the primary leaves and only about 8% was translocated to the shoot in contrast to 20% of translocation for Rb (Table I) (Kannan and Ramani, 1978).

Ca is an important major nutrient element, but it is classified as immobile in plants. Many physiological disorders of fruits are associated with low levels of Ca in the fruits resulting from poor translocation and migration from the leaves at the time of development (Perring, 1979). Since Ca is absorbed directly by the fruits, foliar

Table I. Foliar Absorption and Translocation of MoO_4^{2-} and Rb^+ in Bean Plants[a,b]

	MoO_4^{2-}		Rb^+	
Plant Structure	nmol	Distribution as % of Total	nmol	Distribution as % of Total
Applied primary leaf	604	17.16	383	13.03
Opposite primary leaf	665	18.89	299	10.16
Trifoliate leaf	276	7.84	590	20.05
Stem	1080	30.68	432	14.68
Root	895	25.43	1238	42.08

[a]Source: Kannan and Ramani, 1978. Reproduced with permission of the American Society of Plant Physiologists.

[b]The isotope solution of molybdate/Rb was supplied to a small area of a primary leaf. The amount transported to the remaining part of the applied leaf and other parts is given as absolute amounts.

spraying on the plants to cover the fruit surfaces also is the best method for correcting Ca deficiency. One explanation for the immobility of leaf Ca is that it is transported in the xylem by ion exchange and is concentration dependent (Faust and Shear, 1973). Since the concentration gradient of root-absorbed Ca is toward the leaf, the transport from the leaf to other plant parts is against this gradient. Faust and Shear (1973) reversed the gradient by raising the Ca concentration in the leaf and provided evidence for greater transport of Ca from the leaf.

Perhaps the most widely used nutrient through foliar application is N and the high degree of mobility of this element is a distinct advantage. The mechanism of N transport from the leaf is clearly associated with carbohydrates from the leaf. Okano *et al.* (1983) followed the translocation patterns of $^{13}CO_2$ and $^{15}NO_2$ fed simultaneously to a mature terminal leaf of the rice plant at boot stage and found that the transfer of ^{13}C occurred within a day. Transfer of ^{15}N was rapid in the first few hours and was steady for 8 days. Furthermore, the close similarity in the labeling patterns of the two isotopes in various plant components indicated that (1) ^{13}C and ^{15}N moved together in the bulk stream of the phloem and (2) the relative distribution of C and N is regulated by the characteristics of the sink organs. At the reproductive stage, the roots would compete with other "sinks" like the developing grains, yet a large amount of ^{13}C (19%) and ^{15}N (17%) was exported to the root within a day. The nutrients were retranslocated to the new roots and shoot which seem to function as "sinks."

Tatsumi and Kono (1981) examined the transport of N from different leaves of rice plants. Feeding a single upper (8th) leaf or a lower (5th) leaf with [^{15}N]urea (0.2%) over 7 days resulted in the incorporation into the trichloroacetic acid-insoluble compounds in the shoot and root. About 25% of ^{15}N exported from the upper leaf was fixed in the roots. The percentage of N exported to the roots was larger from the lower leaf than from the upper one. Their results show that the developed leaves are the major sources for supply of N compounds to the growing roots of rice plants. Experiments with sunflower and avocado reveal that only 12% of N supplied as $^{15}NO_2$ to a mature leaf remained in the leaf while the major amount was translocated to young leaves and roots; none was found in the other mature leaves. Zilkah *et al.* (1987) found that foliar-applied urea was translocated basipetally from the current flush of leaves to the developing fruits which acted as strong "sinks" for N.

B. INFLUENCE OF GROWTH REGULATORS

The effects of growth regulators on foliar uptake of nutrient elements were investigated several years ago. Mothes *et al.* (1959) was the first to observe that kinetin applied to a tobacco leaf attracted metabolites from the surrounding regions. Kinetin applied to a detached oat leaf attracted metabolites and delayed the senescence (Gunning and Barkeley, 1963). Definite proof that kinetin attracted inorganic solutes came from the work of Müller and Leopold (1966). They found that the leaf site where kinetin was supplied attracted and accumulated ^{32}P. In an interesting experiment, a competition was created for the mobilization of ^{32}P by applying kinetin at two centers, one at the apex and the other at the base, with the isotope in the middle of the leaf. They found that kinetin could not attract ^{32}P in either direction since the attractive force was equal from both the centers. Further evidence suggested that the action of kinetin is through the conducting vessels, which are parallel in the monocot leaves, since kinetin supplied to the one side of the midrib did not attract ^{32}P placed on the other side.

The effects of several growth regulators on foliar transport in detached leaves of *Pelargonium zonale* were examined using autoradiographic techniques (Penot and Beraud, 1977; Penot, 1978). They found that IAA (indole acetic acid), 2,4-D (dichlorophenoxy acetic acid), NAA (naphthalene acetic acid), and GA_3 (gibberellic acid) influenced the direction of transport by attracting ^{32}P, ^{35}S, and ^{86}Rb to the site of application of growth substances. Penot (1979) noted that IAA, BAP (benzyl aminopurine), and GA_3 applied together showed synergistic action on attracting P, S, and Rb. GA_3 applied alone had a pronounced antisenescence effect and it was concluded that the actions of growth substances on ion migration and senescence are independent effects. The action of ABA (abscisic acid), FC (fusicoccin), BAP, and GA_3 on ion transport was independent on their influence on transpiration (Penot *et al.*, 1981). Penot and Beraud (1985) reported that an increase in the ATP level was paralleled by the enhancement of P uptake by the leaves pretreated with BA_3 (50 μM) for 24 hours.

Growth regulators exert their influence on foliar transport in a number of ways. These could enhance the absorption by the leaf at the site of application, increase the migration within the leaf, and/or stimulate the transport out of the leaf in the acropetal or basipetal direction. Kessler and Moscicki (1958) reported that TIBA (triiodobenzoic acid) and MH (maleic hydrazide) increased the downward transport of ^{45}Ca supplied to the apple leaves. Application of GA_3 or CCC (chloroethyl choline chloride) and kinetin influenced foliar uptake of Fe, and kinetin increased its translocation (Kannan and Mathew, 1970). DMSO (dimethyl sulfoxide), a chemical with multiple effects on plants, increased Fe absorption by maize leaves (Chamel, 1972). These results from short-term experiments support the fact that growth regulators have definite influence on foliar uptake and have the potential for exploitation in the field-grown crops.

V. Environmental/Physiological Factors Affecting Foliar Uptake

A. LIGHT, TEMPERATURE, AND HUMIDITY

The effects of light, atmospheric humidity, and temperature on foliar uptake of nutrients are closely linked with each other. Individually each exerts its influence on photosynthesis, which supplies the energy for absorption. The greater absorption by leaves is favored by low light, optimum temperature, and high humidity. The indirect effects of these factors are on the development processes of the leaf components like wax protrusion, thickness of wax, and the thickness of cuticles. The amount of cuticle, cutin matrix, and cuticular wax on the leaves of *Brassica oleracea* (Macey, 1970), cereal crops (Tribe *et al.*, 1968), and carnation (Reed and Tukey, 1982) is estimated to be higher in those grown under high rather than low light intensities. The wax development varies with the age of the leaf and also with the environmental factors. For example, the wax content of the leaves of 'Williams' banana differed between the field-grown and greenhouse-grown plants (Freeman and Turner, 1985); the wax content of the latter was only 60% of the former. Furthermore, it increased with leaf age but only up to the 4th leaf (Figure 8).

The development of secondary wax structures is also dependent on light intensity. Hull *et al.* (1975) studied in detail the environmental influence on cuticle development of mesquite. They reported the presence of significant amounts of wax structures on the upper surfaces of the youngest leaflets, which were difficult to wet without the use of surfactant. They employed a dye, acridine orange, to record the differences in the permeability of the leaves. It was observed that the dye entered first via both trichomes and stomata and then spread laterally through the epidermal cells in the leaflets of the trees grown outdoors. As the leaves matured, the trichomes continued to absorb the dye, but ceased passing it to the surrounding epidermal cells, indicating the greater cutinization of the trichome bases. As the trees approached full maturity the cuticle became thick and the absorption through the cuticle was the only mode of entry for the dye. After a few months of growth, even cuticular absorption dropped to a very low level. They found that by contrast, the seedlings grown in the controlled environments generally never developed either trichomes or thick cuticles. Thus the entry of the dye was limited to stomatal and

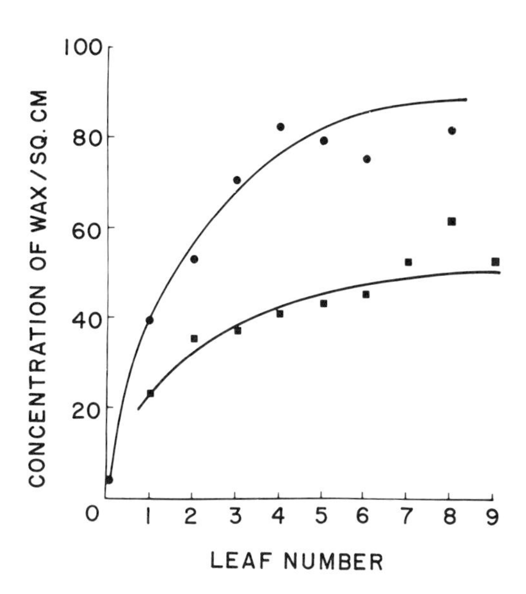

Figure 8. Total leaf wax content in *Musa acuminata* 'Williams' as influenced by age of the leaf (leaf number) and environment. Note that total wax did not alter appreciably on leaves older than 4 in the field (●). Glasshouse-grown leaves (■) produced wax similarly, leveling out by leaf 3 or 4. Total wax on glasshouse-grown leaves was only 60% of that on field-grown leaves. Source: Freeman and Turner, 1985. Reproduced with permission from CSIRO Editorial and Publishing Unit.

cuticular pathways. These findings are important because the development of cuticle and trichomes affects foliar uptake.

Temperature affects foliar absorption in two ways. Increased absorption by leaves at high temperatures has been recorded (Cook and Boynton, 1952). At the same time, high temperature causes drying of the spray drops and thus reduces absorption (Reed and Tukey, 1978). Likewise, the atmospheric humidity exerts its influence partly by keeping the spray droplets from drying. Ca uptake by apple from $CaCl_2$ solution was increased by raising the relative humidity to 87%. However, increasing it to 94% or decreasing to 80% reduced the uptake (Lidster *et al.*, 1977). The thin films of moisture resulting from transpiration also promote foliar absorption and perhaps much more than the water in the spray medium (Thorne, 1958).

B. LEAF AGE, PLANT SPECIES, NUTRITIONAL STATUS, AND PLANT METABOLISM

The capacity of the plant leaves to absorb nutrients greatly differs with age. The young leaves are metabolically more active than the older senescing leaves and therefore are more efficient in absorption. Plant species differ in their ability to absorb foliar-applied nutrients. The reasons for such differences are largely due to the physical conditions of the leaf and the physiological status of the plants. Information on this aspect is severely limited. Swietlik and Faust (1984) report that apple leaves absorbed urea-N three times more than did sour cherry. Van Goor (1973) found that Ca absorption by the leaves of 'Cox Orange Pippin' apple was several times that of 'James Grieve.' The reason for these differences may be the type of leaf surfaces and the degree of wettability. Klein and Weinbaum (1985) examined urea absorption by the leaves of the evergreen plant species *Olea europaea* L. and the deciduous almond fruit tree species *Prunus dulcis* (Mill. D. A. Webb) and found that olive absorbed 15 times more urea than the almond per unit leaf area. Leaf N content of olive was increased to 47% through five successive sprays without causing any toxicity. The initial uptake of urea by olive was very rapid, accounting for 30–40% in the first 2 hours and increasing to nearly 70% within 24 hours. Therefore olive leaves are very efficient in nutrient uptake.

The responses of plants to foliar application of nutrients differ with the conditions of deficiency or sufficiency of the element in the plant. Cook and Boynton (1952) found that apple leaves high in N content absorbed more N from sprays, and this was because of better health of the tree. Furthermore, apple trees with high N status responded quickly to Mg sprays (Forshey, 1963). Interestingly, the influence of one nutrient in the spray on the uptake of the same or additional nutrients from the soil medium has also been observed. Thorne (1955) found that foliar absorption of P and K was significantly more than absorption of N from an NPK mixture when the plants also received N through the soil. However, this effect was obtained when the soil and foliar application were made simultaneously.

The influence of nutrients supplied to the leaf on metabolic processes and photosynthesis has not been fully examined, except for a few reports on the enhancement of photosynthesis by N, P, and K sprays. Root fertilization also should influence photosynthesis once the nutrients are transported to the leaf. Heinicke (1934) was the first to report that N fertilization to apple trees in late summer increased leaf photosynthesis and even prolonged the activity till the late autumn. Likewise, it is possible to obtain enhancement of photosynthesis through supply of N to the leaf at the appropriate time. Urea sprays have been observed to increase (O'Kennedy *et al.*, 1975) or delay leaf abscission (Norton and Childers, 1954) and increase dry matter production in apple (Hansen, 1980). Contrary to these effects, the spraying of complete nutrient solutions reduced photosynthesis in apple, particularly on the day of application (Swietlik *et al.*, 1982), and in maize crops (Harder *et al.*, 1982). However, this effect was temporary. Boote *et al.* (1978) reported that the photosynthetic rates were increased 6 days after spraying of NPKS fertilizers. Foliar application of complete nutrient solutions decreased stomatal conductance, stomatal opening, transpiration, and net photosynthesis in apple seedlings and the effects are attributed to the action of $CaCl_2$, which lowered the rates of transpiration for 2 days after application, especially when the trees were not under drought (Swietlik and Miller, 1987).

C. CHEMICAL FORM AND ORGANIC CARRIERS/ADJUVANTS

The nutrients can be supplied in different forms and the effectiveness is generally the same, provided the amount is equivalent. N supplied in equal amounts as urea, $Ca(NO_3)_2$, or $(NH_4)_2SO_4$ sprays increased the N content and the growth of the tree to the same extent (Boynton *et al.*, 1953). Recently, urea–ammonium nitrate, which provides 50% urea and 25% ammonium and nitrate each, has been found to be very effective for foliar applications of N to barley (Turley and Ching, 1986). Differences in the effectiveness of different forms of phosphatic compounds used for foliar sprays have been well documented. Yogaratnam *et al.* (1981) found greater absorption of P by apple leaves from H_3PO_4 than from potassium, sodium, or calcium phosphate. A number of organic and inorganic phosphate compounds were tested for their suitability in foliar applications, using a technique by which predetermined quantities of P could be applied to a specific leaf area. Barel and Black (1979) obtained comparable data on the absorption and translocation of P from 32 different compounds and found ammonium tripolyphosphate to be the most effective form for maize (Table II). It could be applied at 3.7 g m^{-2} without leaf injury. It was found that 66% of the P applied to the leaf was absorbed, and 87% of it was translocated out of the leaf within 10 days. The leaves of maize and soybean tolerated a concentration of 3.4 and 2.2 g m^{-2}, respectively.

Shafer and Reed (1986) measured the absorption of K from 31 organic and inorganic K compounds by excised leaves of 3-week-old soybean (*Glycine max* L.,

Table II. Absorption and Translocation in 24 Hours after Application of P in Various Forms to Leaves of Corn and Soybeans[a]

	Corn ($\mu g\ cm^{-2}$)			Soybeans ($\mu g\ cm^{-2}$)		
P Compound[b]	P Applied	P Absorbed as % of P Applied	P Translocated as % of P Absorbed	P Applied	P Absorbed as % of P Applied	P Translocated as % of P Absorbed
Orthophosphate	151	78	76	25	83	5
Pyrophosphate	204	16	13	69	30	21
Tripolyphosphate	188	20	55	63	27	26
Tetrapolyphosphate	251	25	54	88	30	34
Trimetaphosphate	210	5	50	62	3	20
Tetrametaphosphate	193	3	46	67	3	20

[a]Source: Barel and Black, 1979. Reproduced with permission of the American Society of Agronomy, Inc.

[b]The phosphate compounds were added in the ammonium form at pH 7.

'Bragg'). The concentration of K salts was kept at 10 m*M* and the pH of the spray solution adjusted to 6.3. Among the 14 inorganic compounds, potassium bicarbonate, nitrate, and a few phosphate forms were promising. Most of the organic forms proved better than the inorganic forms. Monocarboxylic K compounds had a higher percentage of absorption than the di- and tricarboxylic compounds. The absorption of inorganic K depended on the rapidity with which the absorbed K was translocated out of the leaf, i.e., the greater the uptake, the greater the translocation. Smith *et al.* (1987) evaluated the foliar uptake of K from K_2SO_4 and KNO_3, without or with ammonium nitrate or urea, and their results showed that none of the treatments was effective in increasing K content of the leaves of adult pecan trees grown in the orchard.

Compounds which are hydroscopic are likely to be in solution for a long time and, therefore, may be absorbed more effectively. Allen (1970) reported that 90% of $MgCl_2$ was absorbed by apple leaves even at a relative humidity of 30%, while $MgSO_4$ required a relative humidity of 80% for an increase in absorption. This was due to a greater deliquescence of $MgCl_2$ than $MgSO_4$.

Ca deficiency causes many physiological disorders in apples. Bramlage *et al.* (1985) studied the effectiveness of foliar sprays of apple trees with $CaCl_2$, $Ca(H_2PO_4)_2$, and the polyphenolic acid chelate of Ca. They found that technical grades of 77 to 80% $CaCl_2$ flakes consistently increased fruit Ca concentration and reduced senescent breakdown of fruits after storage. $CaCl_2$ and Ca chelate were equally effective in this regard.

The pH of the spray solution is an important factor affecting foliar absorption. Cook and Boynton (1952) found maximum absorption of urea by apple leaves at the pH ranging from 5.4 to 6.6. The effect of pH differed with the chemicals used. Phosphate absorption by chrysanthemum was maximum at pH 3 to 6 for sodium phosphate and pH 7 to 10 for potassium phosphate (Reed and Tukey, 1978).

To increase the effectiveness of foliar uptake, wetting agents are added to the spray liquid. These chemicals are neutral, nonionic compounds which reduce the surface tension and increase the wetting of the surface area of the leaf, thus enabling larger amounts of the solute to be absorbed. However, complete wetting which would reduce the contact angle to zero—a condition recognized as "critical surface tension" (Schönherr and Bukovac, 1972)—is never attained in the field. Triton X-100 or L-77 at 0.1% concentration is used today for urea sprays on fruit trees (Klein and Weinbaum, 1985) and surfactants like Wesco 93X, Multifilm, and X77 are commercially available for different nutrient sprays and crops (Swietlik and Faust, 1984). Stein and Storey (1986) tested the efficacy of a few adjuvants on foliar uptake of N and P by soybean. These were mixed with the fertilizer containing NPKS (12.0 : 1.7 : 3.3 : 0.5) and sprayed on the foliage. The average increases in the nutrient content were 8.9, 2.2, and 2.5% for N and 34, 28, and 21% for P when using the adjuvants glycerol, lecithin, and Pluronic L-121. Glycerol was found to be superior to the others; glycerol and propylene glycol were not phytotoxic at concentrations of up to 10%.

VI. Fertilizer Use Efficiency

A. EFFECTIVENESS OF SOME IMPORTANT ELEMENTS

Among the major nutrients, N in the form of urea has been used successfully as foliar sprays for many crops (Swietlik and Faust, 1984). Foliar nutrition studies were carried out in the 1950s at Rothamsted in the United Kingdom, and application of N sprays to cereals has become a regular practice for increasing the longevity of the leaves (Thorne, 1986). Urea sprays to apple trees during spring or autumn are effective in increasing the shoot growth (Hansen, 1980; Cahoon and Donoho, 1982). However, when the N status of the tree is high, growth increase is not obtained. Urea sprays to supply only 25% of the N as provided through the soil are enough to produce the same effect (Shim *et al.*, 1972). Besides urea, compounds like KNO_3, ammonium sulfate, or nitrate are also used for foliar application to cereals and vegetables (Neumann, 1982; Giskin and Nerson, 1984; Diver *et al.*, 1985). Foliar-applied N is incorporated into a more readily remobilizable N pool than from other sources (Below *et al.*, 1985).

Phosphorus supply through foliage has certain advantages over soil application; however, it is not practicable to supply the entire crop requirement through sprays. On the other hand, P supplied through the leaf becomes readily available to actively growing regions (Bukovac and Wittwer, 1957). The P content of apple fruit appears to be very important for preventing storage breakdown. Spraying of P compounds to the trees during the growing season for apple increases the P content and reduces the breakdown during storage (Yogaratnam and Sharples, 1982). Suwanvesh and Morrill (1986) conducted greenhouse experiments with a late and an early medium-maturing cultivar of Spanish peanut to study their response to KH_2PO_4. They observed that the utilization of foliar P was dependent on root availability, a significant response being obtained when root was limiting. In barley (*Hordeum vulgare* L.), Afridi and Samiullah (1973) recorded more pronounced growth effect when foliar P was given with a basal dressing of a small dose of P rather than foliar P alone. Spraying of 0.1% NaH_2PO_4 was very effective in increasing growth, mineral composition, and carbohydrate content of peanut grown under salinity stress (Malakondaiah and Rajeswara Rao, 1980).

Plants respond better to foliar application of K when soil conditions do not favor root absorption. Potassium is generally included in sprays along with other elements like N and P. The recommended dose for K sprays is about 8.4 g liter^{-1} of water to supply KNO_3 at about 20 kg ha^{-1} (Swietlik and Faust, 1984). A positive correlation exists between K concentration in pecan leaves and oil content in the kernel, but soil application is not effective in increasing leaf K content (Smith *et al.*, 1981). Foliar application of K, especially as KNO_3 applied five times at fortnightly intervals (Diver *et al.*, 1985), distinctly increased the leaf K and the yield of pecan (Gossard and Nevins, 1965). K deficiency is difficult to correct in trees growing in acid clay and silty clay loam soils and response to soil-applied K fertilizers is slow. Foliar application or even trunk injections of K salts are effective for correcting the deficiency in some trees like sweet cherry and prune (Robbins *et al.*, 1982).

Ca and Mg are generally abundant in most soils but their mobility in the plant is significantly much less than that of other elements like N, P, or K. Ca is applied as foliar sprays to prevent nutritional disorders and improve storage quality of fruits. $CaCl_2$ or $Ca(NO_3)_2$ sprays are applied to the trees soon after flowering. There exists a competition between the shoot and fruit for Ca and Ca sprays are effective when the leaves are fully grown (Quinlan, 1969).

Micronutrients are most effective when supplied as foliar sprays. Fe chlorosis in blueberry is easily controlled by spraying even fungicides like Ferbam, which contains Fe (Ritter, 1980). Similarly, Mn and Zn deficiencies in crops are corrected by foliar sprays of $MnSO_4$ or $ZnSO_4$ (Beyers and Terblanche, 1971). Foliar supply of as small an amount of Mn as 0.1 kg ha^{-1} is as effective as soil application at 14 kg ha^{-1} for soybean (Mascagni and Cox, 1985). A single spray application from aircraft to supply Mn at 7 kg ha^{-1} in a low volume of water is very effective in preventing the foliar and seed symptoms of Mn deficiency in lupin (Hannam and Riggs, 1985). Foliar supply of Mn can be superior to soil supply in many crops. Radish is grown as a low-input crop in the Florida Everglades, often following other vegetable crops. Heavy applications of S is widely practiced to lower the soil pH and increase Mn availability. Foliar spraying of Mn is an alternative to soil application of heavy doses of S and has been adopted profitably for celery and radish (Beverly, 1986).

Cu deficiency causes "wither tip" in fruit trees and can be corrected by spraying $CuSO_4$ or Bordeaux mixture (Ritter, 1980). Micronutrients like B and Mo are also applied to crops as foliar sprays to correct their deficiencies. Fruit set in 'Italian' prune trees was increased by B sprays even though the leaf B content was adequate (Hanson and Breen, 1985).

B. UPTAKE DURING CRITICAL PERIODS

The fertilizers supplied to the root or to the leaf are transported to other regions of the plant for utilization. Certain tissues need specific elements at a particular time of growth and thus function as a "sink." Leaves as the center of metabolic activity and photosynthesis functions as the "source" region for carbohydrates which are transported to other regions. However, the leaves act as the "sink" for the inorganic nutrients during the vigorous period. This situation changes with the developmental phase when the fruiting region becomes the "sink" for both organic and inorganic nutrients. Foliar application of nutrients is important at this stage for two reasons: (1) transport of nutrients supplied to the leaf may take place at an enhanced rate and (2) the depletion of nutrients due to partial leaf senescence and migration out of the leaf can be overcome and the vigor of the leaf restored.

The application of N, P, and K to the leaf during autumn, especially prior to leaf fall, is beneficial to fruit trees. Urea-N applied to the foliage of trees is rapidly translocated to the stem and twigs for later retranslocation during the spring growth. Oland (1960) was the first to discover that urea sprays to apple trees prior to leaf fall were the most effective method to supply N. The edible passion fruit (*Passiflora edulis* Sims.), which originates from South America, is now cultivated in tropical and subtropical regions of the world. One of the serious problems limiting fruit production is the low yield in the spring season because of changed climatic conditions in these countries. Vegetative growth is reduced in winter because of a shorter photoperiod and reduced N uptake from the soil at low temperatures. Commercial growers in Australia resort to foliar sprays of urea on passion fruit during winter to provide adequate N (Menzel *et al.*, 1986).

Foliar fertilization is effective when given during the "seed-filling" period in grain crops. The nutrients present in the leaf accumulated during the early growth stages migrate to the grain or fruit during later stages of development. This foliar nutrient depletion affects photosynthesis in the "flag leaf" in particular, which supplies the major portion of the photosynthesis to the grain in a few crops. Batten and Wardlaw (1987) consider that the migration of elements like P from the "low-P" plants during the grain development reduces the photosynthesis. Below *et al.* (1984) conducted experiments with maize to establish whether or not foliar N, P, K, and S at the grain development stage would be profitable. Field studies were conducted to determine if N sprays would delay the remobilization of leaf N and also leaf senescence, help maintain photosynthesis, and increase the grain yield. Only in one of the three hybrids tested, namely, 'FS-854,' did significant increase in growth and grain yield result from N sprays. In a later study, Below *et al.* (1985) found that foliar-applied N was incorporated into a different N pool from that formed by soil-absorbed N, and therefore is not likely to replenish the loss of N in the leaf.

The period between fruit set and seed maturation in crop plants is accompanied by retranslocation of substances to the seed from the senescing leaves. It is not known what mechanism triggers the senescence (Hocking and Pate, 1977). The N requirement of the seeds of legumes far exceeds the supply from the roots, and the deficit is compensated by translocation from the leaves (Sinclair and deWit, 1975). Foliar N sprays are thought to replenish the loss, meet the additional demand from the grain, and rejuvenate the leaves. Extensive work has been carried out with soybeans, which has been classified as a "self-destructive" crop requiring large amounts of N during "seed-fill." Garcia and Hanway (1976) obtained a positive effect from sprays of N, P, K, and S during the R-5 to R-7 stages. Syverud *et al.* (1980) obtained higher yield in soybean from weekly sprays during R-4. Foliar application of polyphosphates also increased the yield of maize and soybeans. Shapiro *et al.* (1987) obtained significant yield increase in soybeans by a combination of foliar sprays of NPKS with irrigation and manuring with compost. Cultivar differences in response to N sprays have also been reported. The 'Ultra' lupin yield was reduced, while that of 'Unicorn' was not affected by N application (Herbert and Dougherty, 1978). Differences were found between determinate and indeterminate soybean cultivars in their response to NPKS sprays (Garcia and Hanway, 1976;

Robertson *et al.*, 1977). Adequate soil moisture is important for good response to foliar application. A few crops like winter wheat (Altman *et al.*, 1983) and rice (Thom *et al.*, 1981) have given consistent yield increase through urea spray applications.

P nutrition is also important for the crops during the "seed-fill" period. Wheat plants grown with a limited supply of P exhibit early senescence of the vegetative tissues and result in fewer and lighter grains (Batten *et al.*, 1986). Low-P plants senesce rapidly. The photosynthetic activity and the inorganic and lipid-bound P content are decreased in the flag leaf. The grain becomes the major sink for P, containing up to 90% of total P in the shoot at the time of maturity. Batten and Wardlaw (1987) support the hypothesis that P exported from the leaves of low-P plants during grain development reduces photosynthesis and grain yield. However, they found that P supply to the flag leaf delayed the leaf senescence.

The nutrition of tea is interesting because the leaf is the economic part of the plant. Annual application of fertilizer N amounts to about 600 kg ha^{-1} in Japan; however, the uptake rates by tea plants are very low. During the growth of new shoots, which are harvested for the production of green tea, a large amount of N must accumulate in the leaves. According to Hoshina (1982), 41% of N in the second crop of green leaves is derived from N stored in the plants by retranslocation, 36% is from the fertilizer applied before sprouting, and the remainder is obtained from the N applied after sprouting. The N required for the formation of new shoots was supplied through urea sprays. Hoshina *et al.* (1978) showed that foliar application of urea increased the chlorophyll content and total N in the shoot growing in summer and the N supplied was readily incorporated in these leaves. Aono *et al.* (1982) found that 1% urea spray increased the amino acid content by about 10-fold and theanine by 50–80%. According to Nakagawa *et al.* (1981), theanine and arginine are the predominant amino acids of the tea leaves and they are responsible for the good quality of green tea. Karasuyama *et al.* (1985) traced the pathway of foliar-applied [^{15}N]urea in tea plants and found that it is assimilated into glutamine first and then to other amino acids. Arginine synthesized from urea-N is accumulated in the leaves, but theanine synthesis is much less. Their studies showed that theanine is largely synthesized in the roots. Furthermore, a high level of ^{15}N labeling in theanine was observed in the roots when the plants were supplied with ^{15}N-labeled ammonium sulfate in the roots. However, theanine is considered to be synthesized partly from the N transported to the roots from the shoot. A small amount of theanine is also synthesized in the leaves. These studies support the general practice of supplying N through sprays, particularly for the second flush of green leaves, since it could be rapidly translocated to the roots for synthesis of theanine which is later retranslocated to the leaves.

C. ABSORPTION OF GASES BY THE LEAVES

The atmosphere is the source of several nutrients present in the gaseous form and plants absorb gases like NH_3 and SO_2 through their leaf surfaces (Erikson, 1952).

Nitrogenous gases are derived from natural and human-made sources, contributing about 6.4 × 10^9 tons per year mainly as NH_3 and 5.7 × 10^7 tons per year as NO, NO_2, and NH_3 (Robinson and Robbins, 1968). Atmospheric N enters the biological cycle through the "wet" and "dry" deposition. Recent evidence suggests that the dry deposition of gaseous NH_3, ammoniated sulfate, and ammoniated nitrate particles on plant surfaces is the major route to the biosystem as opposed to the washout of N compounds by rain into the soil. Hutchinson *et al.* (1972) provided the experimental proof that plant leaves could absorb atmospheric NH_3 in a manner similar to that of CO_2. Using a specially designed growth chamber, they discovered that NH_3 absorption by the leaves was steady during the day, but dropped sharply during the dark period. It again rose up the next day to a slightly higher level than the previous day for some time before reaching the original steady state. Aneja *et al.* (1986) developed a technique for measuring "dry deposition" of atmospheric NH_3 on the vegetative surface and computed the deposition velocities for a few plant species. Accordingly, the velocities for maize, soybean, and oat were 0.3, 0.6, and 1.0 cm sec^{-1}, respectively. The values for SO_2 absorption were 0.1 to 0.8 cm sec^{-1} for a short grass, and 0.2 to 2.0 cm sec^{-1} for tall forest trees. The deposition velocities were 0.3 to 1.3 cm sec^{-1} in the day and decreased by one order of magnitude in the night.

Plant leaves absorb considerable amounts of NO_2 from the atmosphere. Okano and Totsuka (1986) used the ^{15}N dilution method for quantitative assays of NO_2 absorbed by sunflower and maize leaves and studied the influence of nitrate nutrition of the plants on foliar NO_2 absorption. They found that the amount of NO_2 absorbed increased with the concentration of NO_2. Nitrate nutrition of the plants influenced the absorption only at high concentrations of 2 μl $liter^{-1}$ NO_2 and was negligible when the plants were nitrate deficient. No_2 at low concentrations of 0.3 μl $liter^{-1}$ exerted a positive nutritional effect on the nitrate-deficient plants, while at 2 μl $liter^{-1}$ it acted as a toxic air pollutant at all levels of nitrate nutrition. Studies showed that the rate of absorption by sunflower leaves increased during day and decreased during the night in similar fashion as NH_3, but this difference was not significant in spinach (Kaji *et al.*, 1980). It was shown that NO_2-N absorbed by the leaves was translocated to young leaves and roots and converted to nitrate, nitrite, and amino acids.

Plant species differ in their capacity to absorb SO_2 from the atmosphere. The SO_2 fluxes into the leaves of four different plant species, namely, *Pisum sativum* L. (garden pea), *Lycopersicon esculentum* Mill. *flacca* (a mutant of tomato), *Geranium carolinianum* L. (wild geranium), and *Diplacus aurantiacus* (Curtis) Jeps. (a native California shrub), were measured to clarify the relationship among stomatal conductance and internal factors in determining the flux (Olszyk and Tingey, 1985). The most important finding from this study is that the internal leaf conductance was much higher than the stomatal conductance for tomato, indicating that factors inside the leaf play a significant role in regulating SO_2 entry. The absorption of SO_2 from the atmosphere in small amounts by the leaves will be a good source for crops and Fowler and Unsworth (1979) have estimated that 30% of the S content of a wheat crop was derived from SO_2 absorption through stomata.

VII. Summary

The role of the leaves in the uptake of substances from aqueous and gaseous sources is very important not only in plant nutrition but also in terms of the global cycling of the elements which are needed for plant growth. Indirectly foliar nutrition helps in minimizing soil–air pollution within reasonable limits. When the air pollution reaches a very high level, the entire plant ecosystem is affected. The first effect of toxic elements is on the plant surface, which is fully covered by a noncellular cuticular membrane. Once the structure and quality of this membrane are affected by toxic substances from the dry and wet deposition, the leaf cells gradually lose their vitality, affecting primarily photosynthesis. Thus the damage caused by pollutants becomes irreparable. Winner and Atkinson (1986) point out that photosynthesis is affected by air pollutants with a variety of consequences on growth.

The application of fertilizers to the leaves in the form of aqueous solutions is important for crop production. Considerable information is now available on the mechanisms of foliar absorption, transport, and directed movement of nutrients to "sink" regions. Furthermore, the nature of the leaf cuticle, its composition, its development, its permeability, and differences among crop species and cultivars have been well documented.

Foliar nutrition is particularly relevant for timely applications of specific nutrients to the leaf during growth and development. The failure to accomplish the desired benefits from foliar sprays is due to several soil and plant factors. The Diagnosis and Recommendation Integrated System (DRIS), a new method of interpreting plant analyses developed by Beaufils (1971) and subsequently widely employed (Sumner, 1977a,b), has not been followed for foliar fertilization to increase crop yields. Counce and Wells (1986) have attempted to use DRIS for prediction of critical nutrient limits with N and subsequent foliar applications. They obtained yield increases by this method using N sprays in four cases out of ten. Plantation crops have benefited by foliar application, including coffee, tea, cacao, and banana. Foliar application has become economical for large-scale spraying from airplanes.

References

Afridi, M. M. R. K., and Samiullah. 1973. A comparative study of the effect of soil- and leaf-applied phosphorus on the yield of barley (*Hordeum vulgare* L.). *New Phytol.* 72:113–116.

Allen, M. 1970. Uptake from inorganic sprays applied to apple leaves. *Pestic. Sci.* 1:152–155.

Altman, D. W., W. L., McCuistion, and W. E. Kronstad. 1983. Grain protein percentage, kernal hardiness, and grain yield of winter wheat with foliar applied urea. *Agron. J.* 75:87–91.

Aneja, V. P., H. H., Rogers, and E. P. Stahel. 1986. Dry deposition of ammonia at environmental concentrations on selected plant species. *J. Air Pollut. Control Assoc.* 36:1338–1341.

Aono, H., S. Tanaka, and Y. Yanase. 1982. Effect of foliar application of fertilizer on the growth and quality of new shoot of tea plant. *Stud. Tea (Jpn.)* 63:23–32.

Baes, C. F., III, and S. B. McLaughlin. 1984. Trace elements in tree rings: Evidence of recent and historical air pollution. *Science* 224:494–497.

Barel, D., and C. A. Black. 1979. Foliar application of P. II. Yield responses of corn and soybeans sprayed with various condensed phosphates and P-N compounds in greenhouse and field experiments. *Agron. J.* 71:21–24.

Batten, G. D., and I. F. Wardlaw. 1987. Senescence and grain development in wheat plants grown with contrasting phosphorus regimes. *Aust. J. Plant Physiol.* 14:253–265.

Batten, G. D., I. F. Wardlaw, and M. J. Aston. 1986. Growth and the distribution of phosphorus in wheat developed under various phosphorus and temperature regimes. *Aust. J. Agric. Res.* 37:459–469.

Beaufils, E. R. 1971. Physiological diagnosis—A guide to improving maize production based on principles developed for rubber trees. *Fert. Soc. S. Afr. J.* 1:1–30.

Below, F. E., S. J. Crafts-Brandner, J. E., Harper, and R. H. Hageman. 1985. Uptake, distribution and remobilization of ^{15}N-labeled urea applied to maize canopies. *Agron. J.* 77:412–415.

Below, F. E., R. J. Lambertz, and R. H. Hageman. 1984. Foliar applications of nutrients on maize. I. Yield and N content of grain and stover. *Agron. J.* 76:773–776.

Beyers, E., and J. H. Terblanche. 1971. Identification and control of trace element deficiencies. I. Zinc deficiency. *Decid. Fruit Grower* 21:132–137.

Beverly, R. B. 1986. Radish response to foliar nutrient sprays and pH adjustment of an organic soil. *J. Fert. Issues* 3:75–79.

Biddulph, S., O. Biddulph, and R. Cory. 1958. Visual indications of upward movement of foliar applied ^{32}P and ^{14}C in the phloem of the bean stem. *Am. J. Bot.* 45:648–652.

Bohm, J. 1877. Uber die Aufname von Wasser and Kolksalsen durch die Blatter der Feuerbohne. *Landwirtsch Vers.-Stn.* 20:51–59.

Boote, K. J., R. N. Gallaher, W. K. Robertson, K. Hinson, and L. C. Hammond. 1978. Effect of foliar fertilization on photosynthesis, leaf nutrition, and yield of soybeans. *Agron. J.* 70:787–791.

Bowen, J. E. 1969. Absorption of copper, zinc and manganese by sugarcane leaf tissue. *Plant Physiol.* 44:255–261.

Bowen, J. E. 1981. Kinetics of active uptake of boron, zinc, copper and manganese in barley and sugarcane. *J. Plant Nutr.* 3:215–223.

Boynton, D., D. Margolis, and C. R. Gross. 1953. Exploratory studies on N metabolism by McIntosh apple leaves sprayed with urea. *Proc. Am. Soc. Hortic. Sci.* 62:135–146.

Bramlage, W. J., M. Drake, and S. A. Weis. 1985. Comparisons of calcium phosphate, and a calcium chelate as foliar sprays for 'McIntosh' apple trees. *J. Am. Soc. Hortic. Sci.* 110:786–789.

Bukovac, M. J., and S. H. Wittwer. 1957. Absorption and mobility of foliar applied nutrients. *Plant Physiol.* 32:428–435.

Cahoon, G. A., and C. W. Donoho, Jr. 1982. The influence of urea sprays, mulch, and pruning on apple tree decline. *Ohio Agric. Res. Dev. Cent. Res. Circ.* No. 272, pp. 16–19.

Chamel, A. 1972. Pénétration et migration du ^{59}Fe appliqué sur les feuilles de mäis; effect due diméthylsulfoxyde. *Physiol. Plant.* 26:170–174.

Chamel, A. 1980. Penetration du cuivre a travers des cuticules isolées de feuilles de Poirier. *Physiol. Veg.* 18:313–323.

Chamel, A., and J. F. Eloy. 1983. Some applications of the laser probe mass spectrograph in plant biology. *Scanning Electron Microsc.* II:841–851.

Chamel, A., and P. Neumann. 1987. Foliar absorption of nickel: Determination of its cuticular behaviour using isolated cuticles. *J. Plant Nutr.* 10:99–111.

Cook, J. A., and D. Boynton. 1952. Some factors affecting the absorption of urea by McIntosh apple leaves. *Proc. Am. Soc. Hortic. Sci.* 59:82–90.

Counce, P. A., and B. R. Wells. 1986. Rice Y-leaf nutrient analyses and midseason foliar fertilization. *Commun. Soil Sci. Plant Anal.* 17:1071–1087.

Crowdy, S. H., and T. W. Tanton. 1970. Water pathways in higher plants—Free space in wheat leaves. *J. Exp. Bot.* 21:102–111.

Dadykin, V. P. 1951. The application to the aerial parts of plant of nitrogenous fertilizers under soil conditions. *Dokl. Akad. Nauk. SSSR* 79:529–531. (In Russ.)

DeBary, A. 1871. Ueber die Wachsuberzuge der Epidermis. *Bot. Ztg.* 29:129–139.

Diver, S. G., M. W. Smith, and R. W. McNew. 1985. Foliar applications of K_2SO_4, KNO_3, urea and NH_4NO_3 on pecan seedlings. *HortScience* 20:422–425.

Dollard, G. J., M. H. Unsworth, and M. J. Harve. 1983. Pollutant transfer in upland regions by occult precipitation. *Nature (London)* 302:241–243.

Epstein, E. 1973. Mechanisms of ion transport through plant cell membranes. *Int. Rev. Cytol.* 34:123–168.

Erikson, E. 1952. Composition of atmospheric precipitation. *Tellus* 4:215–232.

Erwee, M. G., P. B. Goodwin, and A. J. E. van Bel. 1985. Cell–cell communication in the leaves of *Commelina cyanea* and other plants. *Plant, Cell Environ.* 8:173–178.

Faust, M., and C. B. Shear. 1973. Calcium transport patterns in apples. *Pr. Inst. Sadow. Skierniewice, Ser. E* No. 3, pp. 423–436.

Fitzsimons, P. J., and J. D. B. Weyers. 1986. Potassium ion uptake by swelling *Commelina communis* guard cell protoplasts. *Physiol. Plant.* 66:469–475.

Forshey, C. G. 1963. A comparison of soil nitrogen fertilization and urea sprays as sources of nitrogen for apple trees in sand culture. *Proc. Am. Soc. Hortic. Sci.* 83:32–45.

Fowler, D., and M. H. Unsworth. 1979. Turbulent transfer of sulphur dioxide to a wheat crop. *Q. J. R. Meteorol. Soc.* 105:767–783.

Franceschi, V. R., and R. T. Giaquinta. 1983a. Specialized cellular arrangements in legume leaves in relation to assimilate transport and compartmentation: Comparison of the paraveinal mesophyll. *Planta* 159:415–422.

Franceschi, V. R., and R. T. Giaquinta. 1983b. The paraveinal mesophyll of soybean leaves in relation to assimilate transfer and compartmentation. II. Structural, metabolic and compartment changes during reproductive growth. *Planta* 154:422–431.

Franke, W. 1971. The entry of residues into plants via ectodesmata. *Residue Rev.* 38:81–115.

Franke, W. 1975. Stoffaufnahme durch des Blatt unter besonderer Berücksichtigung der Ektodesmen. *J. Landwirtsch. Forsch.* 26:331–341.

Freeman, B., and D. W. Turner, 1985. The epicuticular waxes on the organs of different varieties of banana (*Musa* spp.) differ in form, chemistry and concentration. *Aust. J. Bot.* 33:393–408.

Garcia, L. R., and J. J. Hanway. 1976. Foliar fertilization of soybeans during the seed-filling period. *Agron. J.* 68:653–657.

Giskin, M., and H. Nerson. 1984. Foliar nutrition of muskmelon. I. Application to seedlings—Greenhouse experiments. *J. Plant Nutr.* 7:1329–1339.

Goldstein, A. H., and A. D. Hunziker. 1985. Phosphate transport across the plasma membrane of wheat leaf protoplasts: Characteristics and inhibitor specificities. *Plant Physiol.* 77:1013–1015.

Gossard, A. C., and R. B. Nevins. 1965. Results of foliar and soil application of potassium and nitrogen to pecan trees. *Proc. Southeast. Pecan Grow. Assoc.* 58:12–20.

Grauke, L. J., J. B. Storey, and E. R. Emino. 1987. Influence of leaf age on the upper and lower leaf surface features of juvenile and adult pecan leaves. *J. Am. Soc. Hortic. Sci.* 112:835–841.

Gris, E. 1844. Nouvelles expériences sur l'action des composés ferrugineux solubles, appliqués à la vegetation et specialement au traitement de la chlorose et à la debilité des plantes. *C. R. Acad. Sci.* 19:1118–1119.

Gunning, B. E. S., and W. K. Barkeley. 1963. Kinetin-induced directed transport and senescence in detached oat leaves. *Nature (London)* 199:262–265.

Haas, K., and J. Schönherr. 1979. Composition of soluble cuticular lipids and water permeability of cuticular membranes from *Citrus* leaves. *Planta* 146:399–403.

Haile-Mariam, S.-N. 1965. Mechanisms of foliar penetration and translocation of mineral ions with special reference to coffee (*Coffea arabica* L.). Ph.D. Thesis, Michigan State Univ., East Lansing.

Hall, D. M., and L. A. Donaldson. 1963. The ultrastructure of wax deposits on plant leaf surfaces. I. Growth of wax on leaves of *Trifolium repens*. *J. Ultrastruct. Res.* 9:259–267.

Hallam, N. D. 1964. Sectioning and electronmicroscopy of *Eucalyptus* leaf waxes. *Aust. J. Biol. Sci.* 17:587–590.

Hallam, N. D. 1970. Growth and regeneration of waxes on the leaves of *Eucalyptus*. *Planta* 93:257–268.
Hamilton, J. M., D. H. Palmiter, and L. C. Anderson. 1943. Preliminary tests with uramon in foliage sprays as a means of regulating the nitrogen supply of apple trees. *Proc. Am. Soc. Hortic. Sci.* 42:123–126.
Hannam, R., and J. L. Riggs. 1985. The effect of manganese applied from an aircraft as a low-volume foliar spray in preventing manganese deficiency in *Lupinus angustifolius* L. *Fert. Res.* 6:149–156.
Hansen, P. 1980. Yield components and fruit development in 'Golden Delicious' apples as affected by the timing of nitrogen supply. *Sci. Hortic.* 12:243–257.
Hanson, E. J., and P. J. Breen. 1985. Effects of fall boron sprays and environmental factors on fruit set and boron accumulation in 'Italian' prune flowers. *J. Am. Soc. Hortic. Sci.* 110:389–392.
Harder, H. J., R. E. Carlson, and R. H. Shaw. 1982. Corn grain yield and nutrient response to foliar fertilizer applied during grain fill. *Agron. J.* 74:106–110.
Hardin, J. W., and L. L. Phillips. 1985. Atlas of foliar surface features in woody plants. VII. *Rhus* subg. *Rhus* (Anacardiaceae) of North America. *Bull. Torrey Bot. Club* 112:1–10.
Harris, N., and N. J. Chaffey. 1985. Plasmatubules in transfer cells of pea (*Pisum sativum* L.). *Planta* 165:191–196.
Heinicke, A. J., 1934. Photosynthesis in apple leaves during late fall and its significance in annual bearing. *Proc. Am. Soc. Hortic. Sci.* 32:77–80.
Herbert, S. J., and C. T. Dougherty. 1978. Influence of irrigation and foliar feeding of N, P, K and S during seed-filling in two lupin species. *N.Z. J. Exp. Agric.* 6:39–42.
Hocking, P. J., and J. S. Pate. 1977. Mobilization of mineral to developing seeds of legumes. *Ann. Bot.* 41:1254–1278.
Holloway, P. J., and E. A. Baker. 1968. Isolation of plant cuticles with zinc chloride–hydrochloric acid solution. *Plant Physiol.* 43:1878–1879.
Hoshina, T. 1982. Uptake and utilization of nitrogen applied to tea plants. JARQ 16:193–197.
Hoshina, T., S. Kozai, and K. Ishigaki. 1978. Examination of emission spectrometric ^{15}N analysis of amino acids and amides separated by thin layer chromatography. *Radioisotopes* 25:448–453.
Hull, H. M., H. L. Morton, and J. R. Wharrie. 1975. Environmental influences on cuticle development and resultant foliar penetration. *Bot. Rev.* 41:421–452.
Hutchinson, G. L., R. J. Millington, and D. B. Peters. 1972. Atmospheric ammonia: Absorption by plant leaves. *Science* 175:771–774.
Jacoby, B. 1975. Light sensitivity of ^{22}Na, ^{86}Rb, and ^{42}K absorption by different tissues of bean leaves. *Plant Physiol.* 55:978–981.
Jacoby, B., and O. E. Plessner. 1971. Sodium export from bean leaves as affected by the mode of application. *Isr. J. Bot.* 20:311–317.
Jyung, W. H., and S. H. Wittwer. 1964. Foliar absorption—An active uptake process. *Am. J. Bot.* 51:437–444.
Jyung, W. H., S. H. Wittwer, and M. J. Bukovac. 1965. Ion uptake by cells enzymically isolated from green tobacco leaves. *Plant Physiol.* 40:410–414.
Kaji, M., T. Yoneyama, T. Tosuka, and H. Iwaki. 1980. Absorption of atmospheric NO_2 by plants and soils. VI. Transformation of NO_2 absorbed in the leaves and transfer of the nitrogen through the plants. In Studies on the Effects of Air Pollutants on Plants and Mechanisms of Phytotoxicity. *Res. Rep. Natl. Inst. Environ. Stud. (Jpn.)* No. 11, pp. 51–58.
Kannan, S. 1969. Penetration of iron and some organic substances through isolated cuticular membranes. *Plant Physiol.* 44:517–521.
Kannan, S. 1970. Course of cation accumulation by leaf tissue in *Phaseolus vulgaris* L. *Experientia* 26:552.
Kannan, S. 1978. Lateral movement of cations in corn leaves. *Plant Physiol.* 61:706–707.
Kannan, S. 1986. Foliar absorption and transport of inorganic nutrients. CRC Crit. Rev. Plant Sci. 4:341–375.
Kannan, S., and H. Keppel. 1976. Differential migration of foliar applied zinc in maize plants. *Z. Naturforsch., C* 31C:195–196.

Kannan, S., and T. Mathew. 1970. Effects of growth substances on the absorption and transport of iron in plants. *Plant Physiol.* 45:206–209.

Kannan, S., and S. Ramani. 1974. Mechanisms of ion absorption by bean leaf slices and transport in intact plants. *Z. Pflanzenphysiol.* 71:220–227.

Kannan, S., and S. Ramani. 1978. Studies on molybdenum absorption and transport in bean and rice. *Plant Physiol.* 62:179–181.

Kannan, S., and S. H. Wittwer. 1965. Effects of chelation and urea on iron absorption by intact leaves and enzymically isolated cells. *Plant Physiol.* 40:Suppl., xii.

Karasuyama, M., T. Yoneyama, and H. Kobayashi. 1985. ^{15}N study on the fate of foliarly applied urea nitrogen in tea plant (*Camellia sinensis* L.). *Soil Sci. Plant Nutr. (Tokyo)* 31:123–131.

Kessler, B., and Z. W. Moscicki. 1958. Effect of triiodobenzoic acid and maleic hydrazide upon the transport of foliar applied calcium and iron. *Plant Physiol.* 33:70–72.

Klein, I., and S. A. Weinbaum. 1985. Foliar application of urea to almond and olive: Leaf retention and kinetics of uptake. *J. Plant. Nutr.* 8:117–129.

Lambertz, P. 1954. Untersuchungen über das Vorkommen von Plasmodesmen in den Epidermi Sauben Wanden. *Planta* 44:147–190.

Leon, J. M., and M. J. Bukovac. 1978. Cuticle development and surface morphology of olive leaves with reference to penetration of foliar-applied chemicals. *J. Am. Soc. Hortic. Sci.* 103:465–472.

Lidster, P. D., S. W. Porrit, and G. W. Eaton. 1977. The effect of storage relative humidity on calcium uptake by 'Spartan' apple. *J. Am. Soc. Hortic. Sci.* 102:394–396.

MacDonald, I. R., and A. E. S. Macklon. 1975. Light-enhanced chloride uptake by wheat laminae: A comparison of chopped and vacuum-infiltrated tissue. *Plant Physiol.* 56:105–108.

Macey, M. J. K. 1970. The effect of light on wax synthesis in leaves of *Brassica oleracea*. *Phytochemistry* 9:757–761.

Madore, M. A., J. W. Oross, and W. J. Lucas. 1986. Symplastic transport in *Ipomoea tricolor* source leaves: Demonstration of functional symplastic connections from mesophyll to minor veins by a novel dye-tracer method. *Plant Physiol.* 82:432–442.

Mahlberg, P. G. 1985. Trichome morphogenesis on leaves of *Cyphomandra betacea*. Sendt. (Solanaceae). *Isr. J. Bot.* 34:253–264.

Malakondaiah, N., and G. Rajeswara Rao. 1980. Changes in nitrogen fractions and carbohydrates by foliar application of phosphorus under salt stress in peanut plants. *Turrialba* 30:197–202.

Mansfield, T. A., and R. Jones. 1971. Effects of abscisic acid on potassium uptake and starch content of stomatal guard cells. *Planta* 101:147–158.

Marshall, C., and I. F. Wardlaw. 1973. A comparative study of distribution and speed of movement of ^{14}C assimilates and foliar-applied ^{32}P-labelled phosphate in wheat. *Aust. J. Biol. Sci.* 26:1–13.

Martin, J. T., and B. E. Juniper. 1970. The Cuticle of Plants. Arnold, London.

Mascagni, H. J., and F. R. Cox. 1985. Effective rates of fertilization for correcting manganese deficiency in soybeans. *Agron. J.* 77:363–366.

Mayberry, B. D., and S. H. Wittwer. 1952. Urea nitrogen applied to the leaves of certain vegetable crops. *Mich. Agric. Exp. Stn., Q. Bull.* No. 34, pp. 365–369.

Mayer, A. 1874. Über die Aufname von Ammoniak durch oberirdische Pflanzenteile. *Landwirtsch Vers.-Stn.* 17:329–340.

Mazliak, P. 1963. La cire cuticulaire des pomme (*Pirus malus* L.). Étude morphologique, biochemique et physiologique. Thése de Doctorat d'État, Univ. de Paris.

McFarlane, J. C., and W. L. Berry. 1974. Cation penetration through isolated leaf cuticles. *Plant Physiol.* 53:723–727.

Menzel, C. M., D. R. Simpson, and G. H. Price. 1986. Effects of foliar-applied nitrogen during winter on growth, nitrogen content and production of passion-fruit. *Sci. Hortic.* 28:339–346.

Meyer, R. E., and S. M. Meola. 1978. Morphological characteristics of leaves and stems of selected Texas woody plants. *U.S. Dep. Agric., Tech. Bull.* No. 1564.

Miller, R. H. 1982. Apple fruit cuticles and the occurrence of pores and transcuticular canals. *Ann. Bot.* 50:355–371.

Miller, R. H. 1984. The multiple epidermis–cuticle complex of Medlar fruit *Mespilus germanica* L. (Rosaceae). *Ann. Bot.* 53:779–792.

Miller, R. H. 1985. The prevalence of pores and canals in leaf cuticular membranes. *Ann. Bot.* 55:459–471.
Miller, R. H. 1986. The morphology and permeability of isolated cuticular membranes of *Hoya carnosa* R. Br. (Asclepiadaceae). *Ann. Bot.* 58:407–416.
Millikan, C. R., and B. C. Hanger. 1969. Movement of foliar-applied ^{45}Ca in Brussels sprouts. *Aust. J. Biol. Sci.* 22:545–558.
Mothes, K., L. Engelbrecht, and O. Kulayeva. 1959. Über die Wirkung des Kinetins auf Stickstoffverteilung und Eiweiss synthese in isolierten Blättern. *Flora (Jena)* 147:445–464.
Müller, K., and A. C. Leopold. 1966. The mechanism of kinetin-induced transport in corn leaves. *Planta* 68:186–205.
Nakagawa, M., T. Anan, and N. Ishima. 1981. The relation of green tea taste with its chemical make-up. *Bull. Natl. Res. Inst. Tea (Kanaya,, Jpn.)* 17:111–118.
Neumann, P. M. 1982. Late-season foliar fertilization with micronutrients—Is there a theoretical basis for increased seed yields? *J. Plant Nutr.* 5:1209–1215.
Nissen, P. 1973. Multiple uptake in plants. II. Mineral cations, chloride and boric acid. *Physiol. Plant.* 29:298–354.
Norton, R. A., and N. F. Childers. 1954. Experiment with urea spray on the peach. *Proc. Am. Soc. Hortic. Sci.* 63:23–31.
Okano, K., J. Tatsumi, T. Yoneyama, Y. Kono, and T. Totsuka. 1983. Investigation on the carbon and nitrogen transfer from a terminal leaf to the root system of rice plant by a double tracer method with ^{13}C and ^{15}N. *Jpn. J. Crop Sci.* 52:331–341.
Okano, K., and T. Totsuka. 1986. Absorption of nitrogen dioxide by sunflower plants grown at various levels of nitrate. *New Phytol.* 102:551–562.
O'Kennedy, B. T., M. J. Hennerty, and J. S. Titus. 1975. The effect of autumn foliar urea sprays on storage-form of nitrogen extracted from bark and wood apple shoots. *J. Hortic. Sci.* 50:331–338.
Oland, K. 1960. Nitrogen feeding of apple trees by postharvest urea sprays. *Nature (London)* 185:857.
Olsen, R. A. 1957. Absorption of sulfur dioxide from the atmosphere by cotton plants. *Soil Sci.* 84:107–111.
Olszyk, D. M. M. and D. T. Tingey. 1985. Interspecific variation in SO_2 flux: Leaf surface versus internal flux and components of leaf conductance. *Plant Physiol.* 79:949–956.
Panić, M., and W. Franke. 1979. Ectodesmata (ectoteichodes) in wheat leaves infected with *Erysiphe graminis tritici*. *Z. Pflanzenkr. Pflanzenschutz* 86:465–471.
Parker, E. R., and R. W. Southwick. 1941. Manganese deficiency in citrus. *Proc. Am. Soc. Hortic. Sci.* 39:51–58.
Penot, M. 1972. Migrations libériennes chez *Tradescantia viridis*. Relation entre les quantites migrées et l'apport initial aux feuilles. *Physiol. Veg.* 10:687–696.
Penot, M. 1978. Hormone-directed transport in detached leaf. Phytohormonal competition. *Acta Hortic.* 80:75–78.
Penot, M. 1979. Some aspects of hormone-directed transport of mineral elements (P, S, Ca, Cl) in plants studied by means of radioactive tracers. *Isotopes and Radiation in Research on Soil–Plant Relationships. Proc. Symp. IAEA. Vienna* SM 235, pp. 527–538.
Penot, M., and J. Beraud. 1977. Phytohormones et transport orienté au niveau de la feuille isolée de *Pelargonium zonale*. Compétition phytohormonale. *Biol. Plant.* 19:430–435.
Penot, M., and J. Beraud. 1985. Hormone directed transport in isolated leaves of *Pelargonium zonale*—Relation with other processes. *Plant Growth Regul.* 3:302–309.
Penot, M., J. Beraud, and D. Poder. 1981. Relationship between hormone-directed transport and transpiration in isolated leaves of *Pelargonium zonale* L. Aiton. *Physiol. Veg.* 19:391–399.
Penot, M., and J. Gallou. 1977. Contribution à l'étude de la physiologie des transports à longue distance du chlore ^{36}Cl dans la plante. *Z. Pflanzenphysiol.* 85:201–214.
Perring, M. A. 1979. The effects of environment and cultural practices on calcium concentration in the apple fruit. *Commun. Soil Sci. Plant Anal.* 10:279–293.
Quinlan, J. D. 1969. Chemical composition of developing and shed fruits of Laxton's Fortune apple. *J. Hortic. Sci.* 44:97–106.

Raese, J. T., and D. C. Staiff. 1988. Chlorosis of 'Anjou' pear trees reduced with foliar sprays of iron compounds. *J. Plant Nutr.* 11:1379–1385.

Rahat, M., and L. Reinhold. 1983. Rb^+ uptake by isolated pea mesophyll protoplasts in light and darkness. *Physiol. Plant.* 59:83–90.

Reed, D. W. 1988. Effects of urea, ammonium and nitrate of foliar absorption of ferric citrate. *J. Plant Nutr.* 11:1429–1437.

Reed, D. W., and H. B. Tukey, Jr. 1978. Effect of pH on foliar absorption of phosphorus compounds by *Chrysanthemum*. *J. Am. Soc. Hortic. Sci.* 103:337–340.

Reed, D. W., and H. B. Tukey, Jr. 1982. Temperature and light intensity effects on epicuticular waxes and internal cuticular ultrastructure of Brussels sprouts and Carnation leaf cuticles. *J. Am. Soc. Hortic. Sci.* 107:417–420.

Ringoet, A., R. V. Rechenmann, and A. J. Gielink. 1971. Calcium accumulation in and possible excretion through guard and accessory cells of stomata and through bulliform cells. *Z. Pflanzenphysiol.* 64:60–64.

Ringoet, A., G. Sauer, and A. J. Gielink. 1968. Phloem transport of calcium in oat leaves. *Planta* 80:15–20.

Ritter, C. M. 1980. Fertilizer recommendations for Pennsylvania orchards. *Pa. Fruit News* 59:19–24.

Robbins, S., M. H. Chaplin, and A. R. Dixon. 1982. The effects of potassium soil amendments, trenching and foliar sprays on the mineral content, growth, yield and fruit quality of Sweet Cherry (*Prunus avium* L.) and *Prunus domestica* L. *Commun. Soil Sci. Plant Anal.* 13:545–560.

Robertson, N. K., K. Hinson, and L. C. Hammond. 1977. Foliar fertilization of soybeans (*Glycine max* L. Merr.) in Florida. *Soil Crop Sci. Am. Proc.* 36:77–79.

Robinson, E., and R. C. Robbins. 1968. Sources, Abundance and Fate of Gaseous Atmospheric Pollutants—Final Report. Project PR 6755. Stanford Res. Inst., Menlo Park, California.

Schönherr, J., and M. J. Bukovac. 1970. Preferential polar pathways in the cuticle and their relationship to ectodesmata. *Planta* 92:189–201.

Schönherr, J., and M. J. Bukovac. 1972. Penetration of stomata by liquids. Dependence on surface tension, wettability and stomatal morphology. *Plant Physiol.* 49:813–819.

Schönherr, J., and R. Huber. 1977. Plant cuticles are polyelectrolytes with isoelectric points around three. *Plant Physiol.* 59:145–150.

Schümacher, W. 1942. Über plasmodesmenartige Strukturen in Epidermisaussenwanden. *Jahrb. Wiss. Bot.* 90:530–545.

Scott, F. M., M. R. Schroeder, and F. M. Turrell. 1948. Development, cell shape, suberization of internal surface and abscission in the leaf of the Valencia orange, *Citrus sinensis*. *Bot. Gaz.* 109:381–411.

Shafer, W. E., and D. W. Reed. 1986. The foliar absorption of potassium from organic and inorganic potassium carriers. *J. Plant Nutr.* 9:143–157.

Shapiro, C. A., T. A. Peterson, and A. D. Flowerday. 1987. Irrigation method, composted manure, variety and foliar fertilizer effect on soybeans. *J. Fert. Issues* 4:122–129.

Shim, K. K., J. S. Titus, and W. E. Splittstoesser. 1972. The utilization of post harvest urea sprays by senescing apple leaves. *J. Am. Soc. Hortic. Sci.* 97:592–596.

Sifton, H. B. 1963. On the hairs and cuticle of Labrador tea leaves. A developmental study. *Can. J. Bot.* 41:199–207.

Sinclair, T. R., and C. T. deWit. 1975. Photosynthate and N requirements for seed production by various crops. *Science* 189:565–567.

Sitte, P., and R. Rennier. 1963. Untersuchungen an cuticularen zellwandschichten. *Planta* 60:19–40.

Smith, F. A., and J. B. Robinson. 1971. Sodium and potassium influx into citrus leaf slices. *Aust. J. Biol. Sci.* 24:861–871.

Smith, M. W., B. C. Cotten, and P. L. Agen. 1987. Foliar potassium sprays on adult pecan trees. *HortScience* 22:82–84.

Smith, M. W., D. Endicott, and N. W. Washmon. 1981. The influence of nitrogen and potassium in pecan. *Proc. Okla. Pecan Grow. Assoc.* 51:52–62.

Smith, R. C., and E. Epstein. 1964. Ion absorption by shoot tissue: Kinetics of potassium and rubidium absorption by corn leaf tissue. *Plant Physiol.* 39:992–996.

Stein, L. A., and J. B. Storey. 1986. Influence of adjuvants on foliar absorption of nitrogen and phosphorus by soybeans. *J. Am. Soc. Hortic. Sci.* 111:829–832.

Sumner, M. E. 1977a. Use of the DRIS system in foliar diagnosis of crops at high yield levels. *Commun. Soil Sci. Plant Anal.* 8:251–268.

Sumner, M. E. 1977b. Effect of corn leaf sampled on N, P, K, Ca, and Mg content and calculated DRIS indices. *Commun. Soil Sci. Plant Anal.* 8:269–280.

Suwanvesh, T., and L. G. Morrill. 1986. Foliar application to peanuts. *Agron. J.* 78:54–58.

Swietlik, D., and M. Faust. 1984. Foliar nutrition of fruit crops. *Hortic. Rev.* 6:287–355.

Swietlik, D., M. Faust, and R. F. Korcak. 1982. Effect of mineral sprays on photosynthesis and stomatal opening of water-stressed and unstressed apple seedlings. I. Complete nutrient sprays. *J. Am. Soc. Hortic. Sci.* 107:563–567.

Swietlik, D., and S. S. Miller. 1987. Effect of foliar-applied calcium and chlorine on stomatal conductance in apple leaves. *HortScience* 22:626–628.

Syverud, T. D., L. M. Walsh, E. S. Oplinger, and K. A. Kelling. 1980. Foliar fertilization of soybeans (*Glycine max* L.). *Commun. Soil Sci. Plant Anal.* 11:637–651.

Tatsumi, J., and Y. Kono. 1981. Translocation of foliar-applied nitrogen to rice roots. *Jpn. J. Crop Sci.* 50:302–310.

Thom, W. O., T. C. Miller, and D. H. Bowman. 1981. Foliar fertilization of rice after midseason. *Agron. J.* 73:411–414.

Thorne, G. N. 1955. Interaction of nitrogen, phosphorus, and potassium supplied in leaf sprays or in fertilizer added to the soil. *J. Exp. Bot.* 6:20–42.

Thorne, G. N. 1958. Factors affecting uptake of radioactive phosphorus by leaves and its translocation to other parts of the plant. *Ann. Bot.* 22:381–398.

Thorne, G. N. 1986. Confessions of a narrow-minded applied biologist, or why do interdisciplinary research? *Ann. Appl. Biol.* 108:205–217.

Tribe, I. S., J. K. Gaunt, and D. W. Parry. 1968. Cuticular lipids in the Gramineae. *Biochem. J.* 109:8P–9P.

Tukey, H. B., R. L. Ticknor, O. N. Hinsvark, and S. H. Wittwer. 1952. Absorption of nutrients by stems and branches of woody plants. *Science* 116:167–168.

Turley, R. H., and T. M. Ching. 1986. Physiological responses of barley leaves to foliar applied urea-ammonium nitrate. *Crop Sci.* 26:987–993.

Turner, J., J. Hemphill, and P. Mahlberg. 1980. Trichomes and cannabinoid content in developing leaves and bracts of *Cannabis sativa* L. (Cannabacaceae). *Am. J. Bot.* 67:1397–1406.

Uphof, J. 1962. Plant hairs. *Handb. Pflanzenanat.* 4:1–206.

Van Goor, B. J. 1973. Penetration of surface-applied ^{45}Ca into apple fruit. *J. Hortic. Sci.* 48:261–270.

Wattendorff, J., and P. J. Holloway. 1980. Studies on the ultrastructure and histochemistry of plant cuticles: The cuticular membranes of *Agave americana* L. *in situ*. *Ann. Bot.* 46:13–28.

Werner, R. 1981. Beiträge zur kenntnis der Teichoden unter Besonderer Berücksichtigung ihrer nachweisbarkeit, der Endoteichoden and der Wurzel Teichoden. Dissertation Dr. Agr. der Höhen Landwirtschaftlichen Fakultät der Rheinischen Friedrick-Wilhelms-Universität zu Bonn

Wetzstin, H. Y., and D. Sparks. 1983. Anatomical indices of cultivar and age-related scab resistance and susceptibility in pecan leaves. *J. Am. Soc. Hortic. Sci.* 108:210–218.

Winner, W. E., and C. J. Atkinson. 1986. Absorption of air pollution by plants and consequences for growth. *Tree* 1:15–18.

Wittwer, S. H., and M. J. Bukovac. 1969. The uptake of nutrients through leaf surfaces. In L. Scharrer and H. Linser (eds.), Handbuch der Pflanzenernährung und Dungung. I: Pflanzenernäshrung. Springer-Verlag, New York. Pp. 235–261.

Yamada, Y., S. H. Wittwer, and M. J. Bukovac. 1964. Penetration of ions through isolated cuticles. *Plant Physiol.* 39:28–32.

Yamada, Y., S. H. Wittwer, and M. J. Bukovac. 1965. Penetration of organic compounds through isolated cuticular membranes with special reference to ^{14}C-urea. *Plant Physiol.* 40:170–175.

Yogaratnam, N., M. Allen, and D. W. P. Greenham. 1981. The phosphorus concentration in apple leaves as affected by foliar application of its compounds. *J. Hortic. Sci.* 56:255–260.

Yogaratnam, N., and R. O. Sharples. 1982. Supplementing the nutrition of Bramley's seedling apple with phosphorus sprays. II. Effects on fruit composition and storage quality. *J. Hortic. Sci.* 57:53–59.

Yoshikawa, E. T. 1988. Correcting of iron deficiency of peach trees. *J. Plant Nutr.* 11:1387–1396.

Zilkah, S., I. Klein, S. Feigenbaum, and S. A. Weinbaum. 1987. Translocation of foliar-applied urea ^{15}N to reproductive and vegetative sinks of avocado and its effect on initial fruit set. *J. Am. Soc. Hortic. Sci.* 112:1061–1065.

Part II

Plant–Soil Interactions in Altering Nutrient Use Efficiency

9

Soil–Plant Interaction on Nutrient Use Efficiency in Plants: An Overview

V. C. BALIGAR, R. R. DUNCAN, and N. K. FAGERIA

Many reviews, chapters, and volumes dealing with genetic control of mineral nutrition have appeared in the literature. These publications give an overview (Vose, 1963; Epstein and Jefferies, 1964; Brown *et al.*, 1972; Clark, 1983; Marschner, 1986), address genetics and breeding (Myers, 1960; Gerloff and Gabelman, 1983; Vose, 1984; Sarić, 1984), and emphasize environmental stress factors (Wright, 1977; Jung, 1978; Clark, 1982; Devine, 1982; Blum, 1988).

Genetics and physiological components of plants have a profound effect on the

Crops as Enhancers of Nutrient Use

plant's ability to absorb and utilize nutrients under various environmental and ecological conditions. In this chapter we will examine the various plant and soil factors that contribute to the nutrient use efficiency of plants; however, we will try not to emphasize the information that has already been presented in the other chapters of this book.

I. Plant Factors

A. GENETICS

Even though crop adaptation is recognized as an important component of crop production, very few breeding programs have been concerned with achieving it directly.

Generally successful breeding efforts require the existence of variability and a means of stable transfer of a character from one individual to another. The choice of an easily identifiable character or marker is highly desirable to simplify the selection procedure.

One of the principal aims in plant breeding has to be to develop cultivars that are not only productive but also have desirable chemical composition from the viewpoint of human and/or animal nutrition. Recent interest in breeding plants for stress environments has centered on breeding plants for tolerance to Fe deficiency, salt, and toxic levels of soil Al and Mn. Recently, attention has been given to plant tolerances to heavy metal toxicities such as Cd, Pb, Cu, Zn, and Ni.

Most Fe-deficient soils contain considerable Fe, but its availability is low due to high soil pH. Improvement in tolerance to Fe deficiency and identification of genotypic differences in sorghum, rice, maize, soybeans, and citrus plants have helped to overcome the Fe deficiency problems (Brown *et al.*, 1972; Graham, 1984; Marschner, 1986). The Fe use efficiency in plants has been associated with excretion of protons (H^+) by roots, release of Fe-reducing compounds from roots, release of Fe-binding ligands such as citrate and other siderophores, and resisting interferences from other elements. Iron-efficient genotypes have a greater ability to absorb Fe from the growth medium than the inefficient plants (Brown *et al.* 1972; Clark, 1982; Marschner, 1986).

Manganese-efficient plants are capable of releasing protons and promoting reduced conditions with or without the aid of rhizosphere organisms, thereby enhancing the solubility of Mn in soil.

Mineral nutritional traits in plants in some cases are under the control of a single gene pair. In most situations more complex genetic systems are involved (Gerloff and Gabelman, 1983; Graham, 1984).

To differentiate efficient and inefficient P users there is need for understanding of the biochemical and physiological mechanisms that control the efficiency of P uptake in plants. Phosphorus use efficiency is largely determined at the soil–root interface. Root extension, root surface area, and internal P utilization are some of the dominating aspects in overall P use efficiency in plants. Phosphorous-efficient cultivars or genotypes could modify the accessibility of the soil P reserves through control of root length, root demand, and solubility of soil P. High efficiency in P uptake is attributed to a plant's ability to release citrate to free P bound by Fe. In some cases a change in rhizosphere pH was responsible for high P use efficiency by plants. Improvement in higher retranslocation of N from vegetative to reproductive parts during maturation results in more grain per unit N taken up. Cultivars that have high efficiency in retranslocation of absorbed N are higher yielders than those having high uptake rate.

Selection for efficient nutrient acquisition and avoidance of ion toxicities may be one approach to adapting plants to unfavorable environments. Species and cultivars or genotypes within species differ greatly in tolerance to mineral toxicities such as Al and Mn (Foy, 1984; Blum, 1988). Differential acid soil (Al, Mn, H) tolerance among cultivars/genotypes within species is often greater than between species (Foy, 1984).

B. PHYSIOLOGY

A close relationship between nutrient uptake (transport) and growth rates of plants has been reported (Pitman, 1972). This relationship raises the possibility that there exists some form of feedback control exerted by the shoot, possibly involving the action of plant growth substances. Roots synthesize hormones such as ethylene, cytokinins, gibberellins, and abscisic acid. These hormones effect root initiation and branching, root permeability, root–shoot ratios, leaf senescence, abscission, and stomatal behavior.

Increased growth induced by the growth regulators brings a higher demand for nutrients in plants. Phytohormones and growth regulators also influence water absorption and long-distance ion transport.

The nutrient status of the plant is known to affect the concentrations of plant growth substances. Low levels of nitrate reduce the concentration of cytokinins. Water deficit, low temperature, and nutrient deficiency increase the abscisic acid levels in plants. Such an enhanced level of abscisic acid has a feedback effect on mineral ion influx into the xylem and hence the translocation to the shoot. Zinc is another element that is involved in N nutrition and synthesis of phytohormones such as indoleacetic acid (IAA). Similarly, Ca also plays an indirect role in IAA and ethylene activities. Higher concentrations of Cu and Fe are known to increase ethylene levels and cause abscission of leaves. On the other hand, elevated amounts of Ca inhibit ethylene production.

Growth regulators are known to influence the uptake, transport, and redistribution of N, K, Ca, and Mg in the plant. The exact role of these phytohormones and growth regulators on nutrient use efficiency when the plant is under some kind of stress is still to be explored. Both synthesis and action of phytohormones are affected by environmental factors such as day length, temperature, and water and nutrient supply. For an extensive coverage of growth regulators (phytohormones) on plant nutrition refer to Marschner (1986).

Most nutrients must move by mass flow and/or diffusion through the soil to the root surface before they are positionally available for absorption into the root; therefore, root length and surface area influence the quantity of nutrient that can reach the root.

Both cultivar and species differences in rooting pattern occur. Rooting patterns in plants are influenced by aerial and soil environmental factors. Cultivar differences in rooting patterns are known to influence yield performance mainly because of differences in nutrient and water use. Genotypic variation in root systems of monocotyledon and dicotyledon species is well documented (O'Toole and Bland, 1987). Root traits such as morphology, rate of growth, and horizontal and vertical spread in soil are under genetic control.

Roots, through their genetic package and interaction with the environment, exert control over whole-plant growth and development by controlling all nutrient and water uptake.

Photosynthetic rates of individual leaves and crop canopies in both C_4 and C_3 species are strongly dependent on environmental conditions such as light, temperature, mineral nutrient content, water status, and concentration of CO_2. Crop species and genotypes within species react differently to environmental conditions, thereby indirectly affecting photosynthetic rates.

A close positive correlation is often observed between the rate of net photosynthesis and the mineral content of leaves. Iron and Mg are essential components of chlorophyll and deficiencies of these elements in plants affect the rate of photosynthesis. Adequate levels of N, P, and K are also essential to maintain appropriate levels of photosynthesis. Plants deficient in minerals usually have smaller leaves and the older leaves die more rapidly, thereby reducing the leaf area indices and photosynthesis. Membranes are probably the first line of defense against adverse environmental changes. Environmental influences such as temperature, light, ion interactions, degree of hydration, and hormonal effects can influence membrane fluidity. Evaluation of environmental effects on membrane fluidity in different crop species and on cultivars within species is an interesting avenue of research.

Allelochemicals are compounds that are released by plants through roots or leaves and affect the performance (usually detrimentally) of other plants. These chemicals can act as phytotoxicants, growth promoters, substrates for microorganisms, enhancers of root exudation, agents for altering soil structure, and predisposers for diseases (Hale and Orcutt, 1987). The role of allelochemicals in growth and nutrient use efficiency of crops grown in mixed cropping systems or in crop rotations is still an unexplored avenue of research.

II. Nutritional Factors

A. DEFICIENCY/TOXICITY OF ELEMENTS IN SOIL GROUPS

Dudal (1976) and Clark (1982) have given a detailed explanation of common stress problems (deficiency/toxicity) that are associated with the major soil groups of the world. Table I lists the major soil groups and associated deficiencies and toxicities of various elements. Many of the mineral stress problems in each soil group are related to the nature of the soil parent material. Soil-forming processes have also contributed to the extent of mineral stress problems. Mineral stress in many of these soils can often be associated with the pH of the soil. Close to 3 billion ha of land area in the world have mineral stress problems, representing 23% of the total world land area (Dudal, 1976). In acid soils, toxicities of Al, Mn, and Fe flooded rice are the major constraints for crop production. In many soil groups, macronutrient deficiency is a major stress factor as well.

B. ESSENTIALITY AND FUNCTIONS OF ELEMENTS

Sixteen elements are known to be essential for green plants. An essential element is one that participates directly as an indispensable requirement for the normal life cycle of a plant. This group consists of C, H, O, N, P, S, K, Ca, Mg, Fe, Mn, Zn, Cu, B, Mo, and Cl. Table II lists the principal forms taken up by plants, concentrations in plant material considered adequate, and the person who demonstrated the essentiality of the element.

Table I. Element Deficiencies and Toxicities Associated with Major Soil Groups[a]

	Element	
Soil Group	Deficiency	Toxicity
Acrisol	N, P, and most other	Al, Mn, Fe
Andosol	Ca, Mg, P, B, Mo	Al
Arenosol	K, Zn, Fe, Cu	—
Chernozem	Zn, Fe, Mn	—
Ferralsol	Ca, Mg, P, Mo	Al, Mn, Fe
Fluvisol	—	Al, Mn, Fe
Gleysol	Mn	Fe, Mo
Histosol	Si, Cu	—
Kastanozem	K, P, Mn, Cu, Zn	Na
Nitosol	P	Mn
Phaeozem	—	Mo
Podzol	Ca, K, N, P, micronutrients	Al
Planosol	Most nutrients	Al
Rendzina	P, Zn, Fe, Mn	—
Solonchak	—	B, Na, Cl
Solonetz	K, N, P, Zn, Cu, Mn, Fe	Na
Vertisol	P, N	S
Xerosol	Mg, K, P, Fe, Zn	Na
Yermosol	Mg, K, P, Fe, Zn	—

[a]Source: Modified from Dudal, 1976; Clark, 1982.

Table II. Essential Nutrients for Plant Growth, Their Principal Forms of Uptake, Concentrations Considered Adequate, and Person(s) Demonstrating the Essentiality of the Element[a]

Nutrient	Chemical Symbol	Principal Forms in which Taken Up	Concentration in Dry Tissue		Person(s) Demonstrating Essentiality
			μmol g⁻¹	mg kg⁻¹	
Carbon	C	CO_2	40 × 10³	45 × 10⁴	J. Sachs (1882)
Hydrogen	H	H_2O	60 × 10³	6 × 10⁴	J. Sachs (1882)
Oxygen	O	H_2O, O_2	30 × 10³	45 × 10⁴	T. De Saussure (1804)
Nitrogen	N	NH_4^+, NO_3^-	1 × 10³	15 × 10³	G. K. Rutherford (1872)
Phosphorus	P	$H_2PO_4^-$, HPO_4^{2-}	60	2 × 10³	Posternak (1903)
Potassium	K	K	250	10 × 10³	A. F. Z. Schimper (1890)
Calcium	Ca	Ca^{2+}	125	5 × 10³	F. Salm-Horstmar (1856)
Magnesium	Mg	Mg^{2+}	80	2 × 10³	Willstatter (1906)
Sulfur	S	SO_4^{2-}, SO_2	30	1 × 10³	Peterson (1911)
Iron	Fe	Fe^{2+}, Fe^{3+}	2	1 × 10²	J. Sachs (1860)
Manganese	Mn	Mn^{2+}	1	50	J. S. McHague (1922)
Boron	B	H_3BO_3, BO_3^-, $B_4O_7^{2-}$	2	20	K. Warington (1923)
Zinc	Zn	Zn^{2+}	0.3	20	A. L. Sommer and C. B. Lipman (1926)
Copper	Cu	Cu^{2+}	0.1	6	C. B. Lipman and G. MacKinney (1931)
Molybdenum	Mo	MoO_4^{2-}	0.001	0.1	D. I. Arnon and P. R. Stout (1938)
Chlorine	Cl	Cl^-	3	1 × 10²	T. C. Broyer *et al.* (1954)

[a]Source: Epstein, 1972; Marschner, 1986.

Mineral nutrients have specific and essential functions in plant metabolism. Table III lists the functions of various essential elements in plants. Plants usually show characteristic symptoms in response to the lack of essential elements. These various elements are constituents of proteins, nucleic acids, organic structures, and enzyme molecules and also play a role in osmoregulation. Lack or excess of any of the essential elements, either in the growth medium and/or in the plant itself, brings changes in overall performance of the plant.

C. NUTRIENT FLUX AT THE SOIL–ROOT INTERFACE

Three mechanisms responsible for much of the ion transport from the soil to the root surface are root interception, mass flow, and diffusion. In the root interception

Table III. Functions of Essential Nutrients in Plants[a]

Nutrient	Function
Carbon	Basic molecular component of carbohydrates, proteins, lipids, and nucleic acids.
Oxygen	Oxygen is somewhat like carbon in that it occurs in virtually all organic compounds of living organisms.
Hydrogen	Hydrogen plays a central role in plant metabolism. Important in ionic balance; acts as main reducing agent, and plays a key role in energy relations of cells.
Nitrogen	Nitrogen is a component of many important organic compounds ranging from proteins to nucleic acids.
Phosphorus	The central role of phosphorus in plants is in energy transfer and protein metabolism.
Potassium	Helps in osmotic and ionic regulation. Potassium functions as a cofactor or activator for many enzymes of carbohydrate and protein metabolism.
Calcium	Calcium is involved in cell division and plays a major role in the maintenance of membrane integrity.
Magnesium	Component of chlorophyll and a cofactor for many enzymatic reactions.
Sulfur	Sulfur is somewhat like phosphorus in that it is involved in plant cell energetics.
Iron	An essential component of many heme and nonheme Fe enzymes and carriers, including the cytochromes (respiratory electron carriers) and the ferredoxins. The latter are involved in key metabolic functions, such as N fixation, photosynthesis, and electron transfer.
Zinc	An essential component of several dehydrogenases, proteinases, and peptidases, including carbonic anhydrase, alcohol dehydrogenase, glutamic dehydrogenase, and malic dehydrogenase.
Manganese	Involved in the O_2-evolving system of photosynthesis and is a component of the enzyme arginase and phosphotransferase.
Copper	A constituent of a number of important enzymes, including cytochrome oxidase, ascorbic acid oxidase, and lactase.
Boron	The specific biochemical function of B is unknown but it may be involved in carbohydrate metabolism and synthesis of cell wall components.
Molybdenum	Required for the normal assimilation of N in plants. An essential component of nitrate reductase, as well as nitrogenase (N_2 fixation enzyme).
Chlorine	Essential for photosynthesis and as an activator of enzymes involved in splitting water. It also functions in osmo-regulation of plants growing on saline soils.

[a]Source: Compiled from Oertli, 1979; Ting, 1982; Stevenson, 1986.

mechanism, soil nutrients at the root surface do not have to move to the interface to be positionally available for absorption. Therefore, the larger the root volume (density), the higher the nutrient interception by the roots and the greater the chances for the root to intercept the ion diffusion paths. Mass flow is a mechanism by which the bulk of the ions in soil solution are transported along the water potential gradient; this process is driven by transpiration. The amount of nutrient movement by mass flow is related to the water used and nutrient concentration in the water. When the supply of ions by root interception and mass flow is not adequate to meet the crop demand, continued uptake reduces the concentration of available nutrients in the soil at the root surface. Such circumstances create a concentration gradient perpendicular to the root surface, and consequently nutrients in the soil diffuse along the gradient toward the root surface.

The relative contributions each of these three mechanisms are largely influenced by soil and plant factors. The relative contribution of mass flow varies with age of the plant and the relative humidity, which affects the transpiration. Plant factors such as root surface area and root density and soil factors such as soil moisture, temperature, density, and texture have an influence on the contribution by diffusion. Species or cultivars within species that have higher root surface area (longer root hairs and root length) are far more efficient in taking up P and K than ones with smaller root surface area.

The potassium supply attributable to different supply mechanisms in maize, onion, and wheat from Raub soil (*Aguic Arguidoll*) is shown in Table IV. Maize plants had larger seeds, grew more rapidly, and produced more roots than wheat and onion. Wheat growth was intermediate between the other two species. The root lengths per gram of shoot were much larger for wheat, followed by maize and onion. As a result, fluxes of K per unit root length were smaller for wheat. Compared to maize, wheat and onions had more root length per gram root, indicating that the roots in these plants were much thinner than those of maize. Root surface area (cm^2 $plant^{-1}$) of maize plants was much greater than that of wheat and onion.

Table IV. Variations in Growth, K Uptake, and K Supply Mechanisms as Influenced by Crop Species[a]

Parameter	Maize	Onion	Wheat
Plant weight, g pot^{-1}	8.0	0.6	2.9
Root length, m pot^{-1}	175.6	9.2	130.8
Root length, m g $root^{-1}$	68.0	98.0	164.0
Root length, m g $shoot^{-1}$	33.0	20.0	64.0
Root surface area, cm^2 $plant^{-1}$	364.0	6.0	104.0
Root density, cm cm^{-3}	12.7	0.7	9.4
Distance between roots, cm	0.2	0.7	0.2
K uptake, m*M* $plant^{-1}$	0.20	0.01	0.97
K flux, p*M* cm^{-1} sec^{-1}	0.15	0.12	0.08
Supply Mechanisms (%)			
Root interception	0.7	0.4	0.5
Mass flow	0.2	60.8	3.9
Diffusion	99.1	38.8	95.6

[a]Source: Baligar, 1985.

Because of lower root density, the mean distance between onion roots was greater than for maize and wheat. About 99% of the K supplied to maize was from the diffusion process, while for onion and wheat, the diffusion mechanism contributed 39 and 96%, respectively. In the case of onion, the mass flow mechanism contributed 61% of the K requirement. In all three species, less than 1% of the absorbed K was due to the root interception process.

Higher root density might induce root competition for the available nutrient pool in the soil. Such effects influence the magnitude of various nutrient flux mechanisms in soil. Various root densities were achieved by changing the number of wheat plants from 1 to 16 per 1.8 kg of soil. The shoot and root growth and root density per plant increased with increasing plant density (Table V). Increasing plant density reduced root surface area per plant, distance between roots, and K uptake and flux rates. Total water uptake increased by threefold with the higher plant density treatment. Increasing plant density from 1 plant per pot to 16 plants per pot increased the diffusion contribution of K from 2 to 63%, whereas the mass flow contribution was reduced from 98 to 35%.

From the preceding results it can be concluded that crop species, their growth parameters, root competition, and nutrient demand greatly influence the percentage of mass flow and diffusion contributions to total nutrient uptake.

Table V. Variations in Growth, K Uptake, and Supply Mechanisms as Influenced by Density of Wheat Plants for a Given Volume of Soil (1.8 kg soil)[a]

	Number of Plants per Pot		
Parameter	1	4	16
Plant weight, g pot^{-1}	0.7	2.0	3.3
Root length, m pot^{-1}	41.4	69.5	126.4
Root length, m g root^{-1}	199.0	126.0	125.0
Root surface, cm^2 plant^{-1}	218.0	125.0	61.0
Root density, cm cm^{-3}	3.0	5.0	9.1
Distance between roots, cm	0.33	0.26	0.19
Water uptake, g pot^{-1}	391.0	1019.0	1469.0
K uptake, m*M* plant^{-1}	0.4	0.3	0.1
K flux, p*M* cm^{-1} sec^{-1}	0.2	0.3	0.1
Supply Mechanisms (%)			
Root interception	1.1	1.1	2.3
Mass flow	97.8	54.5	35.0
Diffusion	1.2	44.5	62.7

[a]Source: Baligar, 1985.

D. NUTRIENT USE EFFICIENCY

For a given genotype, nutrient efficiency is reflected by the ability to produce a high yield in a soil that is limited in one or more mineral nutrients for a standard genotype (Graham, 1984). In a physiological sense, nutrient use efficiency (NUE) is defined as the amount of dry matter or economic yield produced per unit of nutrient in the plant. Equation (1) describes NUE:

$$\text{NUE} = \text{Yield/Nutrient uptake} \quad (1)$$

When comparing efficiency between fertilized plants (F) and unfertilized control (C), the NUE is expressed as shown in Eq. (2). The units of NUE are mg mg^{-1}, g g^{-1}, or kg kg^{-1}.

$$\text{NUE} = (\text{Yield F} - \text{Yield C})/(\text{Nutrient uptake F} - \text{Nutrient uptake C}) \quad (2)$$

Gerloff and Gabelman (1983) have differentiated various strains of plants into most efficient and inefficient for yield expression under a given nutrient stress. They have adapted the nutrient efficiency ratio (NER) as

$$\text{NER} = \text{Units of yield/Unit of element in tissue} \quad (3)$$

where yield is usually grain yield or total biomass in mg, g, or kg and the amount of element present is expressed in mg, g, or kg.

In agronomic terms, fertilizer efficiency is expressed as the amount of economic yield per unit of nutrient applied. Agronomic efficiency, AE (kg kg^{-1}), is calculated as

$$\text{AE} = (\text{Yield F} - \text{Yield C})/(\text{Quantity of nutrient applied}) \quad (4)$$

The apparent nutrient recovery (ANR) can be computed from Eq. (5) and expressed in percentage. The ANR is used to reflect the efficiency of a plant to obtain applied nutrients from the soil.

$$\text{ANR} = [(\text{Nutrient uptake F} - \text{Nutrient uptake C})/(\text{Quantity of nutrient applied})] \times 100 \quad (5)$$

For an extensive coverage of various aspects of mineral nutrient efficiency see Clark in Chapter 4. In Chapter 13, fertilizer efficiency (kg grain kg^{-1} N applied) for N is defined.

Availability of nutrients to the plant is largely influenced by soil factors. Soil physicochemical properties affect the ionic equilibria, thereby influencing the ionic composition of the soil solution, especially at the soil–root interface. The magnitude of diffusion and mass flow fluxes of ions at the soil–root interface are influenced by soil moisture, temperature, ionic properties, organic ions, rate of water absorption by the root, and other root and microbial activities. Root geometry, nutrient status, rate of nutrient and water uptake, and transport and distribution in plants have an effect on the nutrient use efficiency by plants. Mechanisms and processes in soil and plants that affect overall nutrient use efficiency in plants are listed in Table VI. An extensive discussion on these various factors can be found in Gerloff and Gabelman (1983), Barber (1984), and Marschner (1986). Genotypic variations in uptake of nutrients under identical growth conditions could be attributable to the genetic makeup of the plant. The shoot affects transport from the root directly via passive driving forces and indirectly via metabolism-dependent driving forces. Interaction between the shoot and its environment also has an influence on the nutrient uptake efficiency. Species, cultivars, and genotypes within species interact differently with their environment and such variation will result in different growth and nutrient use in plants. Crop species and cultivars and genotypes within species have shown varying responses to micronutrient availability (Brown *et al.*, 1972; Kanwar and Youngdahl, 1985). Genetic improvement of micronutrient efficiency in wheat, sorghum, rice, maize, chick-peas, and soybean has been exploited (Kanwar and Youngdahl, 1985).

Temperature and moisture levels are the two main components of climate that affect micronutrient availability in soil. Early in the growing season, low soil temperatures often lead to Zn and P deficiency, especially where high P levels exacerbate the Zn deficiency. Excess moisture increases the availability and uptake

Table VI. Mechanisms and Processes That Contribute to Nutrient Use Efficiency in Plants

Plant Factors	Soil Factors
Root and Root Hair Geometry	Soil Solution
Number and length	Ionic equilibria
Rate of growth	Solubility, precipitation
Radius	Competing ions
Physiological	Organic ions
Nutrient status	pH
Age and rate of growth	Moisture and temperature
Root/shoot ratio	Diffusion and Mass Flow
Rate of water uptake	Soil moisture
Rate of nutrient influx	Tortuosity
Rate of nutrient efflux	Buffer capacity
Rate of nutrient transport	Ionic properties
Partitioning of nutrient in the plant	Ionic concentrations
Utilization efficiency or low functional nutrient requirement	Other Soil Factors
Environmental Effects	Physicochemical properties of the soil
Intensity and quality of light	
Temperature and moisture	
Soil pH effects	
Rhizosphere	
Nutrient solubility	
Exudates	
Microbial association	

of Fe, Mo, and Mn. Flooding of soil increases the availability of Fe and in some soils the increased Fe levels actually cause toxicity to plants. Drying of soil reduces the moisture content, thereby reducing the rate of diffusion for many of the micronutrients.

Roots induce many modifications in their rhizosphere. Roots preferentially absorb either ions or water, thereby resulting in depletion or accumulation of ions around the root surface. They release H^+ or HCO_3^-, which changes the pH and the consumption or release of O_2 and influences the redox potentials of the rhizosphere soil. Root exudates mobilize mineral nutrients but also enhance microbial growth. The high resultant microbial activities might mobilize or immobilize mineral nutrients. The plant genetic role in these various root activities and their relationship to growth and mineral nutrition is still very much an unexplored area.

Noninfecting microorganisms affect plant mineral nutrition through their influence on the physiology and development of the plant, growth and morphology of roots, and the availability and uptake processes of nutrients (Rovira *et al.*, 1983). The endomycorrhizae are known to enhance the uptake of P, Cu, Zn, and Ni by the roots. The mycorrhizae affect only those nutrients that have a very low availability in soils and that are present in soil solution in very low concentrations.

Sixty-six strains of snap beans and numerous tomato genotypes were subjected to a growth-limiting supply of K, P, N, and Ca (Gerloff and Gabelman, 1983). The efficient and inefficient strains are given in Table VII. Yields of efficient and inefficient strains differed in the utilization of the elements by as much as 44% for N, 72% for P, 47–107% for K, and 169% for Ca.

Table VII. Yield and Nutrient Efficiency Ratios (mg dry wt. per mg element in plant) of Strains of Beans and Tomatoes[a]

Species	Stress Element	Level of Stress Element (mg plant^{-1})	Strain[b]	Dry Matter Yield (g plant^{-1})	Nutrient Efficiency Ratio
Bean	K	11.3	63(I)	6.00	157
			58(E)	8.83	294
Tomato	K	5.0	94(I)	0.95	173
			98(E)	1.97	358
Bean	P	2.0	2(I)	0.87	562
			11(E)	1.50	671
Tomato	N	35.0	51(I)	2.51	83
			63(E)	3.62	118
Tomato	Ca	10.0	139(I)	1.35	381
			39(E)	3.63	434

[a]Source: Gerloff and Gabelman, 1983.
[b]I, most inefficient; E, most efficient.

Twenty-three red clover cultivars were evaluated for their NER to differentiate efficient and inefficient nutrient utilizers at 0 and 50 μM Al levels (Table VIII). Cultivars that have high NER values for essential nutrients, especially for P, Ca, and Mg, might be able to perform well in acidic infertile soils. The red clover cultivars adapted in this study behaved differently for nutrient efficiency ratios. In many instances, the least and the most efficient cultivars for NER differed depending on the presence or absence of Al in the growth medium. The cultivars used in this study showed intraspecific genetic diversity in growth and NER values for the essential elements in the presence or absence of Al. Similar genetic diversity in growth and NER for essential elements was observed in 40 sorghum genotypes grown at 2 and 64% Al saturation (Table VIII) in a Brazilian acid latosol (*Typic haplorthox*) soil with a pH of 4.3. Such findings clearly demonstrate that there is a possibility of selection and breeding for desirable plants adapted to acid infertile soils.

Efficiency of fertilizer use depends on the ability of the plant root to obtain a high proportion of the fertilizer nutrient added to the soil. Knowledge of the variation in ion absorption kinetics of roots among genotypes and its inheritance is important in the development of genotypes that are more efficient in ion absorption from the soil. Six maize genotypes developed in Florida and six genotypes developed in Indiana showed the existence of variations in growth and ion uptake characteristics in a nutrient culture study (Table IX). Florida genotypes had higher average I_{max} (maximal net influx) and K_m (Michaelis–Menten constant; ion concentration in solution, where $I_n = \frac{1}{2}I_{max}$) values for Ca and P. Parental lines of Florida hybrids had higher I_{max} for Ca. Even though Florida and Indiana maize genotypes originated in different geographic regions, they exhibited little difference in their average net influx (I_n) rates for Ca. Florida inbreds with higher I_n for P tend to have hybrids with higher P influx rates. While the differences in this study were not large, investigation of a large number of genotypes is needed to determine the possibility of developing hybrids with roots that are more effective nutrient absorbers.

In higher plants, interactions between nutrients occur when the supply of one nutrient affects the absorption, distribution, or function of another nutrient. Thus, depending on the nutrient supply, interactions between nutrients can modify growth responses and induce either toxicities or deficiencies (Robson and Pitman, 1983). Interactions between cations, anions, micronutrients, and organic ligands might lead to the formation of precipitates or complexes either in the soil or in the plant. Another form of interaction is between ions with sufficiently similar chemical properties. They are known to compete for sites of adsorption, absorption, transport, and function. At a low concentration, SeO_4^{2-}/SO_4^{2-}, K^+/Rb^+, and Ca^{2+}/Sr^{2+} are known to compete with each other for active absorption at the plasmalemma. Fertilizer and other amendments added to soil might change the composition of the soil solution, thereby increasing or decreasing the availability of one or more ions. Such changes might influence the ionic supply pool and may lead to ion interactions both in the soil and in plants. The interaction between essential and nonessential elements may have either positive, negative, or no effect on plant growth.

Table VIII. The Nutrient Efficiency Ratios (NER) for Different Elements of the Most and Least Efficient Red Clover Cultivars (Solution Culture) and Sorghum Genotypes (Soil Culture) with or without Al

Element	Cultivar/Line/Genotype[a]	NER	Cultivar/Line/Genotype	NER
	Red Clover Cultivars[b]			
	0 μ*M* Al		50 μ*M* Al	
P	Redman (E)	1,012	Altaswede (E)	823
	Florie (I)	470	Florie (I)	358
S	Kuhn (E)	970	Altaswede (E)	673
	Redmor (I)	523	Redmor (I)	462
K	Redman (E)	104	Kuhn (E)	55
	K4-183 (I)	61	K4-183 (I)	40
Ca	K4-184 (E)	91	Redland (E)	76
	Florie (I)	53	Norlac (I)	52
Mg	Chesapeake (E)	670	Altaswede (E)	531
	Florie (I)	476	Redmor (I)	359
Mn	K4-184 (E)	37,000	Kenland (E)	25,000
	Prosper I (I)	23,000	K4-184 (I)	16,000
Fe	Altaswede (E)	26,000	Arlington (E)	17,000
	Norlac (I)	14,000	K4-184 (I)	10,000
	Sorghum Genotypes[c]			
	2% Al		64% Al	
N	SC489 (E)	38	SC215 (E)	48
	SC431 (I)	26	PU932204 (I)	32
P	SC489 (E)	1,000	SC237 (E)	1,667
	BR004 (I)	476	SC283 (I)	588
K	BR006R (E)	44	SC308 (E)	164
	SC331 (I)	23	SC283 (I)	42
Ca	Redland B (E)	208	SC215/SC489 (E)	1,250
	SC206/SC207 (I)	123	CMSXS-604 (I)	370
Mg	SC208 (E)	417	SC254 (E)	2,000
	SC226/SC450/SC501 (I)	278	TX430R (I)	1,000
Zn	SC209 (E)	31,000	PU932204 (E)	16,000
	PU932204 (I)	12,000	BR007B (I)	5,000
Fe	BR300 (E)	5,000	SC226 (E)	9,000
	PU932204 (I)	2,000	SC283 (I)	3,000

[a]E, most efficient; I, most inefficient.
[b]Twenty-three red clover cultivars were used (Baligar *et al.*, 1987).
[c]Forty sorghum genotypes were used (Baligar *et al.*, (1989).

Table IX. Growth, Maximum Influx (I_{max}), Michaelis–Menten Constant (K_m), and Net Influx (I_n) for P and Ca in Florida and Indiana Maize Genotypes[a]

			I_{max}		K_m		I_n	
Genotypes	Plant Weight (g plant^{-1})	Root Length (m plant^{-1})	Ca (p*M* cm^{-1} sec^{-1})	P	Ca (m*M*)	P	Ca (p*M* cm^{-1} sec^{-1})	P
Florida								
7B-339-1	0.61	27.4	0.83	0.36	7.3	4.0	0.11	0.36
4I-244-2	0.53	17.0	0.89	0.18	3.6	2.4	0.09	0.35
6B-399-1	0.49	18.7	1.05	0.33	4.2	3.7	0.10	0.57
5I-478-1	0.50	12.9	1.08	0.89	3.0	11.6	0.14	0.68
7B-339-1 × 4I-244-2	1.22	46.6	0.32	0.34	10.4	8.9	0.13	0.31
5I-478-1 × 6B-399-1	1.13	39.0	0.40	0.32	7.7	7.4	0.13	0.38
$\bar{x}$	0.75	26.9	0.76	0.40	6.0	6.3	0.12	0.44
Indiana								
H60	0.36	11.2	0.09	0.12	2.7	1.7	0.14	0.31
W64-A	0.63	18.9	0.11	0.19	2.7	3.3	0.13	0.34
H84	0.62	12.5	0.15	0.09	3.9	1.0	0.11	0.37
H99	0.60	15.4	0.23	0.18	4.5	2.8	0.11	0.44
H60 × W64A	1.05	41.2	0.09	0.14	6.2	1.4	0.11	0.31
H84 × H99	0.91	32.8	0.04	0.17	1.8	1.9	0.04	0.32
$\bar{x}$	0.70	22.0	0.12	0.15	3.6	2.0	0.12	0.35
C1[b]	NS	**	**	**	*	**	NS	**
C2	**	**	**	NS	**	**	NS	**
C3	**	**	NS	NS	NS	NS	*	NS

[a]Source: Baligar and Barber, 1979.

[b]C1, C2, and C3 refer to contrast between Florida and Indiana genotypes, Florida inbreds vs. their hybrids, and Indiana inbreds vs. their hybrids, respectively. Treatment effects were signficant at 5% (*) or 1% (**) levels or nonsignificant (NS).

III. Environmental Factors

A. WATER

Either wet or dry soil conditions affect root growth and functions. A substantial water deficit in the soil, plant, or atmosphere leads to the occurrence of drought stress. In a cultivar improvement program for drought tolerance conditions, one has to use reliable indices such as phenological, morphological, anatomical, physiological, or biochemical characteristics to facilitate selections (Hall, 1981). In plant selection processes for drought tolerance, the morphological drought escape and avoidance and tolerance features in plants need to be harnessed.

Genetic variability for water use efficiency (more yield per unit of water used) is known to exist in various crop species and genotypes. Wheat genotypes with the ability to vary their osmoregulation (osmotic adjustment) mechanisms showed higher growth-maintaining effects under dry conditions (Morgan, 1984). Plants that compensate osmotically for the onset of dryness in soils have a better chance of withstanding drought.

Waterlogging of the soil rapidly and dramatically alters both the physical and biological environment of plant roots. The physiological response to waterlogging affects a wide range of metabolic, hormonal, and developmental processes. Waterlogged conditions in soil lead to cessation of root growth and respiration, which results in a drastic drop in the uptake and transport of nutrients and water. Wilting and epinasty in the plant are related to a decrease in water permeability of roots and an accumulation of ethylene in the shoot. The severity of the effects of waterlogging on the growth and yield of plants depends on the plant species, stages of growth, and soil properties such as pH, organic matter content, and temperature (Marschner, 1986). Waterlogging also inhibits the synthesis and export of cytokinins and gibberellins. Genotypic differences in the Mn tolerance of shoot tissue are closely related to tolerance to waterlogging (Finn *et al.*, 1961).

An increase in mechanical impedance and a decrease in water potential are associated with soil drying, whereas a decrease in aeration and accumulation of phytotoxins are known to be associated with wet soils. Such physical changes in soil are responsible for decreases in root extension and growth. In plants adapted to wetlands, a substantial amount of oxygen is transported from shoot to roots through aerenchyma. The tolerance of plants to oxygen deficiency is improved by adequate mineral nutrition, especially nitrate. Proper soil aeration is also essential for meeting the respiratory requirements of soil microorganisms as well.

B. TEMPERATURE

Among various climatic factors, temperature is the major uncontrollable factor that determines crop production areas and limits crop yields. Plants survive temperature extremes by either avoidance or tolerance of internal heat increases. Optimum temperature varies among species and to some extent within species and tends to be lower for root growth than for shoot growth. High-temperature stress leads to an insufficient supply of carbohydrates to root meristems, whereas low temperature leads to poor or reduced shoot growth due to an insufficient supply of mineral nutrients and water (Marschner, 1986). At low soil temperatures, nutrient availability and uptake are inhibited in addition to reductions in root growth. Root export of cytokinin to the shoot is known to be substantially reduced by low temperatures.

C. SOLAR RADIATION

The quantity and quality of incoming light available to plant leaves will depend on the stand density of the plant, plant height and shape, and orientation of leaves on the culm. Variations in these components eventually affect the capacity of plants to utilize solar radiation. Reflection, absorption, or transmittance of the incident radiation from the sun are influenced by the wavelength of radiation and structure, surface area, and orientation of the leaf. Marked genotypic and cultivar differences in maximum photosynthetic rate per unit leaf area have been reported in many C_3 crops such as soybean, rice, barley, beans, alfalfa, and ryegrass, as well as C_4 crops such as maize, sugarcane, and *Cenchrus ciliaris* (Cooper, 1981). In a few cases, however, correlation with total economic yield suggests that a lower photosynthetic rate per unit leaf area was compensated for by larger but thinner leaves.

Distribution of assimilates within the plant for continued vegetative growth or for accumulation in particular sinks such as seeds, fruits, and storage organs is the most important determinant of economic yield rather than rate of photosynthesis and the supply of assimilates (Cooper, 1981). A reduced rate of photosynthesis in several crop species has been shown to be related to deficiency of N, P, K, S, Mg, Ca, Mo, Mn, B, Fe, and Zn (Moorby and Besford, 1983).

D. pH EXTREMES

Close to 3 billion ha, i.e., nearly 23% of the world's soils, are considered to have some kind of mineral stress. Soil acidity and alkalinity are the two major causes of mineral stress. Reducing the soil chemical constraints for growth by applying amendments is one way of improving yield potentials, but in many instances it is impractical and costly to achieve. Another approach is to breed or select plants that are resistant or tolerant to moderate ecosystem mineral stresses. This appears to be a very fruitful alternative.

In acid soils, lack of adequate levels of Ca is a major limiting factor for proper root growth. Excess levels of Al in soil have toxic effects on root growth and function. Liming of acid soils increases the pH, thereby reducing the activity of Al and Mn and enhancing root growth. Such effects are beneficial to the plant by increasing subsoil penetration by the roots and enhancing utilization of soil nutrients and water. Growth-limiting factors in acid soils include toxicities of Al^{3+}, Mn^{2+}, and other metal ions, low pH (H^+ toxicity), and deficiencies or unavailabilities of certain essential elements, particularly P, Ca, Mg, Fe, and Mo (Foy, 1984). Many of the acid soil factors may also promote or inhibit the survival and function of mycorrhizae, rhizobia, and other microorganisms. Plant species and genotypes within species differ widely in tolerance to various levels of stress in acid soils. The exact physiological mechanisms of Al and Mn tolerances are still debated. Foy (1984) states that Al tolerance in plants has been associated with physiological characteristics (rhizosphere pH changes, resistance to drought, organic acid content, root cation-exchange capacity, root phosphatase activity, and trapping of Al in nonmetabolic sites within plants) and with nutritional characteristics (P and Fe use efficiency, Ca and Mg uptake and transport, and internal concentration of Si, NH_4^+, or NO_3^-).

Manganese tolerance in plants is known to be related to oxidizing power of plant roots, Mn absorption and translocation rates, entrapment in nonmetabolic sites, high internal tolerance to excess Mn, and uptake and distribution of Si and Fe (Foy, 1984). In many species and cultivars, tolerance to Al and Mn is related to nutrient uptake and transport and use efficiency (Foy, 1983, 1984). Cultivars with a high nutrient efficiency ratio may have an advantage in adapting to mineral stress ecosystems. Aluminum tolerance in cultivars of wheat, barley, soybean, snap bean, sorghum, cocoa, tobacco, and tomato is related to Ca and P nutrition, and in wheat, barley, sorghum, potato, maize, and rice to Fe, Mg, Si, or K nutrition (Foy, 1984).

Salinity is a significant limiting factor to agricultural productivity of the world and affects about 1 billion ha of land. Soil salinity problems could be minimized, or to some extent eliminated, by chemical amendments and irrigation. Another approach is introduction and selection or breeding of salt-tolerant (resistant) plants that produce economic yields under conditions of moderate or low levels of salinity. The detrimental effects of salinity on plants are twofold; ions in the external soil solution influence water activity, thereby affecting the water status of the plant, and absorbed ions directly affect the physiological and biochemical functions of the plant. The excess ions in plants lead to (a) turgor reduction, (b) inhibition of membrane function, photosynthesis, and enzyme activities, (c) induction of ion deficiency due to inadequate transport/selective mechanisms, and (d) increased use of metabolic energy for nongrowth processes involved in the maintenance of tolerance (Hasegawa *et al.*, 1986). Differences in salt tolerance exist not only between species but also among genotypes of certain species. Such differences among genotypes are related to restricted Cl translocation into the shoot, preferential accumulation of Na and Cl within the shoot, and high efficiency in the absorption of K. Deficiency of Zn, Fe, P, and to some extent Ca, K, and Mg and toxicity of Na and B are the major mineral constraints for crop production in sodic ($pH > 8$) and saline soils. In saline soils, plant growth is affected mainly by high levels of NaCl and impairment of water balance. In sodic soil, poor physical conditions and poor aeration are the major constraints and are strongly associated with toxicities of B and Na.

Another problem of high-pH soils is the unavailability of some nutrients, and plant species vary in their sensitivity to this problem. Genotypic and cultivar differences in soybean, tomato, oats, Bermuda grass, love grass, and chick-pea have been reported for their high Fe use efficiency in high-pH soils. The efficient types are known to release more protons and/or have a higher root reducing capacity than the inefficient Fe users (Marschner, 1986). Differences in Zn efficiency have been observed for maize, bean, soybean, potato, sugar beets, barley, and wheat in alkaline soils with pH 8.2 (Marschner, 1986).

IV. Microbial Association

Microorganisms present in the rhizosphere may stimulate or inhibit root growth depending on the type of microorganism and on prevailing environmental conditions. Inhibition of root growth is probably due to immobilization of nutrients that are in short supply and the production of phytotoxins. Stimulation of plant growth is attributable to microbial production of phytohormones, mobilization of mineral nutrients, and fixation of atmospheric N into plant-available form. Microorganisms are involved in the enhancement of the availability of K, Ca, P, S, Fe, and other elements through the decomposition of organic compounds and the oxidation or reduction of inorganic compounds. The free-living microbes significantly affect plant growth, pathogens, antagonists, auxin producers, ethylene producers, nitrogen fixers, or phosphate solubilizers.

Noninfective rhizosphere microorganisms can have a large effect on plant nutrition, as does the infective association of mycorrhizal fungi. Rhizosphere microorganisms can affect plant nutrition through their influence on availability of nutrients, nutrient uptake processes, growth and morphology of roots, and physiology and development of plants (Rovira *et al.*, 1983). Nutrient release and immobilization, nitrification and denitrification, nitrogen fixation, P, Mn, and Mo availability, root length, and root hair development are particularly subject to rhizosphere microorganism alterations.

Vesicular-arbuscular mycorrhizae (VAM) form an endotrophic association with higher plants. The VAM can be found associated with virtually all crop plants. Light, temperature, soil moisture, and fertility indirectly affect mycorrhizal morphogenesis largely because they affect root physiology and morphology, the quality and quantity of root exudates, and microbial activity (Curl and Truelove, 1986). Mycorrhizal infection of roots is known to increase the uptake of P, S, Zn, and Cu by plants. Ionic forms of these elements in soils are relatively insoluble, low in concentration, and have low diffusion rates. The growth of fungal hyphae into the soil increases the root surface area, thereby assisting the root in exploring a larger soil volume. Increases in root surface area will increase the probability for the roots to intercept diffusion paths of ions that diffuse short distances. Rovira *et al.* (1983) state that the increase in ion uptake due to endomycorrhizae can have large consequences for nitrogen fixation by legumes and by nonlegumes. VAM are known to increase uptake of Co, Mo, Cu, and Fe, which are involved in nitrogen fixation (Rovira *et al.*, 1983). Crush and Caradus (1980) have reported the lack of plant × mycorrhizae interaction for growth parameters of 10 white clover lines. However, as this study was done with only a very few lines of white clover, a larger number of diversified entries would have given more insight into the interaction effects. Differences in ability to increase nutrient uptake among VAM fungi appear to be due to differences in the rate of formation of extensive fungal hyphae. There is no known case of specificity in any mycorrhizal association such that a single strain of VAM is restricted to a single genotype, cultivar, or provenance of host.

Legumes in symbiotic association with *Rhizobium* fix atmospheric nitrogen. Mineral nutrient deficiency in soil can be a major constraint limiting nitrogen fixation and legume yield. Nitrogen fixation may be specifically limited by low availability of Ca, Co, Cu, and Fe (O'Hara *et al.*, 1988). Deficiency of Ca affects the multiplication of rhizobia and severe deficiency of Co affects nodule initiation. Nodule development is limited by deficiency of B and Fe (O'Hara *et al.*, 1988).

Genotypic differences in symbiotic nitrogen fixation have been observed in soybean. Greater gains may be made in a legume breeding program if both symbionts rather than only the macrosymbiont are considered. The formation and production of a legume/*Rhizobium* symbiosis is essentially an interaction of the two components. Mytton (1975) has reported a highly significant genotypic × *Rhizobium* strain interaction for white clover cultivar yields.

Free-living organisms such as *Azotobacter* and *Azospirillum* are known to fix atmospheric nitrogen and subsequently enhance plant growth (Curl and Truelove, 1986). Enhanced uptake of N, K, and P is a significant factor in increasing yields of *Azospirillum*-inoculated maize and sorghum (Lin *et al.*, 1983). Nitrogen fixation variation in *Azotobacter*-inoculated tropical maize plants was attributed to the genetic differences among host plants (Ela *et al.*, 1982). Further, such nitrogen-fixing traits of maize plants were observed to be inheritable by non-nitrogen-fixing maize plants in crossbreeding programs. The stimulatory effect of rhizosphere bacteria of the genus *Pseudomonas* on growth of plants has been demonstrated for sugar beet, wheat, and rice (Iswandi *et al.*, 1987). Enhanced growth of maize and barley due to seed inoculation with *Rhizopseudomonad* strain 7NSK2 is probably due to the inhibition of deleterious root microorganisms (Iswandi *et al.*, 1987).

V. Soil Management

Soil management practices such as tillage, fertilization, and planting density are influential in modification of plant growth by altering the nutrient use efficiency of any given plant cultivar. The efficiency of added N is 50% or lower and efficiency of added P fertilizer is 10% or less. The K fertilizer efficiency is somewhere around 40%. Nitrogen is lost mostly by leaching, runoff, denitrification, and volatile losses of ammonia. Low efficiency of applied P fertilizer is mainly due to retention of P by soil clay fractions and Fe and Al hydroxides. The loss of K is mainly attributed to leaching and runoff. Reduction in fertilizer efficiency is related to agronomic practices (tillage methods, improper fertilizer placement, time and method of fertilizer application, inadequate irrigation), cropping patterns (delay in sowing, inadequate plant population, inappropriate crop cultivar), weed infestation, and insect and disease attack (Baligar and Bennett, 1986a,b).

Increased crop yields have resulted from improved production practices and improved cultivars or hybrids adapted to the new production practices. During the past 40 to 60 years, improved genotypes have increased crop yields by 100% for peanuts, 60 to 100% for rice, and about 50 to 60% for maize. The newer crop cultivars have higher nutrient efficiency values than the older cultivars. New high-yielding hybrids (cultivars) also respond to higher levels of nutrients. Maintenance of adequate levels of soil-available K and P has improved the N use efficiency in field crops.

VI. Summary

To enhance yield potentials there is a need for understanding the interaction among crop species, soil, and climatic variables. Barber (1976) states that if we can get the crop to absorb a higher proportion of the nutrients added as fertilizer, we will certainly increase fertilizer efficiency, provided other environmental constraints are not seriously limiting. Crops should recover higher portions of added nutrients and, at the same time, produce higher yields per unit of nutrient absorbed (Barber, 1976). Breeding programs to make genetic improvement in nutritional parameters will help to achieve these goals.

References

Baligar, V. C. 1985. Potassium uptake by plant, as characterized by root density, species and K/Rb ratio. *Plant Soil* 85:43–53.

Baligar, V. C., and S. A. Barber. 1979. Genotypic differences of corn for ion uptake. *Agron. J.* 71:870–873.

Baligar, V. C., and O. L. Bennett. 1986a. Outlook on fertilizer use efficiency in the tropics. *Fert. Res.* 10:83–96.

Baligar, V. C., and O. L. Bennett. 1986b. NPK-fertilizer efficiency—A situation analysis for the tropics. *Fert. Res.* 10:147–164.

Baligar, V. C., H. L. dos Santos, G. V. E. Pitta, E. C. Filho, C. A. Vasconcellos, and A. F. de C. Bahia Filho. 1989. Aluminum effects on growth, grain yield and nutrient use efficiency ratios in sorghum genotypes. *Plant Soil* 116:257–264.

Baligar, V. C., R. J. Wright, T. B. Kinraide, C. D. Foy, and J. H. Elgin, Jr. 1987. Aluminum effects on growth, mineral uptake and efficiency ratios in red clover cultivars. *Agron. J.* 79:1038–1044.

Barber, S. A. 1976. Efficient fertilizer use. In F. L. Paterson (ed.), Agronomic Research for Food. *ASA Spec. Publ.* No. 26, pp. 13–29.

Barber, S. A. 1984. Soil Nutrient Bioavailability: A Mechanistic Approach. Wiley (Interscience), New York.

Blum, A. 1988. Plant Breeding for Stress Environments. CRC Press, Boca Raton, Florida.

Brown, J. C., J. E. Ambler, R. L. Chaney, and C. D. Foy. 1972. Differential responses of plant genotypes to micronutrients. In J. J. Mortvedt, P. M. Giordano, and W. L. Lindsay (eds.), Micronutrients in Agriculture. *Soil Sci. Soc. Am.*, Madison, Wisconsin. Pp. 389–418.

Clark, R. B. 1982. Plant response to mineral element toxicity and deficiency. In M. N. Christiansen and C. F. Lewis (eds.), Breeding Plants for Less Favorable Environments. Wiley, New York. Pp. 71–142.

Clark, R. B. 1983. Plant genotype differences in the uptake, translocation, accumulation, and use of mineral elements required for plant growth. *Plant Soil* 72:175–196.

Cooper, J. P. 1981. Physiological constraints to varietal improvements. *Philos. Trans. R. Soc. London Ser. B* 292:431–440.

Crush, J. R., and J. R. Caradus. 1980. Effect of mycorrhizas on growth of some white clovers. *N.Z. J. Agric. Res.* 23:233–237.

Curl, E. A., and B. Truelove. 1986. The Rhizosphere. Springer-Verlag, New York.

Devine, T. E. 1982. Genetic fitting of crops to problem soils. In M. N. Christiansen and C. F. Lewis (eds.), Breeding Plants for Less Favorable Environments. Wiley, New York. Pp. 143–173.

Dudal, R. 1976. Inventory of the major soils of the world with special reference to mineral stress

hazards. In M. J. Wright (ed.), Plant Adaption to Mineral Stress in Problem Soils. Cornell Univ. Press, Ithaca, New York. Pp. 3–13.
Ela, S. W., M. A. Anderson, and W. J. Brill. 1982. Screening and selection of maize to enhance associate bacterial nitrogen fixation. *Plant Physiol.* 70:1564–1567.
Epstein, E. 1972. Mineral Nutrition of Plants: Principles and Perspectives. Wiley, New York.
Epstein, E., and R. L. Jefferies. 1964. The genetic basis of selective ion transport in plants. *Annu. Rev. Plant Physiol.* 15:169–184.
Finn, B. J., S. J. Bourget, K. F. Nielson, and B. K. Dow. 1961. Effects of different soil moisture tensions on grass and legume species. *Can. J. Soil Sci.* 41:16–23.
Foy, C. D. 1983. The physiology of plant adaptation to mineral stress. *Iowa State J. Res.* 57:355–391.
Foy, C. D. 1984. Physiological effects of hydrogen, aluminum, and manganese toxicities in acid soil. In F. Adams (ed.), Soil Acidity and Liming. 2nd ed. American Society of Agronomy. Monograph No. 12. Madison. Pp. 57–97.
Gerloff, G. C., and W. H. Gabelman. 1983. Genetic basis of inorganic plant nutrition. In A. Läuchli and R. L. Bieleski (eds.), Inorganic Plant Nutrition. Encyclopedia of Plant Physiology, New Series, Vol. 15B. Springer-Verlag, New York. Pp. 453–480.
Graham, G. D. 1984. Breeding for nutritional characteristics in cereals. In P. B. Tinker and A. Läuchli (eds.), *Adv. Plant Nutr.* 1:57–102.
Hale, M. G., and D. M. Orcutt. 1987. The Physiology of Plants under Stress. Wiley (Interscience), New York.
Hall, A. E. 1981. Adaptation of annual plants to drought in relation to improvements in cultivars. *Hort. Science* 16:37–38.
Hasegawa, P. M., R. A. Bressan, and A. K. Handa. 1986. Cellular mechanisms of salinity tolerance. *HortScience* 21:1317–1324.
Iswandi, A., P. Bossier, J. Vandenabeele, and W. Verstraete. 1987. Relation between soil microbial activity and the effect of seed inoculation with the *rhizopseudomonad* strain. 7NSK2 on plant growth. *Biol. Fert. Soils* 3:147–151.
Jung, G. A. (ed.). 1978. Crop Tolerance to Suboptimal Land Conditions. *ASA Spec. Publ.* No. 32.
Kanwar, J. S., and L. J. Youngdahl. 1985. Micronutrient needs of tropical food crops. *Fert. Res.* 7:43–67.
Lin, W., Y. Okon, and R. W. F. Hardy. 1983. Enhanced mineral uptake by *Zea mays* and *Sorghum bicolor* roots inoculated with *Azospirillum brasilense*. *Appl. Environ. Microbiol.* 45:1775–1779.
Marschner, H. 1986. Mineral Nutrition of Higher Plants. Academic Press, New York.
Moorby, J., and R. T. Besford. 1983. Mineral nutrition and growth. In A. Läuchli and R. L. Bieleski (eds.), Inorganic Plant Nutrition. Encyclopedia of Plant Physiology, New Series, Vol. 15B. Springer-Verlag. New York. Pp. 481–527.
Morgan, J. M. 1984. Osmoregulation and water stress in higher plants. *Annu. Rev. Plant Physiol.* 35:299–319.
Myers, W. M. 1960. Genetic control of physiological processes: Consideration of differential ion uptake by plants. In R. S. Caldecott and L. A. Snyder (eds.), *A Symposium on Radioisotopes in the Biosphere, University of Minnesota, Minneapolis* pp. 201–226.
Mytton, L. R. 1975. Plant genotype × *Rhizobium* strain interaction in white clover. *Ann. Appl. Biol.* 80:103–107.
Oertli, J. J. 1979. Plant nutrients. In R. W. Fairbridge and C. W. Finkl, Jr. (eds.), The Encyclopedia of Soil Science. Part 1. Dowden, Hutchinson & Ross, Stroudsburg, Pennsylvania. Pp. 382–385.
O'Hara, G. W., N. Boonkerd, and M. J. Dilworth. 1988. Mineral constraints to nitrogen fixation. *Plant Soil* 108:93–110.
O'Toole, J. C., and W. L. Bland. 1987. Genotypic variation in crop plant root systems. *Adv. Agron.* 41:91–145.
Pitman, M. G. 1972. Uptake and transport of ions in barley seedlings. III. Correlation between transport to the shoot and relative growth rate. *Aust. J. Biol. Sci.* 25:243–257.
Robson, A. D., and M. G. Pitman, 1983. Interactions between nutrients in higher plants. In A. Läuchli

and R. L. Bieleski (eds.), Inorganic Plant Nutrition. Encyclopedia of Plant Physiology, New Series, Vol. 15A. Springer-Verlag, New York. Pp. 147–180.

Rovira, A. D., G. D. Bowen, and R. C. Foster. 1983. Significance of rhizosphere microflora and mycorrhizae in plant nutrition. In A. Läuchli and R. L. Bieleski (eds.), Inorganic Plant Nutrition. Encyclopedia of Plant Physiology, New Series, Vol. 15B. Springer-Verlag, New York. Pp. 61–93.

Sarić, M. R. 1984. Genetic improvement of crop yields as related to plant nutrient requirements. In W. Welte and J. Szaboles (eds.), *Proc. 9th CIEC World Fertilizer Congr., Budapest* pp. 115–128.

Stevenson, F. J. 1986. Cycles of Soil Carbon, Nitrogen, Phosphorous, Sulfur, Micronutrients. Wiley, New York.

Ting, I. P. 1982. Plant mineral nutrition and ion uptake. In Plant Physiology. Addison-Wesley, Reading, Massachusetts. Pp. 331–363.

Vose, P. B. 1963. Varietal differences in plant nutrition. *Herb. Abstr.* 33:1–13.

Vose, P. B. 1984. Effects of genetic factors on nutritional requirements of plants. In P. B. Vose and S. G. Blixt (eds.), Crop Breeding: A Contemporary Basis. Pergamon, Oxford. Pp. 67–114.

Wright, M. J. (ed.). 1977. Plant Adaptation to Mineral Stress in Problem Soils. Cornell Univ. Press, Ithaca, New York.

10

Root Microbial Interactions and Rhizosphere Nutrient Dynamics

IAN L. PEPPER and DAVID F. BEZDICEK

Crops as Enhancers of Nutrient Use

I. The Rhizosphere Ecosystem

A. THE SOIL–PLANT–MICROORGANISM SYSTEM

In natural vegetation systems, plant roots are in intimate contact with soil particles. Soil exists as a discontinuous environment with a matrix of organic and inorganic constituents combined under diverse environments, and is therefore a unique environment for many microorganisms. These organisms include viruses, bacteria, actinomycetes, fungi, algae, protozoa, and nematodes. Aerobic and anaerobic microsites exist in close proximity, allowing organic and inorganic substrates to be metabolized by organisms with different modes of nutrition. These conditions permit billions of organisms to coexist in soil. Populations vary with soil, environment, and method of analysis, but reasonable values for "normal" soils are shown in Table I.

Roots are therefore surrounded by organisms and exist as part of the soil–plant–microorganism system, which can be termed the rhizosphere. The complexity of the rhizosphere is shown in Figure 1, where the two inner circles depict, respectively, the early events necessary for colonization and subsequent factors that contribute to rhizosphere competency and the ability to metabolize and reproduce in the rhizosphere in the presence of other organisms. Each component of this system (microorganisms, plant, and soil) interacts with each other, which distinguishes the rhizosphere from the bulk soil. The activity of root microorganisms is affected by soil environmental factors or by environmental factors operating indirectly through the plant. Root microorganisms can affect the plant and plant nutrient uptake, directly by colonizing the root and modifying its structure or indirectly by modifying the soil environment around the root. The spokes of the wheel in Figure 1 depict not only the significant processes that impact plant growth, but those that have potential for enhancement through improved cultural practices, genetic manipulation, and modeling.

Table I. Distribution of Microorganisms in Various Horizons of the Soil Profile[a]

	Organisms per Gram of Soil ($\times 10^3$)				
Depth (cm)	Aerobic Bacteria	Anaerobic Bacteria	Actinomycetes	Fungi	Algae
3–8	7800	1950	2080	119	25
20–25	1800	379	245	50	5
35–40	472	98	49	14	0.5
65–75	10	1	5	6	0.1
135–145	1	0.4	—	3	—

[a]Source: Alexander, 1977. Used with permission of John Wiley & Sons, Inc.

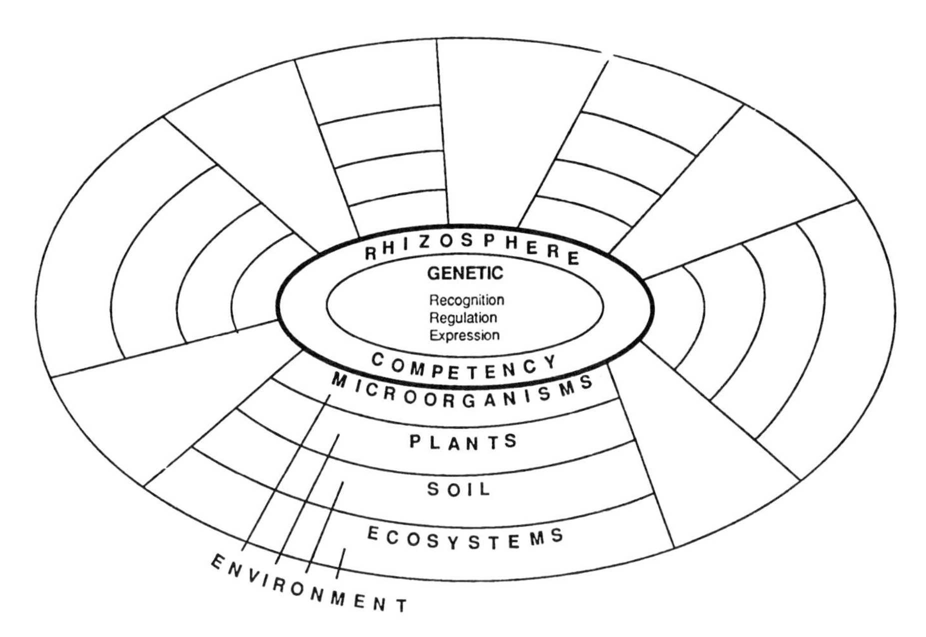

Figure 1. Model of rhizosphere dynamics from the genetic recognition to the ecosystems interaction.

B. STRUCTURE OF THE RHIZOSPHERE

The term rhizosphere was first coined by Hiltner (1904) and was classically used to describe that part of the soil in which legume roots induced high microbial numbers and activity. Now the term includes all plants and has been the subject of numerous reviews (Clark, 1949; Starkey, 1958; Barber, 1968; Rovira, 1979; Bowen, 1980; Curl and Truelove, 1985; Whipps and Lynch, 1986). However, despite intense study, most of our information on the rhizosphere is qualitative rather than quantitative. Originally the rhizosphere effect was thought to extend 2 mm outward from the root surface. Now it is recognized that the rhizosphere can extend to 20 mm as a series of gradients of organic substrate, microorganisms, pH, O_2, CO_2, and H_2O.

Balandreau and Knowles (1978) described three regions of the rhizosphere: the endorhizosphere, rhizoplane, and ectorhizosphere. These are, respectively, the epidermal-cortical region of the root, the interface between the root surface and soil, and the soil around the root. The endorhizosphere is that portion of the rhizosphere found within the root surface. The root surface itself is not continuous as a result of the many holes and wounds in the epidermis. Wounds can be caused by chemical and physical elements, and holes can be bored by bacteria, fungi, and nematodes (Old and Nicholson, 1975). These perforations allow for the entry of organisms into the root cortex and also increase the amount of sloughed cells or "root litter" that are released into the soil. The rhizoplane is a classical term designating the actual root surface. In practice it is difficult to distinguish between the rhizoplane and mucigel. The ectorhizosphere is synonymous with rhizosphere soil and can be defined as that volume of soil in which microorganisms are influenced by the root, typically by increased organic substrate through root exudation.

C. RELEASE OF ORGANIC COMPOUNDS

Substrates released from roots have many origins and have been classified by Rovira *et al.* (1979) as:

1. *exudates*—compounds of low molecular weight that leak nonmetabolically from intact plant cells.
2. *secretions*—compounds metabolically released from active plant cells.
3. *lysates*—compounds released from the autolysis of older cells.
4. *plant mucilages*—polysaccharides from the root cap, root cap cells, primary cell wall, and other cells.
5. *mucigel*—gelatinous material of plant and microbial origin.

The terms "exudates" and "exudation" are sometimes used collectively and perhaps incorrectly to include all of the organic compounds released from roots and most if not all the mechanisms involved in the release of organic compounds. These mechanisms in turn become the driving force behind the term "rhizosphere effect," which usually results in a positive increase in microbial numbers and activity. The less restrictive use of the terms exudates and exudation will be used in this review unless otherwise noted.

1. Exudates, Secretions, and Lysates

The release of soluble organic compounds (loosely known as root exudates) is perhaps the most dominant factor and the driving force behind the "rhizosphere" effect. This refers to the release of soluble compounds from plant roots, which constitutes the only ecosystem in soil where there is a constant supply of available substrate for heterotrophic organisms. Loss of substrates from roots can change the pH and structure of rhizosphere soil and the availability of inorganic nutrients, and can induce toxic or stimulatory effects on soil microorganisms (Hale *et al.*, 1978). The major mechanisms are leakage and secretion. Leakage involves simple diffusion of compounds due to the higher concentrations of compounds within the root as compared to the soil. Secretion can occur against concentration gradients, but requires the expenditure of metabolic energy. Polysaccharides in particular are susceptible to secretion (Hale *et al.*, 1981). Almost any plant metabolite has the potential to be exuded, but most exudates fall into one of several categories: carbohydrates, amino acids, organic acids and lipids, growth factors, enzymes, and miscellaneous compounds (Curl and Truelove, 1985). Of these, the carbohydrates, including sugars, and the amino acids, which also represent a source of nitrogen, are particularly important as a substrate. Organic acids and lipids reduce the pH of the rhizosphere and also have a role in the chelation of metals (Curl and Truelove, 1985). Growth factors, including vitamins, and enzymes stimulate microbial activity and allow growth of organisms with complex heterotrophic requirements. Miscellaneous compounds including volatiles can physiologically stimulate or inhibit organisms (Fries, 1973). When viewed collectively, it is apparent that the rhizosphere is a unique ecosystem in soil that provides a constant supply of substrate and growth factors for organisms.

2. Mucilage and Mucigel

As the root cap extends through soil, plant mucilage or sloughed cells are released into the soil. The amount of sloughed material can be considerable. In solution culture, peanut plants cultured axenically released 0.15% of the plant's carbon, nitrogen, and hydrogen per week (Griffin *et al.*, 1976). One would predict that much more material would be lost in soil because of its abrasive nature. From the root tip to the root hair zone, the root is frequently covered with a mucilaginous layer composed of sloughed root cells and polysaccharides of plant and microbial origin, which is termed mucigel (Miki *et al.*, 1980; Jenny and Grossenbacher, 1963). Plant production of mucilage is particularly predominant at the root tip and consists of highly hydrated chains of glucose, galactose, arabinose, and fucose, as well as galacturonic acid. These plant products are excellent substrates for microbial growth, in particular soil bacteria, which are extremely competitive at metabolizing simple sugars. Thus mucilage is in intimate contact with bacteria that consume the material (Guckert *et al.*, 1975), as well as bacteria that contribute bacterial polysaccharides to the mucilage, resulting in mucigel. The amount of mucigel on a particular root depends on the net production and consumption of the material, so that in some instances parts of the root may have no mucigel. Mucilage may protect the root tip from injury and desiccation as well as play a role in nutrient uptake through its pH-dependent cation-exchange capacity (COO^- groups) (Jenny and Grossenbacher, 1963).

3. Factors Affecting the Release of Organic Compounds

Major factors affecting release of organic compounds include plant species and cultivar (Rovira, 1956; Kraft, 1974); age and stage of plant development (Hamlen *et al.*, 1972); light intensity and temperature (Hale and Moore, 1979); soil factors (Russell, 1973; Reid, 1974; Rittenhouse and Hale, 1971); plant nutrition (Shay and Hale, 1973); plant injury (Barber and Gunn, 1974); and soil microorganisms (Rovira and Davey, 1974).

D. RHIZOSPHERE POPULATIONS

Rhizosphere populations are influenced by many plant, soil, and environmental factors. Crop plant roots tend to have greater rhizosphere populations than tree roots (Dangerfield *et al.*, 1978). Different cultivars of the same plant species may have different rhizosphere populations.

Soils directly affect the growth and vigor of plants and therefore influence shoot growth, photosynthesis, and amounts of exudation into the rhizosphere. The concentration of oxygen in the rhizosphere is usually lower than in nonrhizosphere soil as a result of its utilization by large rhizosphere populations. Hence in heavy-textured soils, oxygen may become limiting and show reduced rhizosphere populations relative to coarser-textured soils.

Inorganic fertilizers and pesticides are important factors governing rhizosphere populations. Applications of inorganic nitrogen or phosphorus fertilizer have been shown to increase the rhizosphere effect (Blair, 1978). Microorganisms also respond to organic fertilizers, including sludges and manures, but nonrhizosphere populations are usually affected more than rhizosphere populations, since organic carbon is usually a limiting factor in nonrhizosphere soil. Nitrates can inhibit specific root microbial associations, including the rhizobia–legume symbiosis and mycorrhizal associations. Pesticides and soil fumigants have also been shown to eradicate mycorrhizal associations (de Bertoldi *et al.*, 1978).

The physical environment around the plant and its roots also affects rhizosphere populations by affecting the amount of organic material release into the soil. Factors such as light, moisture, and temperature can all cause changes in plant metabolism and the rhizosphere effect. In summary, rhizosphere populations are dependent on many diverse interacting factors, and care must be taken when interpreting different studies.

1. Microflora

Bacteria, actinomycetes, fungi, and algae are the predominant microflora. Populations in rhizosphere soil (R) are often compared to those in nonrhizosphere soil (S), giving rise to R : S ratios. The larger the ratio, the larger the rhizosphere effect. Usually bacteria have the largest ratios, followed in decreasing order by actinomycetes, fungi, and algae (Rovira *et al.*, 1974; Papavizas and Davey, 1961; Rouatt *et al.*, 1960). Typical R : S values are 20 : 1, 10 : 1, 5 : 1, and <1 respectively (Rouatt *et al.*, 1960). However, rhizosphere ratios as high as 100 : 1 have been reported for *Pseudomonas* spp. and other gram-negative bacteria (Rovira and Davey, 1974). Growth rate in the rhizosphere may be a better measure of rhizosphere competence than R : S ratios.

Although the fungi are believed to be less numerous in the rhizosphere, they are extremely important since they can be beneficial, as in the case of the mycorrhizal fungi, or harmful when they are pathogenic to plants. Relative respiration was higher from bacteria than fungi in the rhizosphere when specific inhibitors were added (Vancura and Kunc, 1977). The proportion of the root surface occupied by fungi varied considerably, but was generally higher than with bacteria as noted from a survey conducted by Newman *et al.* (1981). Relative biomass estimates of fungi and bacteria in the rhizosphere would be useful, although they are difficult to determine experimentally. Algae are not dominant in the rhizosphere, because of their phototrophic mode of nutrition and dependence on light for an energy source. However, some algae can grow heterotrophically (Alexander, 1977).

2. Microfauna

Most research on the microfauna has centered on protozoa, nematodes, and the microarthropods. Soil protozoa are mostly rhizopods and flagellates, with smaller numbers of ciliates. Protozoan populations tend to mimic those of bacterial populations, since bacteria are their major food supply. Thus the rhizosphere should contain large populations of protozoa. Rouatt *et al.* (1960) reported R : S ratios for protozoa of 2 : 1 in wheat rhizospheres. Darbyshire (1966) reported even higher protozoan populations in ryegrass rhizospheres. Many soil nematodes, including *Heterodera* and *Fylenchus,* are plant parasites that feed on underground roots. Little research has been conducted on nematodes in the rhizosphere, but populations have been reported to be higher in rhizosphere than in nonrhizosphere soil (Henderson and Katznelson, 1961). The Acari (mites) and Collembola (springtails) are important members of soil microarthropods. Mites are predatory on nematodes, and the rhizosphere would be expected to be a favorable habitat for the Acari, but studies in the rhizosphere have been limited in scope. Führer (1961) showed that mites were present in large numbers in some rhizospheres but absent in others. Springtails have been shown to be abundant in cotton rhizospheres with R : S ratios of 4 : 1 in a sandy loam soil (Wiggins *et al.*, 1979), but the reasons for their attraction to roots are not clear.

II. Specific Root Microbial Interactions

A. LEGUME–RHIZOBIA

The legume–rhizobia symbiosis between aerobic, heterotrophic, gram-negative, diazotrophic bacteria of the genus *Rhizobium* and leguminous plants is an intensely studied phenomenon. Rhizobia have the ability to convert atmospheric nitrogen gas into ammonia (biological nitrogen fixation, BNF) in root nodules, which is then utilized by the legume plant. Thus the plant receives a source of nitrogen other than soil nitrogen. In return the plant supplies the rhizobia with simple sugars derived from photosynthates. Thus each symbiotic partner receives a nutrient that is usually limiting its metabolism.

The formation of root nodules is a complex dynamic series of events that requires the intimate association of both plant and bacterium. The process of recognition and nodule formation has been intensely studied and is the subject of several critical reviews (Dazzo and Gardiol, 1984; Rolfe and Shine, 1984; Verma and Nadler, 1984; Halvorson and Stacey, 1986; Long, 1989). Specific steps in the process include recognition of mutual partners and bacterial attachment to the root prior to the actual infection process. New advances in molecular biology techniques are now allowing the symbiosis to be understood at the molecular level, although much remains to be discovered. For example, mutagenesis and gene inactivation experiments are allowing the function of specific genes to be elucidated.

Rhizobia as Inoculants

Lack of nitrogen often limits plant growth and therefore nitrogen fixation by rhizobia is an important process that does not rely on fossil fuel for its production. Unfortunately, indigenous soil rhizobia are often highly competitive and some do not carry out this process efficiently in all soils and environments. Sometimes nitrogen fixation can be increased by inoculating strains of rhizobia directly on the seed just prior to planting. Such inoculation technology requires a competitive and effective strain, a suitable carrier, and proven inoculation procedures. Strains of rhizobia differ in effectiveness of nitrogen fixation (Shoushtari and Pepper, 1985b) and competitiveness (Shoushtari and Pepper, 1985a). Competition in rhizobial ecology refers to the ability of the inoculant strain to override the native population and form nodules. Often inoculant strains dominate the nodules when the native population is low (Materon and Hagedorn, 1984; May and Bohlool, 1983). A classic example of a highly competitive soil population is *Bradyrhizobium japonicum* serogroup 123 in the midwestern United States, which inhabits the majority of nodules in soybean, although it may not dominate the rhizosphere (Moawad *et al.*, 1984). Co-inoculation studies of other *B. japonicum* strains showed that strain 138 was noncompetitive when added equally with strain 110 in the inoculum, but dominated the nodules when inoculated 24 hr previously (Kosslak *et al.*, 1983). These studies show the significance of population dynamics early in the development of the rhizosphere. Differences in the competitiveness of strains of bean rhizobia have also been well documented. KIM-5 is an extremely competitive bean rhizobia and has been known to outcompete other strains even when outnumbered one-million-fold in the inoculant (Josephson and Pepper, 1984).

B. NONLEGUME–*Frankia*

Nonleguminous plants also form symbiotic relations with nitrogen-fixing bacteria. These are usually species of woody shrubs and trees (Vincent, 1974). The bacteria involved in the symbiosis are actinomycetes of the genus *Frankia*. In addition to the *Frankia*, rhizobia and blue-green algae such as *Nostoc* can also fix nitrogen symbiotically with nonleguminous plants (Trinick, 1973; Allen and Allen,

1965). Nonleguminous symbioses have not received the attention that the leguminous associations have and consequently less is known about them. However, the actinomycete symbioses occur mainly in soils that are low in fertility, and they are important colonizers of soils with low nitrogen. The symbiosis occurs in root nodules and, as with leguminous fixation, the infection process involves a series of complex interactions.

C. NONSYMBIOTIC NITROGEN FIXATION

Numerous organisms have been shown to fix nitrogen independently of plants and in nonmutualistic associations in the rhizosphere. The former group includes blue-green algae or *Cyanobacteria* such as *Nostoc* or *Anabaena* and a diverse group of other bacteria. This group also includes the anaerobic *Clostridium pasteurianum* and other aerobic and anaerobic diazotrophic bacteria, e.g., *Azotobacter, Azospirillum, Bacillus, Pseudomonas,* and *Klebsiella.* Some of these, including the azospirilla, are considered to be root associative diazotrophs. These organisms all have a similar nitrogenase enzyme, and all fix nitrogen into ammonia. In the rhizosphere this represents an important source of available nitrogen, but not all nitrogen-fixing bacteria do well in the rhizosphere. *Azotobacter* spp. are often poor rhizosphere competitors (Mulder, 1975). High rates of fixation have been reported for *Azotobacter paspali* (Dobereiner *et al.*, 1972; Dobereiner and Pedrosa, 1987), which form relatively specific associations in the rhizosphere of grasses and cereals. Some of these associations can fix substantial quantities of nitrogen (Dobereiner and Pedrosa, 1987), although quantification can be difficult (Van Berkum and Bohlool, 1980). *Azospirillum* spp. are unique in that they can inhabit the root cortex, where they fix nitrogen. Generally root associative bacteria fix smaller quantities of nitrogen than their mutualistic counterparts. Also, since nitrate or ammonia is known to inhibit the nitrogenase enzyme, associative fixation would likely be more important in natural ecosystems rather than in production agriculture, where nitrogen fertilizers are routinely applied.

D. MYCORRHIZAE

Many soil fungi called mycorrhizae form mutualistic associations with plants. Mycorrhizae are usually divided into three morphologically distinct groups: ectomycorrhizae, endomycorrhizae, and ectendomycorrhizae.

The sheathing or ectomycorrhizae normally infect roots of forest tree species and cause visible morphological changes. These include the development of fungal mycelium around the surface of the host root, which is known as the fungal sheath or mantle (Figure 2). Fungal hyphae radiate from the mantle out into the soil and also penetrate intercellularly to the endodermis of the root. The hyphae do not penetrate into the living cells of the host root. The hyphae within the root are known

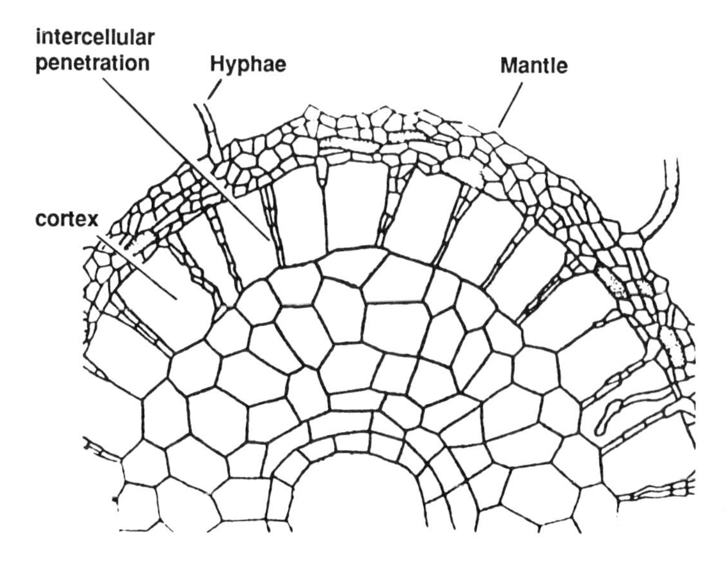

Figure 2. Cross section of ectomycorrhizal rootlet showing the exterior fungal sheath or mantle and intercellular penetration. Source: Harley, 1965. Used with permission of the University of California Press.

as the Hartig net. Most of the ectomycorrhizal fungi are *Basidiomycetes* or *Ascomycetes* (Trappe, 1971). Some fungi are host specific, whereas others have broad host ranges (Meyer, 1974). Some fungi have been used as inoculants, including *Pisolithus tinctorius,* which infects many forest trees (Marx, 1977).

Endomycorrhizal fungi have no fungal mantle and have different internal infection characteristics. Within the host root, fungal hyphae penetrate intracellularly into cortical cells. The most important of the endomycorrhizal fungi are the vesicular-arbuscular (VA) mycorrhizae. These are characterized by the presence of vesicles and arbuscles, which develop intracellularly (Figure 3). Vesicles are storage organs, whereas arbuscles are structures where transfer of nutrients occurs. Other VA mycorrhizae include the ericoid fungi, which are found exclusively in the *Ericaceae,* and the orchidaceous mycorrhizae. The latter group of fungi forms mycorrhizae with orchids and is often pathogenic to other higher plants. The majority of the VA fungi are within the *Zygomycotina,* examples being *Glomus* and *Gigaspora.* The VA fungi form the majority of the mycorrhizal associations, including fruits, vegetables, grain crops, and legumes (Gerdemann, 1968).

Ectendomycorrhizal fungi are intermediate between ecto- and endomycorrhizae. These fungi are normally *Basidiomycetes* and form associations with external mantles as well as inter- and intracellular host root penetration.

The mycorrhizae affect plant growth in numerous ways. The best documented has been increased nutrient uptake through the fungal hyphae acting as an extension of the plant root system. This can be particularly important for phosphorus, which is

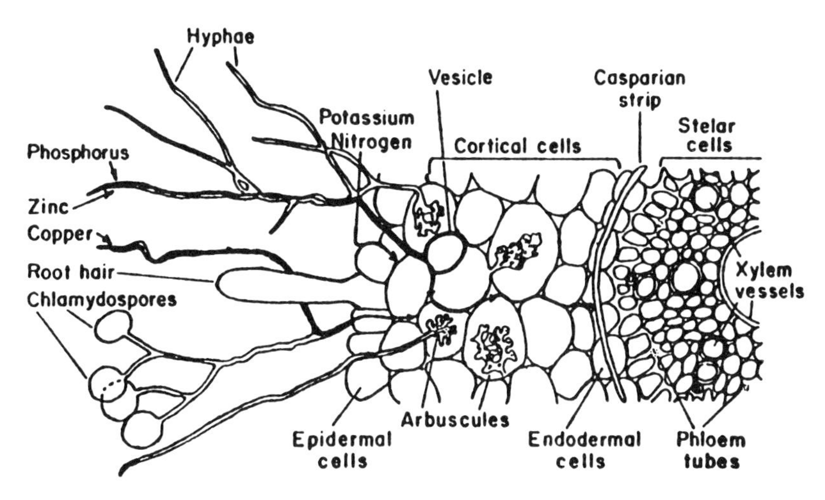

Figure 3. A typical endomycorrhiza showing hyphae extending beyond the root epidermis into the rhizosphere. Intracellular arbuscules are haustoria-like structures. Source: Curl and Truelove, 1985. Used with permission of Springer-Verlag, Inc., New York.

only available in soil solution at low concentrations because of solubility constraints. Mycorrhizal fungi also produce hormones and plant growth regulators that cause rootlets to be more highly branched and with increased absorptive capacity (Curl and Truelove, 1985). The fungi are also important in reducing the incidence of feeder root infection by *Pythium* or *Fusarium* (Marx, 1973). The mechanisms for this could be due to competition between the beneficial and pathogenic fungi, or in the case of the ectomycorrizae to the physical presence of the mantle. Mycorrhizae also interact positively with nitrogen-fixing microorganisms such as rhizobia (Waidyanatha *et al.*, 1979; Redente and Reeves, 1981) and *Azotobacter* (Bagyaradaj and Menge, 1978). Mycorrhizae have only come to light relatively recently, but now it is recognized that they are essential for plant growth and active plant metabolism.

E. GENERAL HETEROTROPHIC RHIZOSPHERE MICROORGANISMS

In addition to the symbiotic associations, other heterotrophic microorganisms interact with plant roots. As noted previously, these are numerous, diverse, and capable of tolerating microaerophylic conditions that arise from the rapid oxidation of root substrates. These organisms mediate numerous important biochemical reactions that will be described later (see Section III). These organisms are actively involved in the form and availability of essential plant nutrients in the rhizosphere. Additionally they can influence the uptake of plant nutrients by altering the morphology and uptake capacity of feeder roots. These changes are caused by microbial production of growth-stimulating or -inhibiting substances. Indole-acetic acid and gibberellins are examples of growth-promoting substances (Riviere, 1963), whereas ethylene has been shown to be microbially produced (Babiker and Pepper, 1984) and is a known inhibitor of plant tissue growth. The production of physiologically active compounds can affect all organisms in the rhizosphere. The complexity of such interactions is difficult to document, but it is undoubtedly important to plant growth.

F. PATHOGENS

Plant pathogens are organisms that cause disease. Pathogens are extremely numerous and diverse, and an in-depth review of these organisms falls outside the scope of this chapter. In the rhizosphere many pathogens cause root disease and consist of bacteria, actinomycetes, fungi, or nematodes. Traditionally rhizosphere studies of pathogens have involved fungi because most root diseases are caused by fungi. Fungal pathogens are prevalent in almost all soils and include *Pythium, Fusarium,* and *Phytophthora* spp. *Agrobacterium tumefaciens* is a bacterial pathogen that causes millions of dollars of damage annually and is particularly damaging to fruit trees. Many rhizosphere bacteria are pathogens such as some of the *Pseudomonas* spp. An example of an actinomycete pathogen is *Streptomyces scabies,* which causes potato scab. Nematodes that are plant pathogens also exist in soil, including *Heterodera* spp.

Since all these organisms coexist in soil, interactions are numerous and complex and can involve the plant host, the soil microbial population, and the pathogen. These interactions have been described by Park (1963) and would be expected to be particularly intense in the rhizosphere. Some of these interactions are described in Section IV.

III. Nutrient Availability and Uptake

A. CARBON FLUX AND ENERGY BALANCE

Early studies carried out under sterile conditions in liquid culture showed that carbon loss from plant roots was 4% or less of carbon equivalents in roots (Rovira and Davey, 1974). Barber and Martin (1976) found that 5 to 10% of photosynthetically fixed carbon was exuded by barley roots under sterile conditions, and 12 to 18% when organisms were introduced. Prikryl and Vancura (1980) showed that root exudation doubled when *Pseudomonas putida* was present in the nutrient solution. Microorganisms tend to collect as colonies or hyphae at the junction of epidermal cells, where energy release is greatest (Rovira, 1965). In soil, plant roots release 5 to 20 times more carbon than that released by similar plants in solution culture.

Pathogen-infected roots can be more leaky than healthy intact roots (Rovira and Campbell, 1975) and root-associated microorganisms may cause permeability changes in the membranes of root cells or accelerate the senescence process of epidermal and cortical cells, which further increase the amount of substrate for fungal growth. Martin (1978) attributed the increase in substrate loss from roots in the presence of microorganisms to microbial lysis of cells. Bowen and Foster (1978), using light microscopy and TEM studies, showed heavily colonized individual senescent cells near the root apex. Bowen (1980) suggests that an organism could have a competitive advantage if it could increase root cell permeability and exudation at a specific point. Increased root permeability has also been attributed to hormone production by microorganisms (Highinbotham, 1968), to drought water stress conditions (Reid and Mexal, 1977; Martin, 1977), and to physical root restriction (Barber and Gunn, 1974).

The preceding studies do not address the loss of photosynthetic carbon through root and microbial respiration. Martin (1977) showed that 39% of carbon translocated into the wheat root was released into the soil. In nonsterile sand, Whipps and Lynch (1986) found that 20 to 25% of the ^{14}C fixed by the plant was lost from the roots of both barley and wheat. They also reported that 12 to 40% of total fixed carbon was lost through roots from soil-grown cereals under a wide variety of growing conditions.

The significance of the loss of assimilates from roots has been addressed by Bowen (1980). Do these losses significantly drain the plant's resources and reduce yield or can the plant compensate? Paul and Kucy (1981) suggest from ^{14}C pulse-labeling studies that faba bean did compensate through increased photosynthesis when mycorrhizal and rhizobial symbionts utilized a portion of the total photosynthate.

B. NITROGEN TRANSFORMATIONS

1. Nitrogen Fixation

Symbiotic organisms such as rhizobia and to a lesser extent nonsymbiotic bacteria such as *Azospirillum* can both augment plant uptake of nitrogen via nitrogen fixation. Nonsymbiotic nitrogen fixation has also been well documented (Jain *et al.*, 1987), but much controversy has centered on its extent and importance. The limiting factor for N_2 fixation is the carbon cost, which has been estimated to be as high as 6.5 g C g^{-1} N fixed (Phillips, 1980). Symbiotic N_2-fixing organisms with direct access to photosynthates can fix significant amounts, but free-living organisms must compete with all other rhizosphere organisms for substrate, and therefore are unlikely to fix large amounts of nitrogen. Early estimates of N_2 fixation by free-living diazotrophs based on acetylene reduction in soil cores were excessive, but more realistic estimates have been in the range of 30 kg N ha^{-1} yr^{-1}. This level of nitrogen input can be significant in natural ecosystems where nitrogen is more limiting, but is unlikely to be important in production agriculture. More recently, Jones and Bangs (1985) estimated N_2 fixation by free-living organisms to be 16.1 kg N ha^{-1} yr^{-1} in an oak forest. Beck and Gilmour (1983) showed that even if 11% of total photosynthate was exuded by wheat, only 14.6 kg N ha^{-1} yr^{-1} could possibly be fixed even if all exudate was utilized by nitrogen-fixing organisms. These estimates were supported by Rennie and Thomas (1987), who found free-living fixation in wheat rhizospheres to average 17.9 kg N ha^{-1} yr^{-1}.

These estimates are in contrast to symbiotic N_2 fixation by legumes of up to 300 kg N ha^{-1} (Bezdicek *et al.*, 1978). The contribution of biologically fixed nitrogen to the nutritional needs of the plant is highly variable and depends on the type of legume, genotype, available soil nitrogen, and climate. After harvest the amount of nitrogen available to subsequent crops also depends on these factors. Yield benefits often observed for crops following legumes in rotation can be attributed both to additional nitrogen and to rotation effects, although rotation effects may be more significant since more nitrogen is often removed in the harvest of seed legumes than what is added from N_2 fixation (Heichel *et al.*, 1981; Bezdicek and Kennedy, 1988).

2. Mineralization, Immobilization, Nitrification, and Denitrification

In addition to supplying plants with fixed nitrogen, rhizosphere organisms greatly influence the quantity and nature of nitrogen compounds available for plant uptake through four key metabolic reactions: mineralization, immobilization, nitrification, and denitrification. Ammonification, which is a specific form of nitrogen mineralization, represents the conversion of amino radicals (plant unavailable) to free ammonium ions (plant available). This reaction is undertaken by numerous heterotrophic bacteria and fungi normally found in rhizosphere populations and is particularly important in the rhizosphere, where amino acids and proteins are constantly supplied via exudates. The R : S ratio for ammonifying bacteria has been reported to be in excess of 50 : 1 (Rouatt *et al.*, 1960). Clarholm (1985) showed that root carbon inputs increased nitrogen mineralization of soil organic matter, and that protozoa bacterial grazers were necessary to liberate such mineralized nitrogen. Thus rhizosphere organisms allow for the reassimilation of nitrogen, which might otherwise have been unavailable to the plant.

Nitrification, which represents the oxidation of ammonium to nitrate in a process most often mediated by aerobic chemoautotrophic bacteria, involves the following organisms and reactions:

$$NH_4^+ \xrightarrow{\textit{Nitrosomonas}\text{ sp.}} NO_2^- \xrightarrow{\textit{Nitrobacter}\text{ sp.}} NO_3^- \qquad (1)$$

The end product nitrate is available to plants, whereas the intermediate nitrite can be toxic to plants, but rarely accumulates in soils. Heterotrophic organisms, including fungi such as *Aspergillus* sp., are also capable of nitrification, but their activities are thought to be insignificant relative to the autotrophic nitrifiers. Although substrate (NH_4^+) is plentiful in the rhizosphere as a result of the activity of the ammonifiers, autotrophic nitrifiers are not strong competitors for ammonium ions with heterotrophic organisms within the rhizosphere. The rhizosphere effect on soil nitrification varies with different plant species, in part due to differences in root exudates, which affect microbial populations as well as microbial toxins and pH changes. The nitrifying bacteria are particularly sensitive to low pH.

The other important nitrogen transformation that can occur in the rhizosphere is denitrification or dissimilatory nitrate reduction. In this reaction nitrate functions as a terminal electron acceptor in the sequence

$$NO_3^- \rightarrow NO_2^- \rightarrow NO \rightarrow N_2O \rightarrow N_2 \quad (2)$$

Most of the organisms capable of denitrification are heterotrophic, gram-negative, nonsporulating bacteria (Payne, 1974), but include a wide variety of genera such as *Pseudomonas, Micrococcus, Bacillus,* and one important autotroph, *Thiobacillus denitrificans*. Since denitrifying organisms are normally heterotrophic aerobes capable of anaerobic respiration and requiring low levels of O_2, the rhizosphere should be an ideal environment for denitrification. Garcia (1973) demonstrated significant populations of denitrifiers in rhizosphere soil with R : S ratios up to 514 : 1 depending on soil type. Denitrification is limited by adequate substrate, and the reaction has been shown to decrease with increasing distance from the root surface of oats (Smith and Tiedje, 1979).

C. PHOSPHORUS NUTRITION

Similarly to nitrogen, the effect of rhizosphere organisms on plant uptake of phosphorus involves both free-living and symbiotic associations.

1. Effect of Free-Living Organisms on Phosphorus Uptake

Phosphorus exists in soil as both organic and inorganic compounds. Typically the availability of phosphate as $H_2PO_4^-$ or HPO_4^{2-} is low and can be deficient due to the low solubility and rate of diffusion of phosphates in soil. In the rhizosphere, exudates frequently contain phosphate radicals that would be subject to microbial modification. Significant microbial processes include mineralization/ immobilization reactions as well as solubilization of inorganic phosphates. Mineralization by soil phosphatase of either plant or microbial origin occurs readily in the rhizosphere, however, much of the phosphate is immediately immobilized because of high general microbial activity (Curl and Truelove, 1985). Cole *et al.* (1978) showed that amoebae in the rhizosphere can be significant in the mineralization of bacterial phosphorus.

A change in rhizosphere pH can influence the availability of soil phosphorus (Grinstead *et al.*, 1982). Maize plants that take up predominantly NH_4^+ decrease the soil pH (Marschner and Römheld, 1983). Other mechanisms also exist that can promote phosphate availability and uptake. Many rhizosphere bacteria are capable of solubilizing phosphate in the immediate root zone, thus allowing increased plant uptake (Louw, 1970; Alagawadi and Gaur, 1988; Laheurte and Berthelin, 1988). A greater number of phosphate-solubilizing bacteria were found in cowpea than in maize rhizosphere (Odunfa and Oso, 1978). Pepper *et al.* (1976) documented solubilization of phosphate from iron and calcium phosphates, and subsequent microbial conversion to polyphosphates. They hypothesized that polyphosphate formation may keep solubilized phosphorus available to plants for longer periods of time. Nye and Kirk (1987) recently developed a physicochemical mechanism for phosphate solubilization in the rhizosphere, showing that significant amounts of phosphate could be solubilized by acid secretion. Additionally, root exudates may contain organic acids capable of releasing phosphates from inorganic compounds. A further mechanism for increased plant uptake of phosphorus is provided through increasing the rate of diffusion of phosphate ions to the root system. Evidence for such a mechanism is limited, but Gardner *et al.* (1983) have suggested that chelating agents released as exudates may be involved.

2. Symbiotic Associations Affecting Phosphorus Uptake

Mycorrhizal fungi strongly influence phosphorus uptake, translocation, and transfer processes. When active, fungi absorb nutrients by processes believed to be similar to those of higher plants. Many studies with different plant species have documented increased phosphorus uptake from mycorrhizal-infected plants as compared to noninfected plants (Hayman, 1983; Gray and Gerdemann, 1969; Mosse *et al.*, 1973). Actual uptake rates by VA fungi have been estimated at between 2×10^{-15} mol cm^{-1} hyphae s^{-1} and 10^{-17} mol cm^{-1} s^{-1} (Cooper and Tinker, 1978). For average hyphae of around 8 μm diameter, this translates to a flux of 10^{-12} mol cm^{-1} s^{-1}, which is similar to P fluxes into roots (Tinker and Gildon, 1983).

The fungi appear to increase phosphorus uptake by more than one mechanism. Mycorrhizal roots seem to be physiologically more efficient in uptake of phosphorus, but also increase the absorbing surface area of the root, with external hyphae being somewhat analogous to root hairs. Diffusion studies on phosphorus uptake by roots in soil have demonstrated that phosphorus diffusion through soil is the rate-limiting step (Nye and Tinker, 1977).

Possible mechanisms for phosphorus uptake from phosphorus-deficient soils are that mycorrhizal roots exude more chelating acids (Tinker, 1975) or alter the pH of the rhizosphere, because of different cation/anion uptake ratios (Buwalda *et al.*, 1983; Smith, 1980). Another mechanism for increased phosphorus uptake from mycorrhizae is the possibility of solubilization of "unavailable soil phosphates" from rock phosphates (Hayman and Mosse, 1972) and iron phosphates (Bolan and Robson, 1987).

Phosphate taken up by mycorrhizae can be supplied directly to the plant in the case of ectomycorrhizae via orthophosphate transfer from the Hartig net to the cortical root cells. For the vesicular-arbuscular fungi, phosphates are usually stored as polyphosphates in arbuscles that sequentially degrade as new arbuscles are formed. During degradation the microbial polyphosphates are hydrolyzed to orthophosphate and taken up by the plant. It is of interest that applications of fertilizer phosphate reduce the degree of mycorrhizal infection. Although the response of plants to infection with vesicular-arbuscular mycorrhizae is usually less than the response to applications of fertilizer phosphorus, the effects of mycorrhizal fungi are economically significant and have been reported to be equivalent to 500 kg P ha^{-1} yr^{-1} for citrus (Menge *et al.*, 1978). In terms of sustainable agriculture, smaller applications of fertilizer phosphate could maximize the effects of the mycorrhizae without sacrificing yield significantly. In terms of carbon requirements for mycorrhizae it has been estimated that 4–10% of total photosynthate can be utilized by the fungi (Kucy and Paul, 1982). Finally, note that mycorrhizae often interact synergistically with symbiotic or free-living nitrogen-fixing organisms, apparently by improving the phosphate nutrition of the organisms (Subba Rao *et al.*, 1985, 1986; Pacovsky *et al.*, 1986).

D. OTHER MACRONUTRIENTS

Fewer studies are available on the rhizosphere effect for plant uptake of potassium, the third major macronutrient. Uptake of potassium by soybeans was enhanced by inoculation with endomycorrhizae (Mojallali and Weed, 1978), confirming the earlier findings of Alexandrov and Zak (1950). More recently, Berthelin and Leyval (1982) have shown that the addition of both *Glomus mosseae* and rhizosphere bacteria of maize can weather micas and promote uptake of several nutrients including potassium. The bacteria in particular were capable of weathering biotite and increasing potassium uptake. Davies (1987) also demonstrated increased potassium uptake due to mycorrhizae while Saif (1987) showed increased calcium and magnesium uptake due to VA fungi.

Jackson and Voigt (1971) demonstrated increased calcium uptake by trees due to calcium silicate-solubilizing rhizosphere bacteria. Barber and Ozanne (1970) used audioradiography to demonstrate increased uptake of calcium by four different plant species and attributed the differences to different root exudate effects on the rhizosphere. Berthelin and Leyval (1982) reported increased plant uptake of calcium and magnesium by maize due to the solubilization of biotite by rhizosphere bacteria. Thus nonspecific solubilization of insoluble elements in the vicinity of the plant root appears to be an important mechanism for increasing plant uptake of phosphorus, potassium, calcium, and magnesium.

Plant uptake of sulfur can also be affected by the rhizosphere. The plant-available form of sulfur (sulfate) is the stable form of inorganic sulfur in aerobic soils. In anaerobic soils sulfur exists as the sulfide ion, which is toxic to plants. Sulfate is reduced to sulfide by anaerobic heterotrophic bacteria such as *Desulfovibrio* sp. or *Desulfotomaculum* sp. Under aerobic conditions sulfide or elemental sulfur can be oxidized to sulfate by autotrophic bacteria such as *Thiobacillus thiooxidans* or by heterotrophic bacteria (Pepper and Miller, 1978). In the rhizosphere, the activity of the sulfate-reducing organisms can be enhanced by organic exudates, particularly in heavy-textured soils with high moisture contents. Under such conditions hydrogen sulfide has resulted in the death of maize plants (Jacq and Dommergues, 1970) and rice (Takai and Kamura, 1966). In the presence of iron in the soil, the hydrogen sulfide is precipitated as iron sulfide in the rhizosphere, thus reducing the toxicity (Takai and Kamura, 1966).

E. MICRONUTRIENTS

In several studies that have evaluated mycorrhizal effects on micronutrient uptake, the results have been contradictory. Zinc uptake was enhanced by mycorrhizal infection (Cooper and Tinker, 1978) and zinc deficiencies occurred following phosphorus fertilization, which suppresses mycorrhizal infection. This contrasts with decreased copper uptake by mycorrhizal versus nonmycorrhizal plants (Timmer and Leyden, 1980). Mycorrhizal infection can also reduce the detrimental effect of heavy metals such as copper and zinc on plant growth (Bradley *et al.*, 1982). In contrast, Killham and Firestone (1983) reported increased uptake of copper, zinc, nickel, lead, iron, and cobalt from mycorrhizal plants. More recently Kucy and Janzen (1987) showed that mycorrhizal fungi increased the uptake of zinc, copper, and iron by field beans, whereas Dueck *et al.* (1986) reported that zinc uptake and toxicity were decreased by mycorrhizal fungi. However, interactions between phosphates and metals frequently occur, and care must be taken when interpreting data, particularly when added phosphate fertilizer can decrease the incidence of mycorrhizal infection as well as precipitate previously available metals such as zinc or manganese.

Bacteria also affect micronutrient uptake by plants, frequently causing decreased uptake. Manganese is less available when oxidized, and certain plant cultivars susceptible to manganese deficiencies have been shown to contain rhizosphere populations with numerous manganese-oxidizing bacteria (Timonin, 1946). Uptake of molybdenum was decreased by rhizosphere organisms, perhaps due to polysaccharides secreted by the organisms (Tan and Loutit, 1976). More recently Stojkovski *et al.* (1986) have implicated glucuronic acid in polysaccharide slimes as instrumental in the binding of molybdenum. Thus when an element is in short supply, its uptake is likely to be restricted by microbial competition, the exception being when mycorrhizal associations enhance nutrient uptake.

F. MICROORGANISM INFLUENCE ON ROOT MORPHOLOGY

Microorganisms on the root surface can directly affect root morphology and influence nutrient uptake. Inhibition of root hair development of pea plants inoculated with a *Pseudomonas* sp. has been shown by Darbyshire and Greaves (1970). In contrast to this, rhizosphere organisms can increase the absorptive capacity of plants by stimulating the development of "proteoid roots" (Malajczuk and Bowen, 1974). These roots are influenced by specific stimuli from rhizosphere organisms.

IV. Rhizosphere Dynamics, Biological Control, and Emerging Technologies

A. MECHANISMS FOR COMPETITIVENESS IN THE RHIZOSPHERE

Microorganisms in the rhizosphere are in intimate contact with the root surface and root hairs, and thus directly influence the microenvironment of the root system. Rhizosphere organisms can be beneficial or detrimental to plants, and humans have always been interested in manipulating rhizosphere populations to ensure that beneficial organisms predominate. One approach to such manipulations is to identify those factors or mechanisms responsible for enhancing the competitiveness of beneficial and detrimental organisms in the rhizosphere. Competitiveness can be defined as the relative ability of an organism to survive, metabolize, and reproduce in the presence of numerous other organisms. Beneficial rhizosphere organisms include:

(a) General heterotrophic rhizosphere organisms that undergo biochemical transformations.
(b) Indigenous symbiotic organisms such as rhizobia, *Frankia* spp., or mycorrhizae.
(c) Inoculant organisms added with seed or sprayed onto soil to manipulate rhizosphere dynamics. These include rhizobia, mycorrhizae, and bacteria such as pseudomonads or *Agrobacterium radiobacter* strain 84, which can biologically control seedling and root diseases.

Since all inoculant organisms compete with the indigenous microflora within the rhizosphere, elucidation of competitiveness mechanisms would allow strain selection for inoculants to be based in part on the competitive nature of the organism, thereby increasing the probability of success of the inoculant strain. For an inoculant strain to be beneficial to a plant, some organisms need only be competitive for a short period of time. For example, biological control of seedling disease is most critical during the first 2–3 weeks of growth following germination. Also infection of plants by rhizobia is usually accomplished within several days of germination.

Several possible mechanisms for enhancing competitiveness in the rhizosphere have been demonstrated, and it is likely that a successful rhizosphere organism may utilize several of these mechanisms concurrently.

1. Recognition

For beneficial organisms to improve plant health, they must first colonize the rhizosphere, which involves some recognition or communication between roots and microorganisms through molecular and genetic mechanisms as depicted in the inner circle in Figure 1. The first step might be chemotaxis, followed by adhesion of the bacterial cell on or near the root, which would place the microorganism in a position to use root exudates and to be transported through the soil as the root grows. Transport of bacteria along the root would be inefficient (Bull, 1987; Bahame and Schroth, 1987), which may also explain why inoculated legumes in rhizobia-free soils are nodulated only on the tap roots (Bezdicek *et al.*, 1978). Mechanisms proposed for early attachment include fimbriae (pili), which function in bacterial binding (Vesper and Bauer, 1986), exopolysaccharides [Dazzo (1980), for rhizobia; Douglas *et al.* (1985) and Thomashow *et al.* (1987), for *Agrobacterium tumefaciens;* Vanderleyden *et al.* (1986), for *Azospirillum brasilense*], and chemotaxis (Currier and Strobel, 1981; Scher *et al.*, 1985).

2. Microbial Antagonism

Once an organism attracts or is attracted to the root surface, the host plant, as well as microbial and abiotic factors, determines which organism dominates the rhizosphere. Antagonism (amensalism) is when one microorganism population produces a substance that is inhibitory to another population (Atlas and Bartha, 1981). A broader concept is often used by plant pathologists (Cook and Baker, 1983) to mean actively expressed opposition, and includes antibiosis, competition and parasitism. The term competition is often strictly defined as two organisms striving for the same resource (Atlas and Bartha, 1981) such as carbohydrates, but it is often used by soil microbiologists (and others) in a broader meaning. Competition between rhizobial strains often means the ability to produce and occupy nodules, which may or may not mean dominance in the rhizosphere by a particular strain. In practice, it is often not known what mechanism(s) is involved when one microorganism is found in larger numbers in the rhizosphere than others.

Antagonism was reported by Kloepper and Schroth (1981) between plant growth-promoting rhizobacteria on potatoes and fungal populations. *Penicillium, Aspergillus,* and *Fusarium* isolates were reduced by 23–64% compared with untreated controls. Such antagonism was strain specific with some rhizobacteria showing an antagonism. Nonfluorescent pseudomonads at low soil temperature have been reported to be inhibitory to winter wheat by production of a toxin (Frederickson and Elliott, 1985). Some beneficial plant rhizobacteria may improve plant growth by specifically reducing populations of subclinical pathogens (Suslow and Schroth, 1982; Schroth and Hancock, 1982; Cook *et al.*, 1987).

Another example is the use of antagonists applied to pruning wounds to provide exclusionary protection against pathogens (Corke, 1978; Rishbeth, 1979). A classic example is the role of resident antagonists in the decline of "take-all" of wheat

(*Gaeumannomyces graminis* var. *tritici*) with monoculture in the Pacific Northwest (Shipton, 1977). Soil organisms responsible for the decline in disease are believed to be *Pseudomonas* spp. (Cook and Rovira, 1976; Weller and Cook, 1983), which provide biological control. Rhizosphere antagonists are implicated (Figure 4) in a 15-year irrigated field study in Washington in which nearly 15 years were required for take-all decline to be expressed, only to have wheat yield decrease from *Rhizoctonia* root rot (Cook, 1988). Perhaps continued wheat culture will control *Rhizoctonia* root rot through other antagonists. Neighboring plots on the Oregon State University Station at Pendleton grown to wheat for 57 consecutive years show no apparent damage by this disease in spite of damage in neighboring fields (R. W. Smiley and R. J. Cook, personal communication). Other examples of disease decline with monoculture have been shown for common scab of potatoes (Weinhold *et al.*, 1964), root rot of sugar beet (Hyakumachi and Ui, 1982), and *Rhizoctonia* damping-off of radish (Chet and Baker, 1981). In some instances, the mechanism of microbial antagonism may be due to unknown toxins, while in other cases, the mechanisms for antagonism are known and arc discussed in the following sections.

3. Antibiotic Production

An antibiotic is a substance produced by one microorganism that, in low concentrations, kills or inhibits other microorganisms. *Pseudomonas* spp. have been implicated as producers of antibiotics (Howell and Stipanovich, 1979), although there are few instances of unequivocal proof. Many other common rhizosphere

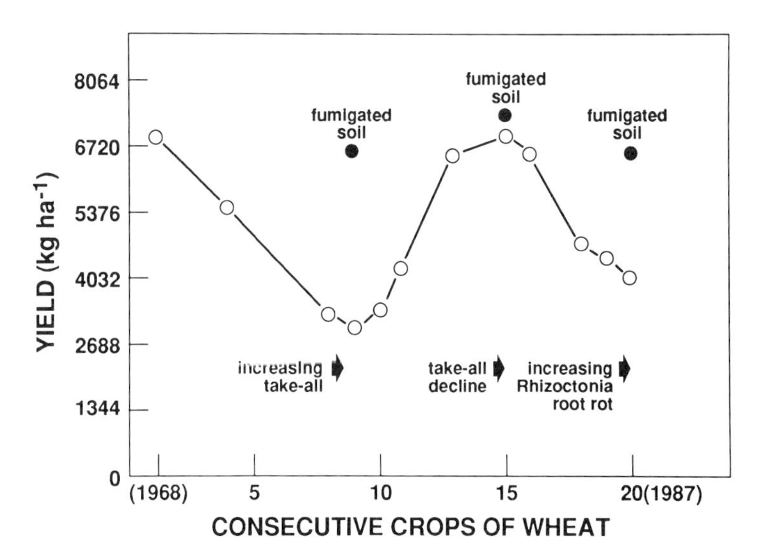

Figure 4. Continuous wheat yield under irrigation and conventional tillage in a Washington State University field study as influenced by take-all, take-all decline, and *Rhizoctonia* root rot. Solid circles represent fumigated plots. Source: After Cook, 1988.

organisms synthesize antibiotics, including *Streptomyces, Penicillium,* and *Aspergillus* (Curl and Truelove, 1985). Thus antibiotic production may be an important mechanism for competitiveness in the rhizosphere. However, resistance to antibiotics may be equally important.

Suppression of plant pathogens by antibiotics has been implicated in only a few instances (Howell and Stipanovic, 1980). Infection of *Pythium ultimum* in cotton was suppressed more by *P. fluorescens* Hv37aR2 than by an isogenic, nonproducing mutant (Howie and Suslow, 1986). More definitive evidence for a role of phenazine antibiotics in inhibition of take-all has been shown (Gurusiddaiah *et al.*, 1986). Through Tn5 mutagenesis of *P. fluorescens* strain 2-79, non-phenazine-producing mutants containing single Tn5 insertions colonized wheat roots equally, but were significantly less suppressive of take-all than the parent strain (Thomashow and Weller, 1988). When phenazine$^-$ strains were genetically restored for phenazine production by DNA complementation from a wild-type genomic library, phenazine was produced and suppression of take-all was similar to that found in the wild type (Thomashow and Weller, 1988; L. S. Pierson and L. S. Thomashow, personal communication). These studies show a definite role for antibiotic production in suppression of a plant pathogen.

Production and resistance are often under the control of extrachromosomal DNA called plasmids. Such plasmids can be transferred horizontally via transconjugation (Broughton *et al.*, 1987). Antibiotic production is highest in soils with a high organic matter content, and the rhizosphere would seem to be a likely place for significant production. However, studies have centered mainly on laboratory cultures, and the role of antibiosis in a natural environment remains controversial (Curl and Truelove, 1985). Undoubtedly molecular biology techniques such as transposon mutagenesis will shed more light on this in the future.

4. Siderophore Production

Siderophores are chemical compounds that can chelate iron and are often produced microbially, under iron-limiting conditions (Neilands, 1981). Certain rhizosphere organisms, particularly *Pseudomonas* sp., produce siderophores that make iron unavailable to other rhizosphere competitors (Kloepper *et al.*, 1980; Teintz *et al.*, 1981). Affinity of bacterial siderophores for Fe^{3+} by pseudomonads is much greater than that of fungal siderophores, which effectively starve fungal pathogens of iron (Leong and Neilands, 1982). Siderophore-producing *Pseudomonas* spp. have also been shown to inhibit fungi, including *Fusarium* sp. (Scher and Baker, 1982). Siderophore production appears to be particularly sensitive to soil pH and is usually produced abundantly in high-pH soils where availability of iron as Fe^{3+} is limiting. Definitive studies proving that siderophore production is the actual mechanism in field soil are needed.

5. Other Factors Enhancing Competitiveness

Garrett (1970) listed several characteristics necessary for an organism to be competitive when inhabiting the same ecological niche as other organisms. These include (a) rapid growth rate, (b) diverse enzyme systems for degradation of substrate, (c) production of antagonistic substances including volatile gases and waste acidic metabolites, and (d) tolerance to inhibitory substances produced by other organisms. Organisms with diverse enzyme systems that can metabolize rapidly are most likely to do well in the rhizosphere. Interestingly, competition for space and water is not considered important, whereas competition for oxygen can cause microbial shifts of aerobic and facultatively anaerobic populations. Competition can also occur between symbiotic microorganisms that infect plants. For example, strains of rhizobia vary in their degree of competitiveness (see Section II,A).

B. BIOLOGICAL CONTROL OF SEEDLING AND ROOT DISEASE

Biological control can be defined in numerous ways. Cook (1988) proposed a broad definition that includes the use of natural or modified organisms, genes, or gene products able to reduce the effects of undesirable organisms (pests) and to favor desirable organisms such as crops, trees, animals, and beneficial insects and microorganisms.

Other definitions have been as simple as the control of one organism by another (Baker, 1985). Plant breeding is considered by many to be a well-proven means of biological control. Although in modern times biological control is sometimes linked to potential advances in recombinant DNA technology, biological control through crop rotation has been a primary means to control pests, weeds, and diseases and will undoubtedly continue to be in the future. In addition to biological methods, chemical (pesticides) and physical (tillage, burning, solar heating) methods are collectively used to control plant pathogens. These approaches can be integrated into a term called holistic plant health (Cook, 1988), which describes the overall welfare of the plant. When the focus is on the pest for control of insects and nematodes, the term integrated pest management (IPM) is often used. Plants with poor overall root health will not utilize nutrients and water efficiently and will often result in poor economic return, especially when the cost of purchased inputs is high. Conversely, healthy roots often found under adequate crop rotation can more fully utilize the nutrient and water resources available and result in a more favorable economic return (Cook, 1988).

General biological control strategies have been categorized into regulation of the pest, exclusionary systems of protection, and systems of self-defense (Cook, 1987). Other strategies of biological control can be categorized into competition, induced resistance/hypovirulence, and predation/parasitism (Whipps and Lynch, 1986), which may be more directly applied to plant pathology. In the competition strategy (exclusionary approach), the biological agent competes for nutrients, time, or space and often uses antibiosis. A classic example is control of *Agrobacterium tumefaciens,* which causes crown gall in stone fruits and ornamental shrubs (Moore and Warren, 1979; Kerr, 1982). Crown gall is controlled by inoculating cut root surfaces with an avirulent antibiotic-producing strain A84 of *A. radiobacter,* which colonizes and protects the roots against the virulent strain. Another commercial biocontrol practice is the inoculation of the fungus *Peniophora gigantea* into stumps of felled pine, which prevents subsequent infection by the virulent root rot fungus *Heterobasidium annosum* (Rishbeth, 1975). Other strategies for control of soil-borne pathogens utilize the inoculation of seed and root pieces with antagonists (Schroth and Hancock, 1982; Kloepper and Schroth, 1981; Cook *et al.*, 1987; Weller and Cook, 1983, 1987).

Hypovirulence and induced resistance are documented mechanisms for control of aerial pathogens (Whipps and Lynch, 1986), although hypovirulent forms of *G. graminis* or cross-protection with an avirulent fungus (*Phialophora graminicola*) has been suggested as one mechanism in take-all decline (Deacon, 1976). However, as shown in Section IV,B, buildup of antagonistic *Pseudomonas* spp. has been suggested as a more plausible explanation (Cook and Rovira, 1976; Weller and Cook, 1983). Hyperparasitism or predation in control of soil-borne pathogens has been implicated but not elucidated in several instances for control of *G. graminis* by parasitic amoebae (Chakraborty, 1983) and *Bacillus mycoides* (Campbell and Faull, 1979).

C. FATE AND DETECTION OF INTRODUCED ORGANISMS

Two of the factors limiting the successful introduction of biocontrol organisms in the rhizosphere are competition from the indigenous soil microflora and a failure to adapt to abiotic factors. There are many possibilities for the fate of a foreign organism introduced into soil containing an indigenous microbial population. Some of these include (a) establishment into the indigenous population, (b) transient survival followed by the death of the organism, and (c) genomic rearrangement and horizontal gene transfer.

1. Incorporation into Indigenous Populations

Competition occurs between introduced and indigenous organisms, and the chance of the introduced organism being stably incorporated into the indigenous population will be enhanced if an ecological niche exists for that organism. Thus if a rhizobial strain is introduced into a rhizobia-free soil, that strain often becomes the indigenous rhizobial strain in that soil. However, free ecological niches are often not available, and introduced strains do not survive indefinitely, although they do have the opportunity to interact with the indigenous organisms.

Andrews (1984) recently classified organisms artificially along an r–k gradient. The r strategists rely on high reproductive rates for survival, whereas the k strategists depend on physiological adaptations to the environment (Andrews and Harris, 1986). Thus r strategists do well in situations where nutrients are not severely limiting, whereas k strategists do better in resource-limited situations. According to these definitions, r strategists would likely do well in the rhizosphere.

2. Transient Survival

Bacterial survival in soil and in the rhizosphere is affected by both biotic and abiotic soil factors, which often impose levels of stress that ultimately lead to the death of introduced organisms. The failure of introduced organisms to adequately colonize the rhizosphere is a critical factor in biological control and in the introduction of rhizobia (Dommergues, 1978; Schmidt, 1979). Seasonal decline of introduced plant growth-promoting organisms is not unusual (Kloepper *et al.*, 1980).

Many studies have shown that organisms added to sterile soil survive for longer periods than when added to nonsterile soil, and biotic factors have been implicated (Vidor and Miller, 1980; Boonkerd and Weaver, 1982). These factors can include competition with similar organisms and predation from other organisms such as protozoa (Danso *et al.*, 1975).

Abiotic factors that influence survival are numerous and diverse. They include soil moisture, temperature, organic matter content, texture, pH, and salinity. Environmental conditions such as low water potential (Bezdicek and Donaldson, 1981), high temperature (Kennedy and Wollum, 1988), high salinity, or extreme soil pH values (Schlinkert-Miller and Pepper, 1988a; Shoushtari and Pepper, 1985a) usually result in a rapid die-off of introduced organisms. However, sometimes organisms become physiologically adapted to specific environments (Schlinkert-Miller and Pepper, 1988b). Survival is usually increased when organisms are added to soils with high organic matter content. The effect of soil texture can vary depending on interactions with other factors, but generally coarse-textured soils are a more hostile environment than fine-textured soils, with extremes of texture adversely affecting survival.

Recently there has been increased interest in the fate of genetically engineered microorganisms (GEM) released into the environment. There are a few published reports on the fate of GEMs deliberately added to soil, and most address the fate of strains with recombinant plasmids. To date such strains appear to behave similarly to wild-type nonrecombinant strains, growing in sterile soil but declining in number when added to nonsterile soil (Devanas and Stotzky, 1986; Walter *et al.*, 1987). Pillai and Pepper (1990) showed that transposon mutants of rhizobia for the most part survived equally as well as the wild type.

3. Genomic Rearrangement and Gene Transfer

During the period of time that an introduced population is viable in a nonsterile soil, genetic alteration can occur within members of the population and genetic interactions with other organisms can take place. Transfer of chromosomally labeled DNA from *Bacillus subtilis* spores to other cells of *B. subtilis* in soil has been demonstrated (Graham and Istock, 1978) after spores were treated with DNAase and heat. Weinberg and Stotzky (1972) showed that F plasmid transfer between *E. coli* strains can occur in soil and was enhanced by the addition of montmorillonite. Plasmid exchange in the rhizosphere and on leaf surfaces can occur at rates of

10^2–10^6 higher than under associated laboratory conditions (Plazinski and Rolfe, 1985; Reanney *et al.*, 1982). The high-density population of the rhizosphere likely encourages genetic exchange between strains, species, and genera (Shapiro, 1985; Slater, 1985).

Recently it has been shown that certain bacteria contain reiterated gene sequences that undergo genomic rearrangements (Martinez *et al.*, 1985; Flores *et al.*, 1988). These rearrangements occur in pure culture under nonstress conditions. It is not known if such rearrangements occur in a natural soil environment.

In addition to internal mutations and rearrangements, there is also the potential for horizontal gene transfer between different species of bacteria (Beringer and Hirsch, 1984). Horizontal gene transfer mediated by naturally occurring or laboratory-derived constructions of self-transmissible plasmids between *Rhizobium, Agrobacterium, Pseudonomas,* and *E. coli* has been documented in pure culture. Less information is available on transfer in soil, but many soil bacteria contain several megaplasmids and a large percentage of their total DNA can be plasmid borne. Some plasmids cannot be mobilized and it is interesting to speculate that essential chromosomal functions may have been mobilized to these plasmids. Other plasmids can be mobilized and transferred to other organisms. Soil populations of *Rhizobium* spp. can transfer genetic material and transfer has been documented in the rhizosphere (Broughton *et al.*, 1987).

4. Detection of Introduced Organisms

To achieve any reasonable accounting of genetic exchange in the environment, suitable methodologies must be developed. Colwell *et al.* (1985) focused on how little is known about genetic exchange in the environment, and how few methods are available to examine this exchange. Methods available to detect and enumerate microorganisms include antibiotic resistance, immunological techniques such as fluorescent antibodies and enzyme-linked immunosorbent assay, plasmid profiles, SDS–PAGE, and DNA : DNA hybridization. Many of these techniques have been reviewed by McCormick (1986). One recent development in the detection and enumeration of specific microorganisms has been based on the expression of cloned *E. coli lacY* and *lacZ* genes in *Pseudomonas fluorescens* (Drahos *et al.*, 1986).

Spontaneous antibiotic-resistant mutants have been used extensively for following the populations of microorganisms in soil and the rhizosphere (Watrud *et al.*, 1985; Fredrickson and Elliott, 1985; Weller and Cook, 1983) and will likely continue to have utility as a selectable marker in microbial ecology. However, the environmental competitiveness of these mutants may be altered as a result of biochemical impairment when compared with the parent (Turco *et al.*, 1986).

More recently innovative molecular approaches have evolved that allow increased sensitivity of detection at the strain level. One approach is the use of transposable elements as a selective marker. Transposons such as Tn5 are mobile DNA sequences that also code for resistance to the amino-glycosidic antibiotics kanamycin and neomycin. Following the insertion of Tn5 into "neutral" DNA, a cell can be detected by plating on antibiotic amended media. Van Elsas *et al.* (1986)

and Fredrickson *et al.* (1988) recently traced Tn5 mutant bacteria in soil by use of such antibiotic resistance. They also used a gene probe specific to the Tn5 DNA sequence to identify the Tn5 bacteria by use of DNA : DNA hybridizations. Pepper *et al.* (1989) also utilized gene probes for detection of rhizobia from root nodules, but their probe was specific to a DNA sequence from a unique megaplasmid. A new method of detection of bacteria involves amplification of target DNA by the use of polymerase chain reaction (PCR). The use of PCR and gene prober produces a very sensitive method of analysis.

D. EMERGING TECHNOLOGIES

The role of the rhizosphere is critical to increasing the efficiency and sustainability of agriculture through the biological control of plant pathogens, insect pests, and weeds and the enhancement of symbiotic processes such as nitrogen fixation and mycorrhizal associations. A more fundamental understanding of the rhizosphere will be needed, especially in the mechanisms of root colonization by microorganisms and the potential for selectively replacing the indigenous population with selected organisms. Collectively, the strategy must be to improve plant health through traditional cultural practices, use of resistant plant cultivars, and through the application of molecular genetics. Healthy plants are able to utilize nutrients and water more efficiently and function more effectively within the biotic and abiotic constraints of the environment. The use of molecular techniques such as transposon mutagenesis is necessary to better understand the basic mechanisms involved in plant–microbe recognition and competition. Furthermore, these and other techniques are already providing new biocontrol organisms and products for potential future use. The future of a more sustainable agriculture will probably depend on how well we are able to merge the proven technologies of the past with the new technologies of the future. The rhizosphere is a key factor since many applications of biotechnology are directed toward the rhizosphere.

References

Alagawadi, A. P., and A. C. Gaur. 1988. Associative effect of *Rhizobium* and phosphate-solubilizing bacteria on the yield and nutrient uptake of chick pea. *Plant Soil* 105:241–246.

Alexander, M. 1977. Introduction to Soil Microbiology. 2nd Ed. Wiley, New York.

Alexandrov, V. G., and G. A. Zak. 1950. Les bactéries destructrices des aluminosilicates. *Mikrobiologiya* 19:97–104.

Allen, E. K., and O. N. Allen. 1965. Nonleguminous plant symbiosis. In C. M. Gilmour and O. N. Allen (eds.), Microbiology and Soil Fertility. Proc. 25th Annu. Biology Colloq., 1964. Oregon State Univ. Press, Corvallis. Pp. 77–106.

Andrews, J. H. 1984. Relevance of r- and k-theory to the ecology of plant pathogens. In M. J. Klug and C. A. Reddy (eds.), Current Perspectives in Microbial Ecology. Am. Soc. Microbiol., Washington, D.C. Pp. 1–7.

Andrews, J. H., and R. F. Harris. 1986. r and k selection and microbial ecology. *Adv. Microb. Ecol.* 9:99–148.

Atlas, R. M., and R. Bartha. 1981. Microbial Ecology, Fundamentals and Applications. Addison-Wesley, Reading, Massachusetts.

Babiker, H. M., and I. L. Pepper. 1984. Microbial production of ethylene in desert soils. *Soil Biol. Biochem.* 16:559–564.

Bagyaradaj, D. J., and J. A. Menge. 1978. Interaction between a VA mycorrhiza and *Azotobacter* and their effects on rhizosphere microflora and plant growth. *New Phytol.* 80:141–145.

Bahme, J. B., and M. N. Schroth. 1987. Spatial–temporal colonization patterns of a rhizobacterium on underground organs of potato. *Phytopathology* 77:1093–1100.

Baker, R. 1985. Biological control of plant pathogens: Definitions. In M. A. Hoy and D. C. Herzog (eds.), Biological Control in Agricultural IPM Systems. Academic Press, New York. Pp. 25–39.

Balandreau, J., and R. Knowles. 1978. The rhizosphere. In Y. R. Dommergues and S. V. Krupa (eds.), Interactions between Non-Pathogenic Soil Microorganisms and Plants. Elsevier, Amsterdam. Pp. 243–268.

Barber, D. A. 1968. Microorganisms and the inorganic nutrition of higher plants. *Annu. Rev. Plant Physiol.* 19:71–88.

Barber, D. A., and K. B. Gunn. 1974. The effect of mechanical forces on the exudation of organic substances by the roots of cereal plants grown under sterile conditions. *New Phytol.* 73:39–45.

Barber, D. A., and J. K. Martin. 1976. The release of organic substances by cereal roots into soil. *New Phytol.* 76:69–80.

Barber, S. A., and P. G. Ozanne. 1970. Audioradiographic evidence for the differential effect of four plant species in altering the calcium content of the rhizosphere soil. *Soil Sci. Soc. Am. Proc.* 34:635–637.

Beck, S. M., and C. M. Gilmour. 1983. Role of wheat root exudates in associative nitrogen fixation. *Soil Biol. Biochem.* 15:33–38.

Beringer, J. E., and P. E. Hirsch. 1984. The role of plasmids in microbial ecology. In M. J. Klug and C. A. Reddy (eds.), Current Perspectives in Microbial Ecology. Am. Soc. Microbiol., Washington, D.C. Pp. 63–70.

Berthelin, J., and C. Leyval. 1982. Ability of symbiotic and non-symbiotic rhizospheric microflora of maize (*Zea mays*) to weather micas and to promote plant growth and plant nutrition. *Plant Soil* 68:369–377.

Bezdicek, D. F., D. W. Evans, B., Abebe, and R. W. Witters. 1978. Evaluation of peat and granular inoculum for soybean yield and N fixation under irrigation. *Agron. J.* 70:865–868.

Bezdicek, D. F., and M. D. Donaldson. 1981. Flocculation of *Rhizobium* from soil colloids for enumeration by immunofluorescence. In R. C. Barkley (ed.), Microbial Adhesion to Surfaces. Harwood, London. Pp. 297–309.

Bezdicek, D. F., and A. C. Kennedy. 1988. Symbiotic nitrogen fixation and nitrogen cycling in terrestrial environments. In J. M. Lynch and J. E. Hubbie (eds.), Micro-Organisms in Action: Concepts and Applications in Microbial Ecology. Blackwell, Oxford. Pp. 241–260.

Blair, W. C. 1978. Interactions of soil fertility and the rhizosphere microflora of seedling cotton in relation to disease caused by *Rhizoctonia solami.* M.S. Thesis, Auburn University, Auburn, Alabama.

Bolan, N. S., and A. D. Robson. 1987. Effects of vesicular-arbuscular mycorrhiza on the availability of iron phosphates to plants. *Plant Soil* 99:401–410.

Boonkerd, N., and R. W. Weaver. 1982. Survival of cowpea rhizobia as affected by soil temperature and moisture. *Appl. Environ. Microbiol.* 43:585–589.

Bowen, G. D. 1980. Misconceptions, concepts and approaches in rhizosphere biology. In D. C. Ellwood, J. N. Hedger, M. J. Ltham, J. M. Lynch, and J. M. Slater (eds.), Contemporary Microbial Ecology. Academic Press, London. Pp. 283–304.

Bowen, G. D., and R. C. Foster. 1978. Dynamics of microbial colonization of plant roots. *Proc. Symp. Soil Microbiology and Plant Nutrition.* Univ. Press, Kuala Lumpur, Malaysia. Pp. 231–256.

Bradley, R., A. J. Burt, and D. J. Read. 1982. The biology of mycorrhiza in the Ericaceae. VIII. The role of mycorrhizal infection in heavy metal resistance. *New Phytol.* 91:197–209.

Broughton, W. J., U. Samrey, and J. Stanley. 1987. Ecological genetics of *Rhizobium melilot:* Symbiotic plasmid transfer in the *Medicago sativa* rhizosphere. *FEMS Microbiol. Lett.* 40:251–255.
Bull, C. T. 1987. Wheat root colonization by disease-suppressive and nonsuppressive bacteria and the effect of population size on severity of take-all caused by *Gaeumannomyces graminis* var. *tritici.* M.S. Thesis, Washington State University, Pullman.
Buwalda, J. G., D. P. Stribley, and P. B. Tinker. 1983. Increased uptake of bromide and chloride by plants infected with vesicular-arbuscular mycorrhizas. *New Phytol.* 93:217–225.
Campbell, R., and Faull, J. L. (1979). Biological control of *Gaeumannomyces graminis:* Field trials and the ultrastructure of the interaction between the fungus and a successful antagonistic bacterium. In B. Schippers and W. Gams (eds.), Soil-Borne Plant Pathogens. Academic Press, London. Pp. 603–609.
Chakraborty, S. 1983. Population dynamics of amoebae in soils suppressive and non-suppressive to wheat take-all. *Soil Biol. Biochem.* 15:661–664.
Chet, I., and R. Baker. 1981. Isolation and biocontrol potential of *Trichoderma hamatum* from soil naturally suppressive to *Rhizoctonia solani. Phytopathology* 71:286–290.
Clarholm, M. 1985. Interactions of bacteria, protozoa and plants leading to mineralization of soil nitrogen. *Soil Biol. Biochem.* 17:181–187.
Clark, F. E. 1949. Soil microorganisms and plant roots. *Adv. Agron.* 1:241–288.
Cole, C. V., E. T. Elliot, H. W. Hunt, and D. C. Coleman. 1978. Trophic interactions in soil as they affect energy and nutrient dynamics. V. Phosphorus transformations. *Microb. Ecol.* 4:381–387.
Colwell, R. R., P. R. Brayton, D. J. Grimes, D. B. Roszakj, S. A. Hua, and L. M. Palmer. 1985. Viable but non-culturable *Vibrio cholerae* and related pathogens in the environment: Implications for release of genetically engineered microorganisms. *Bio/Technology* 3:817–820.
Cook, R. J. 1987. Research Briefing Panel on Biological Control in Managed Ecosystems. Committee on Science, Engineering, and Public Policy. National Academy of Sciences, National Academy of Engineering, and Institute of Medicine. Natl. Acad. Press, Washington, D.C.
Cook, R. J. 1988. Biological control and holistic plant health care in agriculture. *Am. J. Altern. Agric.* **3**, 51–62.
Cook, R. J., and K. F. Baker. 1983. The Nature and Practice of Biological Control of Plant Pathogens. Am. Phytopathol. Soc., St. Paul, Minnesota.
Cook, R. J., and A. D. Rovira. 1976. The role of bacteria in the biological control of *Gaeumannomyces graminis* by suppressive soils. *Soil. Biol. Biochem.* 8:269–273.
Cook, R. J., D. M. Weller, and L. S. Thomashow. 1987. Enhancement of root health and plant growth by rhizobacteria. In Molecular Strategies for Crop Protection. Alan R. Liss, New York. Pp. 125–134.
Cooper, K. M., and P. B. Tinker. 1978. Translocation and transfer of nutrients in vesicular-arbuscular mycorrhizas. II. Uptake and translocation of phosphorus, zinc and sulfur. *New Phytol.* 81:43–52.
Corke, A. T. K. 1978. Interactions between microorganisms. *Ann. Appl. Biol.* 89:89–93.
Curl, E. A., and B. Truelove. 1985. The Rhizosphere. Springer-Verlag, Berlin.
Currier, W. W., and G. A. Strobel. 1981. Characterization and biological activity of trefoil chemotactin. *Plant Sci. Lett.* 21:159–165.
Dangerfield, J. A., D. W. S. Westlake, and F. D. Cook. 1978. Characterization of the bacterial flora associated with root systems of *Pinus contorta* var. *Patifolia. Can. J. Microbiol.* 24:1520–1525.
Danso, S. K. A., S. O. Keya, and M. Alexander. 1975. Protozoa and the decline of Rhizobium populations added to soil. *Can. J. Microbiol.* 21:884–895.
Darbyshire, J. F. 1966. Protozoa in the rhizosphere of *Lolium perrenne* L. *Can. J. Microbiol.* 12:1287–1289.
Darbyshire, J. F., and M. P. Greaves. 1970. An improved method for the study of the interrelationships of soil microorganisms and plant roots. *Soil Biol. Biochem.* 2:63–71.
Davies, F. T., Jr. 1987. Effects of VA-mycorrhizal fungi on growth and nutrient uptake of cuttings of *Rosa multiflora* in two container media with three levels of fertilizer application. *Plant Soil* 104:31–35.
Dazzo, F. B. 1980. Microbial adhesion to plant surfaces. In J. M. Lynch, J. Melling, P. R. Rutter, and B. Vincent (eds.), Microbial Adhesion to Surfaces. Harwood, London. Pp. 311–328.

Dazzo, F. B., and A. E. Gardiol. 1984. Host specificity in *Rhizobium*–legume interactions. In D. P. S. Verma and T. H. Hohn (eds), Genes Involved in Microbe–Plant Interactions. Springer-Verlag, New York. Pp. 3–31.

Deacon, J. W. 1976. Biological control of the take-all fungus *Gaeumannomyces graminis* by *Phialophora radiciola* and similar fungi. *Soil Biol. Biochem.* 8:275–283.

de Bertoldi, M., A. Rambelli, M., Giovannetti, and M. Griselli. 1978. Effects of benomyl and captan on rhizosphere fungi and the growth of *Allium cepa. Soil Biol. Biochem.* 10:265–268.

Devanas, M. A., and G. Stotzky. 1986. Fate in soil of a recombinant plasmid carrying a *Drosophila* gene. *Curr. Microbiol.* 13:279–283.

Dobereiner, J., J. M. Day, and P. J. Dart. 1972. Nitrogenase activity and oxygen sensitivity of the *Paspalum notatum–Azotobacter paspali* association. *J. Gen. Microbiol.* 71:103–116.

Dobereiner, J., and F. O. Pedrosa. 1987. Nitrogen-Fixing Bacteria in Nonleguminous Plants. Sci. Technol. Publ., Madison, Wisconsin.

Dommergues, Y. R. 1978. The plant–microorganism system. In Y. R. Dommergues and S. V. Krupa (eds.), Interactions between Non-Pathogenic Soil Microorganisms and Plants. Elsevier, Amsterdam. Pp. 1–33.

Douglas, C. J., R. J. Staneloni, R. A. Rubin, and E. W. Nester. 1985. Identification and genetic analyses of an *Agrobacterium tumefaciens* chromosomal virulence region. *J. Bacteriol.* 161:850–860.

Drahos, D. J., B. C. Hemming, and S. McPherson. 1986. Tracking recombinant organisms in the environment: β-Galactosidase as a selectable nonantibiotic marker for fluorescent pseudomonads. *Bio/Technology* 4:439–444.

Dueck, T. A., P. Visser, W. H. O. Ernst, and H. Schat. 1986. Vesicular-arbuscular mycorrhizae decrease zinc-toxicity to grasses growing in zinc-polluted soil. *Soil Biol. Biochem.* 18:331–333.

Flores, M., V. Gonzalez, M. A. Pardo, A. Leija, E. Martinez, D. Romero, D. Pinero, G. Davila, and R. Palacios. 1988. Genomic instability in *Rhizobium phaseoli. J. Bacteriol.* 170:1191–1196.

Frederickson, J. K., D. F. Bezdicek, F. J. Brockman, and S. W. Li. 1988. Enumeration of Tn5 mutant bacteria in soil using a most-probable-number-DNA hybridization procedure and antibiotic resistance. *Appl. Environ. Microbiol.* 54:446–453.

Fredrickson, J. K., and L. F. Elliott. 1985. Effect of winter wheat seedling growth by toxin producing rhizobacteria. *Plant Soil* 183:399-409.

Fries, N. 1973. Effects of volatile organic compounds on the growth and development of fungi. *Trans. Br. Mycol. Soc.* 60:1–21.

Führer, von E. 1961. Der Einfluss von Pflanzen-wurzeln auf die Verteilung der Kleinarthropoden in boden, untersucht an *Pseudotritia ardua* (Oribatei). *Pedobiologia* 1:99–112.

Garcia, J. L. 1973. Influence de la rhizosphére du riz sur l'activité dénitrifiante potentielle des sols de riziéres du Senegal. *Oecol. Plant.* 8:315–323.

Gardner, W. K., D. A. Barber, and D. G. Parberry. 1983. The acquisition of phosphorus by *Lupinus albus* L. III. The probable mechanism by which phosphorus movement in the soil/root interface is enhanced. *Plant Soil* 70:107–124.

Garrett, S. D. 1970. Pathogenic Root-Infecting Fungi. Cambridge Univ. Press, London.

Gerdemann, J. W. 1968. Vesicular-arbuscular mycorrhiza and plant growth. *Annu. Rev. Phytopathol.* 6:397–418.

Graham, J. B., and C. A. Istock. 1978. Genetic exchange in *Bacillus subtilis* in soil. *Mol. Gen. Genet.* 66:278–290.

Gray, L. E., and J. W. Gerdemann. 1969. Uptake of phosphorus-32 by vesicular-arbuscular mycorrhizae. *Plant Soil* 30:415–422.

Griffin, G. J., M. G. Hale, and F. J. Shay. 1976. Nature and quantity of sloughed organic matter produced by roots of axenic peanut plants. *Soil Biol. Biochem.* 8:29–32.

Grinstead, M. J., M. J. Hedley, R. E. White, and R. W. Nye. 1982. Plant induced changes in the rhizosphere of rape (*Brassica napus* var. *Emerald*) seedlings. I. pH change and the increase in P concentration in the soil solution. *New Phytol.* 91:19–29.

Guckert, A., H. Breisch, and O. Reisinger. 1975. Interface sol-racine. I. Étude au microscope électronique des relations mucigel-argile-microorganismes. *Soil Biol. Biochem.* 7:241–250.

Gurusiddaiah, S., D. M. Weller, A. Sarkar, and R. J. Cook. 1986. Characterization of an antibiotic produced by a strain of *Pseudomonas fluorescens* inhibitory to *Gaeumannomyces graminis* var. *tritici* and *Pythium* spp. *Antimicrob. Agents Chemother.* 29:488–495.

Hale, M. G., and L. D. Moore. 1979. Factors affecting root exudation. II. 1970–1978. *Adv. Agron.* 31:93–126.

Hale, M. G., L. D. Moore, and G. J. Griffin. 1978. Root exudates and exudation. In Y. R. Dommergues and S. V. Krupa (eds.), Interactions between Non-Pathogenic Soil Microorganisms and Plants. Elsevier, Amsterdam. Pp. 163–203.

Hale, M. G., L. D. Moore, and G. J. Griffin. 1981. Factors affecting root exudation and significance for the rhizosphere ecosystems. *Biological and Chemical Interactions in the Rhizosphere. Symp. Proc. Ecol. Res. Comm. Swed. Natl. Sci. Res. Counc., Stockholm* pp. 43–71.

Halverson, L. J., and G. Stacey. 1986. Signal exchange in plant–microbe interactions. *Microbiol. Rev.* 50:193–225.

Hamlen, R. A., F. L. Lukezic, and J. R. Bloom. 1972. Influence of age and stage of development on the neutral carbohydrate components in root exudates from alfalfa plants grown in gnotobiotic environment. *Can. J. Plant Sci.* 52:633–642.

Harley, J. L. 1965. Mycorrhiza. In K. F. Baker and W. C. Snyder (eds.), Ecology of Soil-Borne Pathogens. Univ. of Calif. Press, Berkeley. Pp. 218–230.

Hayman, D. S. 1983. The physiology of vesicular-arbuscular endomycorrhizal symbiosis. *Can. J. Bot.* 61:944–963.

Hayman, D. S., and B. Mosse. 1972. The role of vesicular-arbuscular mycorrhiza in the removal of phosphorus from soil by plant roots. *Rev. Ecol. Biol. Sol* 9:463–470.

Henderson, V. E., and H. Katznelson. 1961. The effect of plant roots on the nematode population of the soil. *Can. J. Microbiol.* 7:163–167.

Highinbotham, N. 1968. Cell electropotential and ion transport in higher plants. In K. Mathas, F. Muller, A. Nelles, and D. Neumann (eds.), Transport and Distribution of Matter in Cells of Higher Plants. Akademie-Verlag, Berlin. Pp. 167–177.

Hiltner, R. 1904. Uber neurere Erfarhrungen und Probleme auf dem Gebiet der Bodenbakteriologie und unter besonderer Berücksichtigung der Gründüngung und Brache. *Arb. DLG* No. 98, pp. 59–78.

Howell, C. R. and R. D. Stipanovich. 1979. Control of *Rhizoctonia solani* on cotton seedlings with *Pseudomonas fluorescens* and with an antibiotic produced by the bacterium. *Phytopathology* 69:480.

Howell, C. R., and R. D. Stipanovich. 1980. Suppression of *Pythium ultimum*-induced damping-off of cotton seedlings by *Pseudomonas fluorescens* and its antibiotic, pyoluteorin. *Phytopathology* 70:712–715.

Howie, W., and J. Suslow. 1986. Effect of antifungal compound biosynthesis on cotton root colonization and *Pythium* suppression by a strain on *Pseudomonas fluorescens* and its antifungal minus isogenic mutant. *Phytopathology* 76:1069.

Hyakumachi, M., and T. Ui. 1982. Decline of damping-off of sugarbeet seedlings caused by *Rhizoctonia solani* AGZ-Z. *Ann. Phytopathol. Soc. Jpn.* 48:600–606.

Jackson, T. A., and G. K. Voigt. 1971. Biochemical weathering of calcium bearing minerals by rhizosphere micro-organisms and its influence on calcium accumulation in trees. *Plant Soil* 35:655–658.

Jacq, V., and Y. Dommergues. 1970 Influence de l'intensité d'éclairement et de l'age de la plante sur la sulfato-réduction rhizosphérique. *Zentralbl. Bakteriol., Parasitenkd. Infektionskr.* 125:661–669.

Jain, D. K., D. Beyer, and R. J. Rennie. 1987. Dinitrogen fixation (C_2H_2 reduction) by bacterial strains at various temperatures. *Plant Soil* 103:233–237.

Jenny, H., and K. Grossenbacher. 1963. Root–soil boundary zones as seen in the electron microscope. *Soil Sci. Soc. Am. Proc.* 27:273–277.

Jones, K., and D. Bangs. 1985. Nitrogen fixation by free living heterotrophic bacteria in an oak forest: The effect of liming. *Soil Biol. Biochem.* 17:705–709.

Josephson, K. L., and I. L. Pepper. 1984. Competitiveness and effectiveness of strains of *Rhizobium phaseoli* isolated from the Sonoran Desert. *Soil Biol. Biochem.* 16:651–655.

Kennedy, A. C., and A. G. Wollum, II. 1988. A comparison of plate counts, MPN, and FA techniques for enumeration of *Bradyrhizobium japonicum* in soils subjected to high temperatures. *Appl. Environ. Microbiol.* Submitted.

Kerr, A. 1982. Biological control of soil-borne microbial pathogens and nematodes. In N. S. Subba Rao (ed.), Advances in Agricultural Microbiology. Butterworth, London. Pp. 429–463.

Killham, K., and M. K. Firestone. 1983. Vesicular-arbuscular mycorrhizal mediation of grass response to acidic and heavy metal deposition. *Plant Soil* 72:39–48.

Kloepper, J. W., J. Leong, M. Teintz, and M. N. Schroth. 1980. *Pseudomonas* siderophores: A mechanism explaining disease-suppressive soils. *Curr. Microbiol.* 4:317.

Kloepper, J. W., and M. N. Schroth. 1981. Relationship of *in vitro* antibiosis of plant growth-promoting rhizobacteria to plant growth and displacement of root microflora. *Phytopathology* 71:1020–1024.

Kosslak, R. M., B. B. Bohlool, S. Dowdle, and M. J. Sedowsky. 1983. Competition of *Rhizobium japonicum* strains in early stages of soybean nodulation. *Appl. Environ. Microbiol.* 46:870–873.

Kraft, J. M. 1974. The influence of seedling exudates on the resistance of beans to *Fusarium* and *Pythium* root rot. *Phytopathology* 64:190–193.

Kucy, R. M. N., and F. A. Paul. 1982. Carbon flux, photosynthesis and N_2 fixation in mycorrhizal and nodulated fababeans (*Vicia faba* L.). *Soil Biol. Biochem.* 14:407–412.

Kucy, R. M. N., and H. H. Janzen. 1987. Effects of VAM and reduced nutrient availability on growth and phosphorus and micronutrient uptake of wheat and field beans under greenhouse conditions. *Plant Soil* 104:71–78.

Laheurte, F., and J. Berthelin. 1988. Effect of a phosphate solubilizing bacteria on maize growth and root exudation over four levels of labile phosphorus. *Plant Soil* 105:11–17.

Leong, S. A., and J. B. Neilands. 1982. Siderophore production by phytopathogenic microbial species. *Arch. Biochem. Biophys.* 218:351–359.

Long, S. R. 1989. *Rhizobium*–legume nodulation: Life together in the underground. *Cell* 56:203–214.

Louw, H. A. 1970. A study of the phosphate-dissolving bacteria in the root region of wheat and lupin. *Phytophylactica* 2:21–26.

Malajczuk, N., and G. D. Bowen. 1974. Proteoid roots are microbially induced. *Nature (London)* 251:316–317.

Marschner, H., and V. Röomeheld. 1983. *In vivo* measurement of root-induced pH changes at the soil–root interface: Effect of plant species and nitrogen source. *Z. Pflanzenphysiol.* 111:241–251.

Martin, J. K. 1977. Factors influencing the loss of organic carbon from wheat roots. *Soil Biol. Biochem.* 9:1–7.

Martin, J. K. 1978. The variation with plant age of root carbon available to soil microflora. In M. W. Loutit and J. A. R. Miles (eds.), Microbial Ecology. Springer-Verlag, Berlin. Pp. 199–302.

Martinez, E., M. A. Pardo, R. Palacios, and M. A. Cervallos. 1985. Reiteration of nitrogen fixation gene sequences and specificity of *Rhizobium* in nodulation and nitrogen fixation in *Phaseolus vulgaris*. *J. Gen. Microbiol.* 131:1774–1786.

Marx, D. H. 1973. Mycorrhizae and feeder root diseases. In G. C. Marks and T. T. Kozlowski (eds.), Ectomycorrhizae: Their Ecology and Physiology. Academic Press, London. Pp. 351–382.

Marx, D. H. 1977. Tree host range and world distribution of the ectomycorrhizal fungus *Pisolithus tinctorius*. *Can. J. Microbiol.* 23:217–223.

Materon, L. A., and Hagedorn, C. 1984. Competitiveness of *Rhizobium trifolii* strains associated with red clover (*Trifolium pratense* L.) in Mississippi soils. *Appl. Environ. Microbiol.* 44:1096–1101.

May, S. N., and Bohlool, B. B. 1983. Competition among *Rhizobium leguminosarum* strains for nodulation of lentils (*Lens esculenta*). *Appl. Environ. Microbiol.* 45:960–965.

McCormick, D. 1986. Detection technology: The key to environmental biotechnology. *Bio/Technology* 4:419–422.

Menge, J. A., C. K. Labanauskas, E. L. U. Johnson, and R. G. Platt. 1978. Partial substitution of mycorrhizal fungi for phosphorus fertilization in the greenhouse culture of citrus. *Soil Sci. Soc. Am. J.* 42:926–930.

Meyer, F. H. 1974. Physiology of mycorrhiza. *Annu. Rev. Plant Physiol.* 25:567–586.

Miki, N. K., K. J. Clarke, and M. E. McCully. 1980. A histological and histochemical comparison of the mucilage on the root tips of several grasses. *Can. J. Bot.* 58:2581–2593.

Moawad, H. A., W. R. Ellis, and E. L. Schmidt. 1984. Rhizosphere response as a factor in competition among three serogroups of indigenous *Rhizobium japonicum* for nodulation of field grown soybeans. *Appl. Environ. Microbiol.* 47:607–612.

Mojallali, H., and S. B. Weed. 1978. Weathering of micas by mycorrhizal soybean plants. *Soil Sci. Soc. Am. J.* 42:367–372.

Moore, L. W., and G. Warren. 1979. *Agrobacterium radiobacter* strain 84 and biological control of crown gall. *Annu. Rev. Phytopathol.* 17:163–179.

Mosse, B., D. S. Hayman, and D. J. Arnold. 1973. Plant growth responses to vesicular-arbuscular mycorrhiza. V. Phosphate uptake by the plant species from P deficient soils labelled with ^{32}P. *New Phytol.* 72:809–815.

Mulder, E. G. 1975. Physiology and ecology of free living nitrogen-fixing bacteria. In W. D. P. Stewart (ed.), Nitrogen Fixation by Free-Living Microorganisms. Cambridge Univ. Press, London. Pp. 3–28.

Neilands, J. B. 1981. Microbial iron compounds. *Annu. Rev. Biochem.* 50:715–731.

Newman, E. I., A. J. Heap, and R. A. Lawley. 1981. Abundance of mycorrhizas and root–surface microorganisms of *Plantago lanceolata* in relation to soil and vegetation: A multi-variate approach. *New Phytol.* 89:95–108.

Nye, P. H., and G. J. D. Kirk. 1987. The mechanism of rock phosphate solubilization in the rhizosphere. *Plant Soil* 100:127–134.

Nye, P. H., and P. B. Tinker. 1977. Solute Movement in the Soil Root System. Blackwell, Oxford.

Odunfa, V. S. A., and B. A. Oso. 1978. Bacterial populations in the rhizosphere soils of cowpea and sorghum. *Rev. Ecol. Biol. Sol* 15:413–420.

Old, K. M., and T. H. Nicholson. 1975. Electron microscopical studies of the microflora of roots and sand dune grasses. *New Phytol.* 74:51–58.

Pacovsky, R. S., G. Fuller, A. E. Stafford, and E. A. Paul. 1986. Nutrient and growth interactions in soybeans colonized with *Glomus fasciculatum* and *Rhizobium japonicum*. *Plant Soil* 92:37–45.

Papavizas, G. C., and C. B. Davey. 1961. Extent and nature of the rhizosphere of *Lupinus*. *Plant Soil* 14:215–236.

Park, D. 1963. The ecology of soil-borne fungal disease. *Annu. Rev. Phytopathol.* 1:241–258.

Paul, E. A., and R. M. N. Kucey. 1981. Carbon flow in plant microbial associations. *Science* 213:473–474.

Payne, W. J. 1974. Reduction of nitrogenous oxides by microorganisms. *Bacteriol. Rev.* 37:409–452.

Pepper, I. L., K. L. Josephson, C. S. Nautiyal, and D. P. Bourque. 1989. Strain identification of highly competitive bean rhizobia isolated from root nodules: Use of fluorescent antibodies, plasmid profiles and gene probes. *Soil Biol. Biochem.* 21:749–753.

Pepper, I. L., and R. H. Miller. 1978. Comparison of the oxidation of thiosulfate and elemental sulfur by two heterotrophic bacteria and *Thiobacillus thiooxidans*. *Soil Sci.* 126:9–14.

Pepper, I. L., R. H. Miller, and C. P. Ghonsikar. 1976. Microbial inorganic polyphosphates: Factors influencing their accumulation in soil. *Soil Sci. Soc. Am. J.* 40:872–875.

Phillips, D. A. 1980. Efficiency of symbiotic nitrogen fixation in legumes. *Annu. Rev. Plant Physiol.* 31:29–49.

Pillai, S. D., and I. L. Pepper. 1990. Survival of Tn5 mutant bean rhizobia in desert soils: phenotypic expression of Tn5 under moisture stress. *Soil Biol. Biochem.* 22:265–270.

Plazinski, J., and B. G. Rolfe. 1985. *Azospirillum–Rhizobium* interaction leading to a plant growth stimulation without nodule formation. *Can. J. Microbiol.* 31:1026–1030.

Prikryl, Z., and V. Vancura. 1980. Root exudates of plants. VI. Wheat root exudation as dependent on growth, concentration gradient of exudates and the presence of bacteria (*Pseudomonas putida*). *Plant Soil* 57:69–83.

Reanney, D. C., W. J. Kelly, and W. P. Roberts. 1982. Genetic interactions among microbial communities. In A. T. Bull and J. H. Slater (eds.), Microbial Interactions and Communities. Vol. 1. Academic Press, London. Pp. 287–322.

Redente, E. F., and F. B. Reeves. 1981. Interactions between vesicular-arbuscular mycorrhiza and *Rhizobium* and their effect on sweetvetch growth. *Soil Sci.* 132:410–415.

Reid, C. P. P. 1974. Assimilation, distribution and root exudation of ^{14}C by ponderosa pine seedlings under induced water stress. *Plant Physiol.* 54:44–49.
Reid, C. P. P., and J. G. Mexal. 1977. Water stress effects on root exudation of lodgepole pine. *Soil Biol. Biochem.* 9:417–422.
Rennie, R. J., and J. B. Thomas. 1987. ^{15}N determined effect of inoculation with N_2 fixing bacteria on nitrogen assimilation in Western Canadian wheats. *Plant Soil* 100:213–223.
Rishbeth, J. 1975. Stump inoculation: A biological control of *Fomes annosus,* In G. W. Bruehl (ed.), Biology and Control of Soil-Borne Plant Pathogens. Am. Phytopathol. Soc., St. Paul, Minnesota, Pp. 158–162.
Rishbeth, J. 1979. Modern aspects of biological control of *Fomes* and *Amillaria. Eur. J. For. Pathol.* 9:331–340.
Rittenhouse, R. L., and M. G. Hale. 1971. Loss of organic compounds from roots. II. Effect of O_2 and CO_2 tension on the release of sugars from peanut roots under axenic conditions. *Plant Soil* 35:311–321.
Riviere, J. 1963. Rhizosphére et croissance due blé. *Ann. Agron.* 14:619–653.
Rolfe, B. G., and J. Shine. 1984. *Rhizobium–Leguminasae* symbiosis: The bacterial point of view. In D. P. S. Verma and T. H. Hohn (eds), Genes Involved in Microbe–Plant Interactions. Springer-Verlag, New York. Pp. 95–128.
Rouatt, J. W., H. Katznelson, and T. M. B. Payne. 1960. Statistical evaluation of the rhizosphere effect. *Soil Sci. Soc. Am. Proc.* 24:271–273.
Rovira, A. D. 1956. Plant root excretions in relation to the rhizosphere effect. I. The nature of root exudate from oats and peas. *Plant Soil* 7:178–194.
Rovira, A. D. 1965. Plant root exudates and their influence upon soil microorganisms. In K. F. Baker and W. C. Snyder (eds.), Ecology of Soilborne Plant Pathogens: Prelude to Biological Control. Univ. of California Press, Berkeley. Pp. 170–186.
Rovira, A. D. 1979. Biology of the soil–root interface. In J. L. Harley and R. S. Russell (eds.), The Soil–Root Interface. Academic Press, London. Pp. 145–160.
Rovira, A. D., and R. Campbell. 1975. A scanning electron microscope study of interactions between micro-organisms and *Gaeumannomyces graminis* var. *tritici* (syn. *Ophiobolus graminis*) on wheat roots. *Microb. Ecol.* 3:177–185.
Rovira, A. D., and C. B. Davey. 1974. Biology of the rhizosphere. In E. W. Carson (ed.), The Plant Root and Its Environment. Univ. of Virginia Press, Charlottesville. Pp. 153–204.
Rovira, A. D., R. C. Foster, and J. K. Martin. 1979. Note on terminology: Origin, nature and nomenclature of the organic materials in the rhizosphere. In J. L. Harley and R. Scott Russell (eds.), The Soil–Root Interface. Academic Press, London. Pp. 1–4.
Rovira, A. D., E. I. Newman, H. J. Bowen, and R. Campbell. 1974. Quantitative assessment of the rhizoplane microflora by direct microscopy. *Soil Biol. Biochem.* 6:211–216.
Russell, E. W. 1973. Soil Conditions and Plant Growth. 10th Ed. Longman, London.
Saif, S. R. 1987. Growth responses of tropical forage plant species to vesicular-arbuscular mycorrhizae. I. Growth, mineral uptake and mycorrhizal dependency. *Plant Soil* 97:25–35.
Scher, F. M., and R. Baker. 1982. Effect of *Pseudomonas putida* and a synthetic iron chelator on induction of soil suppressiveness to *Fusarium* wilt pathogens. *Phytopathology* 72:1567.
Scher, F. M., J. W. Kloepper, and C. A. Singleton. 1985. Chemotaxis of fluorescent *Pseudomonas* spp. to soybean seed exudates *in vitro* and in soil. *Can. J. Microbiol.* 31:570–574.
Schlinkert-Miller, M., and I. L. Pepper. 1988a. Survival of a fast growing strain of lupine rhizobia in Sonoran Desert soils. *Soil Biol. Biochem.* 20:323–327.
Schlinkert-Miller, M., and I. L. Pepper. 1988b. Physiological and biochemical characteristics of a fast growing strain of lupin rhizobia isolated from the Sonoran Desert. *Soil Biol. Biochem.* 20:319–322.
Schmidt, E. L. 1979. Initiation of plant rock–microbe interactions. *Annu. Rev. Microbiol.* 33:355.
Schroth, M. N., and J. Hancock. 1982. Disease-suppressive soil and root-colonizing bacteria. *Science* 216:1376.
Shapiro, J. A. 1985. Intercellular communication and genetic change in bacterial populations. In H.

Halvorson, (ed.), Engineered Organisms in the Environment: Scientific Issues. Am. Soc. Microbiol., Washington, D.C. Pp. 63–69.

Shay, F. J., and M. G. Hale. 1973. Effect of low levels of calcium on exudation of sugars and sugar derivatives from intact peanut roots under axenic conditions. *Plant Physiol.* 51:1061–1063.

Shipton, P. J. 1977. Monoculture and soilborne plant pathogens. *Annu. Rev. Phytopathol.* 15:387–407.

Shoushtari, N. H., and I. L. Pepper. 1985a. Mesquite rhizobia isolated from the Sonoran Desert: Competitiveness and survival in soil. *Soil Biol. Biochem.* 17:803–806.

Shoushtari, N. H., and I. L. Pepper. 1985b. Mesquite rhizobia isolated from the Sonora Desert: Physiology and effectiveness. *Soil Biol. Biochem.* 17:797–802.

Slater, J. H. 1985. Gene transfer in the microbial communities. In H. O. Halvorson, D. Pramer, and M. Rogul (eds.), Engineered Organisms in the Environment. Am. Soc. Microbiol., Washington, D.C. Pp. 89–98.

Smith, M. S., and J. M. Tiedje. 1979. The effect of roots on soil denitrification. *Soil Sci. Soc. Am. J.* 43:951–955.

Smith, S. E. 1980. Mycorrhizas of autotrophic higher plants. *Biol. Rev. Cambridge Philos. Soc.* 55:475–510.

Starkey, R. L. 1958. Interrelations between microorganisms and plant roots in the rhizosphere. *Bacteriol. Rev.* 22:154–172.

Stojkovski, S., R. Payne, R. J. Magee, and V. A. Stanisich. 1986. Binding of molybdenum to slime produced by *Pseudomonas aeruginosa* PAOI. *Soil Biol. Biochem.* 18:117–118.

Subba Rao, N. S., K. V. B. R. Tilak, and C. S. Singh. 1985. Synergistic effect of vesicular-arbuscular mycorrhizas and *Azospirillum brasilense* on the growth of barley in pots. *Soil Biol. Biochem.* 17:119–121.

Subba Rao, N. S., K. V. B. R. Tilak, and C. S. Singh. 1986. Dual inoculation with *Rhizobium* sp. and *Glomus fasciculatum* enhances nodulation, yield and nitrogen fixation in chick pea (*Cicer arietinum* Linn.). *Plant Soil* 95:351–359.

Suslow, T. V., and M. N. Schroth. 1982b. Role of deleterious rhizobacteria as minor pathogens in reducing crop growth. *Phytopathology* 72:111–115.

Takai, Y., and T. Kamura. 1966. The mechanism of reduction in waterlogged paddy soil. *Folia Microbiol.* 11:304–313.

Tan, E. L., and M. W. Loutit. 1976. Concentration of molbydenum by extracellular material produced by rhizosphere bacteria. *Soil Biol. Biochem.* 8:461–464.

Teintz, M., M. B. Hossain, C. L. Barnis, J. Leong, and D. van der Helm. 1981. Structure of ferric pseudobactin, a siderophore from a plant growth promoting *Pseudomonas*. *Biochemistry* 20:6446.

Thomashow, L. S., and D. M. Weller. 1988. Role of a phenazine antibiotic from *Pseudomonas fluorescens* in biological control of *Gaeumannomyces graminis* var *tritici*. *J. Bacteriol.* 170:3499–3508.

Thomashow, M. F., J. E. Karlinsey, J. R. Marks, and R. E. Hurlbert. 1987. Identification of a new virulence locus in *Agrobacterium tumefaciens* that affects polysaccharide composition and plant cell attachment. *J. Bacteriol.* 169:3209–3216.

Timmer, L. W., and R. F. Leyden. 1980. The relationship of mycorrhizal infection to phosphorus-induced copper deficiency in sour orange seedlings. *New Phytol.* 85:15–23.

Timonin, M. I. 1946. Microflora of the rhizosphere in relation to the manganese deficiency of oats. *Soil Sci. Soc. Am. Proc.* 11:284–292.

Tinker, P. B. 1975. The soil chemistry of phosphorus and mycorrhizal effects on plant growth. In F. E. Sanders, B. Mosse, and P. B. Tinker (eds.), Endomycorrhizae. Academic Press, London. Pp. 353–372.

Tinker, P. B., and A. Gildon. 1983. Mycorrhizal fungi and ion uptake. In D. A. Robb and W. S. Pierpoint (eds.), Metals and Micronutrients: Uptake and Utilization by Plants. Proc. Phytochem. Soc. Symp. Ser. No. 21. Academic Press, New York. Pp. 241–248.

Trappe, J. M. 1971. Mycorrhiza-forming ascomycetes. In E. Hacskaylo (ed.), Mycorrhizae. *Misc. Publ.—U.S. Dep. Agric.* No. 1189, pp. 19–37.

Trinick, M. J. 1973. Symbiosis between *Rhizobium* and the non-legume, *Parasponia*. *Nature (London)* 244:459–460.

Turco, R. F., T. B. Moorman, and D. F. Bezdicek. 1986. Effectiveness and competitiveness of spontaneous antibiotic-resistant mutants of *Rhizobium leguminosarum* and *Rhizobium japonicum*. *Soil Biol. Biochem.* 18:259–262.

Van Berkum, P., and B. B. Bohlool. 1980. Evaluation of nitrogen fixation by bacteria in association with roots of tropical grasses. *Microbial. Rev.* 44:491–517.

Vancura, V., and F. Kunc. 1977. The effect of streptomycin and actidione on respiration in the rhizosphere soil. *Zentralbl. Bakteriol., Parasitenkd. Infektionskr.* 132:472–478.

Vanderleyden, J., C. Vieille, K. Michiels, G. Matassi, and A. Van Gool. 1986. Cloning of DNA sequences from *Azospirillum brasilense,* homologus to *Rhizobium NOD* genes and *Agrobacterium VIR* genes. In B. Lugtenberg (ed.), Recognition in Microbe–Plant Symbiotic and pathogenic Interactions. Springer-Verlag, Berlin. Pp. 215–218.

Van Elsas, H. J. D., A. F. Dijkstra, J. M. Govaert, and J. A. van Veen. 1986. Survival of *Pseudomonas fluorescens* and *Bacillus subtilis* introduced into two soils of different texture in field microplots. *Microb. Ecol.* 15:193–197.

Verma, D. P. S., and K. Nadler. 1984. Legume–*Rhizobium* symbiosis: The hosts' points of view. In D. P. S. Verma and T. H. Hohn (eds.), Genes Involved in Microbe–Plant Interactions. Springer-Verlag, New York. Pp. 57–93.

Vesper, S. J., and W. D. Bauer. 1986. Role of pili (fimbriae) in attachment of *Bradyrhizobium japonicum* to soybean roots. *Appl. Environ. Microbiol.* 52:124–141.

Vidor, C., and R. H. Miller. 1980. Relative saprophytic competence of *Rhizobium japonicum* strains in soils as determined by the quantitative fluorescent antibody technique (FA). *Soil Boil. Biochem.* 12:483–487.

Vincent, J. M. 1974. Root–nodule symbioses with *Rhizobium.* In A. Quispel (ed.), The Biology of Nitrogen Fixation. North-Holland Publ., Amsterdam, Pp. 265–341.

Waidyanatha, D. S., N. Yogaratnam, and W. A. Ariyaratne. 1979. Mycorrhizal infection on growth and nitrogen fixation of *Pueraria* and *Stylosanthes* and uptake of phosphorus from two rock phosphates. *New Phytol.* 82:147–152.

Walter, M. V., K. Barbour, M. McDowell, and R. J. Seidler. 1987. A method to evaluate survival of genetically engineered bacteria in soil extracts. *Curr. Microbiol.* 15:193–197.

Watrud, L. S., F. J. Periak, M. T. Tran, K. Kuspano, E. J. Mayer, M. A. Miller Wideman, M. G. Obukowicz, D. R. Nelson, J. P. Kreitinger, and R. J. Kaufman. 1985. Cloning of *Bacillus thuringiensis* subsp. *kurstaki* deltaendotoxin gene into *Pseudomonas fluorescens:* Molecular biology and ecology of an engineered microbial pesticide. In H. O. Halvorson, D. Pramer, and M. Rogul (eds.), Engineered Organisms in the Environment: Scientific Issues. Am. Soc. Microbiol. Appl. Environ. Microbiol., Washington, D.C. Pp. 40–46.

Weinberg, S. R., and G. Stotsky. 1972. Conjugation and genetic recombination of *Escherichia coli* in soil. *Soil Biol. Biochem.* 4:171–180.

Weinhold, A. R., J. W. Oswald, T. Bowman, J. Bishop, and D. Wright. 1964. Influence of green manures and crop rotation on common scab of potato.*Am. Potato J.* 41:265–273.

Weller, D. M., and R. J. Cook. 1983. Suppression of take-all of wheat by seed treatments with fluorescent pseudomonads. *Phytopathology* 73:463–469.

Weller, D. M., and R. J. Cook. 1987. Increased growth of wheat by seed treatments with fluorescent pseudomonads, and implications of *Pythium* control. *Can. J. Plant Pathol.* 8:328–334.

Whipps, J. M., and J. M. Lynch. 1986. The influence of the rhizosphere on crop productivity. *Adv. Microb. Ecol.* 9:187–243.

Wiggins, E. A., E. A. Curl, and J. D. Harper. 1979. Effects of soil fertility and cotton rhizosphere on populations of Collembola. *Pedobiologia* 19:75–82.

11

Influence of Root System Morphology and Development on the Need for Fertilizers and the Efficiency of Use

D. ATKINSON

A plant's demand for nutrients is related to its biomass and its rate of demand is related to its growth rate; as production increases, then the demand for nutrients,

Crops as Enhancers of Nutrient Use

both the total amount and the intensity of supply required, increases. The link between nutrient demand and plant biomass can be modified by changes in the partitioning of assimilates to the various parts of the plant, i.e., an increase in the relative biomass of leaves will usually increase nutrient demand while an increase in root will normally reduce relative demand. In recent years there have been major increases in the average yields of most crops, and, in many relative increases in the amount of high nutrient content tissue, such as grain, compared to the lower nutrient content straw. As a consequence the quantities of fertilizers being used have increased greatly (Table I) in the last three decades. In 1956 the average rate of N application to wheat was 61 kg ha^{-1}, while in 1985 applications of 140 kg ha^{-1} were being recommended for moderate status soils. In some areas there are now substantial losses of nitrogen to drainage water, which is an increasing environmental problem. Parsons (1987) showed that over the period 1922 to 1983 the average yield of oats in the United Kingdom increased by a factor of 3.8 (from 1.9 to 7.2 Mg ha^{-1}), nitrogen fertilizer by 5.3 (from 26 to 141 kg ha^{-1}), and nitrate losses in drainage by 5.7 (from 11 to 67 kg ha^{-1}).

There are now many reasons to reduce fertilizer applications; however, this needs to be achieved without an excessive reduction in yields. To examine the practicality of this reduction, the need of crops for fertilizer and measures to increase the efficiency of fertilizer use must be assessed. The fertilizer applications to perennial crops are lower than those given to annual crops (SAC, 1985). For similar soils in Scotland, recommended N fertilizer applications are 40 kg ha^{-1} for raspberry and 140 kg ha^{-1} for winter wheat. This is related to the period over which uptake occurs (and the reuse of stored nutrients). In annual crops the rate of nitrogen uptake at least over short periods may exceed the potential of even the most fertile soil to meet demand. Here additional nutrients are needed if growth is not to be limited. Dyson (1986) found that spring barley could absorb around 80% of its total nitrogen

Table I. Comparison of Nitrogen Fertilizer Use by a Number of Crops over the Period 1956–1985

	Nitrogen (kg N ha^{-1})			
Crop	1956[a]	1961[b]	1973[c]	1984[d]
Wheat	61	73	98	140
Barley	54	54	93	140
Oats	55	40	71	—
Potato	125	125	176	150
Swedes	78	91	76	75
Grass, 1 yr	83	76	114	130

[a]Source: Anonymous, 1975.
[b]Source: Anonymous, 1963.
[c]Source: Anonymous, 1975.
[d]Source: SAC, 1985.

content (170 kg ha^{-1}) during a 1-month period (Figure 1). During this period the soil must have provided nitrogen at an average rate of 4.5 kg N day^{-1}. Uptake by winter wheat (250 kg N ha^{-1}) and swede (200 kg ha^{-1}) occurs over a longer period. The higher suggested application rate for barley compared to swede (SAC, 1985) (Table I) emphasizes the importance of the intensity of supply. The longer the period over which nutrients can be supplied, the higher the probability that much of the nitrogen requirement can be met by soil nitrification. Despite this, the ability of the plant to remove nutrients from soil is important both for the use of "natural" nutrients and for the efficiency with which fertilizers are used. The significance of the plant's ability to remove nutrients from soil compared to other factors, which may limit nutrient supply, is discussed in this chapter.

Most nutrient uptake occurs via the roots, but the ability of the root system to influence the supply of nutrients depends both on the root system and on the supply of available nutrients in the rooting medium. Where nutrient supply is abundant, as in solution culture, then the effects of the root characteristics, other than an adequate length of root, will be minimal. Where nutrient supply from the soil is substantial but either diluted through a large soil volume or irregularly distributed, then root system properties will affect performance. In soil-based systems, the root system is always likely to have some influence on the efficiency of nutrient use. This chapter reviews (a) information on variation in root system morphology and functioning, (b) the consequences of changes in these properties for both uptake and the efficiency of fertilizer use, and (c) the contribution which changes to the root system may make to fertilizer requirements and the efficiency of their use. This chapter is not,

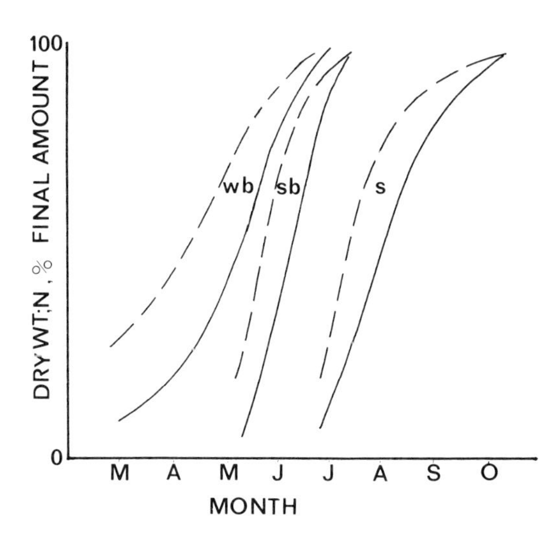

Figure 1. Changes in time in total dry matter (——) and total nitrogen uptake (– – –) expressed as a percentage of final harvest values for winter barley (wb), spring barley (sb), and swedes (s). Source: Dyson, 1986.

however, a review of all the available literature and cited papers are only used to either illustrate general points or as case studies.

I. Crop Nutrient Requirements

A. NUTRIENT DEMAND

The demand for nutrients per unit area varies greatly between crops (Table II). For nitrogen, removal by cereals is lower than by either potato or vegetables and highest in the N-fixing alfalfa and in citrus. Less variation occurs in P removal although relatively similar crops, like potato and tomato, differ greatly. Variation in sulfur removal was small while calcium removal varied by over an order of magnitude. In addition to variation in total nutrient removal, the ratios of nutrients removed also varied. There was major variation in relation to K and Ca with related crops, such as potato and tomato, showing major differences in relative calcium content. This suggests that a range of root adaptations may be needed to optimize the uptake of the nutrients required for satisfactory growth.

In addition to the foregoing there is also variation in the timing and duration of nutrient supply during a single season (Figure 1). Winter wheat, spring barley, and swede absorb at different times. The periodicity of growth means that winter wheat is much more susceptible to low soil temperatures and reduced N-mineralization than swede. Crops with early-season nutrient requirements are most likely to require fertilizers. Crops also vary in the duration of nutrient absorption, which is longer in winter wheat than in spring barley. Any factor which increases the duration of

Table II. Nutrient Removal by a Range of Crops[a]

	Nutrient ($kg\ ha^{-1}$)				
Crop	N	P	K	S	Ca
Barley	68	15	59	10	15
Oats	84	15	94	22	17
Wheat	86	16	48	12	14
Potato	227	9	249	12	59
Turnip	147	12	86	13	56
Alfalfa	249	24	125	18	170
Clover	143	15	91	10	105
Tomato	183	25	284	23	155
Cabbage	182	27	170	—	125
Apple	125	17	136	—	170
Orange	250	24	170	—	227

[a]Source: Derived from Fried and Broeshart, 1967.

nutrient uptake will tend to reduce the required intensity of supply. This effect is clearly shown in woody perennials. Needs are also related to the relative growth of different plant components (Figure 2). In apple, total demand and its duration depend on the balance of carbon investment in root, shoot, leaf, and fruit. Increases in either root or fruit production extend the period of uptake into the autumn. Increases in leaf or shoot growth increase early summer uptake. Choice of an optimum root system, for particular soils or nutrient sources, must therefore take account of both the medium's capacitance and its ability to supply nutrients in a particular month or over periods varying in duration.

B. NUTRIENT USE EFFICIENCY

The efficiency of fertilizer use varies from around 0, total loss, to almost 100% complete recovery and between fertilizers, crops, and soils. It is also influenced by the assessment period. For barley and nitrogen fertilizers, recovery varied with the method of soil management between soil types and spring and winter types (Smith *et al.*, 1984). The effects of tillage varied between years, probably as a result of interactions between rainfall (hence soil water potential) and root distribution. the percentage N derived from N fertilizer increased with increasing rate of application.

Effects due to cultivars were among the largest measured. A study covering a range of barley cultivars (Dick *et al.*, 1985) indicated significant differences in the concentrations of all nutrients (P, K, Na, Ca, Mg, Fe, Mn, Cu, Zn) in both the plant and grain. Differences in the total uptake also occurred and there were significant differences between cultivars with similar grain yields, e.g., Conquest and Gateway.

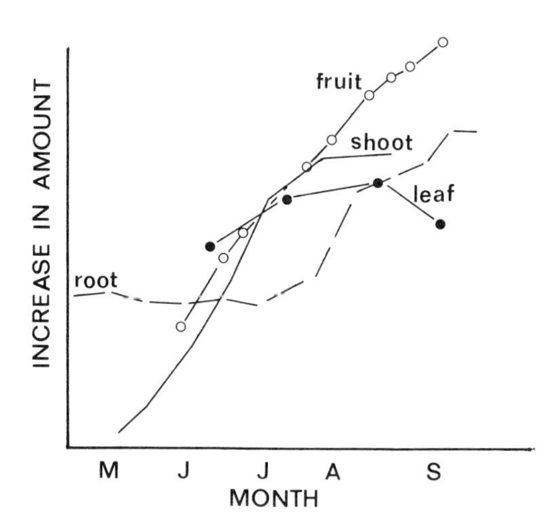

Figure 2. The seasonal periodicity of the growth of root length, extension shoots, leaves, and fruits for young apple trees. Source: Atkinson, 1986a.

These cultivar differences may involve differences in root systems. Chapin (1983) compared the ability of a range of grass and tree species to absorb phosphate from a 10 μmol liter^{-1} solution on a standardized basis (μmol g^{-1} h^{-1}). Uptake varied by almost two orders of magnitude, from 0.2 with *Picea mariana* to 17.6 for *Hordeum leporinum*. The efficiency of nutrient use can vary between both species and genotypes, with root system functioning likely to be at least a component of this variation.

C. NUTRIENT REMOVAL FROM UNFERTILIZED SOIL

Nutrient removal from unfertilized soils varies with both fertility and plant species and many perennial crops regularly grow without fertilizer applications. Atkinson (1978) quantified the uptake of N, P, K, and Ca by apple trees at a range of tree densities. Maximum annual rates of removal were estimated to be as high as 360 and 295 kg ha^{-1} for N and K, respectively. These rates of extraction were related to the extensive nature of the tree root system and the long period over which uptake occurred. Here root system characteristics are clearly important. Similarly the response of forest trees to fertilizers is often limited, even on poor soils (Ballard, 1984), and some growth is always possible in the absence of added nutrients. In the absence of nitrogen, barley produced 55% of the production of plants receiving 120 kg N ha^{-1} (Hansson, 1987). In −N plants the root system represented 23% of total weight, compared with 16% in the +N treatment, and they absorbed 5.8 g N m^{-2}, which was 39% of that with the +N treatment. Interest in more extensive agriculture makes production in the absence of fertilizers of increasing interest. Again changes in rooting strategy may allow further improvements.

II. Root System Characters with Potential for Variation

Plant species vary in their need for nutrients, from both soil and fertilizers, and ability to obtain them from low nutrient content soils. Nutrient supply may be improved by a range of possible changes in the root system. Nutrient uptake depends on (a) the duration of the uptake process, (b) the amount of root present, and (c) the activity of individual root units. Duration is affected by the duration of growth and root survival (Atkinson, 1985). In perennial crops, survival affects active root length as it affects the development of the permanent root system, at least part of which functions in the uptake of both water and nutrients (Atkinson, 1986a). Root length has a major effect on uptake and any given length can be spatially distributed in a range of ways, i.e., distribution with depth and horizontally from the plant stem are both potentially variable. As weight increases so does total root length and the depth of soil through which it is distributed (Greenwood *et al.*,

1982). In annual crops the ratio of root length to plant weight usually follows one of two patterns: it either rises to a maximum and then declines or else declines gradually for much of the period of total production. Although root length and plant weight both increase with time, the relationships between them vary between species but follow a similar trend (Figure 3). Greenwood *et al.* (1982) found that the best fit with available data for a range of vegetable crops was given by

$$\ln L(r) = C_j + b \ln W + M(t) \tag{1}$$

This equation accounted for 90% of observed variation in root length. Here L = root length at time (t), W = plant weight, $M(t)$ = root system losses, C_j is a crop-specific coefficient, and b is a nonspecific coefficient. Values for C_j ranged from 1.5 to 3.7, $b = 1.4$, and $M = -0.035$ day^{-1}.

Any root length can be arranged into different branching structures which will influence the intensity with which soil volumes of a given size within a total volume are exploited. The branching pattern of the root system is one of its more obvious properties. Changes in branching have been modeled by Diggle (1988). The branching of a root system can be simulated if the growth rate (cm day^{-1}) for roots of different orders (the rate decreases from main axes to third and lower orders) and the spacing of the branches of the different orders (spacing increases with root order) are specified.

In addition, there is also scope to vary specific root length (SRL; see Table VIII), which governs the amount of absorbing potential (or soil volume exploited) produced from a given unit of assimilated carbon. This parameter also varies between species and cultivars (Robinson, 1986a; Baligar, 1985a). In many situations, however, nutrient uptake will be determined by the rate of transfer of mineral nutrients to the root surface and so the significance of changes in root length density will vary

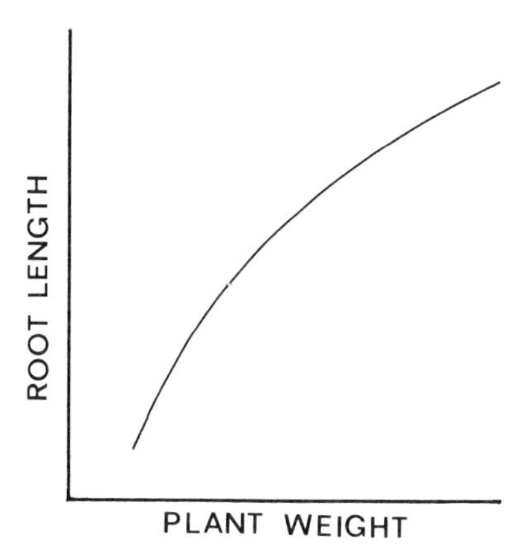

Figure 3. The form of the relationship between the logarithm of root length and the logarithm of plant dry weight based on a number of vegetable species. Source: Greenwood *et al.*, 1982.

between ions (Figure 4). Mobile ions, like nitrate, can be depleted at low rooting densities while for poorly mobile ions, such as phosphate, soil depletion is closely related to root length. In field soils this situation is made more complex by the effects of soil water potential, which normally decreases most with high root densities. This decreases ion mobility, displacing both curves downward, which makes root density important even for mobile ions. In addition, soil drying increases ionic concentration but decreases transfer to the root surface and its ability to absorb. These principles are critical to any appraisal of the effects of root density on potential nutrient uptake.

A range of options is thus available and is summarized in Table III. In the remainder of this chapter, the extent to which variation occurs in root system properties, both within and between species, or can be induced to occur by breeding or means such as the use of growth regulators, as well as the significance of these changes, is discussed. In this consideration the impact of changes in symbiotic relationships, i.e., mycorrhizas, is ignored as these are treated elsewhere (see Chapter 10). Mycorrhizal infection will influence root functioning but mycorrhizas require a root to be present and so most of the considerations outlined here will apply to mycorrhizal and nonmycorrhizal systems.

III. Root System Morphology and Development

A discussion of the potential for changing root systems must be based on an understanding of what is currently known. This is illustrated by case studies of barley and apple roots, which represent a substantial contrast.

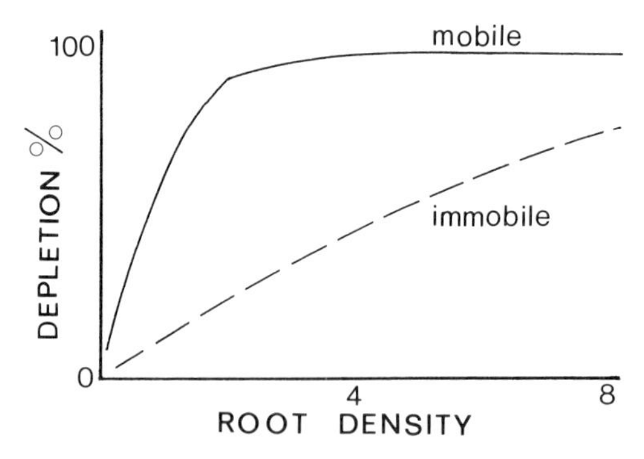

Figure 4. The influence of root length density (in arbitrary units) on the depletion of labile nutrients. (−) Highly mobile ions (e.g., NO_3, SO_4^{2-}), (---) poorly mobile ions (K^+, NH_4, H_2PO_4). Source: Bowen, 1984.

Table III. The Scope for Variation in Plant Root Systems

Strategy	Root Parameter with Potential for Variation
Duration of length	Period of new root production
	Rate of root turnover
	Root survival
	Rate of production of functioning secondarily thickened roots (woody perennial species only)
Root mass	Relative proportion of total assimilate invested in the root system
	Relative investment in roots and mycorrhizas
	Change in specific root length (root diameter)
Root distribution	Variation in root allocation to different depth
	Variation in horizontal root distribution
	Pattern of root branching
Root activity	Ability of a given length of root to absorb nutrients
	Minimum concentration from which roots can absorb nutrient
	Maximum concentration from which roots can absorb
	Ability to produce substances likely to change nutrient supply from the soil
	Speed at which nutrients and water can be removed from root surface
	Nutrient storage potential

A. THE BARLEY ROOT SYSTEM

The barley root system increases to give maximum root length at around 100 days postsowing, 70 days after the start of growth (Welbank *et al.*, 1974). After this, length decreases. The amount of growth was higher in plants receiving 100 kg N ha^{-1} than in those without nitrogen (Figure 5). At their maximum the high-N plants had 66 g of root m^{-2}, a total root length of around 10.2 km m^{-2}, and an *La* (root length per unit soil surface, in cm cm^{-2}) of 102. In this study most of the root system occurred at 0–15 cm, at which depth there were differences between the two nitrogen treatments. The roots had extended to 30 cm depth 60 days postsowing (Figure 6). After 10 July root length decreased presumably because the rate of root loss exceeded new production. Specific root length (SRL) changed with depth and nutrient supply, probably as a result of a changed branching pattern, and was highest near the soil surface and with the zero-nutrient treatment (Table IV).

Rose (1983) modeled changes in the numbers and lengths of different root orders with time. Initially root length was due to the primary axes but with time increasingly lower orders of laterals dominated the available root length (Table V). This dominance of higher-order laterals changes specific root length (Table IV).

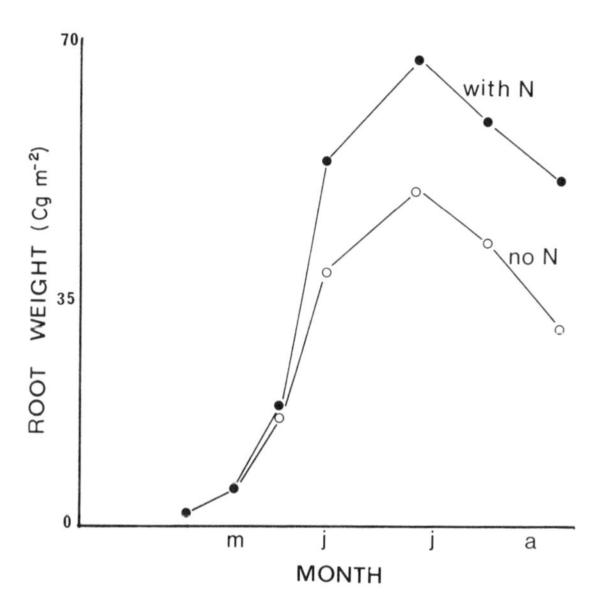

Figure 5. Changes with time in the dry weights of the roots (g m^{-2}) of barley grown with and without nitrogen. Source: Welbank *et al.*, 1974.

Brown *et al.* (1987) found roots to a depth of 120 cm with root length density relatively constant to 45 cm, which contrasts with the data of Welbank *et al.* (1974). Root length density was, however, generally comparable between the two studies. Few data are available on how long individual barley roots survive.

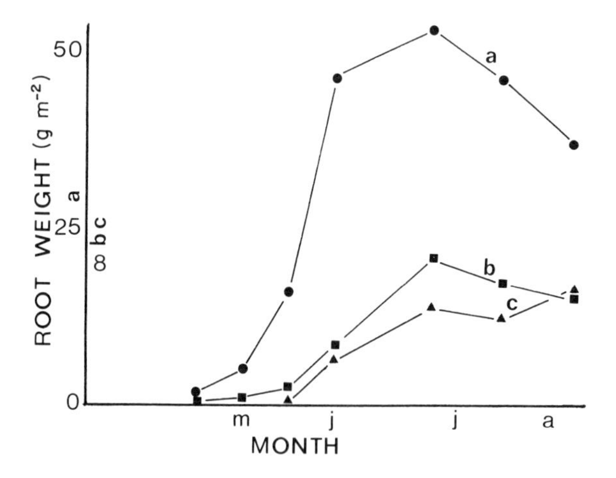

Figure 6. Root dry weight (g m^{-2}) at a range of depths and at intervals over a season. (a) 0–15 cm, (b) 15–30 cm, (c) 30–60 cm depth. Source: Welbank *et al.*, 1974.

Table IV. Variation in Root Weight and Specific Root Length with Soil Depth and Nutrient Treatment[a]

Depth (cm)	Root Weight (g m^{-2}) Nil	Root Weight (g m^{-2}) +K	Specific Root Length (m g^{-1}) Nil	Specific Root Length (m g^{-1}) +K
0–15	13.9	12.9	226	186
15–30	2.6	0.7	197	110
30–60	0.8	0.3	149	93

[a]Source: Welbank *et al.*, 1974.

B. THE APPLE ROOT SYSTEM

The root system of apple is more complex than that of annual crops because of its greater longevity. The system has roots which may vary in age by several years and range from roots with a primary structure, similar to those of barley, to roots with woody and bark tissues. The presence of this range of root types results in a periodicity of white root presence which is very different from that of barley (Figure 7). In the year of planting, root length increased in parallel with shoot growth but in subsequent years root growth occurred at distinct periods, usually late spring and late summer. Normally little growth occurs simultaneously with rapid shoot growth (Figure 7) (Atkinson, 1980, 1983, 1985). During peak growth periods, the total length of functional root is dominated by new white roots. At other times, total length is dominated by roots with secondary tissues.

During the main growing season, new roots took 2–8 weeks to turn brown and shed the tissues external to the pericycle (Atkinson, 1985). This rate of turnover is

Table V. The Effect of Time on the Length of Root (cm) Present as Different Root Orders for Barley[a]

Root Type	Time from Seed Germination (days) 8	13	16	23
Axis	15	22	30	38
First-order lateral	17	89	181	339
Second-order lateral	—	1	32	253

[a]Source: Derived from Rose, 1983.

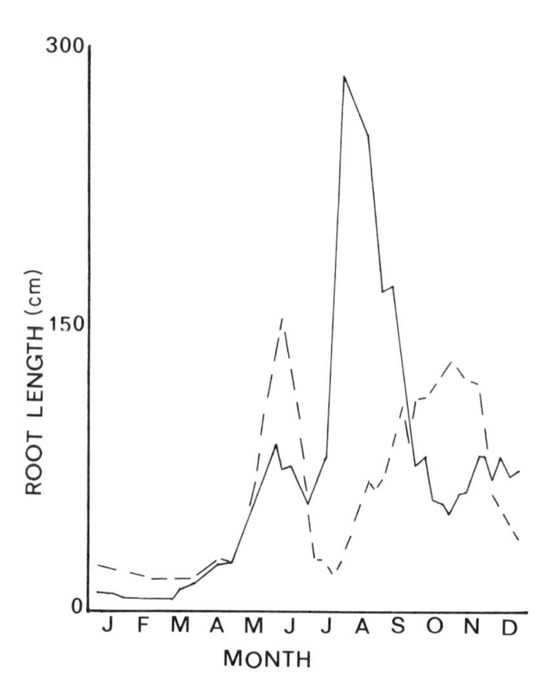

Figure 7. The seasonal periodicity of white apple roots (Worcester/MM104) at two depth (cm per three observation windows). (–) 0.60 cm, (---) 60–120 cm depth. Source: Atkinson, 1985.

quicker than that of apple leaves and strawberry roots (Atkinson, 1985). The combination of new and woody roots (Figure 8) resulted in an adequate root length during a single growing season and from year to year, and also allowed uptake over an extended period, but geared to peaks of nutrient demand. The efficiency of performance is thus influenced by the extent of overlap of the periodicity of new root production and the tree's nutrient demand, the percentage of "new" roots surviving, and the relative activities of the various root types. Considerations which influence the functioning of the cereal root system, i.e., total length and distribution branching, will also be important, but root length density is usually low. In barley, typical values for *La,* root length per unit soil surface, are around 100, although values as high as 1000 have been reported (Newman, 1969). Typical values for apple are in the range 2–10 (Atkinson, 1980). This has major implications for root function.

C. CHANGES IN ROOT SYSTEMS WITH AGE

Attempts to improve root systems must allow for root system changes with age. In annual species, length increases with time and the increasingly thinner roots

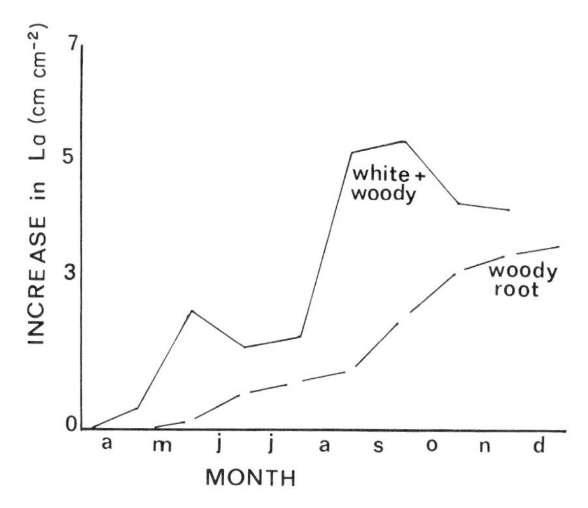

Figure 8. The increase in total root length density (La, cm cm^{-2}) during a single season for an established apple tree (Worcester/MM104). (−) Density due to "white" root plus woody root, (– – –) woody root length. Source: Atkinson, 1986a.

occupy an increasing soil volume. After a time (Figure 5), production decreases and the total length either decreases or remains constant depending on the existing balance of production and loss. Within any given root length, there may simultaneously be depth zones where length is decreasing and others where it is increasing (Welbank *et al.*, 1974). In general there is an increase with time in the depth of soil exploited and the root length density at any given depth (Table VI). The model of the cereal root system developed by Rose (1983) predicts that the root length, for

Table VI. The Length of Roots (km m^{-2}) for Capelle-Desprez (Winter Wheat) on a Number of Dates in a Single Growing Season

	Root Length (km m^{-2})				
Depth	9/12	16/3	14/4	12/5	8/6
0–15	1.75	6.36	9.76	9.63	11.90
0–25	0.23	1.13	2.44	3.13	3.88
0–35	0.06	0.24	0.66	—	1.00
0–45	0.02	0.15	0.31	1.18	0.89
0–55	—	0.06	0.34	—	0.41
0–65	—	0.21	0.23	0.57	0.34
0–75	—	—	0.27	—	0.30
0–100	—	—	0.32	0.42	0.55
Total	2.05	8.12	14.3	14.9	19.3

[a]Source: Welbank *et al.*, 1974.

any given order of laterals, will depend on the properties of the lower orders of laterals and that mean root length will decrease with increasing orders of laterals.

Root diameter usually increases with increasing orders of laterals (Table IV). Branching increases with age and is initially highest in the surface soil. This is important because of the relationships between root diameter and anatomy and functioning. Changes in anatomy have been documented by Miller (1981) for *Zea mays* (Figure 9). The root's external surface and the periphery of the endodermal cells both increased in proportion to root diameter. Roots of larger diameter thus had a proportionally larger external surface area for soil contact and internal (endodermal) surface area to transport material to the stele. The total surface area of the cortical cells, however, increased relative to the increasing root diameter. Relatively larger roots thus have a higher potential for cortical loading of the symplast. This potential should decrease with increasing plant age.

Published descriptions of changes in root system morphology with age for noncultivated species are rare. However, descriptions are available for *Lepidium sativum, Matricaria chamomilla, Valeriana officinalis* (Bockemuhl, 1969), *Taraxacum officinale, Taraxacum laevigatum,* and *Urtica dioica* (Bockemuhl, 1970). *Lepidium sativum* and *V. officinalis* are illustrated in Figure 10. The tap root of *L. sativum* extends rapidly to reach a depth of 110 cm. Subsequent development is by the elongation of first-order laterals and the development of second-order laterals. In contrast, in *V. officinalis* a series of primary axes develop and increases in the depth of rooting are paralleled by increases in horizontal spread. In apple, the amount of new root produced per unit volume seems to be less in later years than during the establishment phase and varies greatly between years (Atkinson, 1985). Hence

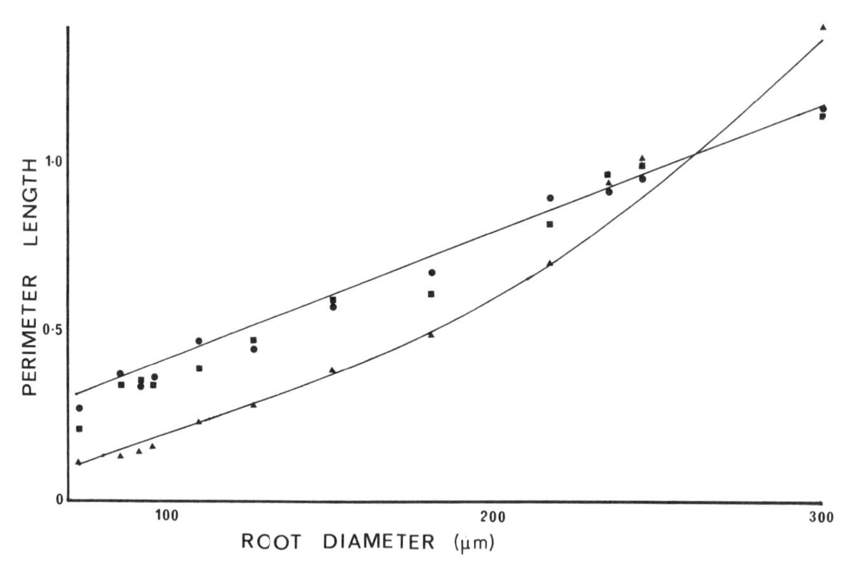

Figure 9. The effect of root diameter (μm) on the external periphery of the root (●) (mm), the total periphery of the endodermis (■) (mm), and the periphery of all the cells in the cortex (▲) (mm × 10^{-1}). Source: Miller, 1981.

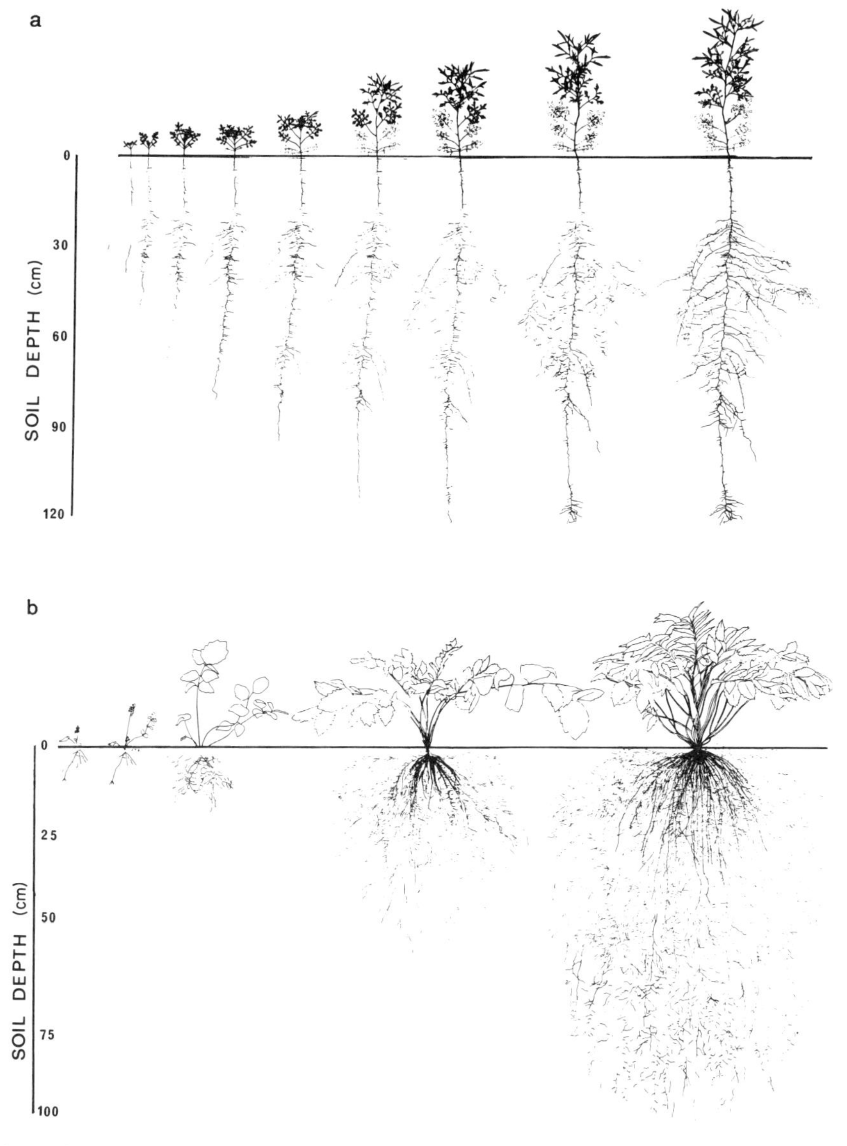

Figure 10. Root system distribution with depth (cm) for developing plants of (a) *Lepidium sativum* and (b) *Valeriana officinalis* with time. Source: Bockemuhl, 1969.

studies of age effects must take account of year-to-year variation and attempts to modify systems by breeding or other methods must allow for developmental changes, otherwise they may undo carefully engineered alterations. In perennial species the continued functioning of older roots gives a clear advantage to a system with a capacity for recycling nutrients.

D. VARIATION IN ROOT SYSTEM MORPHOLOGY BETWEEN SPECIES

Large differences in root morphology also exist between species, e.g., *L. sativum* (Figure 10) differs greatly from *V. officinalis* and both differ from tree root systems (Rogers and Vyvyan, 1934). However, these differences are principally due to differences in branching and, for perennials, root survival. They can therefore be reconciled within a logical framework. The smaller differences which exist between more closely related species are, however, potentially important when explaining the functional significance of particular root properties or when matching a genotype to a soil type. It is therefore useful to examine this range of variation and case studies (see Sections III,A and III,B) illustrate some of the more extreme differences found in cultivated plants. Variation in the root systems of a range of forest trees was documented by Lyr and Hoffmann (1967). The root length produced by 1-year trees varied from 1567 m in *Populus euramericana* to 182 m in *Quercus borealis*. In addition, while 99% of this root length was <1 mm in diameter in *P. euramericana,* only 86% was in this category for *Q. borealis*. Their daily rate of root extension ranged from 6.2 to 2.0 mm day^{-1}, and root depth from 32 cm in *Picea abies* to 210 cm in *Robinia pendula*. In 3-year trees, growth was evenly spread over the period from June to September in *Betula pendula* but concentrated in July in *P. euramericana*. Broeshart and Nethsinghe (1972) compared the uptake of ^{32}P by *Betula* and *Fraxinus*. They found slightly higher uptake by *Betula,* in which the counts were also more variable, suggesting a more uneven distribution of roots within the soil volume. Therefore, within temperate tree species variation occurs in many of the properties previously identified as important (Table III).

Studies have also compared the root systems of annual field crops. Taylor *et al.* (1970) compared the maize and tomato and found that tomato roots penetrated to 200 cm depth more rapidly than maize, which had, however, a higher root length density at most depths. Portas (1973) found that a greater soil volume was used by onion than by lettuce. Root system development in a range of vegetable cultivars was quantified by Greenwood *et al.* (1982), who found the size of the root system to be related to plant size modified by a crop coefficient. Root length per unit area (L) was related to plant size:

$$\ln L = C_j + b_j \ln W - M(t) \tag{2}$$

Here C_j and b_j are crop-dependent coefficients, t is time, W is total plant dry weight (Mg ha^{-1}), and M is a root death coefficient, the same for all crops, which represents a daily root loss of around 3%. Root length density declined with depth for all crops, and in this study varied from 2 m m^{-2} to 15.4 km m^{-2} (La = 154, cm cm^{-2}, comparable to cereals). The soil depth containing most (90%) of the root system and their duration at this depth varied between species. For onion the 90% depth (X_{90}) was 18 cm, and the duration was 80 days, during which L [Eq. (2)] increased from 4.4 to 18.3 cm cm^{-2}. At a root length density of 55 cm cm^{-2}, X_{90} was 30 cm for cauliflower but 90 cm for parsnip. Despite species differences, variation in the depth of rooting was explained simply on the basis of the length of root present. This is a similar situation to that reported for apple (Section I,A), suggesting that promoting growth at depth may simply be a matter of growing more roots.

Closely related species also show differences in the time when they begin to grow. Welbank *et al.* (1974) documented this for a range of cereal species (see Figure 13).

Rapid root growth is often limited by soil physical conditions and so variation in root sensitivity to soil physical condition which varies between species will be important. Dexter (1987) found maximum root growth pressure to vary between 0.24 (*Helianthus annus*) and 1.45 MPa (*Zea mays*). In addition, the penetrometer pressure associated with a halving of the rate of root elongation varied from 0.7 *Gossypium hirumstum*) to 2.03 MPa (*Pisum sativum*). Gooderham (1977) studied pea, ryegrass, and tomato and also found variation between species in their relationship between root growth and soil impedance. Differences between species were, however, smaller than those associated with variation in soil type.

Because the generation of root length in an adequate soil volume is critical to the supply of both water and nutrients, any restriction of the ability of roots to grow in soil will potentially limit this supply.

The significance of variations in root resistance to soil impedance was shown by Ehlers *et al.* (1981). They quantified the root distribution of oats in tilled soil (bulk densities of 1.30 and 1.55 g ml^{-1} at 0–10 and 20–30 cm depth, respectively) and untilled soil (densities of 1.44 and 1.14 g ml^{-1}, respectively). Root length densities were 3.0 and 2.74, 1.49 and 0.81, and 0.72 and 0.62 cm cm^{-3} at 0–10, 10–20, and 20–30 cm depth for tilled and untilled soils, respectively. Similarly, Ellis *et al.* (1977) found that an increase in soil bulk density from 1.32 to 1.50 g ml^{-1} associated with direct drilling rather than ploughing reduced the seminal root growth of young spring barley plants but had little effect in the later stages of growth. Even where the effects of impedance are small, improvement in the ability of seminal roots to penetrate impeded soil zones may be important. Many other soil factors may impede root development, although soil acidity, and the associated increase in soluble aluminum, is among the most common. The effect of aluminum on the root growth of 23 cultivars of red clover was assessed by Baligar *et al.* (1987); 50 μmol Al reduced growth by 16 to 78%. The reduction was poorly related to growth rate but the degree of variation clearly gives scope for selecting genotypes likely to do well in acid soils.

Pasture species also show variation in their root systems (Garwood, 1968). New root production in S23 ryegrass and timothy were higher in late autumn than summer but similar at these times in S24 ryegrass. Evans (1978) found differences in root density and the maximum rooting depth for a range of New Zealand pasture and forage species. Root density, in the surface 20 cm of soil, varied from 418 cm cm^{-2} for *Dactylis glomerata* to 62 cm cm^{-2} for *Triticum aestivum* with *La* (cm cm^{-2}, to 140 cm) being 545 for Cocksfoot and 105 for wheat. The maximum depth of root exploitation was greatest (240 cm) for lucerne.

Large differences clearly exist between closely related species grown in similar conditions. This variation, together with that for cultivars and the effects which can be produced by varying growing conditions, provides the framework against which possible changes (genetic manipulation) must be assessed.

E. VARIATION IN ROOT SYSTEM MORPHOLOGY WITHIN SINGLE SPECIES

While considerable sums of money have been spent on attempts to change the aboveground characteristics of plants, few attempts have been made to alter belowground components. This relates to the inherent difficulties in screening root systems and because links between root morphology and function are still poorly understood. Despite these limitations, a number of studies have documented variation within a single species.

Lupton *et al.* (1974) compared the root development of tall and semidwarf winter wheat cultivars but found little evidence of cultivar differences in root growth, although at depth the roots of semidwarf cultivars were more extensively developed and absorbed more phosphate than did those of taller cultivars. A range of cultivars which varied in shoot weight by 20% varied in root length by only 16%. The semidwarf cultivar TL 365a/34, however, absorbed only 66% of its phosphorus (as ^{32}P) from 20 cm depth compared to 83% in the taller Cappelle-Desprez. The pattern of root system distribution indicated by the distribution of length and the uptake of ^{32}P was relatively similar. Brown *et al.* (1987) compared root system development for two barley cultivars under rain-fed conditions in Syria. Differences between the cultivars 'Arabic' Abiad and Beecher were small and greatest at maturity when La values were 157 and 133 cm cm^{-2}, respectively.

In a study of 25 cultivars of spring barley with relatively similar total production, D. Atkinson (unpublished) found differences in the speed of root penetration to depth, SRL, branching patterns, root density, total root production, and root hair development. Some of these differences are illustrated for four cultivars in Figure 11 (Atkinson, 1987). Estimates of root production at 33 days postplanting varied from 70 to 551 grid intersection units. The differences between cultivars were relatively consistent between dates. Similarly, the depth of the deepest root ranged from 25 to 85 cm at 26 days postplanting. In most cultivars root length increased from planting (late April) to a maximum in early August, after which it decreased. The extent of the decrease varied as did the date when the maximum root length was present (Figure 12). Cultivar differences have also been shown for other cereal crops.

Shukla and Sen (1985) found differences for three rice cultivars. In Pusa 2-21, 93% of the root system was at 0–8 cm depth compared with 57% for IET 1991. In addition, in the shallower cultivar most of the root system was within 12 cm of the plant's center. Similarly in a study of root distribution in soya bean, Raper and Barber (1970) found a mean root length density of 215 mm dm^{-3} in 'Aoda' and 307 in 'Harosoy 63.' In 'Aoda,' root density 75 cm from the plant's center was 35% of that at the center but 64% in 'Harosoy 63.' In contrast, however, in 'Aoda,' root

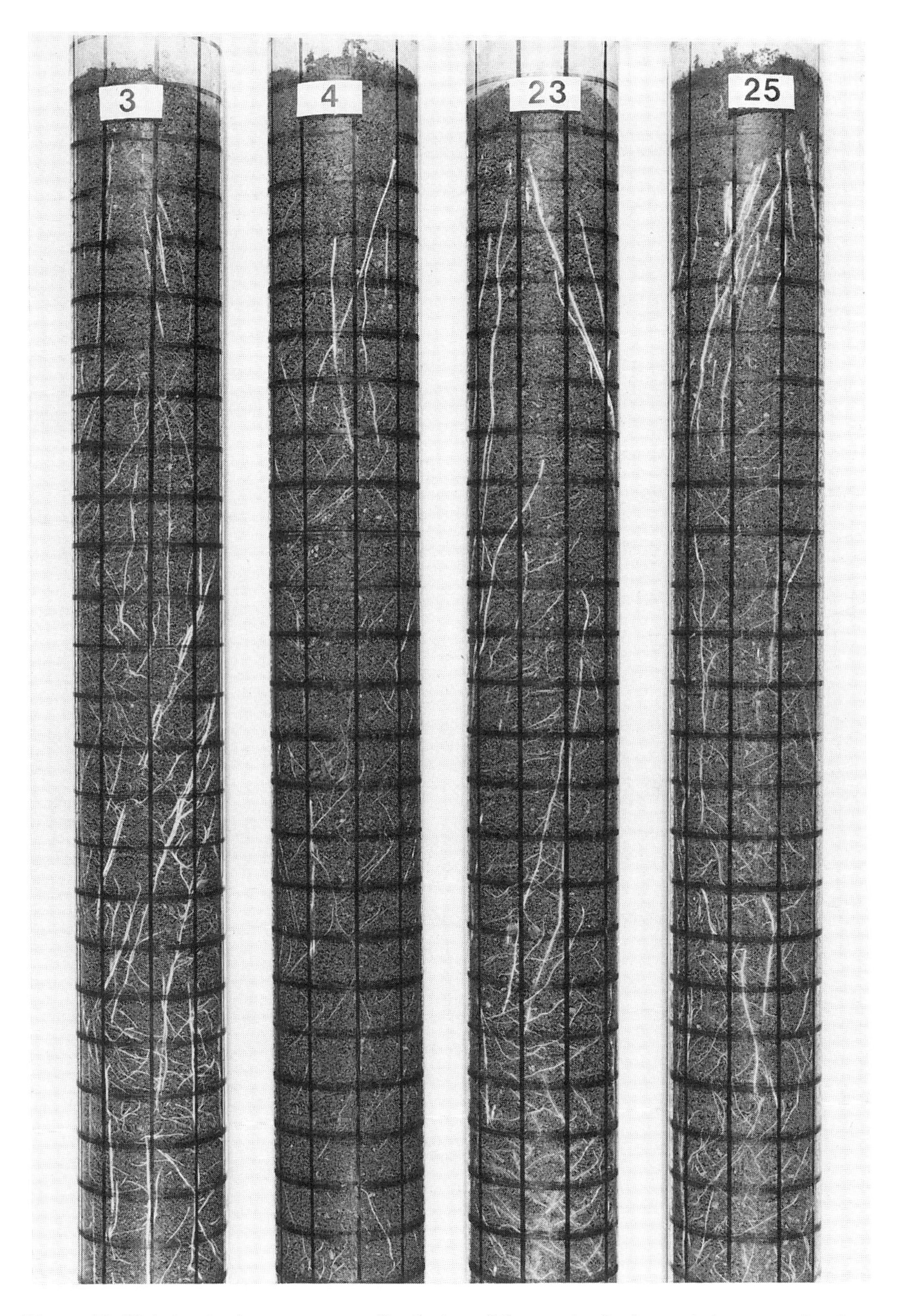

Figure 11. Variation in the root system distribution of four spring barley varieties grown in glass observation tubes: 3 (Plumage), 4 (Keria), 23 (Heriot), 25 (Ayr). The squares have a 2-cm side. Source: Atkinson, 1987.

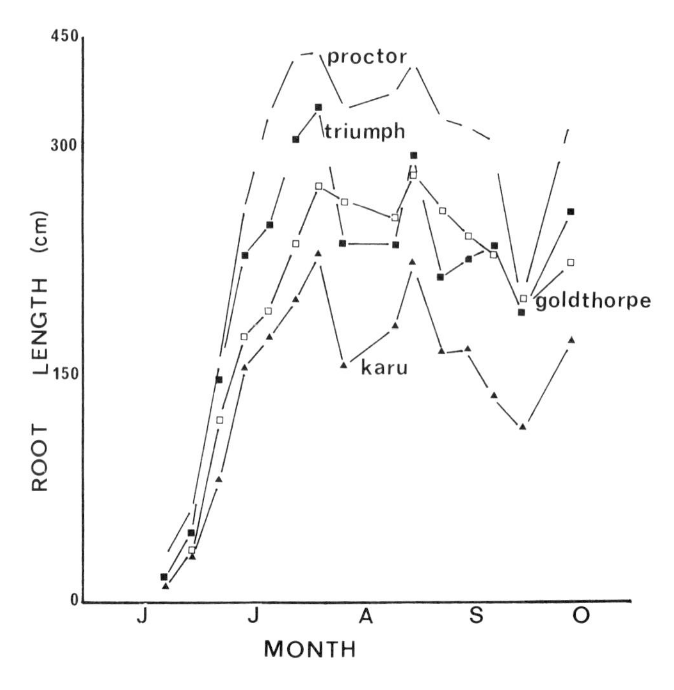

Figure 12. Variation in root length (cm adjacent to tube wall) with time (months) for four barley cultivars grown in glass observation tubes.

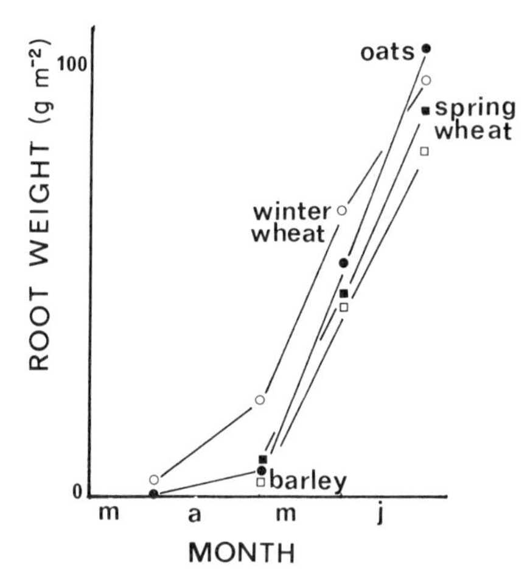

Figure 13. Differences in root weight (g m^{-2}) with time for four cereal species. Source: Welbank *et al.*, 1974.

density at 60–80 cm depth was 60% of that at 0–15 cm depth compared to 35% in 'Harosoy 63.' Similar effects occur in noncultivated species. Bockemuhl (1971) found differences in the number of main vertical roots, the number of orders of roots, the persistence of the root system, and the density of roots for different races of *Senecio vulgaris*.

A number of studies have been made of the root system of fruit trees grown on different rootstocks. Rogers and Vyvyan (1934) compared the root systems of mature trees of Lanes Prince Albert on a range of rootstocks. There were major differences in total weight, spread, and distribution with depth, e.g., on a sandy soil spread varied from 235 to 920 cm. Here interpretations were complicated by large differences in the sizes of the trees and thus in the consequent spacings used. In a study with small rhizatrons, Rogers (1939) found that measured root density varied with tree vigor, from an average of 405 cm for the dwarfing M9 to 998 cm for the more vigorous M16. With M9 more of the new growth was found in the surface 60 cm of soil. The periodicity of new growth was similar for the three rootstocks.

D. Atkinson and R. F. Herbert (unpublished) compared root density for trees on a range of rootstocks. The trees varied in vigor but were planted at a common spacing. Roots were washed from soil monoliths and removed at 50 and 150 cm from the trees (Table VII). The density of fine roots varied with depth, distance from the tree, and rootstock, although there were no significant interactions between those factors. Here observed effects were due to variation in the total amount of root. Despite differences in tree vigor, however, root production by trees on M9 was similar to that with the much more vigorous M26. Clearly there is scope to alter the partitioning of assimilates to the root system.

Table VII. The Effect of Apple Rootstock and Position in the Soil on Fine (<2 mm diameter) Length Root of Density

Rootstock	Distance from Trunk (cm)	Depth (cm) 0–10	10–30	30–50
M9	50	5.2	2.2	0.7
	150	2.5	1.6	0.4
M26	50	5.5	2.0	0.5
	150	1.8	1.4	0.3
M27	50	4.7	2.1	0.5
	150	1.1	1.0	0.3
MM106	50	7.2	3.0	0.9
	150	2.5	2.1	0.5
MM111	50	9.2	3.9	0.9
	150	4.6	2.2	0.7

IV. Variation in Root and Root System Activity

The published literature contains many accounts of studies where excised roots have been assessed for their ability to absorb mineral nutrients. Studies where different cultivars of species have been compared are rare. Nielsen and Schjorring (1983) discussed the kinetic parameters likely to influence the ability of roots to absorb nutrients (Table VIII) and give data for a range of parameters and a number of barley cultivars (Table IX). L varied by 83%, the difference from highest to lowest relative to the lowest, inflow by 75%, K_m by 90%, and C_{min} by 200%. The range of values found was 0.02–0.06 μm for C_{min} and 2.9–5.5 μm for K_m. Baligar (1985b) compared the influx of K into the roots of corn, onion, and wheat and found values of 0.15, 0.12, and 0.08 pmol cm^{-1} sec^{-1}, respectively. For wheat, values for the first 20 days of growth varied from 0.40 to 0.52 pmol cm^{-1} sec^{-1}. For 20- to 30-day-old plants, rates fell to 0.11–0.29, as the length of root per plant increased. Atkinson (1986a) reviewed I_{max} for a range of fruit trees and for P found rates varying from 0.1 to 2.8 pmol cm^{-1} sec^{-1} and for K from 0.3 to 0.9 pmol cm^{-1} sec^{-1}. These rates are much higher than those reported for annual crops. The variation in I_{max} for P (Nielsen and Schjorring, 1983; Atkinson, 1986a) suggests that the ability of roots to absorb nutrients is unlimited under normal soil conditions. Inflow thus represents the balance between plant demand and available root length. Where demand is high and root length low, it will reflect the mobility of the ion in soil and so will more often be of interest for P than for other nutrients. In perennial species, relative root length is low so inflow rates tend to be high. Inflow is thus not a particularly useful selection character. For apple, Bhat (1983) assessed the rela-

Table VIII. Kinetic Parameters Which Are of Importance in Relation to the Efficiency of Nutrient Uptake from the Soil[a]

Parameter	Unit	Common Symbol
Root length per unit mass (root + shoot) of plant	cm g^{-1}	L
Maximal mean net influx per unit length of root	pmol cm^{-1} sec^{-1}	I_{max}
Michaelis–Menten factor for nutrient uptake	μmol	K_m
Minimum concentration in solution at which net influx is zero	μmol	C_{min}
Ion transport to shoot from root	N mol g^{-1} sec^{1}	TS
Root uptake coefficient $I/2\pi r_o C$	cm sec^{-1}	α
Specific root length	m g^{-1}	SRL
Root diameter	mm	r_0
Ion concentration in solution	mol cm^{-3}	C

[a]Source: Nielsen and Schjorring, 1983.

Table IX. Values for a Range of Kinetic Parameters for a Number of Barley Varieties[a]

Variety	Root Length ($m\ g^{-1}$)	Inflow I_{max} (pmol P $cm^{-1}\ sec^{-1}$)	Uptake Capacity K_m (μmol P)	Minimum Effective Concentration C_{min} (μmol P)
Salka	65	0.08	2.9	0.02
Lofa	77	0.08	4.1	0.04
Rupal	46	0.10	3.6	0.04
Nuremberg	68	0.11	3.6	0.06
Mona	42	0.14	5.5	0.05
Zita	57	0.12	4.7	0.03

[a]Source: Nielsen and Schjorring, 1983.

tionship between P inflow and concentration in solution. He found that the P concentration below which P uptake was negligible (C_{min}, Table VIII) was 0.25–0.50 μmol liter^{-1}. P inflow increased linearly to 10 μmol liter^{-1}. These maximum values were high and may represent a useful selection criterion, especially for NO_3, where the range was 1–20 μmol liter^{-1}.

Inflow rates in alfalfa, onion, and corn have been compared in a number of studies (e.g., Baligar, 1985a, 1986, 1987). Rates of influx for Ca, Mg, and Na in maize were lower than in onion, while in contrast the influx of P was slightly higher in maize. K_m values for Ca, Mg, and Na uptake by corn were higher than in onion, suggesting that in maize roots the carriers had a lower affinity for these ions. For Ca, values for maize and onion were, respectively, 0.11 and 0.71 pmol cm^{-1} sec^{-1} for I_{max}, 9.2 and 3.2 μmol for K_m, and 2.0 and 2.0 μmol for C_{min}. For phosphorus the comparable values were 0.75 and 0.60 pmol cm^{-1} sec^{-1}, 2.0 and 2.9 μmol, and 0.1 and 3.0 μmol, respectively (Baligar, 1985a). These values for I_{max} and C_{min} are high compared with those reported (Table IX) for barley. The roots of alfalfa and maize have been compared in relation to their ability to absorb and transport P (Baligar, 1987). With a pH of 5.5 in the medium and a P concentration of 100 μmol, I_{max} was 0.03 and 0.11 for alfalfa and maize, respectively, α was 11.59 and 14.54 cm sec^{-1} × 10^{-6}, while TS was 0.44 and 0.53 nmol g^{-1} sec^{-1}; hence in these species these metabolic parameters are relatively constant. Calculations for the woody (secondarily thickened) and white roots described by Atkinson and Wilson (1979) gave α values of 28.4 and 53.2 cm sec^{-1} × 10^{-6}, higher than the capacity in maize and alfalfa roots (Baligar, 1987). The range of values measured, however, suggests scope for improving these abilities. As with inflow though, it is not clear to what extent α and TS vary as a direct consequence of nutrient demand and root length, which vary greatly in the above species. Consideration of kinetic properties suggests that the scope for improvement, i.e., to increase the ability of a given length of root to absorb, is more limited than the potential for changing root length or distribution. Many derived parameters are directly dependent on root length, although in some situations, and for nutrients like P, uptake may be severely limited by the availability of nutrients at the root surface. This is dominated by diffusion and mass flow rather than the kinetic properties of the root.

V. The Effect on Functioning of Root System Morphology

Despite many experiments where both root length and nutrient uptake have been measured, no clear consensus exists on the functional significance of a range of root system morphological (Table III) and kinetic parameters (Table VIII). Information on the functional significance of these properties is an essential prerequisite to improving performance by breeding for root parameters. Studies in which species or cultivars with contrasting root systems have been compared have often been confounded by the presence of other variables, e.g., that of Rogers and Vyvyan (1934), where the cultivars were planted according to their perceived space requirements. Assessments must be made under realistic conditions, i.e., in rootstock trials the use of unrealistic uniform spacings for all cultivars would not be helpful. Comparisons of plants varying greatly in size or growth stage can also be difficult to interpret. Experiments are needed where treatments have changed the root system and where the impact of the changes have been quantified.

A. EFFECTS OF CHANGES IN ROOT DIAMETER

Root diameter is important because it defines the maximum volume of soil which can be exploited with a given amount of photosynthate. It varies between species and cultivars and changes as plants age (Welbank *et al.*, 1974; Atkinson, 1985). It also affects the relationship between the root surface in contact with soil and the relative amount of cortical cell surface available to load the symplast (Figure 9). Root diameter varies greatly between monocotyledonous and dicotyledonous species (see Table XII). Why dicots appear to invest more carbon in a given length of root than monocots is unclear, but may relate to a need for a greater surface for symplastic loading and a greater need for basic cations (Table II). SRL varies between closely related species (Table X) and was found by Fitter (1985) to be greatest under conditions of poor nutrient supply, i.e., it increased by 19–40% with low P supply. In apple trees grown in competition with grass, and which were thus N deficient, the proportion of new root present as fine laterals increased compared to trees grown with a better nutrient supply (Atkinson, 1983). Welbank *et al.* (1974) also found SRL to be decreased by nutrient additions and reported values of 226, 160, and 186 m g^{-1} for control, +N, and +K treatments. In addition, SRL in winter wheat decreased with increasing depth, i.e., 199 m g^{-1} at 0–15 cm and 62 m g^{-1} at 55–65 cm depth. In a second year on a similar date, comparable values were 262 and 69 m g^{-1}. In contrast, Drew and Saker (1980) found, also with winter

Table X. Specific Root Length (m g^{-1} root) for a Range of Plant Species

Species	Specific Root Length	Reference
Malus pumila	5	Atkinson (1985)
Fragaria × *ananassa*	77	Atkinson (1985)
Phleum pratense	118	Atkinson (1985)
Triticum aestivum	69–151	Rovira (1979)
Poa annua	200–600	Fitter (1985)
Anthoxanthum odoratum	200–400	Fitter (1985)
Holcus lanatus	350–500	Fitter (1985)
Hordeum vulgare	200–350	Fitter (1985)
Deschampsia caespitosa	200–350	Fitter (1985)
Dactylis glomerata	100–400	Fitter (1985)
Phleum pratense	150–550	Fitter (1985)
Briza media	200–700	Fitter (1985)
Thyridolepis mitchelliana	162–192	Christie and Moorby (1975)
Astrebla elymoides	160–205	Christie and Moorby (1975)
Cenchrus ciliaris	136–278	Christie and Moorby (1975)
Lolium perenne	130–750	Robinson and Rorison (1983a,b)
Malus pumila	23–32	Atkinson and Chauhan (1987)
Prunus sp.	37–41	Atkinson and Chauhan (1987)
Triticum aestivum	125–211	Baligar (1985a)
Zea mays	56–68	Baligar (1985a)
Allium cepa	97–98	Baligar (1985a)
Zea mays	150–327	Baligar and Barber (1979)
Medicago sativa	442–529	Baligar (1987)
Zea mays	223–271	Baligar (1987)
Hordeum vulgare	93–226	Welbank *et al.* (1974)
Triticum aestivum	95–175	Welbank *et al.* (1974)
Avena sativa	90–174	Welbank *et al.* (1974)
Hordeum vulgare	71–224	Welbank *et al.* (1974)

wheat, an increase in SRL from 14 at the surface to 500 m g^{-1} at 100 cm depth. Fitter (1985) attributed this effect to a younger age structure with thinner roots at increasing depth.

The consequences of particular values of r_0 (Table VIII) for the inflow of nutrients have been discussed by Robinson (1986a,b). Whether inflow was supplied by mass flow + diffusion or diffusion alone had little influence on the relationship between inflow and root radius for NO_3, K, and NH_4. For H_2PO_4, inflow was always low when supply was by diffusion. For supply by mass flow + diffusion, inflow increased from close to zero for roots with a radius of 0.05 mm to 0.4 p mol cm^{-1} sec^{-1} where the radius was 1 mm. Root radius can thus have an important effect on inflow for phosphate. In addition, the optimum relationship between relative growth rate and SRL can be defined for any given inflow rate of nitrate to the root. To maintain a given growth rate, the ratio of root to shoot needs to be 100 times greater for an inflow rate

Table XI. The Relative Efficiencies of a Range of Possible Adjustments in Partitioning of Resources into and within the Root System[a]

Responses Due to an Increase to	Relative Efficiency	Relative Growth Rate (day^{-1})
Maximal L	0.97	0.07
Maximal K	0.63	0.04
Maximal K and L	0.63	0.16
Balanced K and L	0.63	0.11
Maximum increase in L with a decrease to minimum K	1.14	0.03

[a]Source: Redrawn from Robinson, 1986a.

of 0.1 pmol cm^{-1} sec^{-1} than for one of 10 pmol cm^{-1} sec^{-1} (Robinson, 1986a). With any combination of SRL and root : shoot ratio, however, there is a decrease outside certain limits, i.e., a plant with an SRL of 200 m g^{-1}, a root : shoot ratio of 0.3, and an inflow of 10 pmol cm^{-1} sec^{-1} would have a maximum relative growth rate of 0.37 day^{-1}. Where N inflow was limited to 0.1 pmol cm^{-1} sec^{-1} and with other parameters the same, maximum possible relative growth rate would be 0.019 day^{-1}.

Against this background, the extent to which SRL can compensate for a reduced N inflow rate is limited, but balanced increases in both root : shoot ratio and SRL always give the largest increases in relative growth rate. Increases in root : shoot ratio alone have less effect on growth rate than increases in SRL. The extent to which adjustments in growth can occur just by adjusting the partitioning of resources within the root system, i.e., changes in SRL, is limited. An increase in the mean inflow rate can only occur if the root system gains access to additional areas with a high concentration of nitrogen and this requires changes in root morphology. The relative consequences of the above options are detailed in Table XI, where relative efficiency is calculated as the proportional increase in relative growth rate for a proportional increase in specific root length and root : shoot ratio using the relationship $[Rx/Ra]/[LxKx/LaKa]$, where R is the relative growth rate (day^{-1}), L is the specific root length (m g^{-1}), and K is the root : shoot mass ratio. a is the maximum relative growth rate possible for an inflow rate of 0.1 pmol cm^{-1} sec^{-1}. Values of L and K are as previously described and x is the variable response being tested. This type of analysis can predict the conditions under which a decrease in root diameter and a change in relative partitioning to the root system will affect nutrient supply.

B. EFFECTS OF ROOT LENGTH AND ROOT LENGTH DENSITY

A change in the amount of root possessed by a plant is clearly one of the major options in attempting to improve nutrient supply. It is thus important to assess the relationship between changes in root length and functional properties. Baligar (1985b) assessed the ability of soil-grown wheat, maize, and onion plants to absorb potassium. For wheat, variation in root growth was achieved by varying the planting density, from 1 to 16 plants in a 2-liter pot. The uptake of potassium per pot was closely related to root length, although the relationship changed with age and was different in maize compared to wheat (Figure 14). Effects of spacing on root growth and nutrient uptake were assessed for apple in field studies by Atkinson (1978). He found that for trees at interplant distances of 2.4, 1.2, 0.6, and 0.3 m, root densities were 51, 111, 322, and 933 g m^{-2}, respectively. Over a 5-year period, these densities resulted in the uptake of 1.0, 2.3, 6.3, and 12.7 g K m^{-2}, respectively. For the species used in his study, Baligar (1985b) calculated the percentage of the supply contributed by diffusion and mass flow. For wheat he found that the proportion of potassium supplied by diffusion increased from 1 to 62% as the planting density increased. Nielsen and Schjorring (1983) found that variation in root production between barley cultivars corresponded to P uptake. The ability to cultivars to absorb P in solution culture was related to that under field conditions. Schenk and Barber (1980) established relationships between the root density of a range of maize

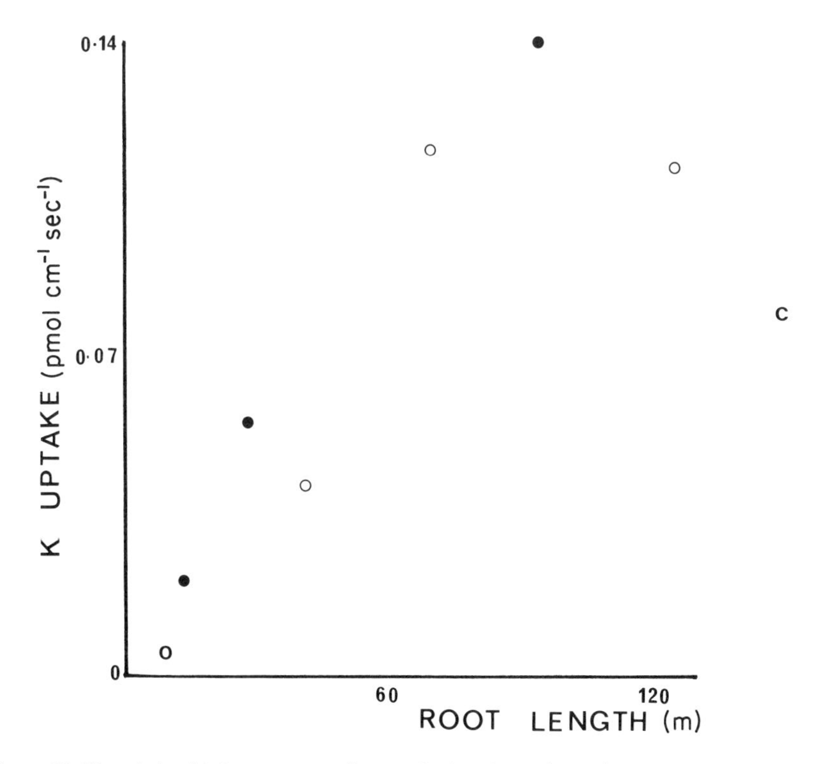

Figure 14. The relationship between potassium uptake (pmol cm^{-1} sec^{-1}) and root length (m) for corn (c), wheat, and onion (O) plants. Solid circles denote 20-day and open circles denote 30-day wheat plants. Source: Baligar, 1985b.

genotypes and their uptake of K and P; an increase in root density from 1 to 4 cm cm^{-3} was associated with an increase in K uptake from 40 to 200 m mol $plant^{-1}$.

Several studies have attempted to relate root density to functioning under field conditions. Mengel (1983) compared root growth and morphology in *Lolium perenne* and *Trifolium pratence* and found that *L. perenne* could use nonexchangeable K^+ which was unavailable to *T. pratence*. He attributed this to an increased root length, which was six times higher in the grass. The relationship between K uptake and root length was steeper with clover than for the grass, while K inflow was three times higher. As a result, clover required a concentration in the soil solution four to five times higher than that needed by the grass. The higher root length density allowed the use of lower concentrations of available K compared to the species with a lower root length.

Evans (1978) compared the root density and distribution of a range of pasture species and related these attributes to their ability to remove water from the soil. Root density at 20–40 cm depth and the maximum soil water potential at this depth were related across thc range of species, as were root density at a range of depths and the associated maximum soil water potential. Castle and Krezdorn (1977) compared root density and water use at a range of soil depths for a variety of citrus rootstocks. They found that water depletion was correlated with the weight of fine roots both within the rooting volume of a single tree and between rootstocks.

Several studies of soil cultivation have involved comparisons of its effects on root development and the consequences of changes for growth and nutrient uptake. Ehlers *et al.* (1981) found that in a tilled soil, a plough layer between 20 and 30 cm depth induced a higher rooting density at 10–20 cm (1.49 compared 0.81 cm cm^{-3}) but a restricted density in the deeper soil layers (0.42 compared to 0.72 cm cm^{-3} at 50 cm depth). These changes corresponded to an increase in water uptake at 10–20 cm depth (45 compared to 31 mm from a 100-mm soil layer) but reduced water use at depth (28 compared to 33 mm from the 200-mm soil layer at 40–60 cm). Water uptake rates were closely related to root length density, i.e., at a water potential of −50 KPa, a density of 2 cm cm^{-3} gave a water uptake rate of 17.5 mm^3 H_2O cm^{-3} soil day^{-1}, a density of 1 cm cm^{-3} gave a rate of 12.7 mm^3 H_2O cm^{-3} day^{-1}, and a density of 0.5 cm cm^{-3} gave a rate of 8.1 mm^3 H_2O cm^{-3} day^{-1}. Similarly, Willatt and Olsson (1982), who assessed effects of irrigation in soya bean, found that at depth the root density limited water uptake, even at high soil water potentials. A root length density of 5 mm cm^{-3} corresponded to a water withdrawal of 0.016 day^{-1} and a density of 1.8 mm cm^{-3} a withdrawal of 0.005 day^{-1}.

The problems involved in making nondestructive measurements of root length have resulted in the development of indirect methods such as the use of radioisotopes. The activity of ^{32}P in the plant, corrected for soil availability, is assumed to be related to root length. In a number of studies (e.g., Kafkafi *et al.*, 1972; Atkinson, 1974; Ellis *et al.*, 1977; Patel and Kabaara, 1975) the relationships between ^{32}P uptake and root length density have been good. For example, in apple (Figure 15), new root production during a season was correlated with the uptake of ^{32}P (Atkinson, 1974). In other studies the relationships have not been as good. Rennie and Halstead (1972) found differences between ^{32}P uptake and the distribution of

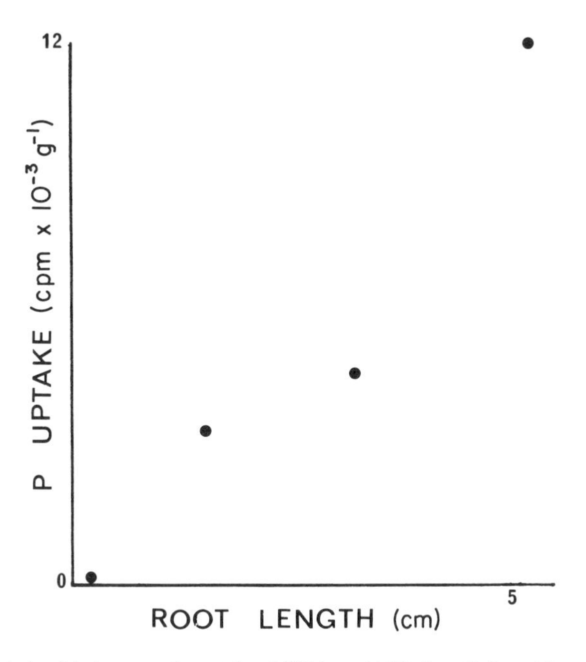

Figure 15. The relationship between the uptake of ^{32}P (cpm × 10^{-3} g^{-1}) from 15 cm depth and white root length (cm per window) as seen in the windows of a root observation laboratory for 2-yr Cox/M9 apple trees. Source: Atkinson, 1974.

root weight, i.e., 80% of "root activity" was found at 0–20 cm depth but most root weight at 30–60 cm depth.

In some crops root pruning is used to stimulate root growth. The effect of this technique has recently been reviewed by Geisler and Ferree (1985), who found its effects to be variable with, in some cases, a decline in nutrient uptake immediately after pruning. However, as the root system regenerates, increases in length (containing more young roots) yield uptake increases. Effects of root pruning will thus depend on the period of assessment. Asamoah and Atkinson (1985) found that a 50% reduction in a cherry root system reduced shoot growth by 40% and water use by 18%.

A wide range of different experiments have thus shown that alterations in root length change the system's ability to function in nutrient uptake. The exact consequences of any given root density, however, will vary between soil types and with plant demand. The precise relationships between field conditions and root length are an important area for future experimentation.

C. EFFECTS OF ROOT BRANCHING AND DISTRIBUTION

Root branching allows the plant to distribute a given root length throughout the soil in a range of different ways. It is the link between length and both the volume and intensity with which a soil is exploited. Fitter (1982, 1985) assessed the effect of root branching on the functioning of the root system. In a study of the effects of fertility on *Poa annua* and *Rumex crispus* he found that first-order roots (the youngest roots in a system) were unresponsive to fertility and root system differences were due to changes in second-order roots (those at the junction of two first-order roots). Root system topology influences both the exploitation and transport characteristics. Systems which are little branched and with few "limbs" along the longest path length in the system seem well adapted to exploit restricted soil volumes, as in soils with high nutrient availability, but are ill-adapted to spatially heterogeneous soils. In a study of the morphology of the barley root system, Hackett (1969) found that although age, cultivar, and nutrition affected root systems properties such as number, length, surface area, and volume, relatively little variation was found between the different dimensions. This seemed to be due mainly to the preponderance of primary laterals with similar root diameters (Hackett, 1971).

Changes in soil cultivation systems have often been reported to result in changes in root distribution and have, in turn, been associated with differences in grain yield or plant nutrition. The differences in root distributions result from alterations in branching pattern within the root system. Boone and Veen (1982) found that a high soil resistance resulted in roots accumulating in the upper soil layers. An increase in resistance measured with a core penetrometer, from 0.9 to 1.6 MPa, reduced root elongation by 50% and changed the direction of root growth. The weight of lateral roots per gram of root mass was unaffected by mechanical resistance although the length of laterals was decreased. The more even distribution of roots with depth and the longer, thinner root system produced in the less compact soil increased nutrient uptake, especially P and K, and as a consequence also increased growth.

Madsen (1985) assessed the effect of soil type on the distribution of spring barley roots. He found similar root densities in the upper part of the soil profile but a decrease in root density with depth which was highest on sandy soils, i.e., a mean maximum root depth of 70 cm compared to 140 cm in loamy soils. Shierlaw and Alston (1984) also found that soil compaction affected plant weight and changed root distribution. In this study using plants grown in pots with compacted central soil zones, ryegrass absorbed twice the amount of phosphorus per unit root length in uncompacted soil compared to maize. P uptake was related to the length of root in that layer, although P uptake per unit root length was higher for compacted soil, especially with ryegrass. In irrigated maize, Aina and Fapohunda (1986) found that differences existed between root length density, specific root length, and root distribution with depth, which influenced both grain yield and water extraction. For the maize cultivars studied, yields were 6.9 Mg ha^{-1} for 'TZESR-W,' 4.2 Mg ha^{-1} for 'TZSR-W,' and 3.7 Mg ha^{-1} for 'FARZ-7.' Associated root length densities were 2.56, 1.88, and 1.70 cm cm^{-3}, specific root lengths were 2.64, 1.93, and 1.62 cm mg^{-1}, and water uptake was 4.2, 3.0, and 2.8 mm day^{-1}, respectively. The improved yield of 'TZESR-W' was attributed to its deeper root system, although grain yield was generally related to root length density. In addition to the effects of total length, spatial distribution in the soil volume will also have major effects. This will be most obvious where the distribution of available plant nutrients is not uniform.

VI. Root System Activity and Functioning

Nutrient uptake must depend on the physiological properties of roots as well as the systems morphology and the amount of root present; e.g., uptake mechanisms which allow the root to remove ions rapidly from the soil solution will be important, at least in the short term. Over longer time periods and for immobile ions, e.g., PO_4, mechanisms which permit the rapid depletion of ions in the soil solution, except where the rate of recharge is high or where roots are in competition, will only advance the time when uptake will depend on movement to the root surface. The ability of roots to influence their own supply of nutrients, the movement of ions from the solid to solution phase, and the environment in which they function are thus important. Moorby *et al.* (1988) assessed the effect of P nutrition on proton efflux from rape roots. They found that increased H^+ efflux began in the terminal portions of the roots within 3 days of the plants becoming P deficient. With increasing time and severity of P deficiency, the length of root producing acid and the amounts of H^+ released increased; the amount of HCO_3 produced and NO_3 uptake decreased. In this study the maximum depression in pH recorded was 0.68 pH unit at 0.2 cm from the root tip. Between 0.5 and 2.0 cm from the tip the fall in pH was 0.3 to 0.45 pH unit. Baligar (1987) assessed the effects of changes in pH on the kinetics of P uptake. In this study, with a low P supply, a reduction in the pH of the medium from 6.5 to 5.5 increased inflow in both alfalfa and maize, in alfalfa from 0.066 to 0.080 pmol cm^{-1} sec^{-1} and in maize from 0.125 to 0.173 pmol cm^{-1} sec^{-1}. In both species it also increased α (cm sec^{-1} $\times$ 10^{-6}) from 9.57 to 11.59 and from 10.5 to 14.54, respectively. K_{min} of corn only increased from 1.17 to 2.24 μmol. In contrast, Baligar (1986) also found that plants of maize and alfalfa increased the pH of their culture solutions, although here plants were grown with a relatively high level of nutrient supply, in contrast to the P-deficient medium used by Moorby *et al.* (1988). Riley and Barber (1971) found that when plants were given NH_4^+ as an N source, the pH fell; but with NO^{3-} as the N source, the pH increased. In the study of Baligar (1986) the N source was NO^{3-}. Root-induced pH changes may also influence the availability of P in the soil, although, as Bache and Crooke (1981) showed, different soils responded differently. In a responsive soil a change in pH from 4.31 to 4.72 corresponded to an increase in available P from 17.4 to 19.5 mg P kg^{-1} soil. A reduction in pH may also result in increased soluble Al, which can be harmful (see Section III,D). Root-induced changes in the pH of the soil immediately adjacent to the root surface are thus likely to affect the supply of nutrients.

The presence of root hairs can also influence the activity of roots. The role of root hairs in *Brassica napus* and *Gossypium hirsutum* was assessed by Misra *et al.*

(1988). Regardless of plant species, P uptake was a function of the soil volume explored by the root hairs. In cotton, similar amounts of P were absorbed within and outside the root hair zone, but in rape, most uptake occurred within the root hair zone.

The ability of roots to supply nutrients to the plant is influenced not only by their ability to absorb nutrients but also their ability to translocate absorbed nutrient within the root system and to the shoot. For Sitka spruce the significance of translocation has been discussed by Coutts and Philipson (1976, 1977) and Philipson and Coutts (1977). Here seedlings were grown with divided root systems and with the two halves of the root system receiving different rates of nutrient supply. Only the side to which nutrients were applied showed increased growth, but nutrient concentrations were increased on both sides of the root system. Root systems thus retain a high degree of plasticity in their responses. This is very important in tree crops, where spatial variation within the root system is usually high, e.g., Reynolds (1970) showed that coefficients of variation between replicate soil cores were as high as 95%. Plasticity is important because many of the treatments applied to tree crops result in a spatially variable water and nutrient distribution within the soil. Levin *et al.* (1979) demonstrated that in trees irrigated so only part of the soil volume is wetted, root density varied greatly within short distances. Large differences in root density associated with the wetted soil areas have also been found by Goode *et al.* (1978).

The supply of nutrients to the plant can thus be influenced by changes in the ability of roots to absorb nutrients or to alter the chemistry of the soil adjacent to the root surface, or by their ability to move nutrients from the root surface to the aboveground parts of the plant.

NUTRIENT INFLOW AND CROP GROWTH

There are many published assessments of nutrient inflow, which is the amount of nutrient absorbed per unit root length per unit time. However, most of these are based on studies in solution culture or in solid media with high rates of nutrient supply. Few studies have been carried out in field soils where nutrient levels vary with depth or over a sufficiently long period to cover the whole production period for a crop. For P, K, and NH_4 transport, rates in the soil are likely to limit inflow. Schenk and Barber (1980) discussed many of these considerations in relation to genotypes of maize. I_{max} for P and K varied between genotypes, e.g., for K from 25 to 42 pmol cm^{-2} sec^{-1}. The genotype with the highest I_{max} for P was, however, not the one with the highest I_{max} for K. For K, the cultivars with the highest and lowest values of inflow also had the highest and lowest values for K_m and C_{min} (i.e., a relatively poor ability to absorb at low soil nutrient levels). For P there was no such correspondence.

In a related field study the cultivar with the highest root surface area per unit of shoot had the best supply of P, although K status was not assessed. Anghinoni *et al.*

(1981) showed that the rate of inflow decreased with time for P and K but not for Ca and Mg. For P, inflow decreased from 0.2 at 5–10 days to 0.03 pmol cm^{-1} sec^{-1} at 36–40 days. Comparable values for K were 0.90 and 0.19 pmol cm^{-1} sec^{-1}. K_m for P changed little for plants at 20 or 40 days, while for K it increased with time. The extent to which inflow is affected by solution concentration also varies for different ions. Baligar (1987) found that at a pH of 5.5, a 10-fold change in P concentration had no effect on inflow. At pH 6.5, inflow decreased from 0.08 to 0.066 pmol cm^{-1} sec^{-1}. Because nutrient inflow is the resultant of a number of processes, it is not a good target for plant breeding and is best optimized by matching root length to nutrient demand.

VII. Prospects for Modifying Plant Root Systems

The root system can thus be modified by breeding or selection, by the use of growth-modifying chemicals, or by cultural practice. The range of changes in root system morphology which are potentially achievable through genetic variation is summarized in Table XII, while the contribution which changes in these properties can make to plant nutrition is summarized in Table XIII. Existing variation within and between species (Table XII) illustrates the scope for changing many of the morphological characteristics of the root system by an amount sufficient to affect functioning. For example, root length density is low in many dicot species and there

Table XII. The Range of Useful Variation in Root System Properties Which Seem to Be Important to Uptake of Nutrients by Field-Grown Crops

Property	Normal Range of Variation	Reference
Root length density	2–1000 (cm cm^{-2})	Atkinson (1980); Newman (1969)
Root depth	40–860 (cm)	Greenwood *et al.* (1982); Atkinson (1980)
Specific root length	5–750 (m g^{-1})	Atkinson (1985); Robinson and Rorison (1983a,b)
Root branching	3.57–11.96 (R_b)	Fitter (1982)
K_m	1–30 (μmol P)	Anghinoni *et al.* (1981)
Inflow P	0.1–2.8 (pmol cm^{-1} sec^{-1})	Atkinson (1986a)
Inflow K	0.3–0.9 (pmol cm^{-1} sec^{-1})	Atkinson (1986a)
Root hairs	0–88 (mm^{-2} root surface)	Shierlaw and Alston (1984)

Table XIII. Possible Consequences of Measured Variation in Root System Properties for Water and Nutrient Uptake

Property	Change in Property	Estimated Consequence for P or K Uptake
Root length density	Increase by 10^3 for a 500-μm-diam. root with 0.275-mm zone of exploitation, which thus increased the exploitable soil volume from 0.067 mm^{-3} cm^{-3} (Lv = 0.1 cm cm^{-3}) to 67 mm^3 cm^{-3} (Lv = 10 cm cm^{-3})	P:access to an additional 45 μg P cm^{-3} soil
	Increase from 2.4 to 5.6 m cm^{-2}	K:possible increased uptake of 1.6 mg cm^{-2}
	Increase from 1 to 3 cm cm^{-3}	K:possible increased uptake of 130 mmol $plant^{-1}$
Root depth	Increase by 10 cm for a single root, diameter + zone exploited as above	P:Access to an additional 0.45 μg P and 20 mm of water
Specific root length	Increase from 100 to 200 m g^{-1} for a root wt. of 1.4 mg cm^{-2}, an increase of La from 14 to 28 cm cm^{-2} in 15 cm of soil depth	K:a potential increase in uptake of 60 mmol $plant^{-1}$ but in perennial species a probable decrease in root survival and so a loss of around 4 cm cm^{-2} yr^{-1} in root length density

is a substantial range of variation in this character, at least 10^3 between species. There is also much variation within species, e.g., from 5.51 to 9.12 km m^{-2} in wheat (Welbank *et al.*, 1974). The depth of root penetration is usually limited by soil characteristics, e.g., impedance at depth, but it can be great. This will increase the volume of soil available to a plant, although decreased nutrient levels at depth and greatly reduced soil temperatures will reduce the significance of increases in rooting volume for nutrient supply. Within a maximum rooting depth the rate or extent of the decrease in root length density, which normally occurs with increasing depth, is variable both with time and between plant species. The consequence of this root length–depth profile will vary with the extent of the variation in the distribution of nutrients down the profile. Its effect will be greatest when the distribution of nutrients with depth is relatively uniform. Even when this is not the case, the access of the plant to an increased water supply may allow sustained growth and more effective use of nutrients previously absorbed.

SRL also varies greatly between species, by an amount in excess of 10^2. An increased SRL allows a larger volume of soil to be exploited with the same mass of root material. The existence of large differences between monocotyledons and dicotyledons and annual and perennial species (Table X) suggests that in some circumstances the low SRL must be of selective value. In woody perennials this is related to root survival (Atkinson, 1985). The potential significance of changes in the physiological indicators discussed, e.g., K_m, will also vary between soils with different transport characteristics and for different nutrient elements. This variation, however, appears to be smaller and perhaps of less significance than that in morphological characters. In addition, I_{max}, a common physiological integrator, is normally a response to the demand of the plant for nutrients and the relationship of this demand to root length. Thus changes in root length density and in root use of soils at depth seem likely to have the greatest effects (Table XIII).

In contrast to genetic variation, the changes which can be induced by cultural or chemical means are limited. As root development is under hormonal control there should be some prospect of modifying root morphology using plant growth regulators. The prospect for doing this has been discussed by Atkinson and Crisp (1983) and Atkinson (1986b). Jackson (1983) reviewed the effect of externally applied growth substances on root growth and morphology. Most of his examples, however, were derived from laboratory studies. Ethylene has been shown, depending on concentration, to either retard or promote root extension. It also influences the development of lower orders of lateral roots and root hairs. Nutrients such as NO_3 can also influence root system morphology. For a range of materials, Jackson (1983) found that root extension increased by 20–147%.

Atkinson and Crisp (1982) assessed the effect of a range of growth regulators on the growth of M25 apple roots and found that a number of growth regulators could increase root weight by more than 20%. New root production was increased by 102% by paclobutrazol. However, not all experiments have given this result (Atkinson, 1986b). Assessments of the effects of the above growth regulator treatments on the uptake of mineral nutrients indicated that both the concentrations of N, P, or K in the plants and the transfer from roots to leaves were affected, e.g., paclobutrazol increased P and decreased K in the plant. These effects probably reflect the measured increases in root mass (increasing the potential for diffusion) and reductions in water movement (reducing mass flow). Root mass in a sward of *Phleum pratense* treated with paclobutrazol showed an increased proportion at 0–7 cm depth and decreased amounts below this depth. These changes were associated with reduced water use. Assessments of the potential for using growth regulators are complicated by the variability of their effects. Paclobutrazol effects (Atkinson, 1986b) seem to be related to its method of application, the plant's developmental stage, and growing conditions. Atkinson and Chauhan (1987) assessed the effect of paclobutrazol on the development of the apple root system under a range of conditions. At 20°C, in contrast to the foregoing, root length was reduced from 136 to 65 m plant^{-1} and specific root length from 32 to 29.5 m g^{-1} in treated plants. These changes were associated with a reduction in the flux of water into the root (Atkinson and Crisp, 1983). Richards and Rowe (1977) found that the cytokinin 1-benzylamino purine (BAP) increased the rate of water uptake per unit length of root. Synthetic growth regulators can therefore modify important root parameters (Table XII), although the magnitude of effects may be less than those which may be achieved by genetic modification. In combination with other methods these materials could be valuable.

Cultural practices, e.g., cultivation, can also affect root properties (Table XII), although under field conditions probably to a smaller extent than is potentially achievable by genetic selection. For example, Ehlers *et al.* (1981) found that tillage compared to no till increased root length density from 2.74 to 3.0 cm cm^{-3}. Major improvements in the efficiency of nutrient use seem likely to require basic changes.

VIII. Conclusions: Possibilities for Improving Nutrient Capture by Modification of the Root System

A plant's demand for nutrients is thus predominantly influenced by the biomass produced. While there is scope for improving the partitioning of resources into the harvested crop and for developing cultivars with lower tissue nutrient concentrations, demands for improved productivity are likely to maintain nutrient need per unit area at least at current levels. This level of nutrient supply can only be met if available nutrients are adequate, the volume of available soil is sufficiently great, and rates of supply are matched to the plant's rate of demand for the whole growing season. Fertilizers are needed when either total nutrient availability or the intensity of supply over a given period of time is inadequate. Where the total nutrient supply is adequate, the main considerations influencing supply are root production, the volume of soil exploited, and the timing of its presence. Root branching and specific root length influence the relationship between total root length and the volume of soil utilized.

The information presented here has shown that all the above parameters vary and that some, e.g., root length density and specific root length, can vary by several orders of magnitude. Within a given species many parameters vary, although usually by a smaller amount than between plants as a whole. The effects of the normal range of field growing conditions, however, seem to result in less variation than that derived from genotypes of given species. Physiological properties seem to vary less than morphological properties. The significance of changes in morphology will vary between the different mineral nutrients and between soils with different transport properties. For example, the mass of root needed to exploit a given soil volume will depend on specific root length and the development of root hairs. Misra *et al.* (1988) showed that root hair length was 0.28 mm in cotton and 0.77 mm in rape, so that a 1-cm length of cotton root would utilize a volume of 19 mm^3 and a similar length of rape root 41 mm^{-3}. Thus for cotton the soil would be completely explored by a root length density of 52 cm cm^{-3} but for rape by a density of only 24 cm cm^{-3}. These values are much higher than are normally found, although values of 50 cm cm^{-3} have been reported and values of 10 cm cm^{-3} are not uncommon. In field soils these values would be modified for some nutrients by the development of mycorrhizal fungi, which would increase the soil volume exploited by a given root length and so reduce the root density needed for complete soil utilization, at least for cotton (rape roots are nonmycorrhizal). Misra *et al.* (1988) also reported that in cotton, substantial amounts of P are removed from outside the volume of the root hairs. For K and NH_4, for which diffusion is appreciable, smaller root length densities would be needed to exploit all available soil K.

The optimum use of soil resources requires plants to be highly plastic in their responses. Jackson (1983) showed that cereals can be highly plastic in their response. For stratified soils, plasticity will be a critical root system characteristic. The variation found in distribution with depth (Table XII) and differences in root density within parts of the root volume of a single individual indicate a potential for high root densities in zones well supplied with nutrients. Plasticity has been given little attention in current studies. The assessments of possible implications which have been made in this paper have all been calculated on the basis of roots alone, but under field conditions most roots of most plants or species will be infected with mycorrhizas, which will greatly extend the effective size of the root system and change its affinity for nutrients. However, all mycorrhizas require the presence of a root and so the considerations advanced in relation to roots will also apply to mycorrhizas and influence their functioning. However, infection varies between roots and species (Atkinson, 1983), which will influence the relative importance of some characteristics and may invalidate physiological measurements made on uninfected roots, e.g., I_{max} may be greatly increased. Where roots are heavily infected with mycorrhizas, then the effect of increasing root length density may be reduced. Where roots are heavily mycorrhizal, increasing length, by increasing specific root length, may have little effect on the volume of soil exploited but may still increase the intensity with which it is colonized.

The discussions in this chapter have been dominated by the functioning of individual plants or of crop plants grown in monoculture. In some agricultural situations, many plants are grown in association with individuals of other species and this is the norm in noncultivated vegetation. Atkinson (1983) showed that when apple trees were grown in association with grass, the form of the root systems, e.g., average root survival and periodicity, was changed. Assessments of the functional consequences of changes in root morphology will also vary depending on whether plants are being grown in monoculture or mixed culture. Where plants grow in mixed culture, rapid uptake may be important to prevent use by another species.

Nutrient applications to crops have been growing and large amounts are being lost to the environment, about which there is increasing public concern. Pollution and the need to control costs argue for more effective fertilizer recovery. The contribution that changes in the root system might make are discussed on the basis of existing variation between root systems and the functional significance of alterations in root morphology and physiology. The significance of potential alterations must be seen in the context of the variation in field soils. It is concluded that alterations in morphological characters, particularly total root length leading to a larger volume of soil being exploited, could increase the efficiency of fertilizer use. It might even allow some reduction in fertilizer rates. Alterations to plant root systems have infrequently been considered by plant breeders; they might with advantage receive emphasis in the future.

References

Aina, P. O., and H. O. Fapohunda. 1986. Root distribution and water uptake pattern of maize cultivars field grown under differential irrigation. *Plant Soil* 94:257–265.

Anghinoni, I., V. C. Baligar, and S. A. Barber. 1981. Growth and uptake rates of P, K, Ca and Mg in wheat. *J. Plant Nutr.* 3:923–933.

Anonymous. 1957. Survey of Fertilizer Practice in Scotland 1956. Rep. No. 3. East Lothian (Lowground), Edinburgh and East of Scotland College of Agriculture and ARC Unit of Statistics, Aberdeen.

Anonymous. 1963. Survey of Fertilizer Practice in Scotland 1961. Rep. No. 33. Mid Lothian and West Lothian (Lowground), Edinburgh and East of Scotland College of Agriculture and ARC Unit of Statistics, Aberdeen.

Anonymous. 1975. Annual Survey of Fertilizer Practice. *Rothamsted Exp. Stn. 1974.*

Asamoah, T. E. O., and D. Atkinson. 1985. The effects of (2*RS*, 3*RS*)-1-(4-chlorophenyl)-4, 4-dimethyl-2-(1*H*-1,2,4 triazol-1-yl)-pentan-3-ol (paclobutrazol: PP 333) and root pruning on the growth, water use and response to drought of colt cherry rootstocks. *Plant Growth Regul.* 3:37–45.

Atkinson, D. 1974. Some observations on the distribution of root activity in apple trees. *Plant Soil* 40:333–342.

Atkinson, D. 1978. The use of soil resources in high density planting systems. *Acta Hortic.* 65:79–90.

Atkinson, D. 1980. The distribution and effectiveness of the roots of tree crops. *Hortic. Rev.* 2:424–490.

Atkinson, D. 1983. The growth, activity and distribution of the fruit tree root system. *Plant Soil* 71:37–48.

Atkinson, D. 1985. Spatial and temporal aspects of root distribution as indicated by the use of a root observation laboratory. In A. H. Fitter, D. Atkinson, D. J. Read, and M. B. Usher (eds.), Ecological Interactions in Soil. Blackwell, Oxford. Pp. 43–66.

Atkinson, D. 1986a. The nutrient requirements of fruit trees: Some current considerations. *Adv. Plant Nutr.* 2:93–128.

Atkinson, D. 1986b. Effects of some plant growth regulators on water use and the uptake of mineral nutrients by tree crops. *Acta Hortic.* 179:395–404.

Atkinson, D. 1987. Variation in root distribution in spring barley. *Rep. Macaulay Inst. Soil Res. 1986* pp. 174–177.

Atkinson, D., and J. S. Chauhan. 1987. The effect of paclobutrazol on the water use of fruit plants at two temperatures. *J. Hortic. Sci.* 62:421–426.

Atkinson, D., and C. M. Crisp. 1982. Prospects for manipulating tree root systems using plant growth regulators: Some preliminary results. *Proc. 1982 British Crop Protection Conf.—Weeds* pp. 593–599.

Atkinson, D., and C. M. Crisp. 1983. The effect of some plant growth regulators and herbicides on root morphology and activity. *Acta Hortic.* 136:21–28.

Atkinson, D., and S. A. Wilson. 1979. The root–soil interface and its significance for fruit tree roots of different ages. In J. L. Harley and R. S. Russell (eds.), The Soil–Root Interface. Academic Press, London. Pp. 259–271.

Bache, B. W., and W. M. Crooke. 1981. Interactions between aluminum, phosphorus and pH in the response of barley to soil acidity. *Plant Soil* 61:365–375.

Baligar, V. C. 1985a. Absorption kinetics of Ca, Mg, Na and P by intact corn and onion roots. *J. Plant Nutr.* 8:543–554.

Baligar, V. C. 1985b. Potassium uptake by plants, as characterized by root density, species and K/Rb ratio. *Plant Soil* 85:43–53.

Baligar, V. C. 1986. Interrelationships between growth and nutrient uptake in alfalfa and corn. *J. Plant Nutr.* 9:1391–1404.

Baligar, V. C. 1987. Phosphorus uptake parameters of alfalfa and corn as influenced by P and pH. *J. Plant. Nutr.* 10:33–46.

Baligar, V. C., and S. A. Barber. 1979. Genotypic differences of corn for ion uptake. *Agron. J.* 71:870–873.
Baligar, V. C., R. J. Wright, T. B. Kinraide, C. D. Foy, and J. H. Elgin. 1987. Aluminum effects on growth, mineral uptake and efficiency ratios in red clover cultivars. *Agron. J.* 79:1038–1044.
Ballard, R. 1984. Fertilization of plantations. In G. D. Bowen and E. K. S. Nambiar (eds.), Nutrition of Plantation Forests. Academic Press, London. Pp. 327–361.
Bhat, K. K. S. 1983. Nutrient inflows into apple roots. *Plant Soil* 71:371–380.
Bockemuhl, J. 1969. Gartenkresse, Kamille, Baldrian Eine neue methode, das wurzelwachstum in erde im verhaltnis zur oberirdischen. Entwicklung der pftanze zu beo bachten. *Elem. Naturwiss.* 11:13–28.
Bockemuhl, J. 1970. Entwicklungsilder zur charakterisierung von lowenzahn und brennessel. *Elem. Naturwiss.* 12:1–14.
Bockemuhl, J. 1971. Beobachtungen am Pflanzenwachstum auf Erden mit Kompostzusatzen aus stadtmull und Klarschlamm. *Elem. Naturwiss.* 15:21–32.
Boone, F. R., and B. W. Veen. 1982. The influence of mechanical resistance and phosphate supply on morphology and function of maize roots. *Neth. J. Agric. Sci.* 30:179–192.
Bowen, G. D. 1984. Tree roots and the use of soil nutrients. In G. D. Bowen and E. K. S. Nambiar (eds.), Nutrition of Plantation Forests. Academic Press, London. Pp. 147–180.
Broeshart, H., and D. A. Nethsinge. 1972. Studies on the pattern of root activity of tree crops using isotope techniques. *Isotopes and Radiation in Soil–Plant Relationships Including Forestry. Proc. Symp. IAEA, Vienna* pp. 453–463.
Brown, S. C., J. D. H. Keatinge, P. J. Gregory, and P. J. M. Cooper. 1987. Effects of fertilizer, variety and location on barley production under rainfed conditions in northern Syria. 1. Root and shoot growth. *Field Crops Res.* 16:53–66.
Castle, W. S., and A. H. Krezdorn. 1977. Soil water use and apparent root efficiencies of citrus trees on four rootstocks. *J. Am. Soc. Hortic. Sci.* 102:403–406.
Chapin, F. S. 1983. Adaptation of selected trees and grasses to low availability of phosphorus. *Plant Soil* 72:283–287.
Christie, E. K., and J. Moorby. 1975. Physiological responses of arid grasses. 1. The influence of phosphorus supply on growth and phosphorus absorption. *Aust. J. Agric. Res.* 26:423–436.
Coutts, M. P., and J. J. Philipson. 1976. The influence of mineral nutrition on the root development of trees. I. The growth of Sitka spruce with individual root systems. *J. Exp. Bot.* 27:1102–1111.
Coutts, M. P., and J. J. Philipson. 1977. The influence of mineral nutrition on the root development of trees. III. Plasticity of root growth in response to changes in the nutrient environment. *J. Exp. Bot.* 28:1071–1075.
Dexter, A. R. 1987. Mechanics of root growth. *Plant Soil* 98:303–312.
Dick, A. C., S. S. Malhi, P. A. Sullivan, and D. R. Walker. 1985. Chemical composition of whole plant and grain yield of nutrients in grain of five barley cultivars. *Plant Soil* 86:257–264.
Diggle, A. J. 1988. ROOTMAP—A model in three dimensional coordinates of the growth and structure of fibrous root systems. *Plant Soil* 105:169–178.
Drew, M. C., and L. R. Saker. 1980. Assessment of a rapid method using soil cones for estimating the amount and distribution of crop roots in the field. *Plant Soil* 55:297–305
Dyson, P. 1986. Timing of nitrogen uptake. *Rep. Macaulay Inst. Soil Res. 1985* pp. 138–139.
Ehlers, W., B. K. Khosla, U. Kopke, R. Stulpnagel, W. Bohm, and K. Baeumer. 1981. Tillage effects on root development, water uptake and growth of oats. *Soil Tillage Res.* 1:19–34.
Ellis, F. B., J. G. Elliot, B. T. Barnes, and K. R. Howse. 1977. Comparison of direct drilling, reduced cultivation and ploughing on the growth of cereals. 2. Spring barley on a sandy loam soil: Soil physical conditions and root growth. *J. Agric. Sci.* 89:631–642.
Evans, P. S. 1978. Plant root distribution and water use patterns of some pasture and crop species. *N.Z. J. Agric. Res.* 21:261–265.
Fitter, A. H. 1982. Morphometric analysis of root systems: Application of the technique and influence of soil fertility on root system development in two herbaceous species. *Plant, Cell Environ.* 5:313–322.
Fitter, A. H. 1985. Functional significance of root morphology and root system architecture. In A. H.

Fitter, D. Atkinson, D. J. Read, and M. B. Usher (eds.), Ecological Interactions in Soil. Blackwell, Oxford. Pp. 87–106.

Fried, M., and H. Broeshart. 1967. The Soil–plant System. Academic Press, New York.

Garwood, E. A. 1968. Some effects of soil water conditions and soil temperature on the roots of grasses and clover. 2. Effects of variation in the soil water content and in soil temperature on root growth. *J. Br. Grassl. Soc.* 23:117–128.

Geisler, D., and D. C. Ferree. 1985. Response of plants to root pruning. *Hortic. Rev.* 6:156–188.

Goode, J. E., K. H. Higgs, and K. J. Hyrycz. 1978. Trickle irrigation of apple trees and the effects of liquid feeding with NO_3^- and K^+ compared with normal manuring. *J. Hortic. Sci.* 53:307–316.

Gooderham, P. T. 1977. Some aspects of soil compaction, root growth and crop yield. *Agric. Progr.* 52:33–44.

Greenwood, D. J., A. Gerwitz, D. A. Stone, and A. Barnes. 1982. Root development of vegetable crops. *Plant Soil* 68:75–96.

Hackett, C. 1969. A study of the root system of barley. II. Relationship between root dimensions and nutrient uptake. *New Phytol.* 68:1023–1030.

Hackett, C. 1971. A study of the root system of barley. III. Branching pattern. *New Phytol.* 70:409–413.

Hansson, A. C. 1987. Roots of Arable Crops: Production, Growth Dynamics and Nitrogen Content. Rep. No. 28. Swedish University of Agricultural Sciences, Uppsala.

Jackson, M. B. 1983. Regulations of root growth and morphology by ethylene and other externally applied growth substances. *Br. Plant Growth Regul. Group Monogr.* 10:103–116.

Kafkafi, U., Z. Karhi, N. Albasal, and N. Roodick. 1972. Root activity of dryland sorghum as measured by radiophosphorus uptake and water consumption. *Isotopes and Radiation in Soil–Plant Relationships Including Forestry. Proc. Symp. IAEA, Vienna* pp. 481–488.

Levin, I., R. Assaf, and B. Bravdo. 1979. Soil moisture and root distribution in an apple orchard irrigated by tricklers. *Plant Soil* 52:31–40.

Lupton, F. G. H., R. H. Oliver, F. B. Ellis, B. T. Barnes, K. R. Howse, P. J. Welbank and P. J. Taylor. 1974. Root and shoot growth of semi-dwarf and tall winter wheats. *Ann. Appl. Biol.* 77:129–144.

Lyr, H., and G. Hoffmann. 1967. Growth rates and growth periodicity of tree roots. *Int. Rev. For. Res.* 2:181–236.

Madsen, H. B. 1985. Distribution of spring barley roots in Danish soils of different texture and under different climatic conditions. *Plant Soil* 88:31–43.

Mengel, K. 1983. Responses of various crop species and cultivars to fertilizer application. *Plant Soil* 72:305–319.

Miller, D. M. 1981. Studies of root function in *Zea Mays*. II. Dimensions of the root systems. *Can. J. Bot.* 59:811–818.

Misra, R. K., A. M. Alston, and A. R. Dexter. 1988. Role of root hairs in phosphorus depletion from a macrostructured soil. *Plant Soil* 107:11–18.

Moorby, H., R. E. White, and P. H. Nye. 1988. The influence of phosphate nutrition on H ion efflux from the roots of young rape plants. *Plant Soil* 105:247–256.

Newman, E. I. 1969. Resistance to water flow in soil and plant. I. Soil resistance in relation to amounts of roots: Theoretical estimates. *J. Appl. Ecol.* 6:1–12.

Nielsen, N. E., and J. K. Schjorring. 1983. Efficiency and kinetics of phosphorus uptake from soil by various barley genotypes. *Plant Soil* 72:225–230.

Parsons, J. W. 1987. The Nitrate Problems in Agriculture. 9th Tom Miller Memorial Lecture. North of Scotland College of Agriculture, Aberdeen.

Patel, R. Z., and A. M., Kabaara. 1975. Isotope studies on the efficient use of P-fertilizers by *Coffea arabica* in Kenya. 1. Uptake and distribution of ^{32}P from labeled KH_2PO_4. *Exp. Agric.* 11:1–11.

Philipson, J. J., and M. P. Coutts. 1977. The influence of mineral nutrition on the root development of trees. II. The effect of specific nutrient elements on the growth of individual roots of Sitka spruce. *J. Exp. Bot.* 28:864–871.

Portas, C. A. M. 1973. Development of root systems during the growth of some vegetable crops. *Plant Soil* 39:507–508.

Raper, C. D., and S. A. Barber. 1970. Rooting systems of soybeans. I. Differences in root morphology among varieties. *Agron. J.* 62:581–584.

Rennie, D. A., and E. H. Halstead. 1972. A ^{32}P injection method for quantitative estimation of the distribution and extent of cereal grain roots. *Isotopes and Radiation in Soil–Plant Relationships Including Forestry. Proc. Symp. IAEA, Vienna* pp. 489–504.

Reynolds, E. R. C. 1970. Root distribution and the cause of its spatial variability in *Pseudotsuga toxifolia* (Poir) Britt. *Plant Soil* 32:501–517.

Richards, D., and R. N. Rowe. 1977. Root–shoot interactions in peach: The function of the root. *Ann. Bot.* 41:1211–1216.

Riley, D., and S. A. Barber. 1971. Effect of ammonium and nitrate fertilization on phosphorus uptake as related to root induced pH changes at the root–soil interface. *Soil Sci. Soc. Am. Proc.* 35:301–306.

Robinson, D. 1986a. Compensatory changes in the partitioning of dry matter in relation to nitrogen uptake and optimum variation in growth. *Ann. Bot.* 58:841–848.

Robinson, D. 1986b. Limits to nutrient inflow rates in roots and root systems. *Physiol. Plant* 68:551–559.

Robinson, D., and I. H. Rorison. 1983a. A comparison of the response of *Lolium perenne* L., *Holcus lanatus* L., *Deschantsia flexuosa* (L.) Trin to a localized supply of nitrogen. *New Phytol.* 94:263–273.

Robinson, D., and I. H. Rorison. 1983b. Relationships between root morphology and nitrogen availability in a recent theoretical model describing nitrogen uptake from soil. *Plant, Cell Environ.* 6:641–647.

Rogers, W. S. 1939. Root studies VIII. Apple root growth in relation to rootstock, soil, seasonal and climatic factors. *J. Pomol. Hortic. Sci.* 17:99–130.

Rogers, W. S., and M. C. Vyvyan. 1934. Root studies V. Rootstock and soil effect on apple root systems. *J. Pomol. Hortic. Sci.* 12:110–150.

Rose, D. A. 1983. The description of the growth of root systems. *Plant Soil* 75:405–415.

Rovira, A. D. 1979. Biology of the soil–root interface. In J. L. Harley and R. S. Russell (eds.), The Soil–Root Interface. Academic Press, London. Pp. 145–160.

SAC. 1985. Fertilizer Recommendations. Publ. No. 160. Scottish Agricultural Colleges, Perth.

Schenk, M. K., and S. A. Barber. 1980. Potassium and phosphorus uptake by corn genotypes grown in the field as influenced by root characteristics. *Plant Soil* 54:65–76.

Shierlaw, J., and A. M. Alston. 1984. Effect of soil compaction on root growth and uptake of phosphorus. *Plant Soil* 77:15–28.

Shukla, L. M., and B. Sen. 1985. Studies on root distribution pattern of some high yielding paddy varieties. *J. Indian Soc. Soil. Sci.* 23:266–267.

Smith, K. A., A. E. Elmes, R. S. Howard, and M. F. Franklin. 1984. The uptake of soil and fertilizer-nitrogen by barley growing under Scottish climatic conditions. *Plant Soil* 76:49–58.

Taylor, H. M., M. G. Huck, B. Klepper, and Z. F. Lund. 1970. Measurement of soil grown roots in a rhizatron. *Agron. J.* 62:807–809.

Welbank, P. J., M. J. Gibb, P. J. Taylor, and E. D. Williams. 1974. Root growth of cereal crops. *Rep. Rothamsted Exp. Stn. 1973* 2:26–66.

Willatt, S. T., and K. A. Olsson. 1982. Root distribution and water uptake by irrigated soybeans on a duplex soil. *Aust. J. Soil Res.* 20:139–146.

12

Role of Moisture Stress in Plant Nutritional Functions

J. F. POWER

In a large majority of agricultural enterprises, water stress at some period during plant development usually limits crop production. This problem is usually frequent and intense under rain-fed agriculture, especially in regions with limited precipitation and/or relatively high evaporation potential. Under arid conditions, and without supplemental irrigation, water deficit is frequently the primary factor limiting crop production. As water availability increases, either through greater precipitation or by irrigation, plant water stress is lessened, and other factors limiting plant growth such as soil nutrient availability assume a greater importance.

Plant stresses resulting from insufficient water or available nutrients appear in many instances to be additive (Power, 1983). Greenwood (1976) developed methodology and equations by which degree of plant growth stress resulting from a

Table I. Percentage Stress in Smooth Bromegrass Resulting from N and Water Deficiencies[a]

	Deficiency of		
Growth Period	N	Water	N + Water
23 May–10 June	3	32	49
11 June–30 June	24	32	54
1 July–22 July	35	10	58

[a]Source: Power, 1971.

nutrient deficiency could be quantified. Power (1971) extended this concept to water stress and found that the reductions in growth of smooth bromegrass resulting from water and nitrogen (N) deficiencies were approximately additive (Table I). When both factors are limiting, the addition of one without the other has very limited benefits on crop production (Power *et al.*, 1973; Ramig and Rhoades, 1963; Smika *et al.*, 1965).

Excess soil water also results in stress for most plant species. Excess water fills the larger-diameter soil pores, thereby reducing oxygen diffusion rates by a factor of 10^4. Since root respiration and a number of other biological processes require O_2, limiting oxygen diffusion rates severely restricts the ability of most plant roots to absorb soil nutrients. Lawton (1945) showed that potassium uptake was particularly restricted in soils containing excess water. By forcing air through soils with high water content, he showed that K absorption by plants would be greatly enhanced, suggesting that lack of O_2 at the root surface is the prime cause for reduced K uptake in wet soils.

The soil nutrients most frequently limiting crop growth and production are N, phosphorus (P), and, in more humid regions, potassium (K). In certain soils, a number of other essential plant nutrients may be involved. However, a thorough discussion of all plant nutrients is beyond the scope of this chapter. Primary emphasis will be placed on water and N stresses, with some mention of P and K problems. Much of the information reported will be from dryland (rain-fed) agriculture in arid and semiarid regions, where water stress is more frequently encountered than in more humid regions.

I. Water and Nutrient Availability

A. SOLUBILITY AND PRECIPITATION

Water affects the availability of many essential plant nutrients by its effect on the solubility and precipitation of salts. This effect is much more pronounced for mineral elements such as P and K than for N. Most N transformations are microbially mediated and few N compounds are precipitated as salts. Soil N transformations as affected by water stress will be discussed in a later section.

Water deficits affect soil chemistry primarily through their effects on the solubility and precipitation of salts and minerals in the soil. As the soil dries, the concentration of soluble salts in the soil solution increases. If this concentration exceeds about 5 dS m^{-1} (2 to 3 dS m^{-1} on oven-dry soil basis), higher soil osmotic potential begins to adversely affect plant water relations and subsequent growth of salt-sensitive crops (Taylor, 1983). Salinity levels of this magnitude are not unusual in dryland soils.

As salt concentrations in the soil solution increase with soil drying, the solubility constant for certain salts is approached, and precipitation occurs. Soluble anions are predominantly bicarbonate (HCO_3^-), sulfate (SO_4^{2-}), and chlorides (Cl^-), along with the cations Ca^{2+}, Mg^{2+}, K^+, and sometimes Na^+. The sulfate salts of these cations generally precipitate first. In many dryland soils, Ca^{2+} is the predominant cation, so one of the first salts to precipitate is gypsum ($CaSO_4 \cdot 2H_2O$). If significant quantities of Ca^{2+} and SO_4^{2+} are present in the soil solution, salinity hazard for such soils is usually relatively low because, when conductivity levels of the saturation extract reach about 2 to 3 dS m^{-1}, gypsum begins to precipitate. As long as there are soluble Ca^{2+} and SO_4^{2-} ions in solution in equilibrium with the precipitated phase, electrical conductivity does not increase regardless of degree of drying. Consequently, in gypsiferous soils, electrical conductivity of the saturation extract will seldom exceed 2 to 3 dS m^{-1}, a level having little adverse affect on many crop species (Bresler *et al.*, 1982).

If sodium (Na^+) exceeds about 6% of the cation-exchange capacity, soil structure and aggregation may begin to disintegrate, resulting in a dispersed soil. Under such conditions, hydraulic conductivity is greatly reduced, soil resistance increases, surface crusting may become severe, and water infiltration becomes restricted; consequently, plant growth is restricted. These adverse effects become acute when the exchangeable sodium percentage reaches about 15%, varying somewhat with soil texture and organic-matter content (Bresler *et al.*, 1982).

In addition to the effects of water deficits on problems associated with soil salinity and sodicity, water availability also affects dissolution rate of soil minerals. In terms of geologic time, dissolution (weathering) of soil minerals proceeds at a much slower rate and at more shallow depths in dryland soils than in those developed in more humid climates. Consequently, feldspars and other K-rich primary minerals often remain intact in dryland soils and serve as a primary source of plant-available K.

Water deficits also directly affect availability of P (Lindsay and Vlek, 1977) and many other nutrients. Except in highly acid soils, soil drying often results in the precipitation of orthophosphate (PO_4^{3-}) as various forms of calcium-phosphate. In temperate-region soils, these precipitated calcium phosphates are predominantly in the form of octacalcium phosphates, a slightly soluble source of plant-available P. However, this mineral form exists in equilibrium not only with soluble PO_4^{3-} ions,

Table II. Hydrolysis of Dicalcium Phosphate Dehydrate (DCPD) to Octacalcium Phosphate (OCP) as Affected by Time and Temperature[a]

Temperature (°C)	Percentage Converted to OCP at			
	1 Month	2 Months	4 Months	10 Months
10	<5	20	20	70
20	<5	40	75	100
30	<5	30	80	100

[a]Source: Sheppard and Racz, 1980.

but also with adsorbed P and other calcium phosphate minerals, with the equilibrium concentrations varying with soil water content, salt concentration, pH, temperature, and other related factors. Sheppard and Racz (1980) showed that hydrolysis of dicalcium-phosphate dihydrate to octacalcium phosphate increased with temperature and time in a western Canadian soil (Table II). Often, much of the P immobilized as octacalcium phosphate can eventually become available to plants (Kissel *et al.*, 1985). In certain acid soils, orthophosphate may precipitate as iron and aluminum phosphates, which are relatively insoluble and unavailable. Stevenson (1986) provides a more thorough discussion of the chemistry and biology associated with soil P transformations.

A third manner in which water deficits may affect chemical solubility of soil nutrients is indirectly through the effects of water deficits on plant root activity (Taylor, 1983). An actively respiring root system releases relatively large amounts of carbon dioxide into the soil solution, decreasing soil solution pH, and thereby altering solubility of many soil minerals. For example, the increased acidity of the soil solution increases P availability in neutral or calcareous dryland soils. Plant root hairs and other root tissue often remain active for only a few days or weeks. After that, they begin to decompose, providing a supply of carbon that enhances microbiological activity within the soil. Taylor and Klepper (1974) found that root growth and activity decreased rapidly as soil water potentials approached or exceeded −0.1 MPa.

B. BIOLOGICAL TRANSFORMATIONS

Microbiological activity in the soil regulates, to a large extent, soil N transformations. This is true to a lesser extent for organic P transformations, as well as for various micronutrients found in soil organic matter (Stevenson, 1986). Soil water availability is, of course, one of the prime factors affecting microbiological activity in the soil. Subsequently, transformations of N are regulated considerably by the influence of water availability on soil microbial activity.

The microbiological population of soils is highly diverse and composed of a wide variety of microbiological species (Paul and Juma, 1981). As related to N transformations, these may be grouped into actinomycetes, fungi, and bacteria. Actinomycetes generally play a very minor role in N cycling and are of occasional concern primarily in arid or highly adverse environments. The fungi comprise a diverse group of microorganisms that account for the major part of the microbial biomass of most soils and are extremely important in decomposing fresh organic matter and converting N to ammonium (NH_4^+) (ammonification). Bacteria are also highly diverse, and certain bacteria (facultative and obligate anaerobes) are metabolically active in the absence of oxygen. These anaerobic organisms often utilize nitrate (NO_3^-) as an alternate electron acceptor, thereby reducing nitrates to various nitrogenous gases (denitrification).

All of the above-mentioned N transformations (ammonification, nitrification, mineralization, denitrification, immobilization) are mediated by microbial activity in soil, and N transformation rates are dependent on the effects of soil water availability on the relative activity of these various soil microbial groups. Some of these relationships will be discussed in more detail later. In general, however, microbial activity is greatly limited in dry soils, resulting in limited dinitrogen fixation and ammonification, and very little nitrification. At the other extreme, in wet (anaerobic) soils, denitrification may dominate. In moist, well-drained soils, rates of mineralization (ammonification and nitrification) and immobilization may be near maximum for the system, while dinitrogen fixation rate is often high but denitrification very low. Readers should consult Mengel (1985) or Stevenson (1986) for a more detailed discussion of mechanisms involved in these N transformations.

C. MASS TRANSPORT, DIFFUSION, AND INTERCEPTION OF NUTRIENTS

Nutrient ions move within the soil by the processes of mass transport and diffusion. The relative importance of these processes in supplying the nutrients needed by a maize (*Zea mays* L.) crop is given in Table III. The processes are discussed by Tisdale *et al.* (1985).

Mass transport of nutrients involves the direct mass flow of dissolved ions in the soil solution toward the plant root as a result of liquid movement in the transpiration stream. Increased evaporation potential of the ambient atmosphere enhances evaporative loss from the soil and plant surfaces, decreasing their water potentials (increasing water tensions). This process results in the flow of water (and its dissolved nutrients) from soil layers with higher water potentials toward soil and plant surfaces with lower potentials. Both the water required to maintain turgor pressure of plant cells and many of the nutrients required for cellular activity and biosynthesis are transported in solution to plant roots as a result of the water potential gradients established by root activity. Nitrates and most other anions used by plants are moved toward the plant roots by this process. Likewise, soluble cations in equilibrium with cations adsorbed on colloidal surfaces are also moved toward root surfaces by mass transport (Mengel, 1985). Because of the preponderance of Ca^{2+}

Table III. Percentage of Nutrients Taken up by Corn Provided by Different Mechanisms[a]

Nutrient	Required for 10 Mg ha^{-1} Corn Yield ($kg\ ha^{-1}$)	Percentage Provided by		
		Root Interception	Mass Flow	Diffusion
Nitrogen	202	1	99	0
Phosphorus	42	3	6	94
Potassium	208	2	20	78
Calcium	42	171	429	0
Magnesium	48	38	250	0
Sulfur	24	5	95	0
Copper	0.10	10	400	0
Zinc	0.30	33	33	33
Boron	0.20	10	350	0
Iron	2.00	11	53	37
Manganese	0.40	33	133	0
Molybdenum	0.01	10	200	0

[a]Source: Barber and Olson, 1968.

and Mg^{2+} ions in the soil solution, the quantities of these cations moved to roots by mass flow often greatly exceed those needed by the crop, leaving a Ca^{2+}- and Mg^{2+}-rich rhizosphere.

Diffusion of ions toward plant roots occurs in response to concentration gradients established by the uptake of ions into the plant root. Diffusion may occur either through the soil solution or by surface migration of ions. Diffusion through the soil solution in water-filled pores is, of course, several orders of magnitude more rapid. For all practical purposes, ions diffuse at best only a few centimeters during a growing season; however, for adsorbed ions, such as various cations adsorbed on clay surfaces or precipitated P compounds, diffusion from these surfaces to the plant root is the primary process by which such ions are made positionally available to plants (Mengel, 1985). Consequently, greater plant rooting density increases the quantity of adsorbed or precipitated ions available for plant uptake. Water deficit would, therefore, reduce diffusion rates by two mechanisms: (1) reduced soil solution volume and mobility of ions and (2) reduced plant root growth and root contact with these ions.

Root interception is another important mechanism by which plants obtain nutrients. This process may include diffusion over very short distances when root hairs come in close proximity to adsorbed or precipitated ions, reducing the distance ions need to move for uptake to fractions of a millimeter. Interception is particularly important for less soluble materials such as Ca (Table III). Again, water stress would reduce ion uptake via interception by reducing root density and frequency of root contact with adsorbed or precipitated ions.

Ellis *et al.* (1985) showed that effective root contact with soil, and subsequent water and P absorption by the plant, was enhanced when the plant was infected with vesicular-arbuscular mycorrhizae (VAM). Hyphae of these small-diameter fungi could act as an extension of the root system by increasing effective soil–root contact. These hyphae explore soil pores less than 30 μm in diameter and extract water, P, and other ions contained in these pores, thereby reducing the impact of water or nutrient deficiencies on plant growth. Fawcett and Quick (1962) concluded that much of the plant-available P in a soil is located in these small-diameter soil pores and is presumably taken up through direct contact with roots or hyphae.

II. Biological Activity

A. WATER–AERATION RELATIONSHIPS

Soil microorganisms exist in the pore space of the soil. While the microbial biomass normally occupies no more than 1 to 2% of the volume of the soil pores, the environment existing within soil pores determines the characteristics of the ecological niches created for the support of soil organisms (Doran and Smith, 1987). The water, oxygen and other gases, and nutrients and other substrates (including soluble C) required for the life activity of these microbes are supplied to them through the soil pores. In a dry soil, any water present exists as thin films around the periphery of pores, resulting in little movement of water, nutrients, and substrates by mass flow and, subsequently, restricted microbial activity. In a wet soil, soil pores are filled or nearly filled with water, so rate of diffusion of oxygen through water-filled pores is reduced by a factor of 10^4 compared to well-aerated soils. Consequently, only those processes that can use ions other than O_2 as the terminal electron acceptor can proceed at an appreciable rate—namely, denitrification and limited ammonification.

These examples illustrate the importance of controlling the balance between the amount of the soil pore volume occupied by water and that occupied by air. This point was nicely illustrated by Linn and Doran (1984) when they showed that most aerobic microbial processes were near optimum when 60% of the soil pore volume was filled with water and the other 40% with air (Figure 1). With less water-filled pore space (WFPS), processes of nutrient and substrate diffusion probably limited biological activity. When WFPS greatly exceeded 60%, oxygen diffusion became restricted and anaerobic processes such as denitrification began to dominate.

It is of interest to note that nitrification (strictly an aerobic process) decreased rapidly when 60% WFPS was exceeded, whereas mineralization processes were reduced but continued to be significant at high WFPS values (Linn and Doran, 1984). This occurred because nitrification results from the activity of only a relatively few bacterial species, all of which are obligate aerobes, while mineralization

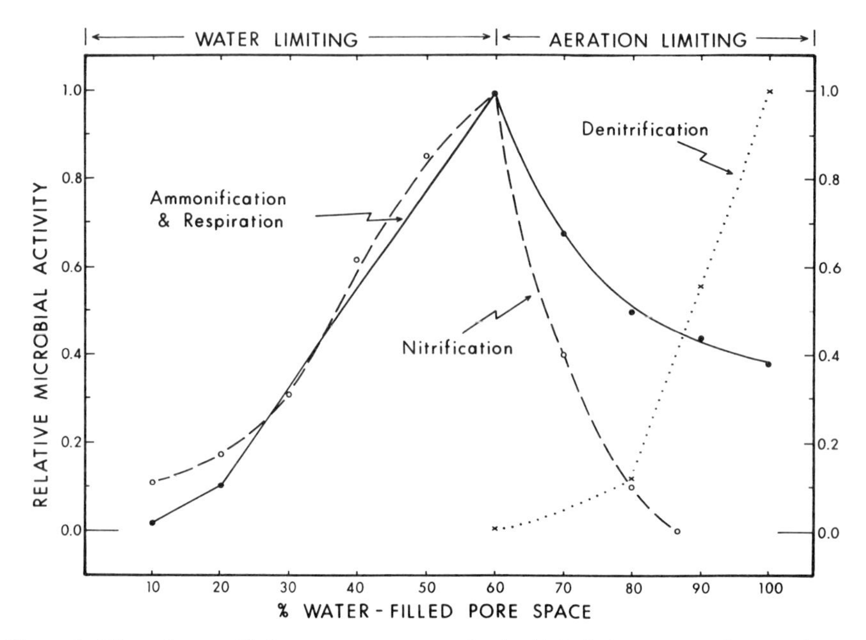

Figure 1. Effect of water-filled pore space on several microbiological processes. Source: Linn and Doran, 1984.

results from activities of a broad group of both bacterial and fungal species that include aerobes, obligate anaerobes, and facultative anaerobes (capable of activity in both the presence and absence of oxygen).

The relationships developed by Linn and Doran (1984) have been more recently tested on a wide range of soils across the United States, and it appears these relationships may be rather universal, at least for temperate-region soils (Doran *et al.*, 1988). These conclusions were also substantiated by recalculating data in a number of other publications (Linn and Doran, 1984). The available literature suggests that WFPS may also apply as a useful index of plant root activity (Grable and Siemers, 1968; Lawton, 1945).

B. NUTRIENT TRANSFORMATIONS

As information presented in Figure 1 suggests, water deficiency stress retards rate of aerobic transformations of soil nutrients. Stanford and Epstein (1974) suggested that rates of N mineralization and other such N transformations vary linearly with water availability between field capacity and wilting point (Table IV). Based on such a relationship, rates of N mineralization have been predicted from data on ambient water and temperature regimes for a number of Oklahoma soils (Smith *et al.*, 1977). Other studies have also shown that, with relatively minor deviations,

Table IV. Linear Regression Equations and Correlation Coefficients Relating N Mineralized (Y in mg kg^{-1}) to Available Soil Water (X in %)[a]

Soil	Regression Equation	r^2
Aastad cl	$Y = -12.9 + 2.3X$	0.96
Amarillo fsl	$Y = -4.3 + 1.7X$	0.93
Barnes l	$Y = -7.7 + 2.9X$	0.96
Bearden sil	$Y = -0.9 + 2.8X$	0.98
Cecil sl	$Y = -11.6 + 1.7X$	0.97
Kranzburg sil	$Y = 8.1 + 1.2X$	0.98
Minidoka sil	$Y = -9.4 + 3.0X$	0.95
Parshall fsl	$Y = 1.5 + 1.7X$	0.99
Pullman sicl	$Y = -0.2 + 2.0X$	0.96

[a]Source: Stanford and Epstein, 1974.

mineralization decreases linearly as soil water decreases from field capacity to the permanent wilting point (Reichman *et al.*, 1966). Most research on this subject shows that, even at water contents below wilting point, a slow rate of mineralization still occurs (Viets, 1972).

Biologically mediated nutrient transformations are controlled to a large extent by the balance between rate of microbiological incorporation of nutrients into microbial biomass (immobilization) and rate of mineralization of the immobilized nutrient occurring when microbial cells die and are decomposed. In addition to immobilization–mineralization balance, nutrient availability to crop plants is also dependent on rate of losses from the soil through leaching and by gaseous diffusion.

The major organic sources of nutrients normally returned to agricultural soils are crop residues and animal wastes. Collectively, the N in these two sources approaches the amount of fertilizer N used annually in the United States (Power and Papendick, 1985). Most of the proteins in these organic materials are normally hydrolyzed and rapidly utilized for biosynthesis of microbial biomass. Often within a matter of days, the soluble carbon in the added organic matter is consumed by the new microbial growth, so subsequent microbial growth rates decline and many of these new microbial cells begin to senesce. As the size of the microbial biomass pool decreases, the N and other nutrients immobilized within senescing cells are mineralized and released into the soil solution for use by the growing crop (Paul and Juma, 1981). Thus, mineralization of N in organic residues added to soils often lags behind immobilization by a few weeks or even months. Relative rates and times of net mineralization depend to a large extent on the C/N ratios of the added organic wastes, with N in low C/N materials being mineralized more rapidly and completely than that in materials with high C/N ratios.

Soil drying reduces rate of microbiological growth, so during drying the rate of senescence of old cells may exceed the rate of proliferation of new cells. As a consequence, more N may be hydrolyzed from senescing cells than is used by new cells, thus increasing net mineralization. Consequently it is common to observe a flush of mineralization with soil drying. Van Veen *et al.* (1987), using C and N isotopes, provided direct proof of this phenomenon.

Fertilizer N added to soils is also subject to immobilization and later mineralization by microbial biomass. Again, quantity and rate of immobilization usually depend on the supply of C available as an energy source for the microorganisms. If fertilizers are broadcast over surface residues, as in a no-till system, immobilization can be relatively rapid because both the fertilizer and the C in crop residues are immediately available to soil organisms near the soil surface. This helps explain why N fertilizer requirements for no-till are sometimes greater than for conventional tillage (Doran and Smith, 1987; Meisinger *et al.*, 1985; Rice and Smith, 1984).

In each generation of microbial biomass through which N passes, a certain percentage of the N is immobilized as cyclic organic compounds that are more resistant to microbial degradation (Clark, 1977). Thus, a variable percentage of the N initially incorporated into microbial biomass is eventually converted to more resistant organic pools. Paul and Juma (1981) indicated that much of the N immobilized in microbial biomass is mineralized in one growing season. Smith and Power (1985) found that, 5 years after application of fertilizer N to semiarid grasslands, 20 to 50% of the applied N resided in slowly mineralized soil organic N pools that, while less labile than the microbial biomass, were still several times more labile than the indigenous soil organic N (Table V).

When stress results from excess water, aerobic processes such as nitrification are inhibited and anaerobic processes such as denitrification can become more dominant. Ammonification can continue at a reduced rate as a result of activity of both obligate and facultative anaerobes. In the anaerobic environment, however, soluble C is utilized much less efficiently, so a smaller microbial population is supported by a given soluble C supply than would occur in an aerobic environment. Consequently, because of a smaller sink for microbial immobilization, ammonium-N may accumulate in greater concentrations under anaerobic than under aerobic conditions. In addition, with excess water the probability of nitrate being lost by leaching is also great.

Nutrient uptake by crops is also impaired when soils are saturated. Lawton

Table V. Effect of Years since Fertilization on Amount of ^{15}N Mineralized (% ^{15}N in total N mineralized) in 24 Weeks[a]

Soil Depth (cm)	Years since ^{15}N Applied					Soil N Reference[b]
	1	2	3	4	5	
0–10	46	39	33	29	28	11
10–30	22	22	22	25	29	6

[a]Source: Smith and Power, 1985.
[b]Percentage of total N mineralized.

(1945) showed that excess water reduced plant concentrations and uptake of calcium, magnesium, K, P, and N in maize. Trought and Drew (1980) published similar results for wheat. With waterlogging, they found that especially N, P, and K were translocated from roots to shoots and from older leaves to newer leaves. This process was accompanied by early senescence of the older tissue. Adverse effects of excess water on ion uptake could be reduced by foliar spraying with urea solutions. Khera and Singh (1975) make similar observations with maize in the Punjab region of northern India.

C. PLANT NUTRIENT UPTAKE

Water stress not only affects the availability of nutrients in the soil, but it also alters physiological processes within the plant, including nutrient uptake and translocation. In this presentation, stress resulting from excess water will be discussed first, followed by a discussion of stress resulting from water deficit.

Kozlowski (1984) recently summarized information on plant responses to flooding. Flooding alters soil pH and the solubility of manganese, iron, P, and a number of other elements, as well as enhances the accumulation of organic compounds such as alcohol, aldehydes, and various types of organic acids. However, Kozlowski concluded that O_2 deficit is the most important cause of flooding injury to plants. Reduced O_2 promotes stomatal closure, which in turn reduces photosynthesis and all processes dependent on the utilization of photosynthate, including mineral uptake (Lawton, 1945; Trought and Drew, 1980).

Length of flooding and growth stage at which flooding occurs are also factors that affect plant nutrient uptake. Leyshon and Sheard (1974) showed that the effects of flooding were more intense on 14-day-old than on 35-day-old barley. After flooding ceased, plants flooded at 14 days old fully recovered whereas those flooded at 35 days failed to completely recover in terms of growth and nutrient uptake. They measured no effect of flooding on Mn solubility or uptake and concluded that O_2 deficiency was the major factor causing reduced plant growth and N, P, and K uptake.

The effects of water deficit on plant growth and nutrient uptake were recently reviewed by Jordan (1983). He and others generally conclude that water deficit also affects stomatal closure and subsequent photosynthesis, thereby affecting energy available for nutrient uptake and transformations. For drought stress, stomatal relationships are complicated somewhat by the relative ability of different cultivars to adjust the osmotic potential of stomatal guard cells. This adjustment in osmotic potential results from translocation of nutrient ions (especially K^+) in and out of the guard cells. Some cultivars were also capable of synthesizing various organic acids that also affect osmotic potential of the guard cells. If stress continues, however, even with closure of stomata, plant water potentials may decline to the extent that cell turgor is lost, thereby greatly restricting the diffusion and translocation of ions.

Hanson and Hitz (1983) discussed the effects of plant water stress on the N economy of plants. Generally, rates of processes such as nitrate reduction and dinitrogen fixation are drastically reduced when leaf water potential falls below 0.2 to 0.5 MPa. If stress continues, plant proteins begin to break down, sometimes accompanied by release of ammonia. For many species, proline is a major product of the disruption of the N metabolism process in water-stressed plants. Betaine may also accumulate in some species when under water stress. Accumulation of such products as proline, betaine, and various organic acids (aconitic, oxalic, and succinic in particular) in water-stressed plants is fairly species specific (Ford and Wilson, 1981).

Raju *et al.* (1987) found that sorghum genotypes differed in their ability to utilize P and other soil minerals. However, when infected with vesicular-arbuscular mycorrhizae, many of these genotype differences disappeared because the VAM hyphae greatly improved the ability of the plant to take up soil P. As indicated earlier, Ellis *et al.* (1985) showed that VAM-infected plants were more drought tolerant because of more efficient utilization of soil water. Thus, it appears that VAM can greatly affect genotype response to both plant nutrient and available water status.

III. Management Practices

A. CROP RESIDUE MANAGEMENT

Crop residues are the major source of organic matter returned to many soils. Factors important in regulating the availability of the nutrients contained in crop residues include quantity, quality, timing, and placement of crop residues.

Bond and Willis (1971) showed that crop residues on the soil surface slow the rates of first- and second-stage evaporation of water from soil surfaces. If surfaces are not rewetted, cumulative evaporation from residue-covered soils will eventually equal that for bare soils. In situations where occasional showers rewet the soil surface after the surface dries, cumulative evaporation after several months for residue-covered soils is less than that for bare soils, and additional water is stored in the soil. However, if no rain is received for several months, differences in cumulative evaporation may not persist, and no effect of crop residues on soil water conservation is shown (Aase and Tanaka, 1987). Increasing the amount of crop residues on the soil surface lengthens the time period during which residue-covered soils conserve more water than bare soils. In water-deficient areas, where crop residues conserve extra soil water, increased crop yields and nutrient uptake may result (Black, 1973; Wilhelm *et al.*, 1986). Power *et al.* (1986) showed that increasing amounts of crop residues on the soil surface increased the availability and uptake of indigenous soil organic N by both maize and soybean. Likewise, recovery of N in crop residues produced by the previous year's soybean crop was also greatly increased by increasing surface crop residue.

Crop residue placement is normally achieved through tillage practices. Residues may either be left predominantly on the soil surface through use of no-till or well-managed subsurface tillage; partially incorporated in the upper few centimeters of soil through many types of stubble-mulch and conservation-tillage systems; or completely incorporated through moldboard plowing or heavy disking. Also, residues may be removed from the soil surface for use as fuel or fodder, or may be burned off. All of these management options have variable effects on the soil environment and, subsequently, on microbial activity and nutrient cycling.

The effects of tillage and other management practices on the soil–plant ecosystem are diagrammatically outlined in Figure 2. For a soil with certain inherent properties and in a given climate, to a large extent, choice of management practices controls the microenvironment existing within the soil pores in which most soil microorganisms are found. The soil environment is characterized by the water, aeration, temperature, and substrate (nutrient) status of these microsites.

Leaving crop residues on the soil surface with no-till, compared to residue incorporation by plowing, results in a surface soil environment that is cooler, more moist (less aerobic), and better supplied with soluble C (Blevins *et al.*, 1984; Broder *et al.*, 1984; Doran, 1987; Mielke *et al.*, 1986). Consequently, in well-drained soils, an environment more conducive for growth of plant roots and aerobic organisms is often created with no-till than with plowing, resulting in a greater microbial biomass (Carter and Rennie, 1984; Doran, 1980, 1987). Subsequently, more N is

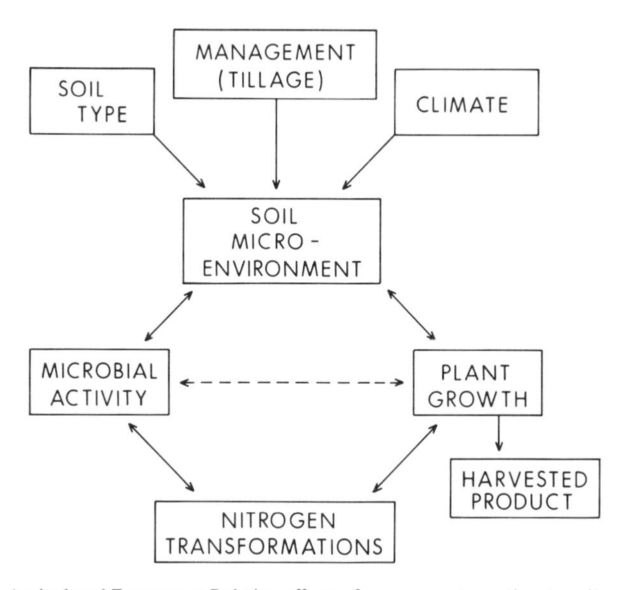

Figure 2. The Agricultural Ecosystem: Relating effects of management practices to soil environment and subsequent soil biology.

Table VI. Effect of Tillage Management on Soil Water, Organic-Matter Contents, Microbial Biomass, and Potentially Mineralized N[a,b]

Location(s) Depth in Soil (cm)	Water Content	Organic Matter	Microbial Biomass	Potentially Mineralizable N
United States (six sites)				
0–7.5	+22	+40	+58	+34
7.5–15.0	+3	−1	−2	−3
15.0–30.0	+5	−7	0	−7
Canada (four sites)				
0–5.0	−4	−11	+11	+59
5.0–10.0	+2	−2	+6	−25
10.0–20.0	—	—	−6	—
England (one site)				
0–5.0	+2	+16	+32	—

[a]Source: Doran and Smith, 1987.
[b]Percentage change for no-till compared to plowing.

temporarily immobilized in the microbial biomass in no-till soils and potentially mineralizable-N is also increased (Table VI).

B. FERTILIZER PRACTICES

Fertilizer management and fertilizer use efficiency are often affected by water stress. Generally, uptake of nutrients applied in fertilizers is greatly reduced as a soil dries below field capacity. Thus, fertilizer practices may be affected by climate and by the use of water-conserving practices such as no-till and stubble-mulch. Much recent research has shown that in dryland regions, with certain management practices (reduced or no-till, snow trapping, etc.), sufficient water may be conserved to enable producers to plant a crop every year or two out of three years, instead of every other year, as occurs with common crop-fallow systems (Black, 1983; Black and Siddoway, 1976; Greb *et al.*, 1967; Wicks, 1976). When more frequent cropping occurs, buildup of available N and P during the fallow period is reduced and crop removal of N and P is increased, escalating the need for N and P fertilization.

A number of studies have shown that fertilizers need to be placed beneath the surface mulch layer of no-till soil to maintain efficiency of fertilizer usage (Doran and Smith, 1987; Meisinger *et al.*, 1985; Mengel, 1985). Reduced efficiency of surface-applied fertilizers on no-till soil may result because of greater microbial immobilization of fertilizer during the decomposition of the crop residues. Compensating for this greater immobilization potential is the fact that the surface of residue-covered soils remains moist for a longer period of time, allowing more microbial recycling of nutrients to occur than in bare soils. Thus, rates of N mineralization may be slower in residue-covered than in bare soil, but total quantity of N mineralized over a growing season could be greater.

In several experiments in Montana and North Dakota, Power and co-workers (Power *et al.*, 1961a,b,c, 1973) showed that both N and P uptake increased for small grain and corn silage crops as water availability increased. Similar data have been published for a number of other crops at many locations (Power, 1983). These results indicate that while uptake of N and P from soil is often increased as available water supply increases, uptake of nutrients from fertilizers may not respond as much additional to water (Table VII). This would suggest that additional water may have a twofold effect on the availability and uptake of soil nutrients: (1) increased water increases the mineralization and availability of nutrients and (2) increased plant root growth and activity resulting from additional water increase the ability of plant roots to find and absorb soil nutrients. The slopes shown for the regression equations in Table VII showed little effect from P fertilization, indicating that fertilization affected uptake of soil nutrients only slightly.

This increased uptake of nutrient with increased water availability usually results from enhanced root activity, plant growth, and nutrient requirement for the crop. With increased water availability, movement of nutrients to and uptake by root surfaces would be enhanced by both greater mass transport and diffusion of nutrients. Also, availability and uptake of indigenous soil N and P are often greater with increased water availability (Power *et al.*, 1961a,b,c, 1986).

Kissel *et al.* (1985) and Nelson (1982) have found that ammonia volatilization from urea-based fertilizers may be enhanced by soil drying, reducing the efficiency of such fertilizers. Harper *et al.* (1983) and Farquhar *et al.* (1979) have also shown that appreciable ammonia may be lost by volatilization from leaf surfaces of well-fertilized crops, especially during periods of drought. These and other mechanisms contribute to reduced efficiency of fertilizers from dry soils.

Table VII. Regression Equation Relating Total N Uptake and Total, Soil, and Fertilizer P Uptake by Spring Wheat (Y in kg ha^{-1}) with Water Use (X in cm)[a]

	P fertilization	
	No P	With P
Total N uptake	$Y = 16.5 + 2.0X$ ($r = 0.94$)	$Y = 21.3 + 2.2X$ ($r = 0.97$)
Total P uptake	$Y = 1.33 + 0.20X$ ($r = 0.95$)	$Y = 2.04 + 0.22X$ ($r = 0.94$)
Soil P uptake	$Y = 1.33 + 0.20X$ ($r = 0.95$)	$Y = 0.95 + 0.21X$ ($r = 0.85$)
Fertilizer uptake	—[b]	NS[c]

[a]Source: Power *et al.*, 1961b,c.
[b]No fertilizer P applied.
[c]NS, not significant.

C. CROPPING SYSTEMS

Cropping systems affect the entire soil–plant ecosystem by altering quantity and composition of crop residues returned to the soil, soil water storage and usage, soil erodibility, and nutrient availability. Frequently, monocultures are utilized because of higher production efficiency and the existence of a defined infrastructure providing various manufactured inputs, as well as defined marketing channels. However, monocultures frequently require an ever-increasing supply of inputs to combat problems associated with buildups in weeds, disease, and insect pests, as well as soil erosion and declining productivity (Power and Follett, 1987). As a result, net return to the producer of monocultures may be limited.

Crop rotations reduce the need for increasing inputs because the biological cycles of many pests are broken by rotating crops. Likewise, if legumes are included in the rotation, the need for fertilizers may be reduced through biological N fixation. Rotations are of many types—long term over a period of many years, or short term where several crops are produced within several years or even within the same year. In regions of limited or uneven distribution of precipitation, fallowing may also be rotated with various crops as a means of conserving stored soil water for later use by crops (Black, 1983).

Frequently, timing and severity of anticipated water stress play major roles in the selection of cropping systems. In semiarid regions such as the western part of the Great Plains of the United States, where growing season precipitation is considerably less than potential evapotranspiration, crops such as wheat and sorghum are routinely rotated with summer-fallow. By using reduced or no tillage, especially when coupled with use of grass barriers or windbreaks to trap snow and enhance soil water conservation, producers may produce two crops every three years or in some instances they may even be able to crop annually (Black, 1983; Black and Siddoway, 1976; Jackson *et al.*, 1982; Wicks, 1976). In such instances, fertilizer practices are altered according to the amount of available nutrients in the soil at planting and anticipated water availability for the crop (Table VIII).

In more humid regions, selection of crops is often closely dependent on water and temperature regimes. If cool night temperatures prevail, sorghum, for instance, would not be a suitable crop. Likewise, in some subhumid regions, maize is a poor choice because precipitation patterns are such that periods of hot, dry weather during anthesis would be expected, greatly reducing grain yield potential. Cotton may be produced under dryland in warmer climates. Suitability of various crops for dryland agriculture is discussed in more detail by Eastin *et al.* (1982).

As mentioned earlier, use of legumes in a cropping system greatly affects fertility relationships. For example, fertilizer N requirements for a maize–soybean rotation are considerably less than those for continuous maize (Table IX), and potential yield level for maize is also often greater when grown in rotation (Voss and Shrader, 1984). This occurs even though, on fertile, highly productive soil, the amount of N removed in the soybean grain may exceed the quantity of N biologically fixed by the soybean (Heichel and Barnes, 1984). Water stress may reduce the amount of N fixed

Table VIII. Relation between Available Soil Water at Seeding and Spring Wheat Response to P Fertilization[a]

Site	$NaHCO_3$ Soluble P (mg kg^{-1})	Available Soil Water (X) (mm)	Grain Increase from Fertilizer P (Y) (kg ha^{-1})
1	33	91	0
2	36	91	50
3	37	119	150
4	42	127	20
5	31	130	220
6	46	132	160
7	37	142	140
8	38	183	440
9	39	190	190

$Y = -235 + 2.77X$, $r = 0.73$

[a]Source: Power *et al.*, 1961a.

biologically, but often this reduction is less affected by stress than is crop growth and N uptake.

Legumes may also be used as a winter cover crop with a monoculture of maize or other crops (Power, 1987). This practice provides some of the benefits of crop rotations such as reduced weed and disease problems and reduced fertilizer N requirement, while maintaining or enhancing grain yields. Hargrove and Frye (1987) report that hairy vetch and crimson clover are among the best legumes for this use (Table X).

It is conceivable that springtime water used by the winter legume could restrict the emergence and growth of the following maize crop. However, in a subhumid climate such as eastern Nebraska, Koerner and Power (1987) did not observe such an effect. The added ground cover provided by the cover crop does reduce water loss by both evaporation and runoff, however.

Table IX. Corn Grain Yields (Mg ha^{-1}) as Affected by Crop Rotation and N Fertilizer[a]

N Rate (kg ha^{-1})	Crop Rotation[b]				
	C–C	*C*SbCO_x	CSb*C*O_x	*C*COM	C*C*OM
0	3.83	7.97	8.22	10.86	7.16
67	6.65	9.73	10.04	11.11	9.73
135	9.29	10.74	10.67	11.17	10.55
200	10.55	10.86	10.86	11.24	10.61

[a]Source: Voss and Shrader, 1984.

[b]C = corn; Sb = soybean; O_x = oats–red clover; M = legume meadow.

Table X. Estimated Nitrogen Contributed by Winter Legumes for No-Till Crop Production[a]

State	Crop	Cover Crop	Fertilizer N Equivalent (kg ha^{-1})
Kentucky	Corn	Hairy vetch	96
		Big flower vetch	50
Georgia	Sorghum	Crimson clover	84
		Hairy vetch	90
		Common vetch	59
		Subterranean clover	57
Alabama	Cotton	Hairy vetch	68
		Crimson clover	68

[a]Source: Hargrove and Frye, 1987.

Legumes are grown in various other types of rotation with grain crops. While this practice is usually beneficial for most situations, Haas and Evans (1957) found that alfalfa grown in rotation with wheat at several Great Plains locations dried the soil to such an extent that the following wheat yields were reduced for several years. This occurred even though the legume increased soil organic matter and organic N levels.

From these results, it is evident that if a legume is to be used as a soil-improving crop in dryland regions, it must be used in such a way that sufficient water is conserved to permit economic production of grain crops. In some regions, it appears that legume cover crops may fit this need. Slinkard *et al.* (1987) in western Canada and Ramig (1987) in Oregon found that lentil, Tangier pea, and field pea were valuable crops in those regions. Likewise, various clovers are grown extensively in rotation with wheat in Australia. However, much additional research is needed to better determine how legumes can be best used in dryland cropping to maintain soil productivity, environmental quality, and net income.

IV. Conclusions

Water stress usually reduces the efficiency with which crops utilize fertilizers (Power, 1983). Water use by plants is controlled mainly by abiotic properties of the environment, especially by radiant energy. Crop growth is the integrated product not only of the abiotic environment, but also of the numerous biologically mediated chemical transformations that occur in soils and plants. Water stress, therefore, has both direct effects on the availability of nutrients in the soil and indirect effects on the physiological processes associated with the growth and development of the crop. Consequently, relationships between water stress and fertilizer use are complex and difficult to generalize.

Rates of N mineralization and nitrification are reduced by increased water deficiency stress. This occurs primarily through the effects of water deficits on the activity of soil microorganisms. Bacterial processes such as nitrification are especially sensitive to water deficits. The effects of water stress and N stress are often additive, that is, correction of one stress without correcting the other often results in only modest increases in plant growth and yield.

Phosphorus is an element affected by both the biological activity and the chemical nature of the soil. Much P is present in crop residues and other forms of organic matter and is subject to the effects of water deficit in a manner similar to the effects of water deficiency on N transformations. However, in many soils a major part of the total P fraction exists in inorganic form and is subject to the effects of water on solubility of these P compounds. Fawcett and Quick (1962) concluded that much of the plant-available P in soils is found in the small-diameter soil pores, many of which were water filled even in relatively dry soil. The availability of a number of the micronutrients probably also parallels that of P and is also dependent on both the organic and inorganic reactions occurring in the soil—Zn, Fe, Mn, B, and others.

Potassium and a few other elements enter into the basic structure of few organic components of the plant or soil, but are usually found in the ionized form in solution or in equilibrium with K absorbed by active organic groups or colloidal surfaces. As such, upon senescence, K is readily released from crop residues into the soil solution. Thus, water stress has relatively little effect on K availability.

Management practices and cropping systems are major factors in controlling nutrient cycling. Use of no tillage rather than plowing leaves nutrients concentrated near the soil surface in a highly active biological zone. Use of monocultures compared to crop rotations greatly alters the soil environment, quantity and quality of crop residues returned, and the entire spectrum of biology of a soil. Water deficiency may be alleviated to some extent by leaving crop residues on the soil surface, but this practice reduces efficiency of utilization of surface-applied fertilizers. Thus, wise choices of management practices and cropping systems employed are needed for efficient nutrient utilization.

References

Aase, J. K., and D. L. Tanaka. 1987. Soil water evaporation comparisons among tillage practices in the Northern Great Plains. *Soil Sci. Soc. Am. J.* 51:436–440.

Barber, S. A., and R. A. Olson. 1968. Fertilizer use on corn. In L. B. Nelson, M. H. McVicker, R. D. Munson, L. F. Seatz, S. L. Tilsdale, and W. C. White (eds.), Changing Patterns in Fertilizer Use. Soil Sci. Soc. Am., Madison, Wisconsin. Pp. 37–63.

Black, A. L. 1973. Soil properties associated with crop residue management in a wheat-fallow rotation. *Soil Sci. Soc. Am. Proc.* 37:943–946.

Black, A. L. 1983. Cropping practices: Northern Great Plains. In H. E. Dregne and W. O. Willis (eds.), Dryland Agriculture. Argon. No. 23. Am. Soc. Agron., Madison, Wisconsin. Pp. 397–406.

Black, A. L., and F. H. Siddoway. 1976. Dryland cropping sequences within a tall wheatgrass barrier system. *J. Soil Water Conserv.* 31:101–105.

Blevins, R. L., M. S. Smith, and G. W. Thomas. 1984. Changes in soil properties under no-tillage. In R. E. Phillips and S. H. Phillips (eds.), No-Tillage Agriculture. Van Nostrand-Reinhold, New York. Pp. 190–230.

Bond, J. J., and W. O. Willis. 1971. Soil water evaporation: Long-term drying as influenced by surface residue and evaporation potential. *Soil Sci. Soc. Am. Proc.* 35:984–987.

Bresler, E., B. L. McNeal, and D. L. Carter. 1982. Saline and Sodic Soils, Principles–Dynamics–Modeling. Springer-Verlag, Berlin.

Broder, M. W., J. W. Doran, G. A. Peterson, and C. R. Fenster. 1984. Fallow tillage influence on spring populations of soil nitrifiers, denitrifiers, and available nitrogen. *Soil Sci. Soc. Am. J.* 48:1060–1067.

Carter, M. R., and D. A. Rennie. 1984. Dynamics of soil microbial biomass N under zero and shallow tillage for spring wheat using ^{15}N urea. *Plant Soil* 76:157–164.

Clark, F. E. 1977. Internal cycling of ^{15}nitrogen in short-grass prairie. *Ecology* 58:1322–1333.

Doran, J. W. 1980. Soil microbial and biochemical changes associated with reduced tillage. *Soil Sci. Soc. Am. J.* 44:765–771.

Doran, J. W. 1987. Microbial biomass and mineralizable nitrogen distributions in no-tillage and plowed soils. *Biol. Fertil. Soils* 5:68–75.

Doran, J. W., L. N. Mielke, and S. Stamatiadis. 1988. Microbial activity and N cycling as regulated by soil water-filled pore space. *Tillage and Traffic in Crop Production. Proc. 11th Int. Conf. Int. Soil Tillage Res. Organ. (ISTRO), Edinburgh* pp. 49–54.

Doran, J. W., and M. S. Smith. 1987. Organic matter management and utilization of soil and fertilizer nutrients. In R. F. Follett, J. W. B. Stewart, and C. V. Cole (eds.), Soil Fertility and Organic Matter as Critical Components of Production Systems. *SSSA Spec. Publ.* No. 19, pp. 53–72.

Eastin, J. D., T. E. Dickinson, D. R. Krieg, and A. B. Mauder. 1982. Crop physiology in dryland agriculture. In H. E. Dregne and W. O. Willis (eds.), Dryland Agriculture. *Agronomy* No. 23, pp. 333–364.

Ellis, J. R., H. J. Larson, and M. G. Boosalis. 1985. Drought resistance of wheat plants inoculated with vesicular-arbuscular mycorrhizae. *Plant Soil* 86:369–378.

Farquhar, G. D., R. Wetselaar, and P. M. Firth. 1979. Ammonia volatilization from senescing leaves of maize. *Science* 203:1257–1258.

Fawcett, R. G., and J. P. Quick. 1962. The effect of soil–water stress on the absorption of soil phosphorus by wheat plants. *Aust. J. Agric. Res.* 13:193–205.

Ford, C. W., and J. R. Wilson. 1981. Changes in levels of solutes during osmotic adjustment to water stress in leaves of four tropical pasture species. *Aust. J. Plant Physiol.* 8:77–91.

Grable, A. R., and E. G. Siemers. 1968. Effects of bulk density, aggregate size, and soil water suction on oxygen diffusion, redox potential, and elongation of corn roots. *Soil Sci. Soc. Am. Proc.* 32:180–186.

Greb, B. W., D. E. Smika, and A. L. Black. 1967. Effect of straw mulch rates on soil water storage during summer fallow in the Great Plains. *Soil Sci. Soc. Am. Proc.* 31:556–559.

Greenwood, E. A. N. 1976. Nitrogen stress in plants. *Adv. Agron.* 28:1–35.

Haas, H. J., and C. E. Evans. 1957. Nitrogen and carbon changes in Great Plains soils as influenced by cropping and soil treatments. *U.S. Dep. Agric., Tech. Bull.* No. 1146.

Hanson, A. D., and W. D. Hitz. 1983. Whole-plant response to water deficit: Water deficits and the nitrogen economy. In H. M. Taylor, W. R. Jordan, and T. R. Sinclair (eds.), Limitations to Efficient Water Use in Crop Production. Am. Soc. Agron., Madison, Wisconsin. Pp. 331–343.

Hargrove, W. L., and W. W. Frye. 1987. The need for legume cover crops in conservation tillage production. In J. F. Power (ed.), The Role of Legumes in Conservation Tillage Systems. Soil Conserv. Soc. Am., Ankeny, Iowa. Pp. 1–5.

Harper, L. A., V. A. Catchpoole, R. Davis, and K. L. Weier. 1983. Ammonia volatilization: Soil, plant, and microclimate effects on diurnal and seasonal fluctuations. *Agron. J.* 75:212–218.

Heichel, G. H., and D. K. Barnes. 1984. Opportunities for meeting crop nitrogen needs from symbiotic nitrogen fixation. In D. F. Bezdicek (ed.), Organic Farming. *ASA Spec. Publ.* No. 46, pp. 49–59.

Jackson, T. L., A. D. Halvorson, and B. B. Tucker. 1982. Soil fertility in dryland agriculture. In H. E. Dregne and W. O. Willis (eds.), Dryland Agriculture. *Agronomy* No. 23, pp. 297–332.

Jordan, W. R. 1983. Whole plant response to water deficit: An overview. In H. M. Taylor, W. R. Jordan, and T. R. Sinclair (eds.), Limitations to Efficient Water Use in Crop Production. Am. Soc. Agron., Madison, Wisconsin. Pp. 289–317.

Khera, K. L., and N. T. Singh. 1975. Fertilizer–aeration interaction in maize (*Zea mays* L.) under temporary flooding. *J. Indian Soil Sci. Soc.* 28:336–343.

Kissel, D. E., D. H. Sander, and R. Ellis, Jr. 1985. Fertilizer–plant interactions in alkaline soils. In O. P. Engelstad (ed.), Fertilizer Technology and Use. Soil Sci. Soc. Am., Madison, Wisconsin. Pp. 153–196.

Koerner, P. T., and J. F. Power. 1987. Hairy vetch winter cover for continuous corn in Nebraska. In J. F. Power (ed.), The Role of Legumes in Conservation Tillage Systems. Soil Conserv. Soc. Am., Ankeny, Iowa. Pp. 57–59.

Kozlowski, T. T. 1984. Plant response to flooding of soil. *BioScience* 34:162–167.

Lawton, K. 1945. The influence of soil aeration on the growth and absorption of nutrients by corn plants. *Soil Sci. Soc. Am. Proc.* 10:263–268.

Leyshon, A. J., and R. W. Sheard. 1974. Influence of short-term flooding on the plant nutrient composition of barley. *Can. J. Soil Sci.* 54:463–473.

Lindsay, W. L., and P. L. G. Vlek. 1977. Phosphate minerals. In J. B. Dixon and S. B. Weed (eds.), Minerals in Soil Environments. Soil Sci. Soc. Am., Madison, Wisconsin. Pp. 639–672.

Linn, D. M., and J. W. Doran. 1984. Effect of water-filled pore space on carbon dioxide and nitrous oxide production in tilled and nontilled soils. *Soil Sci. Soc. Am. J.* 48:1267–1272.

Meisinger, J. J., V. A. Bandel, G. Stanford, and J. O. Legg. 1985. Nitrogen utilization by corn under minimal tillage and moldboard plow tillage. I. Four-years' results using labeled N fertilizer on an Atlantic coastal plain soil. *Agron. J.* 77:602–611.

Mengel, K. 1985. Dynamics and availability of major nutrients in soils. *Adv. Soil. Sci.* 2:65–131.

Mielke, L. N., J. W. Doran, and K. A. Richards. 1986. Physical environment near the surface of plowed and no-tilled soils. *Soil Tillage Res.* 7:355–366.

Nelson, D. W. 1982. Gaseous losses of nitrogen other than through denitrification. In F. J. Stevenson (ed.), Nitrogen in Agricultural Soils. Am. Soc. Agron., Madison, Wisconsin. Pp. 327–363.

Paul, E. A., and N. G. Juma. 1981. Mineralization and immobilization of soil nitrogen by microorganisms. In F. E. Clark and T. Rosswall (eds.), Terrestrial Nitrogen Cycles. *Ecol. Bull.* 33:179–195.

Power, J. F. 1971. The evaluation of water and nitrogen stresses on bromegrass growth. *Agron. J.* 63:726–728.

Power, J. F. 1983. Soil management for efficient water use: Soil fertility. In H. M. Taylor, W. R. Jordan, and T. R. Sinclair (eds.), Limitations to Efficient Water Use in Crop Production. Am. Soc. Agron., Madison, Wisconsin. Pp. 461–470.

Power, J. F. (ed.). 1987. The Role of Legumes in Conservation Tillage Systems. Soil Conserv. Soc. Am., Ankeny, Iowa.

Power, J. F., J. J. Bond, W. A. Sellner, and H. M. Olson. 1973. Effect of supplemental water on barley and corn production in a subhumid region. *Agron. J.* 65:464–467.

Power, J. F., P. L. Brown, T. J. Army, and M. G. Klages. 1961a. Phosphorus responses by dryland spring wheat as influenced by moisture supplies. *Agron. J.* 53:106–108.

Power, J. F., J. W. Doran, and W. W. Wilhelm. 1986. Uptake of nitrogen from soil, fertilizer, and crop residues by no-till corn and soybean. *Soil Sci. Soc. Am. J.* 50:137–142.

Power, J. F., and R. F. Follett. 1987. Monoculture. *Sci. Am.* 256:79–86.

Power, J. F., D. L. Grunes, and G. A. Reichman. 1961b. The influence of phosphorus fertilization and moisture on growth and nutrient absorption by spring wheat. I. Plant growth, N uptake, and moisture use. *Soil Sci. Soc. Am. Proc.* 25:207–210.

Power, J. F., and R. I. Papendick. 1985. Organic sources of nutrients. In O. P. Engelstad (ed.), Fertilizer Technology and Use. 3rd Ed. Soil Sci. Soc. Am., Madison, Wisconsin. Pp. 503–520.

Power, J. F., G. A. Reichman, and D. L. Grunes. 1961c. The influence of phosphorus fertilization and moisture on growth and nutrient absorption by spring wheat. II. Soil and fertilizer P uptake in plants. *Soil Sci. Soc. Am. Proc.* 25:210–213.

Raju, P. S., R. B. Clark, R. J. Ellis, and J. W. Maranville. 1987. Vesicular-arbuscular mycorrhizal infection effects on sorghum growth, phosphorus efficiency, and mineral element uptake. *J. Plant Nutr.* 10:1331–1339.

Ramig, R. E. 1987. Conservation tillage systems for green pea production in the Pacific Northwest. In J. F. Power (ed.), The Role of Legumes in Conservation Tillage Systems. Soil Sci. Soc. Am., Ankeny, Iowa. Pp. 93–94.

Ramig, R. E., and H. F. Rhoades. 1963. Interrelationships of soil moisture level at planting time and nitrogen fertilization on winter wheat production. *Agron. J.* 55:123–127.

Reichman, G. A., D. L. Grunes, and F. G. Viets, Jr. 1966. Effect of soil moisture on ammonification and nitrification in two Northern Plains soils. *Soil Sci. Soc. Am. Proc.* 30:363–366.

Rice, C. W., and M. S. Smith. 1984. Short-term immobilization of fertilizer nitrogen at the surface of no-till and plowed soils. *Soil Sci. Soc. Am. J.* 48:295–297.

Sheppard, S. C., and G. J. Racz. 1980. Phosphorus nutrition of crops as affected by temperature and water supply. *Proc. Western Canada Phosphate Symp., Alberta Soil Feed Test. Lab., Edmonton* pp. 159–199.

Slinkard, A. E., V. O. Biederbeck, L. Bailey, P. Olson, W. Rice, and L. Townley-Smith. 1987. Annual legumes as a fallow substitute in the northern Great Plains in Canada. In J. F. Power (ed.), The Role of Legumes in Conservation Tillage Systems. Soil Sci. Soc. Am., Ankeny, Iowa. Pp. 6–7.

Smika, D. E., H. J. Haas, and J. F. Power. 1965. Effects of moisture and nitrogen fertilizer on growth and water use by native grass. *Agron. J.* 57:483–486.

Smith, S. J., and J. F. Power. 1985. Residual forms of fertilizer nitrogen in a grassland soil. *Soil Sci.* 140:362–367.

Smith, S. J., L. B. Young, and G. E. Miller. 1977. Evaluation of soil nitrogen mineralization potentials under modified field conditions. *Soil Sci. Soc. Am. J.* 41:74–76.

Stanford, G., and E. Epstein. 1974. Nitrogen mineralization–water relations in soils. *Soil Sci. Soc. Am. Proc.* 38:103–107.

Stevenson, F. J. 1986. Cycles of Soil Carbon, Nitrogen, Phosphorus, Sulfur, and Micronutrients. Wiley, New York. Pp. 380.

Taylor, H. M. 1983. Managing root systems for efficient water use: An overview. In H. M. Taylor, W. R. Jordan, and T. R. Sinclair (eds.), Limitations to Efficient Water Use in Crop Production. Am. Soc. Agron., Madison, Wisconsin. Pp. 87–114.

Taylor, H. M., and B. Klepper. 1974. Water relations of cotton. I. Root growth and water use as related to top growth and soil water content. *Agron. J.* 66:584–588.

Tisdale, S. L., W. L. Nelson, and J. D. Beaton. 1985. Soil Fertility and Fertilizers. 4th Ed. Macmillan, New York.

Trought, M. C. T., and M. C. Drew. 1980. The development of waterlogging damage in wheat seedlings (*Triticum aestivum* L.). II. Accumulation and redistribution of nutrients by the shoot. *Plant Soil* 56:187–199.

Van Veen, J. A., J. N. Ladd, J. K. Martin, and M. Amato. 1987. Turnover of carbon, nitrogen, and phosphorus through the microbial biomass in soils incubated with ^{14}C, ^{15}N, and ^{32}P-labelled bacterial cells. *Soil Biol. Biochem.* 19:559–565.

Viets, F. G., Jr. 1972. Water deficits and nutrient availability. In T. T. Kozlowski (ed.), Water Deficits and Plant Growth. Vol. 3. Academic Press, New York. Pp. 217–240.

Voss, R. D., and W. D. Shrader. 1984. Rotation effects and legume sources of nitrogen for corn. In D. F. Bezdicek (ed.), Organic Farming. *ASA Spec. Publ.* No. 46, pp. 61–68.

Wicks, G. A. 1976. Eco-fallow: A reduced tillage system for the Great Plains. *Weeds Today* 7:20–23.

Wilhelm, W. W., J. W. Doran, and J. F. Power. 1986. Corn and soybean yield response to crop residue management under no-tillage production systems. *Agron. J.* 78:184–189.

13

Soil–Plant Nutrient Relationships at Low pH Stress

N. K. FAGERIA, V. C. BALIGAR, and D. G. EDWARDS

Soil pH is a measure of the activity of H^+ ions in the soil solution. The pH of acid soils is less than 7, that of neutral soils is equal to 7, and that of alkaline soils is greater than 7. At neutrality, the concentration of OH^- ions and H^+ ions is the same, since these ions are derived in equal quantities from the ionization of water. The pH of most agricultural soils is in the range 4 to 9. Since most of the plant essential nutrients in a soil reach maximal or near-maximal availability in the pH range 6 to 7 and decrease both above and below this range, soil pH is an indicator of the relative availability of nutrients (McLean, 1973).

Low soil pH stress is a major growth limitation to crop production in many

Crops as Enhancers of Nutrient Use

regions of the world (Sanchez, 1976; Kamprath, 1978; Adams, 1981; Clark, 1982; Foy, 1984). Soil acidity may result from parent materials that were acid and naturally low in the basic cations (Ca^{2+}, Mg^{2+}, K^+, and Na^+) or because these elements have been leached from the soil profile by heavy rains (Kamprath and Foy, 1971). When precipitation exceeds evapotranspiration, leaching occurs. In highly leached soils only iron and aluminum oxides and some of the trace metal oxides, which are highly resistant to weathering, remain from the original parent material (Bohn *et al.*, 1979).

Soil acidity may also develop from exposure to the air of mine spoils containing iron pyrite (FeS_2) or other sulfides. Crop fertilization with ammonia or ammonium fertilizers can result in soil acidification by the microbially mediated reaction (Bohn *et al.*, 1979)

$$NH_4^+ + 2O_2 = NO_3^- + 2H^+ + H_2O \quad (1)$$

Finally, acidity may be produced by the decomposition of plant residues or organic wastes into organic acids. This process is of particular importance in many forest soils.

The factors that constitute acid infertility and govern plant growth in acid soils are complex. In acid mineral soils, a variety of individual chemical constraints and the interactions between them limit plant growth. At low pH, it is not often the H^+ ion activity that limits growth, but the toxicity and/or deficiency of other elements. Plant growth problems associated with poor root growth in acid subsoils include increased drought susceptibility and poor use of subsoil nutrients.

Soils that belong to the orders Oxisols, Ultisols, and Alfisols comprise about 50% of the land surface of the tropics and about 27% of the land area of the world (Lal, 1986). These soils commonly are sufficiently acid to limit plant growth. A preliminary survey indicates that at least 50×10^6 ha of the Cerrado region of Brazil have potential for intensive mechanized agriculture, but development has been limited by soil acidity constraints (Goedert, 1983). Most of these soils are in areas of low to medium population density, present excellent physical conditions, and are located in areas where climatic factors are favorable for crop production (Goedert *et al.*, 1982).

The magnitude of the low pH stress problem in many agricultural areas around the world, and the potential that these areas offer in increasing the production of food and fiber, provide a focus for the objectives of this chapter, i.e., to review the nature, causes, and management of soil acidity in order that crop productivity on acid soils might be improved for the benefit of humankind.

I. Growth-Limiting Factors in Acid Soils

Soil acidity is a complex of factors that may restrict the growth of different plants through different physiological and biochemical pathways (Foy, 1983). In the case of legumes reliant on symbiotic nitrogen fixation, quality and quantity of plant growth are important. Effects of soil acidity on survival and growth of rhizobia, on nodulation per se, and on nodule function are critical to production on acid soils. Considerable research has been conducted to identify soil acidity constraints to crop production, but much still remains to be done. Soil acidity constraints to plant growth include excesses of certain elements, deficiencies or low availabilities of others, and combinations of excesses and deficiencies. At a given soil pH value, the limiting factor may vary with soil type; in a given acid soil, it may vary with plant species or cultivar. A common practical approach to solve the soil acidity problem at a given site is first to identify the growth-limiting factors, and then to tailor both the plant and the soil to the specific soil problem.

On the basis of existing literature, the following yield-limiting factors have been identified in acid soils: (1) toxicities of H, Al, and Mn; (2) deficiencies of N, P, K, Ca, Mg, Mo, and Zn; and (3) reduced activity of beneficial soil microorganisms.

A. LOW pH TOXICITY

Direct toxicity of H^+ ions to plant growth in acid soils is no longer a widely held hypothesis, however, indirect effects of low pH on nutrient availability and on nutrient absorption by plant roots are of importance (Blamey *et al.*, 1987). It is very difficult to determine the direct effects of H^+ ion toxicity on plant growth in acid soils because of the changing interrelationships that occur between pH, Al and Mn concentrations in the soil solution, and the changing availability of essential nutrients including P, Ca, Mg, and Mo. The concentrations of the above elements present in soil solution and their availability to plants as determined by either plant uptake or an empirical soil test vary with soil pH.

Studies have been conducted using both soil and solution culture to determine the effects of pH on plant growth. Data in Figure 1 show the effect of soil pH on the grain yield of upland rice in an Oxisol of central Brazil. Increase in soil pH (1 : 2.5, soil/water ratio) from 5.0 to 5.5 increased grain yield. Since the effects of H^+ are confounded with other factors in acid soils, it is not possible on the basis of the data in Figure 1 to identify the yield-limiting factor. Such difficulties can be avoided by the use of solution cultures in which the solution pH is maintained within relatively narrow limits (Islam *et al.*, 1980; Blamey *et al.*, 1982). Islam *et al.* (1980) grew six species, including cassava, french bean, wheat, and maize, in flowing solution cultures closely controlled at pHs from 3.3 to 8.5 and found that all species attained maximal or near-maximal yields within the pH range of 5.5 to 6.5. Within species differences in growth response to low pH stress have been reported in sunflower (Blamey *et al.*, 1982) and in subterranean clover (Kim, 1985). The critical solution pH, corresponding to 90% of maximum total dry matter yield, for four sunflower cultivars ranged from 4.0 to 5.0 (Blamey *et al.*, 1982). A solution pH of 3.5 was lethal to four sunflower cultivars and also to 11 subterranean clover cultivars studied by Kim (1985). Andrew (1976) used subirrigated sand culture to determine the

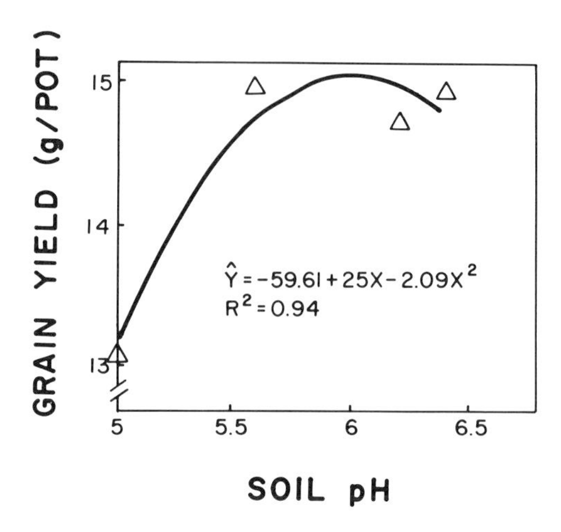

Figure 1. Average of 48 cultivars. Influence of soil pH on grain yield of upland rice grown in an Oxisol of central Brazil. Source: N. K. Fageria, unpublished.

effects of solution pH from 4.0 to 6.0 on growth of nodulated and nitrogen-supplied tropical and temperate pasture legumes. Plants reliant on symbiotic nitrogen fixation were less tolerant of low pH than nitrogen-supplied plants; for example, *Stylosanthes humilis* achieved maximum growth at pH 4.0 when supplied with adequate inorganic nitrogen, but required pH 5.0 to achieve near-maximal growth when dependent on symbiotic nitrogen fixation.

Marked effects of low pH in inhibiting or restricting nodulation of legumes by root nodule bacteria have been reported (Loneragan and Dowling, 1958; Andrew, 1976; Kim, 1985). Nodulation generally appears to be more sensitive than host plant growth to low pH. Root infection is particularly sensitive to low pH (<5) and requires a higher pH (>5) than that necessary for survival of rhizobia (Munns, 1978). Despite the addition of a very high population of *Rhizobium,* alfalfa at pH 4.6 (Munns, 1968) and peas at pH 4.5 (Lie, 1969) failed to nodulate.

Low pH results in increased solubility of certain heavy metals in soils, including Al, Mn, and Fe (Bartuska and Unger, 1980). High concentrations of these elements in acid soils may be toxic to growth of most plants. Hydrogen ions and these metallic cations replace other plant essential cations at the cation-exchange sites; consequently, essential elements such as Ca, Mg, and K are leached from the soil, which becomes base unsaturated.

Mahler and McDole (1987) compiled data from 39 field studies conducted on Mollisols and Alfisols to determine the effect of soil pH on the yield of lentil, spring pea, winter wheat, and spring barley. Lentil and pea were the least tolerant to acid conditions with minimum acceptable pH (1 : 1, soil/water ratio) values required for maximum yield of 5.65 and 5.52, respectively. The cereals were more tolerant, with minimum acceptable pH values of 5.23 for barley and 5.19 for wheat.

Mengel and Kamprath (1978) determined that the critical soil pH for shoot and root growth and nodulation of soybean grown in eight Histosols was in the pH (1 : 3, soil/water ratio) range from 4.6 to 4.8. The lower critical soil pH in organic soils than in mineral soils has been attributed to the considerably lower Al concentration in soil solutions of the former group at the same pH (Evans and Kamprath, 1970). This is due, in part, to the formation of stable Al–organic matter complexes of low solubility (Schnitzer and Skinner, 1973). In addition, solution culture studies have shown that soluble Al–organic complexes are not toxic to plants (Bartlett and Riego, 1972; Hue *et al.*, 1986).

B. ALUMINUM TOXICITY AND TOLERANCE IN CROP PLANTS

Aluminum toxicity is probably the major limiting factor to plant growth and crop production in strongly acid soils (Foy, 1984). Aluminum toxicity reduces growth of both roots and shoots. Symptoms of Al toxicity on shoots are not easily identified, whereas the root system of plants severely affected by Al toxicity is often coralloid in appearance, with many stubby lateral roots and no fine branching (Foy, 1974, 1984). Nevertheless, the extent of the Al toxicity problem in acid soils has been underestimated because of the undue reliance placed on development of the characteristic severe toxicity symptoms on roots. Reduction in root growth, particularly in acid subsoils, limits the plant's ability to absorb water and nutrients, and consequently limits crop yields. Investigations on legumes have shown that the effects of Al toxicity are more severe on nodulation than on growth of the host plant (Carvalho *et al.*, 1981; Murphy *et al.*, 1984; Alva *et al.*, 1987).

Since Al undergoes hydrolysis in aqueous media, including soil solutions of acid soils, the possible hydrolysis species that may be present should be considered (Richburg and Adams, 1970; Marion *et al.*, 1976; Rhodes and Lindsay, 1978; Blamey *et al.*, 1983). The total concentration of monomeric Al in acid solutions is the sum of the concentrations of the various monomeric species, i.e., Al^{3+}, $Al(OH)^{2+}$, $Al(OH)_2^+$, and $Al(OH)_3^0$. With sulfate in solution, an additional species (i.e., $AlSO_4^+$) is also present, while complex ions of Al with SO_4^{2-} and F^- also occur when these anions are present in solution (Lindsay, 1979; Cameron *et al.*, 1986; Tanaka *et al.*, 1987). Marion *et al.* (1976) suggested that the total soluble Al concentration in soil solutions over the pH range 4–6 was best described by

$$[\text{Total Al}] = [Al^{3+}] + [Al(OH)^{2+}] + [Al(OH)_2^+] + [Al(OH)_3^0] + [AlSO_4^+] \quad (2)$$

If the total concentration of monomeric Al in solution (or soil solution), together with the pH, ionic strength, and SO_4^{2-} concentration, is known, the concentrations, activity coefficients, and activities of the individual monomeric species included in Eq. (2) can be calculated. Blamey *et al.* (1983) used this basic approach to describe root elongation of soybean in nutrient solutions in a series of experiments in terms of the sum of the activities of the monomeric Al species ($\Sigma a_{\text{Al mono}}$) included in Eq. (2). This approach was successful, irrespective of the initial (nominal) Al concentration, the concentration of polymeric Al species, ionic strength, pH, P concentration, and solution composition. Alva *et al.* (1986a) have further shown the advantage of using $\Sigma a_{Al\ mono}$ over use of the nominal Al concentration (based on Al added), the measured total Al concentration in solution, or measured monomeric Al concentration in solution.

In Al solutions with low OH : Al and P : Al molar ratios, most of the Al is present as monomeric species; at high OH : Al and P : Al ratios, soluble Al polymers may form. These polymers appear to be readily precipitated by phosphate (Hsu, 1968), although there is some evidence that they remain soluble in dilute solutions (White *et al.*, 1976; Blamey *et al.*, 1983). Monomeric Al species are shown to be phytotoxic (Blamey *et al.*, 1983). However, the relative toxicities of the various monomeric Al species to plant growth remain to be clarified. Pavan and Bingham (1982) suggested that Al^{3+} was the species most toxic to coffee seedlings in solution culture; their assumption that all of the soluble Al was present as monomeric species may not be valid (Bruce *et al.*, 1988). The solution culture experiments of Alva *et al.* (1986a,b, 1987) have shown that the activity of $Al(OH)^{2+}$ or $Al(OH)_2^+$, but never Al^{3+}, correlated best with plant growth parameters. In plant growth experiments where soil solution Al has been measured and activities calculated, the basic assumption was either that all soluble Al was present as Al^{3+} or that all was present as monomeric species. Despite the study of Bruce *et al.* (1988), which showed that the activities of Al^{3+} and $Al(OH)^{2+}$ were the best predictors of Al toxicity limitations to root elongation of soybean in acid soils (mainly subsoils), problems still exist. The methods used to measure monomeric Al (i.e., 15-sec reaction with 8-hydroxyquinoline) before applying speciation and activity calculations fail to discriminate adequately between the phytotoxic monomeric species and the non-phytotoxic, soluble, organically complexed species (Adams and Hathcock, 1984). Kerven *et al.* (1989a,b) have now developed a procedure, based on different reaction rates with the chromogens aluminon and pyrocatechol violet, that makes this discrimination. Hodges (1987) has also approached this problem by using a fluoride electrode technique to quantify the activity of Al^{3+} in soil solutions.

Wright *et al.* (1987a) measured reactive Al in soil solution in 10 limed and unlimed acidic soil horizons using 8-hydroxyquinoline, aluminon, and ferron techniques. They related reactive Al in soil solution to the root and shoot growth of subterranean clover (cv. ‘Mt. Barker’) and switchgrass (cv. ‘Cave-in-Rock’). The amount of Al reacting with 8-hydroxyquinoline and aluminon was significantly related ($P < 0.05$) to root and shoot growth limitations exhibited by both crops. Activities of Al^{3+} calculated from a GEOCHEM program using the Al reacting with 8-hydroxyquinoline and aluminon as the Al inputs were generally the best predictors of root and shoot growth. Predictive capability of regression equations describing root and shoot growth was greatly improved for subterranean clover by including measures of soil or soil solution Ca and pH with Al^{3+} activity. Wright *et al.* (1989) related 4-day root growth of seedlings of three wheat cultivars in an acid soil treated with 10 levels each of $Ca(OH)_2$, $CaSO_4$, and $CaCl_2$ to Al speciation in soil solution. Seedling root length in the $Ca(OH)_2$ treatments was significantly related ($P < 0.01$) to calculated Al^{3+} activity in soil solution. The Al–SO_4 complex in soil solution had a negligible effect on the root growth of 'Hart' wheat, thus confirming the previously reached conclusions concerning the nonphytotoxicity of the Al–SO_4 complex (Cameron *et al.*, 1986).

Solution culture techniques that reproduce, in large part, the properties of soil solutions present in acid soils provide an outstanding opportunity for study of the effects of Al on plant growth and nodulation (Bell and Edwards, 1987). However, this opportunity has not been fully appreciated nor grasped in many past and contemporary studies of Al toxicity. Many solution culture experiments have employed Al concentrations greatly in excess of those commonly found in acid soil solutions, together with high phosphate concentrations and solution pH values that would result in a considerable loss of Al from solution (Blamey *et al.*, 1983). Most studies have related plant growth to the nominal (added) Al concentration only, while some have related plant growth to the total measured Al concentration (Alva *et al.*, 1986a). Very few studies have achieved the rigor that occurs when plant growth is related to the measured concentration of phytotoxic monomeric species or to $\Sigma a_{\mathrm{Al\ mono}}$ (Alva *et al.*, 1986a). Although these various studies have often determined a critical toxicity level for Al corresponding to a 10% decrement from maximum yield, much of this information is of limited utility.

Comparison of a critical nominal Al concentration with a critical $\Sigma a_{\mathrm{Al\ mono}}$, even where the cultivar is common, is not possible. For example, Munns (1965) provided data for subterranean clover cv. Mt. Barker from which a critical Al concentration, based on added Al of about 40 μM, can be derived. The plants were grown at pH 4.1 ± 0.2 in a solution of ionic strength 25.1 m*M*, which contained 5 m*M* calcium and 50 μ*M* phosphorus. Subsequently, Kim (1985) examined the growth response of 14 cultivars of subterranean clover, including Mt. Barker, to Al in an experiment in which the concentration of monomeric Al was regularly measured and adjusted. The solution pH was 4.3 ± 0.1, the ionic strength was 3.28 m*M*, and the solution contained 500 μ*M* calcium and 1 μ*M* phosphorus. In this study, virtually all of the Al remained in solution as monomeric species. The subterranean clover cultivar Mt. Barker was much more sensitive to Al in this study than in that of Munns (1965). A critical monomeric Al concentration of about 3 μ*M* and a critical $\Sigma a_{\mathrm{Al\ mono}}$ of about 2.5 μ*M* were calculated from Kim's (1985) data. A critical nominal Al concentration of 86 μ*M* was determined for sorghum 'Funk G-522DR' (Ohki, 1987). In contrast, Blamey *et al.* (1986b) reported a critical $\Sigma a_{\mathrm{Al\ mono}}$ of about 5 μ*M* for sorghum 'Texas RS610.' These contrasting studies illustrate the difficulties that exist in assessing species and cultivar tolerance to Al toxicity on the basis of nutrient solution studies.

However, when plants are grown in common nutrient solutions or in nutrient solutions in which the activity of monomeric Al species can be measured and controlled, comparisons between species and cultivars are possible. In particular, studies of the comparative tolerance of legume growth and nodulation to Al are very feasible. Thus, Alva *et al.* (1987) showed that the critical $\Sigma a_{\mathrm{Al\ mono}}$ for growth of the roots and shoots of soybean cultivar 'Fitzroy' were 5 and 9 μ*M*, respectively, whereas the critical $\Sigma a_{\text{Al mono}}$ for nodulation was 0.4 μ*M*. Other studies have also shown that nodulation is more sensitive than host plant growth to Al (Carvalho *et al.*, 1982; Kim, 1985; Suthipradit, 1989), while still other reports have shown that the primary Al limitation is on host plant growth and not on nodulation (Munns *et al.*, 1981; Franco and Munns, 1982).

Reports in the literature of toxic Al concentrations associated with yield reduction in a wide range of plant species are quite extensive. Andrew *et al.* (1973) determined the effects of Al concentrations from 0 to 74 μ*M* on growth of 11 pasture legumes supplied with inorganic nitrogen, while Asher (1981) reported that toxic Al concentrations for various species grown in flowing solution culture were: maize, ≥20 μ*M;* cassava, ≥20–30 μ*M;* soybean, ≥40 μ*M;* sweet potato, ≥40–80 μ*M;* ginger, >80 μ*M;* and taro, ≥80 μ*M*. While comparisons can be made among species in studies such as these where all plants are grown in a common solution or similar solutions, comparisons between different studies are difficult. Differences in solution pH, ionic strength, phosphate concentration, and calcium concentration make such comparisons difficult. This difficulty is well illustrated by the data of Alva *et al.* (1986b), who showed that the value of $\Sigma a_{\text{Al mono}}$ necessary to reduce root elongation by 50% varied with increasing solution Ca concentration from 0.5 to 15 m*M;* this variation was from 12 to 17 μ*M* for soybean, <8 to 16 μ*M* for sunflower, <7 to 15 μ*M* for subterranean clover, and 5 to 10 μ*M* for alfalfa. Rigid comparisons of this type between different studies will only become possible when plant growth, and for that matter nodulation, is related to the activity of the monomeric Al species present in nutrient solutions and in soil solutions.

Diagnosis of Al toxicity in soils has, in the past, rested on relationships between plant growth and the amount of exchangeable Al or Al saturation of the effective cation-exchange capacity. Thus, critical soil exchangeable Al levels reported for alfalfa have ranged from 0.2 to 0.9 cmol kg^{-1} (Moschler *et al.*, 1960) or 4 to 19% Al saturation (Foy, 1964). Wright *et al.* (1987b) evaluated acid soil limitations to plant growth in 55 horizons of 14 major Appalachian hill land soils. Aluminum-sensitive 'Romano' and Al-tolerant 'Dade' snap beans were grown for 5 weeks in limed and unlimed conditions. Exchangeable Ca, soil Ca saturation, and total soil solution Ca were positively correlated ($P < 0.01$) with root and shoot growth. Soil Al saturation, total soil solution Al, and soil solution Al reacting in 15 sec with 8-hydroxyquinoline were negatively correlated ($P < 0.01$) with growth. The ratio of Ca/Al in soil solution was more closely related to snap bean growth than the soil solution concentration of any individual element. These findings confirm the importance of considering Ca as well as Al when investigating Al phytotoxicity in acid soils.

Neither exchangeable Al nor Al saturation is suitable for general application to all soils in defining critical soil Al levels, and it is clear that our understanding of Al limitations to growth and nodulation of soil-grown plants will depend on advances in our understanding of the complex chemistry of Al in acid soil solutions (Bell and Edwards, 1987; Bouldin *et al.*, 1987).

Plant species and cultivars within species vary considerably in their ability to grow on Al-toxic soils or media containing Al (Pearson, 1975; Spain *et al.*, 1975; Munns and Fox, 1977; Fageria and Carvalho, 1982; Foy, 1983; Fageria, 1985; Ohki, 1985; Fageria *et al.*, 1987a; Baligar *et al.*, 1987, 1988, 1989). An example of the considerable cultivar differences in Al toxicity that exist between rice cultivars is shown in Figure 2. The exact physiological mechanisms by which certain plants

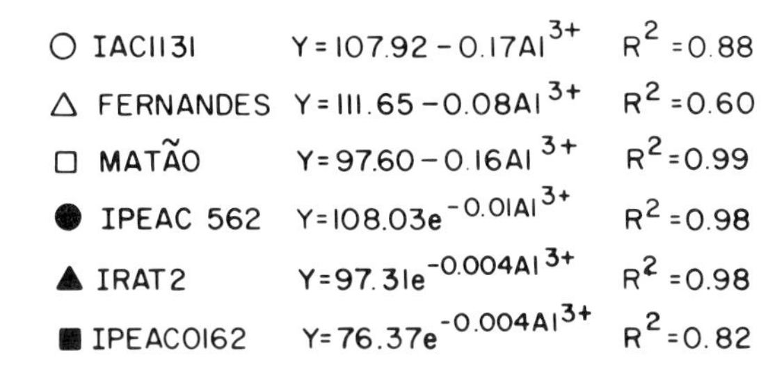

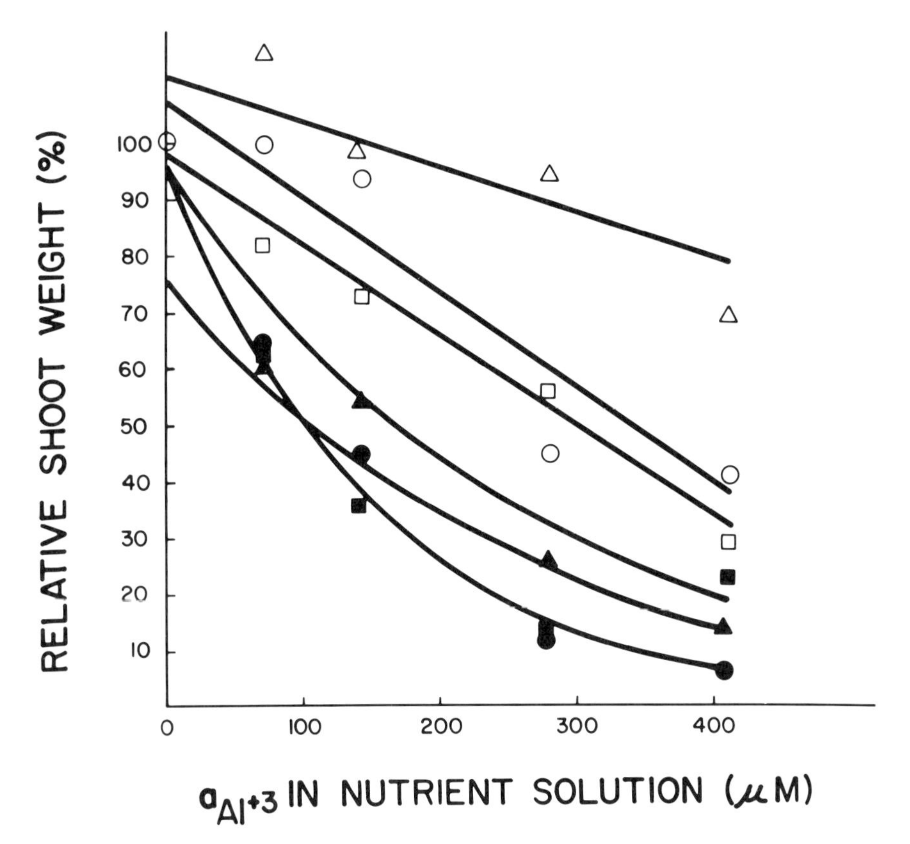

Figure 2. Influence of aluminum activities on relative shoot growth of six rice cultivars. Source: Fageria *et al.*, 1987b.

tolerate high levels of Al in their growth medium are still debated; Foy (1984) and recently Fageria *et al.* (1988) have summarized various proposed hypotheses relating to aluminum tolerance in plants, but considerable research must be done to verify these. Various proposed hypotheses (Foy, 1984; Fageria *et al.*, 1988) state that Al tolerance in plants is associated with pH increases of the growth medium, Al trapping in nonmetabolic sites within plants, greater P, Ca, and Mg uptake and transport and use efficiency, lower root CEC, greater root phosphatase activity, higher internal concentrations of Si, higher NH_4^+ or NO_3^- tolerance or preference, higher organic acid contents, greater Fe efficiency, and increased resistance to drought. Aluminum tolerance in higher plants appears to be due to a combination of both exclusion and internal tolerance mechanisms (Taylor, 1988). The exclusion mechanism refers to the immobilization of Al at the root–soil interface. Binding of Al by cell walls, selective permeability of plasma membranes, plant-induced pH barriers, and exudation of chelate ligands are known to play a role in Al exclusion mechanisms. Chelation of Al by carboxylic acids or Al-binding proteins of the cytosol, compartmentation of Al in the vacuole, and evolution of an Al-tolerant enzyme system have been proposed by Taylor (1988) as possible internal Al tolerance mechanisms of plants.

C. MANGANESE TOXICITY AND TOLERANCE IN CROP PLANTS

Manganese toxicity is another constraint to plant growth in acid soils; it is a common stress for plant growth in Oxisols, but seldom in Ultisols (Pearson, 1975). Although its geographical extent is not known, it is believed to be a less common constraint for crop growth than Al toxicity (Sanchez and Salinas, 1981). Soil pH, the levels of total and easily reducible Mn, organic matter content, aeration, and microbial activity are important factors that govern the availability of Mn to soil-grown plants (Cheng and Ouellette, 1971; Foy, 1973). Manganese toxicity in plants has been associated with a series of deleterious physiological and biochemical effects, including auxin destruction in cotton (Morgan *et al.*, 1966), a reduced number of leaf cells and cell volume in sugar beet (Terry *et al.*, 1975), a loss of control of Mn-activated enzyme systems (Helyar, 1978), and reduced nitrate reductase activity in soybean (Heenan and Campbell, 1981). Plants absorb Mn primarily as the Mn^{2+} ion. Lowering soil pH below about 5.5 increases the concentration of Mn^{2+} ions in the soil solution and, consequently, increases the likelihood of Mn toxicity (Kamprath and Foy, 1985). Manganese toxicity can occur in soils at pH values up to 6.0 (Sanchez, 1976; Kamprath and Foy, 1985). The lime rate commonly needed to raise the pH of Mn-toxic Oxisols to about 6.0 is usually very high. Thus 12 tons ha^{-1} of dolomitic lime was required to increase the soil pH (1 : 2.5; soil/water ratio) of an Oxisol from central Brazil from 5.0 to 6.0 (Fageria, 1984). An alternative or complementary solution to the high rate of lime application is the selection of tolerant plant species or cultivars.

A large number of extraction procedures have been used to evaluate the availability of soil Mn to plants (Cox and Kamprath, 1972; Gambrell and Patrick, 1982). However, few of these methods have been used to predict the likely occurrence of Mn toxicity to crop plants and it has been difficult to find consistent relationships between extractable Mn and toxic effects (Helyar, 1978). Kamprath and Foy (1985) reported that Mn toxicity in alfalfa occurred on several acid soils when the neutral 1 *M* NH_4OAc extractable content of the air-dried soils was >50 mg kg^{-1}. However, Bromfield *et al.* (1983b) found that extraction with neutral NH_4OAc was unsatisfactory because it overestimated the available forms of Mn in acid soils containing high levels of easily reducible Mn. They found that Mn extracted by 0.01 *M* $CaCl_2$ provided a better correlation with Mn concentration in rape and subterranean clover grown in acid soils than soil solution Mn and Mn extracted by neutral NH_4OAc. Growth of rape was significantly depressed when 0.01 *M* $CaCl_2$-extractable Mn was >20 mg kg^{-1} and 0.01 *M* $CaCl_2$-extractable Al was low (<1.8 mg kg^{-1}); however, when extractable Al exceeded 1.8 mg kg^{-1}, Mn toxicity reactions occurred as low as 10 mg kg^{-1} extractable Mn. Manganese toxicity symptoms in the more tolerant subterranean clover were observed only on soils containing >50 mg kg^{-1} $CaCl_2$-extractable Mn. Little information is available on soil solution Mn concentrations that are associated with Mn toxicity.

Wright *et al.* (1987b) estimated soil-extractable Mn by 0.01 *M* $CaCl_2$ and soil solution Mn by immiscible displacement using 1, 1, 2-trichloro-1, 2, 2-trifluoroethane in 55 soil horizons from the 14 major hill land soils of the Appalachian region. They compared soil and solution Mn levels to shoot and root growth of snap beans (cv. Dade and cv. Romano). In general, soil and soil solution Mn were not significantly related to snap bean growth, even though extractable Mn values were as high as 156 mg kg^{-1} of soil. Previous research (Bromfield *et al.*, 1983a) has indicated that Mn levels above 50 mg kg^{-1} soil may pose a Mn toxicity hazard to many crops. Aluminum toxicity and Ca deficiency problems in various soils studied by Wright *et al.* (1987b) might have masked the potential toxic effects of Mn.

Generally, agreement does not exist as to the most appropriate method to estimate plant-available Mn in soils. Wright *et al.* (1988) adopted 1 *M* NH_4OAc (pH 7), 0.01 *M* $CaCl_2$, 0.05 *M* $CaCl_2$, 0.033 *M* H_3PO_4, 0.005 *M* DTPA, 0.2% hydroquinone in 1 *M* NH_4OAc (pH7), and 0.01 *M* $NH_4OH \cdot HCl$ in 0.01 *M* HNO_3 to extract Mn from 11 limed and unlimed acidic subsoil horizons. Further, they also estimated total soil solution Mn and the concentration and activity of Mn^{2+} in soil solution by using the GEOCHEM program. The measured and calculated values of soil solution Mn generally gave the best correlations with subterranean clover and switchgrass Mn concentrations and Mn uptake. Root Mn concentrations were highly correlated with soil solution Mn. The Mn extracted by 0.01 *M* $CaCl_2$ was significantly correlated ($P < 0.01$) with plant Mn concentrations and Mn uptake and proved to be better than the other extractants in estimating plant-available Mn. Although Mn concentrations as high as 1.76 g kg^{-1} (shoots) and 8.49 g kg^{-1} (roots) were noted in subterranean clover, Mn did not appear to be a major growth inhibitor. However, flowing solution culture studies have established that the minimum solution concentration that is toxic to plant growth ranges from as low as 0.3 μM for disc medic and barrel medic

(Robson and Loneragan, 1970) to 51 μM for centro and 65 μM for sunflower (Edwards and Asher, 1982). These latter values are associated with a 10% growth decrement from the yield plateau. The threshold solution concentrations for Mn toxicity obtained from flowing solution culture studies are lower than those reported from earlier studies in which solution concentrations were not constantly maintained (Edwards and Asher, 1982).

The concentration of Mn in plant tissue is often used to confirm a suspected Mn toxicity limitation based on the recognition of toxicity symptoms. A very wide variation exists among plant species in critical internal Mn concentrations associated with Mn toxicity (Andrew and Hegarty, 1969; Helyar, 1978; Edwards and Asher, 1982; Foy, 1984). The Mn concentrations in whole tops of young plants that are sufficient to cause symptoms or depress growth by 10% range from 140 mg kg^{-1} in soybean cultivars 'Custer' and 'Lee' (Heenan and Carter, 1976) to 5300 mg kg^{-1} in sunflower (Edwards and Asher, 1982), a species that is able to localize excess Mn in a metabolically inactive form (Blamey *et al.*, 1986a).

Plant species and cultivars within species differ widely in tolerance to excess soluble Mn (Andrew and Hegarty, 1969; Foy *et al.*, 1969; Robson and Loneragan, 1970; Carter *et al.*, 1975; Heenan and Carter, 1976; Edwards and Asher, 1982; Nelson, 1983; Evans *et al.*, 1987). Manganese tolerance in plants has been associated with oxidizing powers of roots, rates of Mn absorption by roots and transport to tops, Mn entrapment in nonmetabolic centers, a high internal tolerance to excess Mn, and the uptake and distribution of Si and Fe (Foy, 1973). Edwards and Asher (1982) concluded that the differential tolerance to excess Mn among crop and pasture species depended mainly on the ability of the plant tops to tolerate high tissue Mn concentrations, and that low rates of absorption and/or high retention in the roots were important in some species. Tolerance to excessive levels of Mn is not necessarily correlated with tolerance to excessive levels of Al and vice versa in rice (Nelson, 1983). Separate screening for Mn and Al tolerance is therefore necessary, with screening for both required under certain soil conditions.

II. Deficiencies and Uptake Efficiencies of Nutrients in Crops

A. NITROGEN

Nitrogen deficiency is a major limitation to plant growth in acid soils, in both tropical and temperate regions. In tropical America, N deficiency is a major soil constraint over 93% of the region occupied by Oxisols and Ultisols (Sanchez and Salinas, 1981). The main reasons for widespread N deficiency in these and other similar regions in the tropics are a lower rate of N application than that removed in harvested crops or lost by other processes and the decreases in organic matter content with successive cultivations.

Land clearing and cultivation result in degradation of acid tropical soils through a decline in the annual additions of organic matter and an increase in the decomposition rates of organic carbon (Sanchez, 1976). The extent and consequences of this loss of organic matter vary with farming practices, soil type, and climate. The rapid decline in organic matter content of cultivated acid soils in the tropics is a consequence of the continuously high temperature (low moisture) throughout the year. The rate of mineralization of organic matter in tropical soils may be four times greater than that in temperate region soils (Jenkinson and Ayanaba, 1977). Consequently, cultivated soils in the tropics contain lower organic matter levels than cultivated soils in temperate regions. Greenland (1974) reported a substantial decline in organic carbon in an African savanna soil with cultivation; before clearing, the organic carbon ranged from 0.8 to 2.9%, but after 2 years of traditional local cropping, it declined to 0.7 to 2.0%, with a greater drop to 0.6 to 1.6% where intensive cultivation methods were practiced.

The rapid decline in organic matter content reduces the soil's structural stability, renders soils prone to surface crusting, increases soil erosion hazards, and decreases the capacity to hold plant-available water (Lal, 1986). Consequently, upon the decline in organic matter content there is a decrease in effective cation-exchange capacity and the capacity of the soil to retain plant-available nutrients. Organic matter plays an important role in the phytotoxicity of Al in soils; Al released from minerals can be complexed by soluble and solid forms of organic matter. Soil solution concentrations of Al are lower in organic soils than in mineral soils at the same pH (Evans and Kamprath, 1970); they are also low in soils that have received additions or organic matter (Hargrove and Thomas, 1981). Aluminum extracted by 1 *M* KCl was found to be lower as organic matter content increased (0.80 to 5.11%) at any given pH level (Thomas, 1975).

Before discussing N use efficiency in crop production, it is important that the term be defined. The effectiveness of N fertilizer treatments can be assayed by measuring either the ratio of grain yield to N absorbed or the ratio of N absorbed to N applied (Mikkelsen, 1987). Fertilizer efficiency can be described by

$$\begin{array}{c}\text{N fertilizer efficiency}\\ \text{(kg grain kg}^{-1}\text{ N applied)}\end{array} = \begin{array}{c}\text{Percentage of N}\\ \text{recovered}\\ \text{(kg N absorbed kg}^{-1}\\ \text{N applied)}\end{array} \times \begin{array}{c}\text{Efficiency of}\\ \text{utilization}\\ \text{(kg grain kg}^{-1}\\ \text{N absorbed)}\end{array} \quad (3)$$

From an agronomic point of view, efficiency of utilization is more important than efficiency of absorption. Yoshida (1981) reported values of fertilizer use efficiency from 15 to 25 kg grain kg^{-1} applied N for transplanted tropical rice.

Plant recovery of fertilizer N is about 50% on average, but values range from 20 to 80% (Sanchez, 1976; Baligar and Bennett, 1986). Nitrogen losses from soils occur through leaching and through volatilization of various N forms; however, losses through both of these processes are poorly quantified (Greenland, 1977). Nitrogen losses through leaching are a particular problem in the tropics, where light-textured soils are common and high-intensity rainfall events are frequent.

Nitrogen use efficiency can be improved by adopting the following management practices: (1) use of split applications of N adjusted to the N requirement of the crop (Sanchez, 1976; Singh and Sekhon, 1976; Lee and Asher, 1981); (2) use of N-efficient cultivars; (3) incorporation of legumes into cropping systems; (4) incorporation of crop residues after harvest, as opposed to burning (Bartholomew, 1977); (5) use of appropriate forms of fertilizer N; nitrate sources are usually inferior to ammoniacal sources and urea under conditions favoring leaching or denitrification (Sanchez, 1976); (6) deep application to reduce volatilization losses of NH_3; (7) use of green manures to maintain or improve organic matter content of soils; and (8) use of nitrification inhibitors to reduce losses (Prasad *et al.*, 1971).

The role of legumes in agriculture is certain to increase in importance. Pulses are a source of good-quality protein, and forages have a long documented history of supporting livestock on poor soils. Both of these factors are attributed in part to the ability of legumes, in symbiosis with *Rhizobium,* to obtain nitrogen from the air (Larue and Patterson, 1981). Burns and Hardy (1975) averaged many published estimates to arrive at an average figure of 240 kg N fixed per hectare per year of arable land under legume production. Shortly thereafter, that figure was reviewed by a group of scientists attending a conference on nitrogen-fixing microbes; they concluded (Burris, 1978) that a realistic figure would be one-half of the Burns–Hardy calculation. This reassessment, however, was apparently based as much on intuition as on new data. Many legumes perform poorly in acid soils because of the failure to establish an effective symbiotic nitrogen-fixing association with rhizobia. This limits the amount of nitrogen available to succeeding crops.

B. PHOSPHORUS

Phosphorus deficiency is a major limitation to crop production in acid, infertile soils. Thus, Sanchez and Salinas (1981) estimated that P deficiency was a constraint to plant growth over 96% of the total area of acid soils in tropical America (23°N–23°S), while Sanchez (1987) estimated that P deficiency was a constraint over 90% of the total land area of the Amazon Basin. The two major reasons for the occurrence of P deficiency in acid soils are low native soil P content and high P fixation capacity. The P fixation capacity of an Oxisol from central Brazil was studied over a period of 80 days by Fageria and Barbosa Filho (1987) and results are presented in Table I. The amount of P fixed (not recovered by Mehlich 1 extracting solution) increased from 45 to 268 kg P ha^{-1} when the P application rate was increased from 50 to 400 kg P ha^{-1}. It could be suggested that these types of soils require large amounts of fertilizer P for optimum crop production.

The clay fractions of highly weathered Ultisols and Oxisols are dominated by amorphous (noncrystalline) Fe and Al oxides, gibbsite, and kaolinite. These minerals possess a variable surface charge which is highly dependent on pH, concentration of cations, and composition of the solution in contact with those surfaces. In Ultisols and Oxisols, the largest P fixation is by amorphous hydrated oxides of Fe

Table I. Percentage Recovery of P as a Function of Applied P and Reaction Time in an Oxisol of Central Brazil[a]

P Applied ($kg\ ha^{-1}$)	Reaction Time (Days) 0	17	31	45	60	80	Average
0	0	0	0	0	0	0	0
50	13	7	7	6	12	13	10
100	17	16	10	16	12	13	14
150	16	16	19	14	13	14	15
200	15	17	22	21	13	20	18
250	20	17	20	18	17	18	18
300	21	22	18	19	21	15	19
350	23	20	20	19	19	17	20
400	29	20	19	23	24	23	23

[a]Source: Fageria and Barbosa Filho, 1987.

and Al, followed by gibbsite, goethite, and kaolinite (Baligar and Bennett, 1986). The more crystalline the material, the less the fixation of P. Lopes (1977) showed that clay plays a dominant role in P sorption by Cerrado soils of Brazil with P sorption positively correlated with clay content.

Improving P Use Efficiency

In acid soils, P use efficiency can be improved by adopting a series of management practices, including the use of optimum rates, matching the method of application to the site, using rock phosphate in combination with a more soluble source, by liming, and by the use of P-efficient cultivars. Mycorrhizal associations are known to play an important role in P acquisition by plants growing in soils with a limiting P supply.

a. Use of Optimum Rate The optimum rate of P fertilizer can be selected on the basis of a calibration study for a given agroclimatic region and crop. When conducting P calibration experiments, all other growth factors should be at an optimum level to enable the maximum growth response to the different P rates to be obtained. The results of such P calibration experiments will be modified if other factors, including drought, disease, weeds, insects, and deficiency of other nutrients, are not eliminated or controlled.

b. Methods of Application Acid soils are often, but not always, characterized by a high capacity for P fixation. Under these situations, P fertilizer efficiency can be improved through localized application (banding) to reduce P fixation. According to the senior author's experience with the Oxisols of Brazil, the best crop yields are obtained with a combination of broadcast and band application. After building the P level to a certain minimum value that varies with the crop, band application is quite effective in obtaining high yields. A combination of broadcast and banded applications has the additional advantage of a higher yield stability than either an all broadcast or an all banded application.

c. Use of Rock Phosphate with a Soluble P Source Use of rock phosphate in combination with a soluble P source is another strategy that may be used to improve P status of acid soils. Solubility of rock phosphate is greater in soils with low pH and low Ca and P content, conditions that often prevail in acid soils. According to the senior author's experience, rock phosphate alone does not produce good yields of rice and bean in Oxisols of central Brazil. An additional alternative is to broadcast the rock phosphate and apply more soluble P sources in bands to provide an immediate supply while the rock phosphate slowly dissolves.

d. Decrease of P Fixation with Liming Liming decreases the P fixation in acid soils through precipitation of Fe and Al hydroxides (Fageria, 1984). Data in Table II show the effect of lime on the seed yield of common bean grown at different P rates in an Oxisol of central Brazil. Application of lime increased bean yields in the acid soil when low rates of P were applied, but had no effect at the higher rates of P application.

e. Use of P-Efficient Cultivars The possibility of exploiting genotypic differences in the absorption and utilization of P to improve efficiency of P fertilizer use or to obtain higher productivity on P-deficient soils has received considerable attention in recent years (Nielsen and Barber, 1978; Baligar and Barber, 1979; Fageria and Barbosa Filho, 1982).

Crop species of cultivars within species vary widely in their abilities to survive, to grow, and to produce at low levels of P in soils (Loneragan, 1978). The variations among plants result from differences in their abilities to absorb and to utilize P for

Table II. Effects of Lime on Seed Yield (kg/ha) of Irrigated Common Bean Grown at Different P Rates in an Oxisol of Central Brazil[a]

P level (kg ha^{-1})	With Lime (4 tons ha^{-1})					Without Lime				
	R.I.	R.II.	R.III.	Total	$\bar{X}$	R.I.	R.II.	R.III.	Total	$\bar{X}$
0	150	70	110	330	110	40	30	40	110	37
26	980	880	840	2700	900	640	910	780	2330	777
52	1290	1470	860	3620	1207	1460	790	610	2860	953
104	1400	1380	1360	4140	1380	1360	1408	1440	4280	1427
208	1360	1290	1440	4090	1363	1210	1530	1230	3970	1323
Linear					**[b]					**
Quadratic					**					**

[a]Source: Unpublished data of N. K. Fageria.
[b]Significant at the 1% probability level.

their growth and development. According to Loneragan (1978), the differences among plants in their ability to absorb P from soils may be due to at least three distinct root attributes: (1) the physiological ability to absorb P from dilute solutions, (2) metabolic activity resulting in solubilization of sorbed P, and (3) the ability of the root system to explore the soil mass.

f. Mycorrhizal Associations It is well established that several genera and species of vesicular-arbuscular mycorrhizae (VAM) form symbiotic associations with roots of many plant species and consequently increase the uptake of P from soils of low-P status (Bowen, 1973; Tinker, 1975). Mycorrhizae appear to enhance P absorption by extending the physical contact of the root system with the soil rather than by obtaining forms of P from soil that are not available to nonmycorrhizal roots. However, there is some evidence from solution culture studies that mycorrhizal roots are more efficient than nonmycorrhizal roots in absorbing P (Howeler *et al.*, 1982). In general, the relative effects of mycorrhizae are most marked at low and intermediate levels of P availability (Loneragan, 1978); they tend to disappear at extremely low and at high levels of P. Some strains of mycorrhizae are very effective, suggesting that selection of superior strains might provide a powerful technique for increasing P absorption by plants from soils of low P availability (Mosse, 1972).

C. POTASSIUM

Potassium deficiency in acid soils is not as widespread as P deficiency. Thus, Sanchez (1987) estimated that 50% of the Amazon Basin was characterized by soils with low K reserves, whereas 90% of these soils were estimated to be P deficient. Large amounts of K are removed by high-yielding root crops, by forage grasses that are periodically cut, and by sugarcane, while smaller but increasing amounts are removed by cereal crops and food legumes (Sanchez, 1976). The amounts of K removed by the root crops, forage grasses, and sugarcane are much greater than the amounts of the other essential nutrients. In situations where intensive agriculture is practiced, failure to replace K that is removed in the harvested crop can result in K deficiency becoming a limitation to further crop production.

The main factors that influence crop response to K fertilization include available soil K, soil moisture status, soil fertility status, soil temperature, and the particular crop species under consideration (Mengel and Kirkby, 1980). The most widely employed and accepted index to K availability is that extracted by neutral 1 *N* NH_4OAC; this includes the sum of the exchangeable and water-soluble K (Knudsen *et al.*, 1982). Although widely used, this approach has not gained universal acceptance. Mengel and Kirkby (1980) considered that the equilibrium K concentration in the soil solution and the K buffer capacity provide more reliable parameters for assessing K supply to the roots of plants growing in soils.

Table III. Distribution of K in Different Parts of Four Crop Species[a,b]

Crop	Plant Part				
	Roots	Shoots	Grain	Pod Husk	Ear Chaff
Rice	4	78	18	—	—
Wheat	2	72	11	—	15
Common bean	6	43	31	20	—
Cowpea	5	49	36	10	—

[a]Source: Unpublished data of N. K. Fageria.
[b]Values are percentage of total uptake.

Improvement of K Use Efficiency

Three management practices can be used to improve K fertilizer use efficiency by plants growing in acid soils. First, K fertilizers should be applied at an economically feasible rate, bearing in mind the need in the long term to replace K lost through crop removal and leaching. Second, the incorporation of crop residues in the soil after harvest enables a substantial amount of the plant K to be recycled. Data in Table III show that approximately 70 to 80% of the total K content remains in the vegetative shoot of cereals such as wheat and rice, while about 40 to 50% remains in the shoot of legumes such as cowpea and beans. The third practice involves the use of K-efficient cultivars that have an increased K uptake efficiency. Cultivar dif-

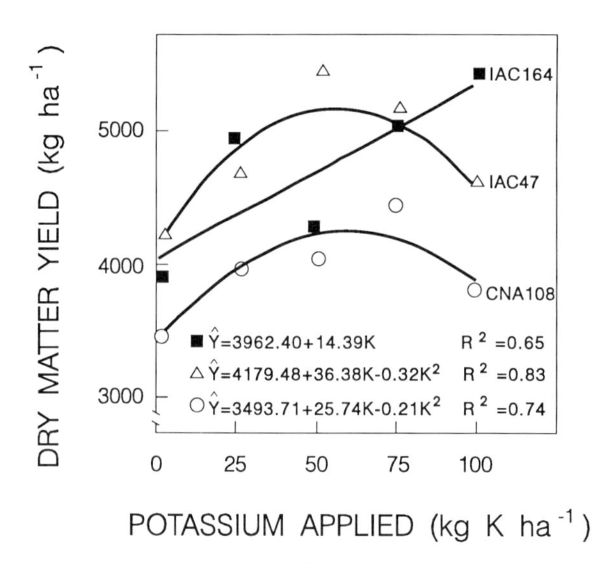

Figure 3. Response of upland rice cultivars to K fertilization in an Oxisol of central Brazil. Source: N. K. Fageria, unpublished.

ferences in K uptake have been reported in a number of crop species (Shea *et al.*, 1967; Glass and Perley, 1980; Siddiqui *et al.*, 1987). The differential response of rice cultivars to K fertilization when grown in an Oxisol of central Brazil is shown in Figure 3.

D. CALCIUM AND MAGNESIUM

Deficiencies of Ca and Mg are also important limitations to plant growth in many acid soils. The correction of Ca and Mg deficiencies is a particular problem in low-input agricultural systems in which plants tolerant to high Al and low available P are grown on soils with a low effective cation-exchange capacity (Sanchez and Salinas, 1981). Bruce *et al.* (1988) have shown that the primary limitations to root elongation in acid soils (mainly subsoils) from southeast Queensland was Ca deficiency and not Al toxicity, in spite of high Al saturations and relatively low soil pH. Crops remove from about 20 to 150 kg Ca ha^{-1} and 10 to 80 kg Mg ha^{-1}, depending on the yield and the particular crop (Doll and Lucas, 1973; Sanchez, 1976). The availability of Ca and Mg to plants growing in soils is influenced by the percentage Ca and Mg saturation, the total Ca and Mg supply, the concentration of Ca and Mg in the soil solution, and by the presence of ions such as Al^{3+} and Mn^{2+}, which inhibit Ca^{2+} and Mg^{2+} absorption (Adams, 1984).

Improving Ca and Mg Use Efficiency

Calcium and Mg use efficiency in acid soils is improved with the application of dolomitic lime. Data in Table IV show an increase in Ca and Mg content of an Oxisol of central Brazil with increasing rates of application of dolomitic lime. In these soils, about 3 cmol (Ca + Mg) kg^{-1} soil is considered to be critical for growth of many food crops (Fageria, 1984). Results in Table IV show that 6 tons of dolomitic lime ha^{-1} is required to maintain Ca + Mg above the critical level in the top 20 cm of the soil profile after harvest of the first crop of upland rice. The critical level can be separated into a critical Mg level of about 1 cmol kg^{-1} and a critical Ca level of about 2 cmol kg^{-1}. Data in Table IV also show that some leaching of Ca + Mg has occurred into the subsoils, particularly at the higher rates of dolomitic lime reaction and after the second harvest of upland rice. Dolomitic lime should be used in those situations where the supply of both Ca and Mg is at best marginal; the use of calcitic lime in such situations may reduce yield through induction or aggravation of Mg deficiency.

The other strategy for increasing Ca and Mg use efficiency lies in the use of efficient species or cultivars. It is widely reported that plant species differ widely in their requirement for Ca (Andrew, 1976; Fageria, 1984) and Mg (Karlen *et al.*, 1978; Gallagher *et al.*, 1981; Clark, 1982; Fageria, 1984). In fact, in an experiment conducted at the International Institute of Tropical Agriculture (IITA), D. G. Edwards and B. T. Kang (unpublished) have shown that lima bean, soybean, and

Table IV. Influence of Dolomitic Lime on Ca + Mg Content at Different Depths in the Soil Profile of an Oxisol of Central Brazil Following the Harvest of Two Consecutive Upland Rice Crops[a]

Lime Applied (tons ha^{-1})	Depth (cm)	After First Harvest	After Second Harvest
		(cmol (Ca + Mg) kg^{-1})	
0	0–20	1.28	1.29
	20–40	0.77	0.97
	40–60	0.64	0.73
	60–80	0.91	0.88
3	0–20	2.01	2.10
	20–40	1.35	1.25
	40–60	0.77	0.92
	60–80	0.82	0.92
6	0–20	3.31	2.76
	20–40	1.33	1.63
	40–60	0.94	1.14
	60–80	0.96	1.11
9	0–20	3.83	3.42
	20–40	1.68	1.89
	40–60	1.08	1.27
	60–80	1.20	1.17
12	0–20	3.99	3.92
	20–40	2.36	2.19
	40–60	1.14	1.50
	60–80	1.12	1.44
	Statistical Significance		
	Lime (L)	**[b]	**
	Depth (D)	**	**
	L × D	**	**

[a]Source: Unpublished data of N. K. Fageria.
[b]Significant at the 1% level of probability.

pigeon pea grew poorly and developed Ca deficiency symptoms in an unlimed Typic Paleudult from southeastern Nigeria, whereas cowpea and winged bean grew much better and did not develop any Ca deficiency symptoms when grown in the same soil. The latter two species achieved much higher Ca concentrations in their tops than the former three species.

E. ZINC

The susceptibility of plants to Zn deficiency varies greatly among species and also cultivars (Viets *et al.*, 1954; Brown *et al.*, 1972). Zinc deficiency has been observed over extensive areas in the United States, Canada, Australia, and many tropical countries (Lindsay, 1972; Sanchez and Salinas, 1981; Kanwar and Youngdahl, 1985; Singh *et al.,* 1987). Low Zn levels also occur in soils in several European countries (Aubert and Pinta, 1977).

Zinc deficiency occurs in a variety of soils, ranging from light-textured sandy soils to fine-textured soils, mucks, and peats (Thorne, 1957). Zinc-deficient soils are often of pH $\geq$ 6.0, but Zn deficiency is common on the Oxisols of central Brazil, which have pH values of 5.0 to 5.5 (Barbosa Filho and Fageria, 1980; Fageria, 1984). Data in Figure 4 show the response of upland rice grown on two Oxisols of central Brazil to application of Zn. The main reasons for Zn deficiency on these soils are naturally low levels of available Zn ($<$1 mg kg^{-1} soil) and management practices such as liming and P fertilization, which sometimes induce Zn deficiency in crops (Fageria and Zimmermann, 1979).

Liming increases soil pH and lowers Zn availability to, and uptake by, plants. It has been suggested that P fertilization induces Zn deficiency through a reaction that occurs at the root surface or within the root (Burleson *et al.*, 1961; Sharma *et al.*, 1968). The P-induced Zn deficiency may be a result of the increased crop growth

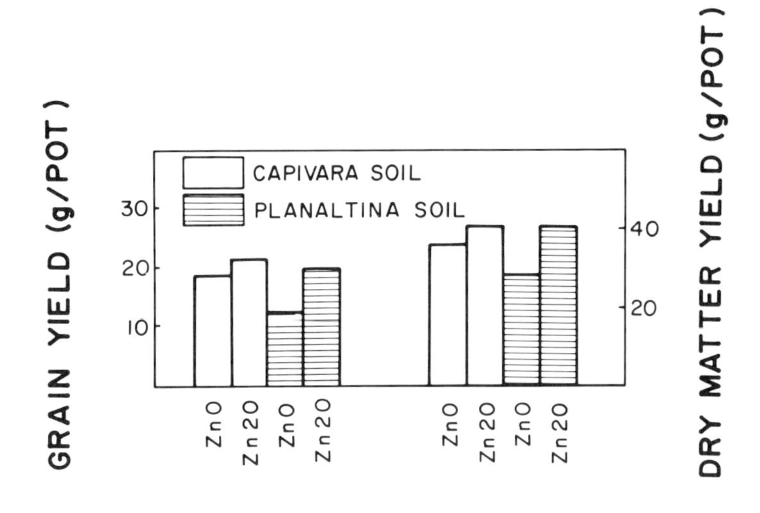

Figure 4. Grain yield of upland rice with and without 20 kg of Zn ha^{-1} when grown in Oxisols of central Brazil. Source: Fageria and Zimmermann, 1979.

due to P fertilization of acid soils in which the available Zn supply is marginal. Recent studies with a number of plant species have shown that Zn deficiency may cause P to accumulate to toxic concentrations in older leaves, thereby inducing or accentuating symptoms resembling those of Zn deficiency (Loneragan *et al.*, 1982; Welch *et al.*, 1982). Zinc deficiency in upland rice can be easily and economically corrected in acid soils of Brazil by application of about 5 kg Zn ha^{-1}, as $ZnSO_4$, at the time of planting (Barbosa Filho and Fageria, 1980). It can also be corrected by foliar application of 0.5% $ZnSO_4$ solution, but soil application is better because residual effects last for 2 to 3 years.

III. Management of Soil Acidity

Crop yield is a function of many soil, climatic, and plant factors, and their interactions. Therefore, when improving crop yields, the whole system must be considered. To achieve higher yields, all these factors should be at an adequate level. In the management of acid soils to increase and/or stabilize crop production, the following strategies can be used to alleviate or reduce low pH soil stress: (1) use of soil amendments and (2) use of plant species and cultivars efficient and/or tolerant to nutrient deficiencies or tolerant to toxicities of nutrients or other elements.

A. USE OF AMENDMENTS

Lime is the most common amendment used for reducing soil acidity problems. Application of a liming material to a soil has a number of direct and indirect effects. The direct effects include (1) increased Ca content and, if dolomitic limestone is used, Mg content in the soil and (2) increased soil pH (Table V). Increased soil pH results in a number of changes that might be considered indirect effects. As soil pH increases, P and Mo become more available. Data in Figure 5 show that extractable P in Oxisols in central Brazil increased by increasing pH from approximately 5 to 6.5 (1 : 2.5 soil/water). Increasing soil pH decreases the amount of exchangeable Al (Table V) and the concentration of Al in soil solution. On the other hand, increasing soil pH decreases the availability of Mn, Cu, Fe, and Zn. If the total amount of any of these elements in the soil is low, increasing the pH through lime application can cause a deficiency in the crop growing on the soil. Zinc deficiency in Brazilian Oxisols is very common with the application of lime (Barbosa Filho and Fageria, 1980).

The lime requirement of an acid soil is the amount of liming material required to neutralize the undissociated and dissociated acidity in the range from the initial acid condition to a selected neutral or less acid condition (McLean, 1973). This practice is mostly applied in the northern temperate regions of the world by liming to increase the soil pH to near neutrality (Sanchez and Salinas, 1981). This strategy is

Table V. Influence of Liming on Soil pH and Exchangeable (Ca + Mg) and Al in the Top 20 cm of an Oxisol of Brazil at 67 Days after Application[a]

Dolomitic Lime Rate (tons ha^{-1})	pH (1 : 2.5 soil : water)	Ca + Mg (cmol kg^{-1})	Al (cmol kg^{-1})
0	5.0	1.49	0.51
3	5.4	2.69	0.26
6	5.6	3.62	0.11
9	5.8	4.29	0.07
12	6.0	5.26	0.04
Statistical Significance	**[b]	**	**

[a]Source: Unpublished data of N. K. Fageria.
[b]Significant at the 1% level of probability.

often not appropriate for Oxisol–Ultisol regions because of differences in crops grown, different chemistry of low-activity clay minerals, and low natural reserves of other nutrients, which often result in yield reduction if such soils are limed or near neutrality (Sanchez and Salinas, 1981). Various methods exist for estimating the lime requirements of soils of tropical America. In one method, levels of soil

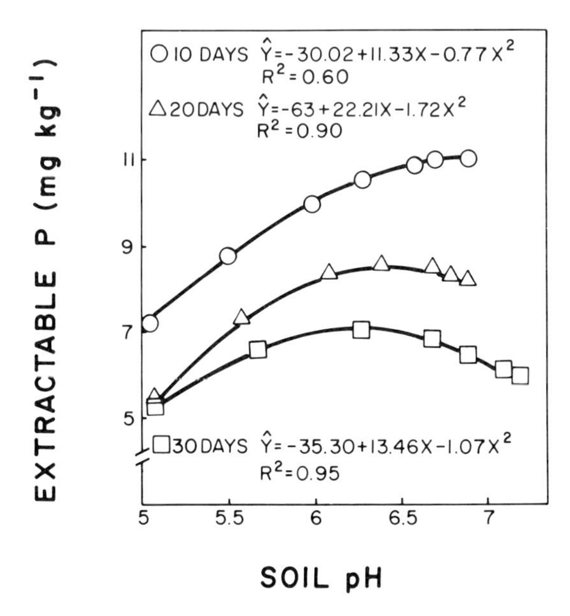

Figure 5. Influence of soil pH on extractable P (Mehlich 1–0.05 *N* HCl + 0.02 N H_2SO_4 extracting solution) in Oxisols of central Brazil. Source: Fageria, 1984.

exchangeable Al, Ca, and Mg are taken into consideration, according to the equation (Olmos and Camargo, 1976)

$$\text{Lime rate (tons ha}^{-1}) = (2 \times \text{Al}) + [2 - (\text{Ca} + \text{Mg})] \tag{4}$$

where Al, Ca, and Mg are in cmol kg^{-1} soil.

In tropical regions, lime recommendations are often based solely on the level of Al extracted with a nonbuffered KC1 solution. This criterion, suggested by Coleman *et al.* (1958) and Kamprath (1970), has been widely used in Brazil (Paula *et al.*, 1987). Shoemaker *et al.* (1961) developed a lime requirement method (known as SMP Method) using a buffer solution that provides good results, principally for soils that have a pH in water of 6.3 or less. According to the same authors, pH buffer capacity is a major consideration when determining the lime requirement of acid soils. At present, in the state of Sao Paulo, Brazil, the common method of recommending a lime application rate is based on percentage base saturation, as seen in

$$\text{Lime rate (tons ha}^{-1}) = \frac{\text{EC }(B_2 - B_1)}{\text{TRNP}} \times \text{df} \tag{5}$$

where

EC = total exchangeable cations ($Ca^{2+} + Mg^{2+} + K^{+} + H^{+} + Al^{3+}$) in cmol kg^{-1}
B_2 = desired optimum base saturation
B_1 = existing base saturation
TRNP = total relative neutralizing power of liming material
df = depth factor, 1 for 20 cm depth and 1.5 for 30 cm depth

For Brazilian Oxisols, the desired optimum base saturation for most of the cereals is considered to be in the range of 50 to 60%, and for legumes it is considered to be in the range of 60 to 70%.

B. USE OF TOLERANT AND EFFICIENT SPECIES/CULTIVARS

Use of tolerant or efficient species or cultivars is another approach to the problem of improving crop production under low pH stress environments. The word "tolerant" is used for nutrient toxicities and the word "efficiency" for deficiencies. If toxicities or deficiencies are very severe in a given soil, use of tolerant or efficient species or cultivars alone is not the most appropriate solution. Under these situations, both soil amendments and tolerant or efficient cultivars should be employed. All crop species or cultivars need minimum nutrient levels for maximal growth and production. If toxicities or deficiencies are not severe, tolerant or efficient cultivars alone can produce desirable yields. Under these situations use of tolerant or efficient crop genotypes should be encouraged. Cultivar differences to toxicities and deficiencies exist, as discussed already under the toxicity and deficiency sections in this chapter and in Chapters 4, 5, 6, and 7.

The possibilities of exploiting genotypic differences in absorption and utilization of nutrients under low pH (or acid soil) stress and improving nutrient use efficiency to obtain higher productivity on acid soils have received considerable attention in recent years (Foy, 1984; Clark, 1984; Gerloff, 1987; Blamey *et al.*, 1987). However, the lack of a clear definition of efficiency can lead to variation in the interpretation of experimental results. The nutrient efficiency of plants can be defined in several ways: (1) plants that accumulate higher concentrations of a nutrient when grown at a given level of fertility are more efficient (Clark and Brown, 1975); (2) production of a large quantity of harvestable dry matter per unit time and area when grown in a medium that has less than sufficient nutrients available for maximum yield under the existing environmental conditions (Fox, 1978); (3) maximum dry matter production per unit of nutrient uptake (Asher and Loneragan, 1967); (4) maximum dry matter production at a constant plant nutrient content (Blair and Cordero, 1978); (5) the ability of a genotype to produce a high yield in a soil that is limiting in that element compared to a standard genotype (Graham, 1984); (6) ability to produce maximum dry matter with a given amount of applied nutrient (Blair and Cordero, 1978; Fox, 1978); and (7) the increase in yield of the harvested portion of the crop per unit of fertilizer nutrient applied where high yields are obtained (Barber, 1976).

It is clear from the above discussion that there is no consistency in the literature as far as nutrient efficiency is concerned. High nutrient efficiency may be due to greater absorption of the nutrient or greater yield per unit of nutrient or greater yield per unit of nutrient absorbed. From the soils viewpoint, the potential for breeding or adapting plants to grow on deficient soils is unlimited. Deficiency, especially for some less mobile nutrients, such as P, is often due to problems of actual availability rather than low total supply.

A number of the plant factors that could be responsible for adaption to nutrient deficiency are summarized by Gerloff (1987). He divided them into four categories associated with (1) nutrient acquisition from the environment; (2) nutrient movement across the root to the xylem; (3) nutrient distribution and remobilization in the shoot; and (4) nutrient utilization in metabolism and growth.

All these plant adaptability factors are related to physiological mechanisms that are summarized by Graham (1978) as (1) better root geometry; (2) faster specific rate of absorption at low concentrations (low K_m); (3) chemical modification of the root–shoot interface to solubilize more of the limiting nutrient; (4) improved interaction and redistribution; and (5) superior utilization, or a lower functional requirement for the nutrient.

In the field, it is difficult to identify the operative mechanism(s) when screening cultivars for nutrient efficiency. However, selection could be more precise if these mechanisms are known. Therefore, generally seed yield, nutrient content, or symptom (deficiency/toxicity) expression are used for screening purposes.

The availability and use of proper screening techniques is a major consideration in the selection of cultivars better adapted to nutrient stress problems in acid soils. In many instances it is very difficult to compare the results of screening experiments because of large experimental variations among various screening methods.

fore, when screening genotypes for some mineral stresses, considerations that should be taken into account include: (1) uniform growth medium (soil or solution culture); (2) uniform ecological conditions; (3) genotypes with the same growth cycle; (4) well-defined evaluation parameters (morphological, physiological, or biochemical); (5) screening techniques that are simple and permit evaluation of a large number of materials with good precision; (6) selection of an appropriate site in the case of field experiments; (7) minimum and maximum levels known in advance; (8) at least three fertility levels considered, i.e., low, medium, and high; (9) greenhouse results verified under field conditions and vice versa; (10) nutrients, other than the variable, present in adequate amounts; (11) efficient and nonefficient cultivars included in the genotype screening; and (12) plant materials genetically uniform.

IV. Summary

Soil pH is an important chemical property that governs nutrient availability to plants. If not in an optimum range, nutrient deficiencies and/or toxicities occur and crop growth and yield decrease. The soil–plant system is a dynamic system, and it is very difficult to define optimum pH values for different plant species. Most food crops grow well in soils if soil pH is around 6. In very acid soils ($pH < 5.5$), deficiencies of N, P, K, Ca, Mg, Mo, and Zn and toxicities of Al and Mn are very common.

Basic principles of management of acid soils are similar for different ecologies, although the technological packages based on these principles vary depending on locally specific constraints and socioeconomic considerations. To obtain high and/or stable yields on acid soils, the following management practices are suggested: (1) use of efficient or tolerant cultivars in combination with soil amendments; (2) adaptation of improved cultural practices to improve nutrient use efficiency; (3) maintaining a regular, adequate supply of organic matter, by use of organic manures encouraged where possibilities exist, especially on small holdings, or use of green manuring on bigger holdings; (4) use of proper crop rotations; and (5) some acid soils very low in native fertility and having a sandy texture should be used for permanent pasture rather than row crop production.

References

Adams, F. 1981. Nutritional imbalances and constraints to plant growth in acid soils. *J. Plant Nutr.* 4:81–87.

Adams, F. 1984. Crop response to lime in the southern United States. In F. Adams (ed.), Soil Acidity and Liming. 2nd Ed. *Agronomy* 12:211–265.

Adams, F., and P. J. Hathcock. 1984. Aluminum toxicity and calcium deficiency in acid subsoil horizons of two Coastal Plains soil series. *Soil Sci. Soc. Am. J.* 48:1305–1309.

Alva, A. K., F. P. C. Blamey, D. G. Edwards, and C. J. Asher. 1986a. An evaluation of aluminum indices to predict aluminum toxicity to plants grown in nutrient solutions. *Commun. Soil Sci. Plant Anal.* 17:1271–1280.

Alva, A. K., D. G. Edwards, C. J. Asher, and F. P. C. Blamey. 1986b. Effects of phosphorus/aluminum molar ratio and calcium concentration on plant response to aluminum toxicity. *Soil Sci. Soc. Am. J.* 50:133–137.

Alva, A. K., D. G. Edwards, C. J. Asher, and S. Suthipradit. 1987. Effects of acid soil infertility factors on growth and nodulation of soybean. *Agron. J.* 79:302–306.

Andrew, C. S. 1976. Effects of calcium, pH, and nitrogen on the growth and chemical composition of some tropical and temperate pasture legumes. I. Nodulation and growth. *Aust. J. Agric. Res.* 27:611–623.

Andrew, C. S., and M. P. Hegarty. 1969. Comparative responses to manganese excess of eight tropical and four temperate pasture legume species. *Aust. J. Agric. Res.* 20:687–696.

Andrew, C. S., A. D. Johnson, and R. L. Sandland. 1973. Effect of aluminium on the growth and chemical composition of some tropical and temperate pasture legumes. *Aust. J. Agric. Res.* 24:325–339.

Asher, C. J. 1981. Limiting external concentrations of trace elements for plant growth: Use of flowing solution culture techniques. *J. Plant Nutr.* 3:163–180.

Asher, C. J., and J. F. Loneragan. 1967. Response of plants to phosphate concentration in solution culture. I. Growth and phosphorus content. *Soil Sci.* 103:225–233.

Aubert, H., and M. Pinta. 1977. Zinc. In Trace Elements in Soils. Elsevier, New York.

Baligar, V. C., and S. A. Barber. 1979. Genotypic differences of corn for ion uptake. *Agron. J.* 71:870–873.

Baligar, V. C., and O. L. Bennett. 1986. Outlook on fertilizer use efficiency in the tropics. *Fert. Res.* 10:83–96.

Baligar, V. C., J. H. Elgin, Jr., and C. D. Foy. 1989. Variability in alfalfa for growth and mineral uptake and efficiency ratios under aluminum stress. *Agron. J.* 81:223–229.

Baligar, V. C., R. J. Wright, N. K. Fageria, and C. D. Foy. 1988. Differential response of forage legumes to aluminum. *J. Plant Nutr.* 11:549–561.

Baligar, V. C., R. J. Wright, T. B. Kinraide, C. D. Foy, and J. H. Elgin, Jr. 1987. Aluminum effects on growth, mineral uptake and efficiency ratios in red clover cultivars. *Agron. J.* 79:1038–1044.

Barber, S. A. 1976. Efficient fertilizer use. In F. L. Patterson (ed.), Agronomic Research for Food. *ASA Spec. Publ.* No. 26, pp. 13–29.

Barbosa Filho, M. P., and N. K. Fageria. 1980. Occurrence, Diagnosis and Correction of Zinc Deficiency in Upland Rice. *EMBRAPA-CNPAF Tech. Circ.* No. 4.

Bartholomew, W. V. 1977. Soil nitrogen changes in farming systems in the humid tropics. In A. Ayanaba and P. J. Dart (eds.), Biological Nitrogen Fixation in Farming Systems of the Tropics. Wiley, London. Pp. 27–42.

Bartlett, R. J., and D. C. Riego. 1972. Effects of chelation on toxicity of aluminum. *Plant Soil* 37:419–423.

Bartuska, A. M., and I. A. Unger. 1980. Elemental concentrations in plant tissues as influenced by low pH soils. *Plant Soil* 55:157–161.

Bell, L. C., and D. G. Edwards. 1987. The role of aluminium in acid soil infertility. In Soil Management under Humid Conditions in Asia (ASIALAND). IBSRAM Proc. No. 5. IBSRAM, Bangkok. Pp. 201–223.

Blair, G. J., and S. Cordero. 1978. The phosphorus efficiency of three annual legumes. *Plant Soil* 50:387–398.

Blamey, F. P. C., C. J. Asher, and D. G. Edwards. 1987. Hydrogen and aluminum tolerance. *Plant Soil* 99:31–37.

Blamey, F. P. C., D. G. Edwards, and C. J. Asher. 1983. Effects of aluminium, OH : Al and P : Al molar ratios, and ionic strength on soybean root elongation in solution culture. *Soil Sci.* 136:197–207.

Blamey, F. P. C., D. G. Edwards, and C. J. Asher. 1986a. Role of trichomes in sunflower tolerance to manganese toxicity. *Plant Soil* 91:171–180.

Blamey, F. P. C., D. G. Edwards, C. J. Asher, and M. K. Kim. 1982. Response of sunflower to low pH. In A. Scaife (ed.), Plant Nutrition 1982. Proc. 9th Int. Plant Nutrition Colloq., Warwick, England. Commonw. Agric. Bur., Slough, England. Pp. 66–71.

Blamey, F. P. C., N. J. Grundon, C. J. Asher, and D. G. Edwards. 1986b. Aluminium toxicity in sorghum and sunflower. *Proc. 1986 Aust. Sorghum Conf.* pp. 6.11–6.18.

Bohn, H. L., B. L. McNeal, and G. E. O'Connor. 1979. Acid soils. In Soil Chemistry. Wiley (Interscience), New York. Pp. 195–216.

Bouldin, D. R., K. D. Ritchey, and E. Lobato. 1987. Management of soil acidity. In Management of Acid Tropical Soils for Sustainable Agriculture. IBSRAM Proc. No. 2. IBSRAM, Bangkok. Pp. 187–203.

Bowen, G. D. 1973. Mineral nutrition of ectomycorrhizae. IN G. C. Marks and T. T. Kozlowski (eds.), Physiology and Ecology of Ectomycorrhizae. Academic Press, New York. Pp. 151–205.

Bromfield, S. M., R. W. Cumming, D. J. David, and C. H. Williams. 1983a. Change in soil pH, manganese and aluminium under subterranean clover pasture. *Aust. J. Exp. Agric. Anim. Husb.* 23:181–191.

Bromfield, S. M., R. W. Cumming, D. J. David, and C. H. Williams. 1983b. The assessment of available manganese and aluminium status in acid soils under subterranean clover pastures of various ages. *Aust. J. Exp. Agric. Anim. Husb.* 23:192–200.

Brown, J. C., J. E. Ambler, R. L. Chaney, and C. D. Foy. 1972. Differential responses of plant genotypes to micronutrients. In J. J. Mortvedt *et al.* (eds.), Micronutrients in Agriculture. Soil Sci. Soc. Am., Madison, Wisconsin. Pp. 389–418.

Bruce, R. C., L. A. Warrell, D. G. Edwards, and L. C. Bell. 1988. Effects of aluminium and calcium in the soil solution of acid soils on root elongation of *Glycine max* cv. Forrest. *Aust. J. Agric. Res.* 39:319–338.

Burleson, C. A., A. D. Dacus, and C. J. Gerard. 1961. The effect of phosphorus fertilization on the zinc nutrition of several irrigated crops. *Soil Sci. Soc. Am. Proc.* 25:365–368.

Burns, R. C., and R. W. F. Hardy. 1975. Nitrogen Fixation in Bacteria and Higher Plants. Springer-Verlag, New York.

Burris, R. H. 1978. Environmental role of nitrogen-fixing blue-green algae and symbiotic bacteria. NFR Editorial Service, Stockholm.

Cameron, R., G. S. P. Ritchie, and A. D. Robson. 1986. The relative toxicities of inorganic aluminum complexes to barley. *Soil Sci. Soc. Am. J.* 50:1231–1236.

Carter, O. G., I. A. Rose, and P. F. Reading. 1975. Variation in susceptibility to manganese toxicity in 30 soybean genotypes. *Crop Sci.* 15:730–732.

Cheng, B. T., and G. J. Ouellette. 1971. Manganese availability in soil. *Soils Fert.* 34:589–595.

Clark, R. B. 1982. Plant response to mineral element toxicity and deficiency. In M. N. Christiansen and C. F. Lewis (eds.), Breeding Plants for Less Favorable Environments. Wiley, New York. Pp. 71–142.

Clark, R. B. 1984. Physiological aspects of calcium, magnesium, and molybdenum deficiencies in plants. In F. Adams (ed.), Soil Acidity and Liming. 2nd Ed. *Agronomy* 12:95–170.

Clark, R. B., and J. C. Brown. 1975. Corn lines differ in mineral efficiency. *Ohio Res.* 60:83–86.

Coleman, N. T., E. J. Kamprath, and S. B. Weed. 1958. Liming. *Adv. Agron.* 10:475–522.

Cox, F. R., and E. J. Kamprath. 1972. Micronutrient soil tests. In J. J. Mortvedt, P. M. Giordano, and W. L. Lindsay (eds.), Micronutrients in Agriculture. Soil Sci. Soc. Am., Madison, Wisconsin. Pp. 289–317.

de Carvalho, M. M., C. S. Andrew, D. G. Edwards, and C. J. Asher. 1981. Aluminum toxicity, nodulation, and growth of *Stylosanthes* species. *Agron. J.* 73:261–265.

de Carvalho, M. M., D. G. Edwards, C. J. Asher, and C. S. Andrew. 1982. Effects of aluminium on nodulation of two *Stylosanthes* species grown in nutrient solution. *Plant Soil* 64:141–152.

Doll, E. C., and R. E. Lucas. 1973. Testing soils for potassium, calcium, and magnesium. In L. M. Walsh and J. D. Beaton (eds.), Soil Testing and Plant Analysis. Soil Sci. Soc. Am., Madison, Wisconsin. Pp. 133–151.

Edwards, D. G., and C. J. Asher. 1982. Tolerance of crop and pasture species to manganese toxicity. In A. Scaife (ed.), Plant Nutrition 1982. Proc. 9th Int. Plant Nutrition Colloq., Warwick, England. Commonwealth Agric. Bur., Slough, England. Pp. 145–150.

Evans, C. E., and E. J. Kamprath. 1970. Lime response as related to percent Al saturation, solution Al and organic matter content. *Soil Sci. Soc. Am. Proc.* 34:893–896.

Evans, J., B. J. Scott, and W. J. Lill. 1987. Manganese tolerance in subterranean clover (*Trifolium subterraneum* L.) genotypes grown with ammonium nitrate or symbiotic nitrogen. *Plant Soil* 97:207–215.

Fageria, N. K. 1984. Response of rice cultivars to liming in Cerrado soils. *Pesqui. Agropec. Bras.* 19:883–889.

Fageria, N. K. 1985. Influence of aluminum in nutrient solutions on chemical composition in two rice cultivars at different growth stages. *Plant Soil* 85:423–479.

Fageria, N. K., V. C. Baligar, and R. J. Wright. 1987a. The effect of aluminum on growth and uptake of Al and P by two rice cultivars. *Agron. Abstr.* p. 201.

Fageria, N. K., V. C. Baligar, and R. J. Wright. 1988. Aluminum toxicity in crop plants. *J. Plant Nutr.* 11:303–319.

Fageria, N. K., and M. P. Barbosa Filho. 1982. Preliminary screening of irrigated rice cultivars for efficient utilization of nitrogen. *Pesqui. Agropec. Bras.* 17:1709–1712.

Fageria, N. K., and M. P. Barbosa Filho. 1987. Phosphorus fixation in oxisol of Central Brazil. *Fert. Agric.* 94:33–37.

Fageria, N. K., and J. R. P. Carvalho. 1982. Influence of aluminum in nutrient solutions on chemical composition in upland rice cultivars. *Plant Soil* 69:31–44.

Fageria, N. K., R. J. Wright, and V. C. Baligar. 1987b. Rice cultivar response to aluminum in nutrient solution. *Commun. Soil Sci. Plant Anal.* 19:1133–1142.

Fageria, N. K., and F. J. P. Zimmermann. 1979. Interaction among P, Zn, and lime in upland rice. *R. Bras. Ci. Solo.* 3:88–92.

Fox, R. H. 1978. Selection for phosphorus efficiency in corn. *Commun. Soil Sci. Plant Anal.* 9:13–37.

Foy, C. D. 1964. Toxic Factors in Acid Soils of the Southern United States as Related to the Response of Alfalfa to Lime. USDA Res. Rep. No. 80. U.S. Gov. Print. Off., Washington, D.C.

Foy, C. D. 1973. Manganese and plants. In Manganese. Natl. Acad. Sci.–Natl. Res. Counc., Washington, D.C. Pp. 51–76.

Foy, C. D. 1974. Effects of aluminum on plant growth. In E. W. Carson (ed.), The Plant Root and Its Environment. Univ. Press of Virginia, Charlottesville. Pp. 601–642.

Foy, C. D. 1983. Plant adaption to mineral stress in problem soils. *Iowa State J. Res.* 57:339–354.

Foy, C. D. 1984. Physiological effects of hydrogen, aluminum and manganese toxicities in acid soils. In F. Adams (ed.), Soil Acidity and Liming. 2nd Ed. *Agronomy* 12:57–97.

Foy, C. D., A. L. Fleming, and W. H. Armiger. 1969. Differential tolerance of cotton varieties to excess manganese. *Agron. J.* 61:690–694.

Franco, A. A., and D. N. Munns. 1982. Acidity and aluminum restraints on nodulation, nitrogen fixation, and growth of *Phaseolus vulgaris* in solution culture. *Soil Sci. Soc. Am. J.* 46:296–301.

Gallagher, R. N., M. D. Jellum, and J. B. Jones, Jr. 1981. Leaf magnesium concentration efficiency versus yield efficiency of corn hybrids. *Commun. Soil Sci. Plant Anal.* 12:345–354.

Gambrell, R. N., and W. H. Patrick, Jr. 1982. Manganese. In A. L. Page, R. H. Miller, and D. R. Kinney (eds.), Methods of Soil Analysis. Part 2: Chemical and Microbiological Properties. 2nd Ed. ASA and SSSA, Madison, Wisconsin. Pp. 313–322.

Gerloff, G. C. 1987. Intact plant screening for tolerance of nutrient-deficiency stress. *Plant Soil* 99:3–16.

Glass, A. D. M., and J. E. Perley. 1980. Varietal differences in potassium uptake by barley. *Plant Physiol.* 65:160–164.

Goedert, W. J. 1983. Management of Cerrado soils of Brazil: A review. *J. Soil Sci.* 34:405–428.

Goedert, W. J., E. Lobato, and M. Resende. 1982. Management of tropical soils and world food prospects. *Proc. 12th Int. Congr. Soil Science, New Delhi* pp. 77–95.

Graham, R. D. 1978. Nutrient efficiency objectives in cereal breeding. In A. R. Ferguson, R. L. Bieleski, and I. B. Ferguson (eds.), Plant Nutrition 1978. Proc. 8th Int. Colloq. Plant Analysis and Fertilizer Problems, Auckland. *Inf. Ser.—N.Z. Dep. Sci. Ind. Res.* No. 134, pp. 165–170.

Graham, R. D. 1984. Breeding for nutritional characteristics in cereals. In P. B. Tinker and A. Läuchli (eds.), *Adv. Plant Nutr.* 1:57–102.

Greenland, D. J. 1974. Intensification of agricultural systems, with special reference to the role of potassium fertilizers. *Potassium Research and Agricultural Production. Proc. 10th Congr. Int. Potash Inst., Budapest* pp. 311–323.

Greenland, D. J. 1977. Contribution of microorganisms to the nitrogen status of tropical soils. In A. Ayanaba and P. J. Dart (eds.), Biological Nitrogen Fixation in Farming Systems of the Tropics. Wiley, London. Pp. 13–25.

Hargrove, W. L., and G. W. Thomas. 1981. Effect of organic matter on exchangeable aluminum and plant growth in acid soils. In R. H. Dowdy, J. A. Ryan, V. V. Volk, and D. E. Baker (eds.), Chemistry in the Soil Environment. *ASA Spec. Publ.* No. 40, pp. 151–165.

Heenan, D. P., and L. C. Campbell. 1981. Soybean nitrate reductase activity influenced by manganese nutrition. *Plant Cell Physiol.* 21:731–736.

Heenan, D. P., and O. G. Carter. 1976. Tolerance of soybean cultivars to manganese toxicity. *Crop Sci.* 16:389–391.

Helyar, K. R. 1978. Effects of aluminium and manganese toxicities on legume growth. In C. S. Andrew and E. J. Kamprath (eds.), Mineral Nutrition of Legumes in Tropical and Subtropical Soils. CSIRO, Melbourne. Pp. 207–231.

Hodges, S. C. 1987. Aluminum speciation: A comparison of five methods. *Soil Sci. Soc. Am. J.* 51:57–64.

Howeler, R. H., C. J. Asher, and D. G. Edwards. 1982. Establishment of an effective endomycorrhizal association on cassava in flowing solution culture and its effects on phosphorus nutrition. *New Phytol.* 90:229–238.

Hsu, P. H. 1968. Interaction between aluminum and phosphate in aqueous solution. *Adv. Chem. Ser.* No. 73, pp. 115–127.

Hue, N. V., G. R. Craddock, and F. Adams. 1986. Effect of organic acids on aluminum toxicity in subsoils. *Soil Sci. Soc. Am. J.* 50:28–34.

Islam, A. K. M. S., D. G. Edwards, and C. J. Asher. 1980. pH optima for plant growth. Results of a flowing solution culture experiment with six species. *Plant Soil* 54:339–357.

Jenkinson, D. S., and A. Ayanaba. 1977. Decomposition of carbon-14 labeled plant material under tropical conditions. *Soil Sci. Soc. Am. J.* 41:912–916.

Kamprath, E. J. 1970. Exchangeable aluminum as a criterion for liming leached mineral soils. *Soil Sci. Soc. Am. Proc.* 34:252–254.

Kamprath, E. J. 1978. Lime in relation to aluminium toxicity in tropical soils. In C. S. Andrew and E. J. Kamprath (eds.), Mineral Nutrition of Legumes in Tropical and Subtropical Soils. CSIRO, Melbourne. Pp. 233–245.

Kamprath, E. J., and C. D. Foy. 1971. Lime–fertilizer–plant interactions in acid soils. In R. W. Olsen, T. J. Army, J. J. Hanway, and V. J. Kilmer (eds.), Fertilizer Technology and Use. 2nd Ed. Soil Sci. Soc. Am., Madison, Wisconsin. Pp. 105–151.

Kamprath, E. J., and C. D. Foy. 1985. Lime–fertilizer–plant interactions in acid soils. In O. P. Engelstad (ed.), Fertilizer Technology and Use. 3rd Ed. Soil Sci. Soc. Am., Madison, Wisconsin. Pp. 91–151.

Kanwar, J. S., and L. J. Youngdahl. 1985. Micronutrient needs of tropical food crops. *Fert. Res.* 7:43–67.

Karlen, D. L., R. Ellis, Jr., D. A. Whitney, and D. L. Grunes. 1978. Influence of soil moisture and plant cultivar on cation uptake by wheat with respect to grass tetany. *Agron. J.* 70:918–921.

Kerven, G. L., D. G. Edwards, C. J. Asher, P. S. Hallman, and S. Kokot. 1989a. Aluminium determination in soil solution. I. Evaluation of existing colorimetric and separation methods for the determination of inorganic monomeric aluminium in the presence of organic acid ligands. *Aust. J. Soil Res.* 27:79–90.

Kerven, G. L., D. G. Edwards, C. J. Asher, P. S. Hallman, and S. Kokot. 1989b. Aluminium determination in soil solution. II. Short-term colorimetric procedures for the measurement of inorganic monomeric aluminium in the presence of organic acid ligands. *Aust. J. Soil Res.* 27:91:102.

Kim, M. K. 1985. Effects of pH and aluminium on nodulation and growth of *Trifolium subterraneum* cultivars. Ph.D. Thesis, University of Queensland, Brisbane, Australia.

Knudsen, D., G. A. Paterson, and P. F. Pratt. 1982. Lithium, sodium, and potassium. In A. L. Page, R. H. Miller, and D. R. Keeney (eds.), Methods of Soil Analysis. Part 2: Chemical and Microbiological Properties. 2nd Ed. ASA and SSSA, Madison, Wisconsin. Pp. 225–246.

Lal, R. 1986. Soil surface management in the tropics for intensive land use and high and sustained production. *Adv. Soil Sci.* 5:1–109.

Larue, T., and T. G. Patterson. 1981. How much nitrogen do legumes fix? *Adv. Agron.* 34:15–38.

Lee, M. T., and C. J. Asher. 1981. Nitrogen nutrition of ginger (*Zingiber officinale*). Effects of sources, rates, and times of nitrogen application. *Plant Soil* 62:23–34.

Lie, T. A. 1969. The effect of low pH on different phases of nodule formation in pea plants. *Plant Soil* 31:391–406.

Lindsay, W. L. 1972. Zinc in soils and plant nutrition. *Adv. Agron.* 24:147–186.

Lindsay, W. L. 1979. Chemical Equilibria in Soils. Wiley, New York.

Loneragan, J. F. 1978. The physiology of plant tolerance to low phosphorus availability. In G. A. Jung (ed.), Crop Tolerance to Suboptimal Land Conditions. *ASA Spec. Publ.* No. 32, pp. 329–343.

Loneragan, J. F., and E. J. Dowling. 1958. The interaction of calcium and hydrogen ions in the nodulation of subterranean clover. *Aust. J. Agric. Res.* 9:464–472.

Loneragan, J. F., D. L. Grunes, R. M. Welch, E. A. Aduayi, A. Tengah, V. A. Lazar, and E. E. Cary. 1982. Phosphorus accumulation and toxicity in leaves in relation to zinc supply. *Soil Sci. Soc. Am. J.* 46:345–352.

Lopes, A. S. 1977. Available water, phosphorus fixation, and zinc levels in Brazilian Cerrado soils in relation to their physical, chemical, and mineralogical properties. Ph.D. Thesis, North Carolina State University, Raleigh.

Mahler, R. L., and R. E. McDole. 1987. Effect of soil pH on crop yield in north Idaho. *Agron. J.* 79:750–755.

Marion, G. M., D. M. Hendricks, G. R. Dutt, and W. H. Fuller. 1976. Aluminum and silica solubility in soils. *Soil Sci.* 121:76–85.

McLean, E. O. 1973. Testing soils for pH and lime requirement. In L. M. Walsh and J. D. Beaton (eds.), Soil Testing and Plant Analysis. Soil Sci. Soc. Am., Madison, Wisconsin. Pp. 77–95.

Mengel, D. B., and E. J. Kamprath. 1978. Effect of soil pH and liming on growth and nodulation of soybeans in Histosols. *Agron. J.* 70:959–963.

Mengel, K., and E. A. Kirkby. 1980. Potassium in crop production. *Adv. Agron.* 33:59–110.

Mikkelsen, D. S. 1987. Nitrogen budgets in flooded soils used for rice production. *Plant Soil* 100:71–97.

Morgan, P. W., H. E. Joham, and J. V. Amin. 1966. Effect of manganese toxicity on the indole-acetic acid oxidase system of cotton. *Plant Physiol.* 41:718–724.

Moschler, W. W., G. D. Jones, and G. W. Thomas. 1960. Lime and soil acidity effects on alfalfa growth in a red-yellow podzolic soil. *Soil Sci. Soc. Am. Proc.* 24:507–509.

Mosse, B. 1972. The influence of soil type and *Endogone* strain on the growth of mycorrhizal plants in phosphate-deficient soils. *Rev. Ecol. Biol. Soil.* 9:537–592.

Munns, D. N. 1965. Soil acidity and growth of a legume. II. Reactions of aluminium and phosphate in solution and effects of aluminium, phosphate, calcium, and pH on *Medicago sativa* L., and *Trifolium subterraneum* L. in solution culture. *Aust. J. Agric. Res.* 16:743–755.

Munns, D. N. 1968. Nodulation of *Medicago sativa* in solution culture. I. Acid-sensitive steps. *Plant Soil* 28:129–146.

Munns, D. N. 1978. Soil acidity and nodulation. In C. S. Andrew and E. J. Kamprath (eds.), Mineral Nutrition of Legumes in Tropical and Subtropical Soils. CSIRO, Melbourne. Pp. 247–263.

Munns, D. N., and R. L. Fox. 1977. Comparative lime requirements of tropical and temperate legumes. *Plant Soil* 46:533–548.

Munns, D. N., J. S. Hohenberg, T. L. Righetti, and D. J. Lauter. 1981. Soil acidity tolerance of symbiotic and nitrogen-fertilized soybeans. *Agron. J.* 73:407–410.
Murphy, H. E., D. G. Edwards, and C. J. Asher. 1984. Effects of aluminium on nodulation and early growth of four tropical pasture legumes. *Aust. J. Agric. Res.* 35:663–673.
Nelson, L. E. 1983. Tolerances of 20 rice cultivars to excess Al and Mn. *Agron. J.* 75:134–138.
Nielsen, N. E., and S. A. Barber. 1978. Differences among genotypes of corn in the kinetics of P uptake. *Agron. J.* 70:695–698.
Ohki, K. 1985. Aluminum toxicity effects on growth and nutrient composition in wheat. *Agron. J.* 77:951–956.
Ohki, K. 1987. Aluminum stress on sorghum growth and nutrient relationships. *Plant Soil* 98:195–202.
Olmos, J. I. L., and M. N. Camargo. 1976. Incidence of aluminum toxicity in Brazilian soils, its characterization and distribution. *Cienc. Cult. (Sao Paulo)* 28:171–180.
Paula, M. B., F. D. Nogueira, H. Andrade, and J. E. Pitts. 1987. Effect of liming on dry matter yield of wheat in pots of low humic gley soil. *Plant Soil* 97:85–91.
Pavan, M. A., and F. T. Bingham. 1982. Toxicity of aluminum to coffee seedlings grown in nutrient solution. *Soil Sci. Soc. Am. J.* 46:993–997.
Pearson, R. W. 1975. Soil Acidity and Liming in the Humid Tropics. Bull. No. 30. Cornell University, Ithaca, New York.
Prasad, R., G. B. Rajale, and B. A. Lakdive. 1971. Nitrification retarders and slow-release nitrogen fertilizers. *Adv. Agron.* 23:337–383.
Rhodes, E. R., and W. L. Lindsay. 1978. Solubility of aluminum in soils of the humid tropics. *J. Soil Sci.* 29:324–330.
Richburg, J. S., and F. Adams. 1970. Solubility and hydrolysis of Al in soil solutions and saturated paste extracts. *Soil Sci. Soc. Am. Proc.* 34:728–734.
Robson, A. D., and J. F. Loneragan. 1970. Sensitivity to annual *Medicago* species to manganese toxicity as affected by calcium and pH. *Aust. J. Agric. Res.* 21:223–232.
Sanchez, P. A. 1976. Properties and Management of Soils in the Tropics. Wiley, New York.
Sanchez, P. A. 1987. Management of acid soils in the humid tropics of Latin America. In Management of Acid Tropical Soils for Sustainable Agriculture. IBSRAM Proc. No. 2. IBSRAM, Bangkok. Pp. 63–107.
Sanchez, P. A., and J. G. Salinas. 1981. Low-input technology for managing Oxisols and Ultisols in tropical America. *Adv. Agron.* 34:279–406.
Schnitzer, M., and S. I. M. Skinner. 1973. Organo-metallic interactions in soils. 3. Properties of iron and aluminum organic matter complexes, prepared in the laboratory and extracted from soil. *Soil Sci.* 98:197–203.
Sharma, K. C., B. A. Krantz, A. L. Brown, and J. Quick. 1968. Interaction of Zn and P in tops and roots of corn and tomato. *Agron. J.* 60:453–456.
Shea, P. F., W. H. Gabelman, and G. C. Gerloff. 1967. The inheritance of efficiency in potassium utilization in strains of snapbean, *Phaseolus vulgaris* L. *Am. Soc. Hortic. Sci. Proc.* 91:286–293.
Shoemaker, H. E., E. O. McLean, and P. F. Pratt. 1961. Buffer methods for determining lime requirements of soil with appreciable amounts of extractable aluminum. *Soil Sci. Soc. Am. Proc.* 25:274–277.
Siddiqui, M. Y., A. D. M. Glass, A. I. Hsiao, and A. N. Minjas. 1987. Genetic differences among wild oat lines in potassium uptake and growth in relation to potassium supply. *Plant Soil* 99:93–105.
Singh, B., and G. D. Sekhon. 1976. Some measures of reducing leaching loss of nitrates beyond potential rooting zone. I. Proper coordination of nitrogen splitting with water management. *Plant Soil* 44:192–200.
Singh, J. P., R. E. Koramnos, and J. W. B. Stewart. 1987. The zinc fertility of Saskatchewan soils. *Can. J. Soil Sci.* 67:103–116.
Spain, J. M., C. A. Francis, R. H. Howeler, and F. Calvo. 1975. Differential species and varietal tolerance to soil acidity in tropical crops and pastures. In E. Bornemisza and A. Alvarado (eds.), Soil Management in Tropical America. Dep. Soil Sci., North Carolina State University, Raleigh. Pp. 308–329.

Suthipradit, S. 1989. Effects of aluminium on growth and nodulation of some tropical crop legumes. Ph.D. Thesis, University of Queensland, Brisbane, Australia.

Tanaka, A., T. Tadano, K. Yamamoto, and N. Kanamura. 1987. Comparison of toxicity to plants among Al^{3+}, $AlSO_4^+$ and Al–F complex ions. *Soil Sci. Plant Nutr.* 33:43–55.

Taylor, G. J. 1988. The physiology of aluminum tolerance in higher plants. *Commun. Soil Sci. Plant Anal.* 19:1179–1194.

Terry, N., P. S. Evans, and D. E. Thomas. 1975. Manganese toxicity effects on leaf cell multiplication and expansion and on dry matter yield of sugar beets. *Crop Sci.* 15:205:208.

Thomas, G. W. 1975. The relationship between organic matter and exchangeable aluminum in acid soil. *Soil Sci. Soc. Am. J.* 39:591–597.

Thorne, W. 1957. Zinc deficiency and its control. *Adv. Agron.* 9:31–65.

Tinker, P. B. 1975. Effects of vesicular-arbuscular mycorrhizas on higher plants. In D. G. Jennings and D. L. Lee (eds.), Symbiosis. *Symp. Soc. Exp. Biol.* 29:325–349.

Viets, F. G., L. C. Boawn, and C. L. Crawford. 1954. Zinc content of bean plants in relation to deficiency symptoms and yield. *Plant Physiol.* 29:76–79.

Welch, R. M., M. J. Webb, and J. F. Loneragan. 1982. Zinc in membrane function and its role in phosphorus toxicity. In A. Scaife (ed.), Plant Nutrition. 1982. Proc. 9th Int. Plant Nutrition Colloq., Warwick, England. Commonwealth Agric. Bur., Slough, England. Pp. 710–715.

White, R. E., L. O. Tiffin, and A. W. Taylor. 1976. The existence of polymeric complexes in dilute solutions of aluminum and orthophosphate. *Plant Soil* 45:521–529.

Wright, R. J., V. C. Baligar, K. D. Ritchey, and S. F. Wright. 1989. Influence of soil solution aluminum on root elongation of wheat seedlings. *Plant Soil* 113:294–298.

Wright, R. J., V. C. Baligar, and S. F. Wright. 1987a. Estimation of phytotoxic aluminum in soil solution using three spectrophotometric methods. *Soil Sci.* 144:224–232.

Wright, R. J., V. C. Baligar, and S. F. Wright. 1987b. The influence of acid soil factors on the growth of snapbeans in major Appalachian soils. *Commun. Soil Sci. Plant Anal.* 18:1235–1252.

Wright, R. J., V. C. Baligar, and S. F. Wright. 1988. Estimation of plant available manganese in acidic subsoil horizons. *Commun. Soil Sci. Plant Anal.* 19:643–662.

Yoshida, S. 1981. Mineral Nutrition of Rice. Int. Rice Res. Inst., Los Banos, Philippines.

14

Plant Nutrient Interactions in Alkaline and Calcareous Soils

D. W. JAMES

The term "fertile," as defined by Webster's, is synonomous with "fruitful" and "prolific." In this context, therefore, a soil may be well supplied with plant nutrient elements in available form and yet be unproductive (infertile) if other growth factors are limiting to plant development. For example, desert soils are generally considered to be highly fertile, but they are not productive unless irrigated. Thus, soil water availability is a factor of soil fertility. Also, all other things being equal, selecting the proper cultivar for a given soil situation may be the difference between success and failure in a crop production enterprise. Accordingly, cultivar, or more specifically crop genetic potential, may be an important factor of soil fertility and productivity.

When plant growth is limited by factors other than plant nutrient availability, fertilizer use efficiency will be depressed. If crop production limitations such as

Crops as Enhancers of Nutrient Use

excess salts or lime-induced chlorosis were reduced or eliminated, then nitrogen, phosphorus, and potassium plant use efficiency (i.e., unit dry weight produced per unit element absorbed) would be enhanced.

This chapter focuses on the soil chemical and fertility aspects of alkaline soils, i.e., soils with pH above 7.0, and on special plant nutritional and crop production problems peculiar to these soils. Attention is focused principally on lime-induced chlorosis or iron deficiency, and the osmolality and specific ion toxicity problems of salt-affected soils. An evaluation is given of the potential role of edaphology (the study of soils as a medium of plant growth) in adapting plants to problem soils. Suggestions are given on how edaphology may become more firmly integrated with physiology and genetics in the development of cultivars that will tolerate specific types of problem soils.

I. Geography and Character of Arid and Semiarid Region Soils

Arid and semiarid regions of the world lie generally in two broad bands between 10 and 40 degrees north and 10 and 40 degrees south latitude, wherein lie the great deserts of the earth. Precipitation in these regions is limited because air masses descend from high altitudes to counterbalance rising air masses in the equatorial regions. Whereas rising air pumps moisture into the tropics, descending air brings aridity to the alternate regions. Rainfall patterns are modified both within and without the two broad dry bands by air flow over mountain ranges and by proximity of land to oceans and lakes (Landsberg and Schloemer, 1967).

Soils that develop within the drier regions seldom, if ever, receive enough natural precipitation for water to pass through the soil profile. These soils are characterized by pH values between 7.0 and 8.3, to as high as 9.5 in sodic soils. They usually have a calcium carbonate accumulation within the profile. This accumulation occurs when calcium and magnesium, after weathering from soil minerals in the surface layers, are translocated into the subsoil by infiltrating water. When water is lost by evapotranspiration, the alkaline earth metals are precipitated as carbonate coatings on soil particles. As the accumulation continues, the precipitates may bridge between soil particles and form caliche, a naturally cemented hardpan consisting of Ca carbonates and silicates. The lime accumulation layer or caliche may occur between 20 and 60 cm in depth, depending on ambient temperature and rainfall conditions.

The pH of calcareous, nonsodic soils, which ranges between 7.2 and 8.3, is not proportional to the amount of lime present. The nature of the carbonate deposition, i.e., coating of soil particles, gives the carbonates an impact on soil chemical and physical properties out of proportion to the amount of lime accumulated. Calcareous soils are well buffered against changes in pH because of the neutralizing effect of carbonate materials on acidifying tendencies.

Of course, soils may be calcareous if they are derived from calcite or dolomite. Therefore, calcareous soils are not limited to arid and semiarid regions. Important areas of calcareous soils are located in the subhumid central and midwestern parts of the United States. Soils derived from limestone may have high total lime content because the soil particles are not merely coated by, but are composed of, carbonates.

Localized areas within arid and semiarid regions may have accumulations of soluble salts, principally the Ca^{2+}, Mg^{2+}, Na^{+}, and K^{+} salts of Cl^{-}, SO_4^{2+}, and occasionally HCO_3^{-}. These salts are collected as water percolates through soil and they are transported to lower parts of the drainage system, such as river flood plains and valley floors. Salts may accumulate in lower physiographic locations as water is lost by evaporation. Artificial salinization of soils, following the onset of irrigation, has been the bane of irrigated agriculture since prehistoric times (James *et al.*, 1982). Reeve and Fireman (1967) indicate that 25 countries have one million acres or more of land under irrigation and that serious salinity problems exist in every one of these.

The proportion of the earth's surface that lies in the arid and semiarid zones is approximately 55% of the total land mass. Within this area lies about 900 million ha of arable land (Wallace and Lunt, 1960; Vose, 1982). Proper management of arid and semiarid soils through fertilization, salinity control, and adaptation of crop cultivars is indispensable to maintenance of world food supplies.

II. Alkaline Soils versus Fe Plant Nutrition

A. LIME-INDUCED CHLOROSIS

Many kinds of plants are prone to be iron deficient when grown on certain types of calcareous soils. This is referred to as lime-induced iron deficiency or iron chlorosis. Iron deficiency is characterized in plants by yellowing of leaves with the tissue immediately adjacent to veins remaining green, hence interveinal chlorosis. As the severity increases the proportion of affected tissue increases and the leaf may become completely chlorotic. In the extreme, chlorosis leads to leaf necrosis and senescence and ultimately death of the plant. Iron deficiency chlorosis in plants within a given field may be persistent over time or it may be ephemeral, since Fe chlorosis may increase or decrease in a plant during the growing season as environmental conditions change. The result of Fe deficiency is reduced crop growth vigor and loss of productivity. Figures 1 and 2 illustrate problems of Fe deficiency frequently encountered on semiarid soils.

Iron is a major component of the earth's crust and of most soils, but it exists as silicates, oxides, and hydroxides and is sparingly soluble in well-aerated soils, even at low pH. Lindsay (1979) showed that Fe^{3+}, the form of Fe utilized by plants, has exceedingly low solubility in soils of pH 7 and above, and that Fe solubility decreases by three orders of magnitude for every unit increase in pH. Römheld and

Figure 1. Maize cultivar trial established on a calcareous soil of pH 7.8 in Utah County, Utah. Commercial seed cultivars were planted in paired rows. Two rows to left of center and two rows to right of center were Fe efficient. Others showed varying degrees of lime-induced chlorosis symptoms. Farmers in the area had done their own cultivar selection based on trial and error. Commercial seed cultivars produced in the intermountain United States area were generally less prone to Fe chlorosis than the commercial seed cultivars imported from the midwestern United States. Source: Nunez-Medina, 1984.

Marschner (1986) state that Fe^{3+} solubility above pH 5 is limiting to plant use and that plants intervene by altering the chemistry of the rhizosphere, mobilizing Fe for plant uptake and utilization. That plant genotypes and cultivars vary markedly in their ability to alter the rhizosphere and to utilize absorbed Fe is the focal point of this section.

Lime in soil is the predisposing factor, but other soil and environmental conditions are associated with Fe deficiency. Wallace and Lunt (1960) and Wallace (1982) listed the following factors, which, they say, operate singly or in combination in contributing to Fe deficiency: bicarbonate in soil solution and irrigation water; excessive soil water; reduced soil aeration; high phosphates or nitrates in soil; high levels of Mn, Zn, and Cu; low or high temperatures; high light intensity; certain organic matter additions to soil; plant virus diseases; and root damage by nematodes and other organisms. Again, it is emphasized, lime in the soil is the predisposing factor as the foregoing factors evidently do not lead to Fe deficiency in noncalcareous soils. An exception to this might be Fe deficiency in citrus growing on acid soils as a result of Cu accumulations following long-term use of Cu-containing fungicides (Smith and Specht, 1952).

Figure 2. A cherry tree growing in Utah County, Utah, showing uniform and severe iron chlorosis. The chlorotic tree was surrounded on all sides by trees that were completely normal in appearance. No soil physical or chemical differences could be discovered in distances between trees in the field. However, the rootstock varied from tree to tree.

Some plant families are, on average, more resistant than others to lime-induced Fe chlorosis, but among most economic plants no one genus or crop type is completely immune. Clark and Gross (1986) presented a review of plant genotype and cultivar variability in Fe stress response. They listed studies of plant genotypes in 45 plant genera ranging from apple to wheat. It is evident from their summary that the preponderant amount of genetics and cultivar selection effort has been focused on four crop types, namely, maize, rice, sorghum, and soybean. This may be a measure of the relative economic significance of Fe deficiency among crop types. Data in Table I give an indication of the extent and intensity of current efforts to improve Fe efficiency in plants through investigations of the mechanisms involved and plant breeding selection for Fe efficiency. The International Symposium on Iron Nutrition and Interactions in Plants, initiated in 1981, is one indication of the recent growth in interest of lime-induced chlorosis and of the potential for improving conditions through plant selection to overcome problem soil conditions (Nelson, 1982; James, 1984; Clark, 1986; Barton, 1988).

As pointed out by Wallace (1982), Fe deficiency has the reputation of being the most difficult to control of any micronutrient deficiency in calcareous soils. It should not be inferred, however, that all calcareous soils induce Fe deficiency, even in all susceptible cultivars. Kuykendall (1956) pointed out that of an estimated 18

Table I. Crops Mentioned in Current Literature as Subjects of Iron Efficiency Research and Improvement

Crop	Reference
Fruits	
Avocado	Chen and Barak (1982); Kadman and Ben-Ya'cov (1982)
Citrus	Hamze *et al.* (1986); Hamze and Nimah (1982)
Grape	Chen and Barak (1982)
Papaya	Kannan (1985)
Tomato	Brown and Ambler (1974)
Grains and Grasses	
Barley	Chen and Barak (1982)
Maize	Brown and Bell (1969); Kovacevic *et al.* (1982); Pierson *et al.* (1984a); Nordquist and Compton (1986)
Kentucky bluegrass	Harivandi and Butler (1982)
Oat	McDaniel and Brown (1982); McDaniel and Murphy (1978); Fehr (1984)
Rice	Kannan (1982)
Sorghum	Clark *et al.* (1982); Kannan (1980); McKenzie *et al.* (1984); Williams *et al.* (1982); Pierson *et al.* (1984a)
Weeping lovegrass	Foy *et al.* (1977)
Wheat	Neikova-Bocheva *et al.* (1986)
Old world bluestems	Berg *et al.* (1986)
Oil and Fiber Crops	
Cotton	Kannan (1982); Brown and Jones (1977b)
Jute	Kannan (1982)
Peanut	Hartzook (1984)
Soybean	Brown and Jones (1977a); Cianzio *et al.* (1979); Coulombe *et al.* (1984); Niebur and Fehr (1981); Pierson *et al.* (1984a)
Sunflower	Alcantara and de la Guardia (1986); Kannan (1984)
Grain Legumes	
Bean	Coyne *et al.* (1982); Pierson *et al.* (1984b)
Chick pea	Singh *et al.* (1986); Hamze *et al.* (1987)
Lentil	Rai *et al.* (1984); Kannan and Ramani (1987)
Pigeon pea	Kannan (1982)

million ha of calcareous soils in the western United States, susceptible crops suffer moderate to severe Fe chlorosis on about one-eighth of this land. In other words, the phenomenon of lime-induced Fe chlorosis occurs only on a portion of calcareous soils. This led Wallace and Lunt (1960) to the assertion that "Fe-chlorosis notwithstanding, calcareous soils are not to be disdained since in general they include many of the most fertile (i.e., productive) soils." They further asserted that calcareous soils are infinitely preferable to acid soils.

Land managers have long been intent on solving Fe deficiency chlorosis with soil amendments and foliar sprays. The persistence of lime-induced chlorosis in many kinds of crops is a general indication of the technical limitations and economic impracticality of such management practices.

There is a growing body of information that indicates that many plant physiological traits associated with Fe deficiency are under genetic control, and that these traits can be selected by routine plant breeding procedures. Therefore, plant breeding, according to Brown *et al.* (1972), represents the possibility of controlling Fe deficiency chlorosis by changing the plant to accommodate peculiar soil problems rather than trying to amend soil chemical properties to satisfy plant needs. Brown and Jones (1977a–c), in a series of papers on soybean, cotton, and sorghum, illustrated the feasibility of this technology.

B. GENETICS OF Fe EFFICIENCY

Several distinctive plant traits are associated with uptake and utilization of Fe under Fe deficiency stress, including both physiological and morphological responses. Excellent reviews of this topic are available (Bennett *et al.*, 1982; Brown, 1977; Chen and Barak, 1982; Clark and Coyne, 1988; Clarkson and Hanson, 1980; Duvick *et al.*, 1981; Olsen *et al.*, 1981; Römheld, 1987; Römheld and Marschner, 1981, 1986). These plant traits, detailed below, are evidently under varying degrees of genetic control.

Iron mobilization in the rhizosphere was described by Römheld and Marschner (1986) as comprising two physiological strategies that are summarized in Table II. Strategy I is more characteristic of dicots, while strategy II is generally limited to the Gramineae. Lime-induced chlorosis is seen in plant genotypes and cultivars where the Fe response mechanisms of Table II are weak or nonexistent. Phenolics in

Table II. Physiological and Morphological Responses to Fe Deficiency Stress in the Root of Plants Growing on Calcareous Soils[a]

Type of Response	Category[b] I	Category[b] II
H^+ release	**[c]	—
Fe(III) reduction at plasma membrane	**	—
Release of phenolics	**	—
Formation of rhizodermal transfer cells	**	—
Release of amino-chelators	—	**

[a]Source: Marschner, 1986.
[b]I is essentially limited to dicotyledonous plants and II to the Gramineae.
[c]**, Highly important, otherwise of limited importance or absent.

Table II include chelates or plant siderophores. Wallace (1982) adds to the three Fe deficiency stress responses of Table II by including the buildup of citric acid within the root, which evidently facilitates transport of Fe to the shoot. Wallace (1982) indicated that these four types of physiological response may occur separately or in combination, for a total of 15 distinctive responses among plant genotypes to Fe deficiency stress.

Data given in Figure 3, adapted from the report of Clark *et al.* (1982), represent an example of Fe stress response in sorghum. As indicated, the nutrient solution pH was lowered under Fe deficiency conditions. This tendency was affected by the form of supplied N: NH_4-N resulted in sharply lowered pH and NO_3-N resulted in slightly increased pH in the nutrient solution culture.

Marschner (1986) listed the morphological changes in roots of dicotyledons associated with Fe stress, including inhibition of root elongation, increased diameter of the root apex, more abundant root hairs, thickened rhizodermal cells, and the formation of rhizodermal transfer cells. The Fe deficiency-induced transfer cells are characterized by high metabolic activity and they are the loci of H^+ and phenolic compound (siderophore) releases. According to Marschner (1986), after the supply of Fe is restored, the transfer cells degenerate and associated metabolic activities return to normal levels while the external pH increases and reducing compounds disappear.

In addition to Fe mobilization in the rhizosphere, there are traits that affect Fe translocation to culms and leaves, and traits that lead to Fe inactivation in the leaves. In the latter phenomenon, total Fe plant composition may be appreciable but a large portion of the contained Fe may be physiologically inactive. Different forms of Fe in plants (e.g., Fe^{3+}, Fe^{2+} total Fe, "physiologically active" Fe) have been analyzed

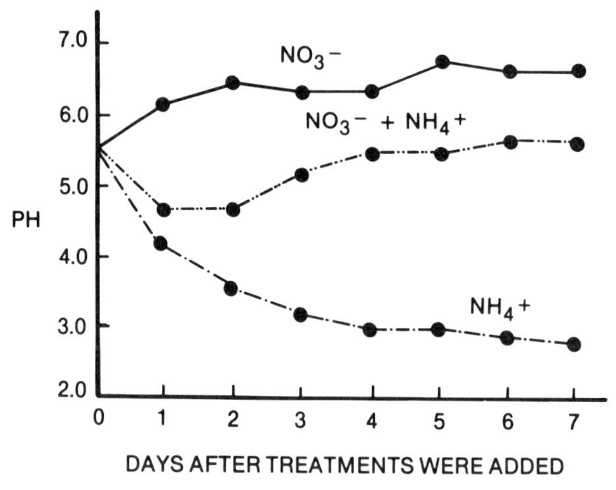

Figure 3. Sorghum effects on pH of nutrient solution as related to time and source of nutrient N. Source: Clark *et al.*, 1982. Reproduced by permission of Marcel Dekker, Inc.

in various plant types as a help in characterizing Fe-efficient genotypes (Pierson *et al.*, 1984a,b; Bell *et al.*, 1958).

C. Fe EFFICIENCY SELECTION PROCEDURES

Wann (1941) and Weiss (1943) did the pioneering work on lime-induced chlorosis with demonstrations that plants could be bred and selected for Fe efficiency. Wann (1941), working with American grape, showed that grafting of scions on selected rootstocks was a solution to an otherwise unmanageable lime-induced chlorosis. Since that time selection of both scions and rootstocks has been done for citrus (Hamze *et al.*, 1986; Hamze and Nimah, 1982; Wutscher *et al.*, 1970) and for avocado (Kadman and Ben-Ya'acov, 1982) with marked success. Avocado production in Israel, a highly successful enterprise, was evidently not feasible before the adaptation of this technology to cultivar improvement. Grafting of soybean scions was used as a research tool by Brown and Tiffin (1960) and by Brown *et al.* (1961). Using this technique, these workers were able to segregate root physiological traits from those of the aboveground parts.

Weiss (1943) was the first to establish that Fe efficiency in soybean was a genotypic trait. He reported that differences in Fe utilization among soybean genotypes were conditioned by a single gene. Weiss' work received little attention initially but Fe-efficient soybean cultivar development is a significant cultivar development success story of recent years. Fehr and associates (Froehlich and Fehr, 1981; Niebur and Fehr, 1981; Hintz *et al.*, 1987), using rapid visual screening as one selection tool, have shown the potential for eliminating lime-induced chlorosis as a limiting factor to soybean production on calcareous soils.

Data in Figure 4, taken from the report of Froehlich and Fehr (1981), show the effect of calcareous soils on Fe chlorosis intensity in soybean and also the striking relationship between chlorosis intensity and yield.

Fehr (1982) mixed the seeds of an Fe-efficient and an Fe-inefficient soybean line and tested them on four calcareous soils. His results are presented in Table III. As indicated, the performance of the Fe-inefficient pure seed line ranged from zero to a relative yield of 76% among sites, averaging 44% across all sites. But the Fe-efficient line alone and in combination with the Fe-inefficient line gave relative yields of 100% at all sites. Fehr (1982) formulated recommendations on the proportion of Fe-efficient and Fe-inefficient seed cultivars to blend depending on the proportion of calcareous and noncalcareous soils present in a given field.

Similar genetic and physiological work has been done for other field crops but the degree of success, though notable, is less marked than for soybean. Examples are sorghum (Brown and Jones, 1977c; Esty *et al.*, 1980; Mikesell *et al.*, 1973), maize (Bell *et al.*, 1958; Clark and Brown, 1974), peanut (Hartzook *et al.*, 1974), and oat (McDaniel and Brown, 1982). There will probably be commercially available Fe-efficient cultivars of these and other crop types as the economic benefits are demonstrated. One direct effect of these efforts would be increased plant use efficiency of the fertilizer elements.

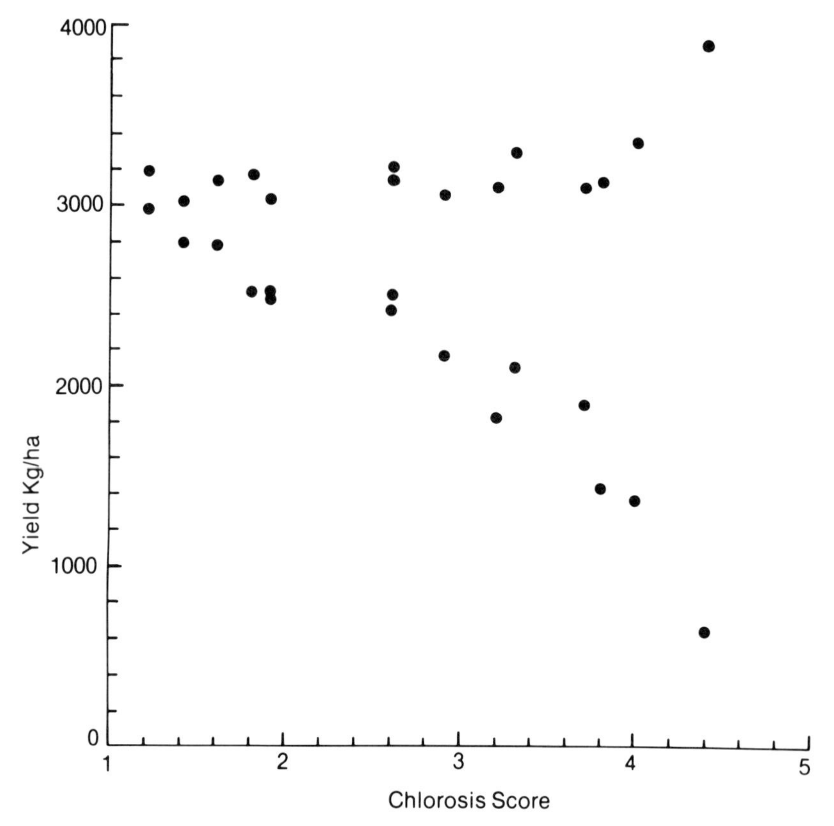

Figure 4. Yield of 15 soybean cultivars and breeding lines produced on noncalcareous (top data set) and calcareous soils (bottom data set) in Iowa as related to visual chlorosis scores (1 = no chlorosis, 5 = highly chlorotic). Each point is the mean of five locations and represents an individual commercial cultivar or breeding line. Each cultivar or line is represented in both the upper and lower data set. Graph produced from data reported by Froehlich and Fehr (1981).

Several approaches have been used in plant screening for Fe-efficient genotypes. Visual scoring of chlorosis in soybean mentioned above has been used in sorghum by McKenzie *et al.* (1984). Pierson *et al.* (1984a,b) measured Fe^{2+} in leaves of sorghum, maize, dry bean, and soybean. Coulombe *et al.* (1984) used the HCO_3 content of the rooting medium in nutrient solution culture of soybean and claimed good success. Alcantara *et al.* (1988) used the HCO_3 technique in selecting sunflower genotypes for Fe efficiency. Hamze *et al.* (1987) used soils varying in calcium carbonate content for selecting chick-pea genotypes. Clark *et al.* (1982) used high PO_4 and NO_3 in the nutrient solution and also calcareous soils in greenhouse pots to induce the Fe response mechanism in identifying Fe-efficient sorghum genotypes. Williams *et al.* (1982) also segregated Fe-efficient sorghum genotypes on the basis of excess NO_3 in the rooting medium.

Table III. Yield (q/ha) of Single and Mixed Seed Sources of Fe-Efficient and Fe-Inefficient Soybean Lines on Iowa Calcareous Soils[a]

Proportion of Mixed Seeds (%)		Location			
A2[b]	S1492[b]	Ames	Humboldt	Knierim	Meservey
100	0	33a	34a	27ab	26a
90	10	33a	34a	27ab	24ab
80	20	31a	33a	29a	25ab
70	30	33a	31a	26abc	25ab
60	40	33a	32a	26abc	24ab
50	50	33a	34a	25bc	25ab
40	60	31a	30a	24bc	20ab
30	70	31a	32a	24bc	24ab
20	80	33a	28a	17d	22ab
10	90	30a	25ab	13e	18ab
0	100	25a	14b	0f	13b

[a]Source: Fehr, 1982. Used by permission of Marcell Dekker, Inc.
[b]A2, a chlorosis-resistant line; S1492, a commercial chlorosis-susceptible cultivar. Values within a column followed by the same letter are not significantly different at the 5% level of probability.

The above factors, alone or in combination with others, may be incorporated into a mass screening program for selection of Fe-efficient genotypes. The intent is to have screening procedures that are efficient in terms of handling large numbers of genotypes and that duplicate field conditions as closely as possible in terms of Fe uptake and utilization.

D. THE POTENTIAL ROLE OF EDAPHOLOGY IN DEVELOPMENT OF Fe-EFFICIENT PLANTS

Chaney and Bell (1987) spoke on the need for interdisciplinary work on solving many of the problems of identifying plant traits associated with plant Fe efficiency interactions on different soils. Ross (1986) and Clarkson and Hanson (1980) addressed similar themes. It is apparent that, to date, edaphology has had a largely tangential role in the overall plant Fe efficiency thrust. The root of the problem is calcareous soils, and herein is where edaphology should have a more direct input in the effort to adapt crops to problem soil conditions. Important soil science contributions could be made in the following areas:

1. Plant physiologists have highlighted HCO_3 in the rooting medium as a factor of plant Fe stress. The soil $H_20–CO_2–CaCO_3$ system should be studied in the field to better understand the wide differences among calcareous soils in inducing Fe

deficiency, which heretofore have not been explained through routine soil chemical and physical measurements. For this purpose, Fe-inefficient plant types or plant indicators (Brown and Wann, 1982) could be used as a device for categorizing Fe-deficient soils. This would help focus on soil chemical characteristics more closely related to plant Fe deficiency.

2. Excess soil moisture and excess PO_4, NO_3, Mn, Zn, and other elements contribute to lime-induced chlorosis. The specific soil chemical and physical conditions associated with these antagonisms need to be better understood, first to help explain plant physiological interactions in Fe stress response, and second to facilitate Fe-efficient genotype screening programs.
3. Edaphology could contribute to plant screening processes by thoroughly characterizing field and greenhouse rooting media, by assuring homogeneous conditions in respect to soil moisture, N, P, and $CaCO_3$ content, and for better reproducibility or prediction of Fe deficiency stress.
4. Edaphologists are usually adept in the benchwork of soil and plant sampling and analysis for inorganic constituents. Soil and plant sampling and analysis procedures could be specifically adapted to large numbers of analyses in a plant screening program. One approach could be to evaluate plant sampling (time and plant part) for this purpose, together with methods of elemental extraction. Procedures not involving complete combustion would increase the efficiency of plant analysis in a genotype screening and cultivar development program.

III. Salt-Affected Soils versus Crop Productivity

Crop selection has long been recognized as an important tool in the management of salt-affected soils and irrigation waters. Ayers and Westcot (1985) stated that there is an 8- to 10-fold range in susceptibility to saline conditions among different crop genera and species. Data in Table IV illustrate their crop salt tolerance classification system. The FAO (1985) gave a broad classification of irrigation water qualities based on total salinity (ECw) and Cl and Na ion toxicities. They also classified water quality on the basis of excess Na (SAR) in relation to soil physical properties. A condensed version of the FAO system is given in Table V. The reports of both Ayers and Westcot (1985) and FAO (1985) are considerably more optimistic than similar classifications and interpretations given earlier (Richards, 1954) in regard to use of salt-affected soils and irrigation waters.

Data like those of Tables IV and V must be interpreted under local conditions because crop sensitivity to salts is modified by farming practices, particularly irrigation amount and frequency, leaching fraction, and drainage (Ayers and Westcot, 1985). In addition, plant sensitivity to salinity varies with growth stage, and there are varying degrees of sensitivity to specific ion toxicities such as Cl and Na. Maas and Hoffman (1977) indicated that, in general, crops tolerate salinity up to a thresh-

Table IV. Salt Tolerance of Representative Crops at 100% and 50% Yield Potential[a]

Crop	ECe[b] (dS/m) at 25°C and Yield Potential of	
	100%	50%
Field		
Barley	8.0	18
Wheat	6.0	13
Soybean	5.0	7.5
Maize	1.7	5.9
Bean	1.0	3.6
Vegetable		
Beet, red	4.0	9.6
Tomato	2.5	7.6
Cabbage	1.8	5.9
Potato	1.7	5.9
Carrot	1.0	4.6
Forage		
Wheatgrass, tall	7.5	19
Ryegrass, perennial	5.6	12
Fescue, tall	3.9	12
Alfalfa	2.0	8.8
Clover, strawberry	1.5	5.7
Fruit		
Date palm	4.0	18
Grapefruit	1.8	4.9
Peach	1.7	4.1
Almond	1.5	4.1
Strawberry	1.0	2.5

[a]Source: Ayers and Westcot, 1985.
[b]ECe is electrial conductivity of soil-saturated paste extract formerly designated as mmhos/cm.

old level that is determined by crop species, after which yield decreases linearly with increasing salt concentrations.

In recent years another factor has been added to those listed above in connection with crop salt tolerance, namely, salt tolerance differences among crop cultivars. The possibility exists that crops bred and selected for salt tolerance could become an important management choice (to be used in conjunction with reclamation procedures such as leaching, drainage, and the application of leaching requirements) in improving the efficiency of salt-affected soils and irrigation waters in crop production. Salt-tolerant cultivars would have the effect of expanding crop production to areas where conventional reclamation procedures are limited economically or technically. It is apparent that salt-tolerant cultivars would have little if any benefit in

Table V. Irrigation Water Quality Classification[a]

Potential Irrigation Water Problem	Degree of Restriction in Use		
	None	Slight to Moderate	Severe
Salinity (dS/m)	<0.7	0.7–3.0	>3.0
Specific ion toxicity[b]			
Sodium (SAR)	<3	3–9	>9
Chloride ($mmol_c$/liter)[c]	<4	4–10	>10
Sodicity[d]			
SAR		ECw	
0–3	>0.7	0.7–0.2	<0.2
3–6	>1.2	1.2–0.3	<0.3
6–12	>1.9	1.9–0.5	<0.5
12–20	>2.9	2.9–1.3	<1.3
20–40	>5.0	5.0–2.9	<2.9

[a]Source: FAO, 1985.
[b]For sensitive crops.
[c]Millimoles of ionic charge per liter, formerly meq/liter.
[d]As related to soil physical properties, evaluated using ECw (electrical conductivity of water expressed as dS/m, formerly mmhos/cm and SAR).

those cases where the limiting factor was simply a matter of high water table. Figure 5 shows the effect on maize of lowered water table and concomitant decrease in salinity in a Dominican Republic reclamation project.

Improved production of crops using salt-tolerant cultivars on marginally saline soils would have a direct effect on fertilizer use efficiency under saline conditions. Dewey (1962) estimated that crested wheatgrass strains selected for salt tolerance could double or triple yield of this forage on mildly saline soils. It is probable that there would be an equivalent improvement in plant nutrient use efficiency.

The following discussion focuses on current knowledge of and potential for breeding improved crop types that are tolerant of salinity.

A. PLANT BREEDING AND SELECTION FOR SALT-TOLERANT PLANTS

The literature of the early 1960s was well aware of plant genetic variability in regard to salt tolerance, ion transport mechanisms, and Cl and Na specific ion toxicities (Huffaker and Wallace, 1959; Epstein, 1963; Epstein and Jefferies, 1964). These authors suggested that increased knowledge of the mechanisms controlling ion transport selectivity in plants could facilitate plant breeding efforts analogous to the thrust provided by the understanding gained earlier on the metabolic pathways.

Figure 5. High water table and saline soil effects on maize near Santiago in the Dominican Republic. Lowering the water table using covered and open drains automatically decreased salt content of the soil. With the exception of rice, plant genotype variability in economic plants to saturated soil conditions has not been demonstrated.

Epstein (1963) presented evidence that several ion transport pumps exist in plants that are distinct from each other, and that they are each under genetic control.

In the interim, Maas and Neiman (1978) reviewed the physiological bases for plant salt tolerance and associated breeding and selection procedures. They emphasized that plant symptoms of salinity stress are vague and, therefore, physiological and biochemical mechanisms of salt tolerance should be identified to provide geneticists with specific selection criteria to facilitate rapid screening of tolerant genotypes.

As evidenced by the number of excellent reviews available on different aspects of the issue, a remarkable amount of progress has been made in recent years on understanding plant salt tolerance (Devine, 1982; Duvick *et al.*, 1981; Epstein *et al.*, 1980; Epstein and Rains, 1987; Epstein, 1985; Flowers and Yeo, 1986; Gerloff and Gabelman, 1983; Greenway and Munns, 1980; Jeschke, 1984; Läuchli, 1984, 1987; Shannon, 1984, 1985; Tal, 1984). Epstein (1985) stated that the genetic approach to crop production on salt-affected soils can be traced to two origins: the European interest in the physiological ecology of halophytes and the need to improve crop production on salt-affected soils in the western United States.

Duvick *et al.* (1981) pointed out that inheritance for salt tolerance can be simple or complex in terms of the number of genes involved. He stated also that progress on developing salt-tolerant cultivars was limited by the lack of reliable and rapid field selection procedures, and that correlations between crop yields and stress tolerance levels were badly needed. The complexity of plant breeding selection for salt tolerance was pointed out by Epstein and Rains (1987), who recognized, with Ayers and Westcot (1985), that plant salt tolerance varies temporally and spatially and with plant growth stage. Gerloff and Gabelman (1983) indicated that, although significant progress had been achieved without detailed knowledge of the physiology of crop salt tolerance, more rapid progress could be made if the mechanisms were better understood.

Plant responses to salt stress are categorized in the following discussion as to morphology, general salinity (osmolality), and excess ion effects. Later, salt stress response variability among plant genotypes will be illustrated. There is duplication among sections and references cited as several authors discuss the general aspects of plant salt stress in connection with detailed evaluations of specific salt tolerance mechanisms. Also, more than one salt tolerance mechanism will usually be operating in a given plant at the same time.

B. MECHANISMS OF PLANT SALT TOLERANCE

Shannon (1979) summarized plant morphological changes engendered by saline conditions as follows: The shape and size of plant organs and cells may change in response to salt stress. This includes increased leaf succulence, decreased leaf size and leaf number, reduced numbers of stomata, thickened cuticle, and deteriorated or undeveloped xylem. Since salinity inhibits top growth more than root growth, there may be a distinctive change in top/root ratio. Visual symptoms (leaf burn) may be evident. Symptomology, however, is apparently not as definitive in diagnosing salt stress as in diagnosing iron chlorosis deficiency stress.

Kramer *et al.* (1977) reported that in *Phaseolus* species xylem parenchyma cells differentiated as transfer cells with well-developed wall protuberances adjacent to the half-bordered pits of the vessels. They said further that the cytoplasm of these transfer cells contains cisternae of rough endoplasmic reticulum, the number of which increased greatly when grown in saline culture. Leopold and Willing (1984) proposed that salt-induced lesions in membranes and subsequent leakage of cell contents could be a distinct effect of ion toxicity.

Osmotic regulation is a plant response to internal water deficits that is characterized by changes in plant water balance and decreased transpiration. Osmotic adjustments include increased delivery of organic solutes to cells such as organic acids, amino acids, and sugars. Salts may be selectively excluded at the root surface or sequestered in leaf cell vacuoles (Flowers and Yeo, 1986). There may be both uni- and bidirectional transport (i.e., phloem transport) of salt (Greenway and Munns, 1980).

It is common to have intraspecies stress variability to more than one salt component. Wyn Jones *et al.* (1984), for example, showed that although Cl and Na were selectively excluded (with separate efficiencies) by wheat, some genotypes showed relatively high tolerance to total salt load in the leaf. Bernstein *et al.* (1956) reported that about one-half of the growth reduction in grape was due to Cl toxicity and the other half was due to high osmotic stress.

Some salinity escape mechanisms, usually more characteristic of the halophytes, include salt accumulation in expanded leaves (while protecting expanding leaves) followed by necrosis of the older leaves. Salt excretion from leaf glands also occurs (Pitman, 1984).

Specific ion effects include ion imbalance and ion toxicities. Plants may react to excess salts by selectively excluding Cl and/or Na from organs or cells. Ion compartmentation or localization within organs or cells has been observed (Eggers and Jeschke, 1983; Flowers and Läuchli, 1983; Lessani and Marschner, 1978; Leigh and Wyn Jones, 1986; Jeschke, 1984; Läuchli, 1984; Wyn Jones, 1981).

The effects of excess Cl and Na, as related to osmolality and direct toxicity, are independent phenomena. Smith *et al.* (1981), working with alfalfa, reported that Cl toxicity effects were independent of salt source (NaCl versus KCl) and also independent of Na per se. Alfalfa tissue Cl concentrations decreased in the order root $<$ culm $<$ leaf. Hajrasuliha (1980) showed that leaf Cl accumulation by bean was dependent on Cl concentration in the rooting medium and that Cl accumulation by the plant was independent of other medium solutes. Flowers and Yeo (1981) reported a negative correlation between Cl and Na accumulation in individual rice plants.

Parker *et al.* (1983, 1987) reported Cl toxicity in soybeans that was induced by KCl fertilization. In their work the overall salt load was apparently lower than that usually associated with saline soils. Another effect of Cl is competition with nitrate uptake under both saline (Torres and Bingham, 1973) and nonsaline (James *et al.*, 1969, 1970) conditions. Some Torres and Bingham (1973) results given in Figure 6 illustrate, among other things, that soil N use efficiency by plants can be sharply reduced by salinity. Furthermore, interpretation of soil and plant N data must take cognizance of salinity in the environment.

It is not clear whether Cl uptake under saline conditions is associated with plant disease suppression as has been reported for nonsaline conditions (Christensen and Brett, 1985; Fixen *et al.*, 1986).

Potassium interacts strongly with Na in the rhizosphere and at cell membranes in salt stress responses. Sodium exclusion is mediated by K storage in the root tip and K retranslocation within the plant. Part of the Na–K interaction, as suggested by Flowers and Läuchli (1983), may be partial substitution by Na for some of the normal plant-K functions. Bogemans and Stassart (1987) reported that salt-tolerant barley retranslocated K more efficiently from older leaves, exchanging it with Na in the expanding leaves. Jeschke (1984) said that the importance of intracellular compartmentation of K and Na is well founded. This occurs by vacuolar Na accumulation and by Na–K exchange in association with preferential K retranslocation. Data in Figure 7 show the reciprocal relationship between K and Na in barley roots grown in Na-salinized solution cultures as shown by Jeschke and Stelter (1976).

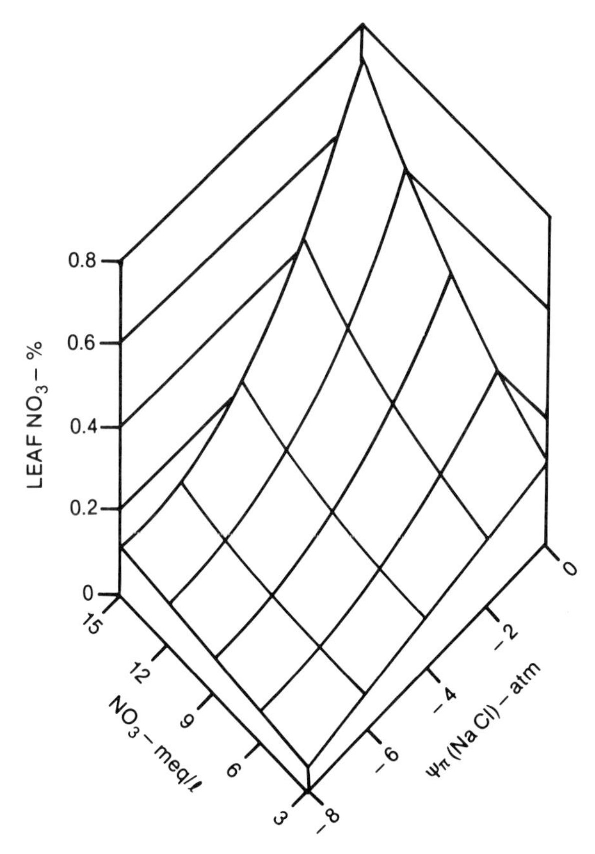

Figure 6. NO_3-N in wheat leaf as related to NO_3 and Cl in the nutrient solution. Source: Adapted from Torres and Bingham, 1973, by permission of the Soil Science Society of America.

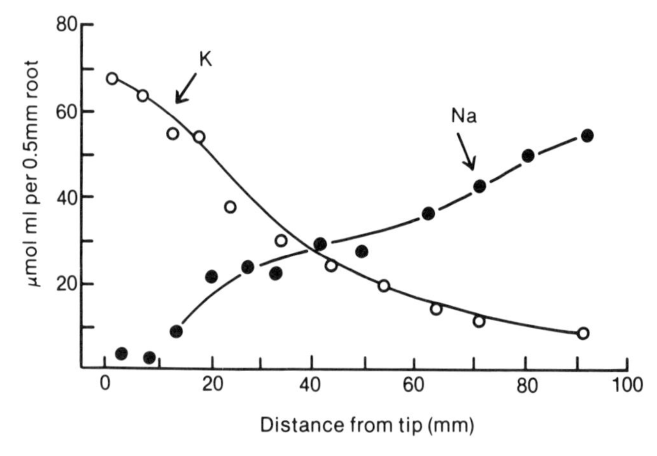

Figure 7. Longitudinal profiles of K and Na content of barley root grown for 4 days in aerated 1 m*M* Na solution. Source: Jeschke and Stelter, 1976. Redrawn by permission of Springer-Verlag.

In studies on alfalfa, Läuchli (1984) said that salt tolerance coincides with K losses and gains to accommodate Na. This is mediated by osmotic adjustment, charge balance by organic acids, and K transport and accumulation. Figdore *et al.* (1987) showed that there was a partial reversal in Na–K roles in tomato wherein, under low-K stress and high Na, Na substituted for some of the normal plant-K functions.

James (1988) described a distinctive K deficiency symptom in alfalfa grown on calcareous low-K, nonsaline, nonsodic soils. In some plants, the classical spotted chlorosis K deficiency symptom was evident. But other alfalfa plants in the same plot area displayed a marginal chlorosis in which the leaves contained Na levels 10-fold higher than normal. Both the spotted and marginal chlorosis symptoms were eliminated by K fertilization and both were described accordingly as K deficiency. But the marginal chlorosis could have been classified as Na toxicity. With K fertilization, leaf K increased and leaf Na decreased to normal levels, i.e., about 2% of K and 0.1% of Na.

In addition to plant K–Na interactions there are evidently significant reductions in Na toxicity by Ca. LaHaye and Epstein (1971) reported that under high NaCl exposure there was a general necrosis of bush bean root tissue at low levels of Ca. But with moderate levels of Ca, root necrosis was prevented. In addition, Na accumulation in leaves was progressively lower with increased levels of Ca.

Maas and Grieve (1987) described NaCl-salinity-induced symptoms in corn as Na-induced Ca deficiency. The symptom was eliminated when $CaCl_2$ was substituted for part of the NaCl osmoticum; a Na/Ca molar ratio of 5.7 in the rooting medium eliminated the symptom and improved plant performance. It is noteworthy that there was a highly significant cultivar effect in their work.

Cramer *et al.* (1987) stated that Ca in the rooting medium was essential for ion absorption, selectivity, and retention, and that Ca counteracts the deleterious effects of high Na levels in cotton. They suggested that high Na displaces Ca in the plasmalemma. Hyder and Greenway (1965) showed that NaCl reduced barley growth much more at low than at high nutrient levels, attributing the differences solely to Ca status of the medium. By contrast, Yeo and Flowers (1982) said that there were no effects on growth and salt concentrations in the shoots of rice plants across a Na/Ca ratio range of 5–25. They concluded that rice responds weakly to Ca under conditions of high Na salinity.

Data in Figure 8, adapted from the report of LaHaye and Epstein (1971), show the striking effect of Ca on Na composition of bean plants growing in NaCl-salinized nutrient culture medium. The growth of bean essentially doubled between zero and 3 mmol of additional Ca added to the nutrient solution. The added Ca effect apparently was not registered where NaCl salinization did not occur. It is evident that the success or failure of a "salt-tolerant" bean line under normal field salinity conditions could rest solely on the relative amount of Ca in the rooting medium during the selection process. In other words, unnatural combinations of salts in the rooting medium during selection raise important questions on the pragmatics of salt-tolerant crop development for saline soil culture.

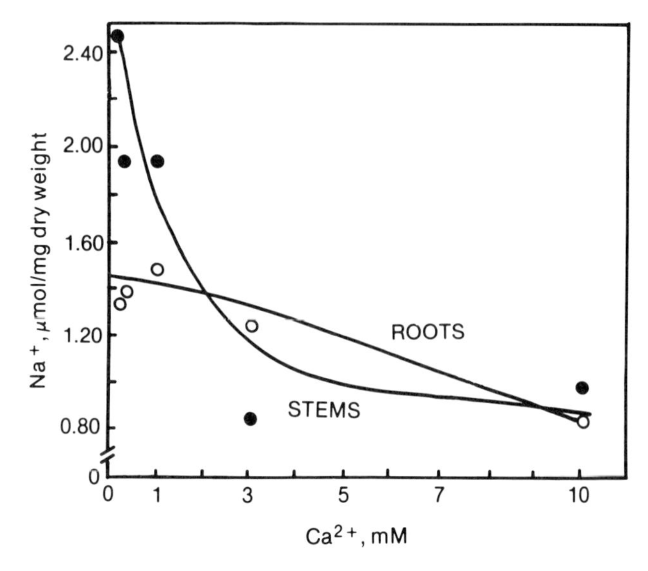

Figure 8. Ca effects on Na composition of bean stem and root. Source: LaHaye and Epstein, 1971. By permission of *Plant Physiology*.

Manufacture of organic solutes, selective ion exclusion or selective ion uptake against concentration gradients, and retranslocation of ions within the plant all require significant expenditures of metabolic energy (Läuchli, 1987; Tal, 1984).

C. PROGRESS ON CROP SALT TOLERANCE DEVELOPMENT

Kingsbury and Epstein (1984) state that there are ample sources of genetic diversity in salt tolerance for most major crop plants. Examples of crop salinity tolerance studies that support this notion are given in Table VI. The number of references in Table VI for each crop type may or may not reflect the overall progress toward development of salinity-tolerant, commercially available cultivars. Data in Table VI show that the research emphasis varies in terms of the different components of salinity stress in crops, i.e., osmolality versus specific ion effects.

All of the fruit studies referenced in Table VI were in connection with rootstock–scion investigations and all reported significant effects of rootstock selection on specific ion effects, mainly Cl. Data in Table VII show some of the results of Ream and Furr (1976), who evaluated salinity tolerance in citrus based on Cl exclusion from tops growing on different rootstocks; there was a six- to sevenfold difference in leaf Cl concentration using the same sweet orange or lemon scion source. Most of the hybrid rootstocks used were less efficient in excluding Cl than was the standard Cleopatra mandarin rootstock.

Two studies reported in Table VI were directed at rhizobial sensitivities (Wilson, 1970; Lauter *et al.*, 1981) and strain variability was demonstrated in both cases.

Relatively few reports were encountered with respect to vegetable crops. Bernstein and Ayers (1953) found all five carrot cultivars they studied to be relatively sensitive to salinity. They reported, however, that within the range of salt tolerance observed, there was a definite relationship between tolerance and the Ca/K ratio of plant tops. Cramer and Spurr (1986a,b) found a large cultivar effect in lettuce for Cl toxicity but little tolerance for salinity per se. Rush and Epstein (1976) found high salinity tolerance in a wild relative of tomato while Figdore *et al.* (1987) found wide ranges in tomato responses to Na under low-K conditions. Tal and Gardi (1976) reported salt tolerance variability between polyploid and tetraploid types of tomato.

Currently the status of breeding and selection for plant salt tolerance seems to be largely limited to identification of physiological mechanisms and correlation of these mechanisms with controlling genetic factors. With the exception of citrus rootstock, the production of commercial cultivars specifically tailored to saline conditions is largely an expectation for the future because little specific progress in that direction is apparent.

Maas and Neiman (1978) and Shannon (1979) emphasized the lack of knowledge on specific mechanisms of plant salt tolerance as a restraint to development of salt-tolerant cultivars. General reactions such as resistance to osmotic shock, plasmolysis, plant rooting, and survival have all been proposed as possible selection criteria but as yet none of these methods have been widely accepted. K/Na ratios in leaves, Cl accumulation, changes in other element concentrations (e.g., PO_4 and NO_3), and plant water use efficiency have also been proposed. Shannon (1979) suggested that a choice could be made as to whether salt exclusion or salt accumulation should be the main selection criterion.

Several narrowly focused procedures have been employed for screening and study of salt tolerance. Ramagopal (1986) used mannitol to generate osmotic stress free of specific ion influences. Isotopically labeled Cl and Na together with autoradiography have been used to elucidate the uptake and transport of these elements (Ramani and Kannan, 1986; Wieneke, 1982).

Noble *et al.* (1984) selected for salt tolerance in lucerne by using leaf damage of less than 10% as the criterion. He said that, coupled with Cl exclusion from shoots, leaf damage indexing could be an efficient scheme for selecting salt-tolerant alfalfa strains.

Yeo and Flowers (1984, 1986) argued that salt tolerance is conferred by no single factor but that it is the net effect of several contributing factors (Cl and Na uptake, preferential accumulation or exclusion from organs and tissues, differences in apoplastic and symplastic salt loads). Therefore they proposed that a "building block" or a "pyramiding approach" should be the basis of action wherein plants with one type of trait be crossed with other plants of different traits, and mass selection be exercised on the progeny.

Shannon (1985) observed that there are two serious constraints to the development of salt-tolerant cultivars: first (as emphasized above), the lack of physiological

Table VI. Crop Types Where Genotype or Cultivar Variability Has Been Demonstrated for Individual or Multiple Salt Parameters

	Salt[a] Component	Reference
Field crops		
Alfalfa	Os, NaCl	Allen *et al.* (1985)
	Cl, Na, Os	Noble *et al.* (1984)
	Cl	Smith *et al.* (1981)
	Os	Stone *et al.* (1979)
Barley	Os	Ayers *et al.* (1952); Ayers (1953)
	Na, Cl, NaCl	Bogemans and Stassart (1987)
	Os	Donovan and Day (1969)
	Os, Cl, Na	Greenway (1965)
	Os, NaCl	Stassart and Bogemans (1987)
Bean	Na	Ayoub (1974)
	Cl	Hajrasuliha (1980)
	Na	LaHaye and Epstein (1971)
	Na	Kramer *et al.* (1977)
Bentgrass	Os, NaCl	Younger *et al.* (1967)
	Cl, Na, Os	Chatelet and Wu (1986)
Chick-pea (rhizobia)	Os, NaCl	Lauter *et al.* (1981)
Maize	Na, Os	Cramer *et al.* (1987)
	Os	Maas *et al.* (1983)
	Na, Os	Maas and Grieve (1987)
	Os, Na	Yeo *et al.* (1977)
Cotton	Os, NaCl	Rathert (1982)
Rice	Na, Cl, NaCl	Flowers and Yeo (1981)
	Cl, Na, NaCl	Ramani and Kannan (1986)
	Na	Yeo and Flowers (1982, 1984)
Sorghum	Os, NaCl	Taylor *et al.* (1975)
Soybean	Os, Cl	Abel and MacKenzie (1964)
	Cl, Os	Abel (1969)
	Cl, Os	Grattan and Maas (1985)
	Cl	Parker *et al.* (1983, 1987)
	Na, Cl	Wieneke (1982)
(rhizobia)	Os	Wilson (1970)
Triticale	Os	Norlyn and Epstein (1984)
Wheat	Os, NaCl	Ayers *et al.* (1952)
	Na	Eggers and Jeschke (1983)
	Cl, Na, Os	Kingsbury *et al.* (1984)
	Cl, Na, Os	Kingsbury and Epstein (1986)
	Cl, Os	Torres and Bingham (1973)
	Cl, Na, Os	Wyn Jones *et al.* (1984)
Wheatgrass	NaCl	McGuire and Dvorak (1981)
	Na, NaCl	Yan and Epstein (1987)
(crested)	Os	Dewey (1960, 1962)
Fruits		
Avocado	Cl	Fenn *et al.* (1970)
Citrus	Cl	Bernstein *et al.* (1956)

Table VI. *(Continued)*

	Salt[a] Component	Reference
Fruits		
	Os	Bhambota and Kanwar (1969)
	Cl, Na	Cooper *et al.* (1952)
	Os	Ream and Furr (1976)
	Cl, NaCl	Sykes (1985)
	Na	Walker (1986)
Grape	Cl, Na	Downton (1977)
	Cl, Os	Bernstein *et al.* (1969)
Stone fruits, almonds	Cl, Os	Bernstein *et al.* (1956)

[a]Os, Osmotics; Cl, Na, specific ion effects of the respective ions (exclusion, compartmentation, toxicity); NaCl, common salt salinization used in screening or testing program. Order of parameters indicates approximate importance as determined by the author's procedure or emphasis.

markers that would facilitate rapid genotype screening, and second, the fact that salinity, as a matter of course, varies qualitatively and quantitatively in the field. He proposed that absolute salt tolerance could best be achieved by improving both absolute yield and relative salt tolerance. The results of Ancalle (1983) would support this notion. He showed that elemental (K, Na, Ca, Mg) composition of 12 alfalfa cultivars tested in six (calcareous, nonsaline) locations was significantly affected by location and cultivar. But most important, there was a highly significant cultivar by location interaction in regard to ionic composition. In other words, the ionic uptake pattern of the alfalfa cultivars was not alike at different locations. Some of Ancalle's results are given in Table VIII. Different genotype responses in different salt-affected fields would be expected for both osmotic and specific ion affects because of uncontrolled or unidentified factors that interact within the soil–plant–water system in the environment.

It has been suggested in several reviews that a serious impediment to salt-tolerant cultivar research and development is the lack of interdisciplinarity in the complexity of issues involved. Kramer (1984), Devine (1982), Duvick *et al.* (1981), Epstein and Rains (1987), and Läuchli (1987) asked for a collaborative, interdisciplinary effort involving a balanced attack from each of the pertinent disciplines, namely, soil science and plant nutrition, physiology and biochemistry, and genetics and plant breeding.

Tal (1985) believed that interdisciplinarity in salt-tolerant crop development has gained some recognition in recent years. He proposed that work should be focused on the amount and stability of yield in different field settings. This would be analogous to the suggestions of Yeo and Flowers (1984, 1986) and Shannon (1985) presented above.

Table VII. Mean Cl Content of Sweet Orange and Lemon Scion Leaves as Related to Hybrid Rootstock and Compared with a Standard Cultivar Rootstock (Cleopatra mandarin)[a,b]

Rootstock[c]	Cl (%)[d]
Sweet Orange Scion	
56-43-14	3.82a
57-149-2	3.75a
61-182-6	2.18b
59-125-3	2.11b
58-209-7	2.00b
58-110-1	1.79bc
56-43-9	1.53bcd
58-208-3	1.52bcd
54-63-27	1.48bcd
Cleopatra mandarin	1.07cde
58-206-3	1.03cde
56-43-7	1.03cde
59-130-19	0.87de
56-43-13	0.74de
61-54-5	0.48e
Lemon Scion	
58-210-5	3.67a
56-146-2	3.50a
58-207-3	2.80b
60-23	2.56b
62-116-11	2.00c
58-143-1	1.43d
58-207-6	1.30d
62-116-7	0.97de
58-210-14	0.55e
Cleopatra mandarin	0.53e

[a]Source: Ream and Furr, 1976. Used by permission of American Society for Horticultural Science.
[b]Plots were irrigated with water salinized with both NaCl and $CaCl_2$.
[c]U.S. Date and Citrus Station accession number.
[d]Means with same letter are not different at 1% level of significance.

Despite the lack of detailed knowledge of salt tolerance mechanisms at the biochemical and enzymatic levels, enough progress has been made to justify a concerted effort along purely empirical lines. In the short term, both production of salt-tolerant cultivars that include other necessary and desirable agronomic traits and an increased understanding of basic phenomena should be forthcoming. The latter would be exploited for long-term objectives.

Table VIII. Location and Cultivar Effects and Location by Cultivar Interactions in Elemental Composition of Alfalfa (summary analyses of variance)[a]

	Mean Squares by Element						
Source of Variation	K	Ca	Mg	Na	Sum of Cations	K/Ca	K/Na
Location	**[b]	**	**	**	**	**	**
Cultivar	**	**	NS[c]	**	**	NS	**
Location × cultivar	**	**	**	**	**	NS	**

[a]Source: Ancalle, 1983.
[b]**, F ratio significant at 1% significance.
[c]NS, not significant.

D. THE POTENTIAL ROLE OF EDAPHOLOGY IN DEVELOPMENT OF SALT-TOLERANT CROPS

As in the case of Fe chlorosis development in plants, edaphology represents a largely unexploited resource in the breeding, selection, and testing of salt-tolerant cultivars. Potentially, edaphology can make several kinds of contributions.

As discussed earlier and shown in Table VI, NaCl-salinized rooting media have been frequently employed in studies of plant salt tolerance. This would be sufficient for fitting plants to seawater irrigation (Epstein and Norlyn, 1977; Epstein *et al.*, 1980). But use of seawater for irrigation, or simple NaCl systems in rooting media, is clearly limited when plant breeding and selection programs are directed toward improved cropping efficiency on saline soils. A comparison of seawater data from Hem (1985) and those of one of Utah's poorer irrigation waters is given in Table IX.

Table IX. Cationic Composition of Seawater and a Saline Irrigation Water

Constituent	Seawater[a]	Irrigation Water[b]
Na ($mmol_c$/liter)	460	14.85
Mg ($mmol_c$/liter)	110	7.51
Ca ($mmol_c$/liter)	20	3.68
K ($mmol_c$/liter)	10	0.36
SAR($mmol_c$/liter)$^{1/2}$	57	6.28
ECw (dS/m)	55	2.51

[a]Computed from data presented by Hem (1985).
[b]Mean values for data collected over several years on the Sevier River at Delta Reservoir in Utah by James and Jurinak (1986).

The Na, K, and Ca ratios in seawater are much different from those normally encountered in salt-affected soils and irrigation waters. The exaggeration may be even more severe in nutrient solutions salinized exclusively with NaCl. Excess Na (SAR) is actually a rare exception, not the rule, in salt-affected soils and waters (Ayers and Westcot, 1985). James and Jurinak (1986), in a survey of Utah surface irrigation water supplies, could not identify Na as a dominating factor in a single instance.

As emphasized earlier, other solutes interact profoundly with Cl and Na in terms of salt effects in plants. These include K and Ca (as related to Na) and, to a lesser extent, SO_4, NO_3, and PO_4 (as related to Cl). Therefore a pragmatic salt-tolerant crop development program would assure the presence of typical levels of K, Ca, and SO_4, together with Cl and Na, in the rooting environment during the selection process.

Some of the potential contributions by edaphologists in plant genotype development are analogous to those suggested earlier for lime-induced Fe chlorosis. Other possible inputs are unique to salt-affected soils and irrigation waters. Edaphologists could:

1. Establish and maintain homogeneous systems in regard to ECe, ECw, and ionic suites to facilitate screening of salt-tolerant genotypes. This would help assure reproducible conditions and aid in verifying plant salt tolerance levels.
2. Adapt tools frequently employed in the study of soil–plant–water relations (matric potential, osmotic potential, transpiration, soil water use efficiency) to expedite screening programs for salt tolerance. This could include adaptation of the thermocouple psychrometer to monitor plant and soil water potentials.
3. Develop or adapt micro ion-specific electrodes for monitoring Cl and Na in plants. This would be aimed at a system of plant sampling that would be nondestructive to the plant, thereby facilitating use of selected plants as sources of genes in the breeding program. This could include identification of the most efficient plant tissue sampling and analysis system for facilitating the rapid screening of salt-tolerant genotypes.

Observations by Läuchli (1987) are apropos in the context of this discussion:

> There is ample opportunity for biotechnology applications in future soil science research. In principle, all organisms are amenable to genetic engineering, including soil organisms and plants. Obviously, the researcher needs to have molecular biology and genetics skills to tackle genetic improvement of organisms. But for the geneticist to succeed in this endeavor, the input by soil and plant scientists is crucial, because geneticists usually do not have the basic understanding of the issues in soil science that are involved, nor do they know how to assess the relative merits of the genetically engineered organism in the specific environment. Biotechnology opens exciting new possibilities in soil science [and] close cooperation between geneticists on one side and soil scientists and plant scientists on the other must be fostered if biotechnology is to be (fully) utilized.

He suggested that high fertilizer use efficiency be considered as a target for genetic improvement by many plant geneticists.

IV. Summary

Alkaline soils, i.e., soils with pH above 7, are characteristic of world regions that are arid or semiarid and of subhumid areas where the soil parent material is calcitic or dolomitic. Many irrigated soils and irrigation waters are affected by excess salts. Certain calcareous soils and all saline soils and irrigation waters represent special problems in regard to improving soil fertility, productivity, and fertilizer use efficiency.

Abundant evidence exists that there are several physiological mechanisms affecting Fe uptake and utilization by plants on one hand, and tolerance of excess salts on the other. These mechanisms are under genetic control; some traits are simply inherited and others are complex.

Considerable progress has been made in development of commercially available Fe-efficient soybean. The same is true for tree fruits, where Fe efficiency is contributed largely by breeding and selection of rootstocks. Iron-efficient cultivars in several other important crop types will continue to become commercially available. In general, progress on development of salt-tolerant cultivars is less marked than is the case for Fe efficiency. Salt tolerance contributed by specially selected rootstocks is being employed in commercial fruit production. With the exception of rice, there is no evidence that plants may be selected for tolerance to saturated soil or high water table conditions, with or without salinity.

Crop tolerance of both lime-induced Fe chlorosis and salinity is complicated by the fact that these problems vary temporally and spatially in the field. Lime in soil, or pH between 7.1 and 8.3, predisposes certain kinds of plants to lime-induced chlorosis but several other factors are involved in the phenomenon and breeding and selection for Fe efficiency must recognize these interactions. Breeding and selection for salinity-tolerant cultivars for field production must take cognizance of the profound interactions among Ca and K versus Na, and among NO_3 and SO_4 versus Cl in overcoming specific ion imbalances and toxicities.

The scientific disciplines mainly involved in adaptation of plants to problem calcareous and saline soils include primarily genetics and plant breeding, and secondarily biochemistry and physiology. Edaphology has had a somewhat indirect role in these efforts to date, but this discipline needs to become more involved if the demonstrated potentials are to be fully realized in adapting plants to problem calcareous and saline soils. Suggestions are given on some direct involvements for edaphologists in these interdisciplinary programs.

The end product of plant adaptation through breeding and selection of desired traits will be greater exploitation of problem alkaline and saline soils in the production of food and fiber. In connection with routine management practices, specially adapted cultivars will expand efficient crop production into marginal soil areas that heretofore have been restrictive of successful economic enterprise. One result of these efforts will be higher plant use efficiency of the fertilizer elements.

References

Abel, G. H. 1969. Inheritance of the capacity for chloride inclusion and chloride exclusion by soybean. *Crop Sci.* 9:697–698.

Abel, G. H., and A. J. MacKenzie. 1964. Salt tolerance of soybean varieties (*Glycine max* L. Merrill) during germination and later growth. *Crop Sci.* 4:157–161.

Alcantara, E., and M. D. de la Guardia. 1986. Variability in the response of sunflower hybrids or parental lines to iron stress. *J. Plant Nutr.* 9:443–451.

Alcantara, E., F. J. Romera, and M. D. de la Guardia. 1988. Genotypic differences in bicarbonate-induced iron chlorosis in sunflower. *J. Plant Nutr.* 11:65–75.

Allen, S. G., A. K. Dobrenz, M. M. Schonhorst, and J. E. Stoner. 1985. Heritability of NaCl tolerance in germinating alfalfa seeds. *Agron. J.* 77:99–101.

Ancalle, R. 1983. P, K, Ca, Mg, and Na composition of alfalfa as related to variety and soil fertility. M.S. Thesis, Utah State University, Logan.

Ayers, A. D. 1953. Germination and emergence of several varieties of barley in salinized soil cultures. *Agron. J.* 45:68–71.

Ayers, A. D., J. W. Brown, and C. H. Wadleigh. 1952. Salt tolerance of barley and wheat in soil plots receiving several salinization regimes. *Agron. J.* 44:307–310.

Ayers, R. S., and D. W. Westcot. 1985. Water Quality for Agriculture. FAO Irrig. Drain. Pap. No. 29, Rev. 1. FAO UN, Rome.

Ayoub, A. T. 1974. Causes of inter-varietal differences in susceptibility to sodium toxicity injury in *Phaseolus vulgaris*. *J. Agric. Sci.* 83:539–543.

Barton, L. L. (ed.). 1988. *Proc. 4th Int. Symp. Iron Nutrition and Interactions in Plants. J. Plant Nutr.* 11:605–626.

Bell, W. D., L. Bogorad, and W. J. McIlrath. 1958. Response of the yellow-stripe maize mutant (ysl) to ferrous and ferric iron. *Bot. Gaz.* 120:36–39.

Bennett, J. H., R. A. Olsen, and R. B. Clark. 1982. Modification of soil fertility by plant roots: Iron stress response mechanism. *What's New Plant Physiol.* (*Plant Physiol.*) 13:1–4.

Berg, W. A., C. L. Dewald, and P. I. Coyne. 1986. Selection of iron-efficient old world bluestems. *J. Plant Nutr.* 9:453–458.

Bernstein, L., and A. D. Ayers. 1953. Salt tolerance of five varieties of carrots. *Proc. Am. Soc. Hortic. Sci.* 61:360–366.

Bernstein, L., J. W. Brown, and H. E. Hayward. 1956. The influence of rootstock on growth and salt accumulation in stone-fruit trees and almonds. *Proc. Am. Soc. Hortic. Soc.* 68:86–95.

Bernstein, L., C. F. Ehlig, and R. A. Clark. 1969. Effect of grape rootstock on chloride accumulation in leaves. *Am. Soc. Hortic. Sci. J.* 94:584–590.

Bhambota, J. R., and J. S. Kanwar. 1969. Salinity tolerance of some rootstocks and scions of citrus species. *Proc. 1st Int. Citrus. Symp.* pp. 1833–1836.

Bogemans, J., and J. M. Stassart. 1987. Ion segregation in different plant parts within different barley cultivars under salt stress. In W. H. Gabelman and B. C. Loughman (eds.), Genetic Aspects of Plant Mineral Nutrition. Nijhoff, The Hague. Pp. 239–246.

Brown, J. C. 1977. Genetically controlled chemical factors involved in absorption and transport of iron by plants. *Adv. Chem. Ser.* No. 162. *Bioinorg. Chem.* 2:93–103.

Brown, J. C., and J. E. Ambler. 1974. Iron stress response in tomato (*Lycopersicon esculentum*). 1. Sites of Fe reduction, absorption and transport. *Physiol. Plant.* 31:221–224.

Brown, J. C., J. E. Ambler, R. L. Chaney, and C. D. Foy. 1972. Differential responses of plant genotypes to micronutrients. In J. J. Mortvedt, P. M. Giordano, and W. L. Lindsay (eds.), Micronutrients in Agriculture. Soil Sci. Soc. Am., Madison, Wisconsin. Pp. 389–418.
Brown, J. C., and W. D. Bell. 1969. Iron uptake dependent upon genotypes of corn. *Soil Sci. Soc. Am. Proc.* 33:99–101.
Brown, J. C., R. S. Holmes, and L. O. Tiffin. 1961. Iron chlorosis in soybeans as related to the genotype of rootstock. 3. Chlorosis susceptibility and reductive capacity at the root. *Soil Sci.* 91:127–132.
Brown, J. C., and W. E. Jones. 1977a. Fitting plants nutritionally to soils. I. Soybeans. *Agron. J.* 69:399–404.
Brown, J. C., and W. E. Jones. 1977b. Fitting plants nutritionally to soils. II. Cotton. *Agron. J.* 69:405–409.
Brown, J. C., and W. E. Jones. 1977c. Fitting plants nutritionally to soils. III. Sorghum. *Agron. J.* 69:410–414.
Brown, J. C., and L. O. Tiffin. 1960. Iron chlorosis in soybeans as related to genotype of rootstock. 2. A relationship between susceptibility to chlorosis and capacity to absorb iron from iron chelate. *Soil Sci.* 89:8–15.
Brown, J. C., and E. V. Wann. 1982. Breeding for Fe efficiency: Use of indicator plants. *J. Plant Nutr.* 5:623–635.
Chaney, R. L., and P. F. Bell. 1987. Complexity of iron nutrition: Lessons for plant–soil interaction research. *J. Plant Nutr.* 10:963–994.
Chatelet, R., and L. Wu. 1986. Contrasting response to salt stress of two salinity-tolerant creeping bentgrass clones. *J. Plant Nutr.* 9:1185–1195.
Chen, Y., and P. Barak. 1982. Iron nutrition of plants in calcareous soils. *Adv. Agron.* 35:217–240.
Christensen, N. W., and M. Brett. 1985. Chloride and liming effects on soil nitrogen form and take-all of wheat. *Agron. J.* 77:157–163.
Cianzio, S. R. de, W. R. Fehr, and I. C. Anderson. 1979. Genotypic evaluation for iron deficiency chlorosis in soybeans by visual scores and chlorophyll concentration. *Crop Sci.* 19:644–646.
Clark, R. B. (ed.). 1986. *Proc. 3rd Int. Symp. Iron Nutrition and Interactions in Plants. J. Plant Nutr.* 9:161–1076.
Clark, R. B., and J. C. Brown. 1974. Differential mineral uptake of maize inbreds. *Commun. Soil Sci. Plant Anal.* 5:213–227.
Clark, R. B., D. P. Coyne, W. M. Ross, and B. E. Johnson. 1989. Genetic aspects of plant resistance to iron deficiency. *Proc. Int. Congr. Plant Physiology, Soc. Plant Physiol. Biochem., New Delhi.*
Clark, R. B., and R. D. Gross. 1986. Plant genotype differences in iron. *J. Plant Nutr.* 9:471–491.
Clark, R. B., Y. Yusuf, W. M. Ross, and J. W. Maranville. 1982. Screening for sorghum differences to iron efficiency. *J. Plant Nutr.* 5:587–604.
Clarkson, D. T., and J. B. Hanson. 1980. The mineral nutrition of higher plants. *Annu. Rev. Plant Physiol.* 31:239–298.
Cooper, W. C., B. S. Gorton, and E. O. Olson. 1952. Ionic accumulation in citrus, as influenced by rootstock and scion and concentrations of salts and boron in the substrate. *Plant Physiol.* 27:191–203.
Coulombe, B. A., R. L. Chaney, and W. J. Weibold. 1984. Use of bicarbonate in screening soybeans for resistance to iron chlorosis. *J. Plant Nutr.* 7:411–426.
Coyne, D. P., S. S. Korban, D. Knudsen, and R. B. Clark. 1982. Inheritance of iron deficiency in crosses of dry beans. (*Phaseolus vulgaris* L.). *J. Plant Nutr.* 5:575–585.
Cramer, G. R., A. Läuchli, and E. Epstein. 1987. Calcium/salinity interactions in plants. *J. Plant Nutr.* 10:1937.
Cramer, G. R., and A. R. Spurr. 1986a. Responses of lettuce to salinity. I. Effects of NaCl and Na_2SO_4 on growth. *J. Plant Nutr.* 9:115–130.
Cramer, G. R., and A. R. Spurr. 1986b. Responses of lettuce to salinity. II. Effects of calcium on growth and mineral status. *J. Plant Nutr.* 9:131–142.
Devine, T. E. 1982. Genetic fitting of crops to problem soils. In M. N. Christiansen and C. F. Lewis (eds.), Breeding Plants for Less Favorable Environments. Wiley, New York. Pp. 143–173.

Dewey, D. R. 1960. Salt tolerance of twenty five strains of *Agropyron*. *Agron. J.* 52:631–635.
Dewey, D. R. 1962. Breeding crested wheatgrass for salt tolerance. *Crop Sci.* 2:403–407.
Donovan, T. J., and A. D. Day. 1969. Some effects of high salinity on germination and emergence of barley (*Hordeum vulgare* L. emend Lam.). *Agron. J.* 61:236–238.
Downton, W. J. S. 1977. Influence of rootstocks on accumulation of chloride, sodium and potassium in grapevines. *Aust. J. Agric. Res.* 28:879–889.
Duvick, D. N., R. A. Kleese, and N. M. Frey. 1981. Breeding for tolerance of nutrient imbalance and constraints to growth in acid, alkaline and saline soils. *J. Plant Nutr.* 4:111–129.
Eggers, H., and W. D. Jeschke. 1983. Comparison of K–Na selectivity mechanisms in roots of *Fagopyrum* and *Triticum*. In M. R. Sarić and B. C. Loughman (eds.), Genetic Aspects of Plant Mineral Nutrition. Nijhoff, The Hague. Pp. 223–228.
Epstein, E. 1963. Selective ion transport in plants and its genetic control. *Proc. Desalinization Research Conf. N. A. S.–N. R. C. Publ.* No. 942.
Epstein, E. 1985. Salt-tolerant crops: Origins, development and prospects of the concept. *Plant Soil* 89:187–198.
Epstein, E., and R. L. Jefferies. 1964. The genetic basis of selective ion transport in plants. *Annu. Rev. Plant Physiol.* 15:169–184.
Epstein, E., and J. D. Norlyn. 1977. Sea water-based crop production: A feasibility study. *Science* 197:249–251.
Epstein, E., J. D. Norlyn, D. W. Rush, R. W. Kingsbury, D. B. Kelley, G. A. Cunningham, and A. F. Wrona. 1980. Saline culture of crops: A genetic approach. *Science* 210:399–404.
Epstein, E., and D. W. Rains. 1987. Advances in salt tolerance. In W. H. Gabelman and B. C. Loughman (eds.), Genetic Aspects of Plant Mineral Nutrition. Nijhoff, The Hague. Pp. 113–125.
Esty, J. A., A. B. Onken, L. R. Hossner, and R. Matheson. 1980. Iron use efficiency in grain sorghum hybrids and parental lines. *Agron. J.* 72:589–592.
FAO. 1985. Water Quality for Agriculture. FAO Irrig. Drain. Pap. No. 29, Rev. 1. FAO UN, Rome.
Fehr, W. R. 1982. Control of iron-deficiency chlorosis in soybeans by plant breeding. *J. Plant Nutr.* 5:611–621.
Fehr, W. R. 1984. Current practices for correcting iron deficiency in plants with emphasis on genetics. *J. Plant Nutr.* 7:347–354.
Fenn, L. B., J. J. Oertli, and F. T. Bingham. 1970. Specific chloride injury in *Persea americana*. *Soil Sci. Soc. Am. Proc.* 34:617–620.
Figdore, S. S., W. H. Gabelman, and G. C. Gerloff. 1987. The accumulation and distribution of sodium in tomato strains differing in potassium efficiency when grown under low-K stress. In W. H. Gabelman and B. C. Loughman (eds.), Genetic Aspects of Plant Mineral Nutrition. Nijhoff, The Hague. Pp. 353–360.
Fixen, P. E., G. W. Buchenau, R. H. Gelderman, T. E. Schumacher, J. R. Gerwing, F. A. Gholick, and B. G. Farber. 1986. Influence of soil and applied chloride on several wheat parameters. *Agron. J.* 78:736–440.
Flowers, T. J., and A. Läuchli. 1983. Sodium versus potassium: Substitution and compartmentation. In A. Läuchli and R. L. Bieleski (eds.), Inorganic Plant Nutrition. Encyclopedia of Plant Physiology, New Series, Vol. 15B. Springer-Verlag, New York. Pp. 651–681.
Flowers, T. J., and A. R. Yeo. 1981. Variability in the resistance of sodium chloride salinity within rice (*Oryza sativa* L.) varieties. *New Phytol.* 88:363–373.
Flowers, T. J., and A. R. Yeo. 1986. Ion relations of plants under drought and salinity. *Aust. J. Plant Physiol.* 13:75–91.
Foy, C. D., P. W. Voigt, and J. W. Schwartz. 1977. Differential susceptibilities of weeping lovegrass strains to an iron-related chlorosis in calcareous soils. *Agron. J.* 69:491–496.
Froehlich, D. M., and W. R. Fehr. 1981. Agronomic performance of soybeans with differing levels of iron deficiency chlorosis on calcareous soils. *Crop Sci.* 21:438–441.
Gerloff, G. C., and W. H. Gabelman. 1983. Genetic basis of inorganic plant nutrition. In A. Läuchli and

R. L. Bieleski (eds.), Inorganic Plant Nutrition. Encyclopedia of Plant Physiology, New Series, Vol. 15B. Springer-Verlag, New York. Pp. 453–480.

Grattan, S. R., and E. V. Maas. 1985. Root control of leaf phosphorus and chlorine accumulation in soybean under salinity stress. *Agron. J.* 77:890–895.

Greenway, J. 1965. Plant response to saline substrates. VII. Growth and ion uptake throughout plant development in varieties of *Hordeum vulgare*. *Aust. J. Biol. Sci.* 18:763–779.

Greenway, H., and R. Munns. 1980. Mechanism of salt tolerance in non-halophytes. *Annu. Rev. Plant Physiol.* 31:149–190.

Hajrasuliha, S. 1980. Accumulation and toxicity of chloride in bean plants. *Plant Soil* 55:133–138.

Hamze, M., and M. Nimah. 1982. Iron content during lime-induced chlorosis with two citrus rootstocks. *J. Plant Nutr.* 5:797–804.

Hamze, M., J. Ryan, R. Middashi, and M. Solh. 1987. Evaluation of chickpea (*Cicer arietinum* L.) genotypes for resistance of lime-induced chlorosis. *J. Plant Nutr.* 10:1031–1040.

Hamze, M., J. Ryan, and M. Zaabout. 1986. Screening of citrus rootstocks for lime-induced chlorosis tolerance. *J. Plant Nutr.* 9:459–469.

Harivandi, M. A., and J. D. Butler. 1982. Factors associated with iron chlorosis of Kentucky bluegrass cultivars. *J. Plant Nutr.* 5:569–573.

Hartzook, A. 1984. The performance of iron absorption-efficient peanut cultivars in calcareous soil in the Lakhish and Beisan valley regions of Israel. *J. Plant Nutr.* 7:407–410.

Hartzook, A., D. Karstadt, M. Naveh, and S. Feldman. 1974. Differential iron absorption efficiency of peanut (*Arachis hypogaea* L.) cultivars grown on calcareous soils. *Agron. J.* 66:114–115.

Hem, J. D. 1985. Study and Interpretation of the Chemical Characteristics of Natural Water. 3rd Ed. *Geol. Surv. Water-Supply Pap. (U.S.)* No. 2254.

Hintz, R. R., Jr., W. R. Fehr, and S. R. Cianzio. 1987. Population development for the selection of high-yielding soybean cultivars with resistance to iron-deficiency chlorosis. *Crop Sci.* 27:707–710.

Huffaker, R. C., and A. Wallace. 1959. Effect of potassium and sodium levels on sodium distribution in some plant species. *Soil Sci.* 88:80–82.

Hyder, S. Z., and H. Greenway. 1965. Effects of Ca on plant sensitivity to high NaCl concentrations. *Plant Soil* 23:258–260.

James, D. W. (ed.). 1984. *Proc. 2nd Int. Symp. Iron Nutrition and Interactions in Plants. J. Plant Nutr.* 7:1–864.

James, D. W. 1988. Leaf margin necrosis in alfalfa: A potassium deficiency symptom associated with high sodium uptake. *Soil Sci.* 145:374–380.

James, D. W., R. J. Hanks, and J. J. Jurinak. 1982. Modern Irrigated Soils. Wiley, New York.

James, D. W., and J. J. Jurinak. 1986. Irrigation Resources in Utah: Water Quality versus Soil Salinity, Soil Fertility and the Environment. *Utah Agric. Exp. Stn., Res. Bull.* No. 514.

James, D. W., D. C. Kidman, W. H. Weaver, and R. L. Reeder. 1970. Factors affecting chloride uptake and implications of the chloride–nitrate antagonism in sugarbeet mineral nutrition. *J. Am. Soc. Sugarbeet Technol.* 15:647–656.

James, D. W., W. H. Weaver, and R. L. Reeder. 1969. Chloride uptake by potatoes and the effects of potassium chloride, nitrogen and phosphorus fertilization. *Soil Sci.* 109:48–52.

Jeschke, W. D. 1984. K–Na exchange at cellular membranes, intracellular compartmentation of cations, and salt tolerance. In R. C. Staples and G. H. Toenniessen (eds.), Salinity Tolerance in Plants: Strategies for Crop Improvement. Wiley, New York. Pp. 37–66.

Jeschke, W. D., and W. Stelter. 1976. Measurement of longitudinal ion profiles in single roots of *Hordeum* and *Atriplix* by use of flameless atomic absorption spectroscopy. *Planta* 128:107–112.

Kadman, A., and A. Ben-Ya'acov. 1982. Selection of avocado rootstocks for calcareous soils. *J. Plant Nutr.* 5:639–643.

Kannan, S. 1980. Differences in iron stress response and iron uptake in some sorghum varieties. *J. Plant Nutr.* 2:347–350.

Kannan, S. 1982. Genotypic differences in iron uptake and utilization in some crop cultivars. *J. Plant Nutr.* 5:531–542.

Kannan, S. 1984. Studies on Fe deficiency stress response in three cultivars of sunflower (*Helianthus annus* L.). *J. Plant Nutr.* 7:1203–1212.
Kannan, W. 1985. Fe-deficiency tolerance in papaya (*Carica papaya* L.): pH reduction and chlorosis recovery response to stress. *J. Plant Nutr.* 8:1191–1197.
Kannan, S., and S. Ramani. 1987. Mechanisms of Fe-deficiency of crop cultivars: Effects of dibutyl phthalate and caffeic acid on Fe-chlorosis recovery. *J. Plant Nutr.* 10:1051–1058.
Kingsbury, R. W., and E. Epstein. 1984. Selection for salt-resistant spring wheat. *Crop Sci.* 24:310–315.
Kingsbury, R. W., and E. Epstein. 1986. Salt sensitivity in wheat: A case for specific ion toxicity. *Plant Physiol.* 80:651–654.
Kingsbury, R. W., E. Epstein, and R. W. Pearcy. 1984. Physiological responses to salinity in selected lines of wheat. *Plant Physiol.* 74:417–423.
Kovacevic, F., B. Bertic, V. Trogrlic, and M. Seput. 1982. Genotype influence on iron concentrations in grain of corn hybrids. *J. Plant Nutr.* 5:605–608.
Kramer, D. 1984. Cytological aspects of salt tolerance in higher plants. In R. C. Staples and G. H. Toenniessen (eds.), Salinity Tolerance in Plants: Strategies for Crop Improvement. Wiley, New York. Pp. 3–15.
Kramer, D., A. Läuchli, and A. R. Yeo. 1977. Transfer cells in roots of *Phaseolus coccines:* Ultrastructure and possible function in exclusion of sodium from the shoot. *Ann. Bot.* 41:1031–1040.
Kuykendall, J. R. 1956. The occurrence of iron chlorosis in the United States with particular respect to the occurrence of lime-induced chlorosis. *Soil Sci.* 84:24–28.
Läuchli, A. 1984. Salt exclusion: An adaptation of legumes for crops and pastures under saline conditions. In R. C. Staples and G. H. Toenniessen (eds.), Salinity Tolerance in Plants: Strategies for Crop Improvement. Wiley, New York. Pp. 171–187.
Läuchli, A. 1987. Soil science in the next twenty-five years: Does biotechnology play a role? *Soil Sci. Soc. Am. J.* 51:1405–1409.
LaHaye, P. A., and E. Epstein. 1971. Calcium and salt tolerance by bean plants. *Physiol. Plant.* 25:213–218.
Landsberg, H. E., and R. W. Schloemer. 1967. World climatic regions in relation to irrigation. In R. M. Hagan, H. R. Haise, and T. W. Edminster (eds.), Irrigation of Agricultural Lands. Monogr. No. 11. Am. Soc. Agron., Madison, Wisconsin. Pp. 25–32.
Lauter, D. J., D. N. Munns, and K. L. Clarkin. 1981. Salt response of chickpea as influenced by N supply. *Agron. J.* 73:961–966.
Leigh, R. A., and R. G. Wyn Jones. 1986. Cellular compartmentation of plant nutrition: The selective cytoplasm and the promiscuous vacuole. In P. B. Tinker and A. Läuchli (eds.), *Adv. Plant Nutr.* 2:249–279.
Leopold, A. C., and R. P. Willing. 1984. Evidence for toxicity effects of salt on membranes. In R. C. Staples and G. H. Toenniessen (eds.), Salinity Tolerance in Plants: Strategies for Crop Improvement. Wiley, New York. Pp. 67–76.
Lessani, H., and H. Marschner. 1978. Relation between salt tolerance and long distance transport of sodium and chloride in various crop species. *Aust. J. Plant Physiol.* 5:27–37.
Lindsay, W. L. 1979. Chemical Equilibria in Soils. Wiley, New York.
Maas, E. V., and C. M. Grieve. 1987. Sodium-induced calcium deficiency in salt-stressed corn. *Plant, Cell Environ.* 10:559–564.
Maas, E. V., and G. J. Hoffman. 1977. Crop salt tolerance—Current assessment. *J. Irrig. Drain. Div., Am. Soc. Civ. Eng.* 103(IR2):115–134.
Maas, E. V., G. J. Hoffman, G. D. Chaba, J. A. Poss, and M. C. Shannon. 1983. Salt sensitivity of corn at various growth stages. *Irrig. Sci.* 4:45–57.
Maas, E. V., and R. H. Nieman. 1978. Physiology of plant tolerance to salinity. In G. A. Jung (ed.), Crop Tolerance to Suboptimal Land Conditions. *ASA Spec. Publ.* No. 32, pp. 277–299.
Marschner, H. 1986. Mineral Nutrition of Higher Plants. Academic Press, Orlando, Florida.
McDaniel, M. E., and J. C. Brown. 1982. Differential iron chlorosis of oat cultivars—A review. *J. Plant Nutr.* 5:545–552.

McDaniel, M. E., and D. J. Murphy. 1978. Differential iron chlorosis of oat cultivars. *Crop Sci.* 18:136–138.

McGuire, P. E., and J. Dvorak. 1981. High salt tolerance potential in wheatgrass. *Crop Sci.* 21:702–705.

McKenzie, D. B., L. R. Hossner, and R. J. Newton. 1984. Sorghum cultivar evaluation for iron chlorosis resistance by visual scores. *J. Plant Nutr.* 7:677–686.

Mikesell, M. E., G. M. Paulsen, R. Ellis, Jr., and A. J. Casady. 1973. Iron utilization by efficient and inefficient sorghum lines. *Agron. J.* 65:77–80.

Neikova-Bocheva, G., B. Bocheve, G. Ganeva, and F. Kaneva. 1986. Chromosomal and cytoplasmic effects on uptake and accumulation of iron, phosphorus, and calcium in *Triticum aestivum* L. *J. Plant Nutr.* 9:493–501.

Nelson, S. D. (ed). 1982. *Proc. 1st Int. Symp. Iron Nutrition and Interactions in Plants. J. Plant Nutr.* 5:229–1001.

Niebur, W. S., and W. R. Fehr. 1981. Agronomic evaluation of soybean genotypes resistant to iron deficiency chlorosis. *Crop Sci.* 21:551–554.

Noble, C. L., G. M. Halloran, and D. W. West. 1984. Identification and selection for salt tolerance in lucerne (*Medicago sativa* L.). *Aust. J. Agric. Res.* 35:239–252.

Nordquist, P. T., and W. A. Compton. 1986. Chlorotic variation of experimental maize (*Zea mays*) hybrids grown on high pH soils. *J. Plant Nutr.* 9:435–442.

Norlyn, J. D., and E. Epstein. 1984. Variability in salt tolerance of four triticale lines at germination and emergence. *Crop Sci.* 24:1090–1092.

Nunez-Medina, R. A. 1984. Iron nutrition of corn as related to variety, soil type and iron soil amendment. M.S. Thesis, Utah State University, Logan.

Olsen, R. A., R. B. Clark, and J. H. Bennett. 1981. The enhancement of soil fertility by plant roots. *Am. Sci.* 69:378–384.

Parker, M. B., G. J. Gascho, and T. P. Gaines. 1983. Cl toxicity of soybeans grown on Atlantic coast Flatwood soils. *Agron. J.* 75:439–443.

Parker, M. B., T. P. Gaines, J. E. Hook, G. J. Gascho, and B. W. Maw. 1987. Chloride and water stress effects on soybean in pot culture. *J. Plant Nutr.* 10:517–538.

Pierson, E. E., R. B. Clark, J. W. Maranville, and D. P. Coyne. 1984a. Plant genotype differences in ferrous and total iron in emerging leaves. I. Sorghum and maize. *J. Plant Nutr.* 7:371–388.

Pierson, E. E., R. B. Clark, D. P. Coyne, and J. W. Maranville. 1984b. Plant genotype differences in ferrous and total iron in emerging leaves. II. Dry beans and soybeans. *J. Plant Nutr.* 7:355–370.

Pitman, M. G. 1984. Transport across the root and shoot/root interactions. In R. C. Staples and G. H. Toenniessen (eds.), Salinity Tolerance in Plants: Strategies for Crop Improvement. Wiley, New York. Pp. 93–123.

Rai, R., V. Prasad, S. K. Choudhury, and N. P. Sinha. 1984. Iron nutrition and symbiotic N_2-fixation of lentil (*Lens culinaris*) genotypes in calcareous soil. *J. Plant Nutr.* 7:399–406.

Ramagopal, S. 1986. Protein synthesis in a maize callus exposed to NaCl and mannitol. *Plant Cell Rep.* 5:430–434.

Ramani, S., and S. Kannan. 1986. Absorption and transport of Na and Cl in rice cultivars differing in their tolerance to salinity: An examination of the effects of ammonium and potassium salts. *J. Plant Nutr.* 9:1553–1564.

Rathert, G. 1982. Increasing salinity stress and carbohydrate metabolism by different salt tolerant cotton genotypes. In A. Scaife (ed.), *Proc. 9th Int. Plant Nutrition Colloq., Warwick University, Warwick, England* pp. 528–544.

Ream, C. L., and J. R. Furr. 1976. Salt tolerance of some citrus species, relatives and hybrids tested as rootstocks. *Am. Soc. Hortic. Sci. J.* 101:265–267.

Reeve, R. C., and M. Fireman. 1967. Salt problems in relation to irrigation. In R. M. Hagan, H. R. Haise, and T. W. Edminster (eds.), Irrigation of Agricultural Lands. Monogr. No. 11. Am. Soc. Agron., Madison, Wisconsin. Pp. 988–1008.

Richards, L. A. (ed.). 1954. Diagnoses and Improvement of Saline and Alkali Soils. *U.S. Dep. Agric., Agric. Handb.* No. 60.

Römheld, V. 1987. Different strategies for iron acquisition in higher plants. *Physiol. Plant.* 70:231–234.
Römheld, V., and H. Marschner. 1981. Effect of Fe stress on utilization of Fe chelates by efficient and inefficient plant species. *J. Plant Nutr.* 3:551–560.
Römheld, V., and H. Marschner. 1986. Mobilization of iron in the rhizosphere of different plant species. In P. B. Tinker and A. Läuchli (eds.), *Adv. Plant Nutr.* 2:155–204.
Ross, W. M. 1986. Improving plants for tolerance to iron deficiency and other mineral nutrition disorders: Breeding and genetic points of view. *J. Plant Nutr.* 9:309–333.
Rush, D. W., and E. Epstein. 1976. Genotypic responses to salinity: Differences between salt-sensitive and salt-tolerant genotypes of the tomato. *Plant Physiol.* 57:162–166.
Shannon, M. C. 1979. In quest of rapid screening for salt tolerance. *Hortic. Sci.* 14:587–589.
Shannon, M. C. 1984. Breeding, selection, and the genetics of salt tolerance. In R. C. Staples and G. H. Toenniessen (eds.), Salinity Tolerance in Plants: Strategies for Crop Improvement. Wiley, New York. Pp. 231–254.
Shannon, M. C. 1985. Principles and strategies in breeding for higher salt tolerance. *Plant Soil* 89:227–241.
Singh, R. A., N. P. Singh, B. P. Singh, and S. G. Sharma. 1986. Reaction of chickpea genotypes to iron deficiency in a calcareous soil. *J. Plant Nutr.* 9:417–422.
Smith, D., A. K. Dobrenz, and M. H. Schonhorst. 1981. Response of alfalfa seedling plants to high levels of chloride salts. *J. Plant Nutr.* 4:143–174.
Smith, P. F., and A. W. Specht. 1952. Heavy metal nutrition in relation to iron chlorosis in citrus seedlings. *Proc. Fla. State Hortic. Soc.* 65:101–108.
Stassart, J. M., and J. Bogemans. 1987. Intervarietal ionic composition changes in barley under salt stress. In W. H. Gabelman and B. C. Loughman (eds.), Genetic Aspects of Plant Mineral Nutrition. Nijhoff, The Hague. Pp. 127–137.
Stone, J. E., D. B. Marx, and A. K. Dobrenz. 1979. Interaction of sodium chloride and temperature on germination of two alfalfa cultivars. *Agron. J.* 71:425–428.
Sykes, S. R. 1985. Effects of seedling age and size on chloride accumulation by juvenile citrus seedlings treated with sodium chloride under glasshouse conditions. *Aust. J. Exp. Agric.* 25:943–953.
Tal, M. 1984. Physiological genetics of salt resistance in higher plants: Studies on the level of the whole plant and isolated organs, tissues, and cells. In R. C. Staples and G. H. Toenniessen (eds.), Salinity Tolerance in Plants: Strategies for Crop Improvement. Wiley, New York. Pp. 301–320.
Tal, M. 1985. Genetics of salt tolerance in higher plants: Theoretical and practical considerations. *Plant Soil* 89:199–226.
Tal, M., and I. Gardi. 1976. Physiology of polyploid plants: Water balance in autotetraploid and diploid tomato under low and high salinity. *Physiol. Plant.* 38:257–261.
Taylor, R. M., E. F. Young, Jr., and R. L. Rivera. 1975. Salt tolerance in cultivars of grain sorghum. *Crop Sci.* 15:734–735.
Torres, C., and F. T. Bingham. 1973. Salt tolerance of Mexican wheat. I. Effect of NO_3 and NaCl on mineral nutrition, growth, and grain production of four wheats. *Soil Sci. Soc. Am. Proc.* 37:711–715.
Vose, P. B. 1982. Iron nutrition in plants: A world overview. *J. Plant Nutr.* 5:233–249.
Walker, R. R. 1986. Sodium exclusion and potassium–sodium selectivity in salt-treated trifoliate orange (*Poncirus trifoliata*) and Cleopatra mandarin (*Citrus reticulata*) plants. *Aust. J. Plant Physiol.* 13:293–303.
Wallace, A. 1982. Historical landmarks in progress relating to iron chlorosis in plants. *J. Plant Nutr.* 5:277–288.
Wallace, A., and O. R. Lunt. 1960. Iron chlorosis in horticultural plants, a review. *Proc. Am. Soc. Hortic. Sci.* 75:819–841.
Wann, F. B. 1941. Control of Chlorosis in American Grapes. *Utah Agric. Exp. Stn., Bull.* No. 299.
Weiss, M. G. 1943. Inheritance and physiology of efficiency in iron utilization in soybeans. *Genetics* 28:253–268.
Wieneke, J. 1982. Application of root zone feeding for evaluation of ion uptake and efflux in soybean

genotypes. In M. R. Sarić and B. C. Loughman (eds.), Genetic Aspects of Plant Mineral Nutrition. Nijhoff, The Hague. Pp. 159–169.
Williams, E. P., R. B. Clark, Y. Yusuf, W. M. Ross, and J. W. Maranville. 1982. Variability of sorghum genotypes to tolerate iron deficiency. *J. Plant Nutr.* 5:553–567.
Wilson, J. R. 1970. Response to salinity of glycine. VI. Some effects of range of short term salt stresses on the growth, nodulation, and nitrogen fixation of *Glycine wightii. Aust. J. Agric. Res.* 21:571–582.
Wutscher, H. K., E. O. Olson, A. V. Shull, and A. Peynado. 1970. Leaf nutrient levels, chlorosis, and growth of young grapefruit trees on 16 rootstocks on calcareous soil. *J. Am. Soc. Hortic. Sci.* 95:259–280.
Wyn Jones, R. G. 1981. Salt tolerance. In C. B. Jonson (ed.), Physiological Processes Limiting Plant Productivity. Butterworth, London. Pp. 271–292.
Wyn Jones, R. G., J. Gorham, and E. McDonnell. 1984. Organic and inorganic solute contents as selection criteria for salt tolerance in the Triticeae. In R. C. Staples and G. H. Toenniessen (eds.), Salinity Tolerance in Plants: Strategies for Crop Improvement. Wiley, New York. Pp. 189–203.
Yan, X., and E. Epstein. 1987. Salt tolerance and transport in two species of wheatgrass, *Elytrigia* spp. *J. Plant Nutr.* 10:1671.
Yeo, A. R., and T. J. Flowers. 1982. Accumulation and localization of sodium ions within the shoots of rice (*Oryza sativa*) varieties differing in salinity resistance. *Physiol Plant* 56:343–348.
Yeo, A. R., and T. J. Flowers. 1984. Mechanisms of salinity resistance in rice and their role as physiological criteria in plant breeding. In R. C. Staples and G. H. Toenniessen (eds.), Salinity Tolerance in Plants: Strategies for Crop Improvement. Wiley, New York. Pp. 151–170.
Yeo, A. R., and T. J. Flowers. 1986. Salinity resistance in rice (*Oryza sativa* L.) and pyramiding approach to breeding varieties for saline soils. *Aust. J. Plant Physiol.* 13:161–173.
Yeo, A. R., D. Kramer, A. Läuchli, and J. Gullasch. 1977. Ion distribution in salt-stressed mature *Zea mays* roots in relation to ultra-structure and retention of sodium. *J. Exp. Bot.* 28:17–29.
Younger, V. B., O. R. Lunt, and F. Nudge. 1967. Salinity tolerance of seven varieties of creeping bentgrass, *Agrostis palustris* Huds. *Agron. J.* 59:335–336.

Glossary of Common and Scientific Names of Plants

COMMON NAME	SCIENTIFIC NAME
A	
Agave	*Agave americana* L.
Alfalfa (lucerne)	*Medicago sativa* L.
Almond	*Prunus dulcis, P. communis*
Apple	*Malus domestica* L., *M. sylvestris* Mill.
Avocado	*Persea americana* Mill.
Azuki bean	*Vigna angularis*
B	
Banana	*Musa acuminata* L.
Barley	*Hordeum vulgare* L., *H. distichon* L.
Barleygrass	*Hordeum leporinum* L.
Barnyard millet	*Echinochloa frumentacea*
Bean (field)	*Phaseolus vulgaris* L.
Beet (red)	*Beta vulgaris* L.
Beggarweed	*Desmondium uncinatum*
Bent grass	*Agrostis palustris, A. solonifera*
Bent grass (Colonial)	*A. tenuis* Sibth.
Bermuda grass	*Cynodon dactylon* (L.) Pers.
Bigfoot trefoil	*Lotus corniculatus* L.
Birch	*Betula* spp.
Blueberry	*Vaccinium* spp.
Broad bean (fababean)	*Vicia faba* L.
Bromegrass (smooth)	*Bromus inermis* Leysser
Browntop (bent grass)	*Agrostis tenuis*
Buckwheat	*Fagopyrum esculentum* Moench
Bufflegrass	*Cenchrus ciliaris*
C	
Cabbage	*Brassica oleracea* L.
Cannabis	*Cannabis* sp.

Carnation	*Dianthus caryophyllus* L.
Carrot	*Daucus carota* L.
Cassava	*Manihot esculenta* Crantz
Celery	*Apium graveolens* L.
Centro	*Centrosema pubescens*
Cherry	*Prunus* spp.
Chick-pea	*Cicer arientinum* L.
Cicer milkvetch	*Astragalus cicer*
Citrus	*Citrus* spp.
Clover (crimson)	*Trifolium incarnatum* L.
Clover (arrowleaf)	*T. vesiculosum* Savi.
Cocoa	*Theobroma cacao* L.
Cotton	*Gossypium hirsutum* L.
Cowpea (southernpea)	*Vigna unguiculata* (L.) Walp.
Crested wheatgrass	*Agropyrum desertorum* (Fisch. ex Link) Shult.
Cucumber	*Cucumis sativus* L.
Cyphomandra	*Cyphomandra* spp.

D

Date palm	*Phoenix dactylifera*
Diplacus shrub	*Diplacus aurantiacus*
Dunegrass	*Elymus mollis, Ammophila arenaria*

E

Eggplant	*Solanum melongena* L.

F

French bean	*Phaseolus vulgaris* L.

G

Geranium	*Geranium carolinianum* L.
Ginger	*Zingiber officinale* Rosc.
Grape	*Vitus vinifera* L.
Greengrass (blackgrass)	*Vigna* spp.
Guinea grass	*Panicum maximum* Jacq.

H

Hairy vetch	*Vicia villosa* Roth.

I

Ipomea	*Ipomea tricolor* L.

J

Jute	*Corchorus capsularis*

K

Kentucky bluegrass	*Poa pratensis* L.

L

Lentil	*Lens culinaris* (L.) Medikus
Lettuce	*Lactuca sativa* L.
Lupine (white)	*Lupinus albus* L.
Lupine (blue)	*Lupinus angustifolius* L.
	Lupinus mutabilis L.

M

Maize (corn)	*Zea mays* L.
Mangrove	*Avicenna marina*
Maple	*Acer pseudoplatanus*
Medic	*Medicago* spp.
Mesquite	*Prosopis* spp.

O

Oak	*Quercus* spp.
Oats	*Avena sativa* L., *A. byzantina* L.
Oldworld bluestem	*Bothriochloa* spp.
Olive	*Olea europaea*
Onion	*Allium cepa* L.
Orange	*Citrus sinensis*
Orchard grass	*Dactylis glomerata* L.

P

Papaya	*Carica papaya*
Passion fruit	*Passiflora edulis* Sims
Pea (field)	*Pisum sativum* L., *P. arvense* L.
Peach	*Prunus persica* L. Sieb. & Zucc.
Peanut (groundnut)	*Arachis hypogaea* L.
Pearl millet	*Pennisetum americanum* (L.) Leeke Schum.
	P. typhoides (Burm. F.) Stapf & C. E. Hubb
	P. glaucum (L.) R. Br.
Pecan	*Carya illinoensis* (Wang.) K. Koch
Perennial ryegrass	*Lolium perenne* L.
Pigeon pea	*Cajanus cajan* (L.) Huth.
Poppy	*Papaver somniferum* L.
Potato	*Solanum tuberosum* L.
Prune	*Prunus domestica* L.

R

Radish	*Raphanus sativus* L.

Rape (winter)	*Brassica napus* L.
Reed canarygrass	*Phalaris arundinacea* L.
Red clover	*Trifolium pratense* L.
Rhus	*Rhus* subq. *Rhus*
Rice	*Oryza sativa* L., *O. indica* L.
Rye	*Secale cereale* L.
Ryegrass (Italian)	*Lolium multiflorum*

S

Sainfoin	*Onobrychis viciifolia*
Sericea lespedeza	*Lespedeza cuneata* (Dument) G. Don
Serradella	*Ornithopus* spp.
Siratro	*Macroptilium atropupureum*
Snap beans	*Phaseolus vulgaris* L.
Sorghum	*Sorghum bicolor* (L.) Moench
Soybean	*Glycine max* (L.) Merr., *G. soya*
Spinach	*Spinacea oleracea* L.
Strawberry clover	*Trifolium fragiferum* L.
Stylosanthes	*Stylosanthes humilis*
Subterranean clover	*Trifolium subterraneum* L.
Sudan grass	*Sorghum bicolor* (L.) Moench sub. *sudanense*
Sugar beet	*Beta vulgaris* L.
Sugarcane	*Saccharum officinarum* L.
Sunflower	*Helianthus annuus* L.
Sweet cherry	*Prunus avium* L.
Sweet potato	*Ipomea batatas* (L.) Lam.
Switchgrass	*Panicum virgatum* L.

T

Tall fescue	*Festuca arundinacea* Schreb.
Tangier pea	*Lathyrus tingitanus* L.
Taro (cocoyam)	*Colocasia esculenta*
Tea	*Cammelia sinensis* L.
Tobacco	*Nicotiana tabaccum* L., *N. plumbaginifolia*
Tomato	*Lycopersicon esculentum* Mill.
Townsville	*Stylosanthes humilis*
Trefoil	*Lotus pedunculatus*
Triticale	X Triticosecale Wittmack
Turnip	*Brassica rapa* (Metzg.) Sinsk.

W

Walnut	*Juglans regia*
Weeping lovegrass	*Eragrostis curvula* (Schrad.) Nees.
Wheat	*Triticum aestivum* (L.) em Thell

Wildoat	*Avena fatua* L.
Wheatgrass (Western)	*Agropyron smithii* Rydb.
White clover	*Trifolium repens* L.

MISCELLANEOUS

Atriplex vesicaria
Chionochloa spp.
Equisetum arvense
Lemna gibba L.
Lemna minor L.
Panicum hylaeicum
Panicum laxam
Panicum milioides
Panicum prionitis
Paoles (family Paoceae)
Trifolium semipilosum

Subject Index

J

K

L

X

Y

Z